Student Solutions Manual
for

Daniel T. Larose's
Discovering Statistics

Third Edition

Christina Morian
Lincoln University

ISBN-13: 978-1-4641-8866-4
ISBN-10: 1-4641-8866-1

Printed in the United States of America.

First Printing 2016

W. H. Freeman and Company
One New York Plaza Suite 4500
New York, NY 10004-1562
www.macmillanhighered.com

Table of Contents

Chapter 1: The Nature of Statistics

Section 1.1

1. The steepest negative slope in the graph is between the years 1994 and 1995.

3.

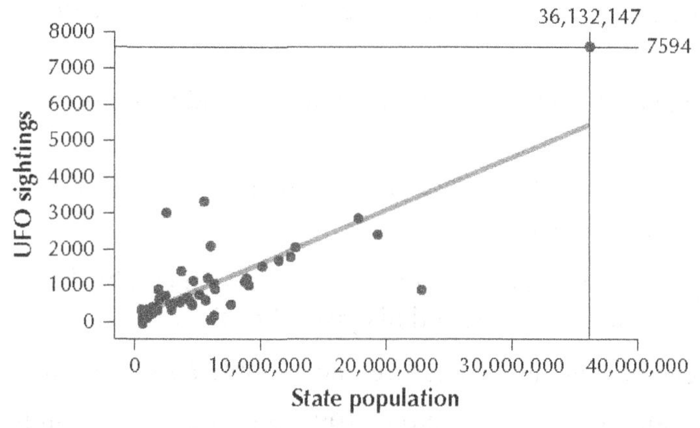

(a) About 36,000,000. Actual answer: 36,132,147.

(b) About 7600. Actual answer: 7594.

5.

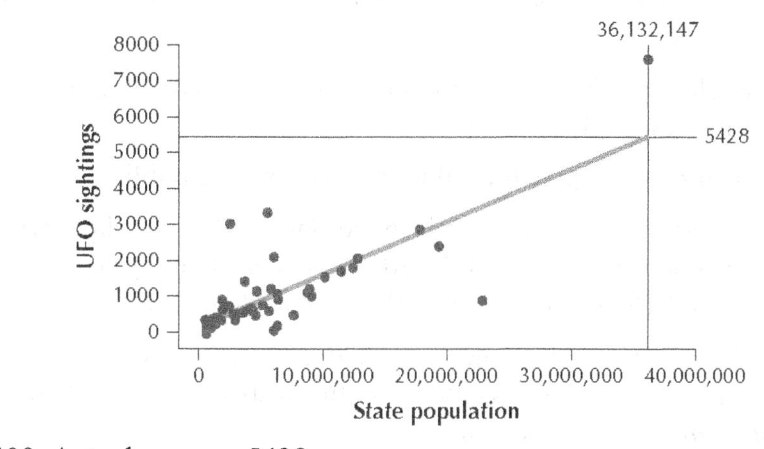

About 5400. Actual answer: 5428.

7. The Eiler fire is the largest. The Junction fire is the smallest.

Section 1.2

1. Statistics is the *art* and *science* of collecting, analyzing, presenting, and interpreting data.

3. A *qualitative variable* is a variable that does not assume a numerical value, but is usually classified into categories. A *quantitative variable* is a variable that takes on numerical values.

5. True.

7. A *statistic* is a characteristic of a sample.

9. A *census* is the collection of data from every element in the population.

11. The elements are the teams: Dragonborn, Sprites, Enchanters, Trolls

13. **(a)** *Captain's gender* can take the values male or female.

(b) The observation for the Sprites is captain's gender is female, 9 wins, rank is 2, and the winning percentage is 0.600.

15. Since the number of wins is counted, the variable *Wins* is discrete:

Since the winning percentage can be any number between 0 and 1 inclusive the variable *Winning percentage* is continuous.

17. The elements are the players: Miguel Cabrera, Michael Cuddyer, Joe Mauer, Michael Trout, Chris Johnson

19. **(a)** The variable *Team* can take the values Detroit Tigers, Colorado Rockies, Minnesota Twins, Los Angeles Angels, and Atlanta Braves.

(b) The observation for Miguel Cabrera is his team is the Detroit Tigers, his batting average is 0.348, the number of hits is 193, his rank is 1, and his year of birth is 1983.

21. The number of hits can be counted. Therefore, the variable *Hits* is discrete.

A player's year of birth can either be 1979 or 1980 or 1981, etc., and nothing in between. Therefore, the variable *Year of birth* is discrete.

A player's batting average can take any value between 0 and 1, inclusive. Therefore, the variable *Batting average* is continuous.

23. The elements are the schools: University of Phoenix, Devry University, ITT Technical Institute, Penn State University, Kaplan University

25. **(a)** The variable *School type* can take the values proprietary and public.

(b) The observation for Penn State University is it is in the state of PA, its school type is public, it had 42,011 federal student loan recipients in the 2013–2014 academic year, and it had a total federal student loan amount $151 million in the 2013–014 academic year.

27. The number of recipients is counted, so the variable *Recipients* is discrete. The total amount of federal student loans for the school is counted to the nearest million dollars, so the variable *Total loan amount ($ millions)* is discrete.

29. **(a)** The values for the variable *year you were born* are numbers that can be ranked or ordered, so the variable is quantitative. Since there are a finite number of years in which you could have been born, the variable is discrete.

(b) There is no natural zero. Division (2000/1990) does not make sense. However, subtraction does make sense. For example, someone born in 1993 is 3 years younger than someone born in 1990 (1993–1990 = 3). Therefore, the variable *year you were born* represents interval data.

31. **(a)** Quantitative. Since the price of tea in China is rounded to the nearest whole unit of currency, it is discrete.

(b) The price of tea in China represents ratio data. There is a natural zero ($0.00 per pound or $0.00 per box). Here, division does make sense. That is, tea that costs $10.00 per pound costs twice as much as tea that costs $5.00 per pound.

33. **(a)** Quantitative. Since the winning score is a whole number, the winning score in next year's Super Bowl is discrete.

(b) The winning score in next year's Super Bowl represents ratio data. There is a natural zero because it is possible for a team to score 0 points. Here division makes sense because a score of 28 points represents twice as many points as a score of 14 points.

35. **(a)** Qualitative.

(b) The rank of the winning Super Bowl team in their division represents ordinal data because the ranks may be arranged in a particular order and no arithmetic may be performed on them.

37. **(a)** Since the possible values of the variable are not numeric but names such as "Bones" or "The Big Bang Theory," the variable is qualitative.

(b) Since there is no natural ordering of the names of television shows the variable represents nominal data.

39. **(a)** Since the possible values of the variable are not numeric, but rather ice cream flavors such as "rocky road" or "strawberry," the variable is qualitative.

(b) Since there is no natural ordering of ice cream flavors, the data represent nominal data.

41. **(a)** Quantitative. Since how old your car is can be any real number greater than or equal to 0, how old your car is continuous.

(b) How old your car is represents ratio data. Since a car that was just purchased is 0 years old, there is a natural zero. A car that is 6 years old is twice as old as a car that is 3 years old, so division makes sense in this case.

43. The 4 teams listed in Table 6 are all of the teams in the intramural league. Therefore, the 4 teams listed in Table 6 represent a population.

45. There are more than 5 universities in the United States. Therefore, the 5 universities in Table 8 represent a sample.

47. Since the data in Table 7 represent a sample, the oldest player is a statistic.

49. Since the data in Table 6 represents a population, descriptive statistics is indicated.

51. The data in Table 8 represent a sample. The result 4 out of 5 (80%) of the universities are proprietary is a statistic. Since this statistic is used to infer that 80% of all universities are proprietary, statistical inference is indicated.

53. The population is all veterans returning from war. The sample is the 20 veterans selected.

55. The population is all older women. The sample is the 10 patients of the physical therapist that she selected.

57. The population is all companies that recently underwent a merger. The sample is the 50 companies that recently underwent a merger that were selected.

59. Statistical inference. A sample of automobile passengers was taken, and the sample proportion of automobile passengers who wear seat belts was calculated. Then this sample proportion was used to make an inference about what percentage of automobile passengers wear seatbelts.

61. Descriptive statistics. The proportion of traffic fatalities in New York that involved alcohol is a descriptive statistic because it describes a sample. But no inference is made regarding a larger population.

63. Statistical inference. A sample of 15- to 18-year-olds was taken, and the sample percentage of 15- to 18-year-olds who use illicit drugs was calculated. Then this percentage was used to make an inference about the percentage of 15- to 18-year-olds who use illicit drugs.

65. **(a)** Elements: Endangered species—pygmy rabbit, Florida panther, Red wolf, and West Indian manatee; variables—*year listed as endangered, estimated number remaining*, and *range*.

(b) Qualitative variables: Since the values of the variable *range* are not numerical, *range* is a qualitative variable. Quantitative variables: Since the values of the variables *year listed as endangered* and *estimated number remaining* are numerical and can be ranked or ordered, the variables *year listed as endangered* and *estimated number remaining* are quantitative variables.

(c) Since there are a finite number of values for the variable *year listed as endangered*, the variable is discrete. Since the values for the variable *estimated number remaining* can be counted, the variable is discrete.

(d) There is no natural zero for the variable *year listed as endangered.* Division of the values of the variable *year listed as endangered* is not possible. However subtraction is possible. For example, a species listed as endangered in 1995 was listed as endangered 7 years before a species listed as endangered in 2002 (2002 – 1995 = 7). Therefore, the variable *year listed as endangered* represents interval data. The variable *estimated number remaining* has a natural zero. Division is possible on the values for the variable *estimated number remaining*. For example, a species with 15 remaining has 2.5 times as many members left as a species with 6 remaining (15/6 = 2.5). Therefore, the variable *estimated number remaining* represents ratio data. There is no natural order for the values of the variable *range*. Therefore, the variable *range* represents nominal data.

(e) 1973, 50, Florida.

67. **(a)** Elements: States—Texas, Missouri, Minnesota, Ohio, and South Dakota; variables—*proportion of GE corn* and *most prevalent type.*

(b) Qualitative variables: Since the values of the variable *most prevalent type* are not numerical, the variable *most prevalent type* is qualitative. Quantitative variables: Since the values of the variable *proportion of GE corn* are numerical and can be ranked or ordered, the variable *proportion of GE corn* is quantitative.

(c) Since the variable *proportion of GE corn* can take on any value between 0 and 1 inclusive, the variable *proportion of GE corn* is continuous.

(d) There is a natural zero for the variable *proportion of GE corn.* Division is also possible for the values of the variable *proportion of GE corn.* For example, a state with 16% GE corn has 2 times the proportion of GE corn as a state with 8% GE corn (16/8 = 2). Therefore, the *variable proportion of GE corn* represents ratio data. Since there is no natural order for the values of the variable *most prevalent type*, the variable *most prevalent type* represents nominal data.

(e) 89%, herbicide-tolerant.

69. **(a)** Elements: Hospitals—Briarcliff Manor, Buchanan, Cortlandt, Croton-on-Hudson, Mount Pleasant, Ossining 1, Ossining 2, Peekskill, Pleasantville, and Sleepy Hollow; variables—*births* and *average maternal age.*

(b) Qualitative variables: There are no qualitative variables. Quantitative variables: The values of the variables *births* and *average maternal age* are numbers that can be ranked or ordered, so both of these variables are quantitative.

(c) The values of the variable *births* represent values that were counted so the variable *births* is discrete. The values of the variable *average maternal age* are calculated from data that were measured and can be any real number between the youngest mother and the oldest mother, so the variable *average maternal age* is continuous.

(d) There is a natural zero for both the *births* and *average maternal age* variables. Division is possible on the values of both the *births* and *average maternal age* variables. For example, a hospital with 80 births had 3.2 times as many births as a hospital that had 25 births (80/25 = 3.2), and a hospital with an average maternal age of 33 has an average maternal age of 1.1 times the average maternal age of a hospital with an average maternal age of 30 (33/30 = 1.1). Therefore, the variables *births* and *average maternal age* represent ratio data.

(e) 134, 29.2.

71. **(a)** Elements are the tornado names: Tri-State, Natchez, St. Louis, Tupelo, Gainesville. Variables: deaths, year

(b) Qualitative variables: There are no qualitative variables. Quantitative variables: Since the values of the variables *deaths* and *year* are numbers that can be ranked or ordered, the variables *deaths* and *year* are quantitative.

(c) Since the values of the variable *deaths* are values that are counted, the variable *deaths* is discrete. Since the values of the variable *year* are whole numbers with no numbers in between, the variable *year* is discrete.

(d) Since it is possible to have 0 tornado deaths in a year, there is a natural 0 for the variable *deaths*. Division is possible on the values of the variable *deaths*. For example, a year with 35 tornado deaths has 7 times as many tornado deaths as a year with 5 tornado deaths (35/5 = 7). Therefore the variable *deaths* represents ratio data. There is an order for the values of the variable *year.* Subtraction can be performed on the values of the variable *year.* However there is no natural 0 for the variable *year* and division cannot be performed on the values of the variable *year.* Therefore, the variable *year* represents interval data.

(e) 255, 1896

73. **(a)** Sample.

(b) This sample could not be considered a random sample of the annual number of tornado deaths of all years. The 5 years selected were not selected randomly, but were selected according to which 5 years had the most tornado deaths.

75. **(a)** This is a statistic because it came from a sample.

(b) An estimate of the average lifetime of all new light bulbs is the average lifetime of the sample of 100 light bulbs, which is 2000 hours. Thus the company can claim that "The average lifetime of this new model of light bulb is 2000 hours."

77. **(a)** Sample.

(b) No, they are the five university campuses with the largest enrollment.

(c) Arizona State University is located in Arizona and in 2014 it had 72,254 students making it the university campus with the second-highest enrollment.

79. The quantitative variables are *sales for week, sales total,* and *weeks on list.*

81. The list in Table 3 represents a sample. Only the 30 best-selling video games are included.

83. Since the values for the variables *platform, studio,* and *type* are qualitative and there is no natural ordering for the values of these variables, they are nominal.

85. The variables *sales for week, sales total,* and *weeks on list* have values that can be divided and have natural zeros Therefore, the variables *sales for week, sales total,* and *weeks on list* are ratio.

87. Descriptive statistics. No attempt was made to use the fact that the Xbox 360 version of *Grand Theft Auto V* outsold the PS3 version of the game during the week of May 17, 2014 to predict that the Xbox 360 version of *Grand Theft Auto V* will outsell the PS3 version of the game during any week after the week of May 17, 2014.

Section 1.3

1. Convenience sampling usually includes only a select group of people. For example, surveying people at a mall on a workday during working hours would probably include few if any people who work full time.

3. The *Literary Digest* could have decreased the bias in their poll by choosing a random sample of houses and apartments and surveying the people door to door. They would have been more likely to include people who were poor or underprivileged by using this method and thus their sample would have been more representative of the population.

5. A random sample is a sample for which every element has an equal chance of being included.

7. Answers will vary.

9. Answers will vary.

11. Illinois, Iowa, Michigan State, Nebraska, Ohio State, Purdue

13. Alabama, Georgia, Mississippi State, Texas A&M

15. Answers will vary.

17. Answers will vary.

19. Answers will vary.

21. No. The sample would likely not be a representative sample of the Southeastern Conference or of all college football teams. This sample will likely not contain at least one of the best teams, at least one of the worst teams, and at least one team in the middle of either the Southeastern Conference or college football.

23. This is cluster sampling because **(a)** the population was divided into clusters (class ranks), **(b)** a random sample of the clusters (class ranks) was taken, and **(c)** all of the students in that class rank (cluster) were selected.

25. This is convenience sampling, since you are choosing a sample that is convenient to you.

27. Target population: all college students; potential population: all students working out at the gymnasium on the Monday night Brandon was there.

29. Target population: all small businesses; potential population: small businesses near the state university.

31. What is meant by “sometimes”? This is vague terminology.

33. This question would only be understood by someone who knows about graduated income taxes, and is neither simple nor clear.

35. **(a)** Observational.

(b) Response variable: how often they attend religious services; predictor variable: whether or not the family is large (at least four children).

37. **(a)** Experimental.

(b) Response variable: performance of the electronics equipment; predictor variable: whether or not a piece of equipment has a new computer processor.

39. Level of insect damage to crops.

41. The new pesticide.

43. LDL cholesterol level in the bloodstream.

45. New medication.

47. **(a)** Randomization is present for the 100 randomly assigned subjects but not for the subjects with high LDL cholesterol levels.

(b) The sample of 100 people is probably enough replication.

49. Experiment

51. **(a)** Answers will vary

(b) No. Every possible sample of 5 video games has the same chance of being selected.

(c) No. Every possible sample of 5 video games has the same chance of being selected. Some of the samples will contain the video game and some won’t.

(d) Answers will vary, answers will vary

53. Answers will vary

55. The poll by Ann Landers was extremely biased. Only people who read Ann Landers’s column and felt strongly about the poll responded to this poll. Further, there was no mechanism to guard against people responding more than once or to keep people who don’t have children from responding. The *Newsday* poll was done professionally; therefore, the sample used was more likely to be representative of the population.

57. The target population is all people living in Chicago, and the potential population is people who have phones and who have their phone number listed in the Chicago phone directory. The potential for selection bias is that many of the people living below the poverty level in Chicago may not have phones. Also, many people may have unlisted numbers. Further, the poverty level is determined by family size as well as income, and this survey does not take that into consideration.

59. The survey question is a leading question.

61. **(a)** No, we do not know what the lowest price in the sample will be before we select the sample. Since the sample is randomly selected, we don't know which stocks will be selected before we select our sample. Different samples may contain different lowest stock prices.

(b) Answers will vary.

(c) No, if we take another sample of size 2, it is not likely to comprise the same two companies. Since the samples are randomly selected, they will probably contain different companies.

(d) Answers will vary.

63. A quantity like "the lowest price in a random sample of stocks" is a variable that may vary from sample to sample.

65. **(a)** Forcing the parents of a treatment group to smoke tobacco would increase the occurrence of respiratory illnesses in their children, which is not very ethical. **(b)** Observational study.

67. **(a)** The control is the placebo bracelet.

(b) The subjects were randomly assigned to wear either the placebo bracelet or the ionized bracelet.

(c) There is replication of data since there are 305 subjects in both the treatment and the control group.

69. This study is an experimental study because the subjects were randomly assigned to either a treatment or a control.

Chapter 1 Review Exercises

1. **(a)** Make/Models: Chevrolet, Corvette, Ferrari 458 Italia, Honda CR-Z, Jaguar F Convertible, Porsche Boxster S

(b) *Cylinders, transmission, combined mileage*

3. The observation for the Chevrolet Corvette is it has 8 cylinders, it has a manual transmission, and its combined city/highway gas mileage is 21 mpg.

5. **(a)** The only way to find out the population average lifetime of all one million light bulbs in the inventory is to turn on all one million light bulbs and leave them all on until they burn out, measuring the time it takes for each light bulb to burn out. All of these lifetimes can then be used to calculate the population average lifetime of all one million light bulbs.

(b) This would require burning out all one million light bulbs that are in stock so that there would be no good light bulbs left to sell. It would be better to take a random sample of the light bulbs, find the average lifetime of the sample, and use the sample average lifetime of the light bulbs to estimate the population average lifetime of the light bulbs.

7. **(a)** You would use an observational study.

(b) Since people are already enrolled in their statistics classes, it would be impractical to randomly reassign people to a statistics class after classes have started.

9. We would use an observational study. It would be impossible to randomly assign a child to come from a single-parent family or a two-parent family.

Chapter 1 Quiz

1. False. Statistical inference consists of methods for estimating and drawing conclusions about *population* characteristics based on the information contained in a *sample*.

3. collecting

5. sample

7. An experimental study is involved.

9. **(a)** The population is all statistics students.

(b) The sample is the random sample of students selected from the statistics class.

(c) The variable is whether the student is left-handed. It is a categorical variable.

(d) The sample proportion is not likely to be exactly the same as the population proportion. But it is not likely to be very far away from the population proportion because which statistics class a person enrolls in is not based on whether the person is left-handed.

Chapter 2: Describing Data Using Graphs and Tables

Section 2.1

1. We use graphical and tabular form to summarize data in order to organize it in a format where we can better assess the information. If we just report the raw data, it may be extremely difficult to extract the information contained in the data.

3. True.

5. The sample size, n.

7. The row totals, the column totals.

9. True

11.

Vote	Frequency
Republican	15
Democrat	2

13.

Size	Frequency
Small	9
Medium	6
Large	2

15.

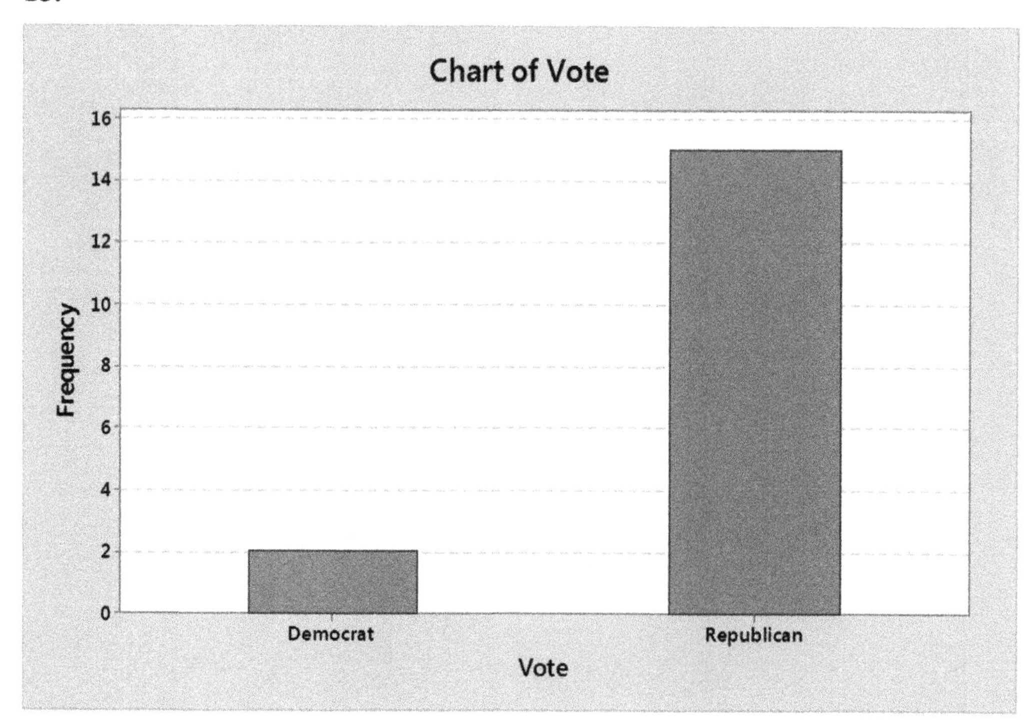

17.

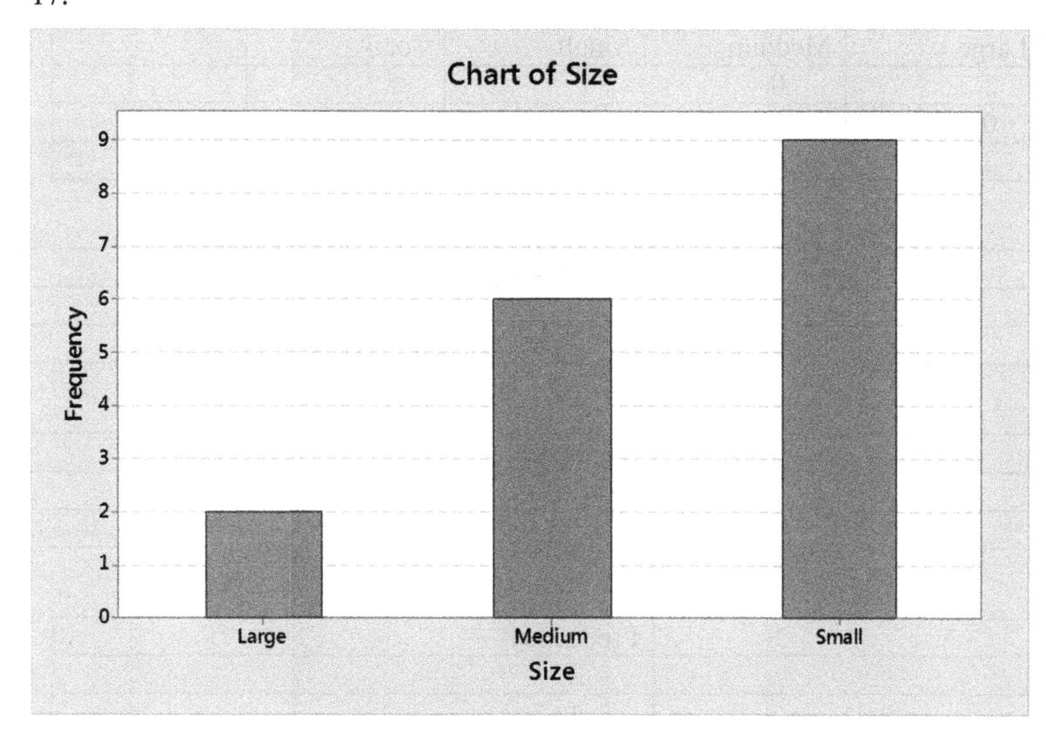
Chart of Size
Frequency
0
1
2
3
4
5
6
7
8
9
Large
Medium
Small
Size

19.

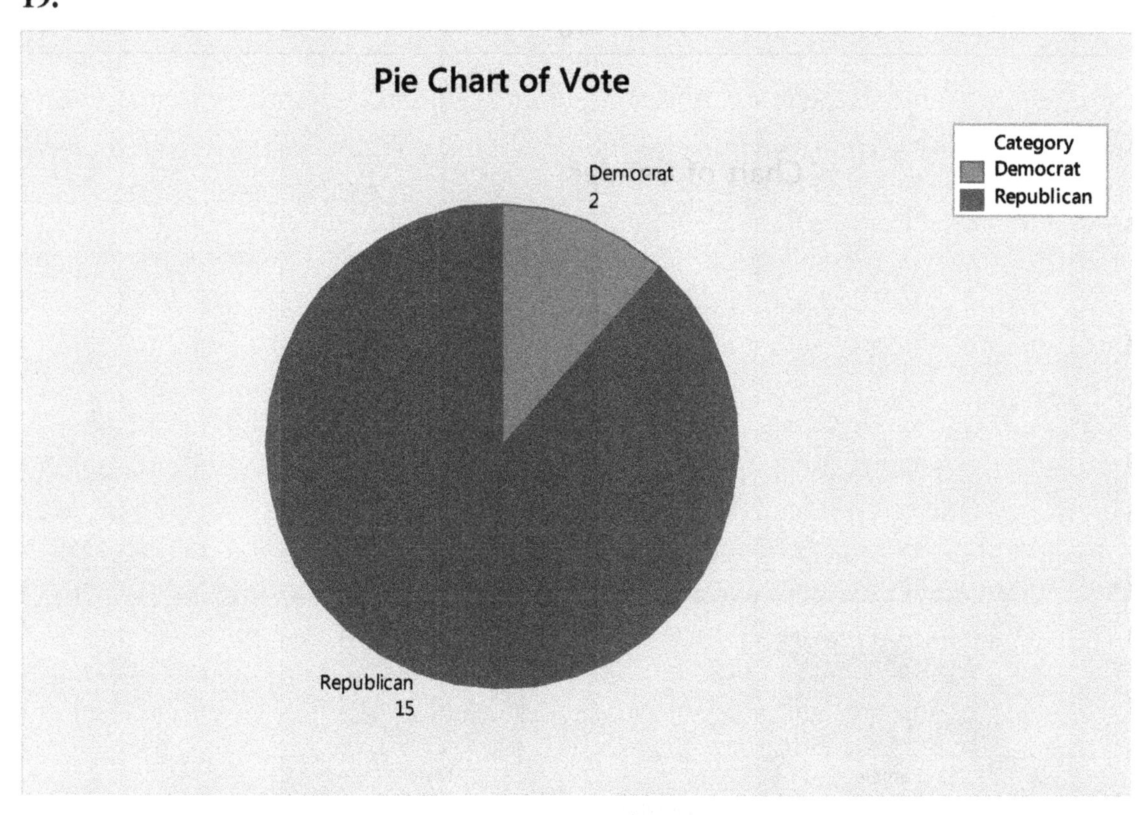
Pie Chart of Vote
Category
Democrat
Republican
Democrat
2
Republican
15

21.

	Large	Medium	Small	Total	
Democrat	2	0	0	2	
Republican	0	6	9	15	
Total	2	6	9	17	

23.

Gender	Frequency
Male	16
Female	4

25.

Survived	Frequency
Yes	9
No	11

27.

Class	Frequency
1st class	2
2nd class	4
3rd class	5
Crew	9

29.

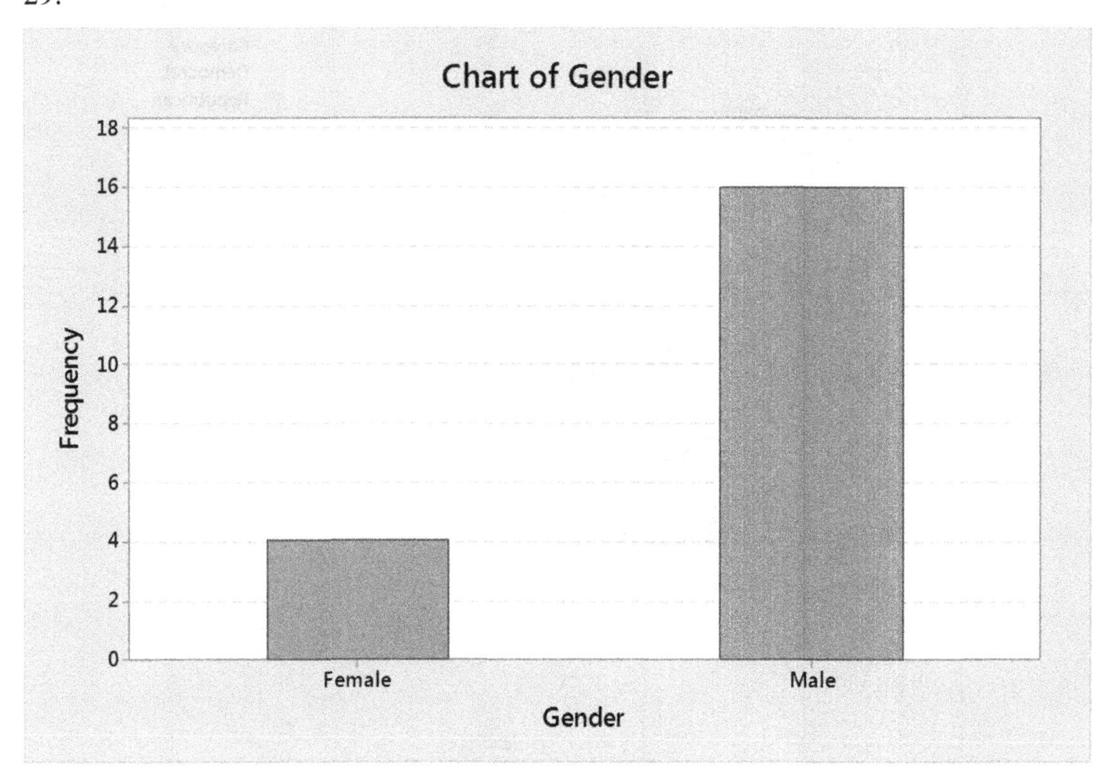

31.

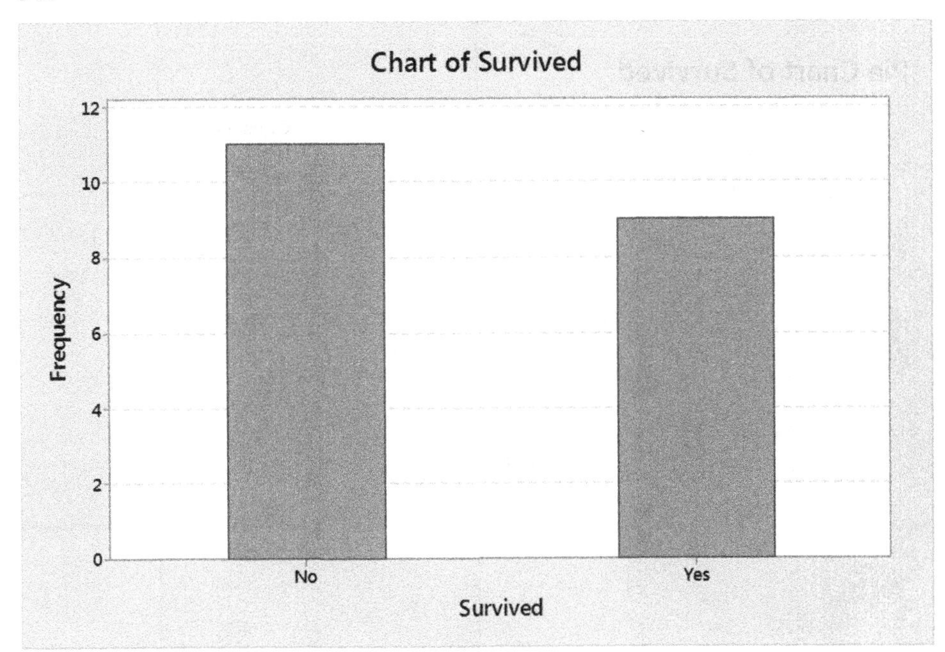
Chart of Survived
Frequency
12
10
8
6
4
2
0
No
Yes
Survived

33.

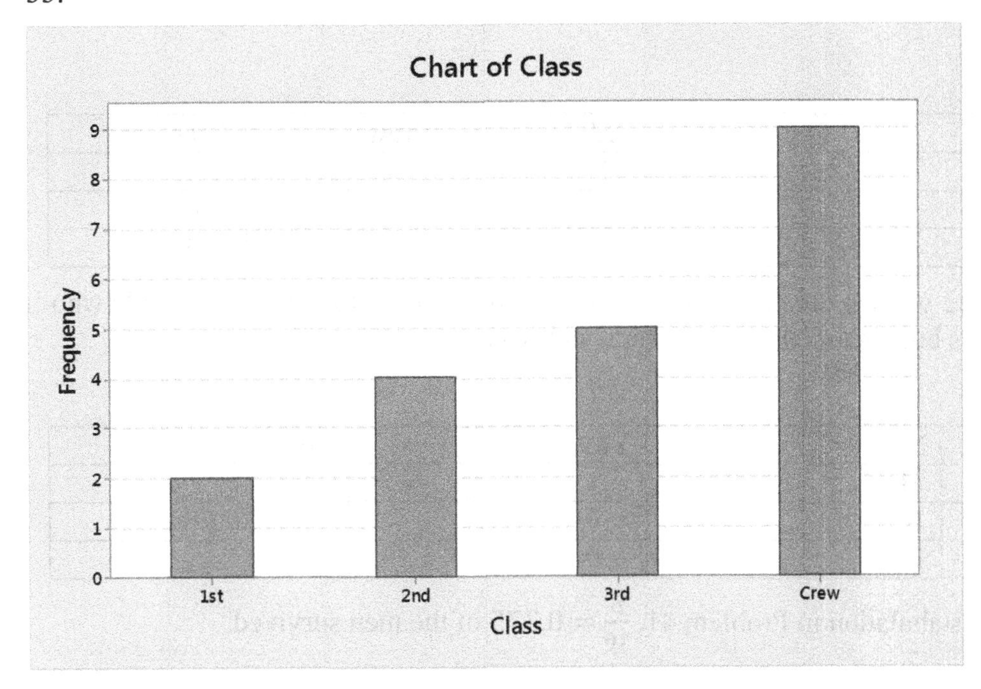
Chart of Class
Frequency
9
8
7
6
5
4
3
2
1
0
1st
2nd
3rd
Crew
Class

35.

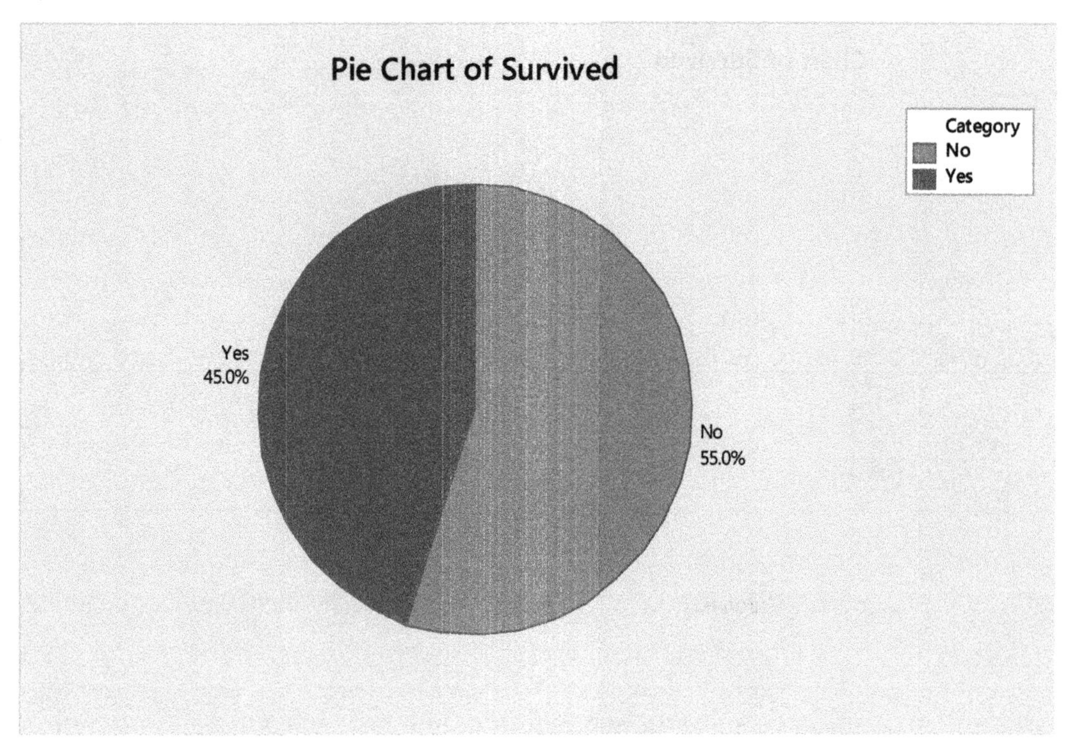

37.

	1st	2nd	3rd	Crew	Total
Female	1	1	2	0	4
Male	1	3	3	9	16
Total	2	4	5	9	20

39. Crew. In 1912 women did not work as crew members on a ship. Only the people who could afford first class bought tickets for both men and women.

41.

	No	Yes	Total
Female	1	3	4
Male	10	6	16
Total	11	9	20

43. From the crosstabulation in Problem 41, $\frac{6}{16} = 0.375$ of the men survived.

45.

	No	Yes	Total
1st	0	2	2
2nd	2	2	4
3rd	2	3	5
Crew	7	2	9
Total	11	9	20

47. From the crosstabulation is Problem 45, $\frac{2}{4} = 0.5$ of the second class passengers survived, $\frac{3}{5} = 0.6$ of the third class passenger survived, and $\frac{2}{9} \approx 0.22$ of the crew survived.

49.

Gender	Frequency
Female	20
Male	21

51.

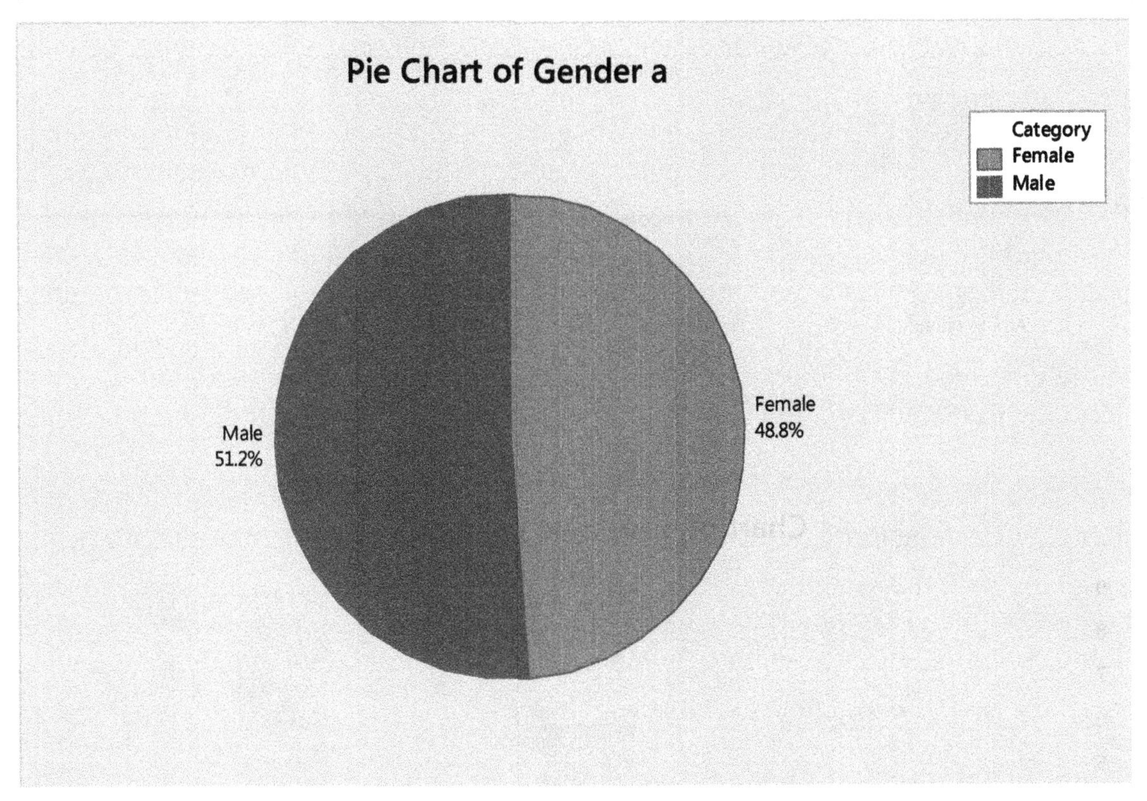

53.

Sport	Frequency
Alpine skiing	21
Figure skating	9
Ice hockey	44

55.

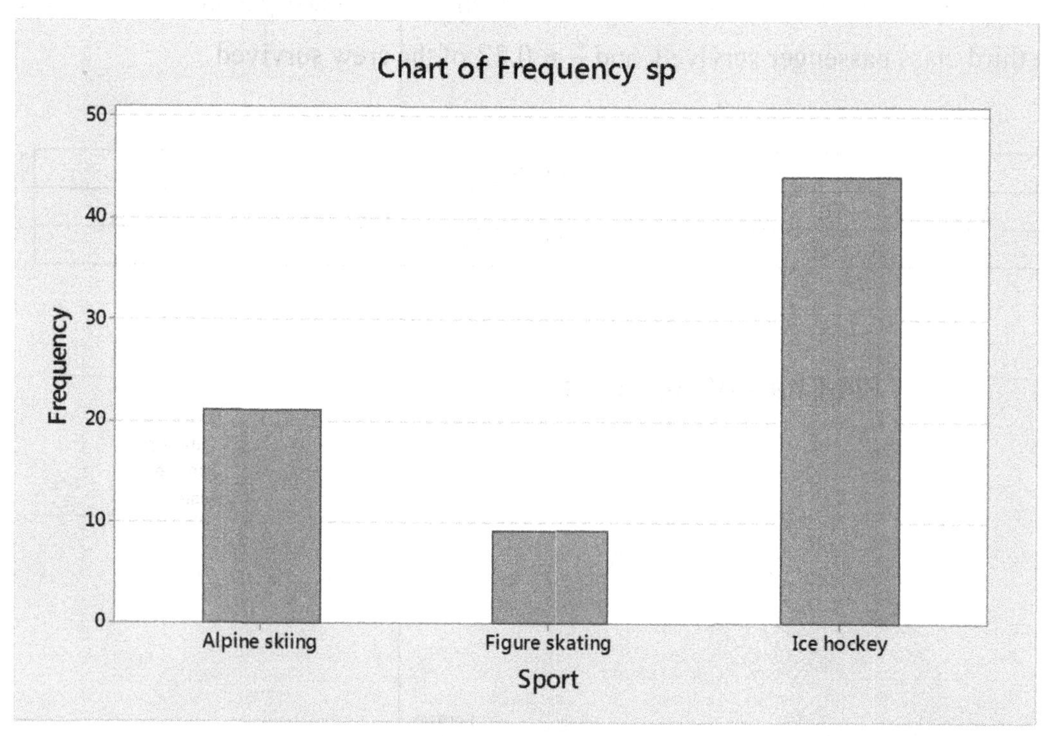
Chart of Frequency sp
Frequency
50
40
30
20
10
0
Alpine skiing
Figure skating
Ice hockey
Sport

57.

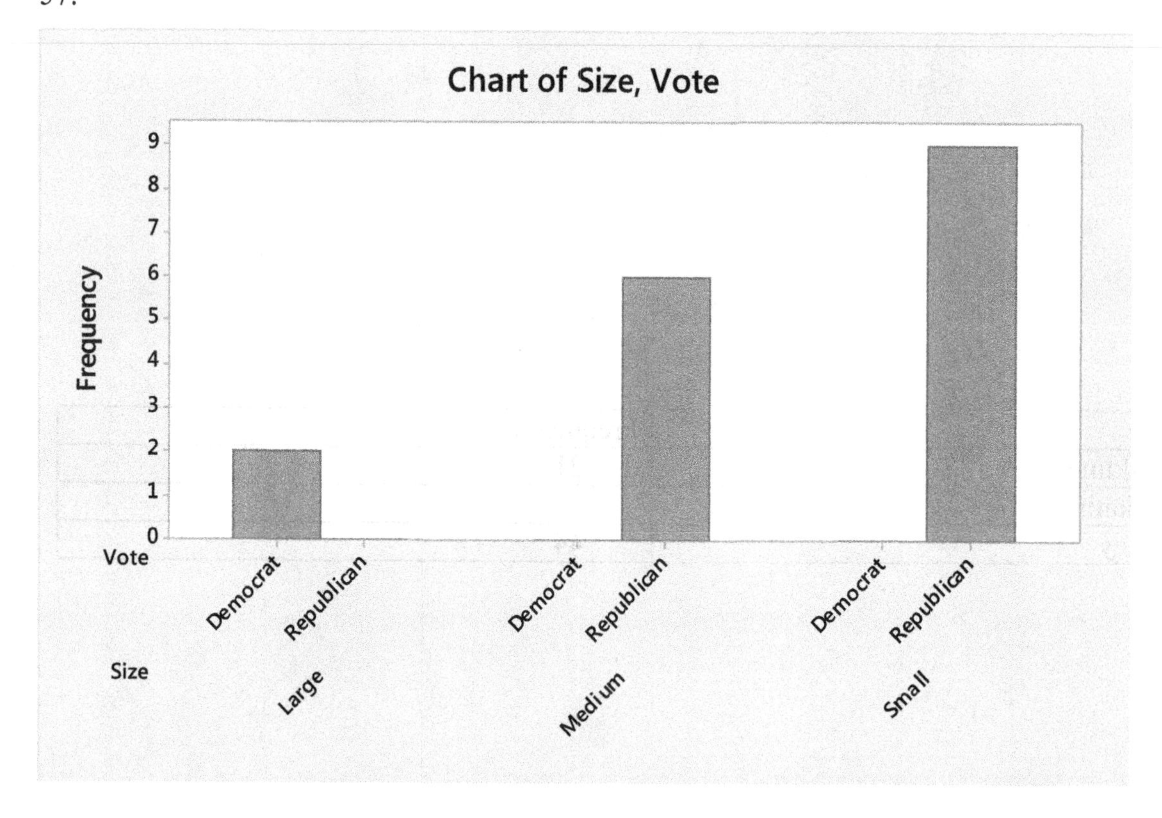
Chart of Size, Vote
Frequency
9
8
7
6
5
4
3
2
1
0
Vote
Democrat
Republican
Democrat
Republican
Democrat
Republican
Size
Large
Medium
Small

59.

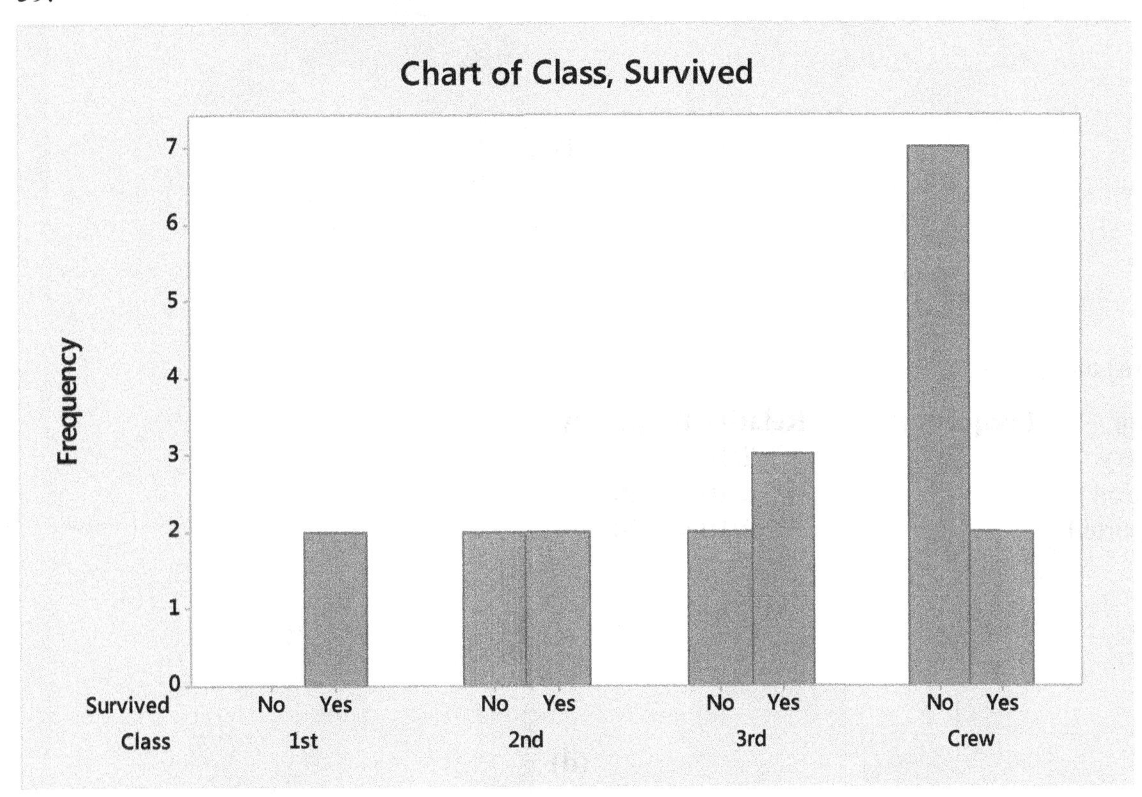

61. **(a)** and **(b)**

Continent	Frequency	Relative frequency
Africa	1	1/10 = 0.10
Asia	5	5/10 = 0.50
Europe	1	1/10 = 0.10
North America	2	2/10 = 0.20
South America	1	1/10 = 0.10

(c)

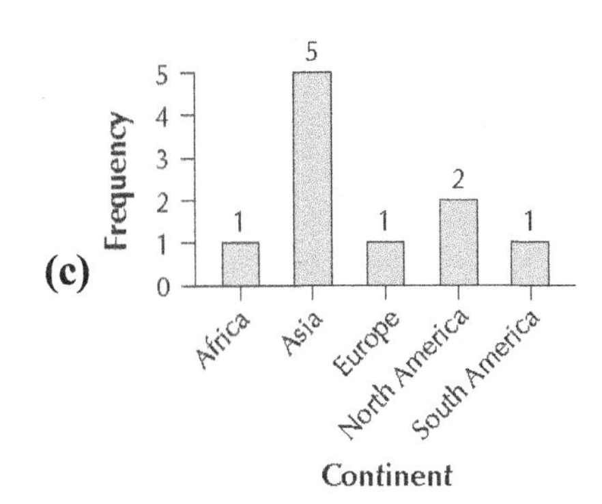

(d)

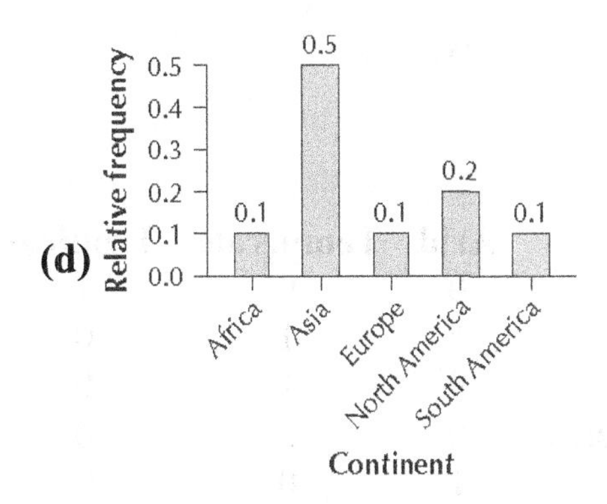

(e)

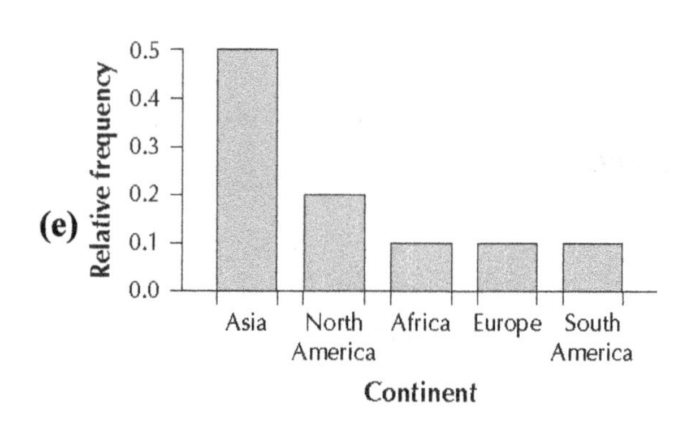

(f)

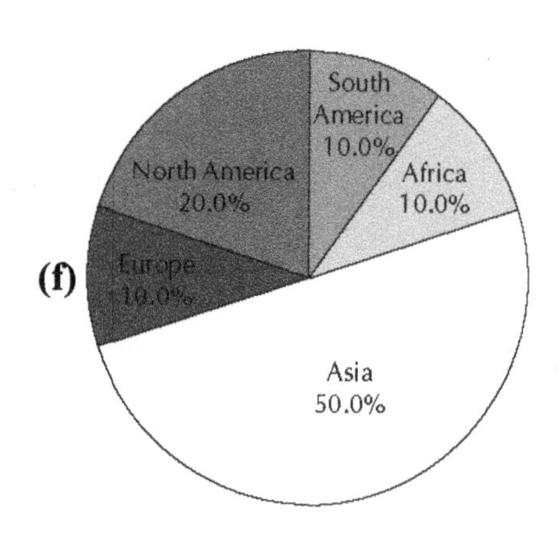

63. **(a)** and **(b)**

Main use	**Frequency**	**Relative frequency**
Industry	2	2/10 = 0.20
Irrigation	6	6/10 = 0.60
Not Reported	2	2/10 = 0.20

(c)

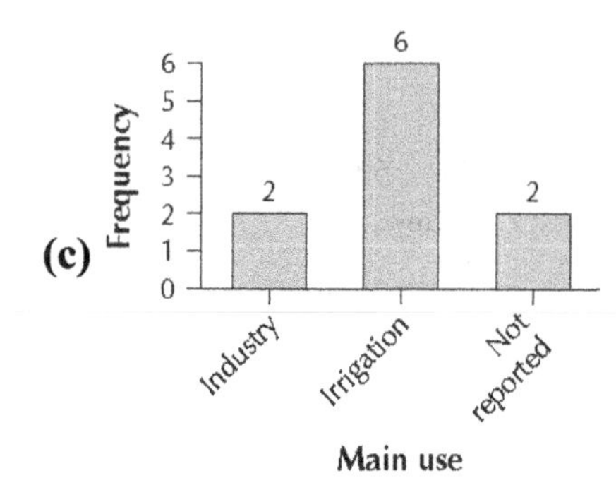

(d)

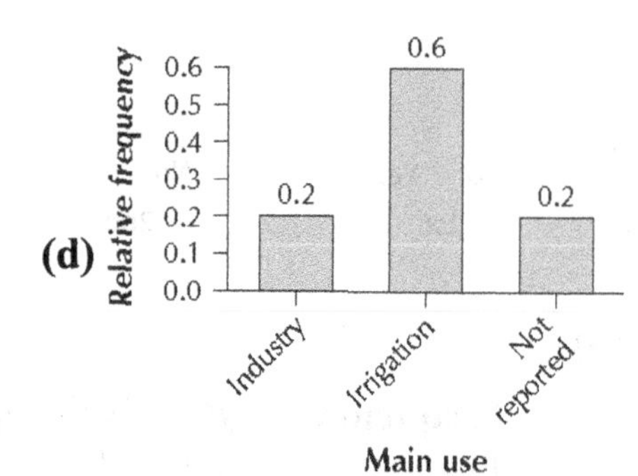

(e)

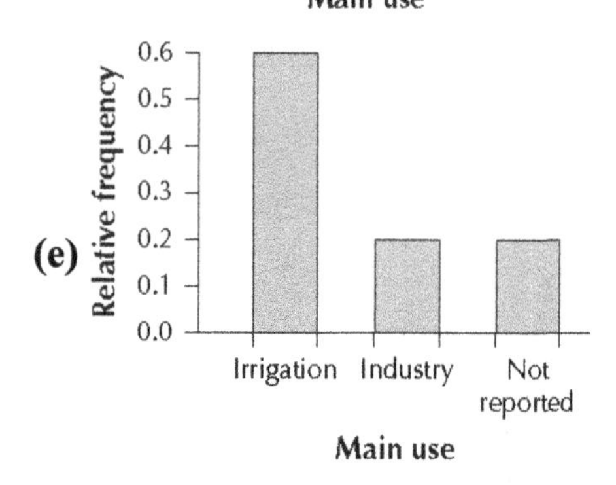

(f)

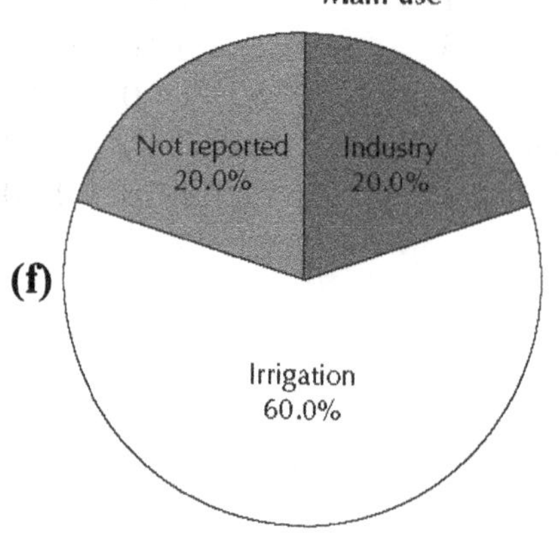

65.

	Arid	**Temperate**	**Tropical**	**Total**
Africa	0	0	1	1
Asia	4	1	0	5
Europe	0	1	0	1
North America	0	2	0	2
South America	1	0	0	1
Total	**5**	**4**	**1**	**10**

67.

	Industry	Irrigation	Not reported	Total
Arid	0	5	0	5
Temperate	2	0	2	4
Tropical	0	1	0	1
Total	**2**	**6**	**2**	**10**

69.

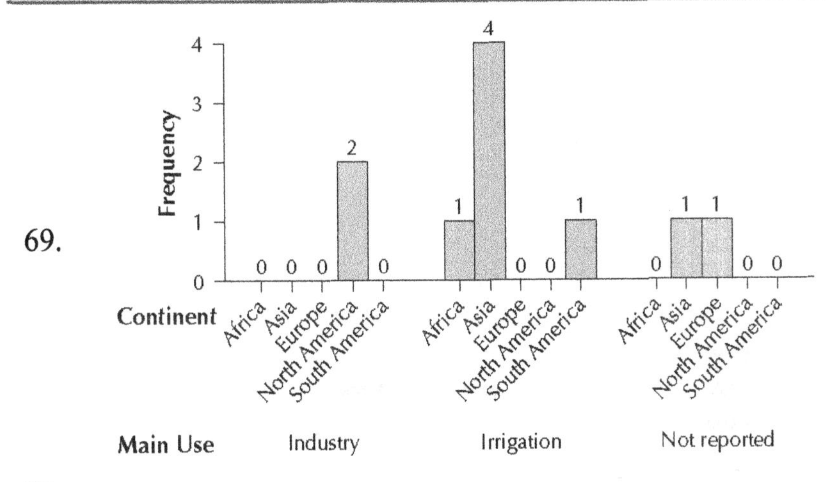

71.

Type	Count	Percent
Action	9	30.00
Adventure	2	6.67
Platform	3	10.00
Racing	2	6.67
Role-Playing	1	3.33
Shooter	10	33.33
Sports	3	10.00
N=	30	

The most common game type is shooter. The least common game type is role-playing.

73.

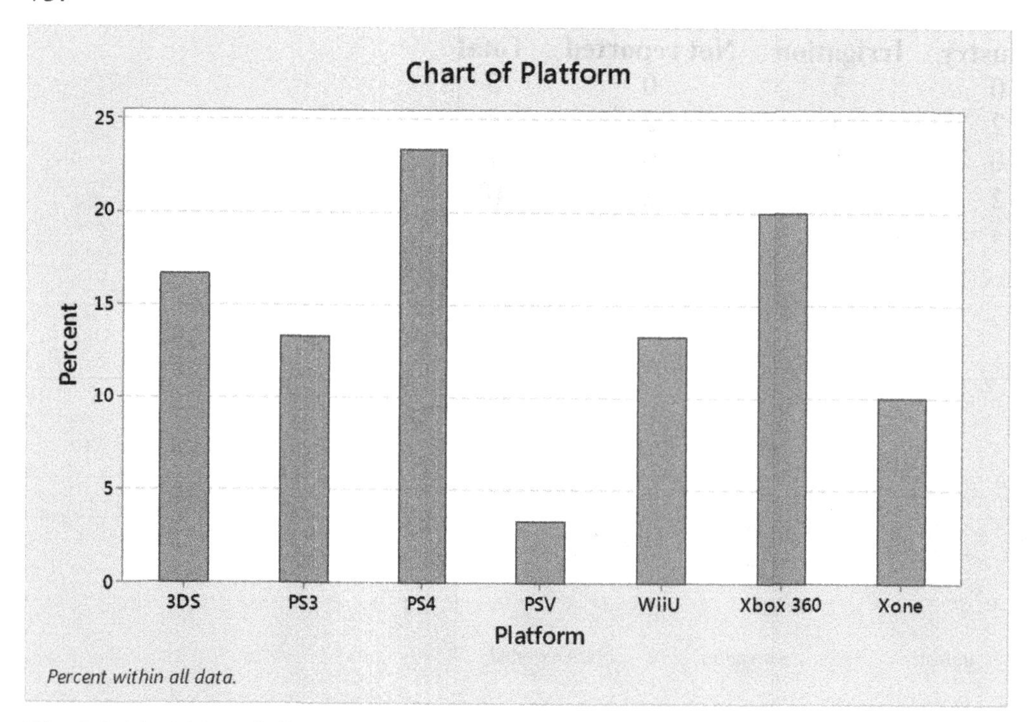

The PS4 is 23% of all platforms.

75.

	3DS	PS3	PS4	PSV	WiiU	Xbox 360	Xone	All
Action	2	1	3	0	2	1	0	9
Adventure	0	1	0	0	0	1	0	2
Platform	1	0	0	0	2	0	0	3
Racing	1	0	0	0	0	0	1	2
Role-Playing	1	0	0	0	0	0	0	1
Shooter	0	2	2	1	0	3	2	10
Sports	0	0	2	0	0	1	0	3
All	5	4	7	1	4	6	3	30

77. No. There are actually two categorical variables—level of education and whether or not the person owns a cell phone. The percents are percents of each category of level of education who own cell phones and not the percent of the whole group who own cell phones.

79. **(a)** Several times a day; 43.4% **(b)** Every few weeks; 5.1%

81. **(a)** Fractures; 26% **(b)** Traumatic brain injury; 9% (c) Yes. It would have to be one of the injuries included in the category "Other injuries."

83. (a) Relative frequency distribution of *vehicle type*

Vehicle type	Relative frequency
SUVs	0.3130
Compact cars	0.1083
Midsize cars	0.1015
Subcompact cars	0.0931
Standard pickup trucks	0.0897
Large cars	0.0643
Station wagons	0.0525
Small pickup trucks	0.0499
Other types	0.1277
Total	1.0000

(b)

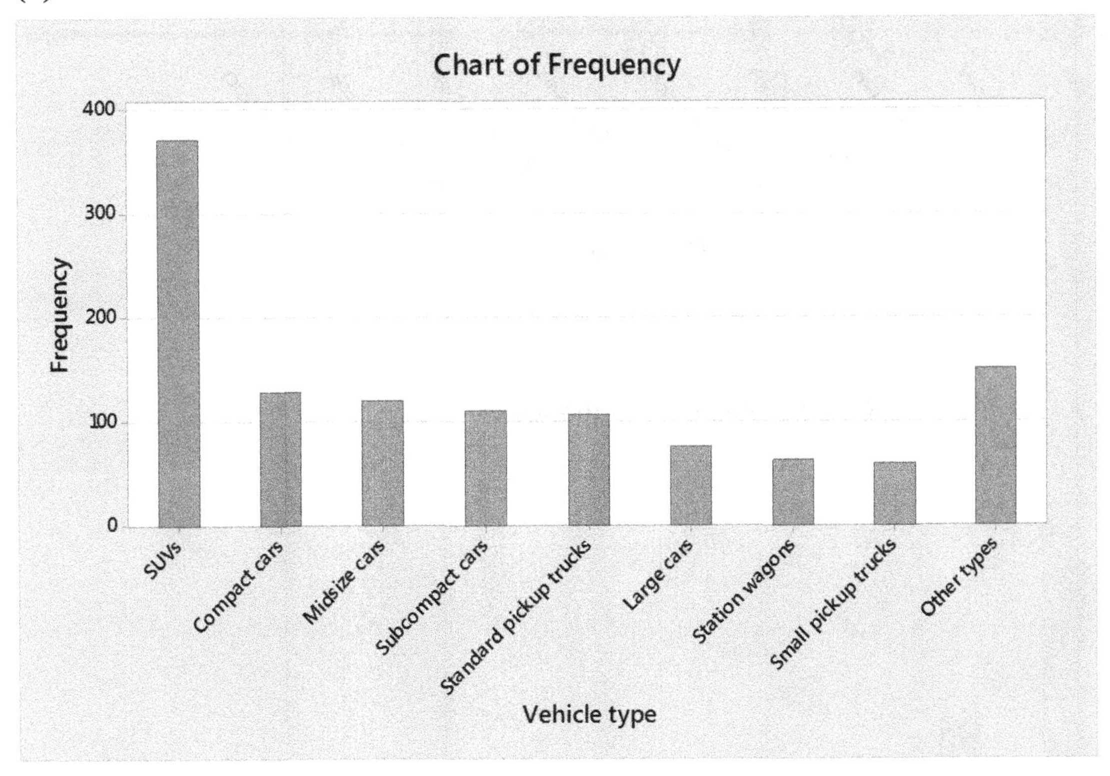

(c)

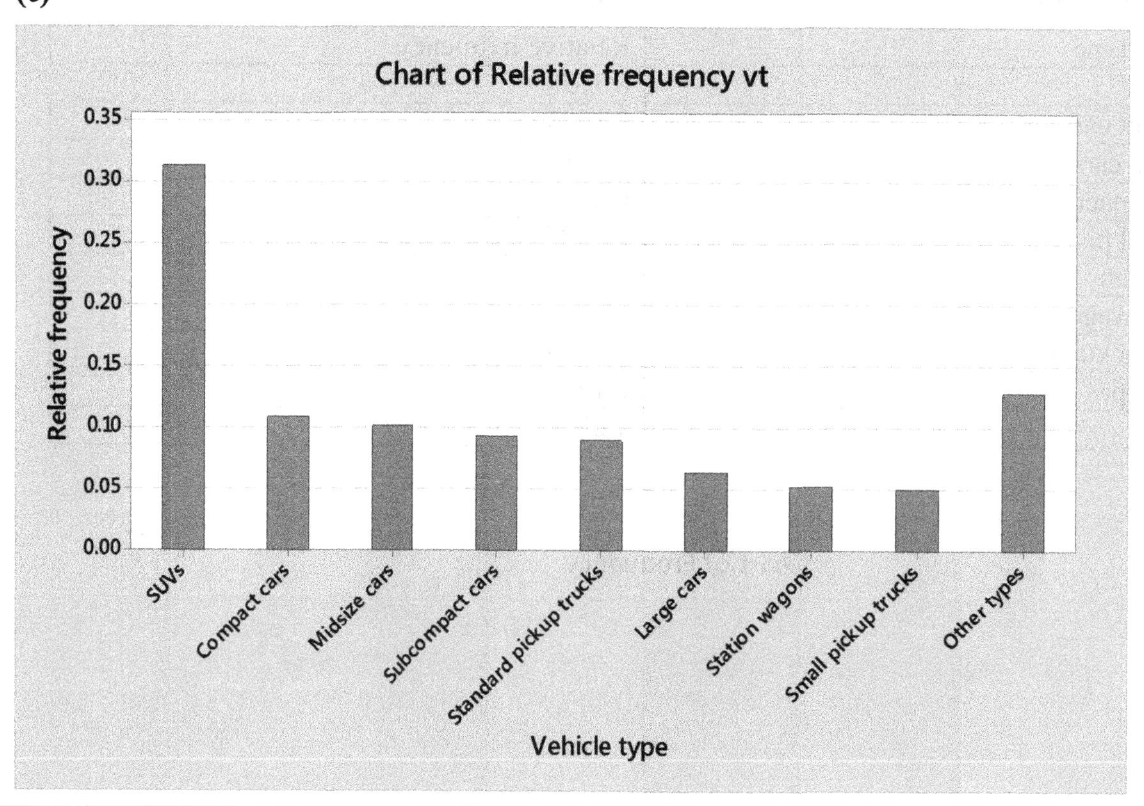
Chart of Relative frequency vt
Relative frequency
0.35
0.30
0.25
0.20
0.15
0.10
0.05
0.00
SUVs
Compact cars
Midsize cars
Subcompact cars
Standard pickup trucks
Large cars
Station wagons
Small pickup trucks
Other types
Vehicle type

(d)

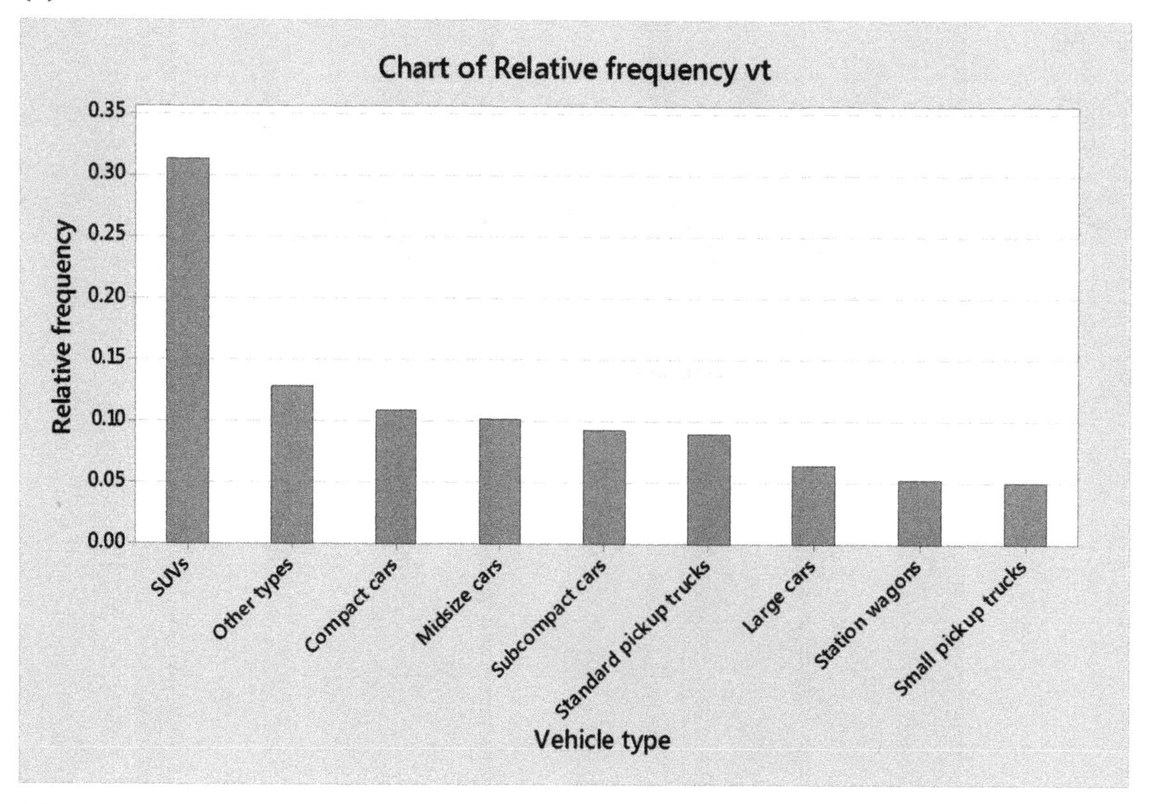
Chart of Relative frequency vt
Relative frequency
0.35
0.30
0.25
0.20
0.15
0.10
0.05
0.00
SUVs
Other types
Compact cars
Midsize cars
Subcompact cars
Standard pickup trucks
Large cars
Station wagons
Small pickup trucks
Vehicle type

(e)

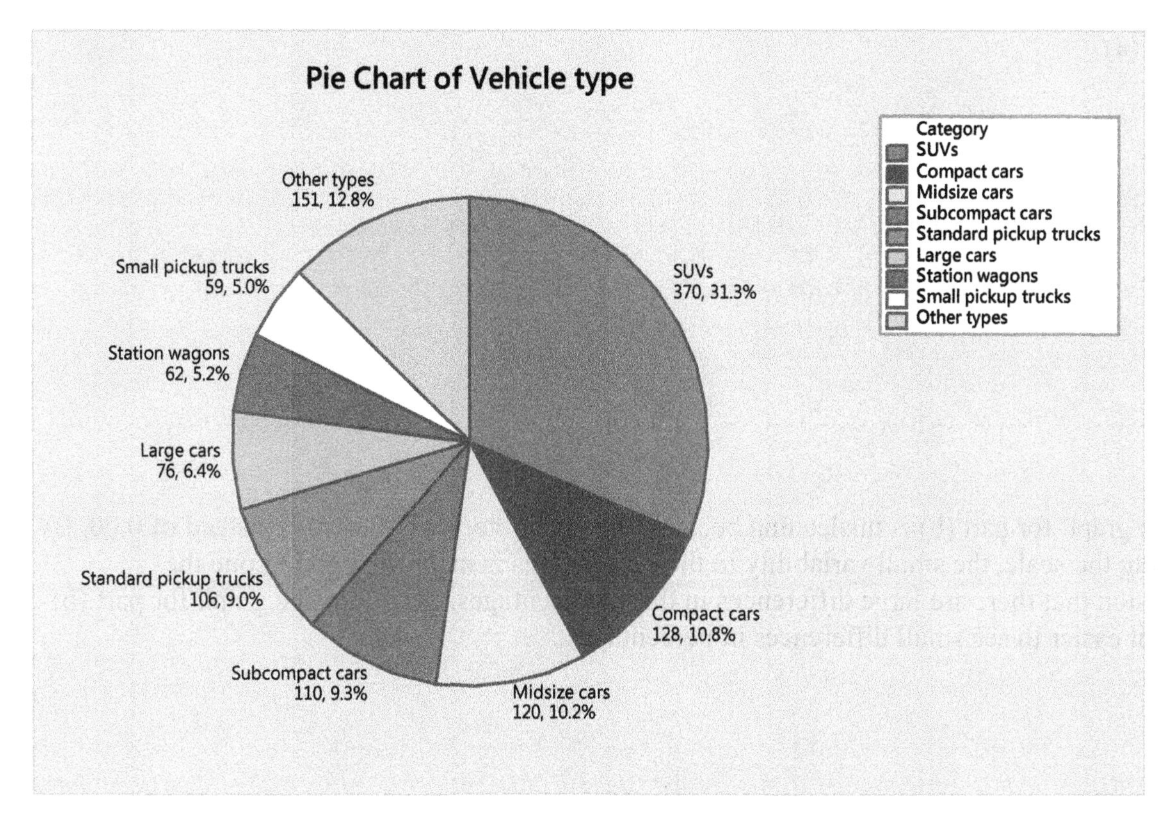

85. A pie chart cannot be constructed for the data since the percentages for the different classes do not add up to 100%. Respondents could select more than one category.

87.

Response	Relative frequency
Yes	$1860/6000 = 0.31$
No	$2700/6000 = 0.45$
I've never thought about it	$1440/6000 = 0.24$

89. NOTE: Relative frequencies are expressed as percentages.

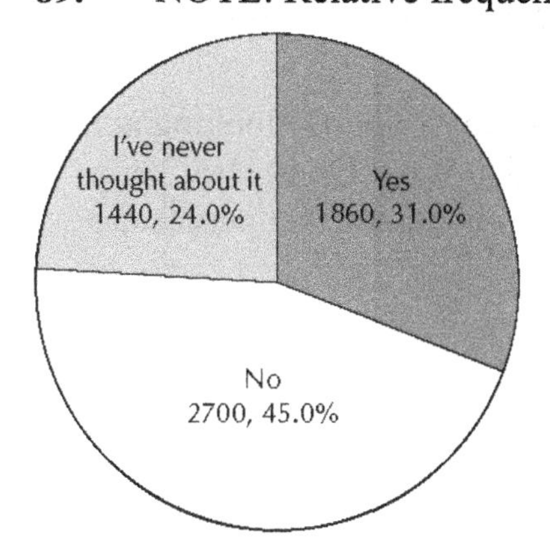

91. **(a)**

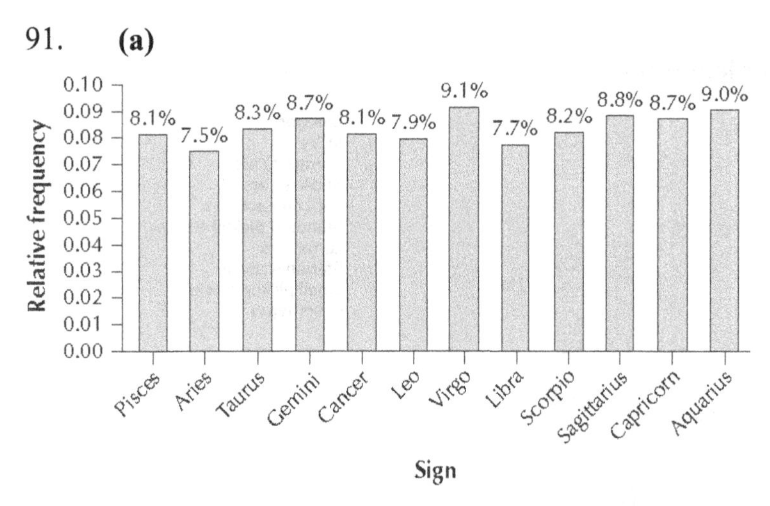

(b) The graph for part (b) is misleading because the y axis starts at 7% (0.07) instead of 0.00. By adjusting the scale, the small variability in the percentages is magnified giving one the impression that there are large differences in these percentages. However, the graph for part (b) makes it easier to see small differences in percentages.

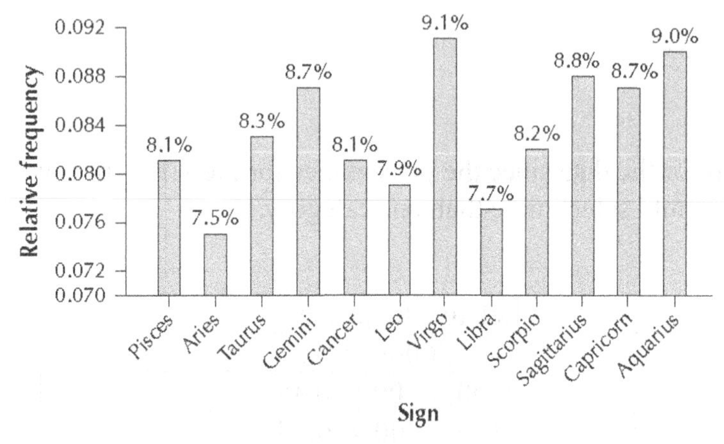

93. **(a)** What is your opinion of the categories? These categories are not quantitative but qualitative. As such, the interpretation of each category could vary from respondent to respondent depending on their own life experiences.

(b) Do you think that everyone interprets these phrases in the same way? No. By looking at the pareto graph one can observe that most of the respondents indicated a "Very Great Deal" or a "Great Deal" of satisfaction with family life. Based on this, one may conclude that people are generally happy with their family life.

95. No.

97.

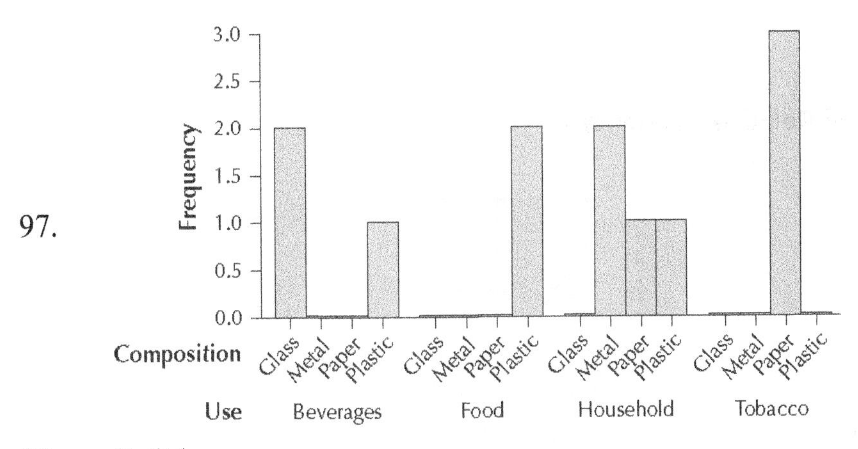

99. Petit larceny

101. Yes. The categories are arranged in order from highest frequency to lowest frequency.

103. **(a)**

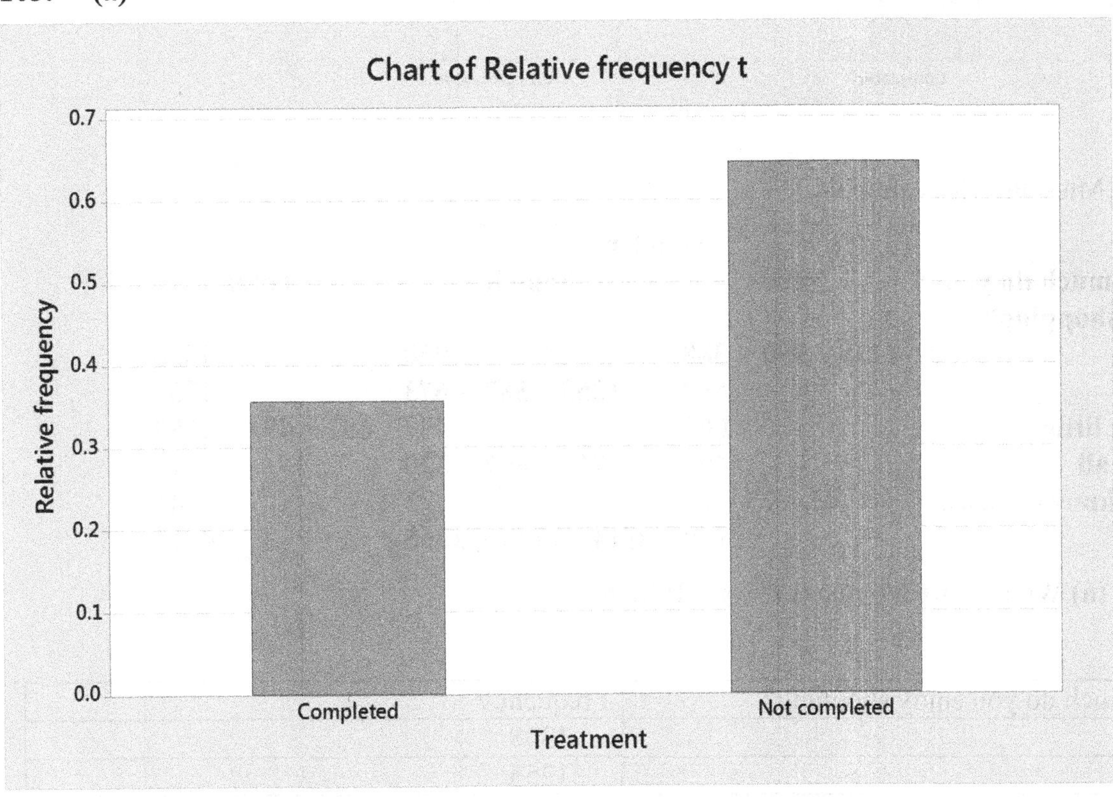

(b)

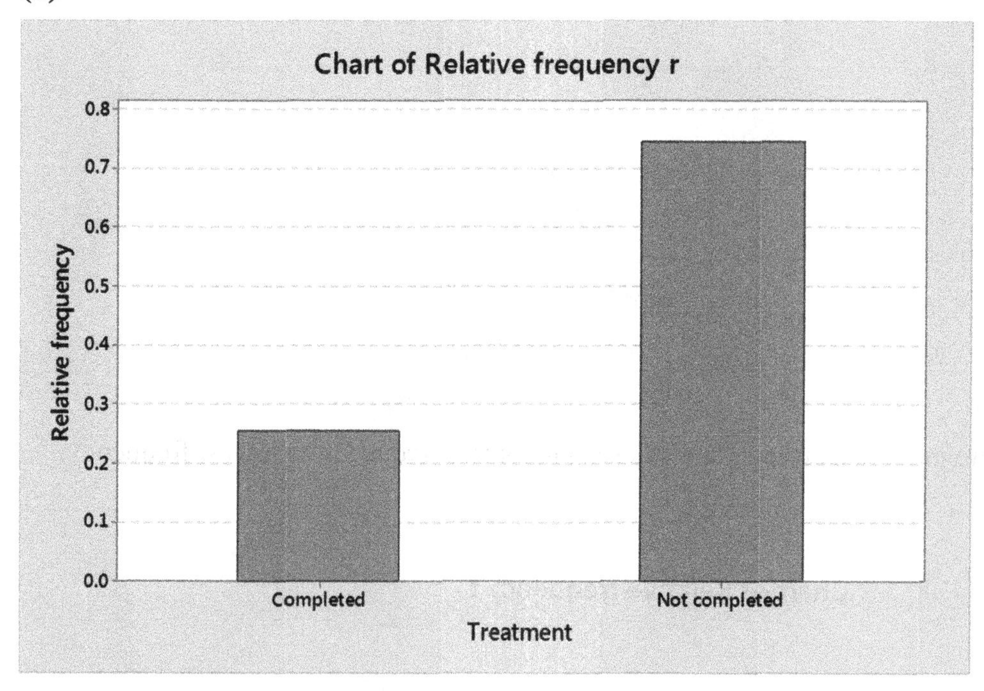

105. Missing values are bold.

	Gender		
"How much do you enjoy shopping?"	**Male**	**Female**	**Total**
A lot	**1338 – 950 = 388**	950	1338
Some	582	**1255 – 582 = 673**	1255
Only a little	662	497	**662 + 497 = 1159**
Not at all	497	**717 – 497 = 220**	717
Don't know/refused	**45 – 25 = 20**	25	45
Total	2149	**4514 – 2149 = 2365**	4514

107. **(a)** Women **(b)** Women **(c)** Men **(d)** Men

109.

How much do you enjoy shopping?	Frequency
A lot	1338
Some	1255
Only a little	1159
Not at all	717
Don't know	45

111.

How much do you enjoy shopping?	Relative frequency
A lot	$1338/4514 \approx 0.2964$
Some	$1255/4514 \approx 0.2780$
Only a little	$1159/4514 \approx 0.2568$
Not at all	$717/4514 \approx 0.1588$
Don't know	$45/4514 \approx 0.0100$

113.

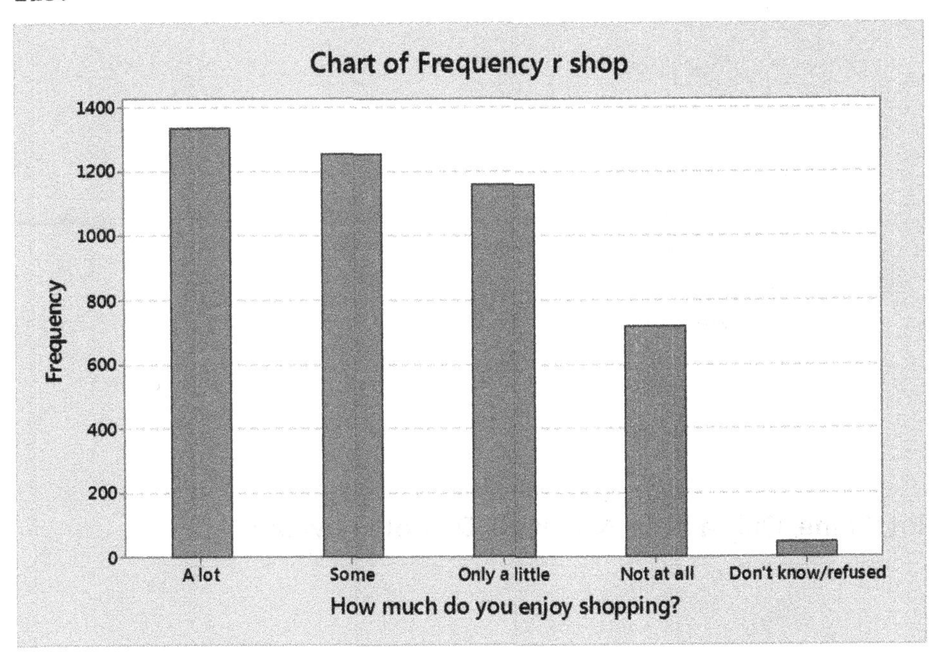

115.

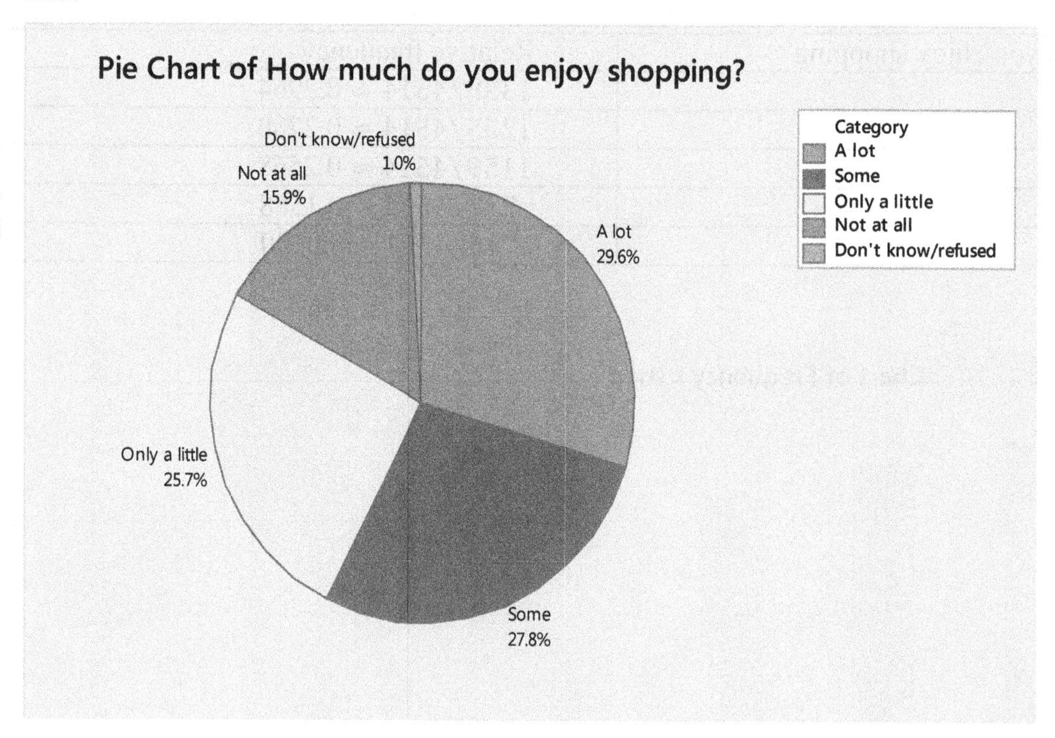

117.

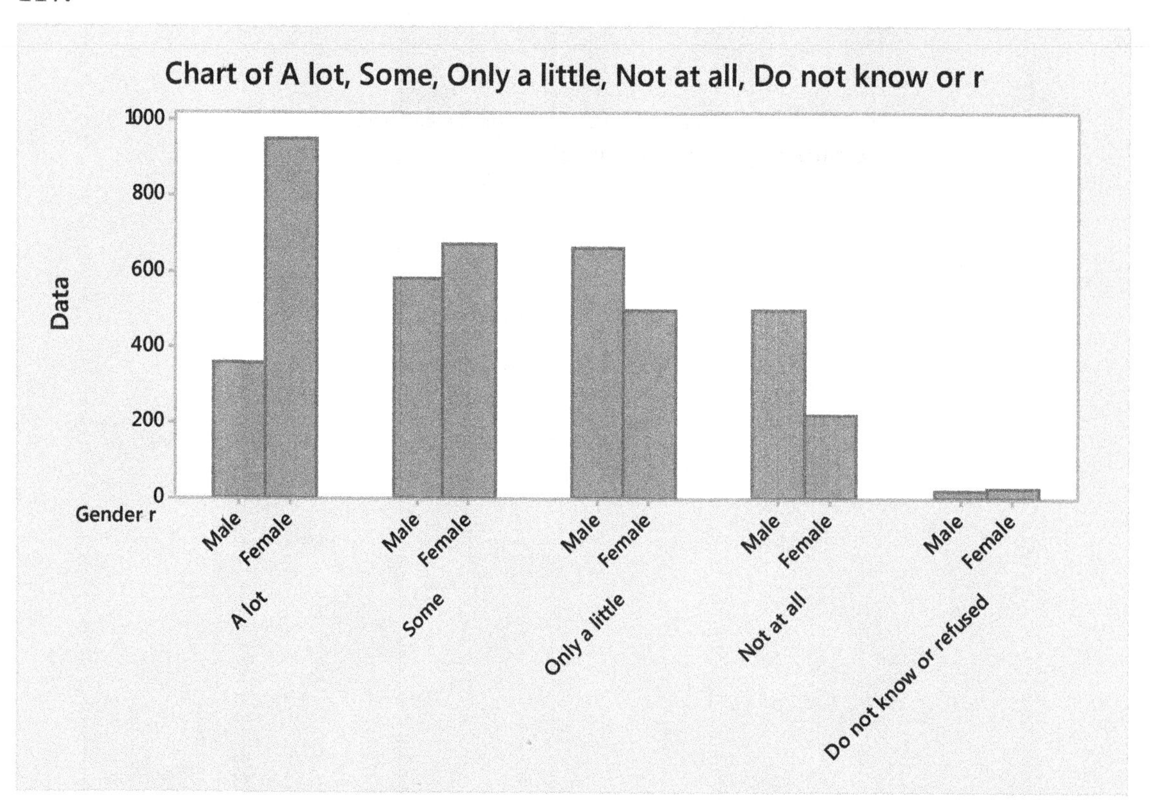

119. **(a)** Since the cell counts are doubled, the total number of men and the total number of women would double. Therefore, the frequencies for men and women would double.

(b) Since the cell counts are doubled, the frequencies for men and women would double. Therefore, the total number of people would also double. Thus, the relative frequency of men would be $\frac{2(\text{original number of men})}{2(\text{original number of people})} = \frac{\text{original number of men}}{\text{original number of people}}$ = original relative frequency of men and the relative frequency of women would be $\frac{2(\text{original number of women})}{2(\text{original number of people})} = \frac{\text{original number of women}}{\text{original number of people}} =$ original relative frequency of women. Therefore, the original relative frequencies of men and women would stay the same.

(c) Since the relative frequencies of men and women would stay the same, the percents of men and women would stay the same. Therefore, the pie chart would stay the same.

121. **(a)**

Tally for Discrete Variables: GENDER

GENDER	Percent
boy	47.49
girl	52.51

(b)

Tally for Discrete Variables: GOALS

GOALS	Percent
Grades	51.67
Popular	29.50
Sports	18.83

123. There is only one qualitative variable: state.

There are seven quantitative variables: Tot_hhld, Fam_tpc, Fam_mpc, Fam_fpc, Nfm-tpc, Nfm_lpc, and Ave_size

125.

(a)

Tally for Discrete Variables: Class

Class	Count	Percent
1st	325	14.77
2nd	285	12.95
3rd	706	32.08
Crew	885	40.21
N=	2201	

(b)

Tally for Discrete Variables: Age

Age	Count	Percent
Adult	2092	95.05
Child	109	4.95
N=	2201	

(c)

Tally for Discrete Variables: Sex

Sex	Count	Percent
Female	470	21.35
Male	1731	78.65
N=	2201	

(d)

Tally for Discrete Variables: Survived_1

Survived_1	Count	Percent
No	1490	67.70
Yes	711	32.30
N=	2201	

127. (a)

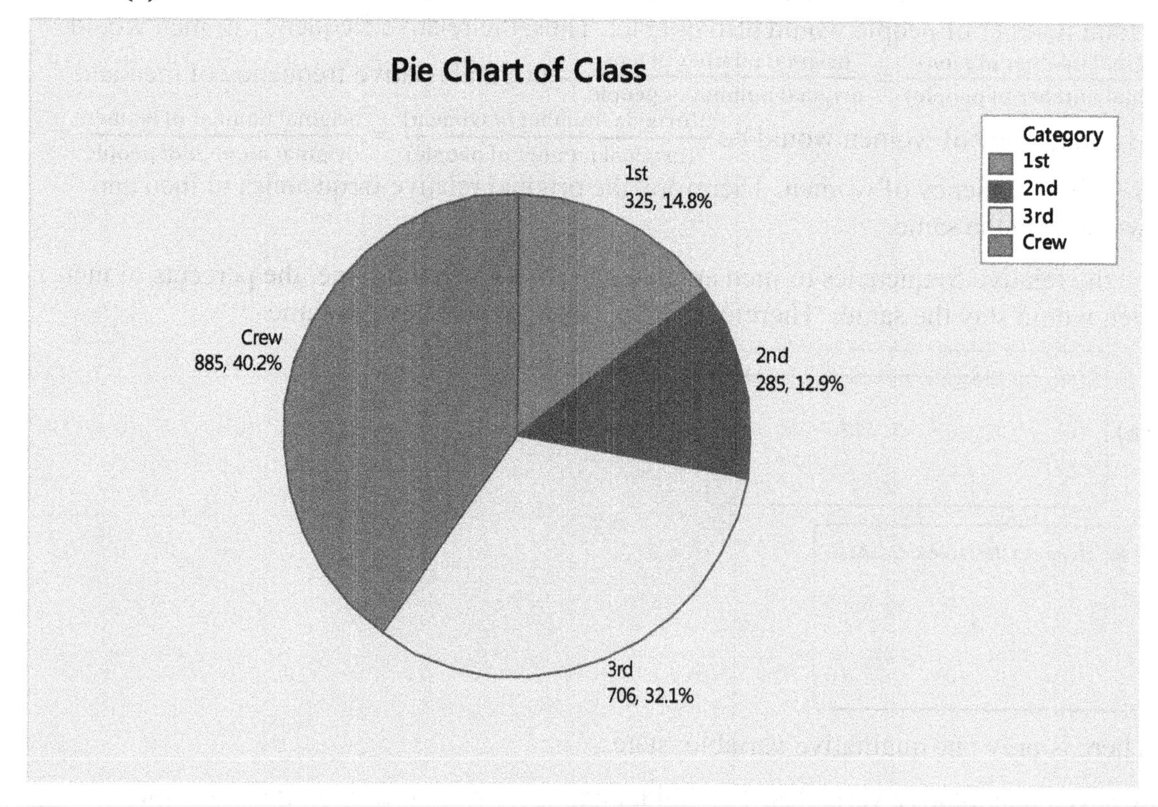
Pie Chart of Class
Category
1st
2nd
3rd
Crew
1st
325, 14.8%
2nd
285, 12.9%
3rd
706, 32.1%
Crew
885, 40.2%

(b)

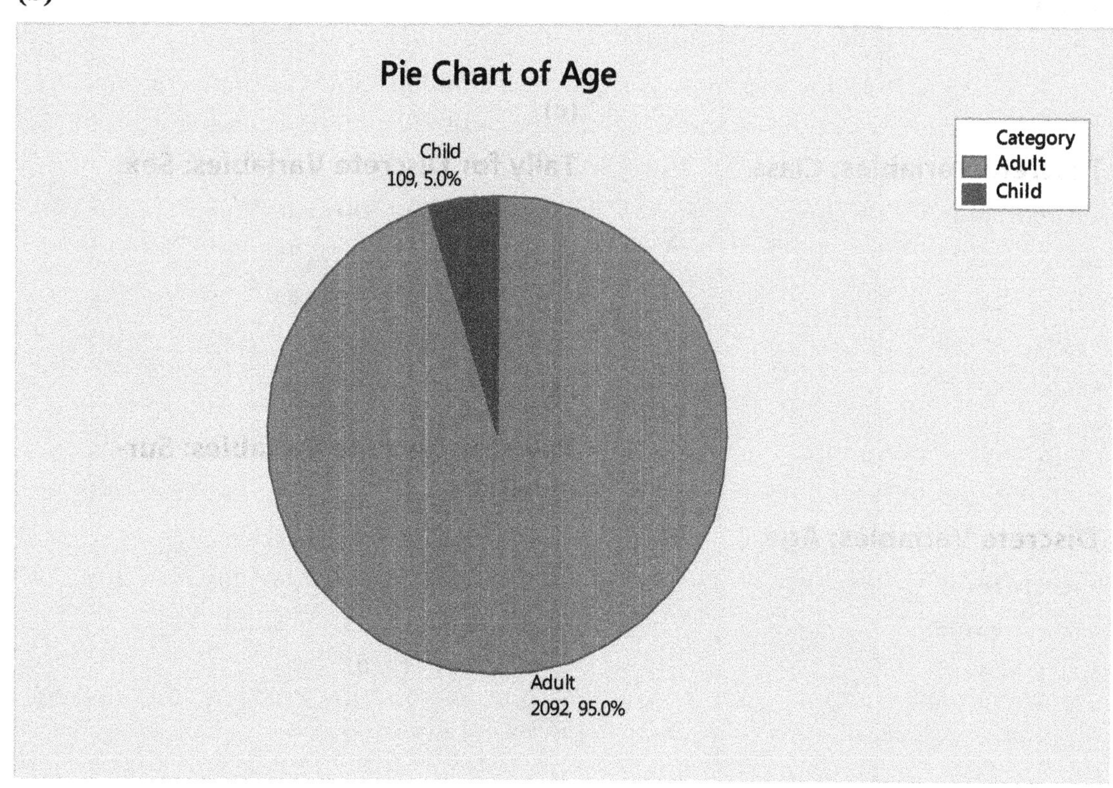
Pie Chart of Age
Category
Adult
Child
Child
109, 5.0%
Adult
2092, 95.0%

(c)

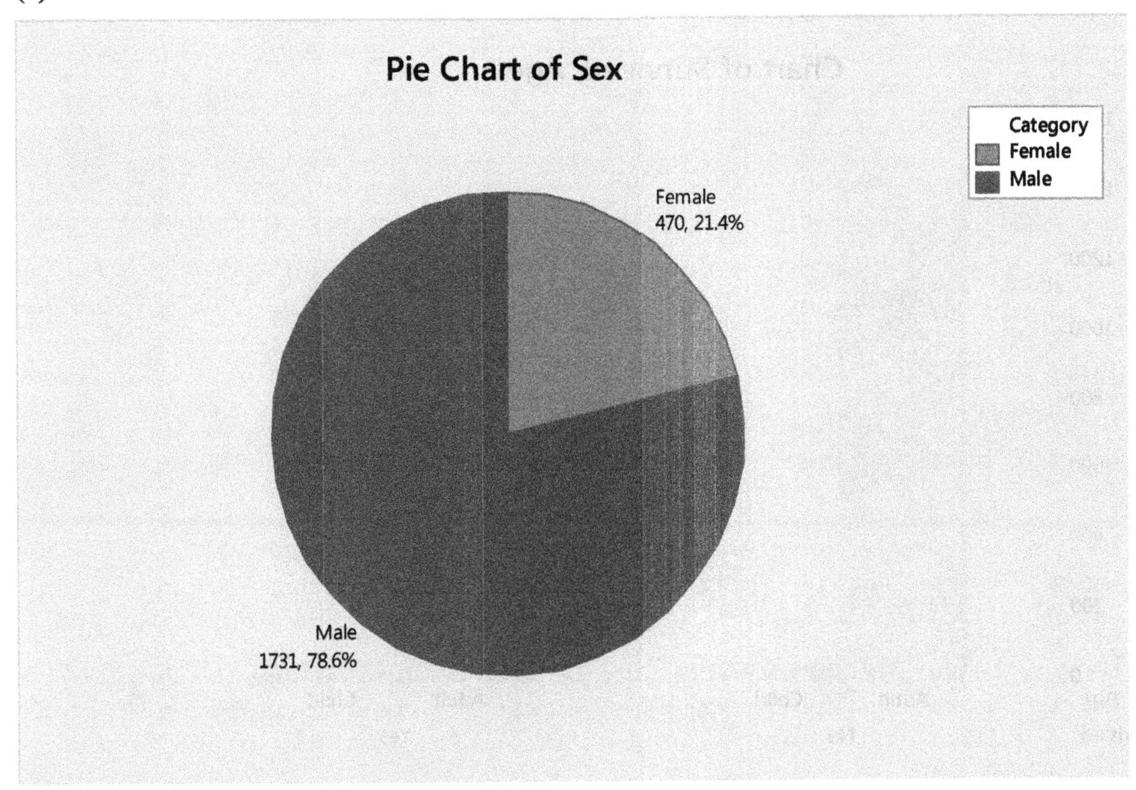
Pie Chart of Sex
Category
Female
Male
Female
470, 21.4%
Male
1731, 78.6%

(d)

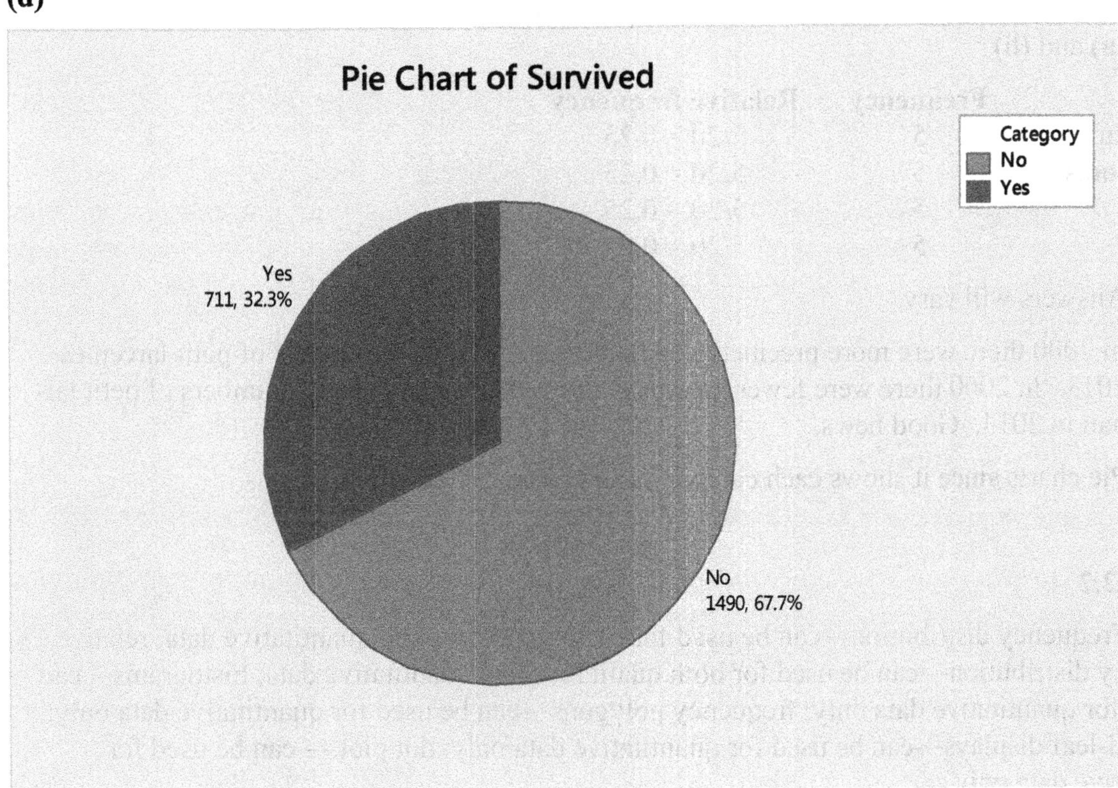
Pie Chart of Survived
Category
No
Yes
Yes
711, 32.3%
No
1490, 67.7%

129.

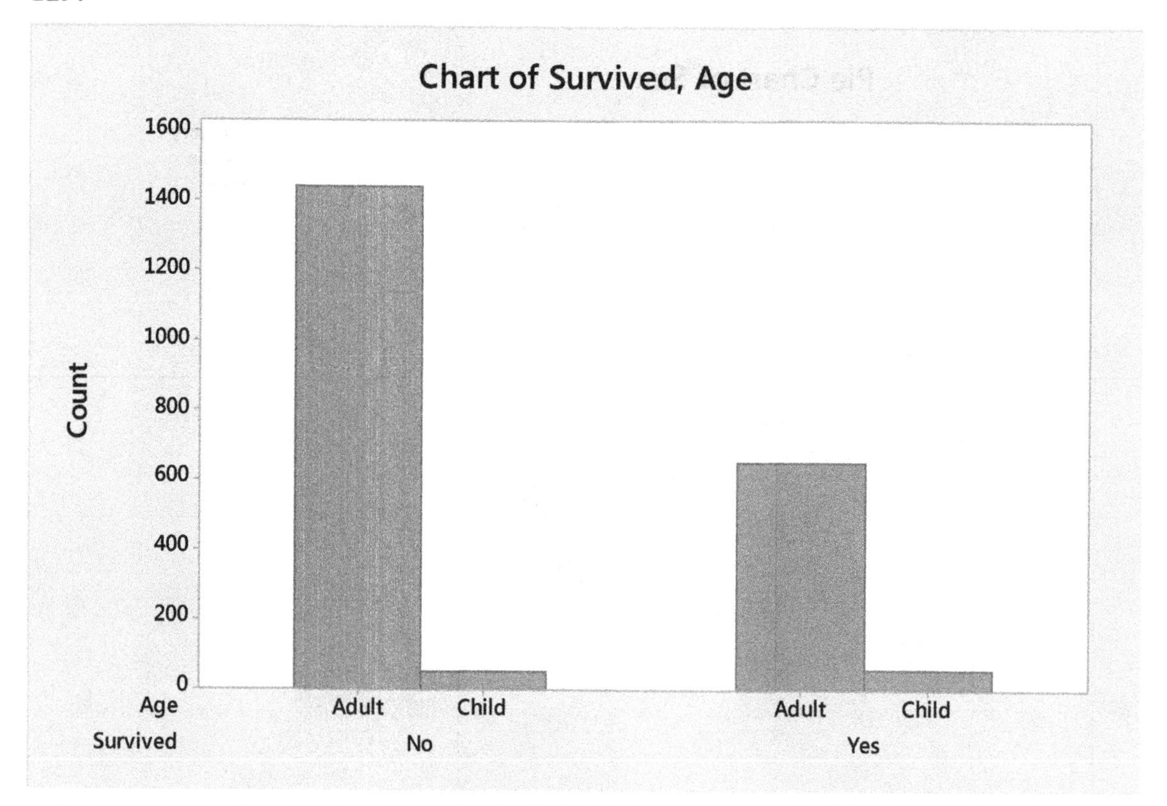

A child.

131. **(a)** and **(b)**

Class	**Frequency**	**Relative frequency**
Freshmen	5	5/20 = 0.25
Sophomores	5	5/20 = 0.25
Juniors	5	5/20 = 0.25
Seniors	5	5/20 = 0.25

133. Answers will vary.

135. In 2000 there were more precincts that had category 3 and 4 numbers of petit larcenies than in 2013. In 2000 there were fewer precincts that had category 1 and 2 numbers of petit larcenies than in 2013. Good news.

137. Pie charts since it shows each category as a percent of the whole.

Section 2.2

1. Frequency distribution—can be used for both qualitative and quantitative data; relative frequency distribution—can be used for both qualitative and quantitative data; histograms—can be used for quantitative data only; frequency polygons—can be used for quantitative data only; stem-and-leaf displays—can be used for quantitative data only; dot plots—can be used for quantitative data only.

3. Between 5 and 20.

5. Dot plots may be useful for comparing two or more variables.

7. Example of a right-skewed distribution—price of precious gems. Example of a left-skewed distribution—lifetime of electrical bulbs.

9.

Number of nominations	Frequency
2	5
3	3
4	1
5	1
6	1
7	1
8	1
Total	**13**

11.

Grand slams	Frequency
14	4
15	2
16	3
17	3
18	2
19	1
21	1
23	1
24	1
Total	**18**

13.

Nominations	Frequency
0 - 3	8
4 - 6	3
7 - 9	2
Total	**13**

15.

Grand slams	Frequency
11 - 15	6
16 - 20	9
21 - 25	3
Total	**18**

17. 1 to < 2, 2 to < 3, 3 to < 4, 4 to < 5, 5 to < 6

19.

Exports to	Frequency
1 to < 2	3
2 to < 3	1
3 to < 4	2
4 to < 5	2
5 to < 6	1
Total	**9**

21. 2 to < 4, 4 to < 6, 6 to < 8, 8 to < 10, 10 to < 12

23.

Imports from	Frequency
2 to < 4	4
4 to < 6	3
6 to < 8	0
8 to < 10	0
10 to < 12	2
Total	**9**

25. 2 to < 3.5, 3.5 to < 5, 5 to < 6.5, 6.5 to < 8, 8 to < 9.5, 9.5 to < 11, 11 to < 12.5

27.

Imports from	Frequency
2 to < 3.5	2
3.5 to < 5	4
5 to < 6.5	1
6.5 to < 8	0
8 to < 9.5	0
9.5 to < 11	1
11 to < 12.5	1
Total	**9**

29. – 6 to < -5, -5 to < -4, -4 to < -3, -3 to < -2, -2 to < -1, -1 to < 0, 0 to < 1, 1 to < 2

31.

Trade balance	Frequency
-6 to < -5	2
-5 to < -4	0
-4 to < -3	0
-3 to < -2	1
-2 to < -1	4
-1 to < 0	0
0 to < 1	1
1 to < 2	1
Total	**9**

33. 0 to < 25, 25 to < 50, 50 to < 75, 75 to < 100, 100 to < 125, 125 to < 150, 150 to < 175, 175 to < 200, 200 to < 225

35.

Motor vehicle theft rate	Frequency
0 to < 25	1
25 to < 50	6
50 to < 75	4
75 to < 100	3
100 to < 125	3
125 to < 150	0
150 to < 175	2
175 to < 200	0
200 to < 225	1
Total	**20**

37. 20 to < 45, 45 to < 70, 70 to < 95, 95 to < 120, 120 to < 145, 145 to < 170, 170 to < 195, 195 to < 220

39.

Motor vehicle theft rate	Frequency
20 to < 45	5
45 to < 70	6
70 to < 95	3
95 to < 120	3
120 to < 145	0
145 to < 170	1
170 to < 195	1
195 to < 220	1
Total	**20**

41.

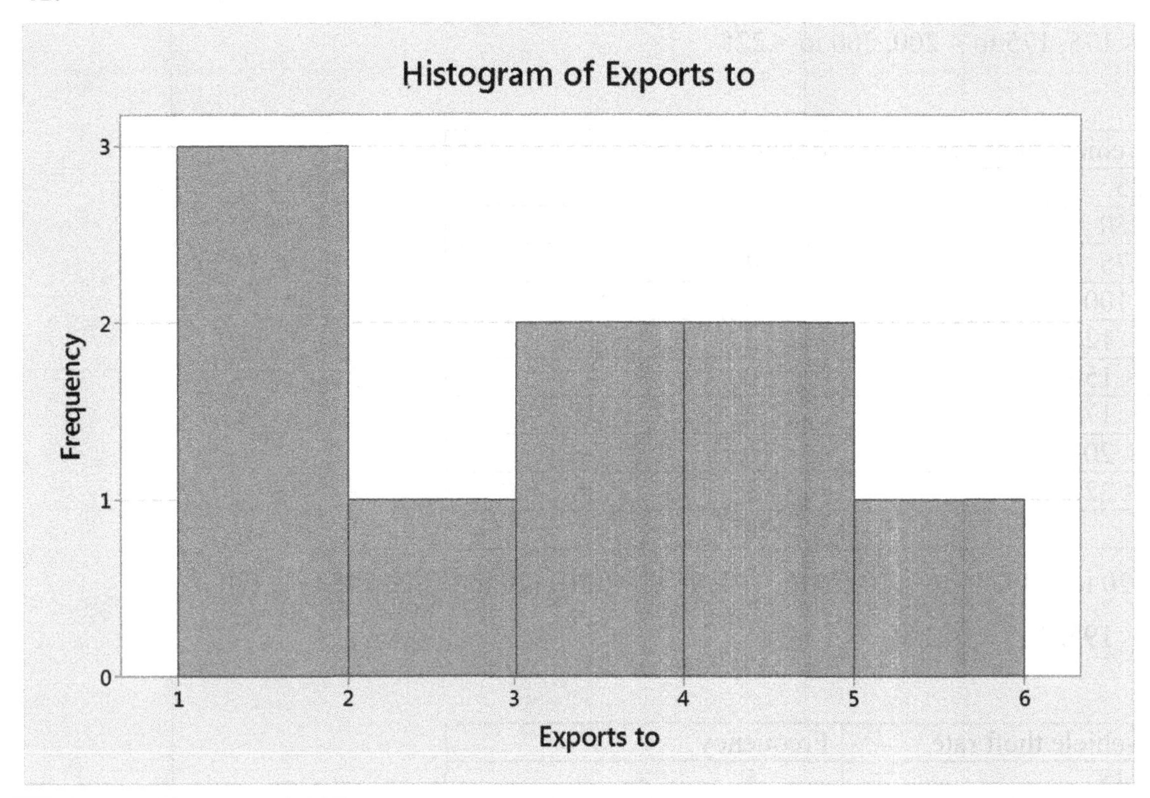
Histogram of Exports to
Frequency
0
1
2
3
1
2
3
4
5
6
Exports to

43.

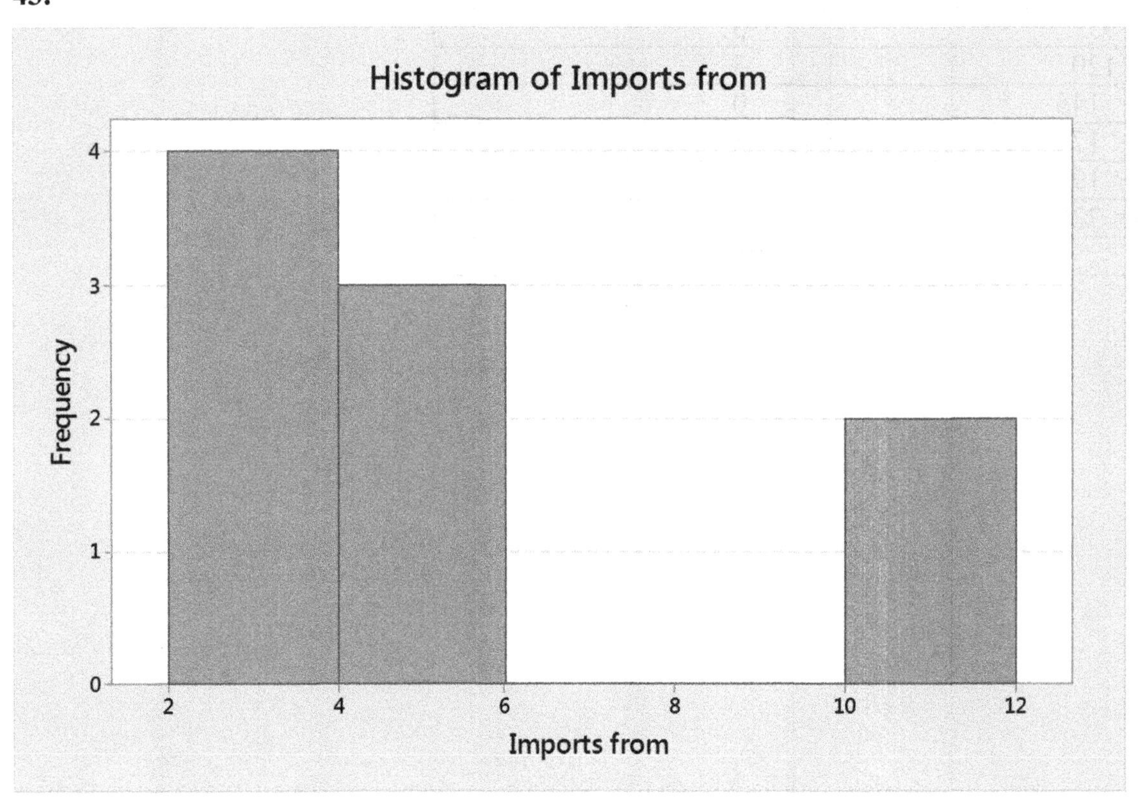
Histogram of Imports from
Frequency
0
1
2
3
4
2
4
6
8
10
12
Imports from

45.

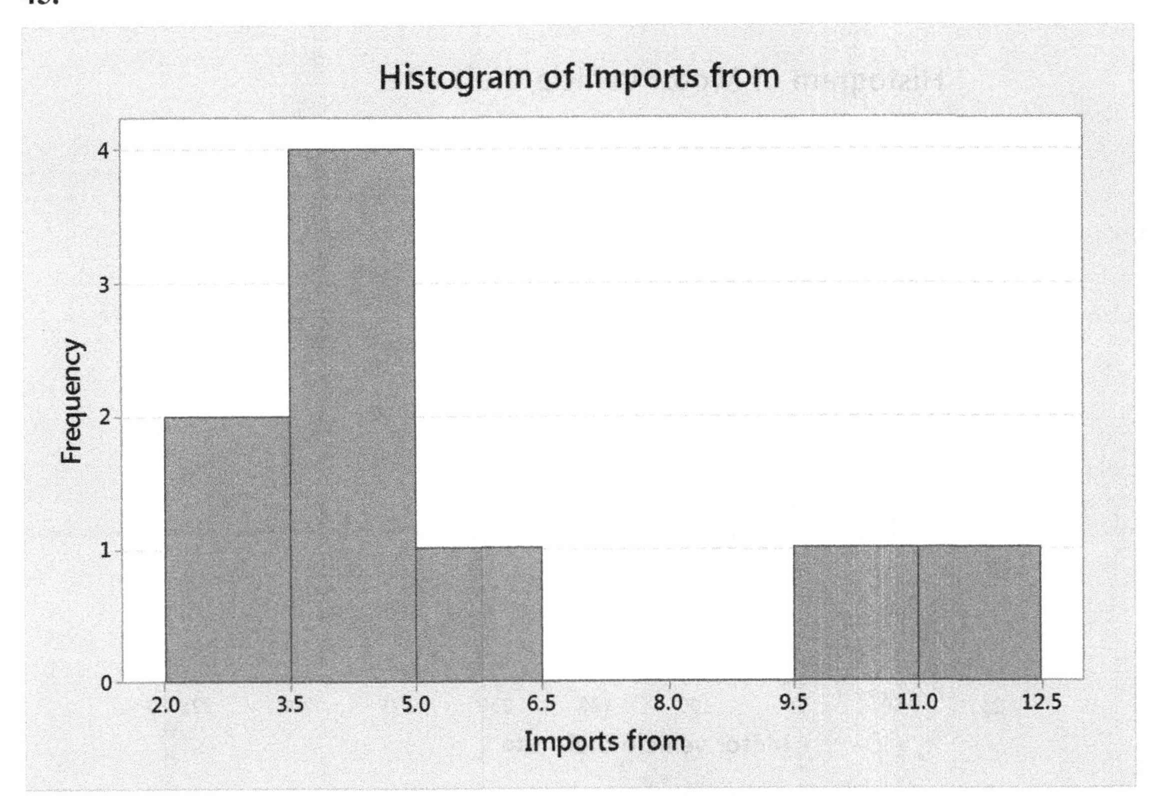
Histogram of Imports from
Frequency
0
1
2
3
4
2.0
3.5
5.0
6.5
8.0
9.5
11.0
12.5
Imports from

47.

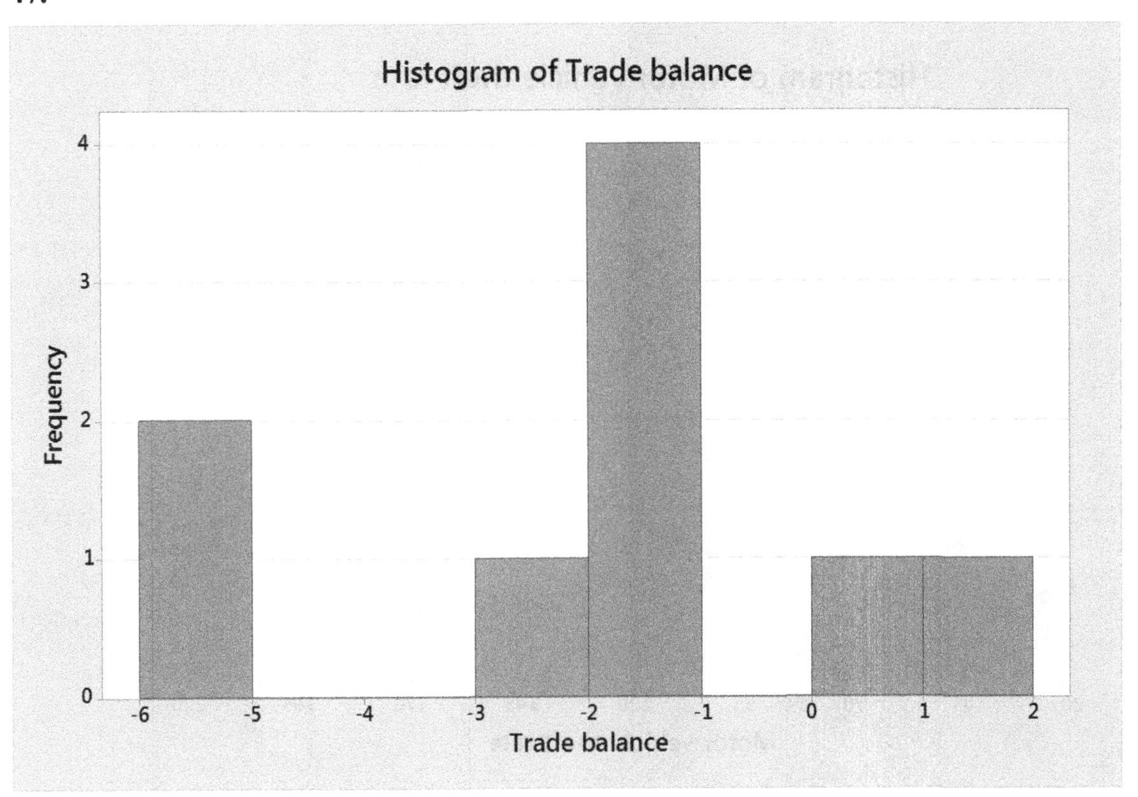
Histogram of Trade balance
Frequency
0
1
2
3
4
-6
-5
-4
-3
-2
-1
0
1
2
Trade balance

49.

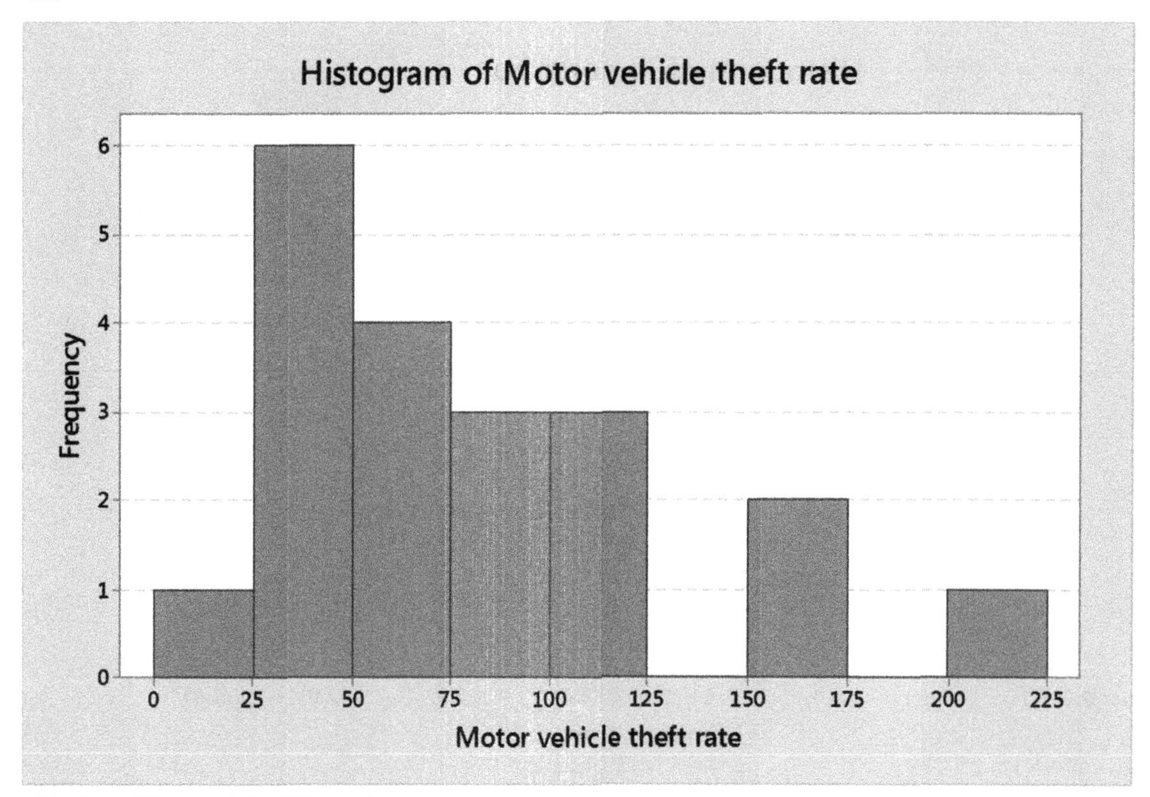
Histogram of Motor vehicle theft rate
Frequency
0
1
2
3
4
5
6
0 25 50 75 100 125 150 175 200 225
Motor vehicle theft rate

51.

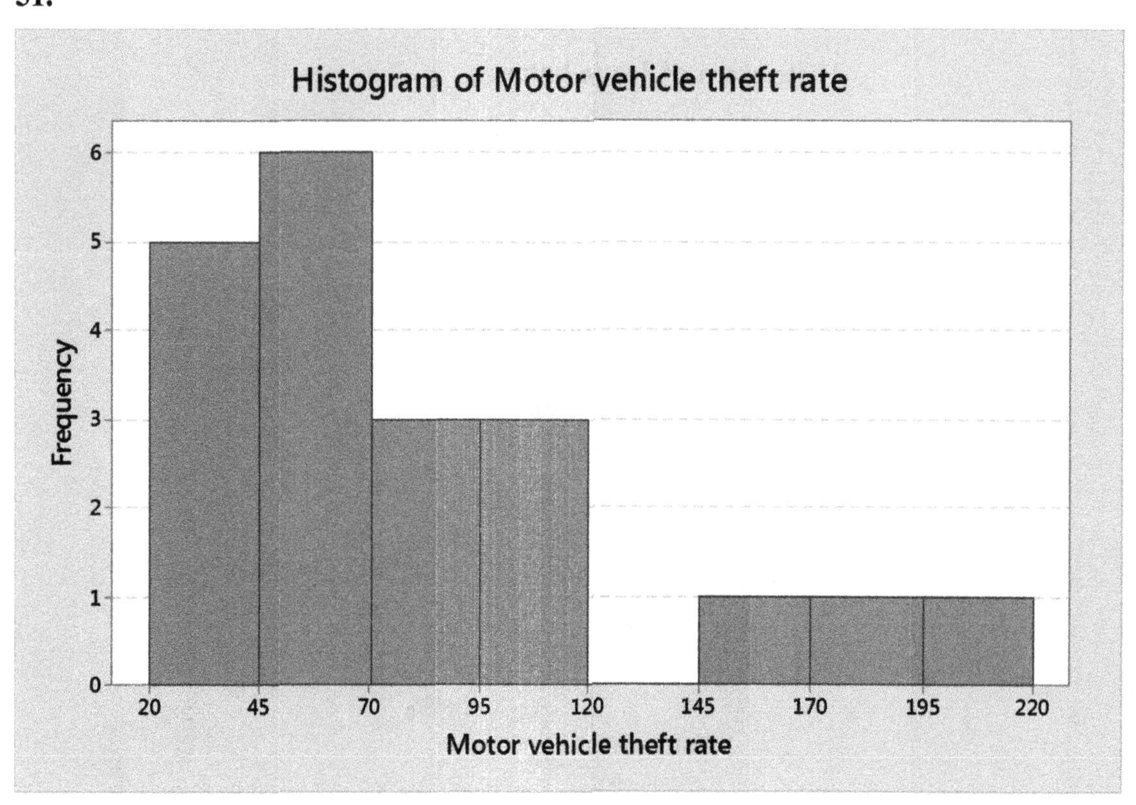
Histogram of Motor vehicle theft rate
Frequency
0
1
2
3
4
5
6
20 45 70 95 120 145 170 195 220
Motor vehicle theft rate

53.

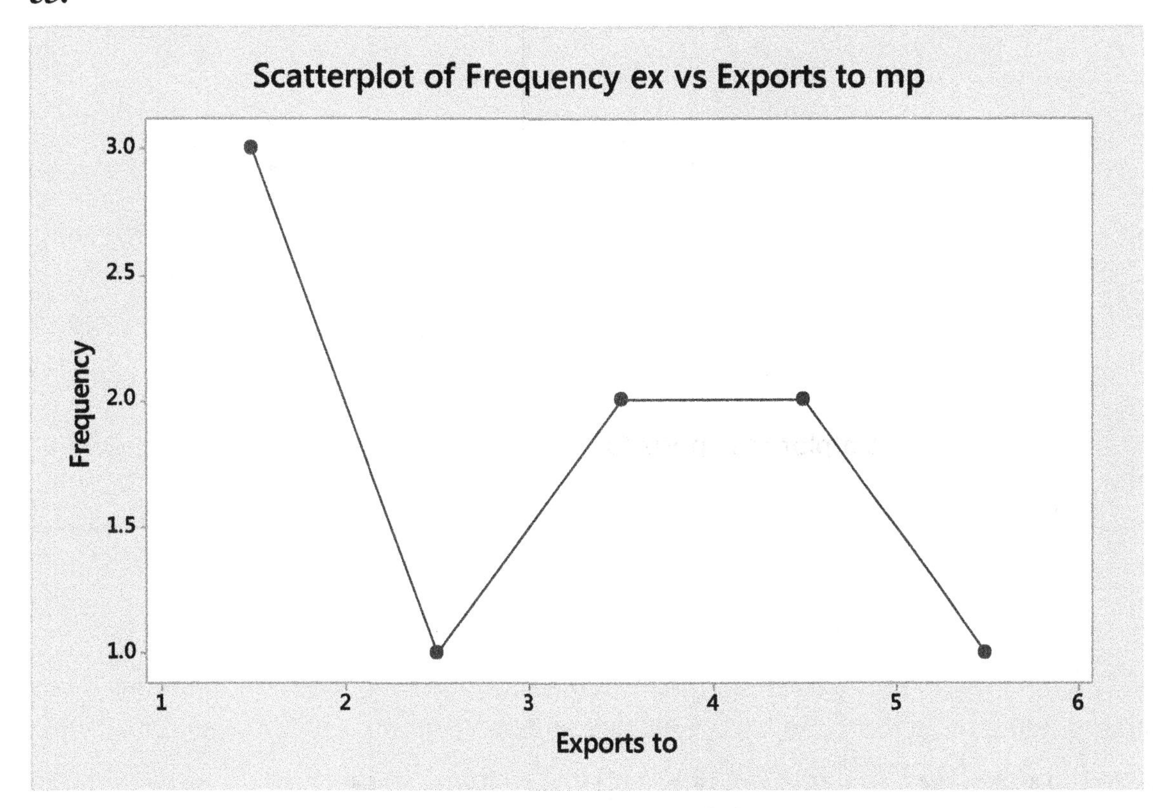
Scatterplot of Frequency ex vs Exports to mp
Frequency
3.0
2.5
2.0
1.5
1.0
1
2
3
4
5
6
Exports to

55.

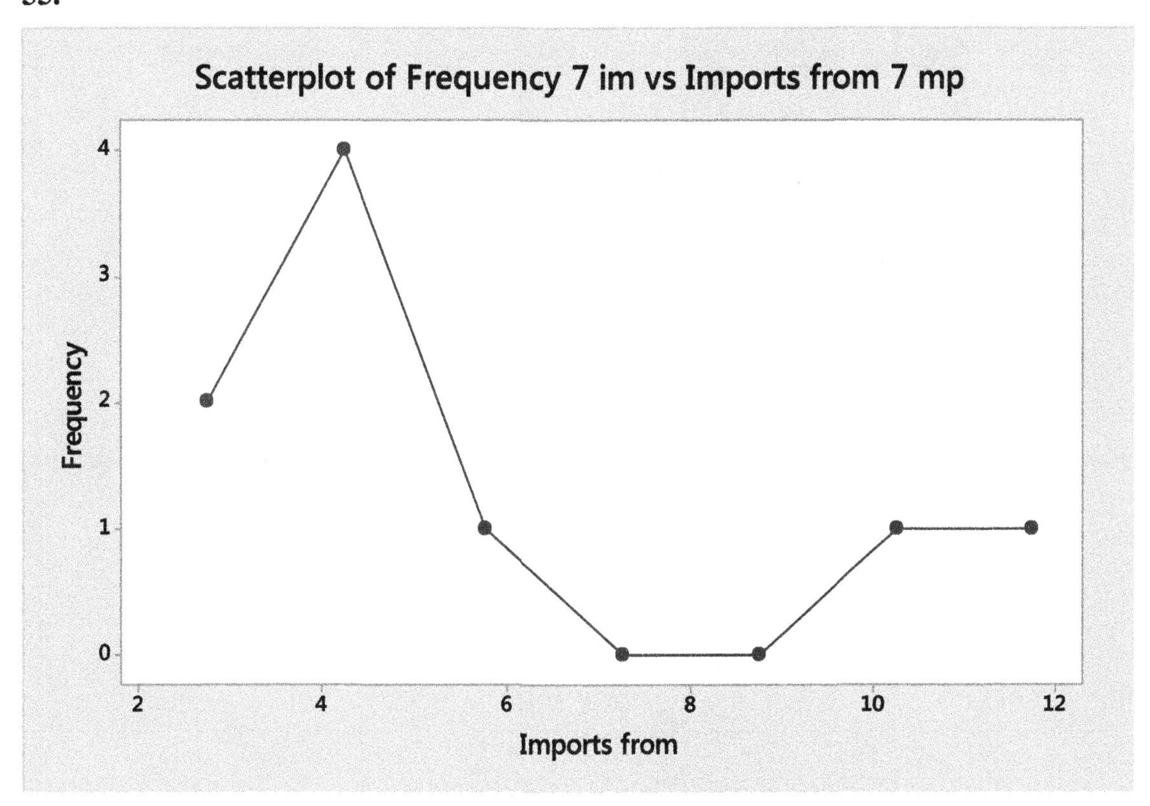
Scatterplot of Frequency 7 im vs Imports from 7 mp
Frequency
4
3
2
1
0
2
4
6
8
10
12
Imports from

57.

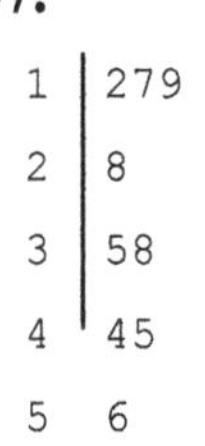

```
1 | 279
2 | 8
3 | 58
4 | 45
5 | 6
```

59.

```
-5 | 66
-4 |
-3 |
-2 | 4
-1 | 8832
 0 | 0
 1 | 0
```

61.

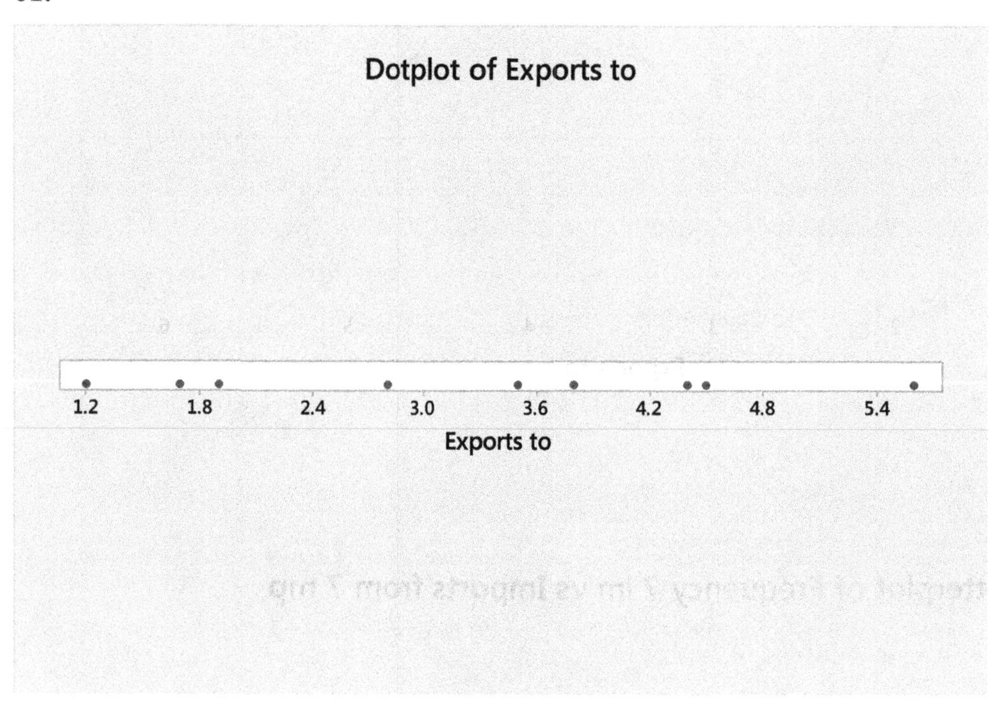

63.

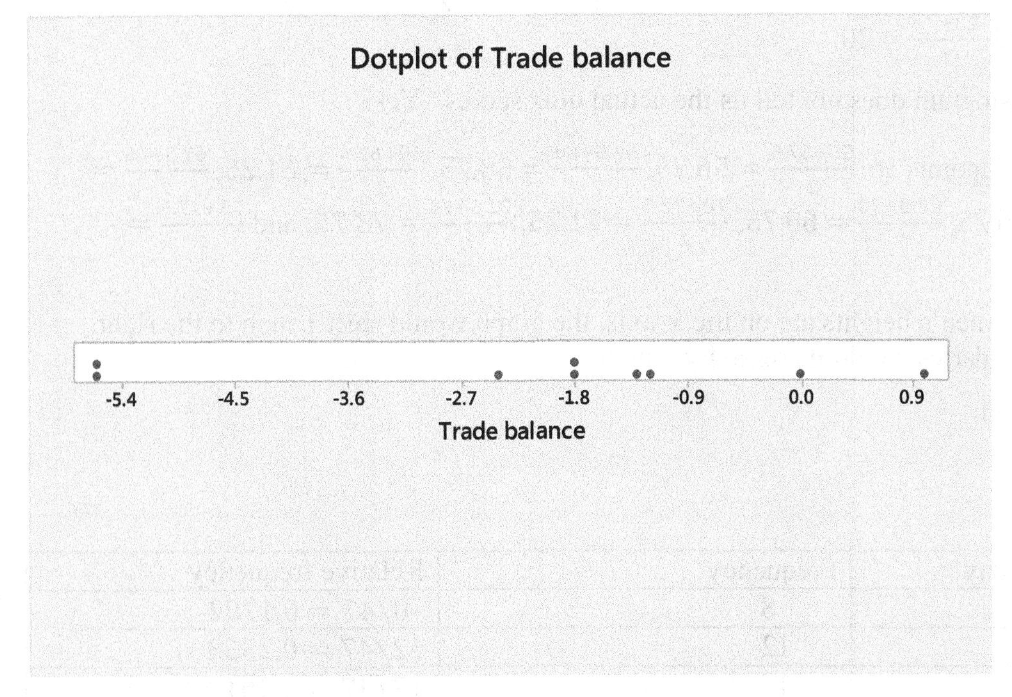

65.

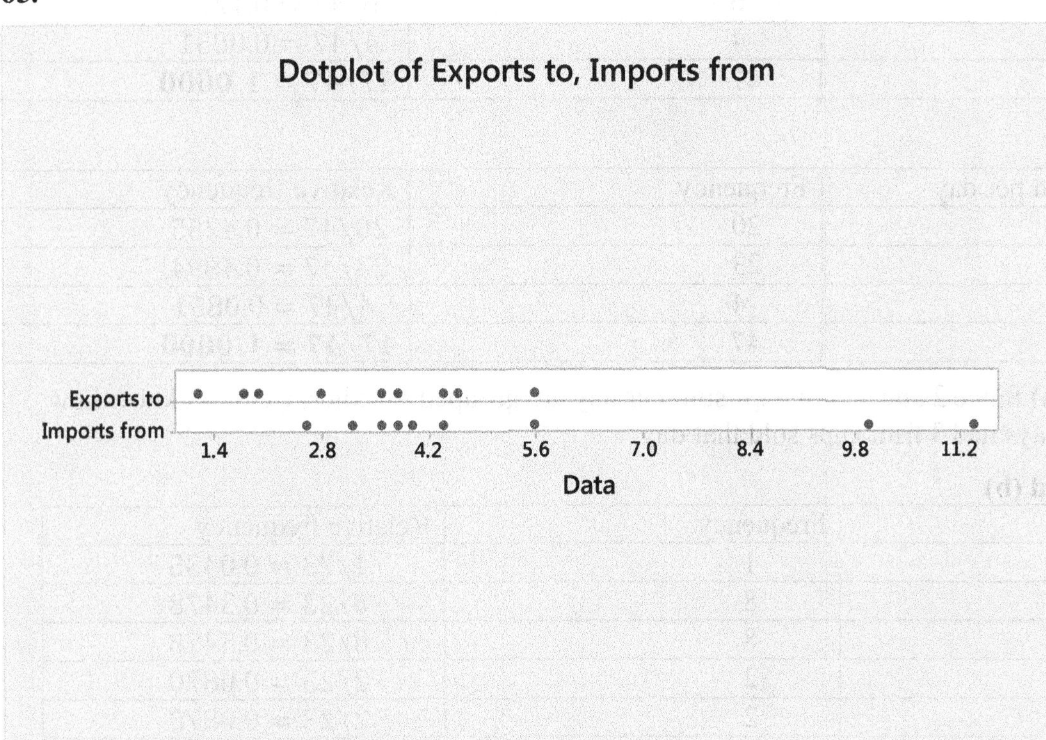

67. The class midpoints are $\frac{0+1.5}{2} = 0.75$, $\frac{1.5+3}{2} = 2.25$, $\frac{3+4.5}{2} = 3.75$, $\frac{4.5+6}{2} = 5.25$, $\frac{6+7.5}{2} = 6.75$, $\frac{7.5+9}{2} = 8.25$, $\frac{9+10.5}{2} = 9.75$, and $\frac{10.5+12}{2} = 11.25$.

69. About 2 students paid $10.50 or more for their last music download.

71. The class midpoints are $\frac{0+4}{2}=2$, $\frac{4+8}{2}=6$, $\frac{8+12}{2}=10$, $\frac{12+16}{2}=14$, $\frac{16+20}{2}=18$, $\frac{20+24}{2}=22$, $\frac{24+28}{2}=26$, and $\frac{28+32}{2}=30$.

73. No. The histogram does not tell us the actual quiz scores. Yes.

75. The class midpoints are $\frac{55+57.5}{2}=56.25$, $\frac{57.5+60}{2}=58.75$, $\frac{60+62.5}{2}=61.25$, $\frac{62.5+65}{2}=63.75$, $\frac{65+67.5}{2}=66.25$, $\frac{67.5+70}{2}=68.75$, $\frac{70+72.5}{2}=71.25$, $\frac{72.5+75}{2}=73.75$, and $\frac{75+77.5}{2}=76.25$.

77. Since the women's heights are on the x-axis, the graph would shift 1 inch to the right. All of the class boundaries would increase by 1 inch.

79. Right-skewed

81. Symmetric

83. **(a)**

Fruit cups sold per day	Frequency	Relative frequency
0	8	$8/47 = 0.1702$
1	12	$12/47 = 0.2553$
2	17	$17/47 = 0.3617$
3	6	$6/47 = 0.1277$
4	4	$4/47 = 0.0851$
Total	**47**	$\mathbf{47/47 = 1.0000}$

(b)

Fruit cups sold per day	Frequency	Relative frequency
0 - 1	20	$20/47 = 0.4255$
2 - 3	23	$23/47 = 0.4894$
4 - 5	4	$4/47 = 0.0851$
Total	**47**	$\mathbf{47/47 = 1.0000}$

(c) 0.2128 **(a)** Since 2 and 3 fruit cups sold per day are grouped together, we don't know how many of the days had 3 fruit cups sold that day.

85. **(a) and (b)**

Frauds	Frequency	Relative frequency
0 to < 40	1	$1/23 = 0.0435$
40 to < 80	8	$8/23 = 0.3478$
80 to < 120	8	$8/23 = 0.3478$
120 to < 160	2	$2/23 = 0.0870$
160 to < 200	2	$2/23 = 0.0870$
200 to < 240	1	$1/23 = 0.0435$
240 to < 280	1	$1/23 = 0.0435$
Total	**23**	$\mathbf{23/23 = 1.0000}$

(c)

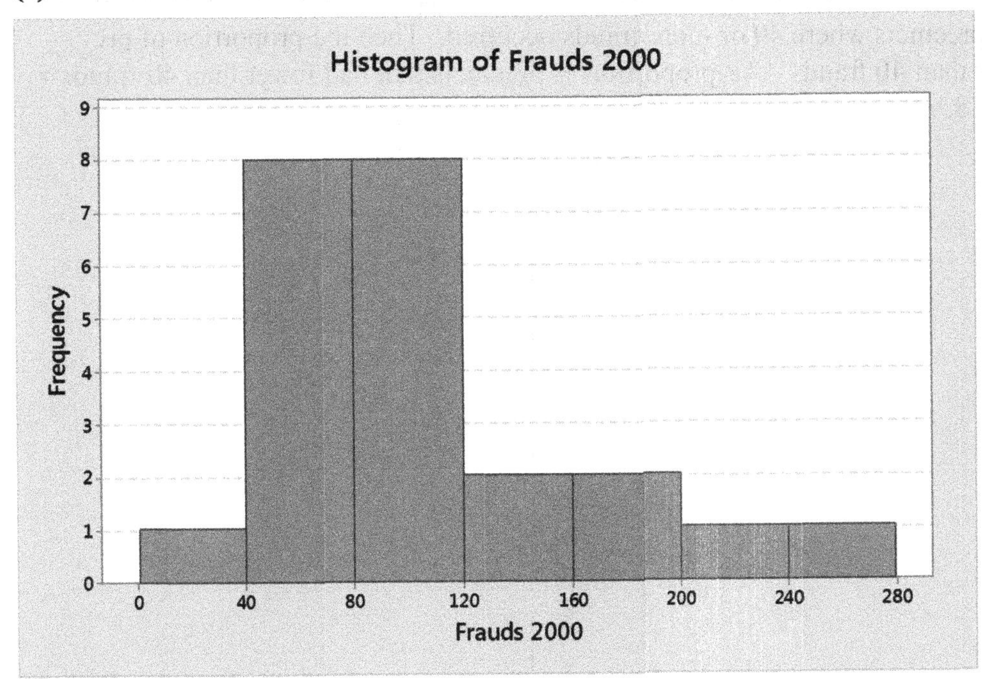

(d)

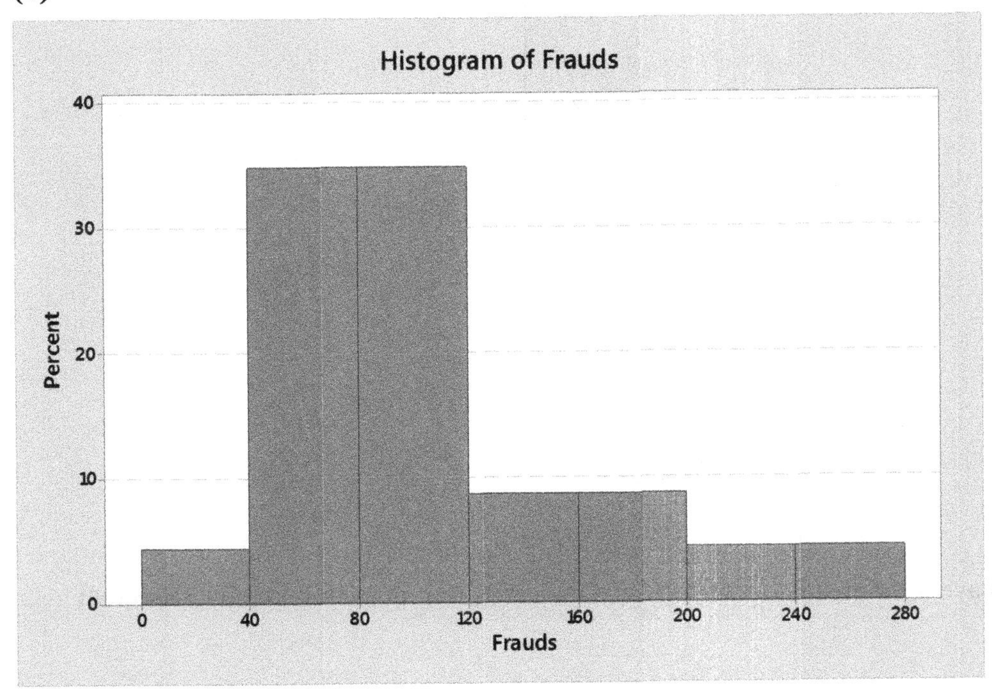

87. **(a)** The relative frequency of precincts where 80 or more frauds occurred is $\frac{2}{23} \approx 0.0870$.

(b) The proportion of precincts where 40 or more frauds occurred is $\frac{15}{23} \approx 0.6522$.

(c) The proportion of precincts that had fewer than 40 frauds is $\frac{8}{23} \approx 0.3478$. The answer in (b) is the proportion of precincts where 40 or more frauds occurred. Then the proportion of precincts that had fewer than 40 frauds = 1 - proportion of precincts that had fewer than 40 frauds= $1 - 0.6522 = 0.3478$.

89.

(a)

Stem	Leaf
1	569
2	137
3	36
4	11248
5	1237
6	38
7	36
8	
9	0
10	
11	
12	
13	3

(b)

Stem	Leaf
1	569
2	13
2	7
3	3
3	6
4	1124
4	8
5	123
5	7
6	3
6	8
7	3
7	6
8	
8	
9	0
9	
10	
10	
11	
11	
12	
12	
13	3

(c) 9. Could not have used the histogram because the histogram does not contain the actual data.

91. (a), (b), (c)

Coffee sold per day	Frequency	Relative frequency
0 to < 8	7	$7/47 = 0.1489$
8 to < 16	7	$7/47 = 0.1489$
16 to < 24	10	$10/47 = 0.2128$
24 to < 32	15	$15/47 = 0.3191$
32 to < 40	6	$6/47 = 0.1277$
40 to < 48	1	$1/47 = 0.0213$
48 to < 56	1	$1/47 = 0.0213$
Total	**47**	$\mathbf{47/47 = 1.0000}$

(d)

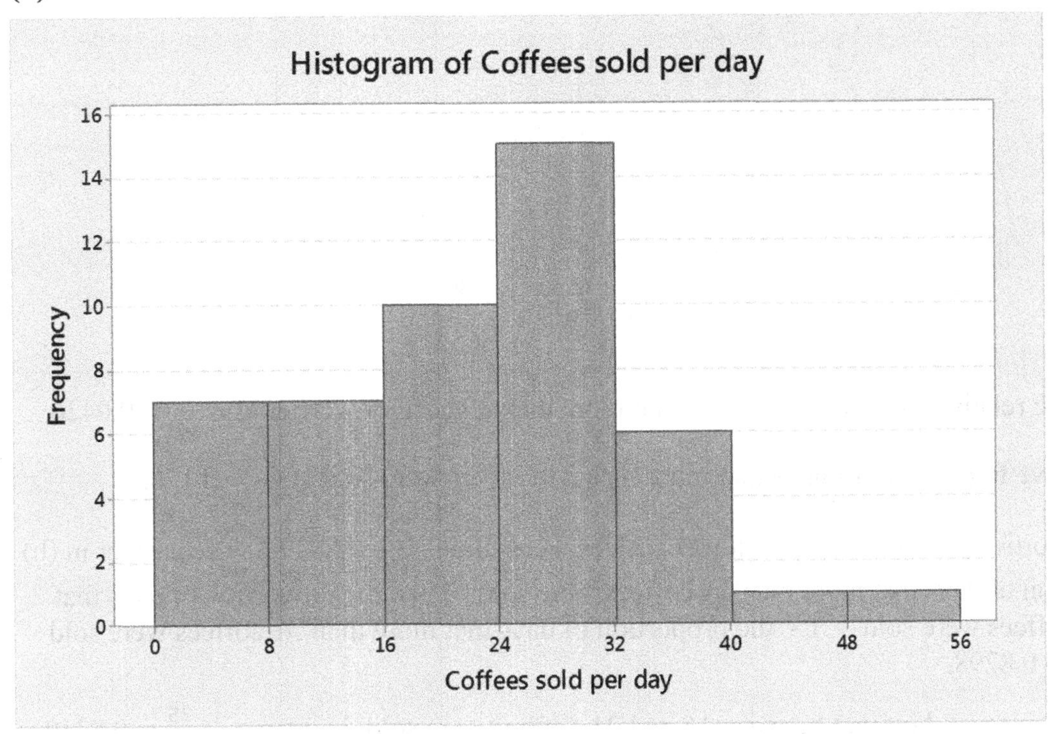

(e)

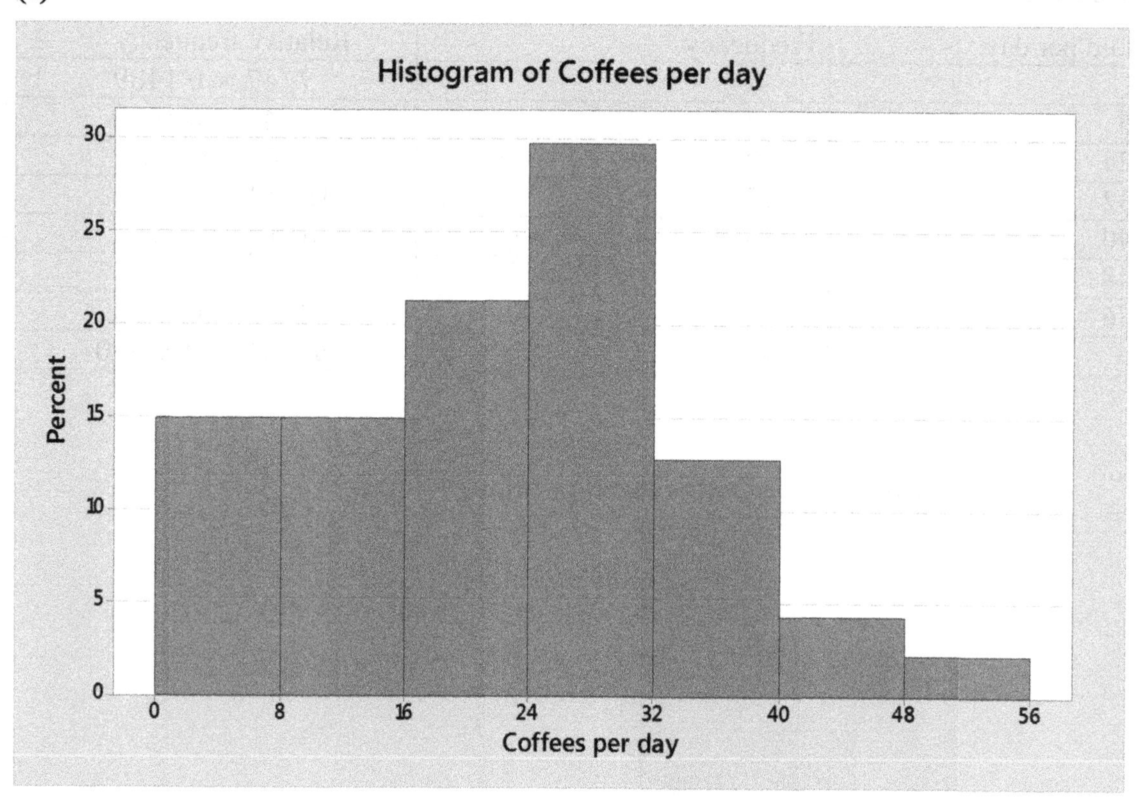

93. **(a)** The relative frequencies of days that more than 39 coffees were sold is $\frac{2}{47} \approx 0.0426$.

(b) The relative frequencies of days that more than 31 coffees were sold is $\frac{8}{47} \approx 0.1702$.

(c) The proportion of days that 31 or fewer coffees were sold is $\frac{39}{47} \approx 0.8298$. The answer in (b) is the proportion of days that more than 31 coffees were sold. Then the proportion of days that 31 or fewer coffees were sold = 1 - the proportion of days that more than 31 coffees were sold $= 1 - 0.1702 = 0.8298$.

(d) The proportion of days that between 16 and 31 coffees were sold, inclusive, is $\frac{25}{47} \approx 0.5319$.

95. **(a)** The class midpoints are $\frac{0+8}{2} = 4$, $\frac{8+16}{2} = 12$, $\frac{16+24}{2} = 20$, $\frac{24+32}{2} = 28$, $\frac{32+40}{2} = 36$, $\frac{40+48}{2} = 44$, and $\frac{48+56}{2} = 52$.

(b)

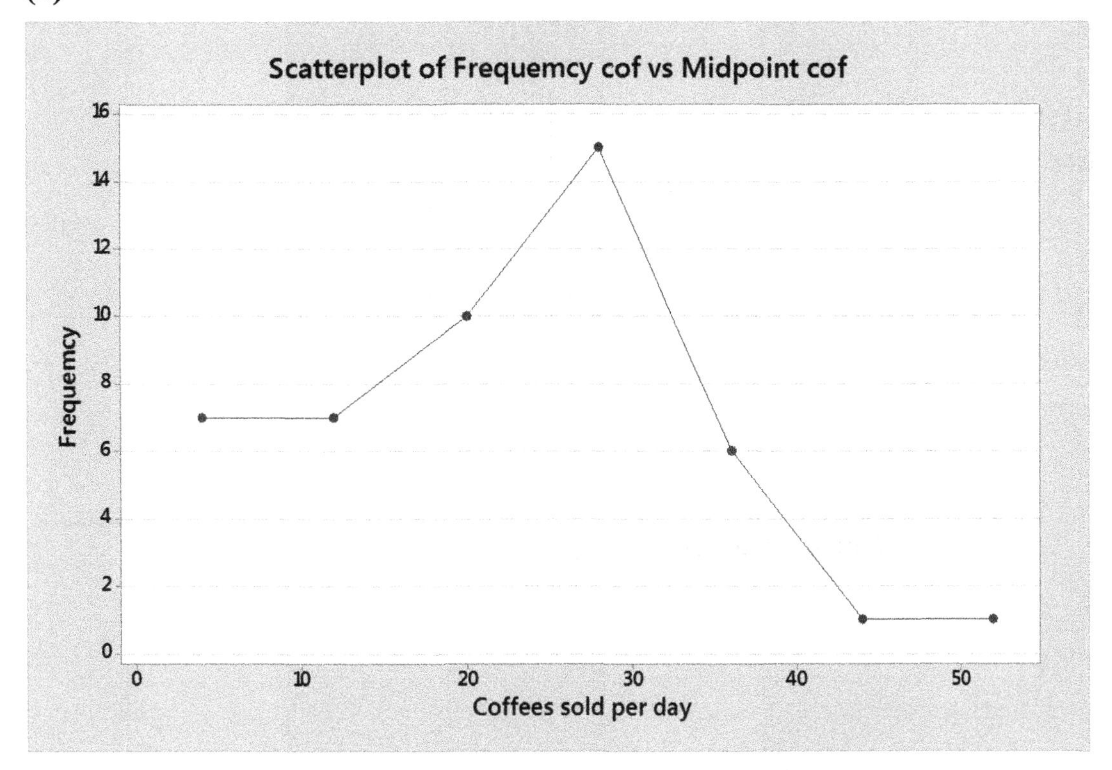

(c)

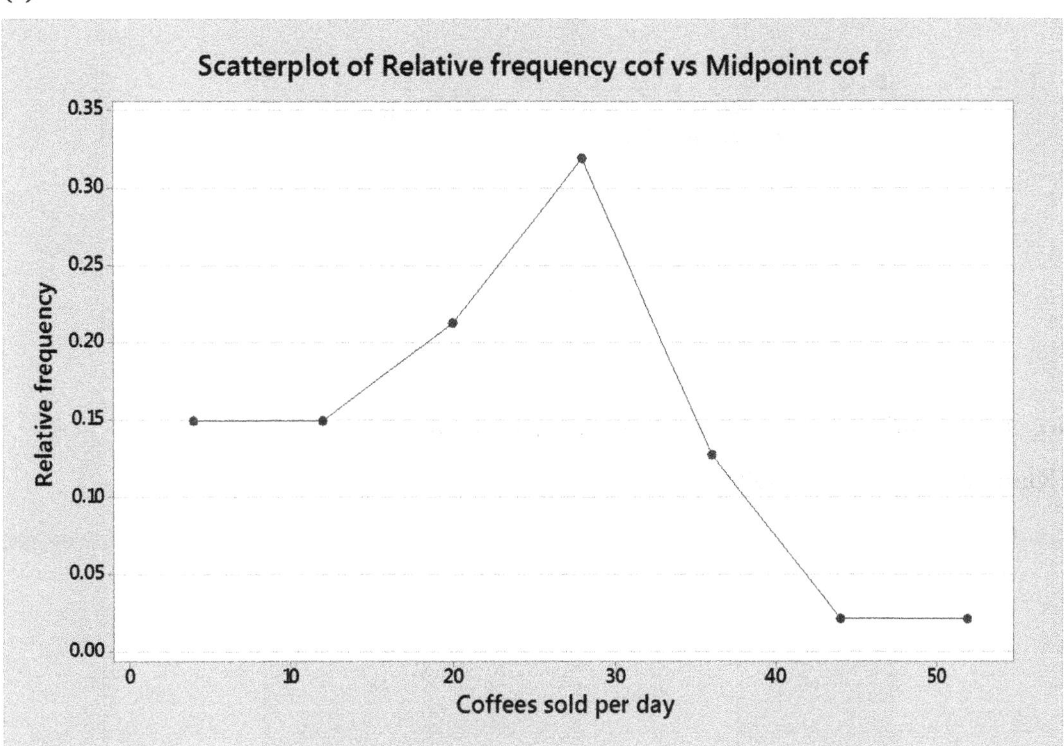

(d) The relative frequency of days that fewer than 17 coffees was sold is $\frac{16}{47} \approx 0.3404$.

97. **(a)** **(b)**

(a)

Stem	Leaves
0	344445688
1	01134668
2	00112334455677789
3	0011233455
4	118

(b)

Stem	Leaves
0	34444
0	5688
1	01134
1	668
2	001123344
2	55677789
3	00112334
3	55
4	11
4	8

(c)

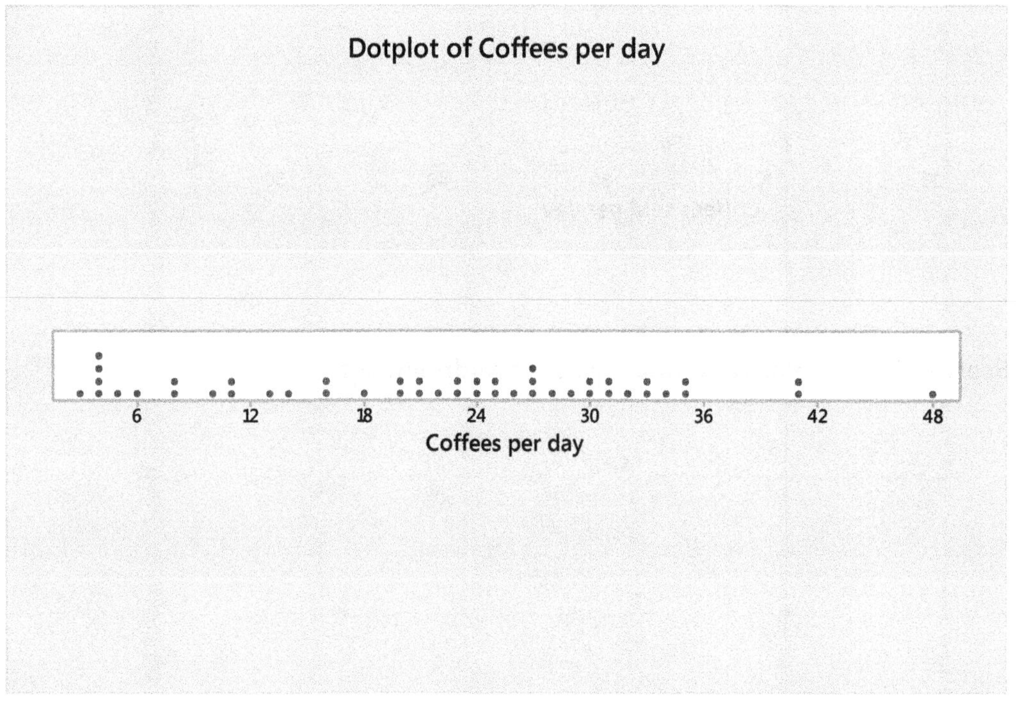

(d) 15. No, stem – and – leaf display contains the actual data.

99. **(a)** Right-skewed, tail on right

(b)

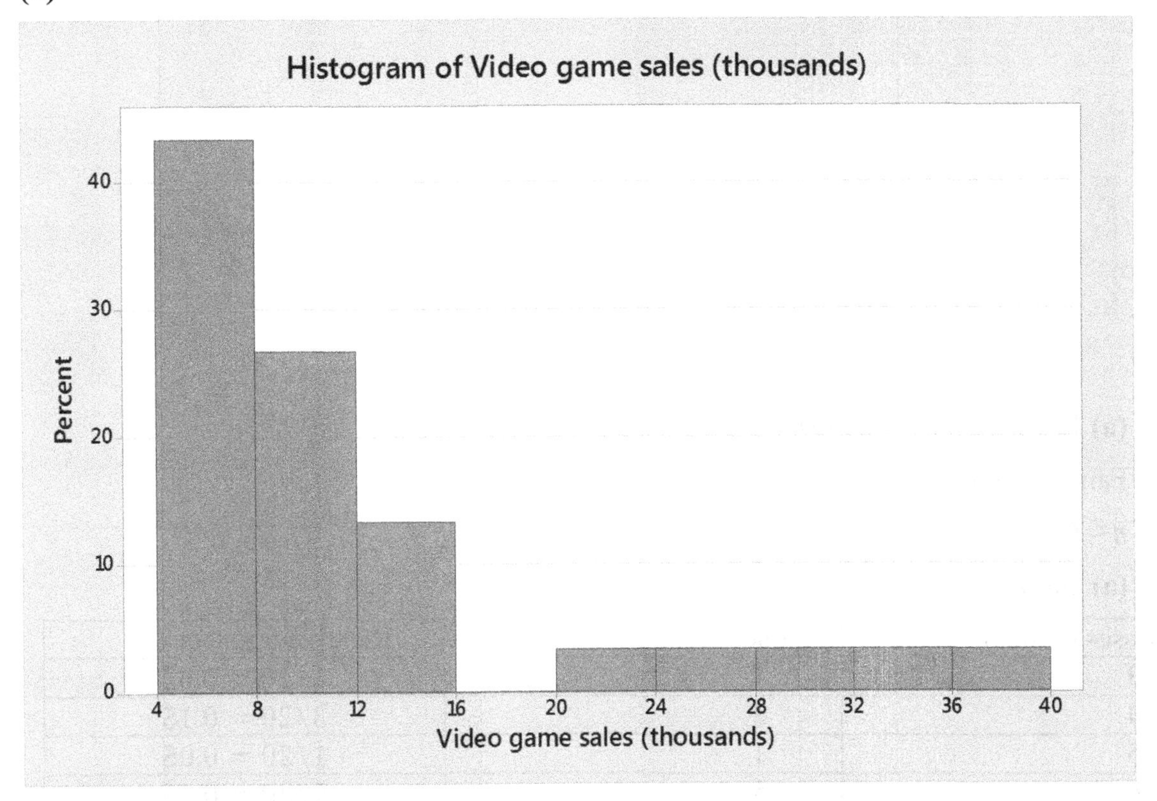

(c) No. Histogram does not contain actual data.

101. **(a)** Approximately 72 **(b)** Approximately 2 **(c)** Approximately 18

103. **(a)** You can change to a relative frequency distribution by dividing the frequency values by the total frequency. The classes will not be affected. **(b)** The relative frequency histogram can be converted into a frequency histogram by changing the scale along the relative frequency axis (vertical axis). This is done by multiplying the relative frequency values by the total frequency. This change in scale will not affect the shape of the distribution. **(c)** The sample size is 19.

105. **(a)** 0 **(b)** 0 **(c)** The class \$25 to \$27.50 has the largest relative frequency. This value is $4/19 \approx 0.2105$. **(d)** 3 **(e)** 0

107. Data set: 23 24 25 26 27 28 28 29 30 31 31 32 32 32 33 35 36 37 39 40

109. Histogram with five classes.

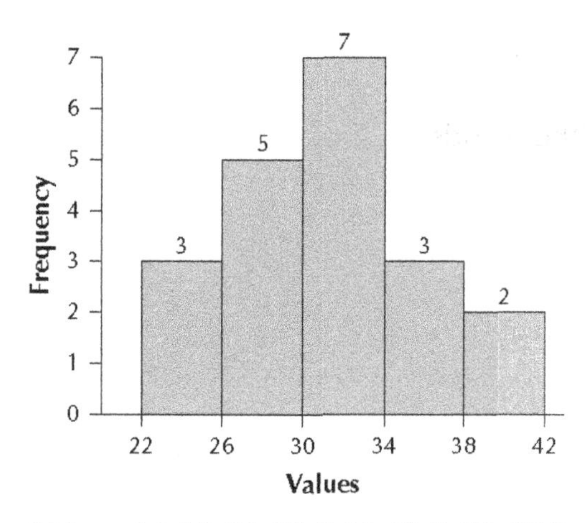

111. **(a)** 15 **(b)** 37.5 **(c)** 52.5 **(d)** 67.5 to 82.5 **(e)** 22.5 to 37.5

113. Fairly symmetrical

115. $n \approx 690$.

117. **(a) and (b)**

Businesses (1000s)	Frequency	Relative frequency
2 to < 3	1	1/20 = 0.05
3 to < 4	3	3/20 = 0.15
4 to < 5	1	1/20 = 0.05
5 to < 6	5	5/20 = 0.25
6 to < 7	2	2/20 = 0.10
7 to < 8	2	2/20 = 0.10
8 to < 9	2	2/20 = 0.10
9 to < 10	2	2/20 = 0.10
10 to < 11	1	1/20 = 0.05
11 to < 12	1	1/20 = 0.05
Total	**20**	**20/20 = 1.00**

(c)

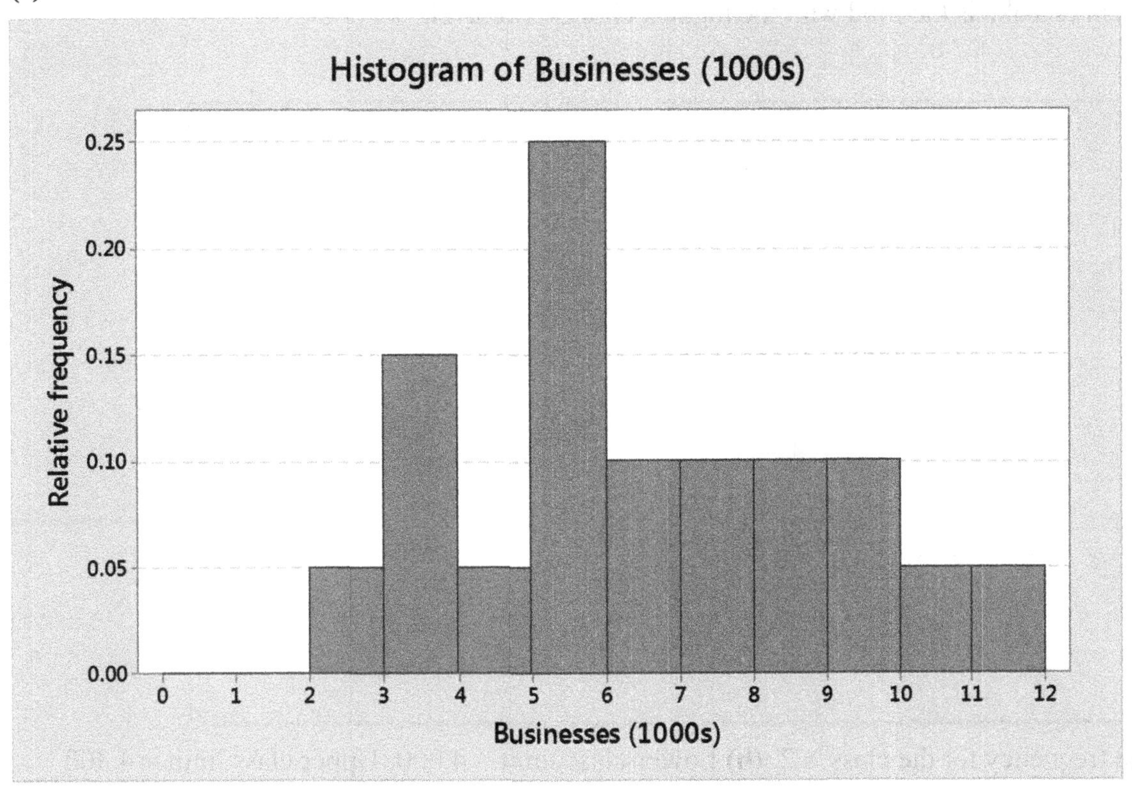

119.

	Dotplot	**Histogram**	**Stem-and-leaf**	**Frequency polygon**
(a) Symmetry and skewness	Appropriate to use for small ranges of data	Appropriate to use	Appropriate to use for small ranges of data	Appropriate to use
(b) Construct using pencil and paper	Easily done for small ranges of data	Easily done for small ranges of data	Easily done for small ranges of data	Easily done for small ranges of data
(c) Retain complete knowledge of the data	Appropriate	Appropriate only if the data are ungrouped	Appropriate	Appropriate only if the data are ungrouped
(d) Presentation in front of non-statisticians	Appropriate	Appropriate	Appropriate	Appropriate

121. There are 961 observations in the data set and there are 25 variables.

123. Yes; fats and oils

125. From Minitab the highest cholesterol value is 2053 grams. From the original data set one whole cheesecake is the food with the highest cholesterol level.

127.

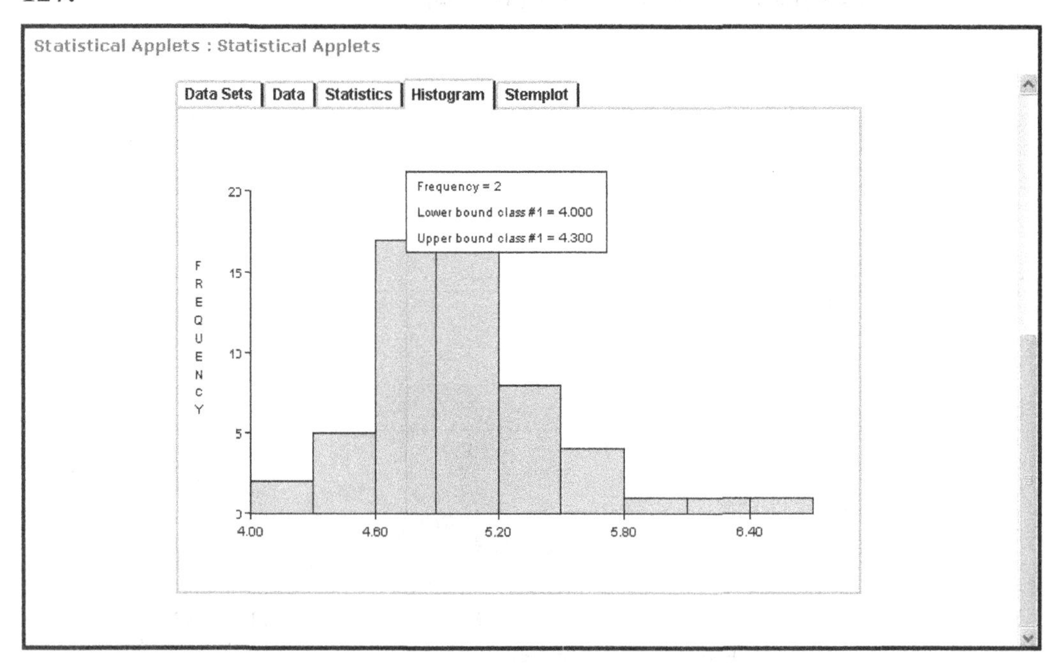

(a) The frequency for the class is 2. **(b)** Lower-class limit = 4.000; Upper class limit = 4.300

129.

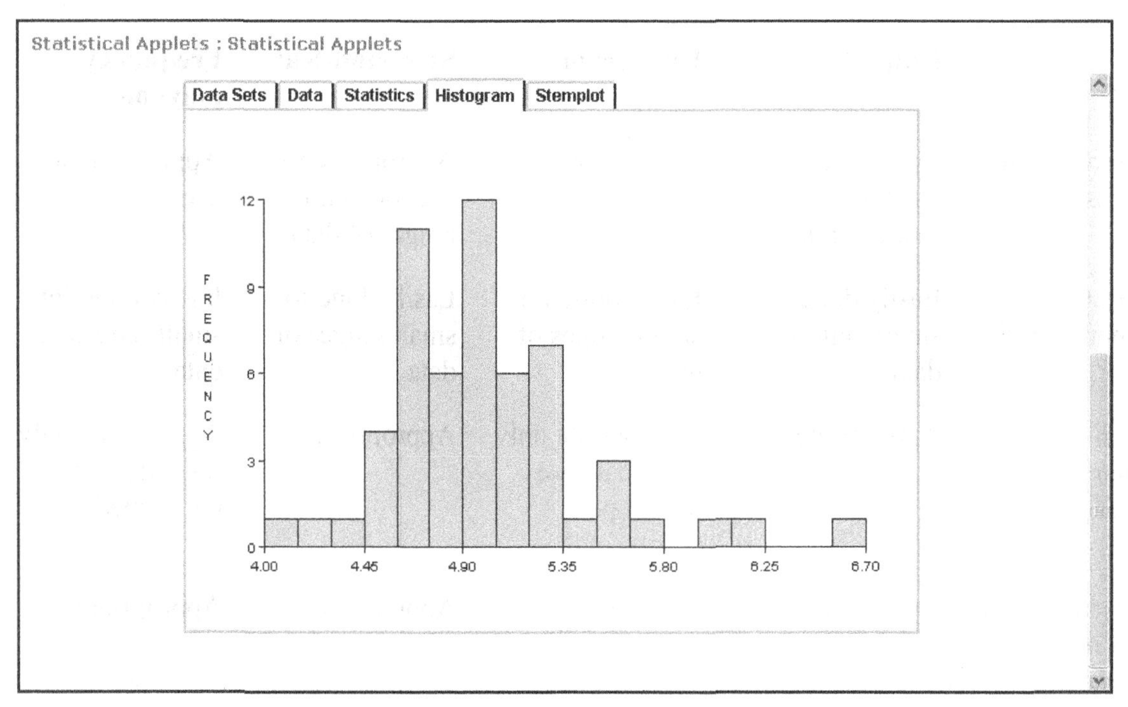

(a) The number of classes increases to 18. **(b)** The class width decreases.

131.

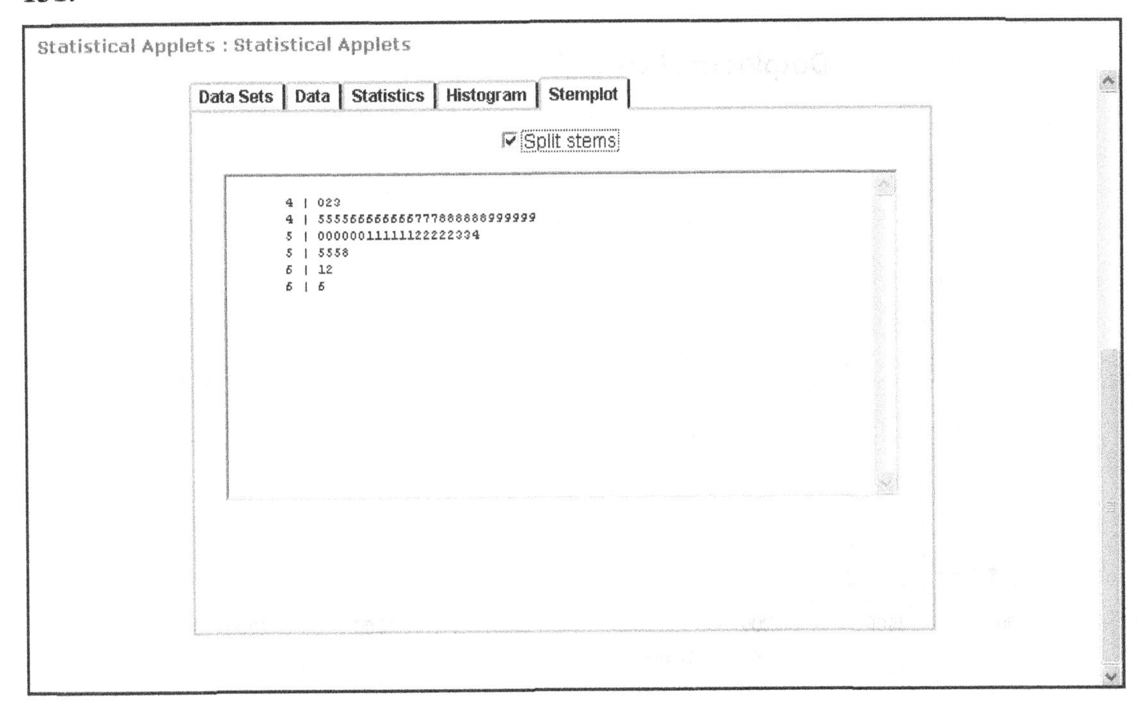

(a) Now there are 6 stems. **(b)** There are still 57 leaves. **(c)** The split stem.

133. Answers will vary.

135.

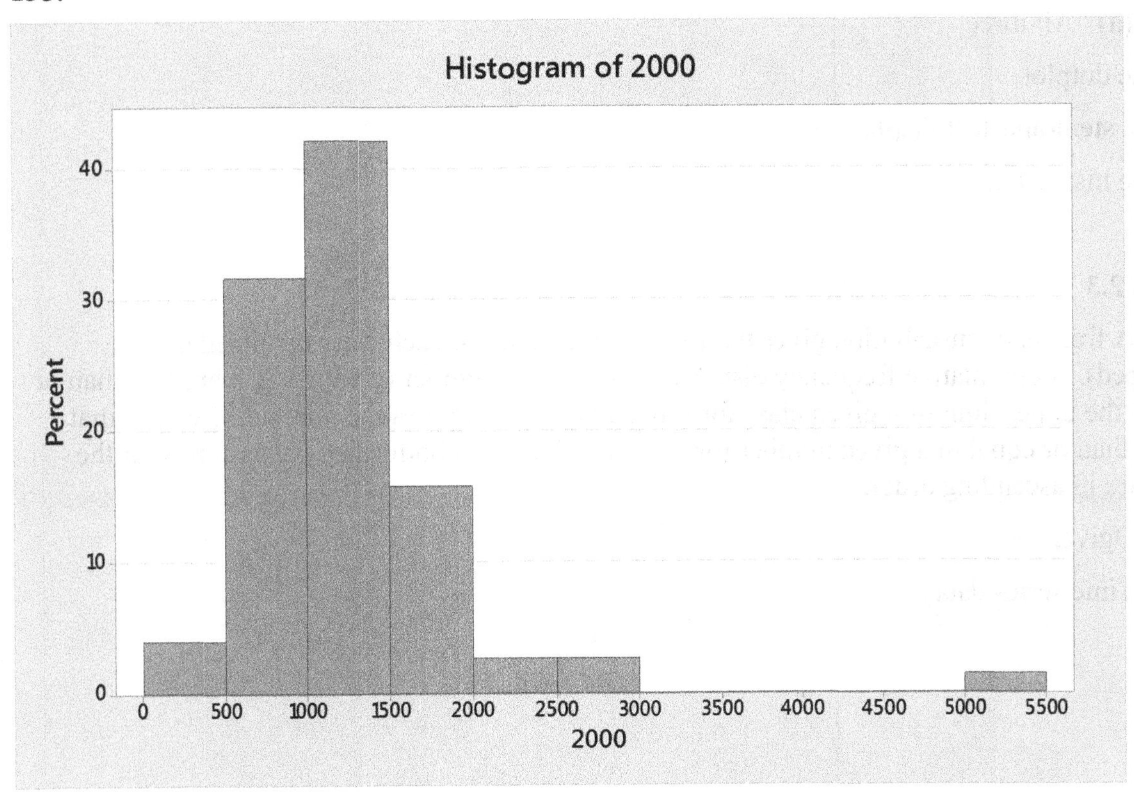

137. The number of petit larceny cases appears to be decreasing. This is good news.

139.

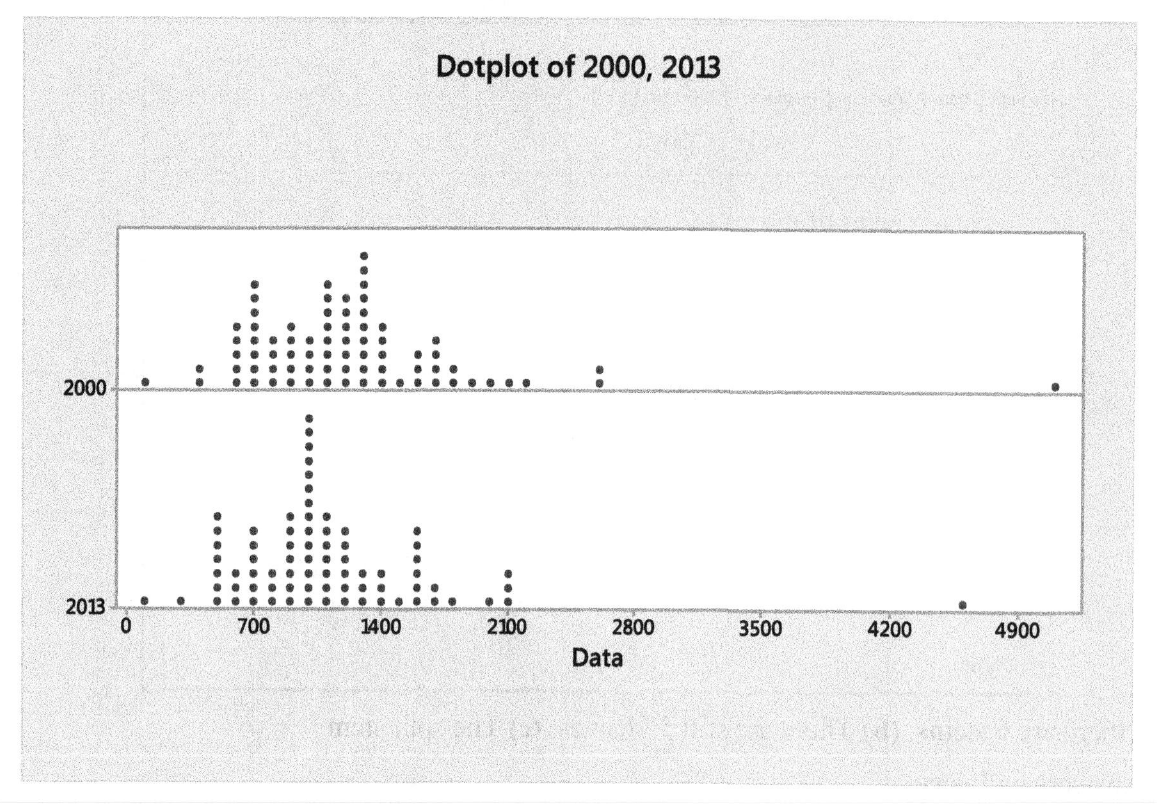

The number of petit larcenies appears to be decreasing.

141. **(a)** All three

(b) The dotplot

(c) The stem-and-leaf display

(d) The histogram

Section 2.3

1. A frequency distribution gives the frequency counts for each class (grouped or ungrouped). A cumulative frequency distribution gives the number of values that are less than or equal to the upper limit of a given class for grouped data, or it gives the number of values that are less than or equal to a given number for ungrouped data. In both cases we assume that the values are in ascending order.

3. Ogive.

5. Time series data.

7.

Age	Frequency (millions)	Relative frequency
$15 \le x < 35$	22.7	$22.7/93.9 \approx 0.24$
$35 \le x < 45$	22.2	$22.2/93.9 \approx 0.24$
$45 \le x < 55$	25.8	$25.8/93.9 \approx 0.27$
$55 \le x < 65$	23.2	$23.2/93.9 \approx 0.25$

9. (a) and (b)

Carbon emissions	Frequency	Relative frequency	Cumulative frequency	Cumulative relative frequency
$50 \le x < 60$	2	$2/20 = 0.10$	2	0.10
$60 \le x < 70$	5	$5/20 = 0.25$	$2 + 5 = 7$	$0.10 + 0.25 = 0.35$
$70 \le x < 80$	2	$2/20 = 0.10$	$2 + 5 + 2 = 9$	$0.10 + 0.25 + 0.10 = 0.45$
$80 \le x < 90$	2	$2/20 = 0.10$	$2 + 5 + 2 + 2 = 11$	$0.10 + 0.25 + 0.10 + 0.10 = 0.55$
$90 \le x < 100$	6	$6/20 = 0.30$	$2 + 5 + 2 + 2 + 6 = 17$	$0.10 + 0.25 + 0.10 + 0.10 + 0.30 = 0.85$
$100 \le x < 110$	2	$2/20 = 0.10$	$2 + 5 + 2 + 2 + 6 + 2 = 19$	$0.10 + 0.25 + 0.10 + 0.10 + 0.30 + 0.10 = 0.95$
$110 \le x < 120$	1	$1/20 = 0.05$	$2 + 5 + 2 + 2 + 6 + 2 + 1 = 20$	$0.10 + 0.25 + 0.10 + 0.10 + 0.30 + 0.10 + 0.05 = 1.00$
Total	**20**	$\mathbf{20/20 = 1.00}$		

11. (a) and (b)

Dangerous weapons cases	Frequency	Relative frequency	Cumulative frequency	Cumulative relative frequency
$0 \le x < 20$	3	$\frac{3}{20} = 0.15$	3	0.15
$20 \le x < 40$	5	$\frac{5}{20} = 0.25$	$3 + 5 = 8$	$0.15 + 0.25 = 0.40$
$40 \le x < 60$	4	$\frac{4}{20} = 0.20$	$3 + 5 + 4 = 12$	$0.15 + 0.25 + 0.20 = 0.60$
$60 \le x < 80$	2	$\frac{2}{20} = 0.10$	$3 + 5 + 4 + 2 = 14$	$0.15 + 0.25 + 0.20 + 0.10 = 0.70$
$80 \le x < 100$	5	$\frac{5}{20} = 0.25$	$3 + 5 + 4 + 2 + 5 = 19$	$0.15 + 0.25 + 0.20 + 0.10 + 0.25 = 0.95$
$100 \le x < 120$	1	$\frac{1}{20} = 0.05$	$3 + 5 + 4 + 2 + 5 + 1 = 20$	$0.15 + 0.25 + 0.20 + 0.10 + 0.25 + 0.05 = 1.00$
Total	**20**	$\frac{20}{20} = \mathbf{1.00}$		

13. (a)

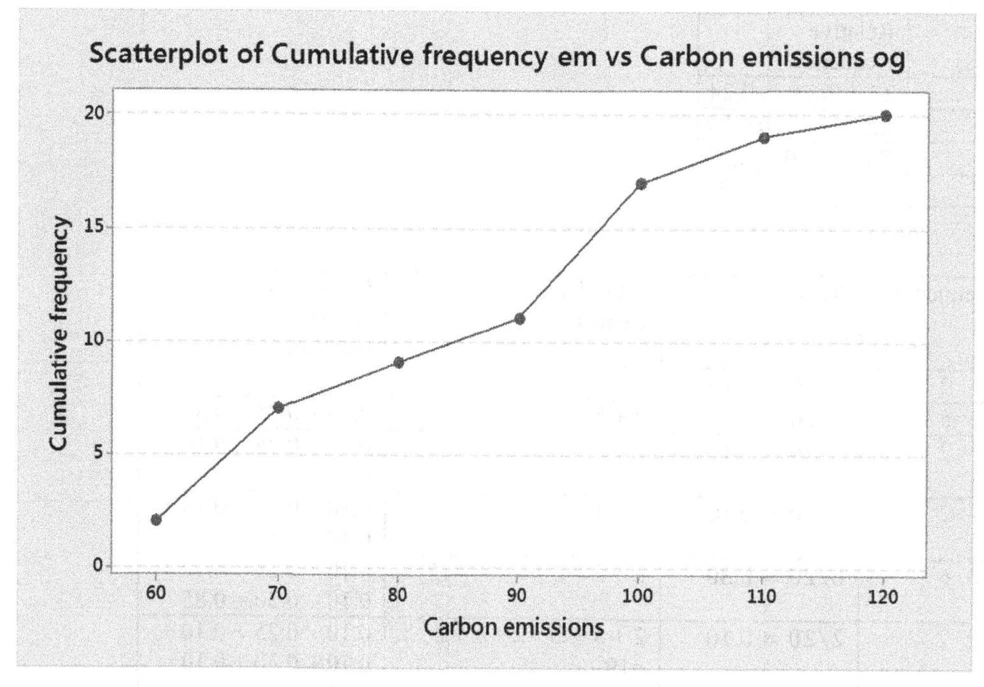
Scatterplot of Cumulative frequency em vs Carbon emissions og
Cumulative frequency
0
5
10
15
20
60
70
80
90
100
110
120
Carbon emissions

(b)

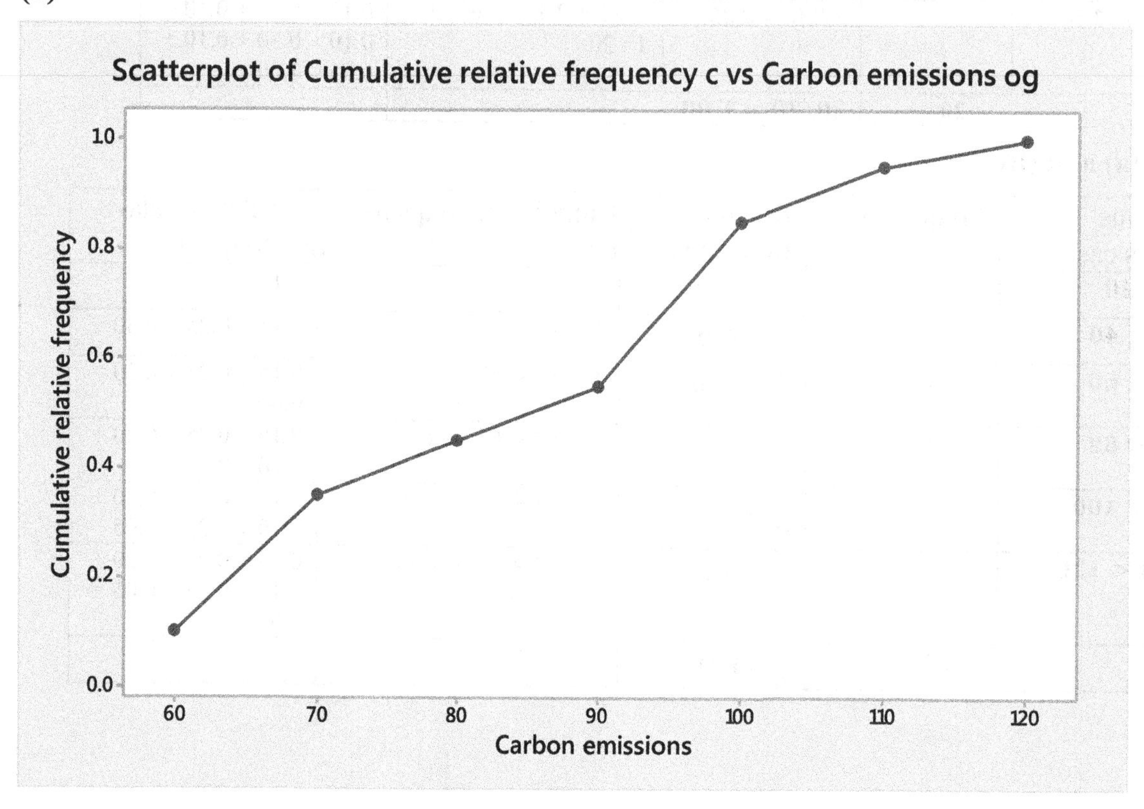
Scatterplot of Cumulative relative frequency c vs Carbon emissions og
Cumulative relative frequency
0.0
0.2
0.4
0.6
0.8
1.0
60
70
80
90
100
110
120
Carbon emissions

15. (a)

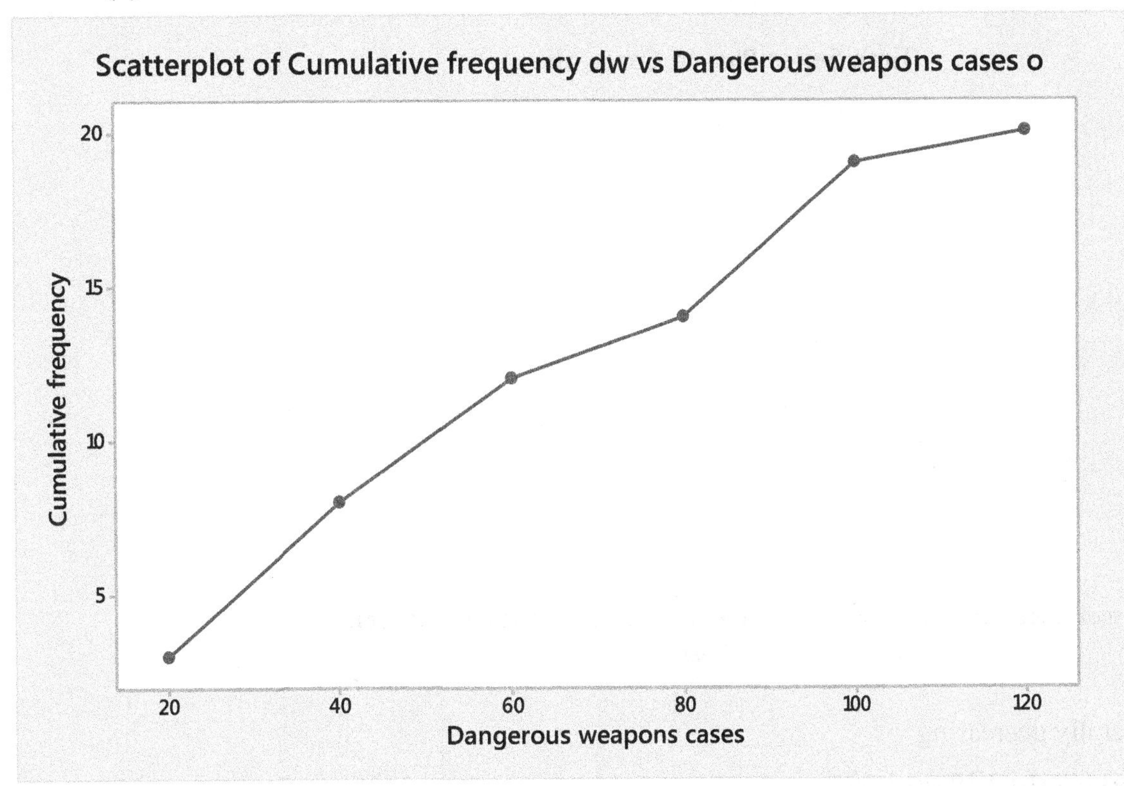
Scatterplot of Cumulative frequency dw vs Dangerous weapons cases o
Cumulative frequency
20
15
10
5
20
40
60
80
100
120
Dangerous weapons cases

(b)

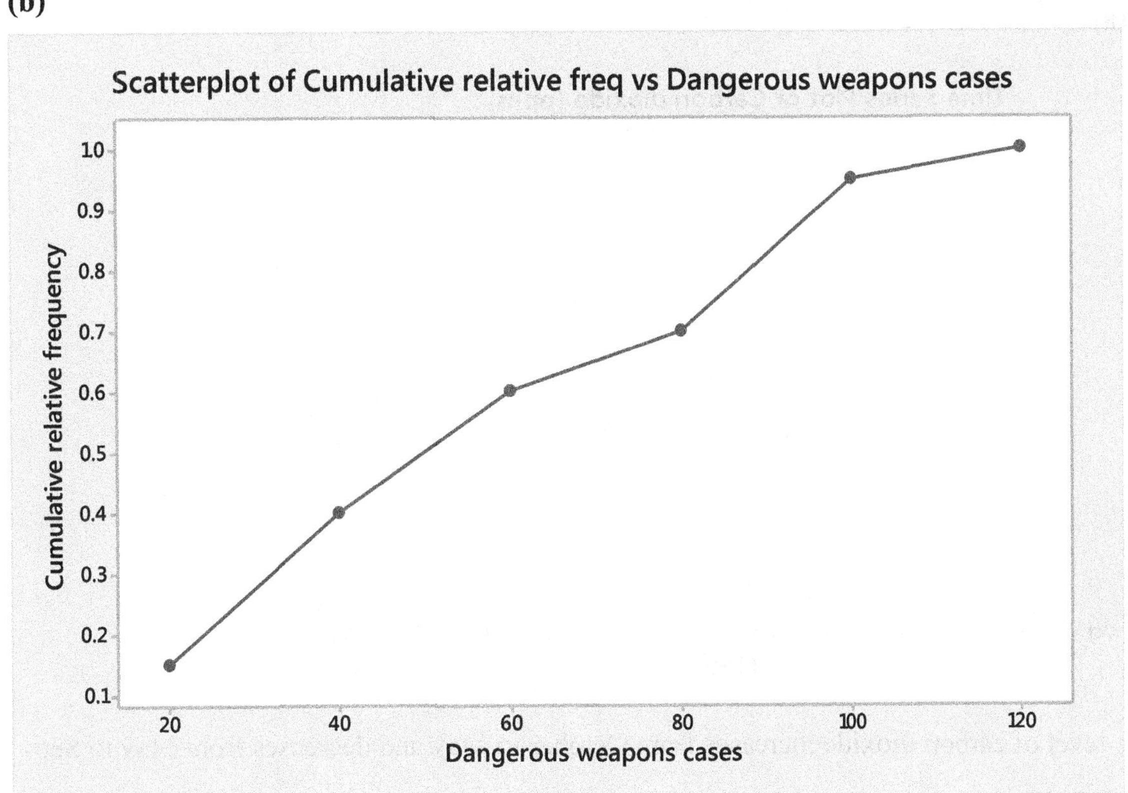
Scatterplot of Cumulative relative freq vs Dangerous weapons cases
Cumulative relative frequency
1.0
0.9
0.8
0.7
0.6
0.5
0.4
0.3
0.2
0.1
20
40
60
80
100
120
Dangerous weapons cases

17. **(a)**

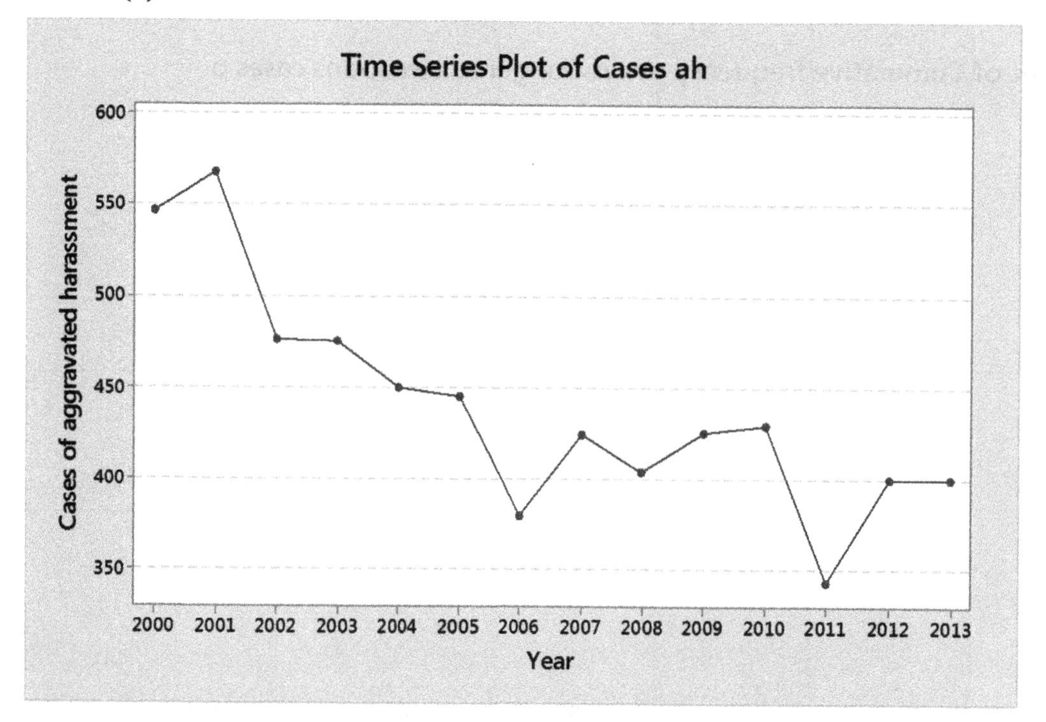

(b) Generally decreasing

19. **(a)** 0.8 **(b)** 2.39 **(c)** 1.99

21. **(a)**

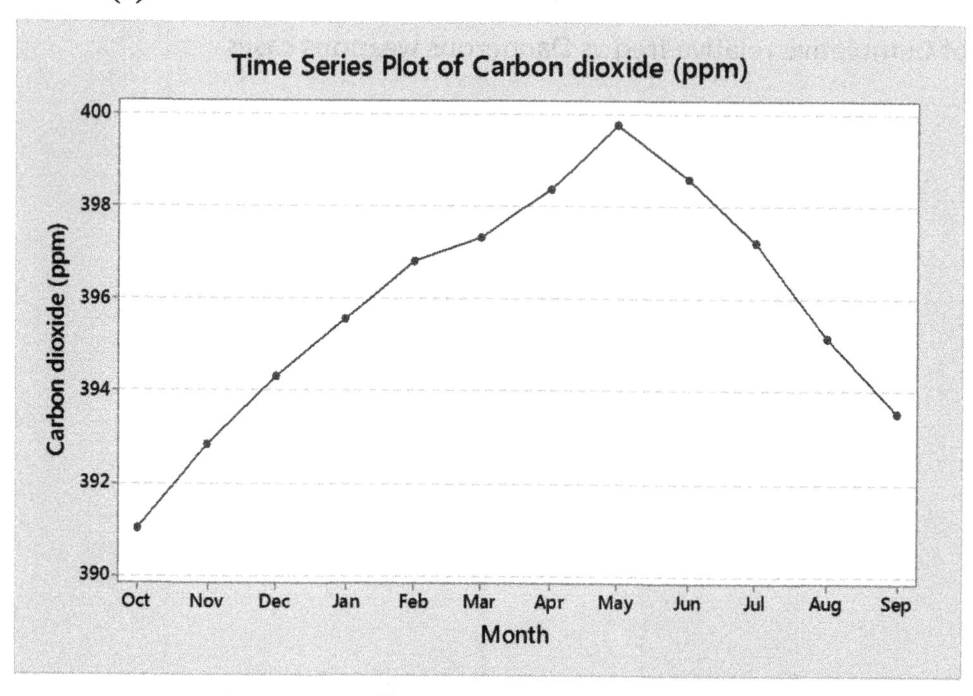

(b) The level of carbon dioxide increases from October to May and decreases from May to September.

23. 2009

25.

Agricultural exports (in billions of dollars)	**Frequency**	**Relative frequency**	**Cumulative relative frequency**
\$0–\$1.9	3	3/20 = 0.15	0.15
\$2.0–\$3.9	9	9/20 = 0.45	0.15 + 0.45 = 0.60
\$4.0–\$5.9	6	6/20 = 0.30	0.15 + 0.45 + 0.30 = 0.90
\$6.0–\$7.9	1	1/20 = 0.05	0.15 + 0.45 + 0.30 + 0.05 = 0.95
\$8.0–\$9.9	0	0/20 = 0	0.15 + 0.45 + 0.30 + 0.05 + 0 = 0.95
\$10.0–\$11.9	0	0/20 = 0	0.15 + 0.45 + 0.30 + 0.05 + 0 + 0 = 0.95
\$12.0–\$13.9	1	1/20 = 0.05	0.15 + 0.45 + 0.30 + 0.05 + 0 + 0 + 0.05 = 1.00
Total	**20**	**1.00**	

(a) 0.60. **(b)** 0.90. **(c)** 0.10.

27. **(a)**

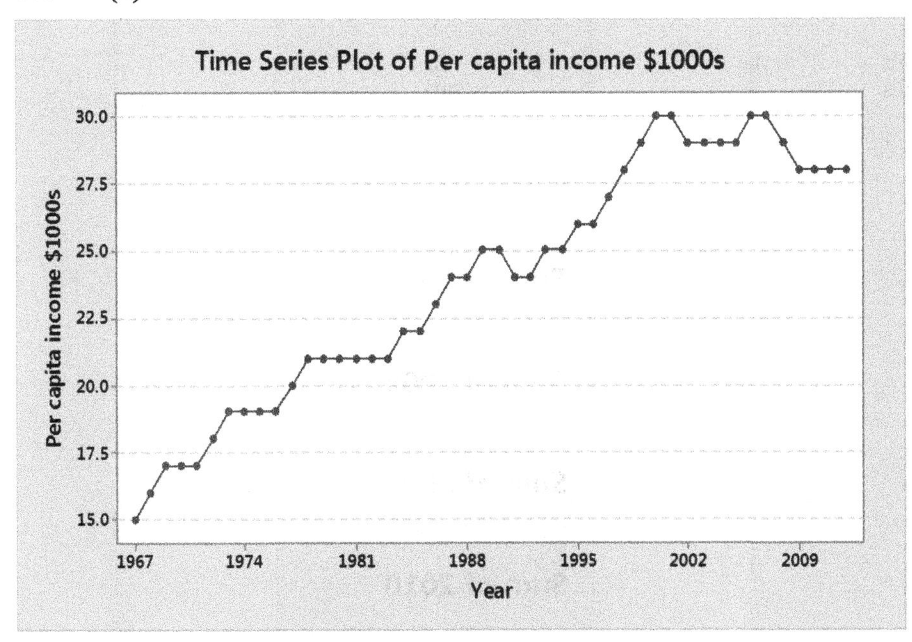

(b) 2008

29. The only change would be that the entire time series graph would shift up 3 units. The horizontal scale would stay the same. The shape of the graph would stay the same.

31.

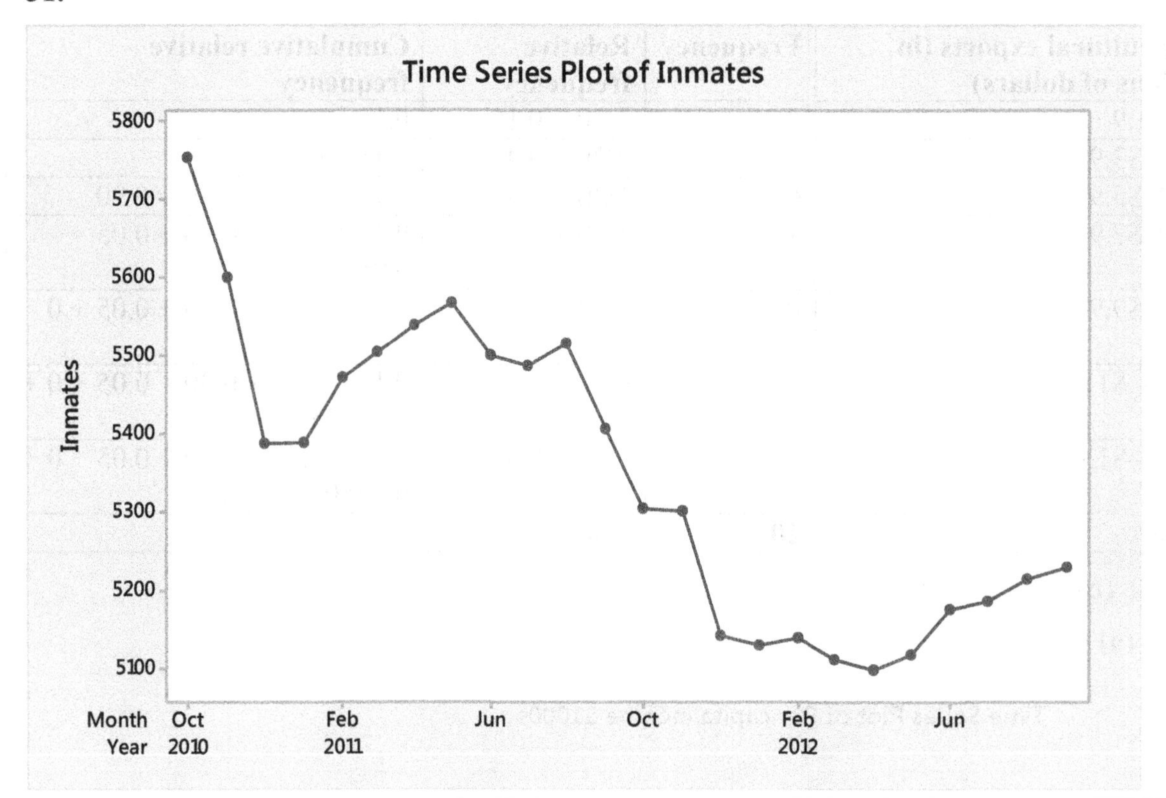

33. 3DS

35.

Sum of 2000

Sum of 2000 = 57304

Sum of 2001

Sum of 2001 = 57753

Sum of 2002

Sum of 2002 = 52469

Sum of 2003

Sum of 2003 = 51298

Sum of 2004

Sum of 2004 = 52158

Sum of 2005

Sum of 2005 = 52408

Sum of 2006

Sum of 2006 = 52169

Sum of 2007

Sum of 2007 = 51429

Sum of 2008

Sum of 2008 = 50310

Sum of 2009

Sum of 2009 = 50216

Sum of 2010

Sum of 2010 = 52716

Sum of 2011

Sum of 2011 = 50972

Sum of 2012

Sum of 2012 = 54495

Sum of 2013

Sum of 2013 = 53445

Year	**Number of third degree assaults**
2000	57,304
2001	57,753
2002	52,469
2003	51,298
2004	52,158
2005	52,408
2006	52,169
2007	51,429
2008	50,310
2009	50,216
2010	52,716
2011	50,972
2012	54,495
2013	53,445

Section 2.4

1. It is important to understand how graphics may be made misleading, confusing, or deceptive because such an understanding enhances our statistical literacy and makes us less prone to being deceived by misleading graphics.

3. Figure 2.54 is more effective at convincing the general public that a problem exists because the graphical representation visually reinforces the magnitude of the differences.

5. The insurance company would prefer to use Table 2.44 rather than Table 2.45. Table 2.44 gives the actual number of cars stolen.

7. **(a)** Biased distortion or embellishment, omitting the zero on the relevant scales, inaccuracy in relative lengths of bars in a bar chart.

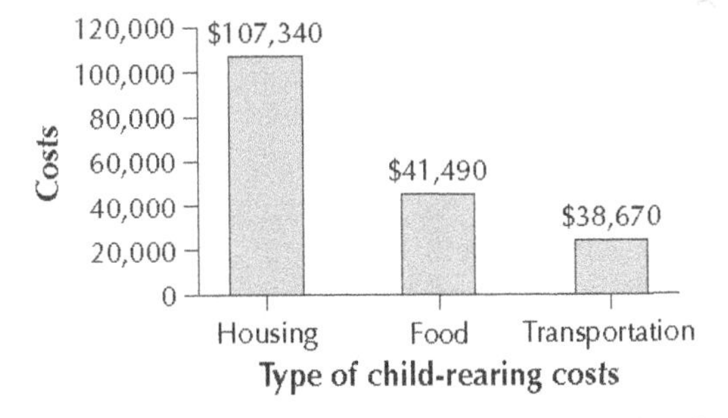

(b) One may use a Pareto chart or pie chart.

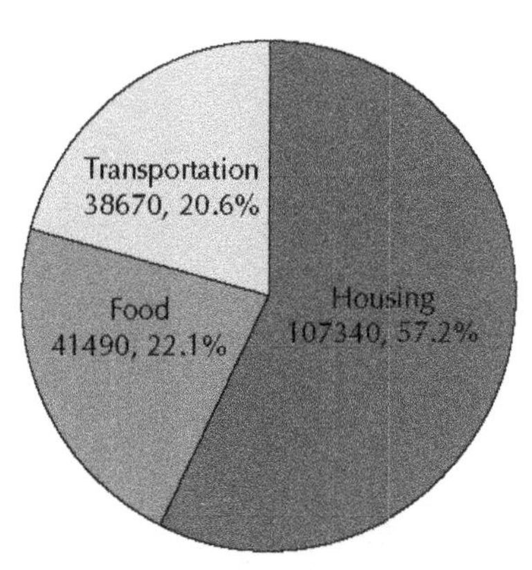

9. **(a)** The number of people living with AIDS is increasing. **(b)** Using two dimensions (area) to emphasize a one-dimensional difference. **(c)** One can use a bar chart.

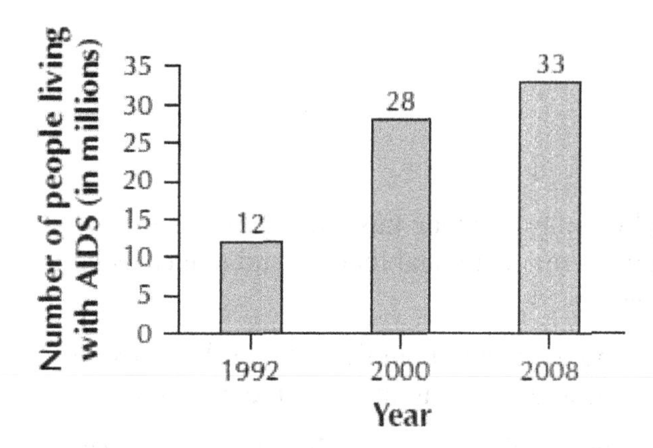

11. **(a)**

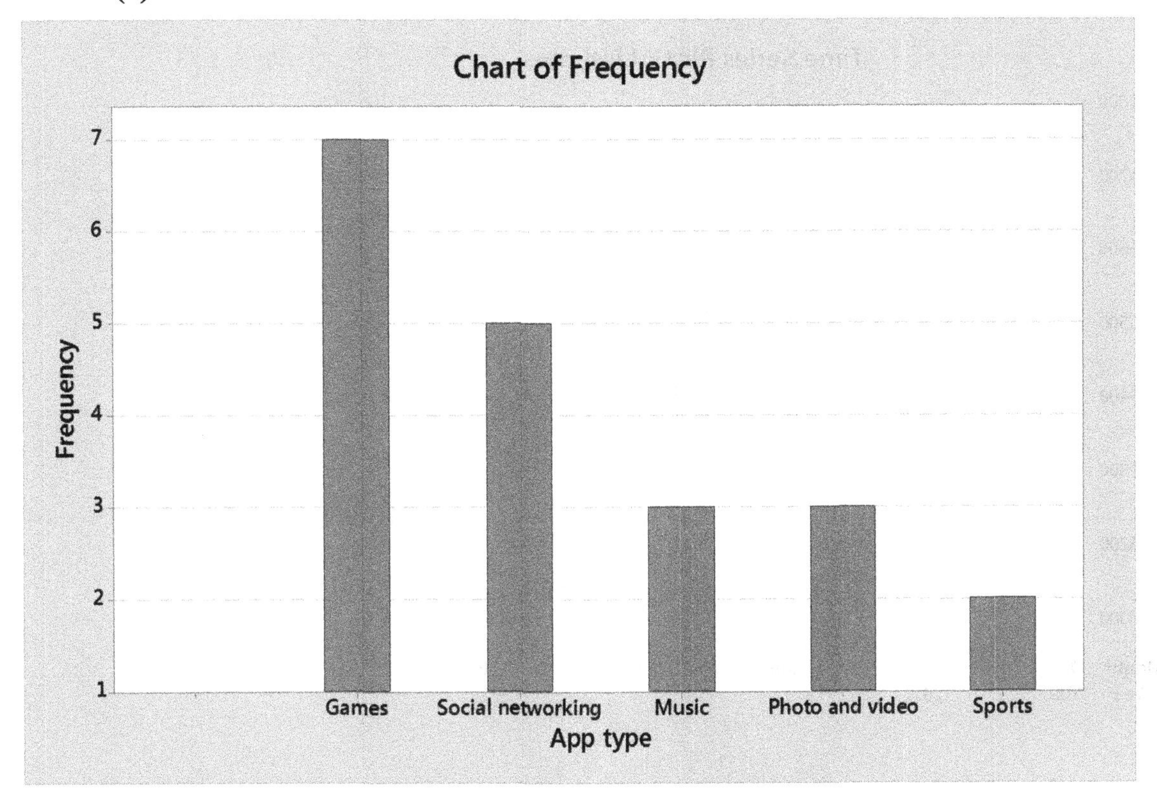

(b) Manipulating the scale, omitting the 0 on the vertical scale

(c)

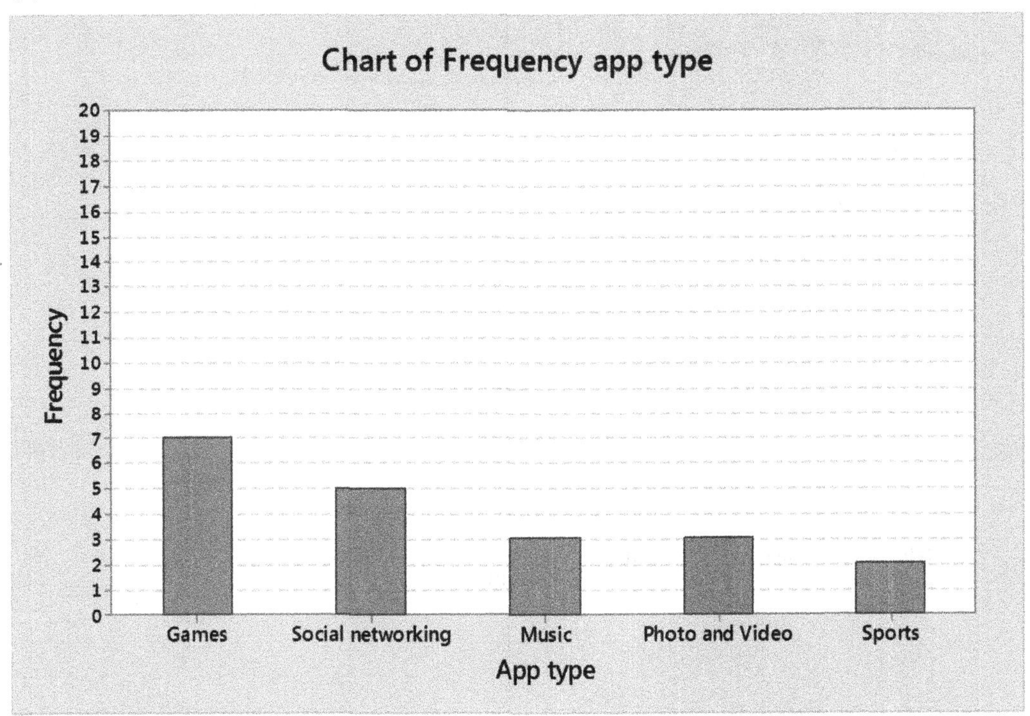

(d) Manipulating the scale

13. **(a)**

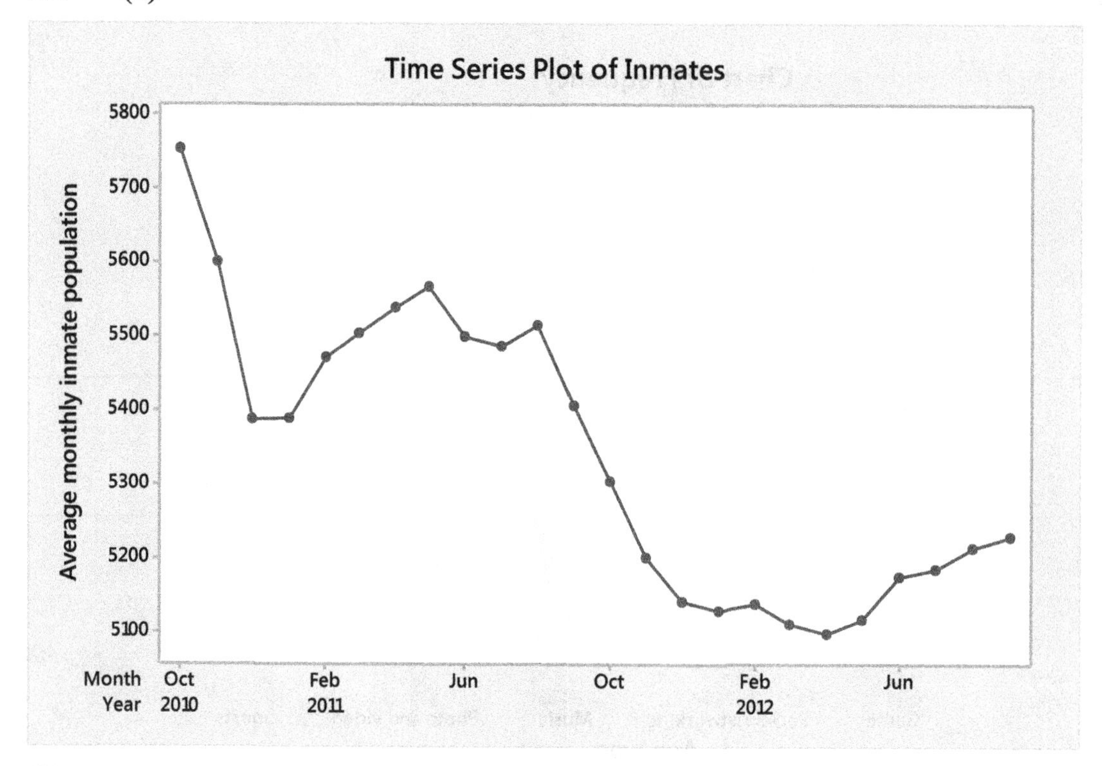
Time Series Plot of Inmates
Average monthly inmate population
5800
5700
5600
5500
5400
5300
5200
5100
Month Oct Feb Jun Oct Feb Jun
Year 2010 2011 2012

(b)

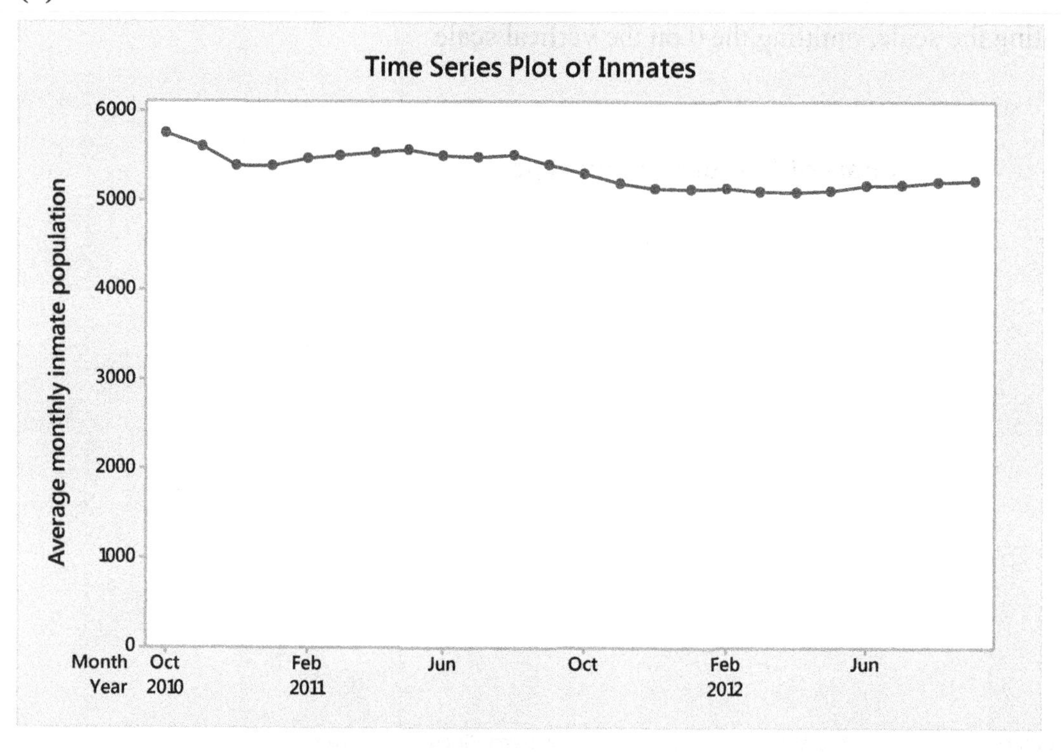
Time Series Plot of Inmates
Average monthly inmate population
6000
5000
4000
3000
2000
1000
0
Month Oct Feb Jun Oct Feb Jun
Year 2010 2011 2012

15.

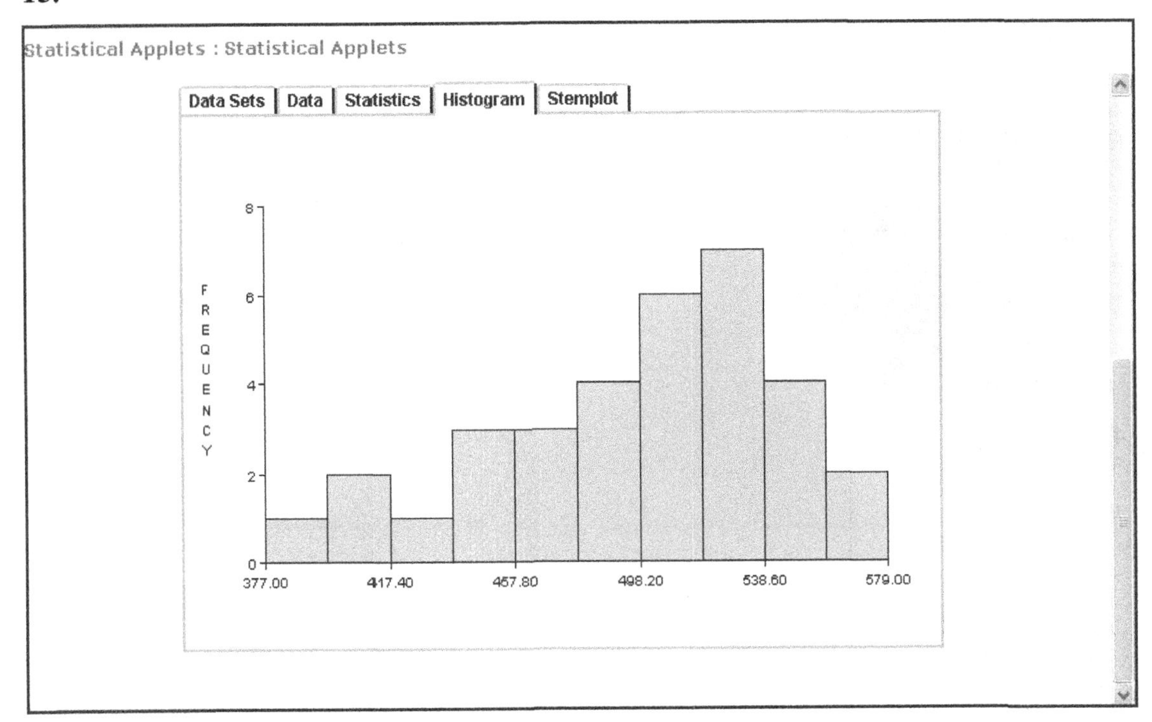

Chapter 2 Review Exercises

1. No, because the variable is categorical.

3.

Part of Speech	Frequency
Adjective	1
Adverb	2
Article	3
Conjunction	3
Preposition	9
Pronoun	7
Verb	6
Total	**31**

5.

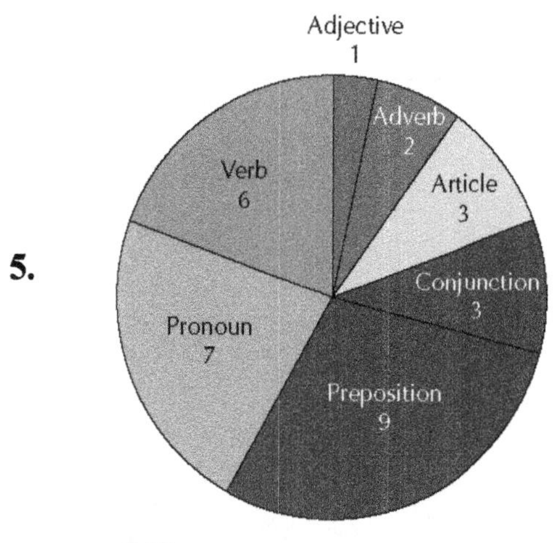

7. $\frac{242}{366} \approx 0.6612$

9. $\frac{9}{366} \approx 0.0246$

11. (a)

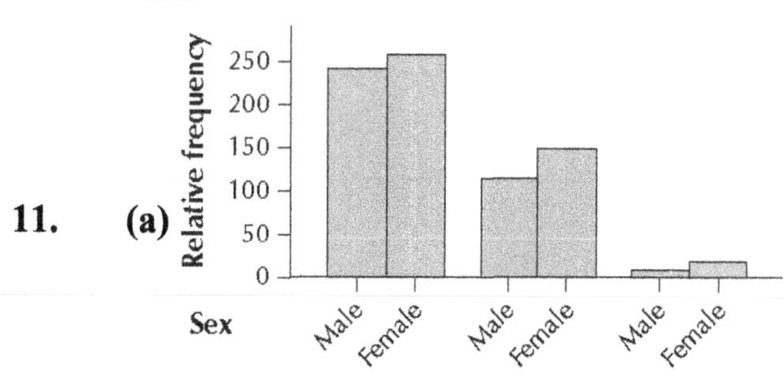

(b)

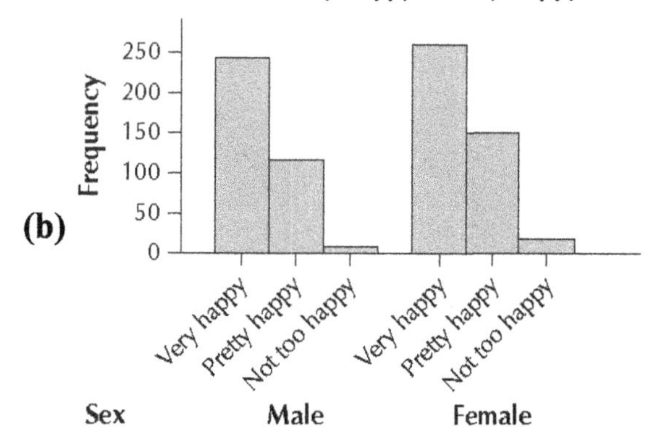

13.

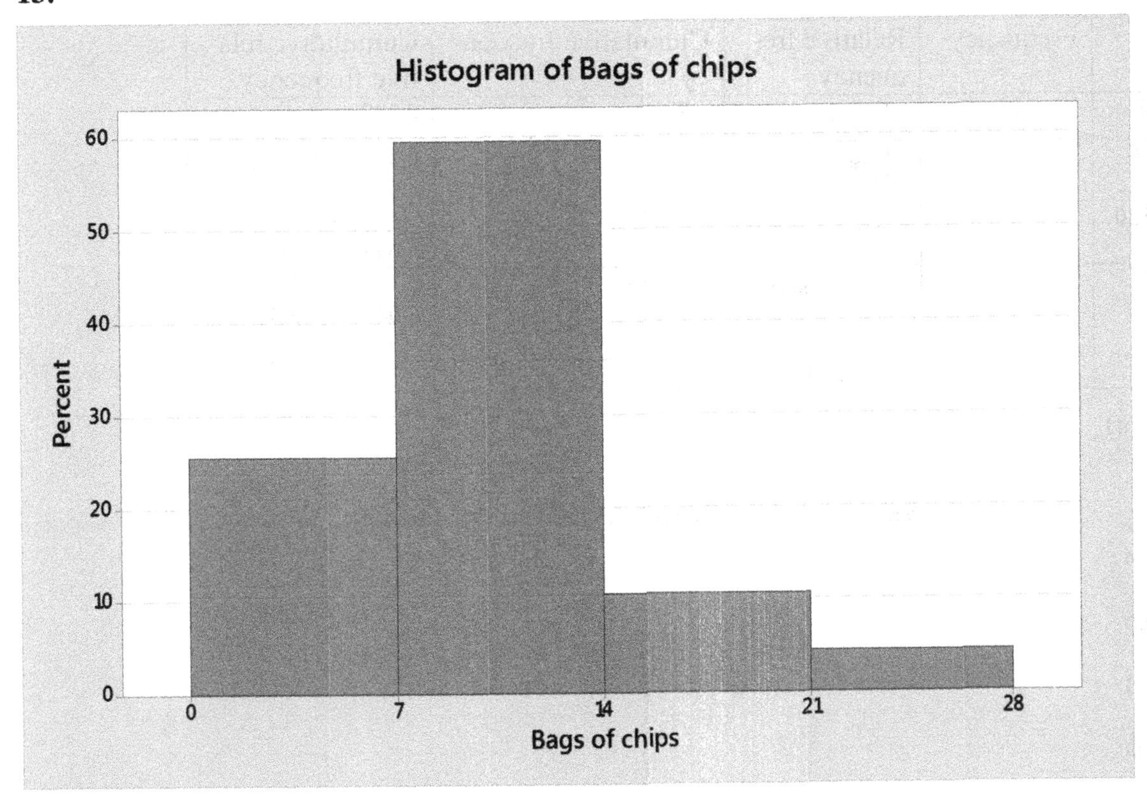

15 and 17.

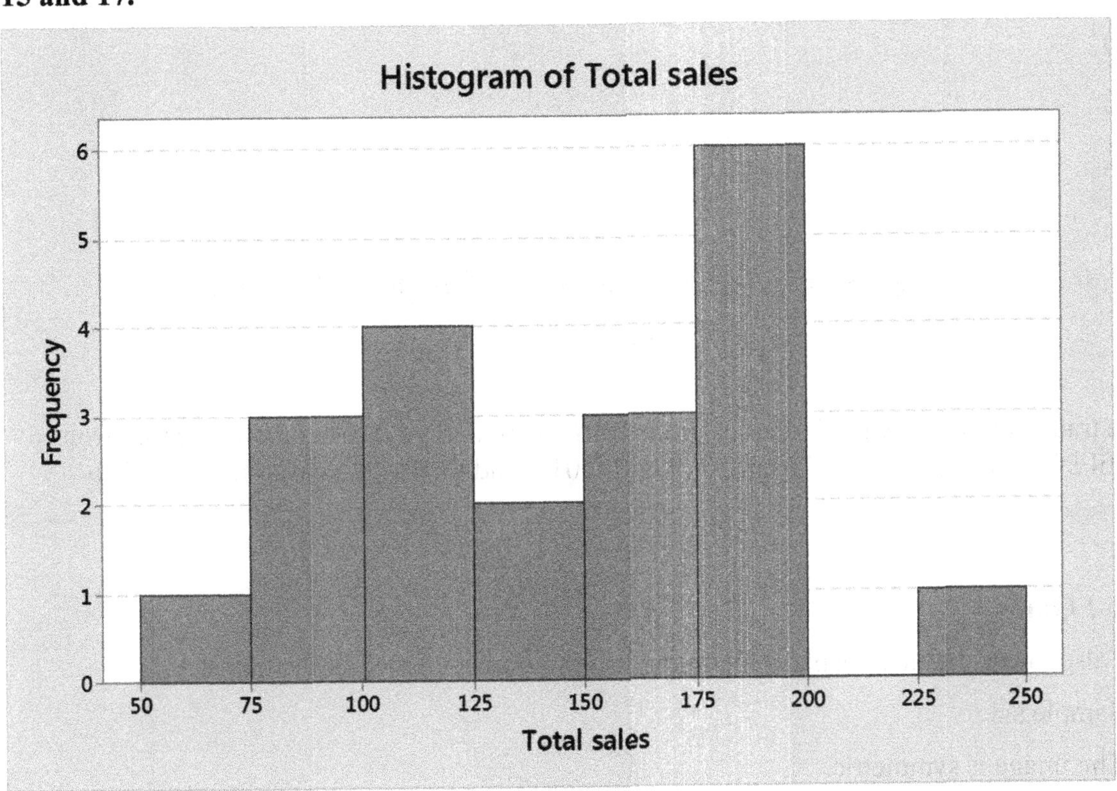

19. $\frac{10}{20} = 0.5$ of the days had sales of at most \$149.99.

21. (a) and (b)

Frauds	Frequency	Relative frequency	Cumulative frequency	Cumulative relative frequency
0 to < 40	8	$\frac{8}{23} \approx 0.3478$	8	0.3478
40 to < 80	13	$\frac{13}{23} \approx 0.5652$	8+13 = 21	0.3478 + 0.5632 = 0.9130
80 to < 120	1	$\frac{1}{23} \approx 0.0435$	8 + 13 + 1 = 22	0.3478 + 0.5632 + 0.0435 = 0.9565
120 to < 160	1	$\frac{1}{23} \approx 0.0435$	8 + 13 + 1 + 1 = 23	0.3478 + 0.5632 + 0.0435 + 0.0435 = 1.0000
Total	**23**	**1.0000**		

23. (a)

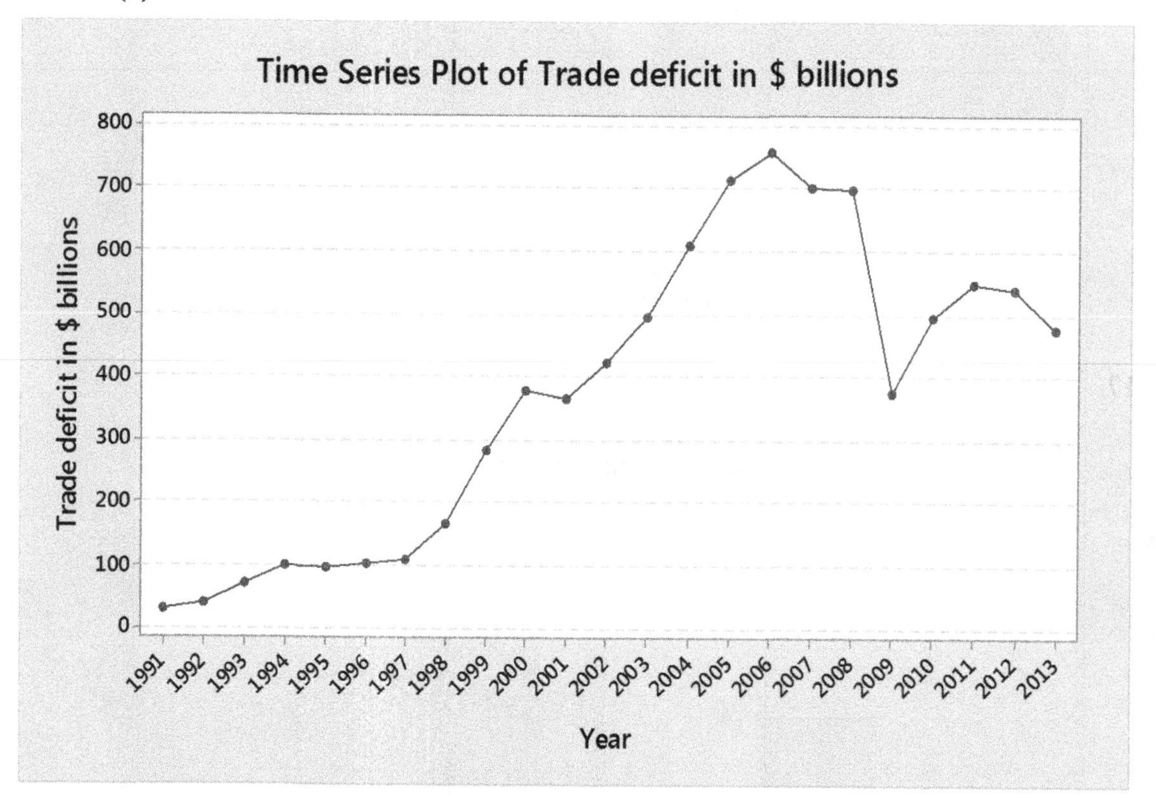

(b) The trade deficit generally increased from 1991 to 2006, then it generally decreased from 2006 until 2009, then it increased from 2009 until 2011, and then it decreased from 2011 to 2013.

Chapter 2 Quiz

1. False. Stem-and-leaf displays retain the information contained in the data set.

3. Sample size

5. The image is symmetric.

7 and 9.

Life Expectancy	Frequency	Cumulative frequency
$40 \leq x < 50$	1	1
$50 \leq x < 60$	0	1 + 0 =1
$60 \leq x < 70$	3	1 + 0 + 3 = 4
$70 \leq x < 80$	3	1 + 0 + 3 + 3 = 7
$80 \leq x < 90$	3	1 + 0 + 3 + 3 + 3 = 10
Total	10	

11.

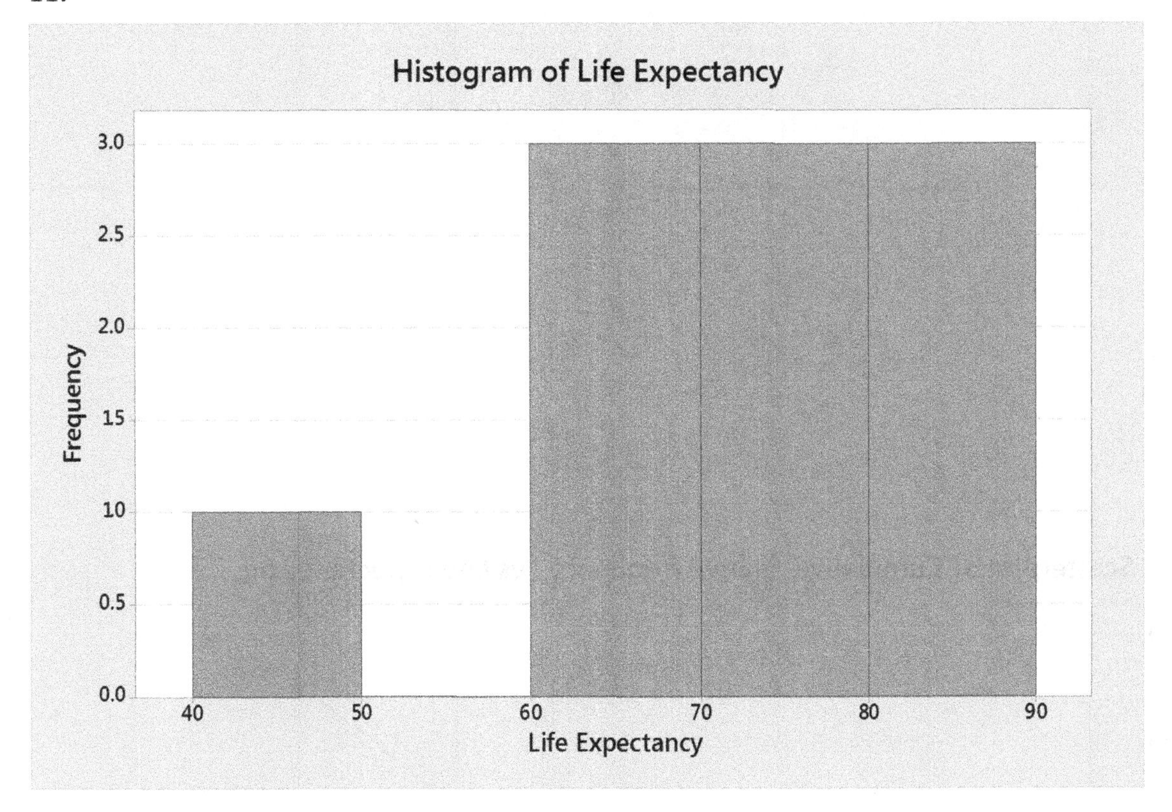

13.

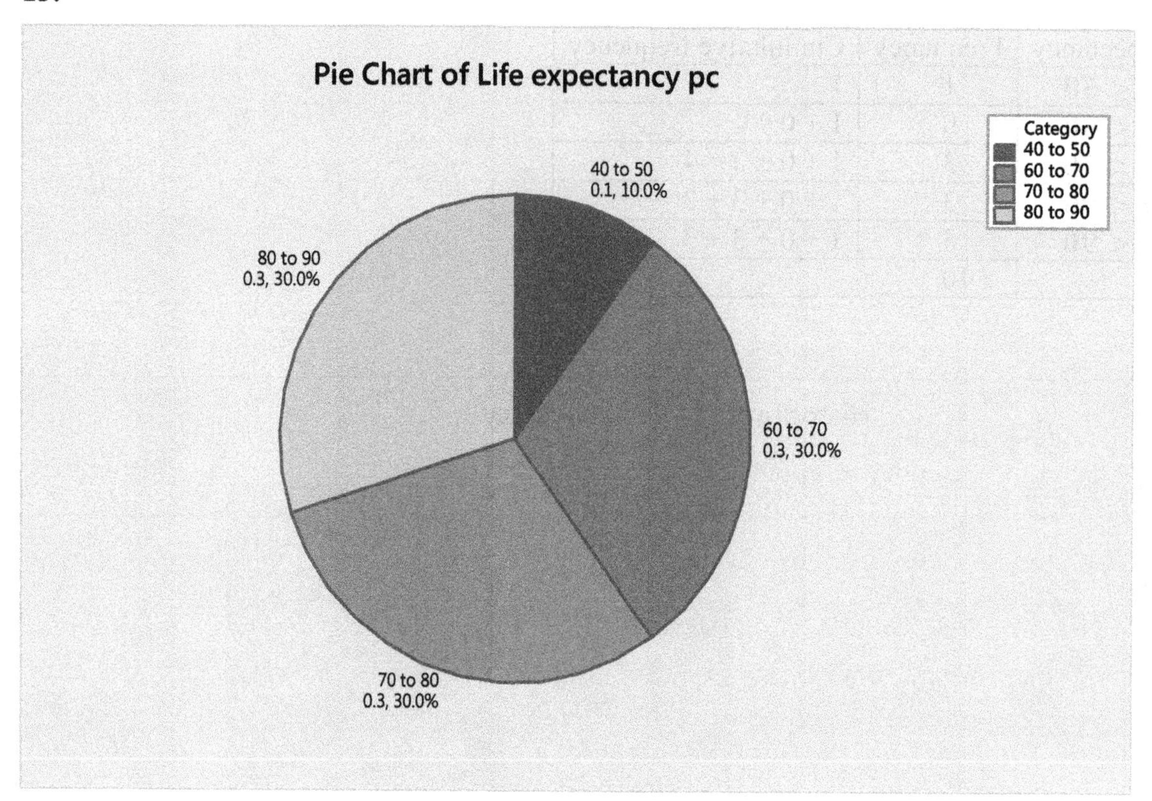

Pie Chart of Life expectancy pc
Category
40 to 50
60 to 70
70 to 80
80 to 90
40 to 50
0.1, 10.0%
80 to 90
0.3, 30.0%
60 to 70
0.3, 30.0%
70 to 80
0.3, 30.0%

15.

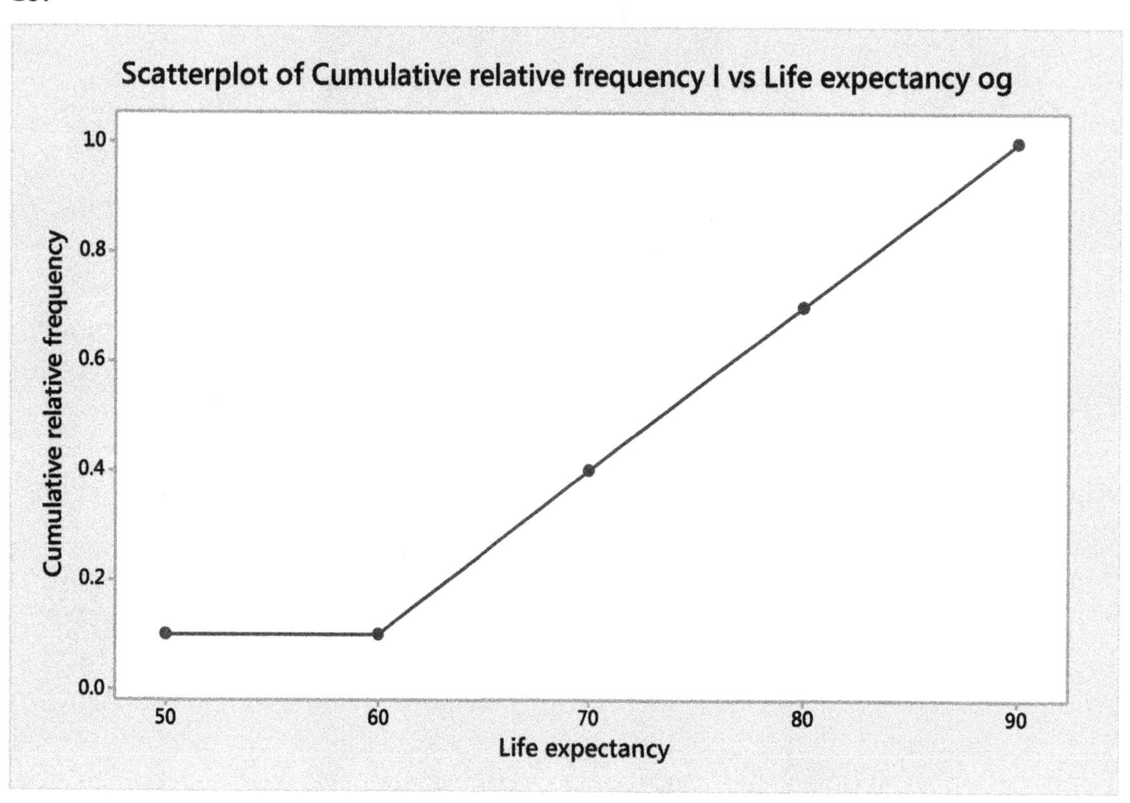

Scatterplot of Cumulative relative frequency l vs Life expectancy og
Cumulative relative frequency
1.0
0.8
0.6
0.4
0.2
0.0
50
60
70
80
90
Life expectancy

Chapter 3: Describing Data Numerically

Section 3.1

1. A measure of center is a value that locates the center of the data set.

3. The mean is sensitive to outliers because each value in the data set is used to compute the mean. If there is an outlier in the data it will skew the value of the mean. The median is not sensitive to outliers because it is the middle value when the data are put in ascending order.

5. The sample size (n).

7. $\sum x$

9. $\bar{x}$

11. The mode.

13. **(a)** $N = 6$ **(b)** $\mu = \frac{\sum x}{N} = \frac{1.4+0.3+2.4+0.4+0.2+0.3}{6} = \frac{5.0}{6} \approx \$0.83$ billion

15. **(a)** $N = 10$ **(b)** $\mu = \frac{\sum x}{N} = \frac{208.0+174.1+167.8+117.2+106.0+100.2+94.1+75.2+75.1+66.0}{10} = \frac{1183.7}{10} = 118.37$ motor vehicles stolen per 100,000 residents

17. **(a)** $N = 10$ **(b)** $\mu = \frac{\sum x}{N} = \frac{2014+1288+1555+584+1607+995+2094+4551+823+2071}{10} = \frac{17{,}582}{10} = 1758.2$ petit larceny cases

19. **(a)** $n = 3$ **(b)** $\bar{x} = \frac{\sum x}{n} = \frac{1.4+2.4+0.2}{3} = \frac{4.0}{3} \approx \$1.33$ **billion**

21. **(a)** $n = 3$ **(b)** $\bar{x} = \frac{\sum x}{n} = \frac{208.0+167.8+100.2}{3} = \frac{476}{3} \approx 158.67$ motor vehicles stolen per 100,000 residents

23. **(a)** $n = 5$ **(b)** $\bar{x} = \frac{\sum x}{n} = \frac{2014+1555+1607+4551+823}{5} = \frac{10{,}550}{5} = 2110$ **petit larceny cases**

25. $\bar{x} = \frac{\sum x}{n} = \frac{1.4+2.4+0.2+10}{4} = \frac{14.0}{4} = \$3.5$ **billion. This is greater than $1.33 billion.**

27. $\bar{x} = \frac{\sum x}{n} = \frac{208.0+167.8+100.2+1000}{4} = \frac{1476}{4} = 369$ motor vehicles stolen per 100,000 residents. **This is larger than 158.67** motor vehicles stolen per 100,000 residents.

29. $\bar{x} = \frac{\sum x}{n} = \frac{2014+1555+1607+4551+823+20{,}000}{6} = \frac{30{,}550}{6} \approx 5091.67$ petit larceny cases. This is larger than **2110 petit larceny cases.**

31. **(a)** First arrange the data in ascending order.

0.2, 1.4, 2.4

Since $n = 3$ is odd, the median is the unique $\left(\frac{n+1}{2}\right)^{th} = \left(\frac{3+1}{2}\right)^{th} = 2$nd value, which is $1.4.billion.

(b) First arrange the data in ascending order.

0.2, 1.4, 2.4, 10

Since $n = 4$ is even, the median is the mean of the two data values that lie on either side of the $\left(\frac{n+1}{2}\right)^{th} = \left(\frac{4+1}{2}\right)^{th} = 2.5$th position. That is, the median is the mean of the 2nd and 3rd data values, 1.4 and 2.4. Splitting the difference between these two we get

median $= \frac{1.4+2.4}{2} = \frac{3.8}{2} = \$1.9$ billion.

Adding the extreme value of $10 billion to the data set only increased the median from $1.4

billion to \$1.9 billion, whereas the mean increased from \$1.33 billion to \$3.5 billion.

33. **(a)** First arrange the data in ascending order.
100.2, 167.8, 208.0

Since $n = 3$ is odd, the median is the unique $\left(\frac{n+1}{2}\right)^{th} = \left(\frac{3+1}{2}\right)^{th} = 2\text{nd}$ value, which is 167.8 motor vehicles stolen per 100,000 residents.

(b) First arrange the data in ascending order.
100.2, 167.8, 208.0, 1000

Since $n = 4$ is even, the median is the mean of the two data values that lie on either side of the $\left(\frac{n+1}{2}\right)^{th} = \left(\frac{4+1}{2}\right)^{th} = 2.5\text{th}$ position. That is, the median is the mean of the 2nd and 3rd data values, 167.8 and 208.0. Splitting the difference between these two we get

$$\text{median} = \frac{167.8+208.0}{2} = \frac{375.8}{2} = 187.9 \text{ motor vehicles stolen per 100,000 residents.}$$

Adding the extreme value of 1000 motor vehicles stolen per 100,000 residents to the data set only increased the median from 167.8 motor vehicles stolen per 100,000 residents.to 187.9 motor vehicles stolen per 100,000 residents, whereas the mean increased from 158.67 motor vehicles stolen per 100,000 residents to 369 motor vehicles stolen per 100,000 residents.

35. **(a)** First arrange the data in ascending order.
823, 1555, 1607, 2014, 4551

Since $n = 5$ is odd, the median is the unique $\left(\frac{n+1}{2}\right)^{th} = \left(\frac{5+1}{2}\right)^{th} = 3\text{rd}$ value, which is 1607 petit larceny cases.

(b) First arrange the data in ascending order.
823, 1555, 1607, 2014, 4551, 20,000

Since $n = 6$ is even, the median is the mean of the two data values that lie on either side of the $\left(\frac{n+1}{2}\right)^{th} = \left(\frac{6+1}{2}\right)^{th} = 3.5\text{th}$ position. That is, the median is the mean of the 3rd and 4th data values, 1607 and 2014. Splitting the difference between these two we get

$$\text{median} = \frac{1607+2014}{2} = \frac{3621}{2} = 1810.5 \text{ petit larceny cases.}$$

Adding the extreme value of 20,000 petit larceny cases.to the data set only increased the median from 1607 petit larceny cases.to 1810.5 petit larceny cases, whereas the mean increased from 2110 petit larceny cases to 5091.67 petit larceny cases.

37. 24 dangerous weapons cases

39. 6.2

41. Mean < Median < Mode

43. Mean = Median = Mode

45. **(a)** $N = 4$ **(b)** $\mu = \frac{\sum x}{N} = \frac{12+11+4+4}{4} = \frac{31}{4} = 7.75$ wins

47. **(a)** $n = 2$ **(b)** $\bar{x} = \frac{\sum x}{n} = \frac{12+4}{2} = \frac{16}{2} = 8$ wins

49. **(a)** $\bar{x} = \frac{\sum x}{n} = \frac{1.7+0.6+0.9+1.2+0.2}{5} = \frac{4.6}{5} = 0.92$ million game units

(b) First arrange the data in ascending order.
0.2, 0.6, 0.9, 1.2, 1.7

Since $n = 5$ is odd, the median is the unique $\left(\frac{n+1}{2}\right)^{th} = \left(\frac{5+1}{2}\right)^{th} = 3\text{rd}$ value, which is 0.9 million game units.

51. **(a)** $\bar{x} = \frac{\sum x}{n} = \frac{(-27.4)+18.7+42.2+(-16.3)+11.2+28.5+1.8+16.9}{8} = \frac{75.6}{8} = 9.45$

(b) First arrange the data in ascending order.

-27.4, -16.3, 1.8, 11.2, 16.9, 18.7, 28.5, 42.2

Since $n = 8$ is even, the median is the mean of the two data values that lie on either side of the $\left(\frac{n+1}{2}\right)^{th} = \left(\frac{8+1}{2}\right)^{th} = 4.5\text{th}$ position. That is, the median is the mean of the 4th and 5th data values, 11.2 and 16.9. Splitting the difference between these two we get

$$\text{median} = \frac{11.2+16.9}{2} = \frac{28.1}{2} = 14.05$$

53. **(a)** $\bar{x} = \frac{\sum x}{n} = \frac{40+28+25+34+26+21+19+24}{8} = \frac{217}{8} = 27.125$ years

(b) First arrange the data in ascending order.

19, 21, 24, 25, 26, 28, 34, 40

Since $n = 8$ is even, the median is the mean of the two data values that lie on either side of the $\left(\frac{n+1}{2}\right)^{th} = \left(\frac{8+1}{2}\right)^{th} = 4.5\text{th}$ position. That is, the median is the mean of the 4th and 5th data values, 25 and 26. Splitting the difference between these two we get

$$\text{median} = \frac{25+26}{2} = \frac{51}{2} = 25.5 \text{ years}$$

55. **(a)** $\bar{x} = \frac{\sum x}{n} = \frac{216+364+149+75+276+375+146+59+67+195}{10} = \frac{1922}{10} = 192.2$ calories

(b) First arrange the data in ascending order.

59, 67, 75, 146, 149, 195, 216, 276, 364, 375

Since $n = 10$ is even, the median is the mean of the two data values that lie on either side of the $\left(\frac{n+1}{2}\right)^{th} = \left(\frac{10+1}{2}\right)^{th} = 5.5\text{th}$ position. That is, the median is the mean of the 5th and 6th data values, 149 and 195. Splitting the difference between these two we get

$$\text{median} = \frac{149+195}{2} = \frac{344}{2} = 172 \text{ calories}$$

57. $n = 9$

59. First arrange the data in ascending order.

−5.6, −5.6, −2.4, −1.8, −1.8, −1.3, −1.2, 0, 1

Since $n = 9$ is odd, the median is the unique $\left(\frac{n+1}{2}\right)^{th} = \left(\frac{9+1}{2}\right)^{th} = 5\text{th}$ value, which is −\$1.8 billion.

61. **(a)** $\bar{x} = \frac{\sum x}{n} = \frac{6+4+6+4+12+4}{6} = \frac{36}{6} = 6$ cylinders

(b) First arrange the data in ascending order.

4, 4, 4, 6, 6, 12

Since $n = 6$ is even, the median is the mean of the two data values that lie on either side of the $\left(\frac{n+1}{2}\right)^{th} = \left(\frac{6+1}{2}\right)^{th} = 3.5\text{th}$ position. That is, the median is the mean of the 3rd and 4th data values, 4 and 6. Splitting the difference between these two we get

$$\text{median} = \frac{4+6}{2} = \frac{10}{2} = 5 \text{ cylinders}$$

(c) 4 cylinders

63. **(a)** $\bar{x} = \frac{\sum x}{n} = \frac{3+2.5+3.5+1.8+6.7+2.4}{6} = \frac{19.9}{6} = 3.317$ liters

(b) First, order the data in ascending order:

1.8, 2.4, 2.5, 3.0, 3.5, 6.7

Since $n = 6$ is even, the median is the mean of the two data values that lie on either side of the $\left(\frac{n+1}{2}\right)^{th} = \left(\frac{6+1}{2}\right)^{th} = 3.5$th position. That is, the median is the mean of the 3rd and 4th data values, 2.5 and 3.0. Splitting the difference between these two, we get

$$\text{median} = \frac{2.5+3.0}{2} = \frac{5.5}{2} = 2.75$$

(c) All of the values occur once. Therefore, this data set has no mode.

65. U.S.A. occurs three times. No other value occurs more than twice. Therefore, the mode is U.S.A.

67. Data set in ascending order: 1, 2, 7, 12, 15, 23, 25, 27, 29, 47

(a) $n = 10$. The sample mean is

$$\bar{x} = \frac{\sum x}{n} = \frac{1+2+7+12+15+23+25+27+29+47}{10} = \frac{188}{10} = 18.8$$

(b) Since $n = 10$ is even, the median is the mean of the two data values that lie on either side of the $\left(\frac{n+1}{2}\right)^{th} = \left(\frac{10+1}{2}\right)^{th} = 5.5$th position. That is, the median is the mean of the 5th and 6th data values, 15 and 23.

$$\text{median} = \frac{15+23}{2} = \frac{38}{2} = 19$$

(c) The mode is the data value that occurs with the greatest frequency. Since all of the data values occur exactly once, there is no mode.

69. The mean: $\bar{x} = \frac{\sum x}{n} = \frac{7.83+7.64+8.48+6.71+8.47}{5} = \frac{39.13}{5} = \$7.826 \approx \$7.83.$

To find the median, first arrange the data in ascending order.

6.71, 7.64, 7.83, 8.47, 8.48

Since $n = 5$ is odd, the median is the unique $\left(\frac{n+1}{2}\right)^{th} = \left(\frac{5+1}{2}\right)^{th} = 3\text{rd}$ value, which is \$7.83

Since all of the prices occur once, there is no mode.

A typical book on the best-sellers list would cost less than \$14.00 since the mean and median are all less than \$14.

71. **(a)** New data set:

$7.83 \times 5 = 39.15, 7.64 \times 5 = 38.20, 8.48 \times 5 = 42.40, 6.71 \times 5 = 33.55, 8.47 \times 5 = 42.35$

The mean of this new data set is $\bar{x} = \frac{\sum x}{n} = \frac{39.15+38.20+42.40+33.55+42.35}{5} = \frac{195.65}{5} = \$39.13$

(b) The new mean is 5 times the original mean.

(c) If each value of a data set is multiplied by a number m, the mean of the resulting data set will be the mean of the original data set times m.

73. It does not make sense to find the mean and the median of the variable author because the variable author is a categorical variable.

75. Manhattan

77. Data set in ascending order:

2011, 2011, 2012, 2012, 2012, 2012, 2013, 2013, 2013, 2014, 2014, 2014, 2014, 2015, 2015, 2015, 2015, 2015, 2015, 2015

The mean:

$$\bar{x} = \frac{\sum x}{n}$$

$$= \frac{2011 + 2011 + 2012 + 2012 + 2012 + 2012 + 2013 + 2013 + 2013 + 2014 + 2014 + 2014 + 2014 + 2015 + 2015 + 2015 + 2015 + 2015 + 2015 + 2015}{20}$$

$$= \frac{40{,}270}{20} = 2013.5$$

Since $n = 20$ is even, the median is the mean of the two data values that lie on either side of the $\left(\frac{n+1}{2}\right)^{th} = \left(\frac{20+1}{2}\right)^{th} = 10.5$th position. That is, the median is the mean of the 10th and 11th data values, 2014 and 2014. Splitting the difference between these two we get

$$\text{median} = \frac{2014+2014}{2} = \frac{4028}{2} = 2014$$

Mode = 2015

79. New data set in descending order:

$2025 - 2011 = 14$, $2025 - 2011 = 14$, $2025 - 2012 = 13$, $2025 - 2012 = 13$, $2025 - 2012 = 13$, $2025 - 2012 = 13$, $2025 - 2013 = 12$, $2025 - 2013 = 12$, $2025 - 2013 = 12$, $2025 - 2014 = 11$, $2025 - 2014 = 11$, $2025 - 2014 = 11$, $2025 - 2014 = 11$, $2025 - 2015 = 10$, $2025 - 2015 = 10$, $2025 - 2015 = 10$, $2025 - 2015 = 10$, $2025 - 2015 = 10$, $2025 - 2015 = 10$, $2025 - 2015 = 10$

The mean:

$$\bar{x} = \frac{\sum x}{n} = \frac{14+14+13+13+13+13+12+12+12+11+11+11+11+10+10+10+10+10+10+10}{20} = \frac{230}{20} = 11.5 \text{ years}$$

Since $n = 20$ is even, the median is the mean of the two data values that lie on either side of the $\left(\frac{n+1}{2}\right)^{th} = \left(\frac{20+1}{2}\right)^{th} = 10.5$th position. That is, the median is the mean of the 10th and 11th data values, 11 and 11. Splitting the difference between these two we get

$$\text{median} = \frac{11+11}{2} = \frac{22}{2} = 11 \text{ years}$$

Since 10 years occurs 7 times and no other age occurs more than 7 times, the mode is 10 years.

81. $\frac{\$84.28}{4} = \$21.07$

83. Let x_4 be the size of the fourth downloaded music file. Then

$$\bar{x} = \frac{\sum x}{n} = \frac{5 + 2 + 3 + x_4}{4} = \frac{10 + x_4}{4} = 3$$

so $10 + x_4 = 4(3) = 12$ Mb. Therefore, $x_4 = 12 - 10 = 2$ Mb.

85. Since $n = 6$ is even, the median is the mean of the two data values that lie on either side of the $\left(\frac{n+1}{2}\right)^{th} = \left(\frac{6+1}{2}\right)^{th} = 3.5th$ position. That is, the median is the mean of the 3rd and 4th data values. The 3rd data value is 14, but the 4th data value x_4 is unknown. Splitting the difference between these two, we get

$$\text{median} = \frac{14+x_4}{2} = 15$$

Then $14 + x_4 = 2(15) = 30$, so $x_4 = 30 - 14 = 16$.

87. Since the mean is larger than the median, the implication is that the distribution is right-skewed.

89. The mean, median, and mode will all be halved. Each statistic will be affected equally.

91. **(a)** Since $n = 65$ is odd, the median is the data value in the $\left(\frac{n+1}{2}\right)^{th} = \left(\frac{65+1}{2}\right)^{th} = 33\text{rd}$ position. Therefore, the median pulse rate for males is 73. **(b)** Since $n = 65$ is odd, the median is the data value in the $\left(\frac{n+1}{2}\right)^{th} = \left(\frac{65+1}{2}\right)^{th} = 33\text{rd}$ position. Therefore, the median pulse rate for females is 76. **(c)** Females have a higher median pulse rate. Yes it does agree.

93. **(a)** Since the mean is $\bar{x} = \frac{\sum x}{n}$, a decrease in the fastest (largest) pulse rate for males would result in a decrease in $\sum x$. Since the number of observations would not change, the mean pulse rate for males would decrease.

(b) Since $n = 65$ is odd, the median is the data value in the $\left(\frac{n+1}{2}\right)^{th} = \left(\frac{65+1}{2}\right)^{th} = 33\text{rd}$ position. Therefore, the median pulse rate for males is 73. If the largest (fastest) pulse rate is decreased so that it is still the largest value, then the positions of all of the numbers will remain unchanged. In this case, since the median pulse rate is the middle value in the data set, a decrease in the fastest pulse rate for males would not affect the median. If the median is decreased so that it is in the 34th or higher position, the number in the 33rd position will remain unchanged. Therefore the median will remain unchanged. Since the numbers in the 31st, 32nd, 33rd, 34th, and 35th positions are 73, a decrease of the fastest (largest) pulse rate so that it is in the 33rd position means that it would have to be 73. Therefore, in this case, the median would remain unchanged. If the fastest (largest) pulse rate was decreased so that it was in a position lower than the 33rd position, then the number in the 32nd position would become the number in the 33rd position, and therefore the new median. Since the number in the 32nd position is 73, the median will remain unchanged. Thus, the median pulse rate for males would be unchanged.

(c) Since the mode is not the fastest pulse rate or the second fastest pulse rate, a decrease in the fastest pulse rate would not affect the mode men's pulse rate. Thus, the modal pulse rate for males would be unchanged.

95. **(a)** For a data set that is right-skewed, the majority of the data values are at the lower end of the distribution, with a few extreme values that are larger than the rest of the values. The smaller $p\%$ of the values that are omitted for a $p\%$ trimmed mean are close to most of the values in the data set and omitting them will not affect the mean very much, but the larger $p\%$ of the values that are omitted for a $p\%$ trimmed mean include some if not all of these extreme high values, causing the $p\%$ trimmed mean to be smaller than the original mean.

(b) For a data set that is left-skewed, the majority of the data values are at the upper end of the distribution, with a few extreme values that are smaller than the rest of the values. The larger $p\%$ of the values that are omitted for a $p\%$ trimmed mean are close to most of the values in the data set and omitting them will not affect the mean very much, but the smaller $p\%$ of the values that are omitted for a $p\%$ trimmed mean include some if not all of these extreme small values, causing the $p\%$ trimmed mean to be larger than the original mean.

(c) For a data set that is symmetric, eliminating the same number of the highest and the lowest values will result in a data set that is symmetric with an axis of symmetry that is approximately the same line as the original axis of symmetry. Therefore, the trimmed mean will be about the same as the original mean.

97. $$\frac{n}{\sum\frac{1}{x}} = \frac{5}{\frac{1}{5}+\frac{1}{4}+\frac{1}{3}+\frac{1}{2}+\frac{1}{1}} = \frac{5}{\frac{12}{60}+\frac{15}{60}+\frac{20}{60}+\frac{30}{60}+\frac{60}{60}} = \frac{5}{\frac{137}{60}} \approx 2.190 \text{ mph}$$

99. Geometric mean $= \sqrt[3]{1.04 \times 1.06 \times 1.10} = \sqrt[3]{1.21264} \approx 1.0664$, 6.64% growth

101. Solutions will vary. The data set should be right-skewed.

103. Solutions will vary. The original data set should be symmetric. Then increasing the largest number in the data set will result in the median remaining the same and the mean increasing.

105. The mean is the sum of the three numbers divided by 3. When the largest number is increased, the sum of the three numbers is increased. Therefore, the mean is increased. However, the median is the middle value of the three numbers, which is unaffected by an increase in the largest value.

107. The mean

109. Mean = 1,562,798, Median = 876,193

111. **(a)** PS4; **(b)** Nintendo; **(c)** Shooter

113. **(a)** Increase by x; **(b)** Increase by x; **(c)** Increase by x

Section 3.2

1. The deviation for a data value is obtained by subtracting the mean of the data set from the value. That is, the deviation for a data value gives the distance the value is from the mean.

3. Benefit—simple to calculate. Drawbacks—quite sensitive to extreme values, does not use all of the data values.

5. Benefit—uses all of the numbers in a data set. Drawback—can be time-consuming to calculate.

7. False. Chebyshev's Rule allows us to find the minimum percentage of data values that lie within a certain interval.

9. When all of the data values are the same.

11. **(a)** Range = largest value − smallest value = \$2.4 - \$0.2 = \$2.2 billion

(b) $\mu = \frac{\sum x}{N} = \frac{1.4+0.3+2.4+0.4+0.2+0.3}{6} = \frac{5}{6} \approx \$0.83$ billion

(c) and (d)

x	$x-\mu$	$(x-\mu)^2$
1.4	1.4 - 0.83 = 0.57	$(0.57)^2 = 0.3249$
0.3	0.3 - 0.83 = - 0.53	$(-0.53)^2 = 0.2809$
2.4	2.4 - 0.83 =1.57	$(1.57)^2 = 2.4649$
0.4	0.4 - 0.83 = - 0.43	$(-0.43)^2 = 0.1849$
0.2	0.2 - 0.83 = - 0.63	$(-0.63)^2 = 0.3969$
0.3	0.3 - 0.83 = - 0.53	$(-0.53)^2 = 0.2809$
Total		$\sum(x-\mu)^2 = 3.9334$

Variance: $\sigma^2 = \frac{\sum(x-\mu)^2}{N} = \frac{3.9334}{6} = 0.6555666667 \approx 0.6556$ billion dollars squared

TI-83/84: $\sigma^2 = 0.6555555556 \approx 0.6556$ billion dollars squared

(e) $\sigma = \sqrt{\sigma^2} \approx \sqrt{0.6556} \approx \$0.8097$ billion

13. **(a)** Range = largest value − smallest value = 208.0 – 66.0 = 142.0 motor vehicles stolen per 100,000 residents

(b) $\mu = \frac{\sum x}{N} = \frac{208.0+174.1+167.8+117.2+106.0+100.2+94.1+75.2+75.1+66}{10} = \frac{1183.7}{10} = 118.37$ motor vehicles stolen per 100,000 residents

(c) and (d)

x	$x - \mu$	$(x-\mu)^2$
208.0	208.0 - 118.37 = 89.63	$(89.63)^2 = 8033.5369$
174.1	174.1 - 118.37 = 55.73	$(55.73)^2 = 3105.8329$
167.8	167.8 - 118,37 = 49.43	$(49.43)^2 = 2443.3249$
117.2	117.2 - 118.37 = - 1.17	$(-1.17)^2 = 1.3689$
106	106.0 - 118. 37= - 12.37	$(-12.37)^2 = 153.0169$
100.2	100.2 - 118.37 = -18.17	$(-18.17)^2 = 330.1489$
94.1	94.1 - 118.37 = - 24.27	$(-24.27)^2 = 589.0329$
75.2	75.2 - 118.37 = -43.17	$(-43.17)^2 = 1863.6489$
75.1	75.1 - 118.37 = - 43.27	$(-43.27)^2 = 1872.2929$
66	66.0 - 118.37 = - 52.37	$(-52.37)^2 = 2742.6169$
Total		$\sum(x-\mu)^2 = 21{,}134.821$

Variance: $\sigma^2 = \frac{\sum(x-\mu)^2}{N} = \frac{21{,}134.821}{10} = 2{,}113.4821$ motor vehicles stolen per 100,000 residents squared

(e) $\sigma = \sqrt{\sigma^2} = \sqrt{2113.4821} \approx 45.97$ motor vehicles stolen per 100,000 residents

15. **(a)** Range = largest value − smallest value = 4551 – 584 = 3967 petit larceny cases

(b) $\mu = \frac{\sum x}{N} = \frac{2014+1288+1555+584+1607+995+2094+4551+823+2071}{10} = \frac{17582}{10} = 1758.2$ petit larceny cases

(c) and (d)

x	$x - \mu$	$(x-\mu)^2$
2014	2014 - 1758.2 = 255.8	$(255.8)^2 = 65{,}433.64$
1288	1288 - 1758.2 = - 470.2	$(-470.2)^2 = 221{,}088.04$
1555	1555 - 1758.2 = - 203.2	$(-203.2)^2 = 41{,}290.24$
584	584 - 1758.2 = - 1174.2	$(-1174.2)^2 = 1{,}378{,}745.64$
1607	1607 - 1758.2 = - 151.2	$(-151.2)^2 = 22{,}861.44$
995	995 - 1758.2 = - 763.2	$(-763.2)^2 = 582{,}474.24$
2094	2094 - 1758.2 = 335.8	$(335.8)^2 = 112{,}761.64$
4551	4551 - 1758.2 = 2792.8	$(2792.8)^2 = 7{,}799{,}731.84$
823	823 - 1758.2= - 935.2	$(-935.2)^2 = 874{,}599.04$
2071	2071 - 1758.2 = 312.8	$(312.8)^2 = 97{,}843.84$
Total		$\sum(x-\mu)^2 = 11{,}196{,}829.6$

Variance: $\sigma^2 = \frac{\sum(x-\mu)^2}{N} = \frac{11{,}196{,}829.6}{10} = 1{,}119{,}682.96$ petit larceny cases squared

(e) $\sigma = \sqrt{\sigma^2} = \sqrt{1{,}119{,}682.96} \approx 1058.15$ petit larceny cases

17. **(a)** First find the sample mean.

$\bar{x} = \frac{\sum x}{n} = \frac{1.4+2.4+0.2}{3} = \frac{4}{3} \approx \$1.33$ billion

x	$x - \bar{x}$	$(x - \bar{x})^2$
1.4	1.4 – 1.33 = 0.07	$(0.07)^2 = 0.0049$
2.4	2.4 – 1.33 = 1.07	$(1.07)^2 = 1.1449$
0.2	0.2 – 1.33 = -1.13	$(-1.13)^2 = 1.2769$
		$\sum(x - \bar{x})^2 = 2.4267$

$s^2 = \frac{\sum(x-\bar{x})^2}{n-1} = \frac{2.4267}{3-1} = 1.21335$ billion dollars squared (TI-83/84: 1.2133 billion dollars squared)

(b) $s = \sqrt{s^2} = \sqrt{1.2133} \approx \$1.10$ billion

(c) For this sample of state export data, the typical difference between a state's export amount and the mean export amount is $1.10 billion.

19. **(a)** First find the sample mean.

$\bar{x} = \frac{\sum x}{n} = \frac{208.0+167.8+100.2}{3} = \frac{476}{3} \approx 158.67$ motor vehicles stolen per 100,000 people

x	$x - \bar{x}$	$(x - \bar{x})^2$
208.0	208.0 – 158.67 = 49.33	$(49.33)^2 = 2433.4489$
167.8	167.8 – 158.67 = 9.13	$(9.13)^2 = 83.3569$
100.2	100.2 – 158.67 = -58.47	$(-58.47)^2 = 3418.7409$
		$\sum(x - \bar{x})^2 = 5935.5467$

$s^2 = \frac{\sum(x-\bar{x})^2}{n-1} \approx \frac{5935.5467}{3-1} = 2967.7735$ motor vehicles stolen per 100,000 people squared (TI-83/84: 2967.7733 motor vehicles stolen per 100,000 people squared)

(b) $s = \sqrt{s^2} \approx \sqrt{2967.7733} \approx 54.48$

(c) For this sample of motor vehicle theft data, the typical difference between a country's motor vehicle theft rate and the mean motor vehicle theft rate is 54.48 motor vehicles stolen per 100,000 people.

21. **(a)** First find the sample mean.

$\bar{x} = \frac{\sum x}{n} = \frac{2014+1555+1607+4551+823}{5} = \frac{10{,}550}{5} = 2110$ petit larceny cases

x	$x - \bar{x}$	$(x - \bar{x})^2$
2014	2014 – 2110 = - 96	$(-96)^2 = 9216$
1555	1555 – 2110 = -555	$(-555)^2 = 308{,}025$
1607	1607 – 2110 = -503	$(-503)^2 = 253{,}009$
4551	4551 – 2110 = 2441	$(2441)^2 = 5{,}958{,}481$
823	823 – 2110 = -1287	$(-1287)^2 = 1{,}656{,}369$
		$\sum(x - \bar{x})^2 = 8{,}185{,}100$

$s^2 = \frac{\sum(x-\bar{x})^2}{n-1} = \frac{8{,}185{,}100}{5-1} = 2{,}046{,}275$ petit larceny cases squared

(b) $s = \sqrt{s^2} = \sqrt{2{,}046{,}275} \approx 1430.48$ petit larceny cases

(c) For this sample of New York precincts, the typical difference between a precinct's number of petit larceny cases and the mean number of petit larceny cases is 1430.48 petit larceny cases.

23. First we need to find k.

$k = \left|\frac{\text{data value} - \text{mean}}{\text{standard deviation}}\right| = \left|\frac{45-50}{5}\right| = \left|\frac{55-50}{5}\right| = 1$

The Empirical Rule tells us that about 68% of the data are between 45 and 55.

25. First we need to find k.

$k = \left|\frac{\text{data value} - \text{mean}}{\text{standard deviation}}\right| = \left|\frac{35-50}{5}\right| = \left|\frac{65-50}{5}\right| = 3$

The Empirical Rule tells us that about 99.7% of the data are between 35 and 65.

27. Here $k = 1$. The Empirical Rule tells us that about 68% of the data are between –1 and 1

29. Here $k = 2$. The Empirical Rule tells us that about 95% of the data are between –2 and 2. Therefore, about 100% - 95% = 5% of the data are less than –2 or greater than 2. Since the distribution is bell-shaped, the distribution is symmetric. Therefore, about $\left(\frac{1}{2}\right) \times 5\% = 2.5\%$ of the data are less than –2.

31. First we need to find k.

$k = \left|\frac{\text{data value}-\text{mean}}{\text{standard deviation}}\right| = \left|\frac{16-20}{2}\right| = \left|\frac{24-20}{2}\right| = 2$.

Chebyshev's Rule tells us that at least $\left(1 - \frac{1}{(k)^2}\right)100\% = \left(1 - \frac{1}{(2)^2}\right)100\% = 75\%$ of the data are between 16 and 24.

33. First we need to find k.

$k = \left|\frac{\text{data value}-\text{mean}}{\text{standard deviation}}\right| = \left|\frac{12-20}{2}\right| = \left|\frac{28-20}{2}\right| = 4$.

Chebyshev's Rule tells us that at least $\left(1 - \frac{1}{(k)^2}\right)100\% = \left(1 - \frac{1}{(4)^2}\right)100\% = 93.75\%$ of the data are between 12 and 28.

35. First we need to find k.

$k = \left|\frac{\text{data value}-\text{mean}}{\text{standard deviation}}\right| = \left|\frac{0-20}{5}\right| = \left|\frac{40-20}{5}\right| = 4$.

Chebyshev's Rule tells us that at least $\left(1 - \frac{1}{(k)^2}\right)100\% = \left(1 - \frac{1}{(4)^2}\right)100\% = 93.75\%$ of the data are between 0 and 40.

37. First we need to find k.

$k = \left|\frac{\text{data value}-\text{mean}}{\text{standard deviation}}\right| = \left|\frac{12.5-20}{5}\right| = \left|\frac{27.5-20}{5}\right| = 1.5$.

Chebyshev's Rule tells us that at least $\left(1 - \frac{1}{(k)^2}\right)100\% = \left(1 - \frac{1}{(1.5)^2}\right)100\% \approx 55.56\%$ of the data are between 12.5 and 27.5.

39. All of the data values in graph **(a)** are less than 75, so the mean of the data set in graph **(a)** is less than 75. The middle of the histogram is about 50 and the value with the highest frequency is 50. Therefore, the mean is 50.

The middle of the data set in graph **(b)** is somewhere between 70 and 80. The value with the highest frequency is 80 and the value with the second highest frequency is 75. Therefore, the mean is 75.

The majority of the data values in graph **(c)** are less than 75 so the mean of the data set in graph **(c)** is less than 75. The majority of the data values are between 30 and 75, inclusive. Therefore, the mean is between 30 and 75, so the mean is 50.

The middle of the data values in graph **(d)** is around 75–80. The data value with the highest frequency is 75. The mean is 75.

The data sets in graphs **(b)** and **(d)** both have a mean of 75. Therefore, (i) and (ii) match up with graphs **(b)** and **(d)**. The range of the data values in graph **(b)** is 105 − 55 = 50 and the range of the data values in graph **(d)** is 140 − 35 = 105, so the data set in graph **(d)** is more spread out than the data set in graph **(b)**. Therefore, the data set in graph **(d)** has a larger standard deviation than the data set in graph **(b)**. Thus, (i) corresponds to graph **(d)** and (ii) corresponds to graph **(b)**.

The data sets in graphs **(a)** and **(c)** both have a mean of 50. The range of the data values in graph **(a)** is 70 − 25 = 45 and the range of the data set in graph **(c)** is 105 − (−30) = 135. Therefore, the spread of the data set in graph **(c)** is more spread out than the data set in graph **(a)**, so the data set in graph **(c)** has a larger standard deviation than the data set in graph **(a)**. Thus, (iii) corresponds to graph **(c)**, and (iv) corresponds to graph **(a)**.

(i) **(d)**
(ii) **(b)**
(iii) **(c)**
(iv) **(a)**

41. **(a)** Range = largest number − smallest number = 1.7 − 0.2 = 1.5 million game units

(b) First find the sample mean.

$$\bar{x} = \frac{\sum x}{n} = \frac{1.7+0.6+0.9+1.2+0.2}{5} = \frac{4.6}{5} = 0.92 \text{ million game units}$$

x	$x - \bar{x}$	$(x - \bar{x})^2$
1.7	1.7 – 0.92 = 0.78	$(0.78)^2 = 0.6084$
0.6	0.6 – 0.92 = -0.32	$(-0.32)^2 = 0.1024$
0.9	0.9 – 0.92 = -0.02	$(-0.02)^2 = 0.0004$
1.2	1.2 – 0.92 = 0.28	$(0.28)^2 = 0.0784$
0.2	0.2 – 0.92 = -0.72	$(-0.72)^2 = 0.5184$
		$\sum(x - \bar{x})^2 = 1.308$

$s^2 = \frac{\sum(x-\bar{x})^2}{n-1} = \frac{1.308}{5-1} = 0.327$ million game units squared

(c) $s = \sqrt{s^2} = \sqrt{0.327} \approx 0.5718$ million game units

43. **(a)** Range = largest number − smallest number = 42.2 − (−27.4) = 69.6

(b) First find the sample mean.

$$\bar{x} = \frac{\sum x}{n} = \frac{-27.4+18.7+42.2+(-16.3)+11.2+28.5+1.8+16.9}{8} = \frac{75.6}{8} = 9.45$$

x	$x-\bar{x}$	$(x-\bar{x})^2$
-27.4	-27.4 – 9.45 = -36.85	$(-36.85)^2 = 1357.9225$
18.7	18.7– 9.45 = 9.25	$(9.25)^2 = 85.5625$
42.2	42.2 – 9.45 = 32.75	$(32.75)^2 = 1072.5625$
-16.3	-16.3 – 9.45 = -25.75	$(-25.75)^2 = 663.0625$
11.2	11.2 – 9.45 = 1.75	$(1.75)^2 = 3.0625$
28.5	28.5 – 9.45 = 19.05	$(19.05)^2 = 362.9025$
1.8	1.8 - 9.45 = -7.65	$(-7.65)^2 = 58.5225$
16.9	16.9 – 9.45 = 7.45	$(7.45)^2 = 55.5025$
		$\sum(x-\bar{x})^2 = 3659.1$

$s^2 = \frac{\sum(x-\bar{x})^2}{n-1} = \frac{3659.1}{8-1} \approx 522.7286$

(c) $s = \sqrt{s^2} \approx \sqrt{522.7286} \approx 22.86$

45. **(a)** Range = largest number - smallest number = 40 - 19 = 21 years

(b) First find the sample mean.

$\bar{x} = \frac{\sum x}{n} = \frac{40+28+25+34+26+21+19+24}{8} = \frac{217}{8} = 27.125$ years

x	$x-\bar{x}$	$(x-\bar{x})^2$
40	40 – 27.125 = 12.875	$(12.875)^2 = 165.765625$
28	28– 27.125 = 0.875	$(0.875)^2 = 0.765625$
25	25 – 27.125 = -2.125	$(-2.125)^2 = 4.515625$
34	34 – 27.125 = 6.875	$(6.675)^2 = 47.265625$
26	26 – 27.125 = -1.125	$(-1.125)^2 = 1.265625$
21	21 – 27.125 = -6.125	$(-6.125)^2 = 37.515625$
19	19 – 27 125 = -8.125	$(-8.125)^2 = 66.015625$
24	24 – 27.125= -3.125	$(-3.125)^2 = 9.765625$
		$\sum(x-\bar{x})^2 = 332.875$

$s^2 = \frac{\sum(x-\bar{x})^2}{n-1} = \frac{332.875}{8-1} \approx 47.5536$ years squared

(c) $s = \sqrt{s^2} \approx \sqrt{47.5536} \approx 6.90$ years

47. **(a)** Range = largest number - smallest number = 375 - 59 = 316 calories

(b) First find the sample mean.

$\bar{x} = \frac{\sum x}{n} = \frac{216+364+149+75+276+375+146+59+67+195}{10} = \frac{1922}{10} = 192.2$ calories

x	$x-\bar{x}$	$(x-\bar{x})^2$
216	216 – 192.2 = 23.8	$(23.8)^2 = 566.44$
364	364 – 192.2 = 171.8	$(171.8)^2 = 29.515.24$
149	149 – 192.2 = - 43.2	$(-43.2)^2 = 1866.24$
75	75 – 192.2 = - 117.2	$(-117.2)^2 = 13{,}735.84$
276	276 – 192.2 = 83.8	$(83.8)^2 = 7{,}022.44$
375	375 – 192.2 = 182.8	$(182.8)^2 = 33{,}415.84$
146	146 – 192.2 = - 46.2	$(-46.2)^2 = 2{,}134.44$
59	59 – 192.2 = - 133.2	$(-133.2)^2 = 17{,}742.24$
67	67 – 192.2 = - 125.2	$(-125.2)^2 = 15{,}675.04$
195	195 – 192.2 = 2.8	$(2.8)^2 = 7.84$
		$\sum(x-\bar{x})^2 = 121{,}681.6$

$s^2 = \frac{\sum(x-\bar{x})^2}{n-1} = \frac{121{,}681.6}{10-1} \approx 13{,}520.1778$ calories squared

(c) $s = \sqrt{s^2} \approx \sqrt{13{,}520.1778} \approx 116.28$ calories

49. No, standard deviation

51. From Exercise 41, $\bar{x} = 0.92$ million game units and $s \approx 0.5718$ million game units. Because 0.0048 lies 1.600559636 standard deviations below the mean $\bar{x} - 1.600559636s \approx 0.92 - 1.600559636(0.5718) \approx 0.0048$ and 1.8352 lies 1.600559636 standard deviations above the mean $\bar{x} + 1.600559636s \approx 0.92 + 1.600559636(0.5718) \approx 1.8532$, the question is really asking what is the minimum percentage that lies within $k = 1.600559636$ standard deviations from the mean. From Chebyshev's Rule, the minimum percentage is $\left(1 - \frac{1}{(k)^2}\right)100\% = \left(1 - \frac{1}{(1.600559636)^2}\right)100\% \approx 60.96\%$. Thus, at least 60.96% of total sales lie between 0.0048 million game units and 1.8352 million game units.

53. From Exercises 43 and 44:

Darts: Range = 69.6, Variance: $s^2 \approx 522.7286$, Standard deviation: $s \approx 22.86$

DJIA: Range = 28.6, Variance: $s^2 \approx 115.2393$, Standard deviation: $s \approx 10.73$

Since all three measures of spread are larger for the darts than for the DJIA, the stock market return that reflects more variability is the darts.

55. From Exercise 43, $\bar{x} = 9.45$ and $s \approx 22.86$. Because -13.41 lies 1 standard deviation below the mean $\bar{x} - 1s \approx 9.45 - 1(22.86) = -13.41$ and 32.31 lies 1 standard deviation above the mean $\bar{x} + 1s \approx 9.45 + 1(22.86) = 32.31$, the question is really asking what is the percentage that lies within $k = 1$ standard deviation from the mean. From the Empirical Rule, about 68% of darts stock market returns lie between -13.41 and 32.31.

57. **(a)** No; **(b)** No; **(c)** Yes

59. **(a)** Yes; **(b)** Yes; **(c)** Yes

61. **(a)** No; **(b)** No; **(c)** Yes

63. **(a)** Range = largest data value − smallest data value = 12 cylinders − 4 cylinders = 8 cylinders

(b) $n = 6$,

x	x^2
6	36
4	16
6	36
4	16
12	144
4	16
$\Sigma x = 36$	$\Sigma x^2 = 264$

$$s^2 = \frac{\sum x^2 - \left(\sum x\right)^2/n}{n-1} = \frac{264 - (36)^2/6}{6-1} = \frac{264 - 1296/6}{5} = \frac{264 - 216}{5} = \frac{48}{5} = 9.6 \text{ cylinders}^2$$

(c) $s = \sqrt{s^2} = \sqrt{9.6} \approx 3.098$ cylinders

65. **(a)** Range = largest data value − smallest data value = 41 mpg − 11 mpg = 30 mpg

(b) $n = 6$,

x	x^2
18	324
41	1681
18	324
25	625
11	121
31	961
$\Sigma x = 144$	$\Sigma x^2 = 4036$

$$s^2 = \frac{\sum x^2 - \left(\sum x\right)^2/n}{n-1} = \frac{4036 - (144)^2/6}{6-1} = \frac{4036 - 20{,}736/6}{5}$$

$$= \frac{4036 - 3456}{5} = \frac{580}{5} = 116 \text{ mpg}^2$$

(c) $s = \sqrt{s^2} = \sqrt{116} \approx 10.770$ mpg

67. <u>**Colony A**</u>

First find the sample mean.

$$\bar{x} = \frac{\Sigma x}{n} = \frac{109+120+94+61+72+134+94+113+111+106}{10} = \frac{1014}{10} = 101.4 \text{ milligrams}$$

x	$x - \bar{x}$	$(x - \bar{x})^2$
109	109 – 101.4 = 7.6	$(7.6)^2 =$ 57.76
120	120 – 101.4 = 18.6	$(18.6)^2 =$ 345.96
94	94 – 101.4 = - 7.4	$(-7.4)^2 =$ 54.76
61	61 – 101.4 = -40.4	$(-40.4)^2 =$ 1632.16
72	72 – 101.4 = -29.4	$(-29.4)^2 =$ 864.36
134	134 – 101.4 = 32.6	$(32.6)^2 =$ 1062.76
94	94 – 101.4 = -7.4	$(-7.4)^2 =$ 54.76
113	113 – 101.4 = 11.6	$(11.6)^2 =$ 134.56
111	111 – 101.4 = 9.6	$(9.6)^2 =$ 92.16
106	106 – 101.4 = 4.6	$(4.6)^2 =$ 21.16
		$\sum(x - \bar{x})^2 = 4320.4$

Standard deviation: $s = \sqrt{\frac{\sum(x-\bar{x})^2}{n-1}} = \sqrt{\frac{4320.4}{10-1}} \approx 21.91$ milligrams

Colony B

First find the sample mean.

$\bar{x} = \frac{\sum x}{n} = \frac{148+110+110+97+136+115+101+158+67+114}{10} = \frac{1156}{10} = 115.6$ milligrams

x	$x - \bar{x}$	$(x - \bar{x})^2$
148	148 – 115.6 = 32.4	$(32.4)^2 =$ 1049.76
110	110 – 115.6 = -5.6	$(-5.6)^2 =$ 31.36
110	110 – 115.6 = - 5.6	$(-5.6)^2 =$ 31.36
97	97 – 115.6 = -18.6	$(-18.6)^2 =$ 345.96
136	136 – 115.6 = 20.4	$(20.4)^2 =$ 416.16
115	115 – 115.6 = - 0.6	$(-0.6)^2 =$ 0.36
101	101 – 115.6 = -14.6	$(-14.6)^2 =$ 213.16
158	158 – 115.6 = 42.4	$(42.4)^2 =$ 1797.76
67	67 – 115.6 = -48.6	$(-48.6)^2 =$ 2361.96
114	114 – 115.6 = -1.6	$(-1.6)^2 =$ 2.56
		$\sum(x - \bar{x})^2 = 6250.4$

Standard deviation: $s = \sqrt{\frac{\sum(x-\bar{x})^2}{n-1}} = \sqrt{\frac{6250.4}{10-1}} \approx 26.35$ milligrams

(a) Colony B has the greatest standard deviation.
(b) Colony B has the greatest standard deviation, so it has the greater variability. This concurs with the answer to Exercise 66.
(c) Since Colony B has the greatest standard deviation and since the variance is the standard deviation squared, Colony B will have the greatest variance.

69. **(a)** $n = 3$

x	x^2
85	7225
92	8464
85	7225
$\sum x = 262$	$\sum x^2 = 22.914$

$s^2 = \frac{\sum x^2 - (\sum x)^2/n}{n-1} = \frac{22{,}914-(262)^2/3}{3-1} = \frac{22{,}914-68{,}644/3}{2} = \frac{22{,}914-22{,}881.33333}{2} = \frac{32.66667}{2} \approx 16.33$ wins squared

(b) $s = \sqrt{s^2} \approx \sqrt{16.33} \approx 4.04$ wins

(c) For this sample of baseball teams in the American League East Division, the typical difference between a team's number of wins and the mean number of wins is 4.04 wins.

71. $\mu = 63°\text{F}$ and $\sigma = 3°\text{F}$.

Since the distribution is bell-shaped, we may use the Empirical Rule.

(a) First we need to find k.

$$k = \left|\frac{\text{data value} - \text{mean}}{\text{standard deviation}}\right| = \left|\frac{60-63}{3}\right| = \left|\frac{66-63}{3}\right| = 1$$

The Empirical Rule tells us that about 68% of the October days in Santa Monica, California have temperatures that are between 60 and 66 degrees Fahrenheit.

(b) First we need to find k.

$$k = \left|\frac{\text{data value} - \text{mean}}{\text{standard deviation}}\right| = \left|\frac{57-63}{3}\right| = \left|\frac{69-63}{3}\right| = 2$$

The Empirical Rule tells us that about 95% of the October days in Santa Monica, California have temperatures that are between 57 and 69 degrees Fahrenheit.

(c) First we need to find k.

$$k = \left|\frac{\text{data value} - \text{mean}}{\text{standard deviation}}\right| = \left|\frac{55-63}{3}\right| = \left|\frac{71-63}{3}\right| \approx 2.67.$$

Can't do; the Empirical Rule does not say what percent of the data values lie within 2.67 standard deviations of the mean.

73. $\mu = 63°\text{F}$ and $\sigma = 3°\text{F}$.

Since the distribution is unknown, we may not use the Empirical Rule. Therefore, we will use Chebyshev's Rule.

(a) First we need to find k.

$$k = \left|\frac{\text{data value} - \text{mean}}{\text{standard deviation}}\right| = \left|\frac{60-63}{3}\right| = \left|\frac{66-63}{3}\right| = 1$$

Since $k = 1$, we can't use Chebyshev's Rule.

(b) First we need to find k.

$$k = \left|\frac{\text{data value} - \text{mean}}{\text{standard deviation}}\right| = \left|\frac{57-63}{3}\right| = \left|\frac{69-63}{3}\right| = 2$$

From Chebyshev's Rule, the minimum percentage is $\left(1 - \frac{1}{(k)^2}\right)100\% = \left(1 - \frac{1}{(2)^2}\right)100\% = 75\%$. Thus, at least 75% of the October days in Santa Monica,

California have temperatures that are between 57 and 69 degrees Fahrenheit.

(c) First we need to find k.

$k = \left|\frac{\text{data value}-\text{mean}}{\text{standard deviation}}\right| = \left|\frac{55-63}{3}\right| = \left|\frac{71-63}{3}\right| \approx 2.666666667$.

From Chebyshev's Rule, the minimum percentage is $\left(1-\frac{1}{(k)^2}\right)100\% \approx$ $\left(1-\frac{1}{(2.6666666677)^2}\right)100\% \approx 85.94\%$. Thus, at least 85.94% of the October days in Santa Monica, California have temperatures that are between 55 and 71 degrees Fahrenheit.

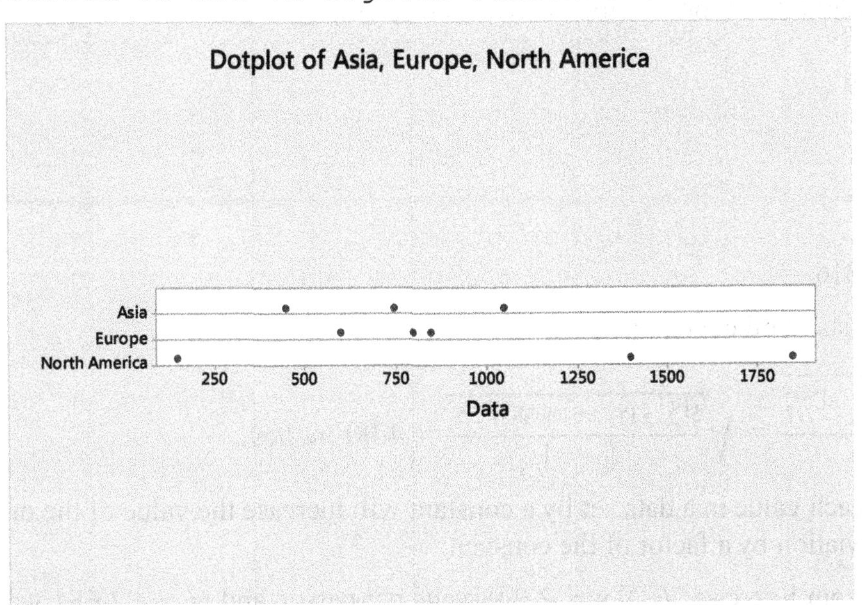

75.
North America. The dots for North America are more spread out than the dots for Asia and Europe.

77. **(a)** Since the standard deviation is the square root of the variance and since North America has the largest variance, North America will have the largest standard deviation.
(b) Since the standard deviation is the square root of the variance and since Europe has the smallest variance, Europe will have the smallest standard deviation.

79. **(a)** The standard deviation will increase because the contribution of a value of 84 to the computation is greater than the contribution of a value of 72. **(b)** Original standard deviation = 2.45 inches

New standard deviation:
To calculate the standard deviation we first need to find $\sum x$ and $\sum x^2$.

x	x^2
66	4356
67	4489
70	4900
70	4900
84	7056
$\sum x = 357$	$\sum x^2 = 25{,}701$

$n = 5$. Therefore, the standard deviation is

$$s = \sqrt{\frac{\sum x^2 - \left(\sum x\right)^2/n}{n-1}} = \sqrt{\frac{25{,}701 - (357)^2/5}{5-1}} \approx 7.27 \text{ inches.}$$

Yes it is correct.

81. Data set in ascending order after each of the heights has been doubled:

132, 134, 140, 140, 144 $n = 5$

(a) Range = largest value − smallest value = 144 − 132 = 12

To calculate the standard deviation we first need to find $\sum x$ and $\sum x^2$.

x	x^2
132	17,424
134	17,956
140	19,600
140	19,600
144	20,736
$\sum x = 690$	$\sum x^2 = 95{,}316$

$n = 5$. Therefore, the standard deviation is

$$s = \sqrt{\frac{\sum x^2 - \left(\sum x\right)^2/n}{n-1}} = \sqrt{\frac{95{,}316 - (690)^2/5}{5-1}} \approx 4.90 \text{ inches.}$$

(b) Multiplying each value in a data set by a constant will increase the value of the original range or standard deviation by a factor of the constant.

83. **(a)** Asia: From Exercise 76. $\sum x = 2259$ watts per person and $s^2 = 87{,}651$ watts per person squared. Therefore, $\bar{x} = \frac{\sum x}{n} = \frac{2259}{3} = 753$ watts per person, $s = \sqrt{s^2} = \sqrt{87{,}651} =$ 296.0591157 watts per person.

$\text{CV} = \frac{\text{standard deviation}}{\text{mean}} \cdot 100\% = \frac{296.0591157}{753} \cdot 100\% \approx 39.32\%$

Europe: From Exercise 76. $\sum x = 2287$ watts per person and $s^2 = 15{,}582.33333$ watts per person squared. Therefore, $\bar{x} = \frac{\sum x}{n} = \frac{2287}{3} = 752.3333333$ watts per person and $s = \sqrt{s^2} =$ $\sqrt{15{,}582.33333} = 124.8292167$ watts per person,

$\text{CV} = \frac{\text{standard deviation}}{\text{mean}} \cdot 100\% = \frac{124.8292167}{752.3333333} \cdot 100\% \approx 16.37\%$

North America: From Exercise 76. $\sum x = 3404$ watts per person and $s^2 = 810{,}500.3333$ watts per person squared. Therefore, $\bar{x} = \frac{\sum x}{n} = \frac{3404}{3} = 1134.666667$ watts per person and $s = \sqrt{s^2} =$ $\sqrt{810{,}500.3333} = 900.2779201$ watts per person.

$\text{CV} = \frac{\text{standard deviation}}{\text{mean}} \cdot 100\% = \frac{900.2779201}{1134.666667} \cdot 100\% \approx 79.34\%$

(b) North America has the largest coefficient of variation, so it has the greatest variability. This agrees with the measures of spread in Exercise 36.

85. **(a)** Asia: From Exercise 83, $\bar{x} = 753$ watts per person.

x_i	$\lvert x_i - \bar{x}\rvert$
447	$\lvert 447 - 753\rvert = \lvert -306\rvert = 306$
774	$\lvert 774 - 753\rvert = \lvert 21\rvert = 21$
1038	$\lvert 1038 - 753\rvert = \lvert 285\rvert = 285$
	$\sum\lvert x_i - \bar{x}\rvert = 612$

MAD $= \frac{\sum\lvert x_i - \bar{x}\rvert}{n} = \frac{612}{3} = 204$ watts per person

Europe: From Exercise 83, $\bar{x} = 762.3333333$ watts per person

x_i	$\lvert x_i - \bar{x}\rvert$
861	$\lvert 861 - 762.3333333\rvert = \lvert 98.6666666667\rvert = 98.6666666667$
804	$\lvert 804 - 762.3333333\rvert = \lvert 41.6666666667\rvert = 41.6666666667$
622	$\lvert 662 - 762.3333333\rvert = \lvert -140.33333333333\rvert = 140.33333333333$
	$\sum\lvert x_i - \bar{x}\rvert = 280.66666666667$

MAD $= \frac{\sum\lvert x_i - \bar{x}\rvert}{n} = \frac{280.66666666667}{3} \approx 93.56$ watts per person

North America: From Exercise 83, $\bar{x} = 1134.666667$ watts per person.

x_i	$\lvert x_i - \bar{x}\rvert$
1402	$\lvert 1402 - 1134.666667\rvert = \lvert 267.33333333333\rvert = 267.3333333333$
1871	$\lvert 1871 - 1134.666667\rvert = \lvert 736.3333333333\rvert = 736.3333333333$
131	$\lvert 131 - 1134.666667\rvert = \lvert -1003.6666666667\rvert = 1003.6666666667$
	$\sum\lvert x_i - \bar{x}\rvert = 2007.3333333333$

MAD $= \frac{\sum\lvert x_i - \bar{x}\rvert}{n} = \frac{2007.3333333333}{3} \approx 669.11$ watts per person

(b) North America has the largest mean average deviation, so it has the greatest variability. This agrees with the measures of spread in Exercise 36.

87. If the mean equals the median, then mean − median = 0. Therefore,

$$\text{Skewness} = \frac{3(\text{mean} - \text{median})}{\text{standard deviation}} = \frac{3(0)}{\text{standard deviation}} = \frac{0}{\text{standard deviation}} = 0.$$

89. **(a)**

	Range	**Sample variance**	**Sample standard deviation**	**Coefficient of variation**	**Mean absolute deviation**
Cylinders	8	9.6	3.098	51.64%	2
Engine size	4.9	3.078	1.754	52.89%	1.189
City mpg	30	116	10.770	44.88%	8.333

(b) City mpg: Range, Sample variance, Sample standard deviation, Mean absolute deviation
Engine size: Coefficient of variation
Cylinders: None of them

91. Answers will vary.
93. Answers will vary.
95. Answers will vary.

97.

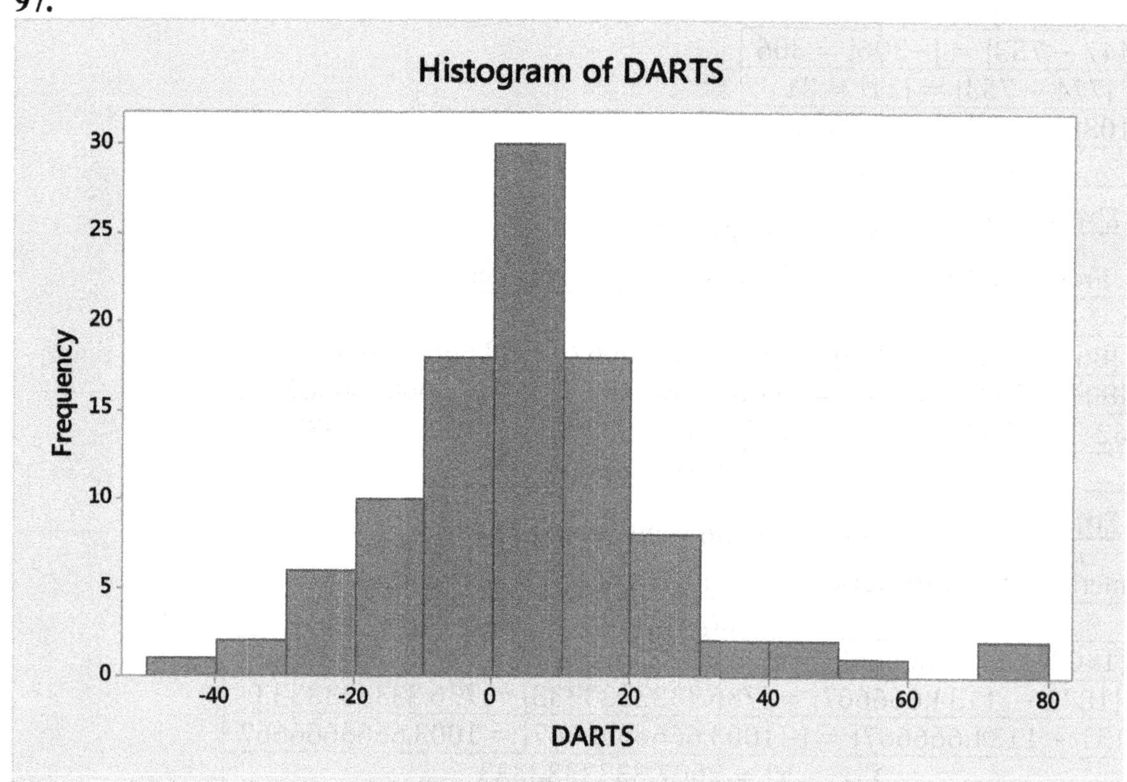
Histogram of DARTS
Frequency
DARTS
0
5
10
15
20
25
30
-40
-20
0
20
40
60
80

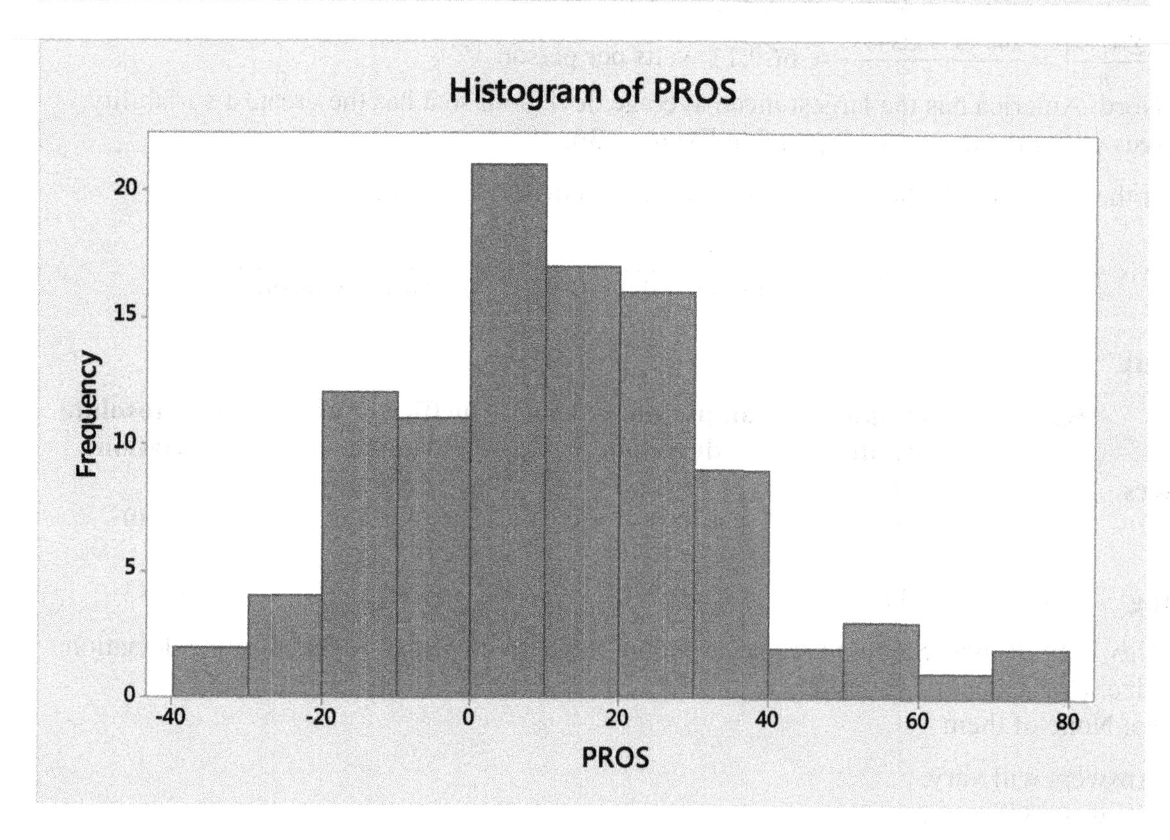
Histogram of PROS
Frequency
PROS
0
5
10
15
20
-40
-20
0
20
40
60
80

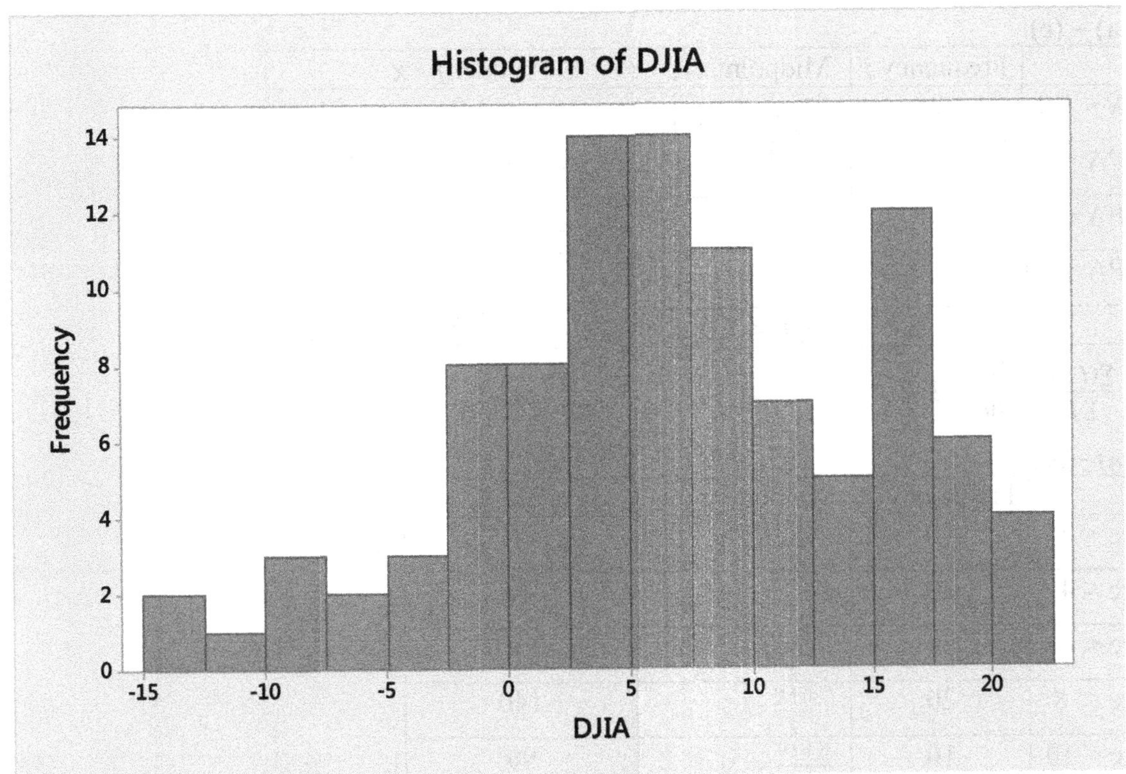

99. **(a)** $\mu - 1\sigma = 4.52 - 1 \cdot 19.39 = -14.87, \mu + 1\sigma = 4.52 + 1 \cdot 19.39 = 23.91$
$\mu - 2\sigma = 4.52 - 2 \cdot 19.39 = -34.26, \mu + 2\sigma = 4.52 + 2 \cdot 19.39 = 43.3$
$\mu - 3\sigma = 4.52 - 3 \cdot 19.39 = -53.65, \mu + 3\sigma = 4.52 + 3 \cdot 19.39 = 62.69$
(b) About 68% of the stock returns lie between -14.87 and 23.91.
About 95% of the stock returns lie between -34.26 and 43.3.
About 99.7% of the stock returns lie between -53.65 and 62.69.
(c) 76% of the stock returns lie between -14.87 and 23.91.
94% of the stock returns lie between -34.26 and 43.3.
98% of the stock returns lie between -53.65 and 62.69.
(d) The actual percentages that lie within each interval are close to the approximate percentages given by the Empirical Rule.

Section 3.3

1. These formulas will only provide estimates because we will not know the exact data values. We will estimate the data values in each class with the class midpoint. These midpoints will then be used in the computations.

3. The mean

5. $\bar{x} = \frac{\sum(w \cdot x)}{\sum w} = \frac{(0.25)(100)+(0.40)(60)+(0.35)(90)}{0.25+0.40+0.35} = \frac{25+24+31.5}{1} = \frac{80.5}{1} = 80.5$

7. $\bar{x} = \frac{\sum(w \cdot x)}{\sum w} = \frac{(3)(2.0)+(3)(3.5)+(3)(2.5)+(3)(3.0)+(8)(2.0)}{3+3+3+3+8} = \frac{6+10.5+7.5+9+16}{20} = \frac{49}{20} = 2.45$

9. **(a) – (c)**

Class	Frequency f	Midpoint x	Product $\boldsymbol{f \cdot x}$
$0 \le$ GPA < 1.0	2	$\frac{0+1.0}{2} = 0.5$	1.0
$1.0 \le$ GPA < 2.0	10	$\frac{1.0+2.0}{2} = 1.5$	15.0
$2.0 \le$ GPA < 3.0	13	$\frac{2.0+3.0}{2} = 2.5$	32.5
$3.0 \le$ GPA < 4.0	5	$\frac{3.0+4.0}{2} = 3.5$	17.5
Total	$\sum f = 30$		$\sum(f \cdot x) = 66$

(d) $\bar{x} = \frac{\sum(f \cdot x)}{\sum f} = \frac{66}{30} = 2.2$

11. **(a) – (c)**

Class	Frequency f	Midpoint x	Product $\boldsymbol{f \cdot x}$
$0 \le$ score < 2	10	$\frac{0+2}{2} = 1$	10
$2 \le$ score < 4	20	$\frac{2+4}{2} = 3$	60
$4 \le$ score < 6	30	$\frac{4+6}{2} = 5$	150
$6 \le$ score < 8	20	$\frac{6+8}{2} = 7$	140
$8 \le$ score < 10	10	$\frac{8+10}{2} = 9$	90
Total	$\sum f = 90$		$\sum(f \cdot x) = 450$

(d) $\bar{x} = \frac{\sum(f \cdot x)}{\sum f} = \frac{450}{90} = 5$

13. **(a) – (c)**

Class	Frequency f	Midpoint x	Product $\boldsymbol{f \cdot x}$
$0 \le$ cost < 5	100	$\frac{0+5}{2} = 2.5$	250
$5 \le$ cost < 10	150	$\frac{5+10}{2} = 7.5$	1,125
$10 \le$ cost < 15	200	$\frac{10+15}{2} = 12.5$	2,500
$15 \le$ cost < 20	250	$\frac{15+20}{2} = 17.5$	4,375
$20 \le$ cost < 30	300	$\frac{20+30}{2} = 25$	7,500
$30 \le$ cost < 50	350	$\frac{30+50}{2} = 40$	14,000
$50 \le$ cost < 100	400	$\frac{50+100}{2} = 75$	30,000
$100 \le$ cost < 200	450	$\frac{100+200}{2} = 150$	67,500
Total	$\sum f = 2200$		$\sum(f \cdot x) = 127{,}250$

(d) $\bar{x} = \frac{\sum(f \cdot x)}{\sum f} = \frac{127{,}250}{2200} \approx 57.84$

15. From Exercise 9, $\bar{x} = 2.2$.

Class	Frequency f	Midpoint x	$\bar{x}$	$x - \bar{x}$	$(x - \bar{x})^2 \cdot f$
$0 \leq$ GPA < 1.0	2	$\frac{0+1.0}{2} = 0.5$	2.2	$0.5 - 2.2 = -1.7$	$(-1.7)^2 \cdot 2 = 5.78$
$1.0 \leq$ GPA < 2.0	10	$\frac{1.0+2.0}{2} = 1.5$	2.2	$1.5 - 2.2 = -0.7$	$(-0.7)^2 \cdot 10 = 4.9$
$2.0 \leq$ GPA < 3.0	13	$\frac{2.0+3.0}{2} = 2.5$	2.2	$2.5 - 2.2 = 0.3$	$(0.3)^2 \cdot 13 = 1.17$
$3.0 \leq$ GPA < 4.0	5	$\frac{3.0+4.0}{2} = 3.5$	2.2	$3.5 - 2.2 = 1.3$	$(1.3)^2 \cdot 5 = 8.45$
Total	$\sum f = 30$				$\sum(x - \bar{x})^2 \cdot f = 20.3$

Variance: $s^2 = \frac{\sum(x-\bar{x})^2 \cdot f}{\sum f} = \frac{20.3}{30} \approx 0.6767$

Standard deviation: $s = \sqrt{s^2} \approx \sqrt{0.6767} \approx 0.8226$

17. From Exercise 11, $\bar{x} = 5$.

Class	Frequency f	Midpoint x	$\bar{x}$	$x - \bar{x}$	$(x - \bar{x})^2 \cdot f$
$0 \leq$ score < 2	10	$\frac{0+2}{2} = 1$	5	$1 - 5 = -4$	$(-4)^2 \cdot 10 = 160$
$2 \leq$ score < 4	20	$\frac{2+4}{2} = 3$	5	$3 - 5 = -2$	$(-2)^2 \cdot 20 = 80$
$4 \leq$ score < 6	30	$\frac{4+6}{2} = 5$	5	$5 - 5 = 0$	$(0)^2 \cdot 30 = 0$
$6 \leq$ score < 8	20	$\frac{6+8}{2} = 7$	5	$7 - 5 = 2$	$(2)^2 \cdot 20 = 80$
$8 \leq$ score < 10	10	$\frac{8+10}{2} = 9$	5	$9 - 5 = 4$	$(4)^2 \cdot 10 = 160$
Total	$\sum f = 90$				$\sum(x - \bar{x})^2 \cdot f = 480$

Variance: $s^2 = \frac{\sum(x-\bar{x})^2 \cdot f}{\sum f} = \frac{480}{90} \approx 5.3333$

Standard deviation: $s = \sqrt{s^2} \approx \sqrt{5.3333} \approx 2.3094$

19. From Exercise 13, $\bar{x} = \frac{\sum(f \cdot x)}{\sum f} = \frac{127{,}250}{2200} = 57.8409$.

Class	Frequency f	Midpoint x	$\bar{x}$	$x - \bar{x}$	$(x - \bar{x})^2 \cdot f$
$0 \le \text{cost} < 5$	100	$\frac{0+5}{2} = 2.5$	57.8409	$2.5 - 57.8409 = -53.3409$	$(-55.3409)^2 \cdot 100 = 306{,}261.5213$
$5 \le \text{cost} < 10$	150	$\frac{5+10}{2} = 7.5$	57.8409	$7.5 - 57.8409 = -50.3409$	$(-50.3409)^2 \cdot 150 = 380{,}130.9319$
$10 \le \text{cost} < 15$	200	$\frac{10+15}{2} = 12.5$	57.8409	$12.5 - 57.8409 = -45.3409$	$(-45.3409)^2 \cdot 200 = 411{,}159.4426$
$15 \le \text{cost} < 20$	250	$\frac{15+20}{2} = 17.5$	57.8409	$17.5 - 57.8409 = -40.3409$	$(-40.3409)^2 \cdot 250 = 406{,}847.0532$
$20 \le \text{cost} < 30$	300	$\frac{20+30}{2} = 25$	57.8409	$25 - 57.8409 = -32.8409$	$(-32.8409)^2 \cdot 300 = 323{,}557.4138$
$30 \le \text{cost} < 50$	350	$\frac{30+50}{2} = 40$	57.8409	$40 - 57.8409 = -17.8409$	$(-17.8409)^2 \cdot 350 = 111{,}404.1995$
$50 \le \text{cost} < 100$	400	$\frac{50+100}{2} = 75$	57.8409	$75 - 57.8409 = 17.1591$	$(17.1591)^2 \cdot 400 = 117{,}773.8851$
$100 \le \text{cost} < 200$	450	$\frac{100+200}{2} = 150$	57 57.8409	$150 - 57.8409 = 92.1591$	$(92.1591)^2 \cdot 450 = 3{,}821{,}984.871$
Total	$\sum f = 2200$				$\sum (x - \bar{x})^2 \cdot f = 5{,}879{,}119.318$

Variance: $s^2 = \frac{\Sigma(x-\bar{x})^2 \cdot f}{\Sigma f} \approx \frac{5{,}879{,}119.318}{2200} \approx 2672.3270$

Standard deviation: $s = \sqrt{s^2} \approx \sqrt{2672.3270} \approx 51.6946$

21. (a) and (b)

Class	**Frequency f**	**Midpoint x**	**Product $f \cdot x$**
$0 \le \text{age} < 5$	63,422	$\frac{0+5}{2} = 2.5$	$(63{,}422)(2.5) = 158{,}555$
$5 \le \text{age} < 18$	240,629	$\frac{5+18}{2} = 11.5$	$(240{,}629)(11.5) = 2.767{,}233.5$
$18 \le \text{age} < 65$	540,949	$\frac{18+65}{2} = 41.5$	$(540{,}949)(41.5) = 22{,}449.383.5$
	$\sum f = 845{,}000$		$\sum (f \cdot x) = 25{,}375{,}172$

$\bar{x} = \frac{\Sigma(f \cdot x)}{\Sigma f} = \frac{23{,}375{,}172}{845{,}000} \approx 30.0298$ years

(c)

Class	**Frequency f**	**Midpoint x**	$\bar{x}$	$x - \bar{x}$	$(x - \bar{x})^2 \cdot$
$0 \le \text{age} < 5$	63,422	$\frac{0+5}{2} = 2.5$	30.0298	$2.5 - 30.0298 = -27.5298$	$(-27.5298)^2 \cdot 63{,}422 = 48{,}066{,}892.48$
$5 \le \text{age} < 18$	240,629	$\frac{5+18}{2} = 11.5$	30.0298	$11.5 - 30.0298 = -18.5298$	$(-18.5298)^2 \cdot 240{,}629 = 82{,}620{,}806.47$
$18 \le \text{age} < 65$	540,949	$\frac{18+65}{2} = 41.5$	30.0298	$41.5 - 30.0298 = 11.4702$	$(11.4702)^2 \cdot 540{,}949 = 71{,}170{,}219.19$
	$\sum f = 845{,}000$				$\sum (x - \bar{x})^2 = 201{,}857{,}918.14$

Variance: $s^2 = \frac{\Sigma(x-\bar{x})^2 \cdot f}{\Sigma f} \approx \frac{201{,}857{,}918.14}{845{,}000} \approx 238.8851$ years squared

Standard deviation: $s = \sqrt{s^2} \approx \sqrt{238.8851} \approx 15.4559$ years

23.

Class	Years f	Midpoints x	$f \cdot x$
20 ≤ value < 60	13	$\frac{20+60}{2} = 40$	$(13)(40) = 520$
60 ≤ value < 100	21	$\frac{60+100}{2} = 80$	$(21)(80) = 1680$
100 ≤ value < 140	10	$\frac{100+140}{2} = 120$	$(10)(120) = 1200$
140 ≤ value < 180	6	$\frac{140+180}{2} = 160$	$(6)(160) = 960$
180 ≤ value < 260	10	$\frac{180+260}{2} = 220$	$(10)(220) = 2200$
260 ≤ value < 460	7	$\frac{260+460}{2} = 360$	$(7)(360) = 2520$
	$\sum f = 67$		$\sum f \cdot x = 9080$

$\bar{x} = \frac{\sum(f \cdot x)}{\sum f} = \frac{9080}{67} \approx 135.5224$ deaths per year

Class	Years f	Midpoints x	$\bar{x}$	$x - \bar{x}$	$(x - \bar{x})^2 \cdot f$
20 ≤ value < 60	13	$\frac{20+60}{2} = 40$	135.5224	$40 - 135.5224 = -95.5224$	$(-95.5224)^2 \cdot 13 = 118{,}618.87572288$
60 ≤ value < 100	21	$\frac{60+100}{2} = 80$	135.5224	$80 - 135.5224 = -55.5224$	$(-55.5224)^2 \cdot 21 = 64{,}737.47493696$
100 ≤ value < 140	10	$\frac{100+140}{2} = 120$	135.5224	$120 - 135.5224 = -15.5224$	$(-15.5224)^2 \cdot 10 = 2{,}409.4490176$
140 ≤ value < 180	6	$\frac{140+180}{2} = 160$	135.5224	$160 - 135.5224 = 24.4776$	$(24.4776)^2 \cdot 6 = 3594.91741056$
180 ≤ value < 260	10	$\frac{180+260}{2} = 220$	135.5224	$220 - 135.5224 = 84.4776$	$(84.4776)^2 \cdot 10 = 71{,}364.6490176$
260 ≤ value < 460	7	$\frac{260+460}{2} = 360$	135.5224	$360 - 135.5224 = 224.4776$	$(224.4776)^2 \cdot 7 = 352{,}731.35031232$
	$\sum f = 67$				$\sum (x - \bar{x})^2 \cdot f = 613{,}456.7164$

Variance: $s^2 = \frac{\sum(x-\bar{x})^2 \cdot f}{\sum f} \approx \frac{613{,}456.7164}{67} \approx 9{,}156{,}0704$ deaths per year squared

Standard deviation: $s = \sqrt{s^2} \approx \sqrt{9{,}156.0704} \approx 95.6874$ deaths per year

25. $\bar{x} = \frac{\sum wx}{\sum w} = \frac{w_1x_1+w_2x_2+w_3x_3+w_4x_4+w_5x_5}{w_1+w_2+w_3+w_4+w_5}$

$= \frac{(12{,}380)(\$60.32) + (18{,}580)(\$60.25) + (9{,}540)(\$59.39) + (35{,}550)(\$57.98) + (10{,}130)(\$55.95)}{12{,}380 + 18{,}580 + 9{,}540 + 35{,}550 + 10{,}130}$

$= \frac{\$5{,}060{,}749.70}{1086{,}1800} \approx \$58.72$

Weighted mean wage across the five states = \$58.72.

27. If $w_i = 1$ for all i, then the weighted mean formula will be equivalent to the formula for the sample mean.

29.

Delay (minutes)	Frequency f	Midpoint x	Product $\boldsymbol{f \cdot x}$
0 to < 16	665	8	5320
16 to < 31	551	23.5	12,948.5
31 to < 46	497	38.5	19,134.5
46 to < 61	399	53.5	21,346.5
61 to < 91	355	76	26,980
91 to < 120	27	105.5	2848.5
Total	$\sum f = 2494$		$\sum(f \cdot x) = 88{,}578$

31.

Delay (minutes)	Midpoint x	Frequency f	$\bar{x}$	$x - \bar{x}$	$(x - \bar{x})^2 \cdot f$
0 to < 16	8	665	35.52	−27.52	503,638.016
16 to < 31	23.5	551	35.52	−12.02	79,608.7004
31 to < 46	38.5	497	35.52	2.98	4,413.5588
46 to < 61	53.5	399	35.52	17.98	128,988.8796
61 to < 91	76	355	35.52	40.48	581,713.792
91 to < 120	105.5	27	35.52	69.98	132,224.4108
Total		$\sum f$ = 2494			$\sum (x - x)^2 \cdot f$ = 1,430,587.3576

33. $s = \sqrt{s^2} \approx \sqrt{573.6116} \approx 23.9502$

35.

Professionals

Class	Frequency f	Midpoint x	$f \cdot x$
-50 ≤ price change < -25	3	$\frac{(-50)+(-25)}{2} = -37.5$	$(3)(-37.5) = -112.5$
-25 ≤ price change < 0	26	$\frac{(-25)+0}{2} = -12.5$	$(26)(-12.5) = -325$
0 ≤ price change < 25	43	$\frac{0+25}{2} = 12.5$	$(43)(12.5) = 537.5$
25 ≤ price change < 50	22	$\frac{25+50}{2} = 37.5$	$(22)(37.5) = 825$
50 ≤ price change < 75	5	$\frac{50+75}{2} = 62.5$	$(5)(62.5) = 312.5$
75 ≤ price change < 100	1	$\frac{75+100}{2} = 87.5$	$(1)(87.5) = 87.5$
Total	$\sum f = 100$		$\sum(f \cdot x) = 1325$

$$\bar{x} = \frac{\sum(f \cdot x)}{\sum f} = \frac{1325}{100} = 13.25$$

Darts

Class	Frequency	Midpoint x	$f \cdot x$
-50 ≤ price change < -25	4	$\frac{(-50)+(-25)}{2} = -37.5$	$(4)(-37.5) = -150$
-25 ≤ price change < 0	33	$\frac{(-25)+0}{2} = -12.5$	$(33)(-12.5) = -412.5$
0 ≤ price change < 25	53	$\frac{0+25}{2} = 12.5$	$(53)(12.5) = 662.5$
25 ≤ price change < 50	7	$\frac{25+50}{2} = 37.5$	$(7)(37.5) = 262.5$
50 ≤ price change < 75	3	$\frac{50+75}{2} = 62.5$	$(3)(62.5) = 187.5$
75 ≤ price change < 100	0	$\frac{75+100}{2} = 87.5$	$(0)(87.5) = 0$
Total	100		$\sum(f \cdot x) = 550$

$$\bar{x} = \frac{\sum(f \cdot x)}{\sum f} = \frac{550}{100} = 5.5$$

DJIA

Class	Frequency f	Midpoint x	$f \cdot x$
-50 ≤ price change < -25	0	$\frac{(-50)+(-25)}{2} = -37.5$	$(0)(-37.5) = 0$
-25 ≤ price change < 0	19	$\frac{(-25)+0}{2} = -12.5$	$(19)(-12.5) = -237.5$
0 ≤ price change < 25	81	$\frac{0+25}{2} = 12.5$	$(81)(12.5) = 1012.5$
25 ≤ price change < 50	0	$\frac{25+50}{2} = 37.5$	$(0)(37.5) = 0$
50 ≤ price change < 75	0	$\frac{50+75}{2} = 62.5$	$(0)(62.5) = 0$
75 ≤ price change < 100	0	$\frac{75+100}{2} = 87.5$	$(0)(87.5) = 0$
Total	100		$\sum(f \cdot x) = 775$

$$\bar{x} = \frac{\Sigma(f \cdot x)}{\Sigma f} = \frac{775}{100} = 7.75$$

37.

```
Variable    Mean   StDev  Variance
PROS       10.95   22.25    494.91
Variable   Mean   StDev  Variance
DARTS      4.52   19.39    375.91
Variable    Mean   StDev  Variance
DJIA       6.793   8.031    64.505
```

Professionals:
Mean: 10.95 – 13.25 = -2.3
Variance: 494.91 - 555.69 = -60.78
Standard deviation: 22.25 - 23.57 = -1.32
Darts:
Mean: 4.52 – 5.5 = -0.98
Variance: 375.91 – 376 = -0.09
Standard deviation: 19.39 – 19.39 = 0
DJIA:
Mean: 6.793 – 7.75 = -0.957
Variance: 64.505 - 96.188 = -31.683
Standard deviation: 8.031 - 9.808 = -1.777

39.
Using Minitab,

Sum of Frequency

```
Sum of Frequency = 308692174
```

Sum of Age*Frequency

```
Sum of Age*Frequency = 11514388721
```

Then $\bar{x} = \frac{\Sigma(f \cdot x)}{\Sigma f} = \frac{11,514,388,721}{308,692,174} \approx 30.70$ years.

41. The Empirical Rule tells us that about 68% of the observations lie within 1 standard deviation of the mean. Then Mean – 1 standard deviation = 37.30 – 1(22.70) = 14.6 years and Mean + 1 standard deviation = 37.30 + 1(22.70) = 60 years.
Therefore about 68% of the ages lie between 14.6 years and 60 years old, inclusive.

Section 3.4

1. A positive *z*-score indicates that the data value is above the mean. A negative *z*-score indicates that the data value is below the mean. A *z*-score of zero indicates that the data value is equal to the mean.

3. There can't be more than 100% of the data values less than or equal to any value in the data set.

5. A percentile is a data value, while a percentile rank is a percentage.

For Exercises 7-10, $\mu = 130$ Facebook friends and $\sigma = 30$ Facebook friends.

7 $x = 190$ Facebook friends, so $z = \frac{x-\mu}{\sigma} = \frac{190-130}{30} = 2$.

9. $x = 100$ Facebook friends, so $z = \frac{x-\mu}{\sigma} = \frac{100-130}{30} = -1$.

For Exercises 11-14, $\mu = 100$ hours and $\sigma = 25$ hours.

11. $x = 125$ hours, so $z = \frac{x-\mu}{\sigma} = \frac{125-100}{25} = 1$.

13. $x = 200$ hours, so $z = \frac{x-\mu}{\sigma} = \frac{200-100}{25} = 4$.

For Exercises 15–18, $\mu = 100$ mg/dl and $\sigma = 10$ mg/dl.

15. $x = 90$ mg/dl, so $z = \frac{x-\mu}{\sigma} = \frac{90-100}{10} = -1$. Therefore, Alyssa's blood sugar level is 1 standard deviation below the mean.

17. **(a)** The standard deviation, which is 10 mg/dl. **(b)** $x = 125$ mg/dl, so

$z = \frac{x-\mu}{\sigma} = \frac{125-100}{10} = 2.5$.

(c) Chelsea's blood sugar level lies 2.5 standard deviations above the mean blood sugar level of 100 mg/dl.

For Exercises 19 - 22, $\mu = 130$ Facebook friends and $\sigma = 30$ Facebook friends.

19. For a z-score of -1.0, we have

$x = z\text{-score} \cdot \sigma + \mu = (-1.0) \cdot (30) + 130 = 100$ Facebook friends.

21. For a z-score of 0.0, we have

$x = z\text{-score} \cdot \sigma + \mu = (0.0) \cdot (30) + 130 = 130$ Facebook friends.

For Exercises 23 - 26, $\mu = 100$ hours and $\sigma = 25$ hours.

23. For a z-score of 2.0, we have

$x = z\text{-score} \cdot \sigma + \mu = (2.0) \cdot (25) + 100 = 150$ hours.

25. For a z-score of -0.5, we have

$x = z\text{-score} \cdot \sigma + \mu = (-0.5) \cdot (25) + 100 = 87.5$ hours.

For Exercises 27–30, $\mu = 100$ mg/dl and $\sigma = 10$ mg/dl.

27. For a z-score of 1.96, we have

$x = z\text{-score} \cdot \sigma + \mu = (1.96) \cdot (10) + 100 = 119.6$ mg/dl.

29. For a z-score of -1.96, we have

$x = z\text{-score} \cdot \sigma + \mu = (-1.96) \cdot (10) + 100 = 80.4$ mg/dl.

31. Elizabeth: $x = 85, \bar{x} = 70, s = 15$, so $z = \frac{x-\bar{x}}{s} = \frac{85-70}{15} = 1$

Fiona: $x = 85, \bar{x} = 75, s = 5$, so $z = \frac{x-\bar{x}}{s} = \frac{85-75}{5} = 2$

Since the $z-score$ for Fiona's statistics quiz is larger than the *zscore* for Elizabeth's statistics quiz, Fiona did better relative to her class.

33. The z-score for 190 Facebook friends is 2, which lies in the range $2 \le x < 3$. Therefore, a person with 190 Facebook friends may be considered moderately unusual.

35. The z-score for 100 Facebook friends is -1, which lies in the range $-2 < x < 2$. Therefore, a person with 100 Facebook friends is not considered unusual.

37. The z-score for 125 hours is 1, which lies in the range $-2 < x < 2$. Therefore, 125 hours of video uploaded to YouTube every minute is not considered unusual.

39. The z-score for 200 hours is 4, which lies in the range $x \ge 3$. Therefore, 200 hours of video uploaded to YouTube every minute may be considered an outlier.

41. Since $-2 < z\text{-score} = -1 < 2$, Alyssa's blood sugar level is not an outlier.

43. Since $2 \le z\text{-score} = 2.5 < 3$, Chelsea's blood sugar level is moderately unusual.

Data is ascending order for Exercises 45, 45, 47, and 49..

20, 21, 23, 24, 25, 25, 26, 26, 27, 29, 30, 31

45. $i = \left(\frac{p}{100}\right)n = \left(\frac{75}{100}\right)12 = 9.$

Here, $i = 9$ is an integer, so the 75th percentile is the mean of the values in positions 9 and 10.

Position	1	2	3	4	5	6	7	8	9	10	11	12
Mpg	20	21	23	24	25	25	26	26	27	29	30	31

From the chart, the data value in position 9 is 27 mpg and the data value in position 10 is 29 mpg. The mean of these two values is

$\frac{27+29}{2} = \frac{56}{2} = 28$ mpg. Therefore, the data value corresponding to the 75th percentile is 28 mpg.

47. $i = \left(\frac{p}{100}\right)n = \left(\frac{95}{100}\right)12 = 11.4.$

Here, $i = 11.4$ is not an integer, so we round i up to the nearest integer, which is $i = 12$. Therefore, the 95th percentile is the value in position 12.

Position	1	2	3	4	5	6	7	8	9	10	11	12
Mpg	20	21	23	24	25	25	26	26	27	29	30	31

From the chart, the data value in position 12 is 31 mpg. Therefore, the data value corresponding to the 95th percentile is 31 mpg.

49. $i = \left(\frac{p}{100}\right)n = \left(\frac{10}{100}\right)12 = 1.2.$

Here, $i = 1.2$ is not an integer, so we round i up to the nearest integer, which is $i = 2$. Therefore, the 10th percentile is the value in position 2.

Position	1	2	3	4	5	6	7	8	9	10	11	12
Mpg	20	21	23	24	25	25	26	26	27	29	30	31

From the chart, the data value in position 2 is 21 mpg. Therefore, the data value corresponding to the 10th percentile is 21 mpg.

Data arranged in ascending order for Problems 51, 53, and 55.

80, 125, 125, 135, 140, 150, 180, 180, 200, 210, 210, 220, 230, 240, 260, 290

51. $i = \left(\frac{p}{100}\right)n = \left(\frac{75}{100}\right)16 = 12.$

Here, $i = 12$ is an integer, so the 75th percentile is the mean of the values in positions 12 and 13.

Position	1	2	3	4	5	6	7	8	9	10	11	12	13	14	15	16
Calories	80	125	125	135	140	150	180	180	200	210	210	220	230	240	260	290

From the chart, the data value in position 12 is 220 mg per serving and the data value in position 13 is 230 mg per serving. The mean of these two values is $\frac{220+230}{2} = \frac{450}{2} = 225$ mg per serving. Therefore, the data value corresponding to the 75th percentile is 225 mg per serving.

53. $i = \left(\frac{p}{100}\right)n = \left(\frac{90}{100}\right)16 = 14.4.$

Here, $i = 14.4$ is not an integer, so we round i up to the next integer, which is $i = 15$. Therefore, the 90th percentile is the value in position 15.

Position	1	2	3	4	5	6	7	8	9	10	11	12	13	14	15	16
Calories	80	125	125	135	140	150	180	180	200	210	210	220	230	240	260	290

From the chart, the data value in position 15 is 260 mg per serving. Therefore, the data value corresponding to the 90th percentile is 260 mg per serving.

55. $i = \left(\frac{p}{100}\right)n = \left(\frac{5}{100}\right)16 = 0.8.$

Here, $i = 0.8$ is not an integer, so we round i up to the next integer, which is $i = 1$. Therefore, the 5th percentile is the value in position 1.

Position	1	2	3	4	5	6	7	8	9	10	11	12	13	14	15	16
Calories	80	125	125	135	140	150	180	180	200	210	210	220	230	240	260	290

From the chart, the data value in position 1 is 80 mg per serving. Therefore, the data value corresponding to the 5th percentile is 80 mg per serving.

Data is ascending order for Exercises 57, 59, and 61.

20, 21, 23, 24, 25, 25, 26, 26, 27, 29, 30, 31

57. Here, $x = 30$. Eleven cars have x-values at or below 30 mpg, so the percentile rank of a car with a combined mpg of 30 mpg is percentile rank of data value $(x = 30) = \frac{\text{number of values in data set} \leq 30}{\text{total number ofvalues in data set}} \cdot 100 = \frac{11}{12} \cdot 100 \approx 92\%$. Therefore, 30 mpg represents the 92nd percentile of combined gas mileage for cars.

59. Here, $x = 20$. One car has an x-value at or below 20 mpg, so the percentile rank of a car with a combined mpg of 20 mpg is percentile rank of data value $(x = 20) = \frac{\text{number of values in data set} \leq 20}{\text{total number ofvalues in data set}} \cdot 100 = \frac{1}{12} \cdot 100 \approx 8\%$. Therefore, 20 mpg represents the 8th percentile of combined gas mileage for cars.

61. Here, $x = 27$. Nine cars have x-values at or below 27 mpg, so the percentile rank of a car with a combined mpg of 27 mpg is percentile rank of data value $(x = 27) = \frac{\text{number of values in data set} \leq 27}{\text{total number ofvalues in data set}} \cdot 100 = \frac{9}{12} \cdot 100 = 75\%$. Therefore, 27 mpg represents the 75th percentile of combined gas mileage for cars.

63. Here, $x = 80$. One breakfast cereal has an x-value at or below 80 mg per serving, so the percentile rank of a breakfast cereal with 80 mg per serving is percentile rank of data value $(x = 80) = \frac{\text{number of values in data set} \leq 80}{\text{total number ofvalues in data set}} \cdot 100 = \frac{1}{16} \cdot 100 \approx 6\%$. Therefore, 80 mg per serving represents the 6th percentile of mg of sodium per serving for breakfast cereals.

65. Here, $x = 260$. Fifteen breakfast cereals have x-values at or below 260 mg per serving, so the percentile rank of a breakfast cereal with 260 mg per serving is percentile rank of data value $(x = 260) = \frac{\text{number of values in data set} \leq 260}{\text{total number ofvalues in data set}} \cdot 100 = \frac{15}{16} \cdot 100\% \approx 94\%$. Therefore, 260 mg per serving represents the 94th percentile of mg of sodium per serving for breakfast cereals.

67. Here, $x = 230$. Thirteen breakfast cereals have x-values at or below 230 mg per serving, so the percentile rank of a breakfast cereal with 230 mg per serving is percentile rank of data value $(x = 230) = \frac{\text{number of values in data set} \leq 230}{\text{total number ofvalues in data set}} \cdot 100 = \frac{13}{16} \cdot 100\% \approx 81\%$. Therefore, 230 mg per serving represents the 81st percentile of mg of sodium per serving for breakfast cereals.

69. Q1 is the 25th percentile.

Then, $i = \left(\frac{p}{100}\right)n = \left(\frac{25}{100}\right)12 = 3$. Here, $i = 3$ is an integer, so the 25th percentile is the mean of the values in positions 3 and 4.

Position	1	2	3	4	5	6	7	8	9	10	11	12
Mpg	20	21	23	24	25	25	26	26	27	29	30	31

From the chart, the data value in position 3 is 23 mpg and the data value in position 4 is 24 mpg.

The mean of these two values is $\frac{23+24}{2} = \frac{47}{2} = 23.5$ mpg. Therefore, the data value corresponding to the 25th percentile is 23.5 mpg.

71. Q3 is the 75th percentile.

Then, $i = \left(\frac{p}{100}\right)n = \left(\frac{75}{100}\right)12 = 9$. Here, $i = 9$ is an integer, so the 75th percentile is the mean of the values in positions 9 and 10.

Position	1	2	3	4	5	6	7	8	9	10	11	12
Mpg	20	21	23	24	25	25	26	26	27	29	30	31

From the chart, the data value in position 9 is 27 mpg and the data value in position 10 is 29 mpg. The mean of these two values is $\frac{27+29}{2} = \frac{56}{2} = 28$ mpg. Therefore, the data value corresponding to the 75th percentile is 28 mpg.

73. Q1 is the 25th percentile.

Then, $i = \left(\frac{p}{100}\right)n = \left(\frac{25}{100}\right)16 = 4$. Here, $i = 4$ is an integer, so the 25th percentile is the mean of the values in positions 4 and 5.

Position	1	2	3	4	5	6	7	8	9	10	11	12	13	14	15	16
Calories	80	125	125	135	140	150	180	180	200	210	210	220	230	240	260	290

From the chart, the data value in position 4 is 135 mg per serving and the data value in position 5 is 140 mg per serving. The mean of these two values is $\frac{135+140}{2} = \frac{275}{2} = 137.5$ mg per serving. Therefore, the data value corresponding to the 25th percentile is 137.5 mg of sodium per serving.

75. Q3 is the 75th percentile.

Then, $i = \left(\frac{p}{100}\right)n = \left(\frac{75}{100}\right)16 = 12$. Here, $i = 12$ is an integer, so the 75th percentile is the mean of the values in positions 12 and 13.

Position	1	2	3	4	5	6	7	8	9	10	11	12	13	14	15	16
Calories	80	125	125	135	140	150	180	180	200	210	210	220	230	240	260	290

From the chart, the data value in position 12 is 220 mg per serving and the data value in position 13 is 230 mg per serving. The mean of these two values is $\frac{220+230}{2} = \frac{450}{2} = 225$ mg per serving. Therefore, the data value corresponding to the 75th percentile is 225 mg of sodium per serving.

77. IQR = Q3 – Q1 = 28 – 23.5 = 4.5. The middle 50%, or middle half, of the highway MPG data ranged over 4.5 miles per gallon.

79. The sample mean is

$$\bar{x} = \frac{\sum x}{n} = \frac{90 + 90 + 100 + 110 + 110 + 110 + 110 + 110 + 110 + 120 + 120 + 130}{12}$$

$$= \frac{1310}{12} \approx 109.1667 \text{ calories.}$$

To calculate the standard deviation we first need to find $\sum x$ and $\sum x^2$.

x	x^2
90	8,100
90	8,100
100	10,000
110	12,100
110	12,100
110	12,100
110	12,100
110	12,100
110	12,100
120	14,400
120	14,400
130	16,900
$\Sigma x = 1{,}310$	$\Sigma x^2 = 144{,}500$

$n = 12$. Therefore, the standard deviation is

$$s = \sqrt{\frac{\sum x^2 - \left(\sum x\right)^2/n}{n-1}} = \sqrt{\frac{144{,}500 - (1{,}310)^2/12}{12-1}} \approx 11.6450 \text{ calories.}$$

(a) For Corn Flakes, $x = 100$ calories. Thus the z-score for the Corn Flakes is

$$z = \frac{x - \bar{x}}{s} = \frac{100 - 109.1667}{11.6450} = -0.79.$$

(b) For Basic 4, $x = 130$ calories. Thus the z-score for the Basic 4 is

$$z = \frac{x - \bar{x}}{s} = \frac{130 - 109.1667}{11.6450} = 1.79.$$

(c) For Bran Flakes, $x = 90$ calories. Thus the z-score for the Bran Flakes is

$$z = \frac{x - \bar{x}}{s} = \frac{90 - 109.1667}{11.6450} = -1.65.$$

(d) For Cap'n Crunch, $x = 120$ calories. Thus the z-score for the Cap'n Crunch is

$$z = \frac{x - \bar{x}}{s} = \frac{120 - 109.1667}{11.6450} = 0.93.$$

81. For cereals with $x = 90$ calories, $= \frac{x - \bar{x}}{s} \approx \frac{90 - 109.1555557}{11.64500153} \approx -0.176.$

$-2 < z\text{-score} \approx -0.176 < 2$, so a cereal with 90 calories per serving is not an outlier.

For cereals with $x = 100$ calories, $= \frac{x - \bar{x}}{s} \approx \frac{100 - 109.1555557}{11.64500153} \approx -0.084.$

$-2 < z\text{-score} \approx -0.084 < 2$, so a cereal with 100 calories per serving is not an outlier.

For cereals with $x = 110$ calories, $= \frac{x - \bar{x}}{s} \approx \frac{110 - 109.1555557}{11.64500153} \approx 0.008.$

$-2 < z\text{-score} \approx 0.008 < 2$, so a cereal with 110 calories per serving is not an outlier.

For cereals with $x = 120$ calories, $= \frac{x - \bar{x}}{s} \approx \frac{120 - 109.1555557}{11.64500153} \approx 0.099.$

$-2 < z\text{-score} \approx 0.099 < 2$, so a cereal with 120 calories per serving is not an outlier.

For cereals with $x = 130$ calories, $= \frac{x - \bar{x}}{s} \approx \frac{130 - 109.1555557}{11.64500153} \approx 0.191.$

$-2 < z\text{-score} \approx 0.191 < 2$, so a cereal with 130 calories per serving is not an outlier.
Therefore, this data set has no outliers.

83. First, arrange the data in ascending order:

90 90 100 110 110 110 110 110 110 120 120 130

(a) Here, $x = 90$ calories.

$$\text{percentile rank of data value } x = \frac{\text{number of data values} \le x = 90}{\text{total number of values in data set}} \cdot 100 = \frac{2}{12} \cdot 100 \approx 17\%$$

(b) Here, $x = 120$ calories.

$$\text{percentile rank of data value } x = \frac{\text{number of data values} \le x = 120}{\text{total number of values in data set}} \cdot 100 = \frac{11}{12} \cdot 100 \approx 92\%$$

(c) Here, $x = 110$ calories.

$$\text{percentile rank of data value } x = \frac{\text{number of data values} \le x = 110}{\text{total number of values in data set}} \cdot 100 = \frac{9}{12} \cdot 100 = 75\%$$

(d) Here, $x = 100$ calories.

$$\text{percentile rank of data value } x = \frac{\text{number of data values} \le x = 100}{\text{total number of values in data set}} \cdot 100 = \frac{3}{12} \cdot 100 = 25\%$$

85. The middle 50%, or half, of the number of calories in 12 breakfast cereals, ranges over 10 calories.

87. From the TI-83/84, $\bar{x} \approx 5.073333333$ and $s \approx 3.359308319$.

(a) $x = 14.7$ million American adults,

$$z = \frac{x - \bar{x}}{s} \approx \frac{14.7 - 5.073333333}{3.359308319} \approx 2.87.$$

(b) $x = 2.0$ million American adults,

$$z = \frac{x - \bar{x}}{s} \approx \frac{2.0 - 5.073333333}{3.359308319} \approx -0.91.$$

(c) For Valerian, $x = 2.1$ million. Thus the z-score for Valerian is

$$z = \frac{x - \bar{x}}{s} = \frac{2.1 - 5.0733}{3.3593} = -0.89.$$

(d) For Ginseng, $x = 8.8$ million. Thus the z-score for Ginseng is

$$z = \frac{x - \bar{x}}{s} = \frac{8.8 - 5.0733}{3.3593} = 1.11.$$

89. Echinacea: $x = 14.7$ million American adults,

$$z = \frac{x - \bar{x}}{s} \approx \frac{14.7 - 5.073333333}{3.359308319} \approx 2.87$$

$2 \le z\text{-score} \approx 2.87 < 3$, so Echinacea with 14.7 million users is moderately unusual.
Ginseng: $x = 8.8$ million American adults,

$$z = \frac{x - \bar{x}}{s} \approx \frac{8.8 - 5.073333333}{3.359308319} \approx 1.11$$

$-2 < z\text{-score} \approx 1.11 < 2$, so Ginseng with 8.8 million users is not an outlier.
Ginkgo biloba: $x = 7.7$ million American adults,

$$z = \frac{x - \bar{x}}{s} \approx \frac{7.7 - 5.073333333}{3.359308319} \approx 0.78$$

$-2 < z\text{-score} \approx 0.78 < 2$, so Ginkgo biloba with 7.7 million users is not an outlier.
Garlic: $x = 7.1$ million American adults,

$$z = \frac{x - \bar{x}}{s} \approx \frac{7.1 - 5.073333333}{3.359308319} \approx 0.60$$

$-2 < z\text{-score} \approx 0.60 < 2$, so Garlic with 7.1 million users is not an outlier.
Glucosamine: $x = 5.2$ million American adults,

$$z = \frac{x - \bar{x}}{s} \approx \frac{5.2 - 5.073333333}{3.359308319} \approx 0.04$$

$-2 < z\text{-score} \approx 0.04 < 2$, so Glucosamine with 5.2 million users is not an outlier.
St. John's wort: $x = 4.4$ million American adults,

$$z = \frac{x - \bar{x}}{s} \approx \frac{4.4 - 5.073333333}{3.359308319} \approx -0.20$$

$-2 < z\text{-score} \approx -0.20 < 2$, so St. John's wort with 4.4 million users is not an outlier.
Peppermint: $x = 4.3$ million American adults,

$$z = \frac{x - \bar{x}}{s} \approx \frac{4.3 - 5.073333333}{3.359308319} \approx -0.23$$

$-2 < z\text{-score} \approx -0.23 < 2$, so Peppermint with 4.3 million users is not an outlier.
Fish oil: $x = 4.2$ million American adults,

$$z = \frac{x - \bar{x}}{s} \approx \frac{4.2 - 5.073333333}{3.359308319} \approx -0.26$$

$-2 < z\text{-score} \approx -0.26 < 2$, so fish oil with 4.2 million users is not an outlier.
Ginger: $x = 3.8$ million American adults,

$$z = \frac{x - \bar{x}}{s} \approx \frac{3.8 - 5.073333333}{3.359308319} \approx -0.38$$

$-2 < z$-score $\approx -0.38 < 2$, so Ginger with 3.8 million users is not an outlier.
Soy: $x = 3.5$ million American adults,

$$z = \frac{x - \bar{x}}{s} \approx \frac{3.5 - 5.073333333}{3.359308319} \approx -0.47$$

$-2 < z$-score $\approx -0.47 < 2$, so Soy with 3.5 million users is not an outlier.
Chamomile: $x = 3.1$ million American adults,

$$z = \frac{x - \bar{x}}{s} \approx \frac{3.1 - 5.073333333}{3.359308319} \approx -0.59$$

$-2 < z$-score $\approx -0.59 < 2$, so Chamomile with 3.1 million users is not an outlier.
Bee pollen: $x = 2.8$ million American adults,

$$z = \frac{x - \bar{x}}{s} \approx \frac{2.8 - 5.073333333}{3.359308319} \approx -0.68$$

$-2 < z$-score $\approx -0.68 < 2$, so Bee pollen with 2.8 million users is not an outlier.
Kava kava: $x = 2.4$ million American adults,

$$z = \frac{x - \bar{x}}{s} \approx \frac{2.4 - 5.073333333}{3.359308319} \approx -0.80$$

$-2 < z$-score $\approx -0.80 < 2$, so Kava kava with 2.4 million users is not an outlier.
Valerian: $x = 2.1$ million American adults,

$$z = \frac{x - \bar{x}}{s} \approx \frac{2.1 - 5.073333333}{3.359308319} \approx -0.89$$

$-2 < z$-score $\approx -0.89 < 2$, so Valerian with 2.1 million users is not an outlier.
Saw palmetto: $x = 2.0$ million American adults,

$$z = \frac{x - \bar{x}}{s} \approx \frac{2.0 - 5.073333333}{3.359308319} \approx -0.91$$

$-2 < z$-score $\approx -0.91 < 2$, so Saw palmetto with 2.0 million users is not an outlier.

91. **(a)** Here, $x = 14.7$ million.

$$\text{percentile rank of data value } x = \frac{\text{number of data values} \le x = 14.7}{\text{total number of values in data set}} \cdot 100 = \frac{15}{15} \cdot 100 = 100\%$$

(b) Here, $x = 2.0$ million.

$$\text{percentile rank of data value } x = \frac{\text{number of data values} \le x = 2.0}{\text{total number of values in data set}} \cdot 100 = \frac{1}{15} \cdot 100 \approx 7\%$$

(c) Here, $x = 8.8$ million.

$$\text{percentile rank of data value } x = \frac{\text{number of data values} \le x = 8.8}{\text{total number of values in data set}} \cdot 100 = \frac{14}{15} \cdot 100 \approx 93\%$$

(d) Here, $x = 2.1$ million.

$$\text{percentile rank of data value } x = \frac{\text{number of data values} \le x = 100}{\text{total number of values in data set}} \cdot 100 = \frac{2}{15} \cdot 100 = 13\%$$

93. The middle 50%, or half of the usage of dietary supplements, ranges over 4.3 million.
95. **(a)** The difference between the 5th percentile and the 50th percentile is $|8998-6381| = 2617$ and the difference between the 95th percentile and the 50th percentile is $|17{,}188-8998| = 8190$. Since the 50th percentile is closer to the 5th percentile than it is to the 95th percentile the distribution of expenditures is right-skewed. **(b)** Since the distribution of expenditures it right-skewed, we would expect the mean expenditure per pupil to be greater than the 50th percentile which is $8998. (c) Answers will vary.

97. The data set would have to be right-skewed with a few values much larger than the rest of the data set. The boxplot will have the line for the median closer to the line for Q3 than the line for Q1.

Example: Data −4, −3, −2, −1, −1, −1, 0, 20

−5 0 5 10 15 20

99. Not possible. Q_1, the 25th percentile, will always be less than or equal to Q_3, the 75th percentile. Thus, the $IQR = Q_3 - Q_1$ is always greater than or equal to zero.

101. From the table, 30% of Twitter accounts have 19 or fewer followers and 10% of Twitter accounts have 3 or fewer followers. Therefore, 30%−10% = 20% of Twitter accounts have between 3 and 19 followers.
103. From the table, 99.9% of Twitter accounts have 24,964 or fewer followers. Therefore, 100%− 99.9% = 0.1% of Twitter accounts have more than 24,964 followers.
105. 95th percentile
107.

```
Variable    Mean   StDev   Median
PROS       10.95   22.25     9.60
```

$$z\text{-score} = \frac{\textbf{median-mean}}{\textbf{standard deviation}} = \frac{\textbf{9.60-10.95}}{\textbf{22.25}} \approx \mathbf{-0.0607}$$

```
Variable   Mean   StDev   Median
DARTS      4.52   19.39     3.25
```

$$z\text{-score} = \frac{\textbf{median-mean}}{\textbf{standard deviation}} = \frac{\textbf{3.25-4.52}}{\textbf{19.39}} \approx \mathbf{-0.0655}$$

```
Variable    Mean   StDev   Median
DJIA       6.793   8.031    7.000
```

$$z\text{-score} = \frac{\text{median-mean}}{\text{standard deviation}} = \frac{7.000-6.793}{8.031} \approx 0.0258$$

For the pros and the darts data the median is below the mean. For the DJIA data the median is above the mean.
109. From the Empirical Rule, about 95%.
111.
Pros: IQR = Q3 – Q1 = 26.55 – (-5.85) = 32.4
For the professionals, the middle 50%, or middle half, of the change in stocks data ranged over 32.4%.
Darts: IQR = Q3 – Q1 = 14.35 – (-6.25) = 20.60

For using darts, the middle 50%, or middle half, of the change in stocks data ranged over 20.60%.
DJIA: IQR= Q3 – Q1 = 13.2 – 1.55 = 11.65
For the DJIA, the middle 50%, or middle half, of the change in stocks data ranged over 11.65%.

113. $\bar{x} = \frac{\sum x}{n} = \frac{0.38+0.9+1.25+1.64+1.66+2.57}{6} = \frac{8.4}{6} = 1.4$ fatalities per 100,000 people

x	$x - \bar{x}$	$(x - \bar{x})^2$
0.38	$0.38 - 1.4 = -1.02$	$(-1.02)^2 = 1.0404$
0.9	$0.9 - 1.4 = -0.5$	$(-0.5)^2 = 0.25$
1.25	$1.25 - 1.4 = -0.15$	$(-0.15)^2 = 0.0225$
1.64	$1.64 - 1.4 = 0.24$	$(0.24)^2 = 0.0576$
1.66	$1.66 - 1.4 = 0.26$	$(0.26)^2 = 0.0676$
2.57	$2.57 - 1.4 = 1.17$	$(1.17)^2 = 1.3689$
		$\sum(x - \bar{x})^2 = 2.807$

Standard deviation: $s = \sqrt{\frac{\sum(x-\bar{x})^2}{n-1}} = \sqrt{\frac{2.807}{6-1}} = \sqrt{0.5614} \approx 0.7493$ fatality per 100,000 people

(a) For Ohio, $x = 0.90$ fatality per 100,000 people. Therefore, $z = \frac{x-\mu}{\sigma} \approx \frac{0.90-1.4}{0.7493} \approx -0.6673$.
(b) For Texas, $x = 1.64$ fatalities per 100,000 people. Therefore, $z = \frac{x-\mu}{\sigma} \approx \frac{1.64-1.4}{0.7493} \approx -0.3203$.

(c) For Florida, $x = 2.57$ fatalities per 100,000 people. Therefore, $z = \frac{x-\mu}{\sigma} \approx \frac{2.57-1.4}{0.7493} \approx 1.5615$.

115. For Nebraska, $x = 0.38$ fatality per 100,000 people. Therefore, $z = \frac{x-\mu}{\sigma} \approx \frac{0.38-1.4}{0.7493} \approx -1.3613$.

For Ohio, $x = 0.90$ fatality per 100,000 people. Therefore, $z = \frac{x-\mu}{\sigma} \approx \frac{0.90-1.4}{0.7493} \approx -0.6673$.
For Tennessee, $x = 1.25$ fatalities per 100,000 people. Therefore, $z = \frac{x-\mu}{\sigma} \approx \frac{1{,}25-1.4}{0.7493} \approx -0.2002$.
For Texas, $x = 1.64$ fatalities per 100,000 people. Therefore, $z = \frac{x-\mu}{\sigma} \approx \frac{1.64-1.4}{0.7493} \approx 0.3203$.
For California $x = 1.66$ fatalities per 100,000 people. Therefore, $z = \frac{x-\mu}{\sigma} \approx \frac{1.66-1.4}{0.7493} \approx 0.3470$.
For Florida, $x = 2.57$ fatalities per 100,000 people. Therefore, $z = \frac{x-\mu}{\sigma} \approx \frac{2.57-1.4}{0.7493} \approx 1.5615$.

Since all of the z-scores lie in the interval $-2 < z\text{-score} < 2$, there are no outliers.

117. **(a)** $i = \left(\frac{p}{100}\right)n = \left(\frac{50}{100}\right)6 = 3$. Here, $i = 3$ is an integer, so the 50th percentile is the mean of the values in positions 3 and 4.

Position	1	2	3	4	5	6
Pedestrian fatality rate	0.38	0.90	1.25	1.64	1.66	2.57

From the chart, the data value in position 3 is 1.25 fatalities per 100,000 people and the data value in position 4 is 1.64 fatalities per 100,000 people. The mean of these two values is $\frac{1.25+1.64}{2} = \frac{2.89}{2} = 1.445$ fatalities per 100,000 people. Therefore, the data value corresponding to the 50th percentile is 1.445 fatalities per 100,000 people.

(b) $i = \left(\frac{p}{100}\right)n = \left(\frac{75}{100}\right)6 = 4.5$. Here, $i = 4.5$ is not an integer so round i up to the next whole number $i = 5$. Therefore, the 75th percentile is the value in positions 5.

Position	1	2	3	4	5	6
Pedestrian fatality rate	0.38	0.90	1.25	1.64	1.66	2.57

From the chart, the data value in position 5 is 1.66 fatalities per 100,000 people. Therefore, the data value corresponding to the 75th percentile is 1.66 fatalities per 100,000 people.

(c) $i = \left(\frac{p}{100}\right)n = \left(\frac{25}{100}\right)6 = 1.5$. Here, $i = 1.5$ is not an integer so round i up to the next whole number $i = 2$. Therefore, the 25th percentile is the value in positions 2.

Position	1	2	3	4	5	6
Pedestrian fatality rate	0.38	0.90	1.25	1.64	1.66	2.57

From the chart, the data value in position 2 is 0.9 fatality per 100,000 people. Therefore, the data value corresponding to the 25th percentile is 0.90 fatality per 100,000 people.

119. **(a)** Q1 is the 25th percentile.

Then, $i = \left(\frac{p}{100}\right)n = \left(\frac{25}{100}\right)6 = 1.5$. Here, $i = 1.5$ is not an integer so round i up to the next whole number $i = 2$. Therefore, the 25th percentile is the value in positions 2.

Position	1	2	3	4	5	6
Pedestrian fatality rate	0.38	0.90	1.25	1.64	1.66	2.57

From the chart, the data value in position 2 is 0.9 fatality per 100,000 people. Therefore, the data value corresponding to the 25th percentile is 0.90 fatality per 100,000 people.

(b) Q2 is the 50th percentile.

Then, $i = \left(\frac{p}{100}\right)n = \left(\frac{50}{100}\right)6 = 3$. Here, $i = 3$ is an integer, so the 50th percentile is the mean of the values in positions 3 and 4.

Position	1	2	3	4	5	6
Pedestrian fatality rate	0.38	0.90	1.25	1.64	1.66	2.57

From the chart, the data value in position 3 is 1.25 fatalities per 100,000 people and the data value in position 4 is 1.64 fatalities per 100,000 people. The mean of these two values is $\frac{1.25+1.64}{2} = \frac{2.89}{2} = 1.445$ fatalities per 100,000 people. Therefore, the data value corresponding to the 50th percentile is 1.445 fatalities per 100,000 people.

(c) Q3 is the 75th percentile.

Then, $i = \left(\frac{p}{100}\right)n = \left(\frac{75}{100}\right)6 = 4.5$. Here, $i = 4.5$ is not an integer so round i up to the next whole number $i = 5$. Therefore, the 75th percentile is the value in positions 5.

Position	1	2	3	4	5	6
Pedestrian fatality rate	0.38	0.90	1.25	1.64	1.66	2.57

From the chart, the data value in position 5 is 1.66 fatalities per 100,000 people. Therefore, the data value corresponding to the 75th percentile is 1.66 fatalities per 100,000 people.

(d) IQR = Q3 – Q1 = 1.66 – 0.90 = 0.76 fatality per 100,000 people

Section 3.5

1. False. The five-number summary consists of the following: Minimum, Q1, *Median*, Q3,

Maximum.

3. **(a)** The median will be about the same distance from Q1 and Q3 and the upper and lower whiskers will be about the same length. **(b)** The median is closer to Q1 than to Q3 and the upper whisker is much longer than the lower whisker. **(c)** The median is closer to Q3 than to Q1 and the lower whisker is much longer than the upper whisker.

5. Any data values located 1.5(IQR) or more below Q1 or 1.5(IQR) or more above Q3 is considered an outlier.

7. Data arranged in ascending order: $200, $200, $200, $200, $250, $300, $600, $800
Q1 is the 25th percentile.
Then, $i = \left(\frac{p}{100}\right)n = \left(\frac{25}{100}\right)8 = 2$. Here, $i = 2$ is an integer, so the 25th percentile is the mean of the values in positions 2 and 3.

Position	1	2	3	4	5	6	7	8
Cell phone prices	$200	$200	$200	$200	$250	$300	$600	$800

From the chart, the data value in position 2 is $200 and the data value in position 3 is $200. The mean of these two values is $\frac{\$200+\$200}{2} = \frac{\$400}{2} = \$200$. Therefore, the data value corresponding to the 25th percentile is $200.
Q2 is the 50th percentile.

Then, $i = \left(\frac{p}{100}\right)n = \left(\frac{50}{100}\right)8 = 4$. Here, $i = 4$ is an integer, so the 50th percentile is the mean of the values in positions 4 and 5.

Position	1	2	3	4	5	6	7	8
Cell phone prices	$200	$200	$200	$200	$250	$300	$600	$800

From the chart, the data value in position 4 is $200 and the data value in position 4 is $250. The mean of these two values is $\frac{\$200+\$250}{2} = \frac{\$450}{2} = \$225$. Therefore, the data value corresponding to the 50th percentile is $225.
Q3 is the 75th percentile.
Then, $i = \left(\frac{p}{100}\right)n = \left(\frac{75}{100}\right)8 = 6$. Here, $i = 6$ is an integer, so the 75th percentile is the mean of the values in positions 6 and 7.

Position	1	2	3	4	5	6	7	8
Cell phone prices	$200	$200	$200	$200	$250	$300	$600	$800

From the chart, the data value in position 6 is $300 and the data value in position 7 is $600. The mean of these two values is $\frac{\$300+\$600}{2} = \frac{\$900}{2} = \$450$. Therefore, the data value corresponding to the 75th percentile is $450.

9. IQR = Q3 - Q1 = $450 - $250 = $250

11. Q1 – 1.5(IQR) = $200 – 1.5($250) = $200 - $375 = - $175
Q3 + 1.5(IQR) = $450 + 1.5($250) = $450 + $375 = $825
Since $200 is not -$175 or less and $200 is not $825 or more, $200 is not an outlier.

13. Data set in ascending order:
1750, 2150, 2300, 2350, 2400, 2500, 2500, 2550, 2750, 2800, 2950, 3100
Q1 is the 25th percentile.
Then, $i = \left(\frac{p}{100}\right)n = \left(\frac{25}{100}\right)12 = 3$. Here, $i = 3$ is an integer, so the 25th percentile is the mean of the values in positions 3 and 4.

Position	1	2	3	4	5	6	7	8	9	10	11	12
Annual fuel cost	$1750	$2150	$2300	$2350	$2400	$2500	$2500	$2550	$2750	$2800	$2950	$3100

From the chart, the data value in position 3 is $2300 and the data value in position 4 is $2350. The mean of these two values is $\frac{\$2300+\$2350}{2} = \frac{\$4650}{2} = \$2325$. Therefore, the data value corresponding to the 25th percentile is $2325.

Q2 is the 50th percentile.

Then, $i = \left(\frac{p}{100}\right)n = \left(\frac{50}{100}\right)12 = 6$. Here, $i = 6$ is an integer, so the 50th percentile is the mean of the values in positions 6 and 7.

Position	1	2	3	4	5	6	7	8	9	10	11	12
Annual fuel cost	$1750	$2150	$2300	$2350	$2400	$2500	$2500	$2550	$2750	$2800	$2950	$3100

From the chart, the data value in position 6 is $2500 and the data value in position 7 is $2500. The mean of these two values is $\frac{\$2500+\$2500}{2} = \frac{\$5000}{2} = \$2500$. Therefore, the data value corresponding to the 50th percentile is $2500.

Q3 is the 75th percentile.

Then, $i = \left(\frac{p}{100}\right)n = \left(\frac{75}{100}\right)12 = 9$.

Here, $i = 9$ is an integer, so the 75th percentile is the mean of the values in positions 9 and 10.

Position	1	2	3	4	5	6	7	8	9	10	11	12
Annual fuel cost	$1750	$2150	$2300	$2350	$2400	$2500	$2500	$2550	$2750	$2800	$2950	$3100

From the chart, the data value in position 9 is $2750 and the data value in position 10 is $2800. The mean of these two values is $\frac{\$2750+\$2800}{2} = \frac{\$5550}{2} = \$2775$. Therefore, the data value corresponding to the 75th percentile is $2775.

15. IQR = Q3 – Q1 = $2775 - $2325 = $450

17. Q1 – 1.5(IQR) = $2325 – 1.5($450) = $2325 - $675 = $1650

Q3 + 1.5(IQR) = $2775 + 1.5($450) = $2775 + $675 = $3450

Since $1750 is not $1650 or less and $1750 is not $3450 or more, $1750 is not an outlier.

19. Q1 is the 25th percentile.

Then, $i = \left(\frac{p}{100}\right)n = \left(\frac{25}{100}\right)23 = 5.75$. Here, $i = 5.75$ is not an integer, so we round i up to the next integer $i = 6$. Then the 25th percentile is the value in positions 6.

Position	Criminal Trespass Cases
1	32
2	41
3	48
4	49
5	55
6	55
7	67
8	68
9	88
10	98
11	101
12	101
13	111
14	111
15	131
16	145
17	150
18	166
19	190
20	223
21	258
22	363
23	451

From the chart, the data value in position 6 is 55 criminal trespass cases. Therefore, the data value corresponding to the 25th percentile is 55 criminal trespass cases.
Q2 is the 50th percentile.

Then, $i = \left(\frac{p}{100}\right)n = \left(\frac{50}{100}\right)23 = 11.5$. Here, $i = 11.5$ is not an integer, so we round i up to the next integer $i = 12$. Then the 50th percentile is the value in positions 12.

Position	Criminal Trespass Cases
1	32
2	41
3	48
4	49
5	55
6	55
7	67
8	68
9	88
10	98
11	101
12	101
13	111
14	111
15	131
16	145
17	150
18	166
19	190
20	223
21	258
22	363
23	451

From the chart, the data value in position 12 is 101 criminal trespass cases. Therefore, the data value corresponding to the 50th percentile is 101 criminal trespass cases.
Q3 is the 75th percentile.

Then, $i = \left(\frac{p}{100}\right)n = \left(\frac{75}{100}\right)23 = 17.25$. Here, $i = 17.25$ is not an integer, so we round i up to the next integer $i = 18$. Then the 75th percentile is the value in positions 18.

Position	Criminal Trespass Cases
1	32
2	41
3	48
4	49
5	55
6	55
7	67
8	68
9	88
10	98
11	101
12	101
13	111
14	111
15	131
16	145
17	150
18	166
19	190
20	223
21	258
22	363
23	451

From the chart, the data value in position 18 is 166 criminal trespass cases. Therefore, the data value corresponding to the 75th percentile is 188 criminal trespass cases.

21. IQR = Q3 – Q1 = 166 – 55 = 111 criminal trespass cases

23. Q1 – 1.5(IQR) = 55 – 1.5(111) = 55 – 166.5 = -111.5 criminal trespass cases
Q3 + 1.5(IQR) = 166 + 1.5(111) = 166 + 166.5 = 332.5 criminal trespass cases
Since 32 criminal trespass cases is not -111.5 criminal trespass cases or less and 32 criminal trespass cases is not 332.5 criminal trespass cases or more, 32 criminal trespass cases is not an outlier.

25. **(a)** The left whisker is shorter than the right whisker, and the line for the median is closer to the line for Q1 than to the line for Q3. Therefore, the distribution is right-skewed.
(b) Minimum = 0, Q1 = 1, Q2 = Median = 3, Q3 = 7.5, Maximum = 12.

27. **(a)** The left whisker is shorter than the right whisker, and the line for the median is closer to the line for Q1 than to the line for Q3. Therefore, the distribution is right-skewed.
(b) Minimum = 5, Q1 = 10, Q2 = Median = 15, Q3 = 25, Maximum = 45.

29. The IQR for variable x is Q3 − Q1 = 25 − 10 = 15.
The IQR for variable y is Q3 − Q1 = 30 − 20 = 10.
Since the IQR for variable x is greater than the IQR for variable y, variable x has greater variability.

31. Q1 is the 25th percentile.
Then, $i = \left(\frac{p}{100}\right)n = \left(\frac{25}{100}\right)10 = 2.5$. Here, $i = 2.5$ is not an integer, so we round i up to the

next integer $i = 3$. Then the 25th percentile is the value in position 3.

Position	1	2	3	4	5	6	7	8	9	10
Price	3.38	9.66	25.28	31.25	35.02	41.54	51.43	65.28	94.75	95.18

From the chart, the data value in position 3 is $25.28. Therefore, the data value corresponding to the 25th percentile is $25.28.

Q2 is the 50th percentile.

Then, $i = \left(\frac{p}{100}\right)n = \left(\frac{50}{100}\right)10 = 5$. Here, $i = 5$ is an integer so the 50th percentile is the mean of the values in positions 5 and 6.

Position	1	2	3	4	5	6	7	8	9	10
Price	3.38	9.66	25.28	31.25	35.02	41.54	51.43	65.28	94.75	95.18

From the chart, the data value in position 5 is $35.02 and the data value in position 6 is $41.54. The mean of these two values is $\frac{\$35.02+\$41.54}{2} = \frac{\$76.56}{2} = \$38.28$. Therefore, the data value corresponding to the 50th percentile is $38.28.

Q3 is the 75th percentile.

Then, $i = \left(\frac{p}{100}\right)n = \left(\frac{75}{100}\right)10 = 7.5$. Here, $i = 7.5$ is not an integer, so we round i up to the next integer $i = 8$. Then the 75th percentile is the value in position 8.

Position	1	2	3	4	5	6	7	8	9	10
Price	3.38	9.66	25.28	31.25	35.02	41.54	51.43	65.28	94.75	95.18

From the chart, the data value in position 8 is $65.28. Therefore, the data value corresponding to the 75th percentile is $65.28.

The five-number summary is:

Minimum = $3.38, Q1 = $25.28, Q2= median = $38.28, Q3 = $65.28, maximum = $95.18

33. Q1 – 1.5(IQR) = $25.28 – 1.5($40) = $25.28 – $60 = -$34.72

Q3 + 1.5(IQR) = $65.28 + 1.5($40) = $65.28 + $60 = $125.28

Thus, for this data set, a data value would be an outlier if it was - $34.72 or less or $125.28 or more. No data values are $-$34.72 or less or $125.28 or more. Therefore, there are no outliers in this data set.

35. Q1 is the 25th percentile.

Then, $i = \left(\frac{p}{100}\right)n = \left(\frac{25}{100}\right)10 = 2.5$. Here, $i = 2.5$ is not an integer, so we round i up to the next integer $i = 3$. Then the 25th percentile is the value in position 3.

Position	1	2	3	4	5	6	7	8	9	10
Change	−0.36	−0.15	−0.01	−0.01	0.09	0.09	0.14	0.15	0.41	1.09

From the chart, the data value in position 3 is – 0.01. Therefore, the data value corresponding to the 25th percentile is -0.01.

Q2 is the 50th percentile.

Then, $i = \left(\frac{p}{100}\right)n = \left(\frac{50}{100}\right)10 = 5$. Here, $i = 5$ is an integer so the 50th percentile is the mean of the values in positions 5 and 6.

Position	1	2	3	4	5	6	7	8	9	10
Change	−0.36	−0.15	−0.01	−0.01	0.09	0.09	0.14	0.15	0.41	1.09

From the chart, the data value in position 5 is 0.09 and the data value in position 6 is 0.09. The mean of these two values is $\frac{0.09+0.09}{2} = \frac{0.18}{2} = 0.09$. Therefore, the data value corresponding to the 50th percentile is 0.09.

Q3 is the 75th percentile.
Then, $i = \left(\frac{p}{100}\right)n = \left(\frac{75}{100}\right)10 = 7.5$. Here, $i = 7.5$ is not an integer, so we round i up to the next integer $i = 8$. Then the 75th percentile is the value in position 8.

Position	1	2	3	4	5	6	7	8	9	10
Change	−0.36	−0.15	−0.01	−0.01	0.09	0.09	0.14	0.15	0.41	1.09

From the chart, the data value in position 8 is 0.15. Therefore, the data value corresponding to the 75th percentile is 0.15.
The five-number summary is:
Minimum = -0.36, Q1 = -0.01, Q2 = median = 0.09, Q3 = 0.15, maximum = 1.09

37. Q1 – 1.5(IQR) = (-0.01) – 1.5(0.16) = (-0.01) – 0.24 = - 0.25
Q3 + 1.5(IQR) = 0.15 + 1.5(0.16) = 0.15 + 0.24 = 039
Thus, for this data set, a data value would be an outlier if it was – 0.25 or less or 0.39 or more. -0.36 is less than -0.25 and 0.41 and 1.09 are more than 0.39. Therefore, -0.36, 0.41, and 1.09 are outliers.

39. Data set for *usage* arranged in ascending order (in millions):
2.0, 2.1, 2.4, 2.8, 3.1, 3.5, 3.8, 4.2, 4.3, 4.4, 5.2, 7.1, 7.7, 8.8, 14.7 $n = 15$

Here, Q1 is the 25th percentile, so $p = 25$. Therefore,

$$i = \left(\frac{p}{100}\right)n = \left(\frac{25}{100}\right)15 = 3.75.$$

Since i is not an integer, round i up to 4. We know that Q1 is the number in the 4th position, so Q1 = 2,800,000.
Here, Q2 is the 50th percentile, so $p = 50$. Therefore,

$$i = \left(\frac{p}{100}\right)n = \left(\frac{50}{100}\right)15 = 7.5.$$

Since i is not an integer, round i up to 8. We know that Q2 is the number in the 8th position, so Q2 = 4,200,000.
Here, Q3 is the 75th percentile, so $p = 75$. Therefore,

$$i = \left(\frac{p}{100}\right)n = \left(\frac{75}{100}\right)15 = 11.25.$$

Since i is not an integer, round i up to 12. We know that Q3 is the number in the 12th position, so Q3 = 7,100,000.
Therefore, the five-number summary for *usage* is:
Minimum = 2,000,000, Q1 = 2,800,000, median = 4,200,000, Q3 = 7,100,000, maximum = 14,700,000

41. Lower fence = Q1 − 1.5 · IQR = 2,800,000 − 1.5(4,300,000) = −3,650,000

Upper fence = Q3 + 1.5 * IQR = 7,100,000 + 1.5(4,300,000) = 13,550,000
The only value that lies outside of this interval is 14,700,000. Thus 14,700,000 is the only outlier.

43. The sample mean is

$$\bar{x} = \frac{\sum x}{n} = \frac{2.0 + 2.1 + 2.4 + 2.8 + 3.1 + 3.5 + 3.8 + 4.2 + 4.3 + 4.4 + 5.2 + 7.1 + 7.7 + 8.8 + 14.7}{15}$$
$$= \frac{76.1}{15} \approx 5.0733 \text{ million} = 5,073,300.$$

To calculate the standard deviation we first need to find $\sum x$ and $\sum x^2$.

x	x^2
2.0	4.00
2.1	4.41
2.4	5.76
2.8	7.84
3.1	9.61
3.5	12.25
3.8	14.44
4.2	17.64
4.3	18.49
4.4	19.36
5.2	27.04
7.1	50.41
7.7	59.29
8.8	77.44
14.7	216.09
$\sum x = 76.1$	$\sum x^2 = 544.07$

$n = 15$
Therefore, the standard deviation is

$$s = \sqrt{\frac{\sum x^2 - \left(\sum x\right)^2/n}{n-1}} = \sqrt{\frac{544.07 - (76.1)^2/15}{15-1}} \approx 3.3593 \text{ million} = 3,359,300.$$

45. Honda:

Q1 is the 25th percentile.
Then, $i = \left(\frac{p}{100}\right)n = \left(\frac{25}{100}\right)7 = 1.75$. Here, $i = 1.75$ is not an integer, so we round i up to the next integer $i = 2$. Then the 25th percentile is the value in position 2.

Position	1	2	3	4	5	6	7
mpg	17	18	21	23	24	31	42

From the chart, the data value in position 2 is 18 mpg. Therefore, the data value corresponding to the 25th percentile is 18 mpg.

Q2 is the 50th percentile.
Then, $i = \left(\frac{p}{100}\right)n = \left(\frac{50}{100}\right)7 = 3.5$. Here, $i = 3.5$ is not an integer, so we round i up to the next integer $i = 4$. Then the 50th percentile is the value in position 4.

Position	1	2	3	4	5	6	7
mpg	17	18	21	23	24	31	42

From the chart, the data value in position 4 is 23 mpg. Therefore, the data value corresponding to the 50th percentile is 23 mpg.

Q3 is the 75th percentile.
Then, $i = \left(\frac{p}{100}\right)n = \left(\frac{75}{100}\right)7 = 5.25$. Here, $i = 5.25$ is not an integer, so we round i up to the next integer $i = 6$. Then the 75th percentile is the value in position 6.

Position	1	2	3	4	5	6	7
mpg	17	18	21	23	24	31	42

From the chart, the data value in position 6 is 31 mpg. Therefore, the data value corresponding to the 75th percentile is 31 mpg.
The five-number summary is:
Minimum = 17 mpg, Q1 = 18 mpg, Q2 = median = 23 mpg, Q3 = 31 mpg, maximum = 42 mpg

Lexus:
Q1 is the 25th percentile.
Then, $i = \left(\frac{p}{100}\right)n = \left(\frac{25}{100}\right)6 = 1.5$. Here, $i = 1.5$ is not an integer, so we round i up to the next integer $i = 2$. Then the 25th percentile is the value in position 2.

Position	1	2	3	4	5	6
mpg	15	18	19	20	23	24

From the chart, the data value in position 2 is 18 mpg. Therefore, the data value corresponding to the 25th percentile is 18 mpg.
Q2 is the 50th percentile.

Then, $i = \left(\frac{p}{100}\right)n = \left(\frac{50}{100}\right)6 = 3$. Here, $i = 3$ is an integer so the 50th percentile is the mean of the values in positions 3 and 4.

Position	1	2	3	4	5	6
mpg	15	18	19	20	23	24

From the chart, the data value in position 3 is 19 mpg and the data value in position 4 is 20 mpg. The mean of these two values is $\frac{19+20}{2} = \frac{39}{2} = 19.5$ mpg. Therefore, the data value corresponding to the 50th percentile is 19.5 mpg.

Q3 is the 75th percentile.
Then, $i = \left(\frac{p}{100}\right)n = \left(\frac{75}{100}\right)6 = 4.5$. Here, $i = 4.5$ is not an integer, so we round i up to the next integer $i = 5$. Then the 75th percentile is the value in position 5.

Position	1	2	3	4	5	6
mpg	15	18	19	20	23	24

From the chart, the data value in position 5 is 23 mpg. Therefore, the data value corresponding to the 75th percentile is 23 mpg.
The five-number summary is: Minimum = 15 mpg, Q1 = 18 mpg, Q2 = median = 19.5 mpg, Q3 = 23 mpg, Maximum = 24 mpg

47. The distribution of the mpg for the Honda cars is right-skewed.
The distribution for the mpg for the Lexus cars is right-skewed.

49. Honda: Mean $= \frac{\sum x}{n} = \frac{17+18+21+23+24+31+42}{7} = \frac{176}{7} \approx 25.14$ mpg

Lexus: Mean $= \frac{\sum x}{n} = \frac{15+18+19+20+23+24}{6} = \frac{119}{6} \approx 19.83$ mpg

Yes. For the Honda cars, the mean of 25.14 mpg is greater than the median of 23 mpg. For the Lexus cars, the mean of 19.83 mpg is greater than the median of 19.5 mpg.

51. Honda: IQR = Q3 – Q1 = 31 mpg – 18 mpg = 13 mpg

Lexus: IQR = Q3 – Q1 = 23 mpg – 18 mpg = 5 mpg
The IQR for the Honda cars is larger than the IQR for the Lexus cars. This indicates that the data for the Honda cars has a larger spread than the data for the Lexus cars.

53. Honda: Q1 – 1.5(IQR) = 18 – 1.5(13) = 18 – 19.5 = -1.5 mpg

Q3 + 1.5(IQR) = 31 + 1.5(13) = 31 + 19.5 = 50.5 mpg
Thus, for this data set, a data value would be an outlier if it was 1.5 mpg or less or 50.5 or more. No data values are -1.5 mpg or less or 50.5 mpg or more. Therefore, there are no outliers in this data set.
Lexus: Q1 – 1.5(IQR) = 18 – 1.5(5) = 18 – 7.5 = 10.5 mpg
Q3 + 1.5(IQR) = 23 + 1.5(5) = 23 + 7.5 = 30.5 mpg
Thus, for this data set, a data value would be an outlier if it was 10.5 mpg or less or 30.5 or more. No data values are = 10 mpg or less or 30.5 mpg or more. Therefore, there are no outliers in this data set.

55. From Minitab:

Descriptive Statistics: IRON

Variable	N	N*	Mean	SE Mean	StDev	Minimum	Q1	Median	Q3
IRON	961	0	1.784	0.101	3.138	0.000	0.300	0.800	1.700

Variable	Maximum
IRON	37.600

Mean = 1.784 mg, standard deviation = 3.138 mg, Min = 0.000 mg, Q1 = 0.300 mg, median = 0.800 mg, Q3 = 1.700 mg, Max = 37.600 mg
Range = Maximum – minimum = 37.600 mg – 0.000 mg = 37.600 mg
IQR = Q3 – Q1 = 1.700 mg – 0.300 mg = 1.400 mg

57.

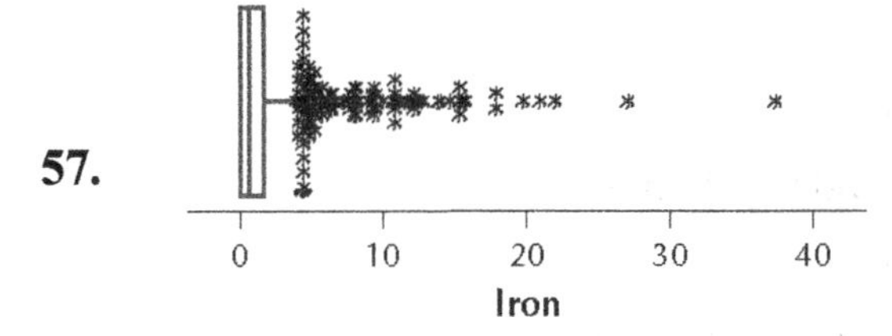

The boxplot is very right-skewed.

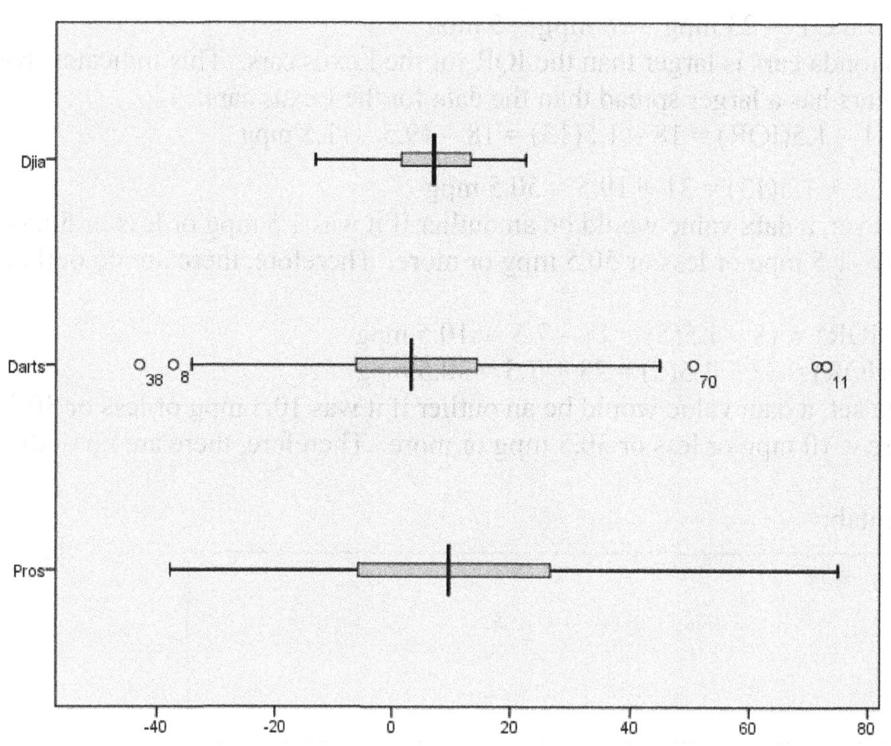

59.
PROS, DJIA
61. Pros and DJIA
63.

```
Variable  Mean  StDev
DARTS     4.52  19.39
```

z-score for $x = -43.0$: $z = \frac{x-\bar{x}}{s} \approx \frac{-43.0-4.42}{10.38} \approx$ -2.4507
Since $-3 < z$-score ≤ -2, $x = -43.0$ is considered moderately unusual.
z-score for $x = -37.3$: $z = \frac{x-\bar{x}}{s} \approx \frac{-37.3-4.42}{10.38} \approx$ -2.1568
Since $-3 < z$-score ≤ -2, $x = -37.3$ is considered moderately unusual.
z-score for $x = 72.9$: $z = \frac{x-\bar{x}}{s} \approx \frac{72.9-4.42}{10.38} \approx 3.5266$
Since z-score ≥ 3, $x = 72.9$ is an outlier.
z-score for $x = 71.3$: $z = \frac{x-\bar{x}}{s} \approx \frac{71.3-4.42}{10.38} \approx 3.4440$
Since z-score ≥ 3, $x = 71.3$ is an outlier.
z-score for $x = 50.5$: $z = \frac{x-\bar{x}}{s} \approx \frac{50.5-4.42}{10.38} \approx 2.3713$
Since $2 \leq z -$ score < 3, for $x = 50.5$ is considered moderately unusual.

Chapter 3 Review Exercises

1. $\mu = \frac{\sum x}{N} = \frac{7.8+1.9+14.9+1.5+2.7+1.6}{6} = \frac{30.4}{6} \approx \$5.07$ million

3. Mean: $\mu = \frac{\sum x}{N} = \frac{7.8+1.9+14.9+1.5+2.7+1.6+62.1}{7} = \frac{92.5}{7} \approx \$13.21$ million
Median: First arrange the data in ascending order: 1.5, 1.6, 1.9, 2.7, 7.8, 14.9, 62.1

Since $N = 7$ is odd, the median is the unique $\left(\frac{N+1}{2}\right)^{th} = \left(\frac{7+1}{2}\right)^{th} = 4$th value, which is \$2.7 million.

The mean increased from \$5.07 million to \$13.21 million, whereas the median increased \$2.3 million to \$2.7 million. Therefore, the mean was more affected by the presence of California. The mean is affected by extreme values more than the median is.

5. First arrange the data in ascending order:

90, 90, 100, 110, 110, 110, 110, 110, 110, 120, 120, 130

Since $n = 12$ is even, the median is the mean of the two data values that lie on either side of the $\left(\frac{n+1}{2}\right)^{th} = \left(\frac{12+1}{2}\right)^{th} = 6.5$th position. That is, the median is the mean of the 6th and 7th data values, 110 and 110. Splitting the difference between these two we get

$$\text{median} = \frac{110+110}{2} = \frac{220}{2} = 110 \text{ calories}$$

7. The mode will not be affected because the value with the largest frequency is unaffected by the deletion of values 90 or less calories.

9. Range = largest number in the data set – smallest number in the data set = \$14.9 million−\$1.5 million = \$13.4 million

11. $\sum(x-\mu) \approx 2.73 + (-3.17) + 9.83 + (-3.57) + (-2.37) + (-3.47) = -\$0.02$ million. The average deviation $= \frac{\sum(x-\mu)}{N} \approx \frac{-0.02}{6} \approx -\$0.0033$ million $\approx \$0$ million. The average deviation is not a good measure of spread because it is always 0.

13. $\sigma = \sqrt{\sigma^2} \approx \sqrt{24.0889} \approx \$4.9080$ million.

15. To calculate the standard deviation we first need to find $\sum x$ and $\sum x^2$.

x	x^2
90	8,100
90	8,100
100	10,000
110	12,100
110	12,100
110	12,100
110	12,100
110	12,100
110	12,100
120	14,400
120	14,400
130	16,900
$\sum x = 1{,}310$	$\sum x^2 = 144{,}500$

$n = 12$. Therefore, the standard deviation is

$$s = \sqrt{\frac{\sum x^2 - \left(\sum x\right)^2/n}{n-1}} = \sqrt{\frac{144{,}500 - (1{,}310)^2/12}{12-1}} \approx 11.6450 \text{ calories.}$$

17. The Empirical Rule tells us that about 95% of data values lie within 2 standard deviations of the mean. The values that are 2 standard deviations from the mean are $\bar{x} - k \cdot s \approx 109.17 - 2(11.6450) = 109.17 - 23.29 = 85.88$ calories and $\bar{x} + k \cdot s \approx 109.17 + 2(11.6450) = 109.17 + 23.29 = 132.46$ calories. Therefore, about 95% of breakfast cereals have calories per serving values between 85.88 calories and 132.46 calories.

19. About 300.

21. To calculate the variance we first need to find □x and □x².

x	x^2
462	213,444
621	385,641
104	10,816
907	822,649
293	85,849
186	34,596
470	220,900
136	18,496
675	455,625
114	12,996
$\Sigma x = 3968$	$\Sigma x^2 = 2{,}261{,}012$

$n = 10$. Therefore, the standard deviation is

$$s = \sqrt{\frac{\sum x^2 - \left(\sum x\right)^2/n}{n-1}} = \sqrt{\frac{2{,}261{,}012 - (3968)^2/10}{10-1}} \approx 276.2.$$

(a) Difference = 300 − 276.2 = 23.8. **(b)** The frequency counts for the syllables typically differ from the mean of 396.8 by 276.2.

23.

Age x	Numbers (millions) f	$x - \bar{x}$	$(x - \bar{x})^2 \cdot f$
20	4.5	$20 - 24.45 = -4.45$	$(-4.45)^2 \cdot 4.5 = 89.11125$
21	4.4	$21 - 24.45 = -3.45$	$(-3.45)^2 \cdot 4.4 = 52.371$
22	4.3	$22 - 24.45 = -2.45$	$(-2.45)^2 \cdot 4.3 = 25.81075$
23	4.2	$23 - 24.45 = -1.45$	$(-1.45)^2 \cdot 4.2 = 8.8305$
24	4.2	$24 - 24.45 = -0.45$	$(-0.45)^2 \cdot 4.2 = 0.8505$
25	4.3	$25 - 24.45 = 0.55$	$(0.55)^2 \cdot 4.3 = 1.30075$
26	4.2	$26 - 24.45 = 1.55$	$(1.55)^2 \cdot 4.2 = 10.0905$
27	4.2	$27 - 24.45 = 2.55$	$(2.55)^2 \cdot 4.2 = 27.3105$
28	4.2	$28 - 24.45 = 3.55$	$(3.55)^2 \cdot 4.2 = 52.9305$
29	4.2	$29 - 24.45 = 4.55$	$(4.55)^2 \cdot 4.2 = 86.8505$
Total	$\Sigma f = 42.7$		$\Sigma(x - \bar{x})^2 \cdot f = 355.55675$

$$s = \sqrt{s^2} = \sqrt{\frac{\Sigma(x-\bar{x})^2 \cdot f}{\Sigma f}} \approx \sqrt{\frac{355.55675}{42.7}} \approx \sqrt{8.326855972} \approx 2.89 \text{ years}$$

25. The actual percentage of people in their twenties who are between 21.56 and 27.34 years old is $\frac{4.3+4.2+4.2+4.3+4.2+4.2}{42.7} \cdot 100\% = \frac{25.4}{42.7} \cdot 100\% \approx 59.48\%$, Since 59.48% is significantly less than the 68% from the Empirical Rule, this tells us that the distribution is not bell-shaped.

27. $i = \left(\frac{p}{100}\right) n = \left(\frac{50}{100}\right) 10 = 5$

Here, $i = 5$ is an integer so the 50th percentile is the mean of the values in positions 5 and 6.

Position	1	2	3	4	5	6	7	8	9	10
Ragweed pollen index	8	25	25	26	31	38	43	48	59	60

From the chart, the data value in position 5 is 31 and the data value in position 6 is 38. The mean of these two values is $\frac{31+38}{2} = \frac{69}{2} = 34.5$. Therefore, the data value corresponding to the 50th percentile is 34.5.

29. For Albany, $x = 48$. Then the z-score is

$z = \frac{x-\bar{x}}{s} \approx \frac{48-36.3}{16.5062} \approx 0.7088$

31. For Tupper Lake, $x = 8$. Then the z-score is

$z = \frac{x-\bar{x}}{s} \approx \frac{8-36.3}{16.5062} \approx -1.7145$

For Exercises 33-35: Data set arranged in ascending order:

8, 25, 25, 26, 31, 38, 43, 48, 59, 60

33. Here, $x = 25$. Three cities have x-values at or below 25, so the percentile rank of a city with a ragweed pollen index of 25 is percentile rank of data value $(x = 25) = \frac{\text{number of values in data set} \leq 25}{\text{total number ofvalues in data set}} \cdot 100 = \frac{3}{10} \cdot 100 = 30\%$. Therefore, 25 represents the 30th percentile of ragweed pollen indices for cities in New York.

35. Here, $x = 48$. Eight cities have x-values at or below 48, so the percentile rank of a city with a ragweed pollen index of 48 is percentile rank of data value $(x = 48) = \frac{\text{number of values in data set} \leq 48}{\text{total number ofvalues in data set}} \cdot 100 = \frac{8}{10} \cdot 100 = 80\%$. Therefore, 48 represents the 80th percentile of ragweed pollen indices for cities in New York.

37. IQR = Q3 − Q1 = 48 − 25 = 23. The spread of the middle 50% of the data set is 23.

39. **(a)** Minimum = 8, Q1 = 25, Median = 34.5, Q3 = 48, Maximum = 60

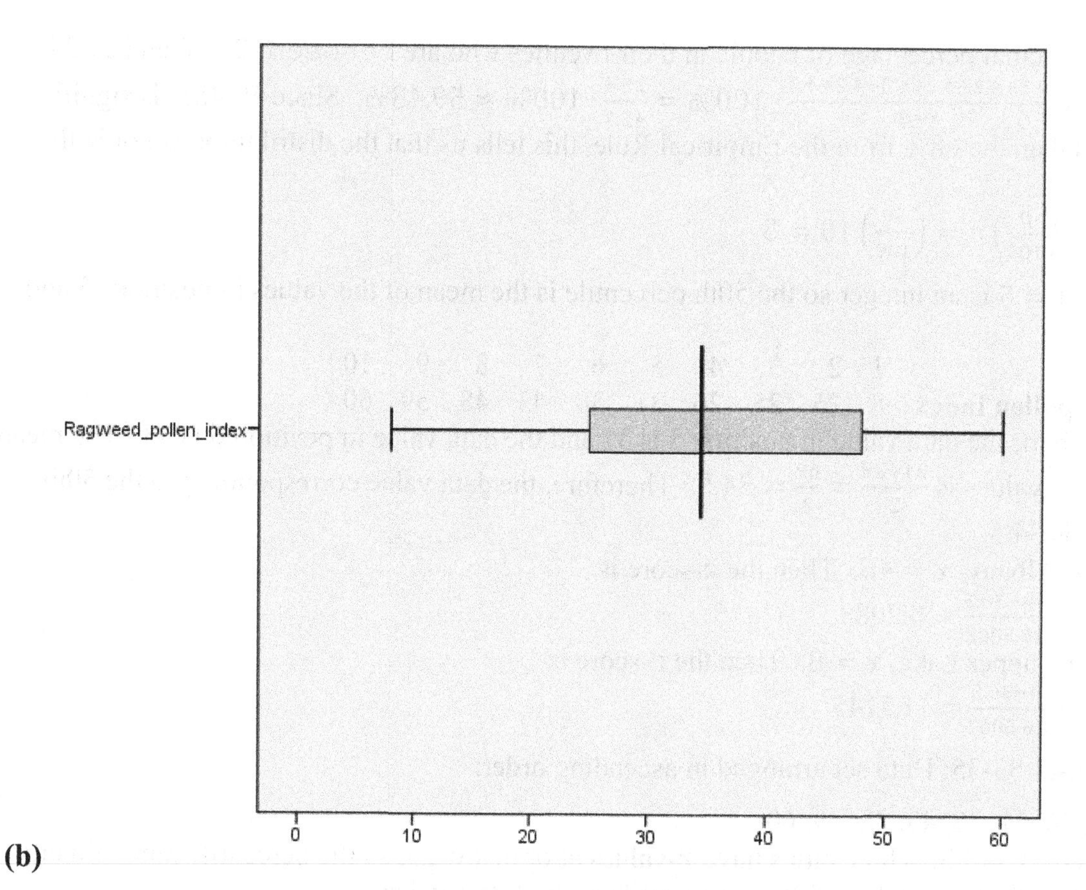

(b)

(c) The data set is close to symmetric. It is slightly right-skewed.

(d) The mean should be close to the median or a little above the median.

(e) The sample mean is

$$\bar{x} = \frac{\sum x}{n} = \frac{8+25+25+26+31+38+43+48+59+60}{10} = \frac{363}{10} = 36.30.$$

To calculate the standard deviation we first need to find $\sum x$ and $\sum x^2$.

x	x^2
8	64
25	625
25	625
26	676
31	961
38	1444
43	1849
48	2304
59	3481
60	3600
$\sum x = 363$	$\sum x^2 = 15{,}629$

$n = 10$

Therefore, the standard deviation is

$$s = \sqrt{\frac{\sum x^2 - \left(\sum x\right)^2/n}{n-1}} = \sqrt{\frac{15{,}629 - (363)^2/10}{10-1}} \approx 16.51.$$

The value of the mean is slightly above the median of 34.50, so the data is slightly right-skewed.

41. Since none of the quartiles are equal to the largest value, increasing the largest value would leave all of the quartiles and the IQR unchanged. This is an example of robustness.

Chapter 3 Quiz

1. False. For example let data set A be 1, 6, 6, 6, 16 and data set B be 4, 6, 6, 6, 13. Then both data sets have a mean of 7, a median of 6, and a mode of 6, but they are not the same data sets.

3. False. The Empirical Rule only applies to data sets with bell-shaped distributions.

5. center

7. robust measures

9. zero

11. Angelina's grade point average

$$\bar{x} = \frac{\sum wx}{\sum w} = \frac{w_1x_1 + w_2x_2 + w_3x_3 + w_4x_4 + w_5x_5}{w_1 + w_2 + w_3 + w_4 + w_5}$$

$$= \frac{(4)(4.0) + (3)(3.7) + (3)(3.3) + (3)(2.7) + (2)(2.3)}{4 + 3 + 3 + 3 + 2} = \frac{49.7}{15} \approx 3.133$$

13. **(a)** Since the distribution is bell-shaped it is symmetric with the median = mean = 60 gallons.

(b) Because 20 gallons lies 1 standard deviation below the mean

$$\mu - 1\sigma = 60 - 1(40) = 20$$

and 100 lies 1 standard deviation above the mean

$$\mu + 1\sigma = 60 + 1(40) = 100$$

the question is really asking what is the percentage that lies within $k = 1$ standard deviation from the mean. From the Empirical Rule, about 68% of Americans consume between 20 and 100 gallons of carbonated beverages per year.

(c) No, Chebyshev's Rule can't be used for $k = 1$.

(d) From (b), about 68% of Americans consume between 20 and 100 gallons of carbonated beverages per year.

Therefore, about 100%− 68% = 32% of Americans consume less than 20 or more than 100 gallons of carbonated beverages per year.

Since the distribution is bell-shaped it is symmetric. Therefore, about $\left(\frac{1}{2}\right)32\% = 16\%$ of Americans consume more than 100 gallons of carbonated beverages per year.

15. IQR = Q3 − Q1 = 518 − 501.5 = 16.5

17. Lower fence = Q1 − 1.5 * IQR = 501.5 − 1.5(16.5) = 476.75

Upper fence = Q3 + 1.5 * IQR = 518 + 1.5(16.5) = 542.75

All of the SAT scores lie between 476.75 and 542.75, so there are no outliers.

Chapter 4: Correlation and Regression

Section 4.1

1. Scatterplot.

3. Between –1 and 1, inclusive. Thus $-1 \leq r \leq 1$.

5. Often, the value of the x variable can be used to predict or estimate the value of the y variable.

7. They decrease.

9. A women's weight depends in part on her height. We can use a women's height to predict her weight. Thus, a women's height is the predictor (x) variable, and a women's weight is the response (y) variable.

11. The cost of a repair job depends in part on the number of hours spent on the repair. We can use the number of hours spent on a repair job to predict the cost of the repair job. Thus, the number of hours spent on the repair is the predictor (x) variable, and the cost of a repair job is the response (y) variable.

13. **(a)**

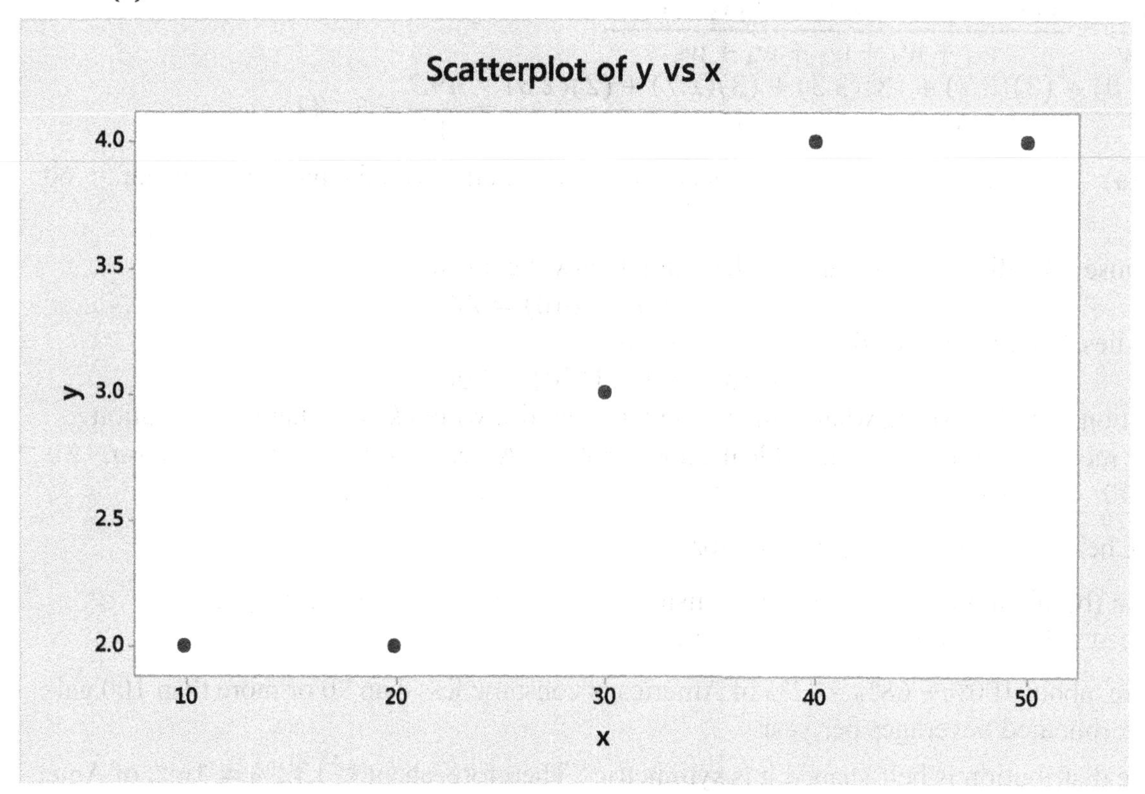

(b) The variables x and y have a positive linear relationship.

(c) First, we calculate the respective means, $\bar{x}$ and $\bar{y}$.

$$\bar{x} = \frac{\sum x}{n} = \frac{10+20+30+40+50}{5} = \frac{150}{5} = 30 \quad \text{and } \bar{y} = \frac{\sum y}{n} = \frac{2+2+3+4+4}{5} = \frac{15}{5} = 3$$

x	y	$(x-\bar{x})$	$(x-\bar{x})^2$	$(y-\bar{y})$	$(y-\bar{y})^2$	$(x-\bar{x})(y-\bar{y})$
10	2	−20	400	−1	1	20
20	2	−10	100	−1	1	10
30	3	0	0	0	0	0
40	4	10	100	1	1	10
50	4	20	400	1	1	20
			$\sum(x-\bar{x})^2 = 1000$		$\sum(y-\bar{y})^2 = 4$	$\sum(x-\bar{x})(y-\bar{y}) = 60$

Next, calculate the respective standard deviations, s_x and s_y.

$s_x = \sqrt{\frac{\Sigma(x-\bar{x})^2}{n-1}} = \sqrt{\frac{1000}{5-1}} = \sqrt{250} \approx 15.8113883$ and

$$s_y = \sqrt{\frac{\Sigma(y-\bar{y})^2}{n-1}} = \sqrt{\frac{4}{5-1}} = \sqrt{1} = 1$$

Put these values all together in the formula for the correlation coefficient r:

$$r = \frac{\Sigma(x-\bar{x})(y-\bar{y})}{(n-1)s_x s_y} = \frac{60}{(5-1)(15.8112883)(1)} = 0.9486832981 \approx 0.9487$$

(d) This value of r is very close to the maximum value $r = 1$. We would therefore say that x and y are positively correlated. As x increases, y also tends to increase.

15. **(a)**

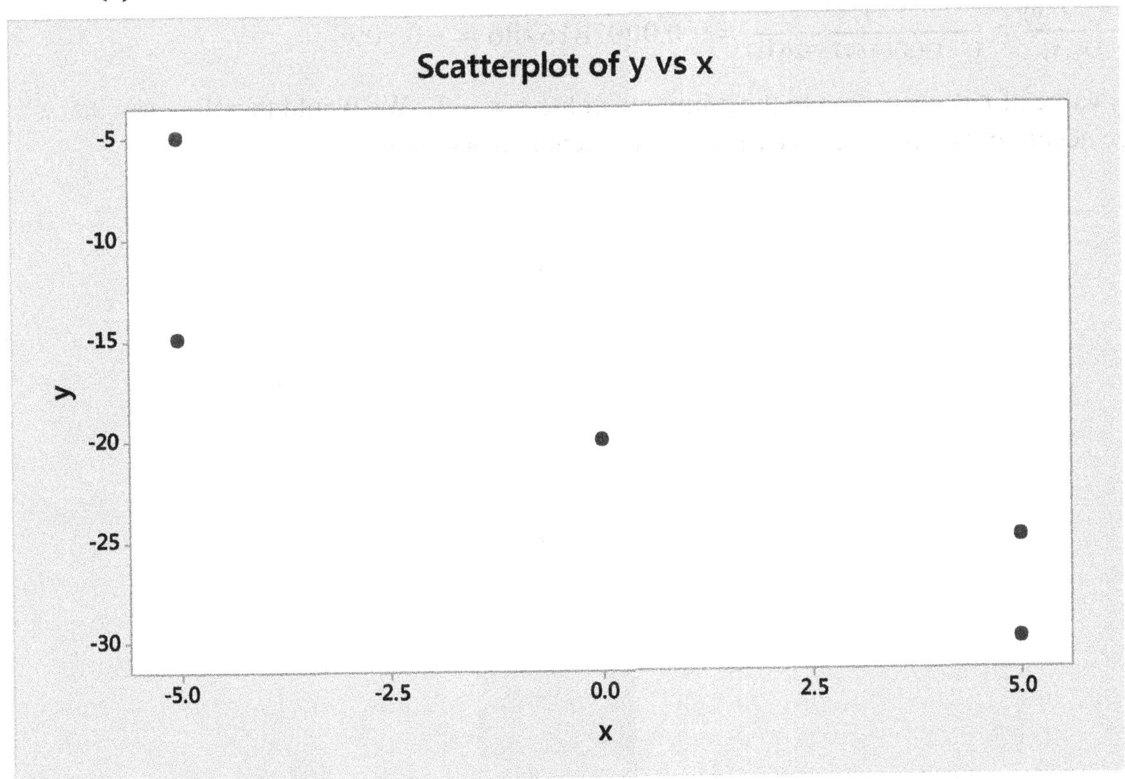

(b) The variables x and y have a negative linear relationship.

(c) First, we calculate the respective means, $\bar{x}$ and $\bar{y}$.

$\bar{x} = \frac{\sum x}{n} = \frac{(-5)+(-5)+0+5+5}{5} = \frac{0}{5} = 0$ and

$\bar{y} = \frac{\sum y}{n} = \frac{(-5)+(-15)+(-20)+(-25)+(-30)}{5} = \frac{-95}{5} = -19$

x	y	$(x - \bar{x})$	$(x - \bar{x})^2$	$(y - \bar{y})$	$(y - \bar{y})^2$	$(x - \bar{x})(y - \bar{y})$
−5	−5	−5	25	14	196	−70
−5	−15	−5	25	4	16	−20
0	−20	0	0	−1	1	0
5	−25	5	25	−6	36	−30
5	−30	5	25	−11	121	−55
			$\sum(x-\bar{x})^2 = 100$		$\sum(y-\bar{y})^2 = 370$	$\sum(x-\bar{x})(y-\bar{y}) = -175$

Next, calculate the respective standard deviations, s_x and s_y.

$s_x = \sqrt{\frac{\sum(x-\bar{x})^2}{n-1}} = \sqrt{\frac{100}{5-1}} = \sqrt{25} = 5$ and

$$s_y = \sqrt{\frac{\sum(y-\bar{y})^2}{n-1}} = \sqrt{\frac{370}{5-1}} = \sqrt{92.5} \approx 9.617692031$$

Put these values all together in the formula for the correlation coefficient r:

$r = \frac{\sum(x-\bar{x})(y-\bar{y})}{(n-1)s_x s_y} = \frac{-175}{(5-1)(5)(9.617692031)} = -0.9097816786 \approx -0.9098$

(d) This value of r is very close to the minimum value $r = -1$. We would therefore say that x and y are negatively correlated. As x increases, y tends to decrease.

17. **(a)**

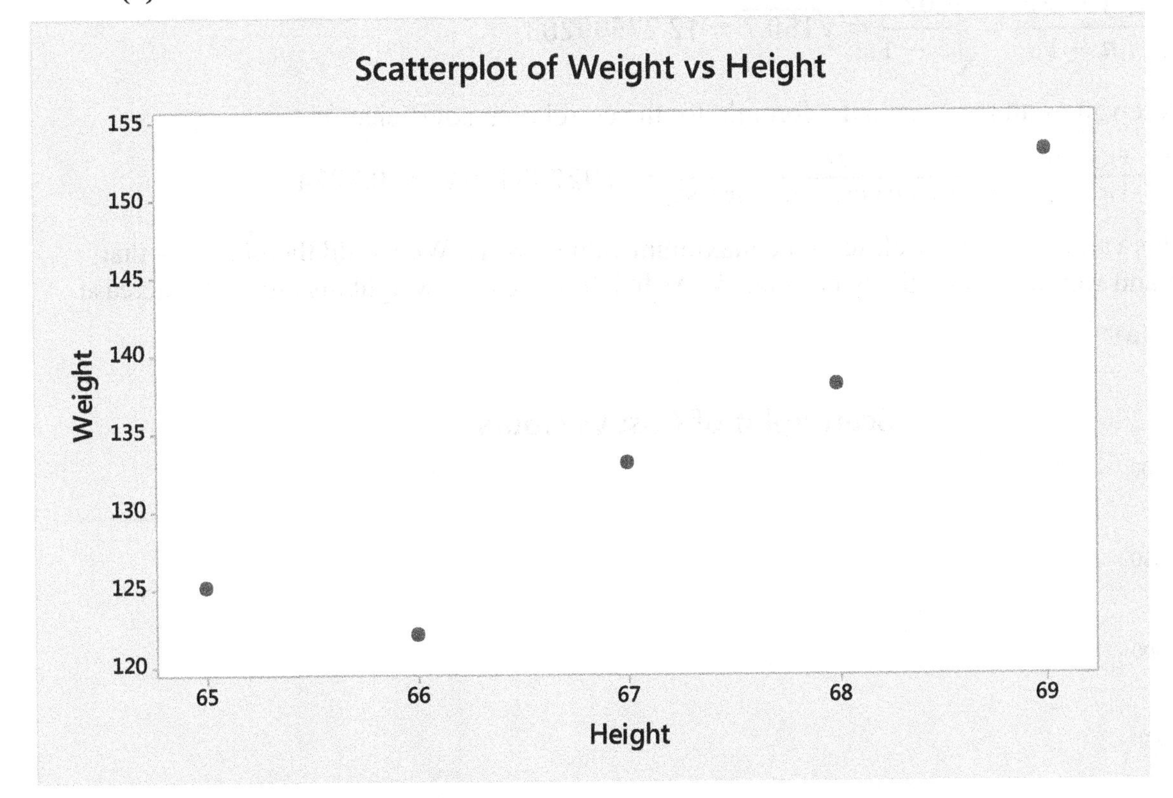

(b) The variables height and weight have a positive linear relationship.

(c) First, we calculate the respective means, $\bar{x}$ and $\bar{y}$.

$\bar{x} = \frac{\sum x}{n} = \frac{66+67+69+68+65}{5} = \frac{335}{5} = 67$ and

$$\bar{y} = \frac{\sum y}{n} = \frac{122+133+153+138+125}{5} = \frac{671}{5} = 134.2$$

x	y	$(x - \bar{x})$	$(x-\bar{x})^2$	$(y - \bar{y})$	$(y-\bar{y})^2$	$(x-\bar{x})(y-\bar{y})$
66	122	−1	1	−12.2	148.84	12.2
67	133	0	0	−1.2	1.44	0
69	153	2	4	18.8	353.44	37.6
68	138	1	1	3.8	14.44	3.8
65	125	−2	4	−9.2	84.64	18.4
			$\sum(x-\bar{x})^2 = 10$		$\sum(y-\bar{y})^2 = 602.8$	$\sum(x-\bar{x})(y-\bar{y}) = 72$

Next, calculate the respective standard deviations, s_x and s_y.

$s_x = \sqrt{\frac{\sum(x-\bar{x})^2}{n-1}} = \sqrt{\frac{10}{5-1}} = \sqrt{2.5} \approx 1.58113883$ and

$$s_y = \sqrt{\frac{\Sigma(y-\bar{y})^2}{n-1}} = \sqrt{\frac{602.8}{5-1}} = \sqrt{150.7} = 12.27599283$$

Put these values all together in the formula for the correlation coefficient r:

$$r = \frac{\Sigma(x-\bar{x})(y-\bar{y})}{(n-1)s_x s_y} = \frac{72}{(5-1)(1.58113883)(12.27599283)} = 0.9273546941 \approx 0.9274$$

(d) This value of r is very close to the maximum value $r = 1$. We would therefore say that height and weight are positively correlated. As height increases, weight also tends to increase.

19. **(a)**

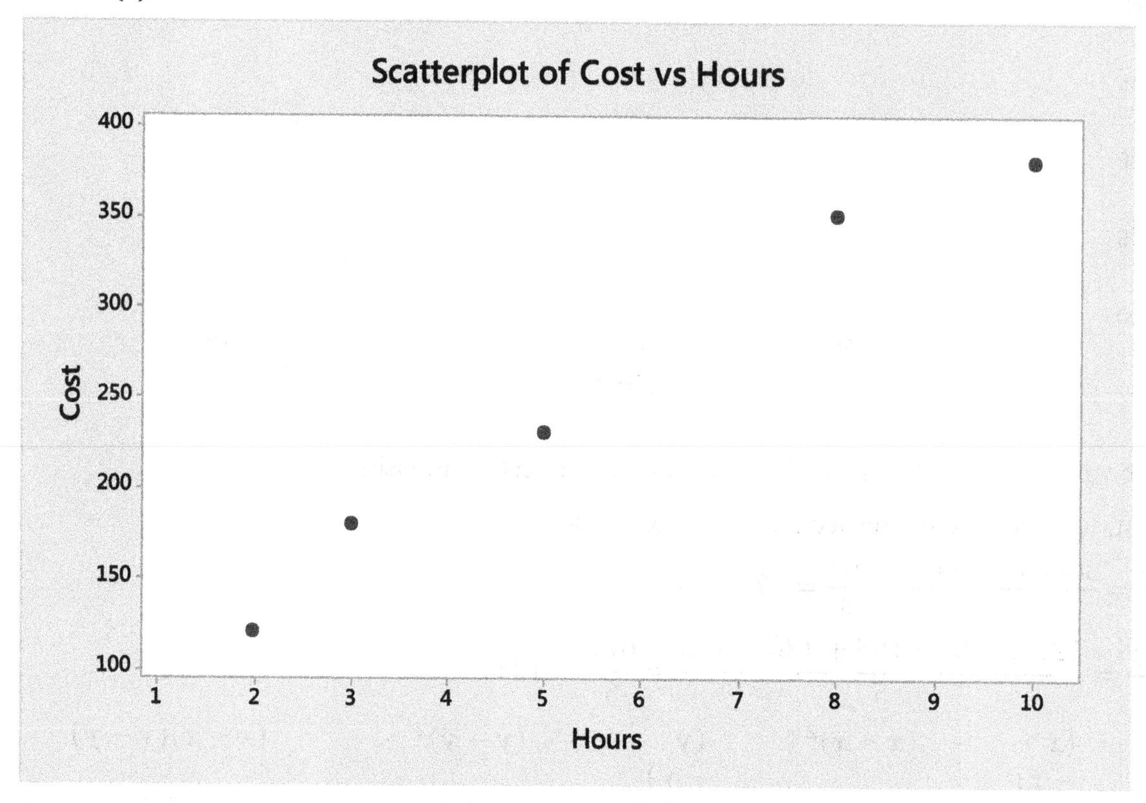

(b) The hours a repair job takes and the cost of the repair job have a positive linear relationship.

(c) First, we calculate the respective means, $\bar{x}$ and $\bar{y}$.

$$\bar{x} = \frac{\Sigma x}{n} = \frac{2+3+5+8+10}{5} = \frac{28}{5} = 5.6 \text{ and } \bar{y} = \frac{\Sigma y}{n} = \frac{120+180+230+350+380}{5} = \frac{1260}{5} = 252$$

x	y	$(x-\bar{x})$	$(x-\bar{x})^2$	$(y-\bar{y})$	$(y-\bar{y})^2$	$(x-\bar{x})(y-\bar{y})$
2	120	−3.6	12.96	−132	17,424	475.2
3	180	−2.6	6.76	−72	5,184	187.2
5	230	−0.6	0.36	−22	484	13.2
8	350	2.4	5.76	98	9,604	235.2
10	380	4.4	19.36	128	16,384	563.2
			$\sum(x-\bar{x})^2 = 45.2$		$\sum(y-\bar{y})^2 = 49{,}080$	$\sum(x-\bar{x})(y-\bar{y}) = 1474$

Next, calculate the respective standard deviations, s_x and s_y.

$s_x = \sqrt{\frac{\Sigma(x-\bar{x})^2}{n-1}} = \sqrt{\frac{45.2}{5-1}} = \sqrt{11.3} \approx 3.361547263$ and $s_y = \sqrt{\frac{\Sigma(y-\bar{y})^2}{n-1}} = \sqrt{\frac{49{,}080}{5-1}} = \sqrt{12{,}270} \approx$ 110.770032

Put these values all together in the formula for the correlation coefficient r:

$$r = \frac{\Sigma(x-\bar{x})(y-\bar{y})}{(n-1)s_x s_y} = \frac{1474}{(5-1)(3.361547263)(110.770032)} = 0.9896371485 \approx 0.9896$$

(d) This value of r is very close to the maximum value $r = 1$. We would therefore say that hours worked and cost are positively correlated. As the number of hours worked increases, the cost also tends to increase.

21. **(a)** Strong negative linear relationship. **(b)** They decrease.

23. **(a)** Moderate positive linear relationship. **(b)** They increase.

25. **(a)** Perfect negative linear relationship. **(b)** They decrease.

27. A value of r near 1 means that the points in the scatterplot lie close to a straight line with a positive slope. Thus the answer is graph **i**.

29. A value of r near –0.5 means that the points in the scatterplot lie somewhat close to a straight line with a negative slope. Thus the answer is graph **iii**.

31. **(a)** (1,1), (2,3), (3,3), (4,4), (5,6), (6,6), (7,7), (8,7), (9,9), (10,11) **(b)** From Minitab:

Correlations: x, y

```
Pearson correlation of x and y = 0.978
```

From the TI-83/84: r =0.9781316853.

Using the formula:

$$\bar{x} = \frac{\sum x}{n} = \frac{1+2+3+4+5+6+7+8+9+10}{10} = \frac{55}{10} = 5.5$$

$$\bar{y} = \frac{\sum y}{n} = \frac{1+3+3+4+6+6+7+7+9+11}{10} = \frac{57}{10} = 5.7$$

x	y	$x-\bar{x}$	$(x-\bar{x})^2$	$y-\bar{y}$	$(y-\bar{y})^2$	$(x-\bar{x})(y-\bar{y})$
1	1	-4.5	20.25	-4.7	22.09	21.15
2	3	-3.5	12.25	-2.7	7.29	9.45
3	3	-2.5	6.25	-2.7	7.29	6.75
4	4	-1.5	2.25	-1.7	2.89	2.55
5	6	-0.5	0.25	0.3	0.09	-0.15
6	6	0.5	0.25	0.3	0.09	0.15
7	7	1.5	2.25	1.3	1.69	1.95
8	7	2.5	6.25	1.3	1.69	3.25
9	9	3.5	12.25	3.3	10.89	11.55
10	11	4.5	20.25	5.3	28.09	23.85
			$\sum(x-\bar{x})^2 = 82.5$		$\sum(y-\bar{y})^2 = 82.1$	$\sum(x-\bar{x})(y-\bar{y}) = 80.5$

$$s_x = \sqrt{\frac{\sum(x-\bar{x})^2}{n-1}} = \sqrt{\frac{82.5}{10-1}} = 3.027650354$$

$$s_y = \sqrt{\frac{\sum(y-\bar{y})^2}{n-1}} = \sqrt{\frac{82.1}{10-1}} = 3.020301677$$

$$r = \frac{\sum(x-\bar{x})(y-\bar{y})}{(n-1)s_x s_y} = \frac{80.5}{(10-1)(3.027650354)(3.020301677)} = 0.9781316854$$

33. **(a)** (1,7), (2,8), (3,7), (4,6), (5,6), (6,5), (7,6), (8,5), (9,7), (10,6)

(b) From Minitab:

Correlations: x, y

```
Pearson correlation of x and y = -0.522
```

From the TI-83/84: $r = -0.5222329679$.

Using the formula:

$$\bar{x} = \frac{\sum x}{n} = \frac{1+2+3+4+5+6+7+8+9+10}{10} = \frac{55}{10} = 5.5$$

$$\bar{y} = \frac{\sum y}{n} = \frac{7+8+7+6+6+5+6+5+7+6}{10} = \frac{63}{10} = 6.3$$

x	y	$x-\bar{x}$	$(x-\bar{x})^2$	$y-\bar{y}$	$(y-\bar{y})^2$	$(x-\bar{x})(y-\bar{y})$
1	7	-4.5	20.25	0.7	0.49	-3.15
2	8	-3.5	12.25	1.7	2.89	-5.95
3	7	-2.5	6.25	0.7	0.49	-1.75
4	6	-1.5	2.25	-0.3	0.09	0.45
5	6	-0.5	0.25	-0.3	0.09	0.15
6	5	0.5	0.25	-1.3	1.69	-0.65
7	6	1.5	2.25	-0.3	0.09	-0.45
8	5	2.5	6.25	-1.3	1.69	-3.25
9	7	3.5	12.25	0.7	0.49	2.45
10	6	4.5	20.25	-0.3	0.09	-1.35
			$\Sigma(x-\bar{x})^2 = 82.5$		$\Sigma(y-\bar{y})^2 = 8.1$	$\Sigma(x-\bar{x})(y-\bar{y}) = -13.5$

$$s_x = \sqrt{\frac{\sum(x-\bar{x})^2}{n-1}} = \sqrt{\frac{82.5}{10-1}} = 3.027650354$$

$$s_y = \sqrt{\frac{\sum(y-\bar{y})^2}{n-1}} = \sqrt{\frac{8.1}{10-1}} = 0.9486832981$$

$$r = \frac{\sum(x-\bar{x})(y-\bar{y})}{(n-1)s_x s_y} = \frac{-13.5}{(10-1)(3.027650354)(0.9486832981)} = -0.5222329679$$

35. **(a)**

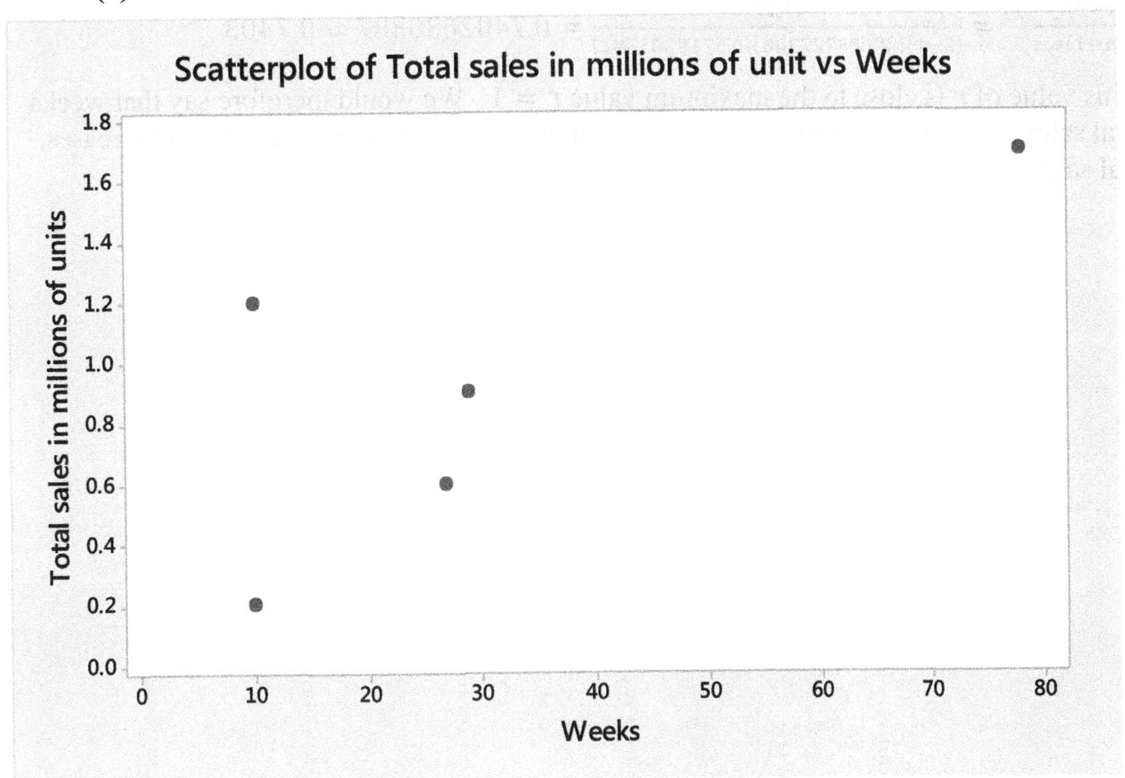

(b) The number of weeks on the top 30 list and the total sales in millions of game units have a positive linear relationship.

(c) First, we calculate the respective means, $\bar{x}$ and $\bar{y}$.

$\bar{x} = \frac{\sum x}{n} = \frac{78+27+29+10+10}{5} = \frac{154}{5} = 30.8$ and $\bar{y} = \frac{\sum y}{n} = \frac{1.7+0.6++0.9+1.2+0.2}{5} = \frac{4.6}{5} = 0.92$

x	y	$(x - \bar{x})$	$(x - \bar{x})^2$	$(y - \bar{y})$	$(y - \bar{y})^2$	$(x - \bar{x})(y - \bar{y})$
78	1.7	47.2	2227.84	0.78	0.6084	36.816
27	0.6	−3.8	14.44	−0.32	0.1024	1.216
29	0.9	−1.8	3.24	−0.02	0.0004	0.036
10	1.2	−20.8	432.64	0.28	0.0784	−5.824
10	0.2	−20.8	432.64	−0.72	0.5184	14.976
			$\sum(x - \bar{x})^2 = 3110.8$		$\sum(y - \bar{y})^2 = 1.308$	$\sum(x - \bar{x})(y - \bar{y}) = 47.22$

Next, calculate the respective standard deviations, s_x and s_y.

$$s_x = \sqrt{\frac{\sum(x-\bar{x})^2}{n-1}} = \sqrt{\frac{3110.8}{5-1}} = \sqrt{777.7} \approx 27.88727308 \text{ and}$$

$$s_y = \sqrt{\frac{\sum(y-\bar{y})^2}{n-1}} = \sqrt{\frac{1.308}{5-1}} = \sqrt{0.327} \approx 0.5718391382$$

Put these values all together in the formula for the correlation coefficient r:

$$r = \frac{\sum(x-\bar{x})(y-\bar{y})}{(n-1)s_x s_y} = \frac{47.22}{(5-1)(27.88727308)(0.5718391382)} = 0.7402630387 \approx 0.7403$$

(d) This value of r is close to the maximum value $r = 1$. We would therefore say that weeks and total sales are positively correlated. As the number of weeks on the top 30 lists increases, the total sales also tends to increase.

37. **(a)**

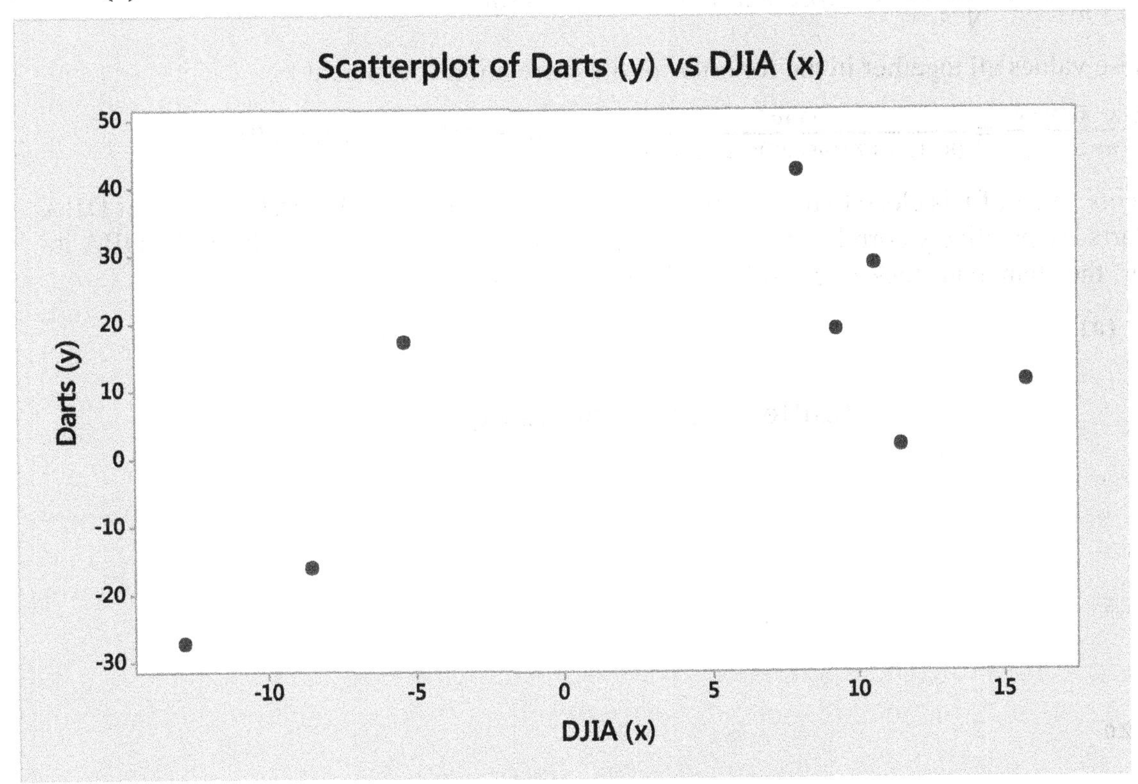

(b) The change in price of the stocks in the DJIA and the change in price of the stocks in the portfolio selected by the darts have a positive linear relationship.

(c) First, we calculate the respective means, $\bar{x}$ and $\bar{y}$.

$$\bar{x} = \frac{\sum x}{n} = \frac{(-12.8)+9.3+8.0+(-8.5)+15.8+10.6+11.5+(-5.3)}{8} = \frac{28.6}{8} = 3.575 \text{ and}$$

$$\bar{y} = \frac{\sum y}{n} = \frac{(-27.4)+18.7+42.2+(-16.3)+11.2+28.5+1.8+16.9}{8} = \frac{75.6}{8} = 9.45$$

x	y	$(x-\bar{x})$	$(x-\bar{x})^2$	$(y-\bar{y})$	$(y-\bar{y})^2$	$(x-\bar{x})(y-\bar{y})$
−12.8	−27.4	−16.375	268.140625	−36.85	1357.9225	603.41875
9.3	18.7	5.725	32.775625	9.25	85.5625	52.95625
8.0	42.2	4.425	19.580625	32.75	1072.5625	144.91875
−8.5	−16.3	−12.075	145.805625	−25.75	663.0625	310.93125
15.8	11.2	12.225	149.450625	1.75	3.0625	21.39375
10.6	28.5	7.025	49.350625	19.05	362.9025	133.82625
11.5	1.8	7.925	62.805625	−7.65	58.5225	−60.62625
−5.3	16.9	−8.875	78.765625	7.45	55.5025	−66,11875
			$\sum(x-\bar{x})^2 = 806.675$		$\sum(y-\bar{y})^2 = 3659.1$	$\sum(x-\bar{x})(y-\bar{y}) = 1140.7$

Next, calculate the respective standard deviations, s_x and s_y.

$$s_x = \sqrt{\frac{\sum(x-\bar{x})^2}{n-1}} = \sqrt{\frac{806.675}{8-1}} = \sqrt{115.2392857} \approx 10.73495625 \text{ and}$$

$$s_y = \sqrt{\frac{\Sigma(y-\bar{y})^2}{n-1}} = \sqrt{\frac{3659.1}{8-1}} = \sqrt{522.7285714} \approx 22.86325811$$

Put these values all together in the formula for the correlation coefficient r:

$$r = \frac{\Sigma(x-\bar{x})(y-\bar{y})}{(n-1)s_x s_y} = \frac{1140.7}{(8-1)(10.73495625)(22.86325811)} \approx 0.6639494316 \approx 0.6639$$

(d) This value of r is close to the maximum value $r = 1$. We would therefore say that DJIA and Darts are positively correlated. As the change in stock prices of the stocks in the DJIA increases, the change in stock prices selected by darts also tends to increase.

39. **(a)**

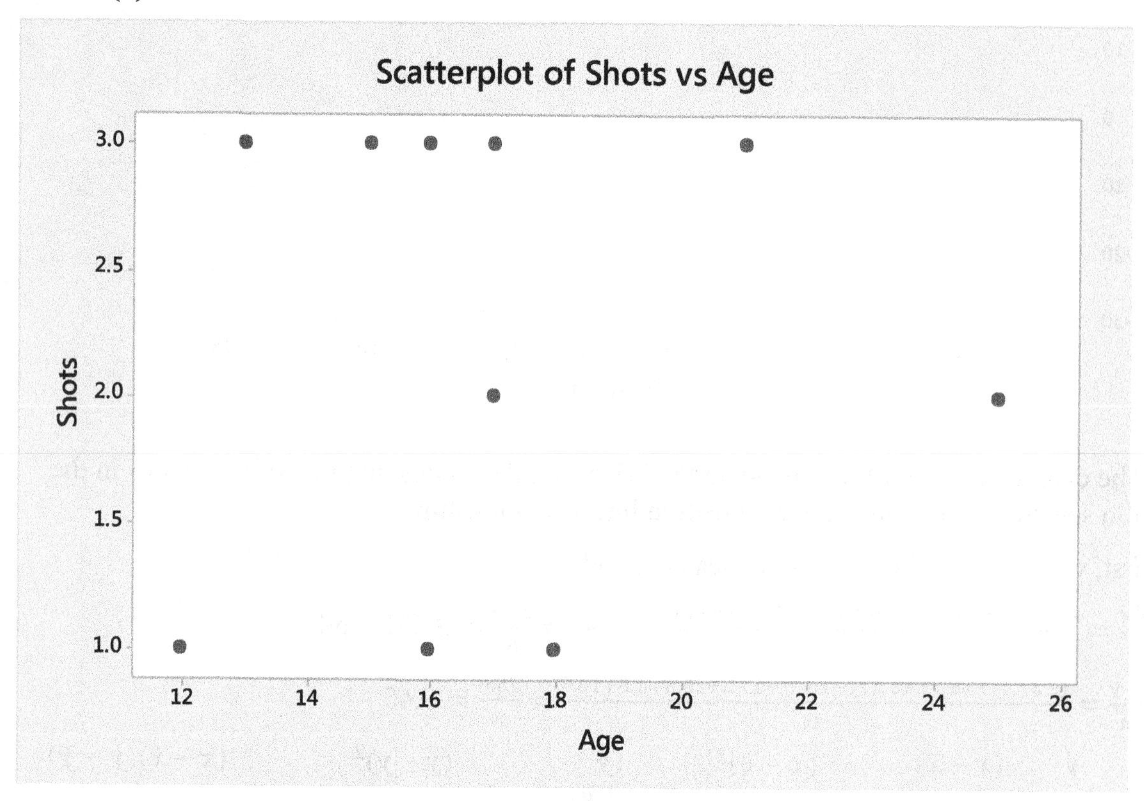

(b) Age and number of shots have no apparent relationship.

(c) First, we calculate the respective means, $\bar{x}$ and $\bar{y}$.

$$\bar{x} = \frac{\Sigma x}{n} = \frac{13+21+16+17+17+18+25+15+12+16}{10} = \frac{170}{10} = 17 \text{ and}$$

$$\bar{y} = \frac{\Sigma y}{n} = \frac{3+3+3+2+3+1+2+3+1+1}{10} = \frac{22}{10} = 2.2$$

x	y	$(x-\bar{x})$	$(x-\bar{x})^2$	$(y-\bar{y})$	$(y-\bar{y})^2$	$(x-\bar{x})(y-\bar{y})$
13	3	-−4	16	0.8	0.64	−3.2
21	3	4	16	0.8	0.64	3.2
16	3	− 1	1	0.8	0.64	−0.8
17	2	0	0	− 0.2	0.04	0
17	3	0	0	0.8	0.64	0
18	1	1	1	−1.2	1.44	−1.2
25	2	8	64	−0.2	0.04	−1.6
15	3	−2	4	0.8	0.64	−1.6
12	1	−5	25	−1.2	1.44	6
16	1	−1	1	−1.2	1.44	1.2

$\sum(x-\bar{x})^2 = 128$ $\quad\sum(y-\bar{y})^2 = 7.6$ $\quad\sum(x-\bar{x})(y-\bar{y}) = 2$

Next, calculate the respective standard deviations, s_x and s_y.

$$s_x = \sqrt{\frac{\Sigma(x-\bar{x})^2}{n-1}} = \sqrt{\frac{128}{10-1}} = \sqrt{14.22222222} \approx 3.771236166 \text{ and}$$

$$s_y = \sqrt{\frac{\Sigma(y-\bar{y})^2}{n-1}} = \sqrt{\frac{7.6}{10-1}} = \sqrt{0.8444444444} \approx 0.9189365835$$

Put these values all together in the formula for the correlation coefficient r:

$$r = \frac{\Sigma(x-\bar{x})(y-\bar{y})}{(n-1)s_x s_y} = \frac{2}{(10-1)(3.771236166)(0.9189365835)} \approx 0.0641$$

(d) This value of r is very close to the value $r = 0$. We would therefore say that there is no linear relationship between age and shots.

41. **(a)**

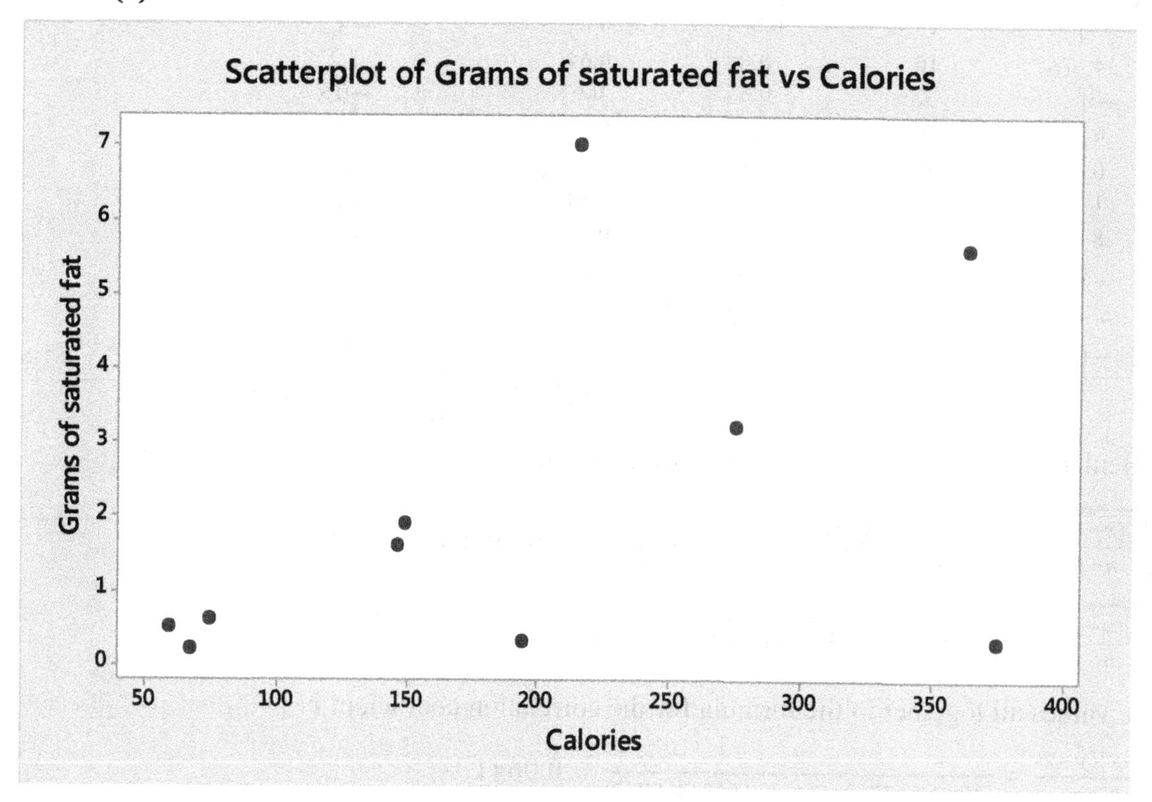

(b) The number of calories in a serving of food and the grams of saturated fat in a serving of food have a positive linear relationship.

(c) First, we calculate the respective means, $\bar{x}$ and $\bar{y}$.

$$\bar{x} = \frac{\sum x}{n} = \frac{216+364+149+75+276+375+146+59+67+195}{10} = \frac{1922}{10} = 192.2 \text{ and}$$

$$\bar{y} = \frac{\sum y}{n} = \frac{7+5.6+1.9+0.6+3.2+0.3+1.6+0.5+0.2+03}{10} = \frac{21.2}{10} = 2.12$$

x	y	$(x-\bar{x})$	$(x-\bar{x})^2$	$(y-\bar{y})$	$(y-\bar{y})^2$	$(x-\bar{x})(y-\bar{y})$
216	7	23.8	566.44	25 4.88	23.8144	116.144
364	5.6	171.8	29,515.24	3.48	12.1104	597.864
149	1.9	−43.2	1,866.24	−0.22	0.0484	9.504
75	0.6	−117.2	13,735.84	−1.52	2.3104	178.144
276	3.2	83.8	7.022.44	1.08	1.1664	90.504
375	0.3	182.8	33,415.84	−1.82	3.3124	−332.696
146	1.6	−46.2	2,134.44	−0.52	0.2704	24.024
59	0.5	−133.2	17,742.24	−1.62	2.6244	215.784
67	0.2	−125.2	15,675.04	−1.92	3.6864	240.384
195	0.3	2.8	7.84	−1.82	3.3124	−5.096
			$\sum(x-\bar{x})^2 = 121{,}681.6$		$\sum(y-\bar{y})^2 = 52.656$	$\sum(x-\bar{x})(y-\bar{y}) = 1134.56$

Next, calculate the respective standard deviations, s_x and s_y.

$$s_x = \sqrt{\frac{\Sigma(x-\bar{x})^2}{n-1}} = \sqrt{\frac{121{,}681.6}{10-1}} = \sqrt{13{,}520.17778} \approx 116.2762993 \text{ and}$$

$$s_y = \sqrt{\frac{\Sigma(y-\bar{y})^2}{n-1}} = \sqrt{\frac{52.656}{10-1}} = \sqrt{5.850666667} \approx 2.418815137$$

Put these values all together in the formula for the correlation coefficient r:

$$r = \frac{\Sigma(x-\bar{x})(y-\bar{y})}{(n-1)s_x s_y} = \frac{1134.56}{(10-1)(116.2762993)(2.418815137)} \approx 0.4482$$

(d) This value of r is positive. We would therefore say that the number of calories in a serving of food and the grams of saturated fat in a serving of food are positively correlated. As the number of calories in a serving of food increases, the grams of saturated fat in a serving of food also tends to increase.

43. (a)

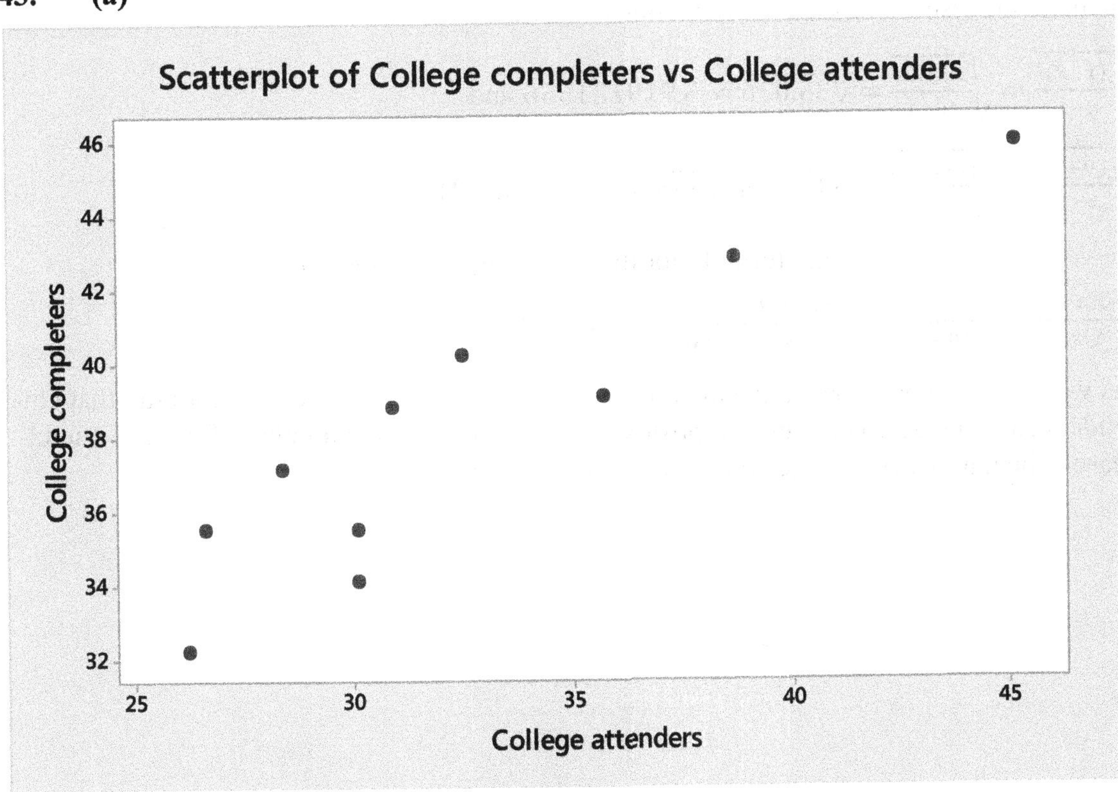

(b) The percent of residents of a state who attend college and the percent of residents of a state who graduate from college have a positive linear relationship.

(c) First, we calculate the respective means, $\bar{x}$ and $\bar{y}$.

$$\bar{x} = \frac{\Sigma x}{n} = \frac{30.9+26.6+30.1+35.7+45.2+38.7+30.1+28.4+32.5+26.2}{10} = \frac{324.4}{10} = 32.44 \text{ and}$$

$$\bar{y} = \frac{\Sigma y}{n} = \frac{38.8+35.5+34.1++39.1+45.9+42.8+35.5+37.1+40.2+32.2}{10} = \frac{381.2}{10} = 38.12$$

x	y	$(x-\bar{x})$	$(x-\bar{x})^2$	$(y-\bar{y})$	$(y-\bar{y})^2$	$(x-\bar{x})(y-\bar{y})$
30.9	38.8	−1.54	2.3716	0.68	0.4624	−1.0472
26.6	35.5	−5.84	34.1056	−2.62	6.8644	15.3008
30.1	34.1	−2.34	5.4756	−4.02	16.1604	9.4068
35.7	39.1	3.26	10.6276	0.98	0.9604	3.1948
45.2	45.9	12.76	162.8176	7.78	60.5284	99.2728
38.7	42.8	6.26	39.1876	4.68	21.9024	29.2968
30.1	35.5	−2.34	5.4756	−2.62	6.8644	6.1308
28.4	37.1	−4.04	16.3216	−1.02	1.0404	4.1208
32.5	40.2	0.06	0.0036	2.08	4.3264	0.1248
26.2	32.2	− 6.24	38.9376	−5.92	35.0464	36.9408
			$\sum(x-\bar{x})^2 = 315.324$		$\sum(y-\bar{y})^2 = 154.156$	$\sum(x-\bar{x})(y-\bar{y}) = 202.742$

Next, calculate the respective standard deviations, s_x and s_y.

$$s_x = \sqrt{\frac{\Sigma(x-\bar{x})^2}{n-1}} = \sqrt{\frac{315.324}{10-1}} = \sqrt{35.036} \approx 5.919121556 \text{ and}$$

$$s_y = \sqrt{\frac{\Sigma(y-\bar{y})^2}{n-1}} = \sqrt{\frac{154.156}{10-1}} = \sqrt{17.12844444} \approx 4.138652491$$

Put these values all together in the formula for the correlation coefficient r:

$$r = \frac{\Sigma(x-\bar{x})(y-\bar{y})}{(n-1)s_x s_y} = \frac{202.742}{(10-1)(5.919121556)(4.138652491)} \approx 0.9196$$

(d) This value of r is very close to the maximum value $r = 1$. We would therefore say that college attenders and college completers are positively correlated. As the number of college attenders increases, the number of college completers also tends to increase.

45. **(a)**

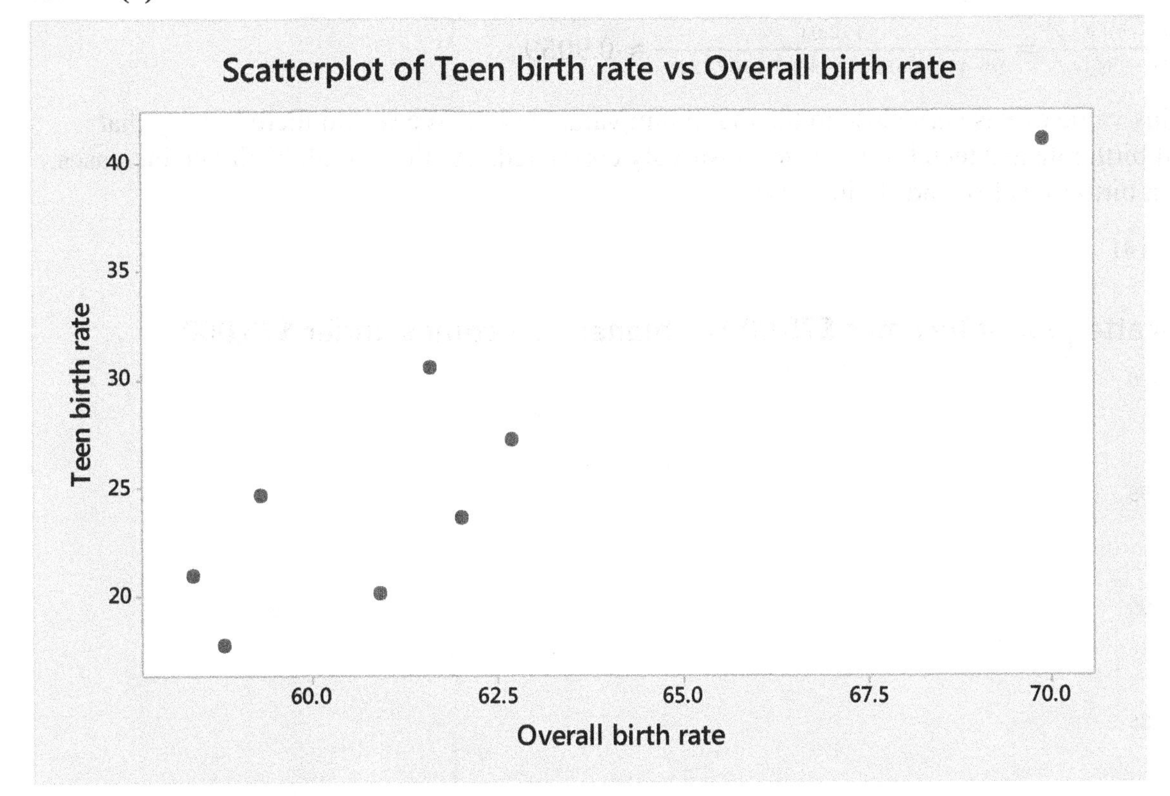

(b) The overall birth rate and the teenage birth rate have a positive linear relationship.

(c) First, we calculate the respective means, $\bar{x}$ and $\bar{y}$.

$$\bar{x} = \frac{\sum x}{n} = \frac{62+59.3+61.6+58.8+62.7+58.4+69.9+60.9}{8} = \frac{493.6}{8} = 61.7 \text{ and}$$

$$\bar{y} = \frac{\sum y}{n} = \frac{23.6+24.6+30.5+17.7+27.2+20.9+41+20.1}{8} = \frac{205.6}{8} = 25.7$$

x	y	$(x-\bar{x})$	$(x-\bar{x})^2$	$(y-\bar{y})$	$(y-\bar{y})^2$	$(x-\bar{x})(y-\bar{y})$
62	23.6	0.3	0.09	−2.1	4.41	−0.63
59.3	24.6	−2.4	5.76	−1.1	1.21	2.64
61.6	30.5	−0.1	0.01	4.8	23.04	−0.48
58.8	17.7	−2.9	8.41	−8.0	64.00	23.20
62.7	27.2	1.0	1.00	1.5	2.25	1.50
58.4	20.9	−3.3	10.89	−4.8	23.04	15.84
69.9	41.0	8.2	67.24	15.3	234.09	125.46
60.9	20.1	−0.8	0.64	−5.6	31.36	4.48
			$\sum(x-\bar{x})^2 = 94.04$		$\sum(y-\bar{y})^2 = 383.4$	$\sum(x-\bar{x})(y-\bar{y}) = 172.01$

Next, calculate the respective standard deviations, s_x and s_y.

$$s_x = \sqrt{\frac{\sum(x-\bar{x})^2}{n-1}} = \sqrt{\frac{94.04}{8-1}} = \sqrt{13.43428571} \approx 3.665281124 \text{ and}$$

$$s_y = \sqrt{\frac{\sum(y-\bar{y})^2}{n-1}} = \sqrt{\frac{383.4}{8-1}} = \sqrt{54.77142857} \approx 7.40077216$$

Put these values all together in the formula for the correlation coefficient r:

$$r = \frac{\Sigma(x-\bar{x})(y-\bar{y})}{(n-1)s_x s_y} = \frac{172.01}{(8-1)(3.665281124)(7.40077216)} \approx 0.9059$$

(d) This value of r is very close to the maximum value $r = 1$. We would therefore say that overall birth rate and teen birth rate are positively correlated. As the overall birth rate increases, the teen birth rate also tends to increase.

47. (a)

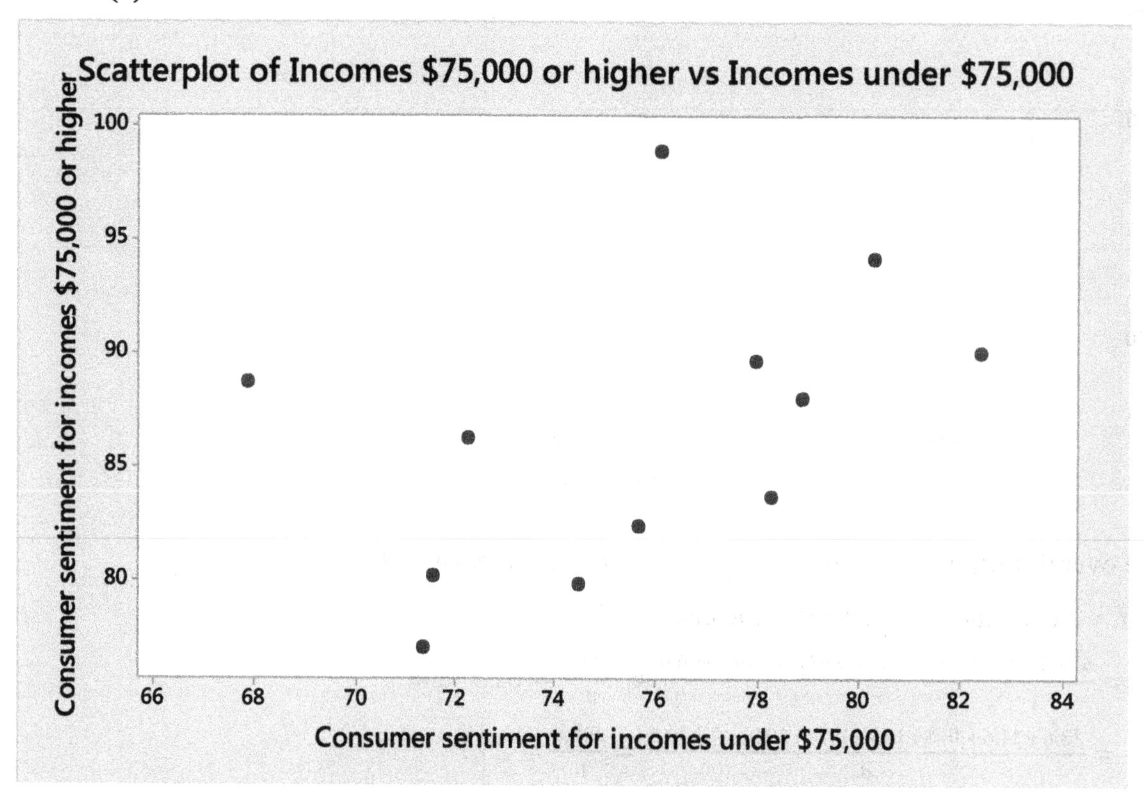

(b) The consumer sentiment for incomes under $75,000 and the consumer sentiment for incomes $75,000 or higher have a positive linear relationship.

(c) First, we calculate the respective means, $\bar{x}$ and $\bar{y}$.

$$\bar{x} = \frac{\Sigma x}{n} = \frac{71.6+75.7+78.3+74.5+80.3+76.1+82.4+78+72.3+71.4+67.9+78.9}{12} = \frac{907.4}{12} \approx 75.6167 \text{ and}$$

$$\bar{y} = \frac{\Sigma y}{n} = \frac{80.2+82.4+83.7+79.8+84.1+98.9+90.0+89.6+86.2+77.0+88.7+88.0}{12} = \frac{1038.6}{12} = 86.55$$

x	y	$(x-\bar{x})$	$(x-\bar{x})^2$	$(y-\bar{y})$	$(y-\bar{y})^2$	$(x-\bar{x})(y-\bar{y})$
71.6	80.2	−4.0167	16.13387889	−6.35	40.3225	25.506045
75.7	82.4	0.0833	0.00693889	−4.15	17.2225	−0.345695
78.3	83.7	2.6833	7.20009889	−2.85	8.1225	−7.647405
74.5	79.8	−1.1167	1.24701889	−6.75	45.5625	7.537725
80.3	94.1	4.6833	21.93329889	7.55	57.0025	35.358915
76.1	98.9	0.4833	0.22357889	12.35	152.5225	5.968755
82.4	90.0	6.7833	46.01315889	3.45	11.9025	23.402385
78.0	89.6	2.3833	5.68011889	3.05	9.3025	7.269065
72.3	86.2	−3.3167	11.00049889	−0.35	0.1225	1.160845
71.4	77.0	−4.2167	17.78055889	−9.55	91.2025	40.269485
67.9	88.7	−7.7167	59.54745889	2.15	4.6225	−16.590905
78.9	88.0	3.2833	10.78005889	1.45	2.1025	4.760785
			$\sum(x-\bar{x})^2 = 197.5566667$		$\sum(y-\bar{y})^2 = 440.01$	$\sum(x-\bar{x})(y-\bar{y}) = 126.65$

Next, calculate the respective standard deviations, s_x and s_y.

$$s_x = \sqrt{\frac{\Sigma(x-\bar{x})^2}{n-1}} = \sqrt{\frac{197.5566667}{12-1}} = \sqrt{17.95969697} \approx 4.237888268 \text{ and}$$

$$s_y = \sqrt{\frac{\Sigma(y-\bar{y})^2}{n-1}} = \sqrt{\frac{440.01}{12-1}} = \sqrt{40.00090909} \approx 6.32462719$$

Put these values all together in the formula for the correlation coefficient r:

$$r = \frac{\Sigma(x-\bar{x})(y-\bar{y})}{(n-1)s_x s_y} = \frac{126.65}{(12-1)(4.237888268)(6.32462719)} \approx 0.4296$$

(d) This value of r is positive. We would therefore say that the consumer sentiment for incomes under \$75,000 and the consumer sentiment for incomes \$75,000 or higher are positively correlated. As the consumer sentiment for incomes under \$75,000 increases, the consumer sentiment for incomes \$75,000 or higher also tends to increase.

49. **(a)**

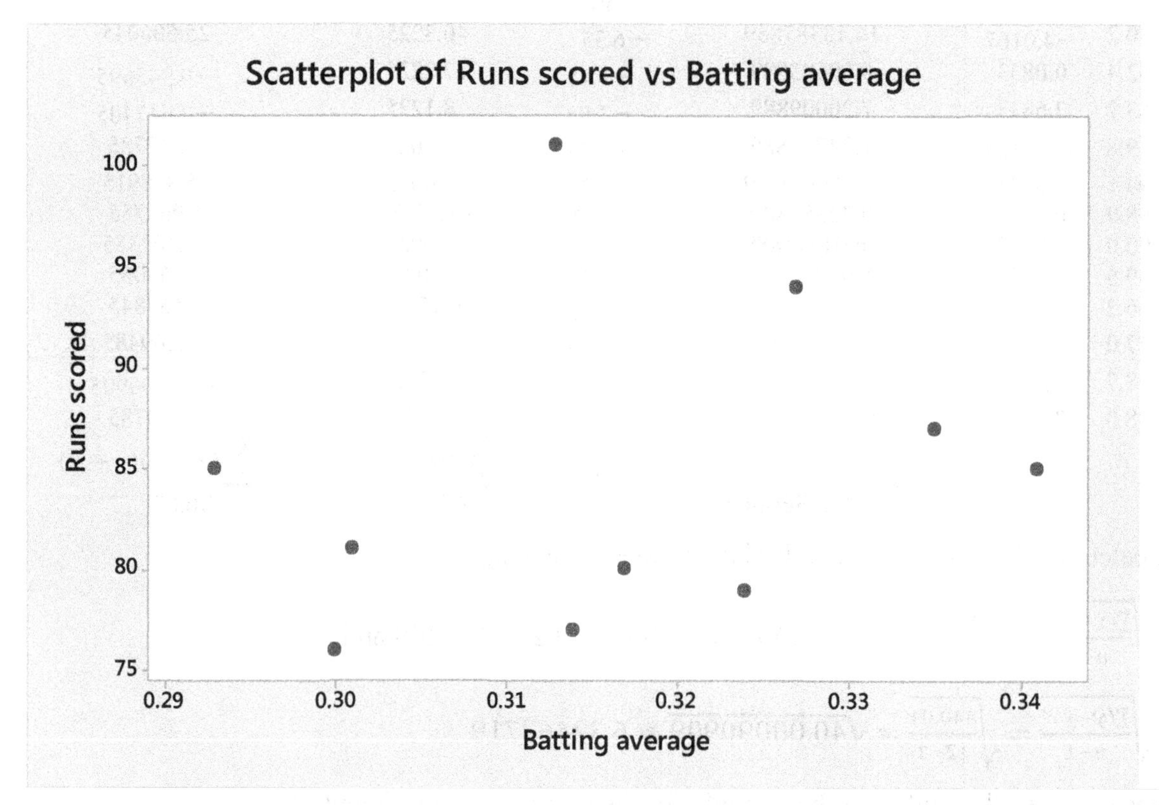

(b) The batting average of a baseball player and the number of runs scored by a baseball player have no apparent relationship.

(c) First, we calculate the respective means, $\bar{x}$ and $\bar{y}$.

$\bar{x} = \frac{\sum x}{n} = \frac{0.341+0.335+0.327+0.324+0.317+0.314+0.313+0.301+0.300+0.293}{10} = \frac{3.165}{10} = 0.3165$ and

$\bar{y} = \frac{\sum y}{n} = \frac{85+87+94+79+80+77+101+81+76+85}{10} = \frac{845}{10} = 84.5$

x	y	$(x-\bar{x})$	$(x-\bar{x})^2$	$(y-\bar{y})$	$(y-\bar{y})^2$	$(x-\bar{x})(y-\bar{y})$
0.341	85	0.0245	0.00060025	0.5	0.25	0.01225
0.335	87	0.0185	0.00034225	2.5	6.25	0.04625
0.327	94	0.0105	0.00011025	9.5	90.25	0.09975
0.324	79	0.0075	0.00005625	−5.5	30.25	−0.04125
0.317	80	0.0005	0.00000025	−4.5	20.25	−0.00225
0.314	77	−0.0025	0.00000625	−7.5	56.25	0.01875
0.313	101	−0.0035	0.00001225	16.5	272.25	−0.05775
0.301	81	−0.0155	0.00024025	−3.5	12.25	0.05425
0.300	76	−0.0165	0.00027225	−8.5	72.25	0.14025
0.293	85	− 0.0235	0.00055225	0.5	0.25	−0.01175
			$\sum(x-\bar{x})^2 = 0.0021925$		$\sum(y-\bar{y})^2 = 560.5$	$\sum(x-\bar{x})(y-\bar{y}) = 0.2585$

Next, calculate the respective standard deviations, s_x and s_y.

$$s_x = \sqrt{\frac{\Sigma(x-\bar{x})^2}{n-1}} = \sqrt{\frac{0.0021925}{10-1}} = \sqrt{0.0002436111111} \approx 0.0156080464 \text{ and}$$

$$s_y = \sqrt{\frac{\Sigma(y-\bar{y})^2}{n-1}} = \sqrt{\frac{560.5}{10-1}} = \sqrt{62.27777778} \approx 7.891627068$$

Put these values all together in the formula for the correlation coefficient r:

$$r = \frac{\Sigma(x-\bar{x})(y-\bar{y})}{(n-1)s_x s_y} = \frac{0.2585}{(10-1)(0.0156080464)(7.891627068)} \approx 0.2332$$

(d) This value of r is very close to the value $r = 0$. We would therefore say that there is no linear relationship between batting average and runs scored.

51. First, we calculate the respective means, $\bar{x}$ and $\bar{y}$.

$$\bar{x} = \frac{\Sigma x}{n} = \frac{77.9+53.1+33.7+53.9+40.5}{5} = \frac{259.1}{5} = 51.82 \text{ and}$$

$$\bar{y} = \frac{\Sigma y}{n} = \frac{38.3+11.4+28+26+4.5}{5} = \frac{108.2}{5} = 21.64$$

x	y	$(x-\bar{x})$	$(x-\bar{x})^2$	$(y-\bar{y})$	$(y-\bar{y})^2$	$(x-\bar{x})(y-\bar{y})$
77.9	38.3	26.08	680.1664	16.66	277.5556	494.4928
53.1	11.4	1.28	1.6384	−10.24	104.8576	−13.1072
33.7	28.0	−18.12	328.3344	6.36	40.4496	−115.2432
53.9	26.0	2.08	4.3264	4.36	19.0096	9.0688
40.5	4.5	−11.32	128.1424	−17.14	293.7796	194.0248
			$\sum(x-\bar{x})^2 = 1142.608$		$\sum(y-\bar{y})^2 = 735.652$	$\sum(x-\bar{x})(y-\bar{y}) = 509.236$

Next, calculate the respective standard deviations, s_x and s_y.

$$s_x = \sqrt{\frac{\Sigma(x-\bar{x})^2}{n-1}} = \sqrt{\frac{1142.608}{5-1}} = \sqrt{285.652} \approx 16.90124256 \text{ and}$$

$$s_y = \sqrt{\frac{\Sigma(y-\bar{y})^2}{n-1}} = \sqrt{\frac{735.652}{5-1}} = \sqrt{183.913} \approx 13.56145272$$

Put these values all together in the formula for the correlation coefficient r:

$$r = \frac{\Sigma(x-\bar{x})(y-\bar{y})}{(n-1)s_x s_y} = \frac{509.236}{(5-1)(16.90124256)(13.56145272)} \approx 0.5554$$

This value of r is positive. We would therefore say that assets and liabilities are positively correlated. As assets increase, liabilities also tend to increase.

53. First, we calculate the respective means, $\bar{x}$ and $\bar{y}$.

$$\bar{x} = \frac{\Sigma x}{n} = \frac{12.51+18.44+10.95+24.57+18.87}{5} = \frac{85.34}{5} = 17.068 \text{ and}$$

$$\bar{y} = \frac{\sum y}{n} = \frac{1.82+2.79+1.28+1.88+10.62}{5} = \frac{18.39}{5} = 3.678$$

x	y	$(x-\bar{x})$	$(x-\bar{x})^2$	$(y-\bar{y})$	$(y-\bar{y})^2$	$(x-\bar{x})(y-\bar{y})$
12.51	1.82	−4.558	20.775364	−1.858	3.452164	8.468764
18.44	2.79	1.372	1.882384	−0.888	0.788544	−1.218336
10.95	1.28	−6.118	37.429924	−2.398	5.750404	14.670964
24.57	1.88	7.502	56.280004	−1.798	3.232804	−13.488596
18.87	10.62	1.802	3.247204	6.942	48.191364	12.509484
			$\sum(x-\bar{x})^2 = 119.61488$		$\sum(y-\bar{y})^2 = 61.41528$	$\sum(x-\bar{x})(y-\bar{y}) = 20.94228$

Next, calculate the respective standard deviations, s_x and s_y.

$$s_x = \sqrt{\frac{\sum(x-\bar{x})^2}{n-1}} = \sqrt{\frac{119.61488}{5-1}} = \sqrt{29.90372} \approx 5.468429391 \text{ and}$$

$$s_y = \sqrt{\frac{\sum(y-\bar{y})^2}{n-1}} = \sqrt{\frac{61.41528}{5-1}} = \sqrt{15.35382} \approx 3.91839508$$

Put these values all together in the formula for the correlation coefficient r:

$$r = \frac{\sum(x-\bar{x})(y-\bar{y})}{(n-1)s_x s_y} = \frac{20.94228}{(5-1)(5.468429391)(3.91839508)} \approx 0.2443$$

This value of r is positive. We would therefore say that assets and liabilities are positively correlated. As assets increase, liabilities also tend to increase.

This value of r is very close to the value $r = 0$. We would therefore say that there is no linear relationship between price-earnings ratio and current ratio.

55.

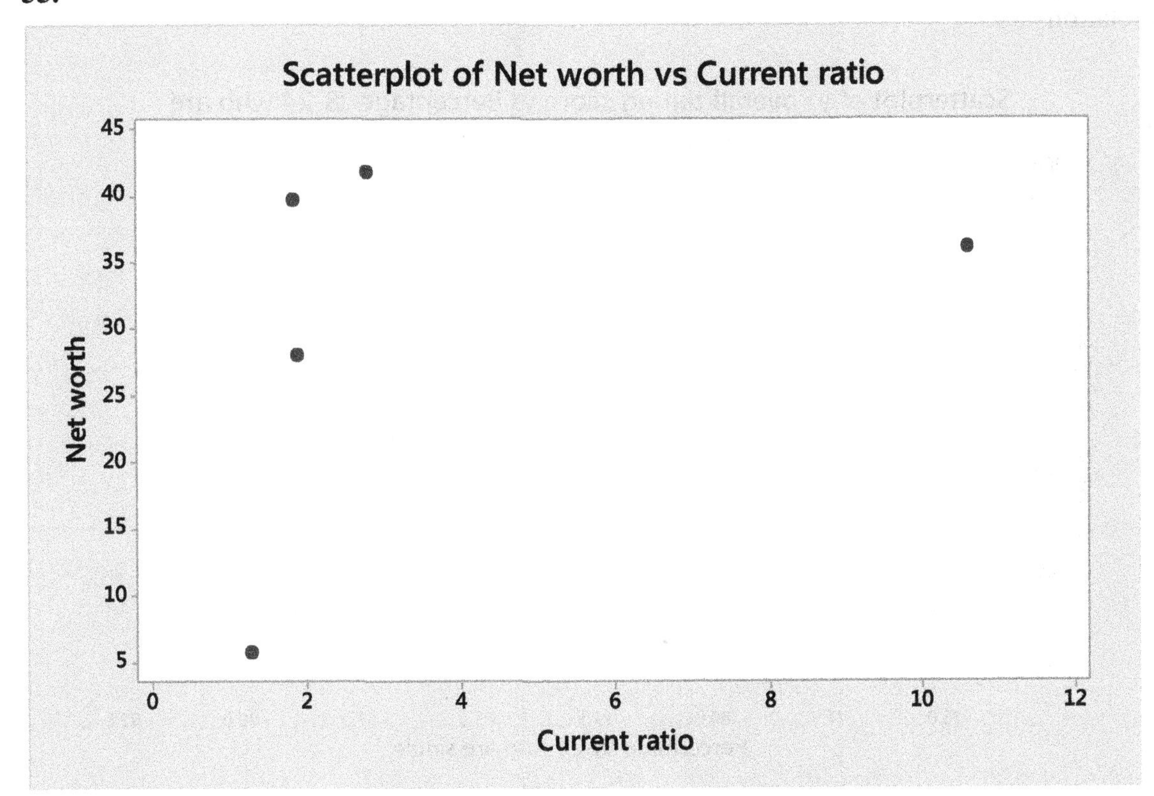

The current ratio of a company and the company's net worth have a positive linear relationship.

57. (a)

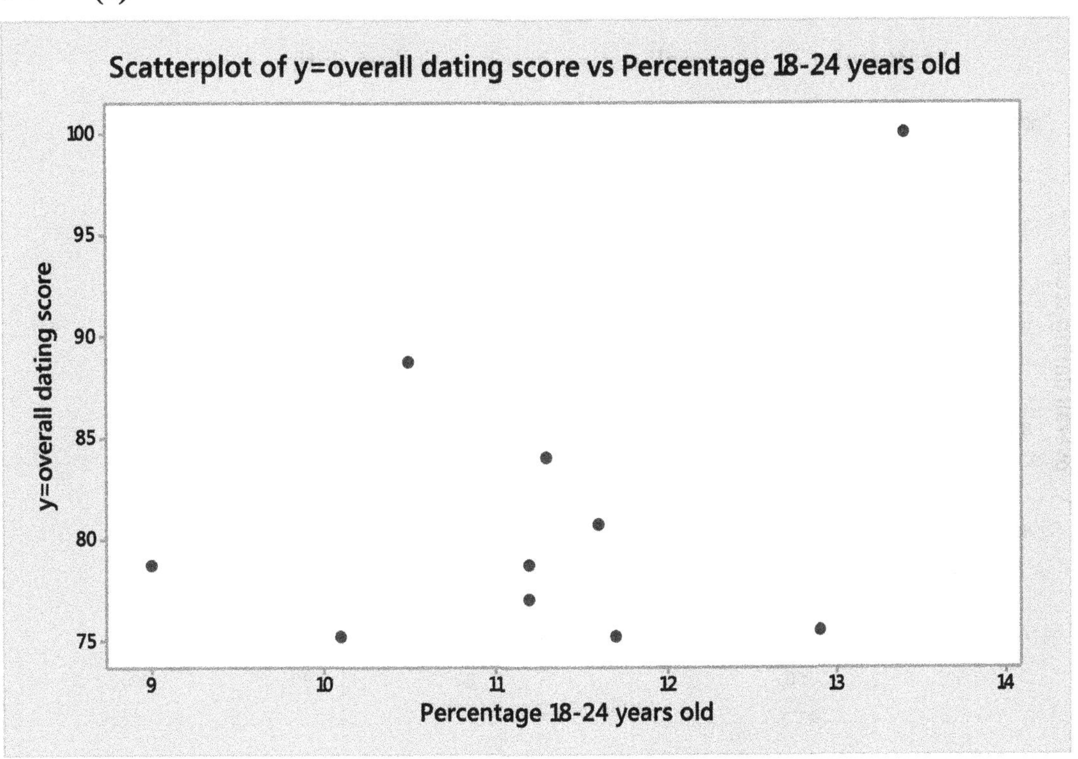

The percentage of 18–24 year-olds in a city and the overall dating score in a city have a positive linear relationship.

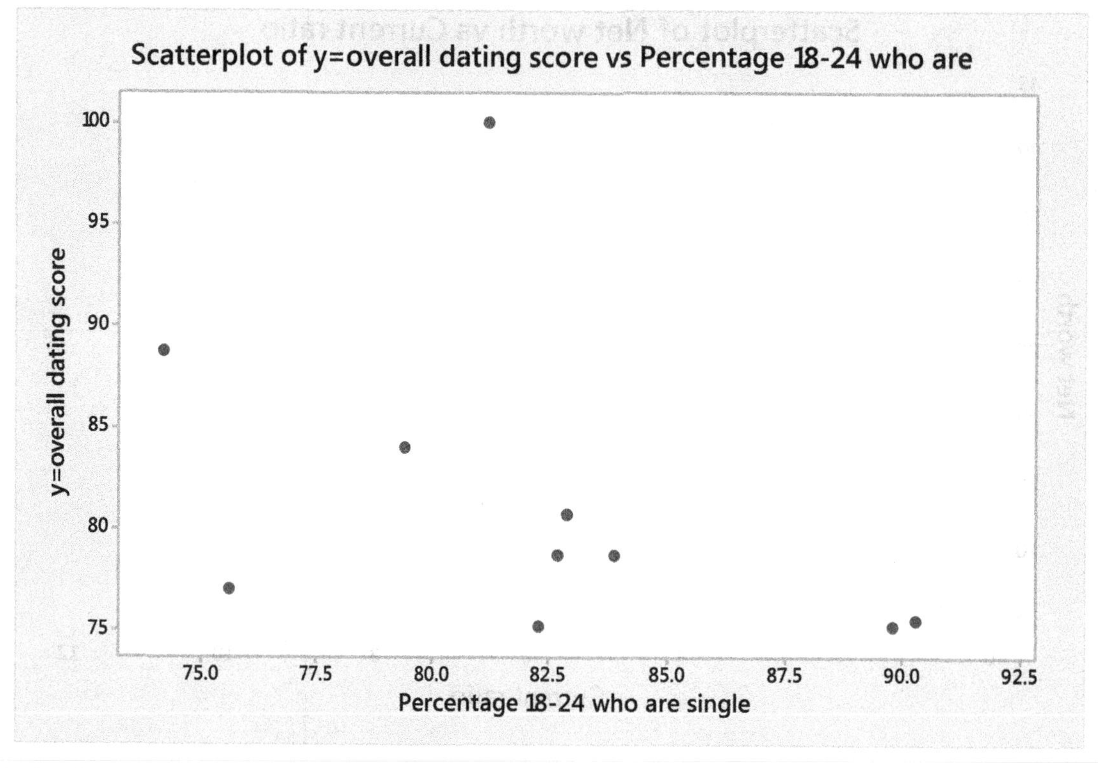

(b)

The percentage of a city's residents who are 18–24 years old and single and the overall dating score of a city have a negative linear relationship.

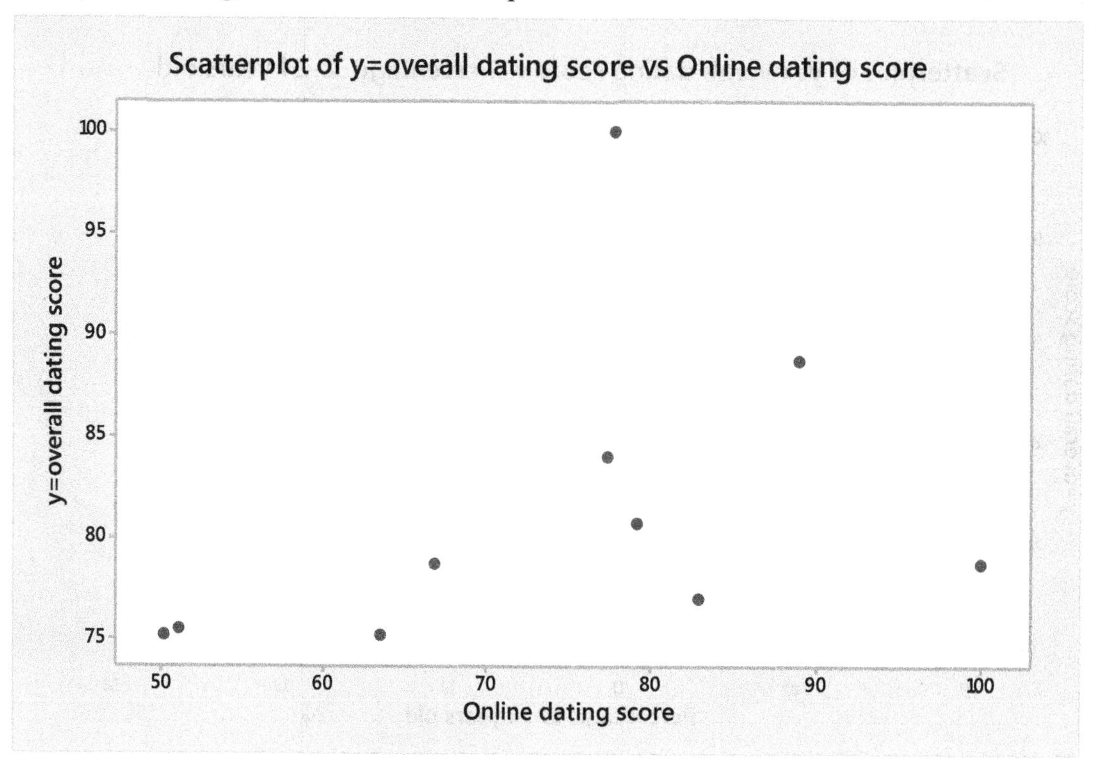

(c)

The online dating score of a city and the overall dating score of a city have a positive linear relationship.

59. Of the three regressions in Exercise 58, the one with the largest absolute value of the correlation coefficient is the percentage of people 18–24 years old who are single. Therefore, the predictor variable that is the best indicator of the overall dating score is the percentage of people 18–24 years old who are single.

61. Positive. The scatterplot indicates a positive linear relationship between cooling degree-days and average January temperature.

63.

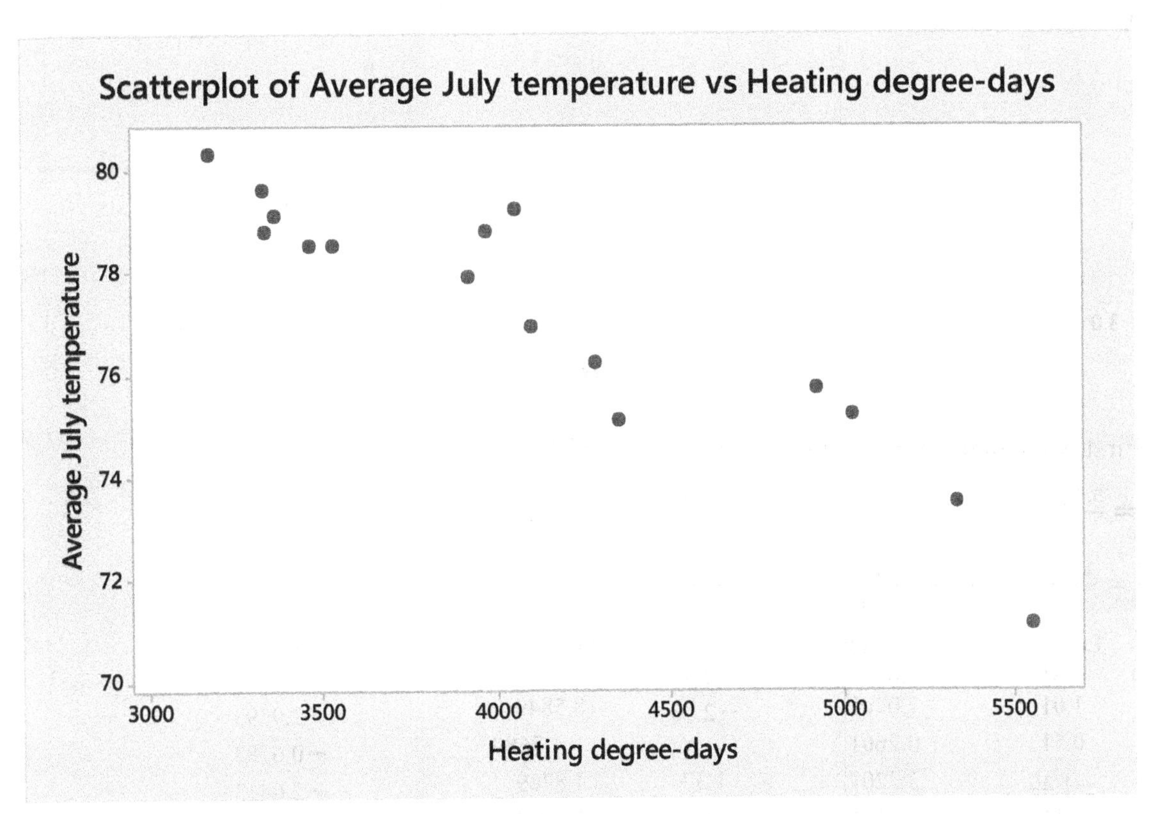

Heating degree-days and average July temperature have a negative linear relationship.

65. The relationship between the average July temperatures and the heating degree-days has the largest absolute value of the correlation coefficient, so it is the stronger relationship.

67. No.

69.

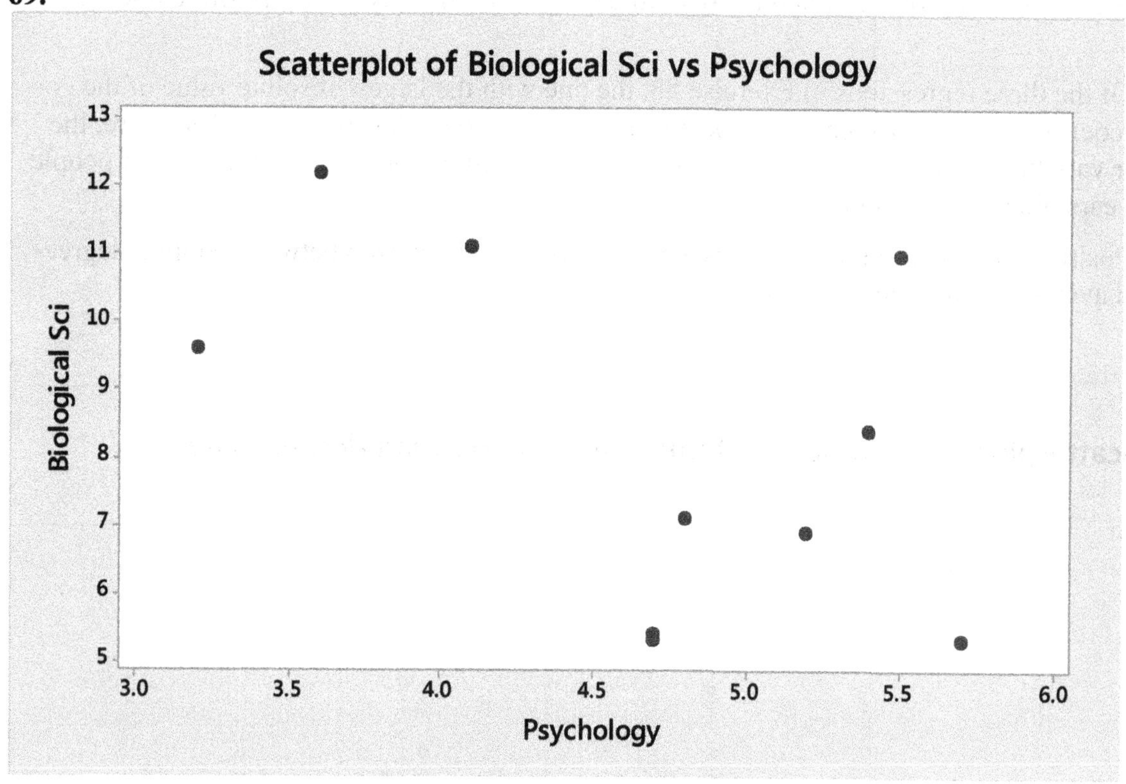

71. First, we calculate the respective means, $\bar{x}$ and $\bar{y}$.

$$\bar{x} = \frac{\sum x}{n} = \frac{5.5+5.7+5.2+3.2+4.8+4.1+4.7+5.4+4.7+3.6}{10} = \frac{46.9}{10} = 4.69 \text{ and}$$

$$\bar{y} = \frac{\sum y}{n} = \frac{11.0+5.3+6.9+9.6+7.1+11.1+5.3+8.4+5.4+12.2}{10} = \frac{82.3}{10} = 8.23$$

x	y	$(x-\bar{x})$	$(x-\bar{x})^2$	$(y-\bar{y})$	$(y-\bar{y})^2$	$(x-\bar{x})(y-\bar{y})$
5.5	11.0	0.81	0.6561	2.77	7.6729	2.2437
5.7	5.3	1.01	1.0201	−2.93	8.5849	−2.9593
5.2	6.9	0.51	0.2601	−1.33	1.7689	−0.6783
3.2	9.6	−1.49	2.2201	1.37	1.8769	−2.0413
4.8	7.1	0.11	0.0121	−1.13	1.2769	−0.1243
4.1	11.1	−0.59	0.3481	2.87	8.2369	−1.6933
4.7	5.3	0.01	0.0001	−2.93	8.5849	−0.0293
5.4	8.4	0.71	0.5041	0.17	0.0289	0.1207
4.7	5.4	0.01	0.0001	−2.83	8.0089	−0.0283
3.6	12.2	−1.09	1.1881	3.97	15.7609	−4.3273
			$\sum(x-\bar{x})^2 = 6.209$		$-\bar{y})^2 = 61.801$	$\sum(x-\bar{x})(y-\bar{y}) = -9.517$

Next, calculate the respective standard deviations, s_x and s_y.

$$s_x = \sqrt{\frac{\sum(x-\bar{x})^2}{n-1}} = \sqrt{\frac{6.209}{10-1}} = \sqrt{0.6898888889} \approx 0.8305955026 \text{ and}$$

$$s_y = \sqrt{\frac{\sum(y-\bar{y})^2}{n-1}} = \sqrt{\frac{61.801}{10-1}} = \sqrt{6.866777778} \approx 2.620453735$$

Put these values all together in the formula for the correlation coefficient r:

$$r = \frac{\Sigma(x-\bar{x})(y-\bar{y})}{(n-1)s_x s_y} = \frac{-9.517}{(10-1)(0.8305955026\,)(2.620453735\,)} \approx -0.4858$$

This value of r is negative. We would therefore say that Psychology majors and Biological science majors are negatively correlated. As the percent of Psychology majors at a school increases, the percent of Biological science majors tends to decrease.

This confirms what we observed in the scatterplot in Exercise 69.

73. First, we calculate the respective means, $\bar{x}$ and $\bar{y}$.

$$\bar{x} = \frac{\Sigma x}{n} = \frac{71.5+68.5+61.9+50+69+42+41.5+73+72.9+77.8}{10} = \frac{628.1}{10} = 62.81 \text{ and}$$

$$\bar{y} = \frac{\Sigma y}{n} = \frac{91.5+96.7+91.1+66.2+94.0+74.7+66.9+96.7+92.8+89.4}{10} = \frac{860}{10} = 86$$

x	y	$(x-\bar{x})$	$(x-\bar{x})^2$	$(y-\bar{y})$	$(y-\bar{y})^2$	$(x-\bar{x})(y-\bar{y})$
71.5	91.5	8.69	75.5161	5.5	30.25	47.795
68.5	96.7	5.69	32.3761	10.7	114.49	60.833
61.9	91.1	−0.91	0.8281	5.1	26.01	−4.641
50.0	66.2	−12.81	164.0961	−19.8	392.04	253.638
69.0	94.0	6.19	38.3161	8.0	64.00	49.52
42.0	74.7	−20.81	433.0561	−11.3	127.69	235.153
41.5	66. 9	−21.31	454.1161	−19.1	364.81	407.021
73.0	96.7	10.19	103.8361	10.7	114.49	109.033
72.9	92.8	10.09	101.8081	6.8	46.24	68.612
77.8	89.4	14.99	224.7001	3.4	11.56	50.966
			$\sum(x-\bar{x})^2 = 1628.649$		$-\bar{y})^2 = 1291.58$	$\sum(x-\bar{x})(y-\bar{y}) = 1277.98$

Next, calculate the respective standard deviations, s_x and s_y.

$$s_x = \sqrt{\frac{\Sigma(x-\bar{x})^2}{n-1}} = \sqrt{\frac{1628.649}{10-1}} = \sqrt{180.961} \approx 13.45217455 \text{ and}$$

$$s_y = \sqrt{\frac{\Sigma(y-\bar{y})^2}{n-1}} = \sqrt{\frac{1291.58}{10-1}} = \sqrt{143.5088889} \approx 11.97951956$$

Put these values all together in the formula for the correlation coefficient r:

$$r = \frac{\Sigma(x-\bar{x})(y-\bar{y})}{(n-1)s_x s_y} = \frac{1277.98}{(10-1)(13.45217455)(11.97951956)} \approx 0.8811$$

This value of r is very close to the maximum value $r = 1$. We would therefore say that economy and literacy are positively correlated. As the economy increases, literacy also tends to increase.

75. Yes.

First, we calculate the respective means, $\bar{x}$ and $\bar{y}$.

$$\bar{x} = \frac{\Sigma x}{n} = \frac{91.5+96.7+91.1+66.2+94.0+74.7+66.9+96.7+92.8+89.4}{10} = \frac{860}{10} = 86 \text{ and}$$

$$\bar{y} = \frac{\Sigma y}{n} = \frac{71.5+68.5+61.9+50+69+42+41.5+73+72.9+77.8}{10} = \frac{628.1}{10} = 62.81$$

x	y	$(x-\bar{x})$	$(x-\bar{x})^2$	$(y-\bar{y})$	$(y-\bar{y})^2$	$(x-\bar{x})(y-\bar{y})$

91.5	71.5	5.5	30.25	8.69	75.5161	47.795
96.7	68.5	10.7	114.49	5.69	32.3761	60.833
91.1	61.9	5.1	26.01	−0.91	0.8281	−4.641
66.2	50.0	−19.8	392.04	−12.81	164.0961	253.638
94.0	69.0	8.0	64.00	6.19	38.3161	49.52
74.7	42.0	−11.3	127.69	−20.81	433.0561	235.153
66.9	41.5	−19.1	364.81	−21.31	454.1161	407.021
96.7	73.0	10.7	114.49	10.19	103.8361	109.033
92.8	72.9	6.8	46.24	10.09	101.8081	68.612
89.4	77.8	3.4	11.56	14.99	224.7001	50.966
			$\sum(x-\bar{x})^2 = 1291.58$		$, -\bar{y})^2 = 1628.649$	$\sum(x-\bar{x})(y-\bar{y}) = 1277.98$

Next, calculate the respective standard deviations, s_x and s_y.

$$s_x = \sqrt{\frac{\Sigma(x-\bar{x})^2}{n-1}} = \sqrt{\frac{1291.58}{10-1}} = \sqrt{143.5088889} \approx 11.97951956$$

and $s_y = \sqrt{\frac{\Sigma(y-\bar{y})^2}{n-1}} = \sqrt{\frac{1628.649}{10-1}} = \sqrt{180.961} \approx 13.45217455$

Put these values all together in the formula for the correlation coefficient r:

$$r = \frac{\Sigma(x-\bar{x})(y-\bar{y})}{(n-1)s_x s_y} = \frac{1277.98}{(10-1)(11.97951956)(13.45217455\,)} \approx 0.8811$$

Yes.

77. First, we calculate the respective means, $\bar{x}$ and $\bar{y}$.

$$\bar{x} = \frac{\Sigma x}{n} = \frac{95.2+92.8+92.8+51.7+100.0+78.3+49.3+87.9+90.3+85.5}{10} = \frac{823.8}{10} = 82.38 \text{ and}$$

$$\bar{y} = \frac{\Sigma y}{n} = \frac{91.5+96.7+91.1+66.2+94.0+74.7+66.9+96.7+92.8+89.4}{10} = \frac{860}{10} = 86$$

x	y	$(x-\bar{x})$	$(x-\bar{x})^2$	$(y-\bar{y})$	$(y-\bar{y})^2$	$(x-\bar{x})(y-\bar{y})$
95.2	91.5	12.82	164.3524	5.5	30.25	70.51
92.8	96.7	10.42	108.5764	10.7	114.49	111.494
92.8	91.1	10.42	108.5764	5.1	26.01	53.142
51.7	66.2	−30.68	941.2624	−319.8	392.04	607.464
100.0	94.0	17.62	310.4644	8.0	64.00	140.96
78.3	74.7	−34.08	16.6464	−311.3	127.69	46.104
49.3	66.9	−333.08	1094.286	−319.1	364.81	631.828
87.9	96.7	5.52	30.4704	10.7	114.49	59.064
90.3	92.8	7.92	62.7264	6.8	46.24	53.856
85.5	89.4	3.12	9.7344	3,4	11.56	10.608
			$\sum(x-\bar{x})^2 = 2847.096$		$\sum(y-\bar{y})^2$ 91.58	$\sum(x-\bar{x})(y-\bar{y}) = 1785.03$

Next, calculate the respective standard deviations, s_x and s_y.

$$s_x = \sqrt{\frac{\Sigma(x-\bar{x})^2}{n-1}} = \sqrt{\frac{2847.096}{10-1}} = \sqrt{316.344} \approx 17.78606196 \text{ and}$$

$$s_y = \sqrt{\frac{\Sigma(y-\bar{y})^2}{n-1}} = \sqrt{\frac{1291.58}{10-1}} = \sqrt{143.5088889} \approx 11.97951956$$

Put these values all together in the formula for the correlation coefficient r:

$$r = \frac{\Sigma(x-\bar{x})(y-\bar{y})}{(n-1)s_x s_y} = \frac{1785.03}{(10-1)(17.78606196)(11.97951956)} \approx 0.9309$$

This value of r is very close to the maximum value $r = 1$. We would therefore say that health and literacy are positively correlated. As health increases, literacy also tends to increase.

79.

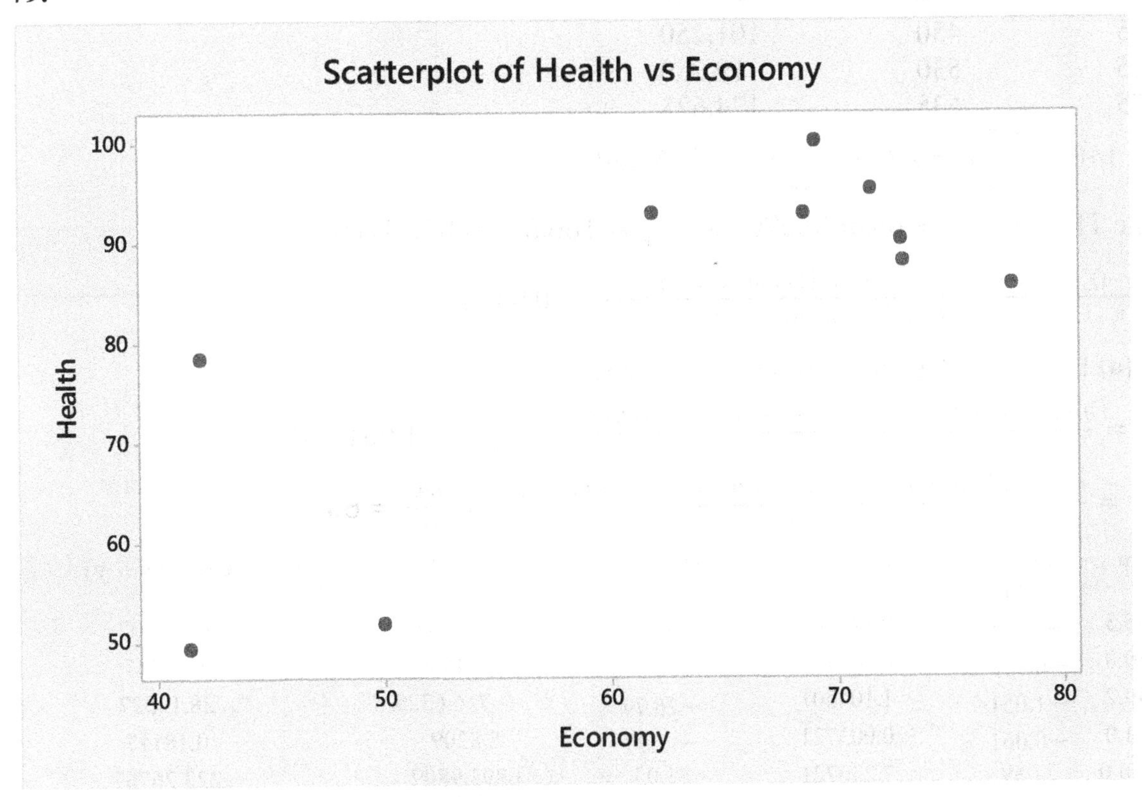

Economy and health have a positive linear relationship.

81. The correlation coefficient r for the linear relationship between x = health and y = economy equals the correlation coefficient r for the linear relationship between x = economy and y = health. From Exercise 80, $r \approx 0.7598$.

83.

x	y	xy
75	155	11,625
125	210	26,250
125	290	36,250
175	360	63,000
175	250	43,750
225	450	101,250
225	530	119,250
275	635	174,625
$\sum x = 1400$	$\sum y = 2880$	$\sum xy = 576{,}000$

From the TI-83/84: $s_x \approx 65.46536707$ and $s_y \approx 166.5404284$. Then

$$r = \frac{\sum xy - (\sum x \sum y)/n}{(n-1)s_x s_y} = \frac{576{,}000 - ((1400)(2880))/8}{(8-1)(65.46536707)(166.5404284)} \approx 0.9434$$

85. **(a)** First, we calculate the respective means, $\bar{x}$ and $\bar{y}$.

$$\bar{x} = \frac{\sum x}{n} = \frac{0.73+1.14+0.6+1.59+4.34+1.98+3.12+1.8+0.65+0.56}{10} = \frac{16.51}{10} = 1.651 \text{ and}$$

$$\bar{y} = \frac{\sum y}{n} = \frac{45.3+50.8+40.2+64.0+150.0+106.0+90.8+58.8+35.4+28.4}{10} = \frac{669.7}{10} = 66.97$$

x	y	$(x - \bar{x})$	$(x-\bar{x})^2$	$(y - \bar{y})$	$(y-\bar{y})^2$	$(x-\bar{x})(y-\bar{y})$
0.73	45.3	−0.921	0.848241	−21.67	469.5889	19.95807
1.14	50.8	−0.511	0.261121	−16.17	261.4689	8.26287
0.60	40.2	−1.051	1.104601	−26.77	716.6329	28.13527
1.59	64.0	−0.061	0.003721	−-2.97	8.8209	0.18117
4.34	150.0	2.689	7.230721	83.03	6893.9809	223.26767
1.98	106.0	0.329	0.108241	39.03	1523.3409	12.84087
3.12	90.8	1.469	2.157961	23.83	567.8689	35.00627
1.80	58.8	0.149	0.022201	−8.17	66.7489	−1.21733
0.65	35.4	−1.001	1.002001	−31.57	996.6649	31.60157
0.56	28.4	−1.091	1.190281	−38.57	1487.6449	42.07987
			$\sum(x-\bar{x})^2 = 13.92909$		$\sum(y-\bar{y})^2 = 12{,}992.761$	$\sum(x-\bar{x})(y-\bar{y}) = 400.1163$

Next, calculate the respective standard deviations, s_x and s_y.

$$s_x = \sqrt{\frac{\sum(x-\bar{x})^2}{n-1}} = \sqrt{\frac{13.92909}{10-1}} = \sqrt{1.547676667} \approx 1.244056537 \text{ and}$$

$$s_y = \sqrt{\frac{\sum(y-\bar{y})^2}{n-1}} = \sqrt{\frac{12{,}992.761}{10-1}} = \sqrt{1443.640111} \approx 37.99526432$$

Put these values all together in the formula for the correlation coefficient r:

$$r = \frac{\sum(x-\bar{x})(y-\bar{y})}{(n-1)s_x s_y} = \frac{400.1163}{(10-1)(1.244056537)(37.99526432)} \approx 0.9405$$

(b) Yes.

(c) This value of r is very close to the maximum value $r = 1$. We would therefore say that lead and zinc are positively correlated. As the lead content in the fish increases, the zinc content in the fish also tends to increase.

87. **(a)**

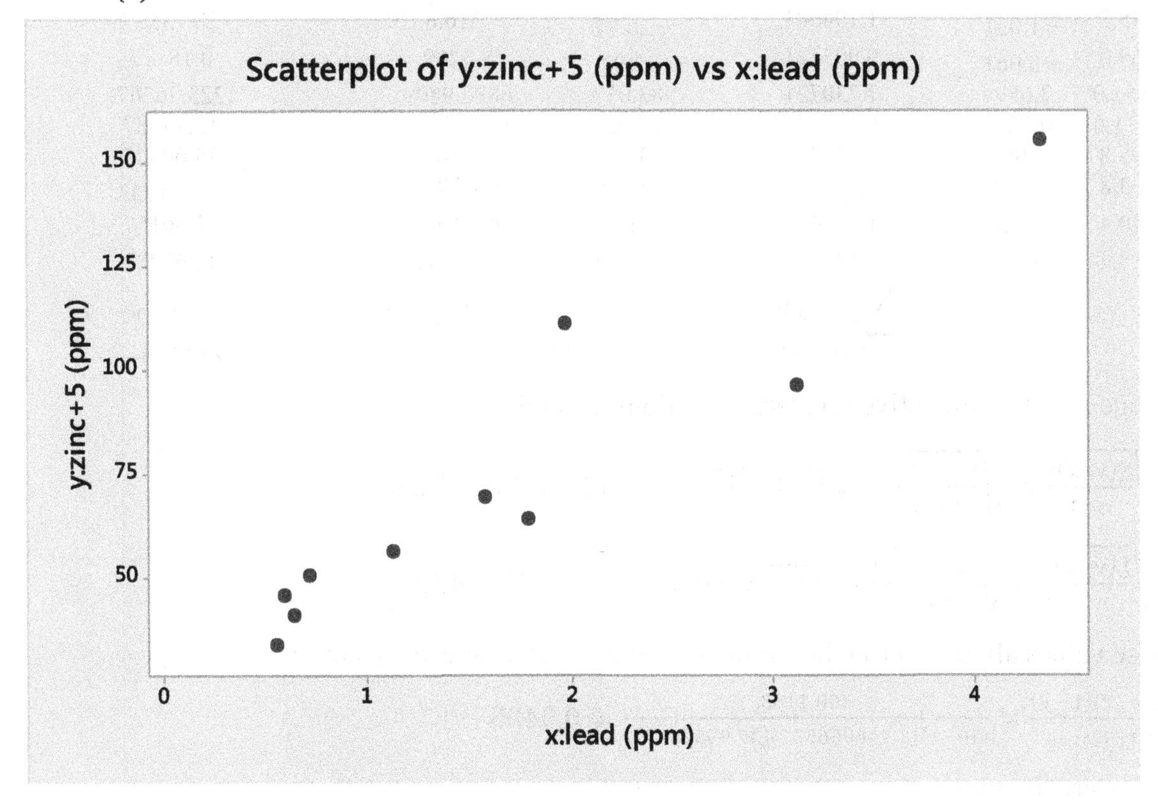

Everything is the same except the dots are shifted up 5 ppm.

(b) First, we calculate the respective means, $\bar{x}$ and $\bar{y}$.

$$\bar{x} = \frac{\sum x}{n} = \frac{0.73+1.14+0.6+1.59+4.34+1.98+3.12+1.8+0.65+0.56}{10} = \frac{16.51}{10} = 1.651 \text{ and}$$

$$\bar{y} = \frac{\sum y}{n} = \frac{50.3+55.8+45.2+69.0+155.0+111.0+95.8+63.8+40.4+33.4}{10} = \frac{719.7}{10} = 71.97$$

x	y	$(x - \bar{x})$	$(x-\bar{x})^2$	$(y-\bar{y})$	$(y-\bar{y})^2$	$(x-\bar{x})(y-\bar{y})$
0.73	50.3	−0.921	0.848241	-21 −21.67	469.5889	19.95807
1.14	55.8	−0.511	0.261121	−16.17	261.4689	8.26287
0.60	45.2	−1.051	1.104601	−26.77	716.6329	28.13527
1.59	69.0	−0.061	0.003721	−2.97	8.8209	0.18117
4.34	155.0	2.689	7.230721	83.03	6893.9809	223.26767
1.98	111.0	0.329	0.108241	39.03	1523.3409	12.84087
3.12	95.8	1.469	2.157961	23.83	567.8689	35.00627
1.80	63.8	0.149	0.022201	−8.17	66.7489	−1.21733
0.65	40.4	−1.001	1.002001	−31.57	996.6649	31.60157
0.56	33.4	−1.091	1.190281	−38.57	1487.6449	42.07987
			$\sum(x-\bar{x})^2$ = 13.92909		$\sum(y-\bar{y})^2$,992.761	$\sum(x-\bar{x})(y-\bar{y})$ = 400.1163

Next, calculate the respective standard deviations, s_x and s_y.

$$s_x = \sqrt{\frac{\sum(x-\bar{x})^2}{n-1}} = \sqrt{\frac{13.92909}{10-1}} = \sqrt{1.547676667} \approx 1.244056537 \text{ and}$$

$$s_y = \sqrt{\frac{\sum(y-\bar{y})^2}{n-1}} = \sqrt{\frac{12{,}992.761}{10-1}} = \sqrt{1443.640111} \approx 37.99526432$$

Put these values all together in the formula for the correlation coefficient r:

$$r = \frac{\sum(x-\bar{x})(y-\bar{y})}{(n-1)s_x s_y} = \frac{400.1163}{(10-1)(1.244056537)(37.99526432)} \approx 0.9405$$

(c) They are the same.

(d) When a constant is added to each y-data value the correlation coefficient stays the same.

89. Answers will vary.

91. **(a)** Answers will vary. One possible answer:

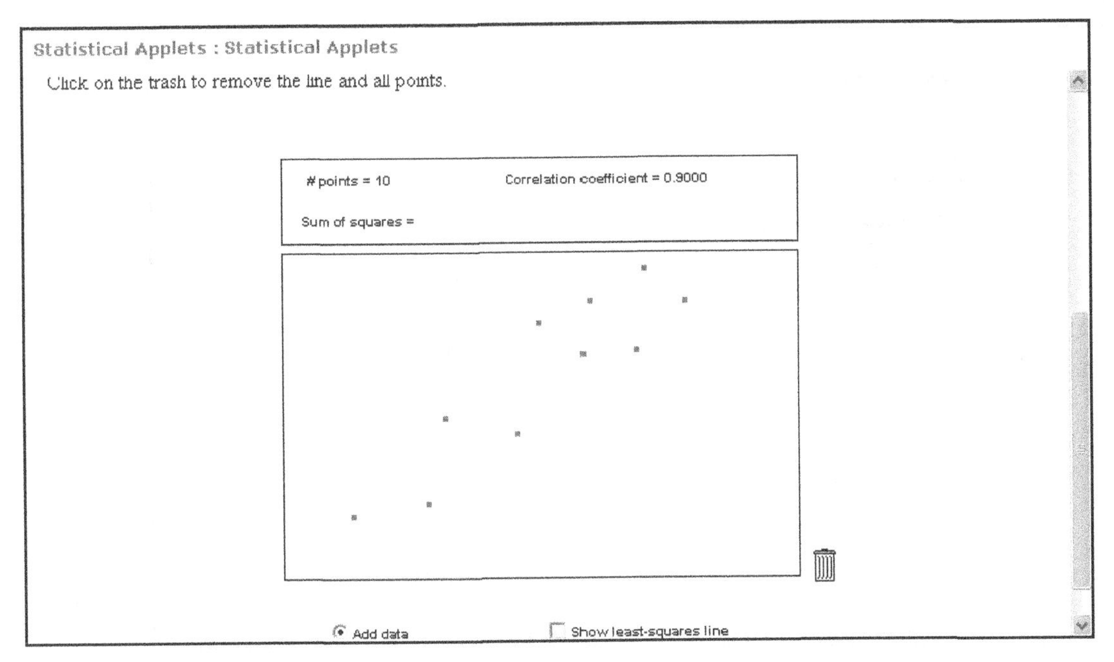

(b) Answers will vary. One possible answer:

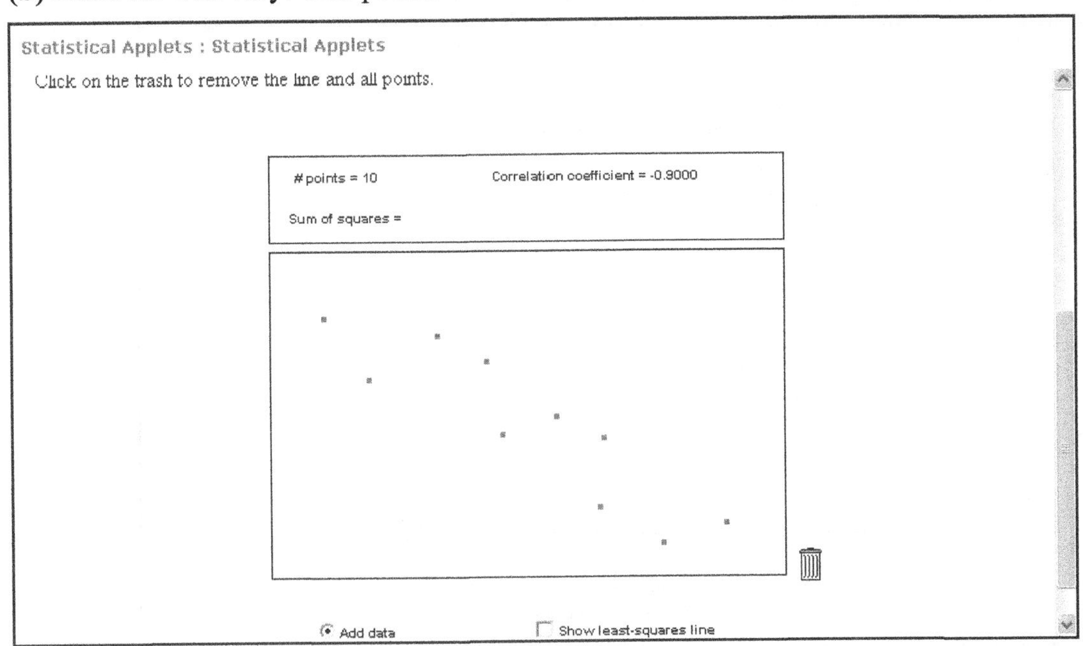

(c) Answers will vary. One possible answer:

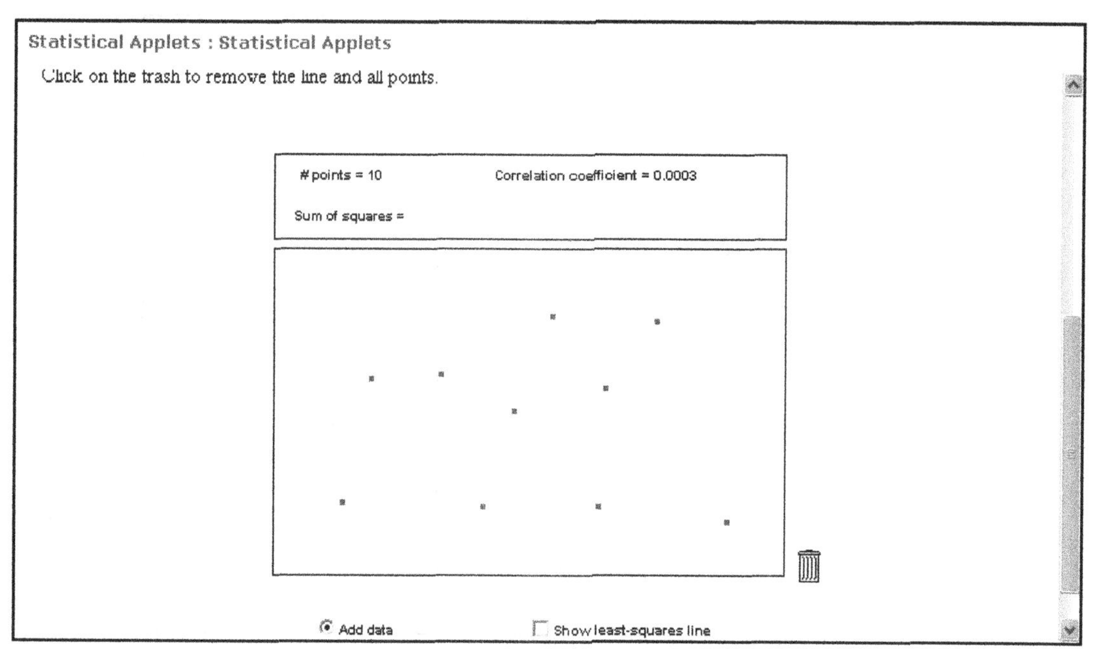

93. **(a)** Answers will vary. One possible answer:

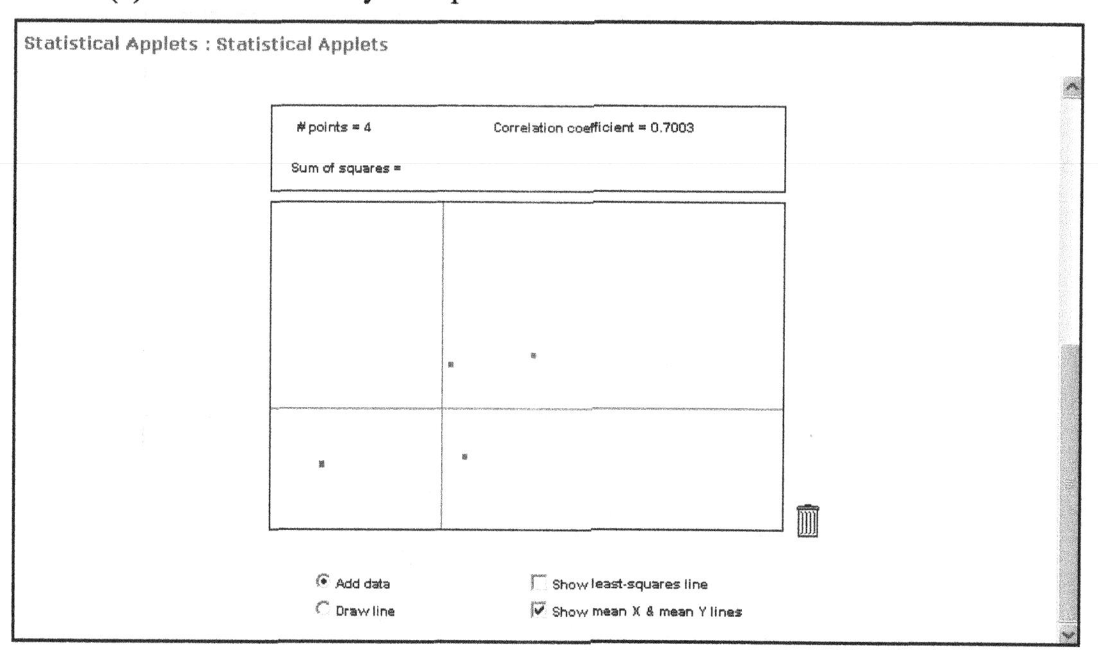

(b) Answers will vary. One possible answer:

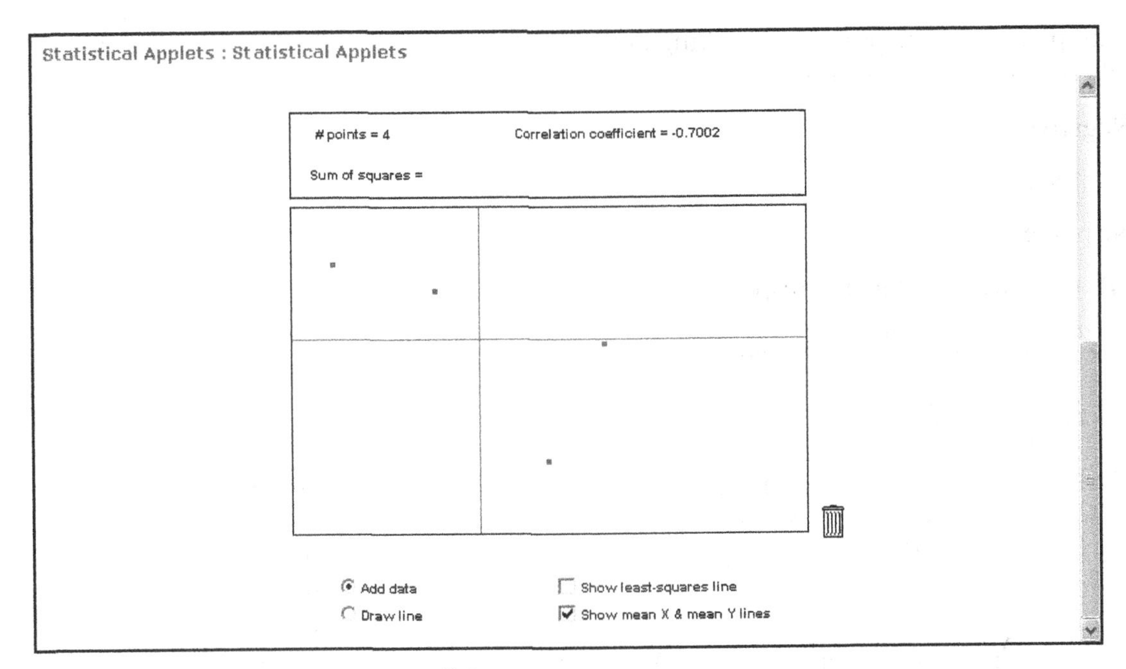

(c) Answers will vary. One possible answer:

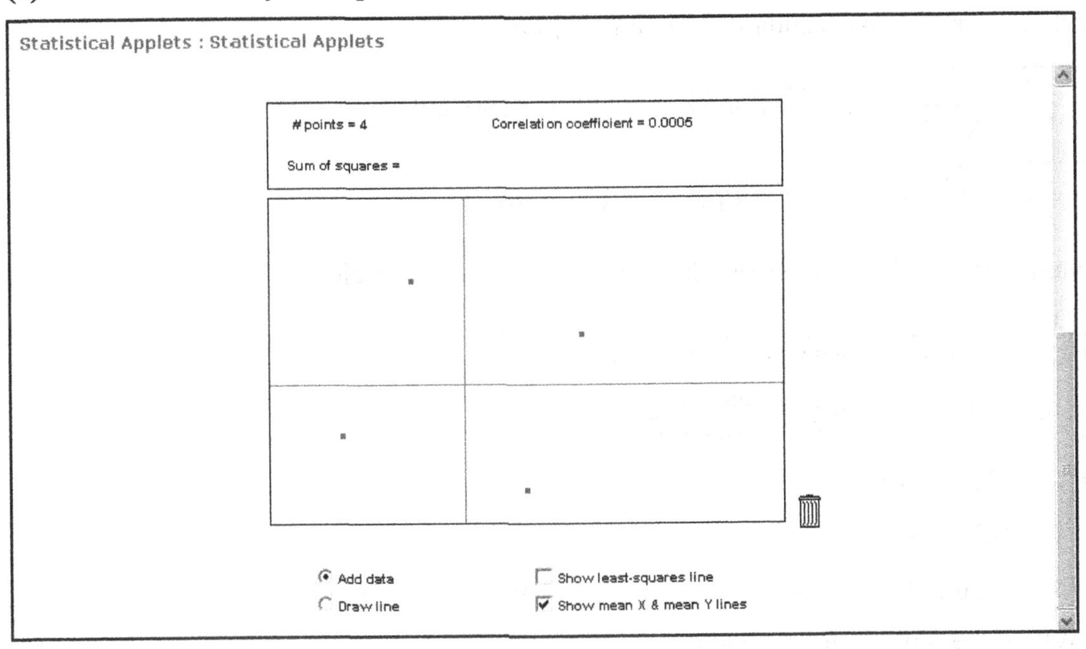

95. Positive

97. This value of r is positive. We would therefore say that height and weight are positively correlated. As height increases, weight also tends to increase.

Section 4.2

1. The objective of regression analysis is to approximate the relationship between two numerical variables using the regression line and the regression equation.

3. We can find the predicted value of y by plugging a given value of x into the regression equation and simplifying.

5. Extrapolation is the process of making predictions based on x-values that are beyond the range of the x-values in our data set.

7. Negative.

9. Positive.

11. Negative.

13. **(a)** First, we calculate the respective means, $\bar{x}$ and $\bar{y}$.

$$\bar{x} = \frac{\sum x}{n} = \frac{10+20+30+40}{4} = \frac{100}{4} = 25 \text{ and}$$

$$\bar{y} = \frac{\sum y}{n} = \frac{2+5+9+12}{4} = \frac{28}{4} = 7$$

x	y	$(x-\bar{x})$	$(x-\bar{x})^2$	$(y-\bar{y})$	$(y-\bar{y})^2$	$(x-\bar{x})(y-\bar{y})$
10	2	−15	225	5 −5	25	75
20	5	−5	25	−2	4	10
30	9	5	25	2	4	10
40	12	15	225	5	25	75

$$\sum(x-\bar{x})^2 = 500 \qquad \sum(y-\bar{y})^2 = 58 \quad \sum(x-\bar{x})(y-\bar{y}) = 170$$

Next, calculate the respective standard deviations, s_x and s_y.

$$s_x = \sqrt{\frac{\sum(x-\bar{x})^2}{n-1}} = \sqrt{\frac{500}{4-1}} = \sqrt{166.6666667} \approx 12.90994449 \text{ and}$$

$$s_y = \sqrt{\frac{\sum(y-\bar{y})^2}{n-1}} = \sqrt{\frac{58}{4-1}} = \sqrt{19.33333333} \approx 4.396968653$$

Put these values all together in the formula for the correlation coefficient r:

$$r = \frac{\sum(x-\bar{x})(y-\bar{y})}{(n-1)s_x s_y} = \frac{170}{(4-1)(12.90994449)(4.396968653)} \approx 0.9982743729$$

$$b_1 = r \cdot \frac{s_y}{s_x} = (0.9982743729) \cdot \left(\frac{4.396968653}{12.90994449}\right) = 0.34,\ b_0 = \bar{y} - (b_1 \cdot \bar{x})$$

$$= 7 - (0.34 \cdot 25) = -1.5$$

Therefore, $\hat{y} = 0.34x - 1.5$.

(b) For each increase of 1 unit, the estimated value of y increases by 0.34 unit. When $x = 0$ the estimated value of y is -1.5.

15. **(a)** First, we calculate the respective means, $\bar{x}$ and $\bar{y}$.

$$\bar{x} = \frac{\sum x}{n} = \frac{(-5)+(-4)+(-3)+(-2)+(-1)}{5} = \frac{-15}{5} = -3 \text{ and}$$

$$\bar{y} = \frac{\sum y}{n} = \frac{10+18+18+26+26}{5} = \frac{98}{5} = 19.6$$

x	y	$(x-\bar{x})$	$(x-\bar{x})^2$	$(y-\bar{y})$	$(y-\bar{y})^2$	$(x-\bar{x})(y-\bar{y})$
−5	10	−2	4	−9.6	92.16	19.2
−4	18	−1	1	−1.6	2.56	1.6
−3	18	0	0	−1.6	2.56	0
−2	26	1	1	6.4	40.96	6.4
−1	26	2	4	6.4	40.96	12.8

$\sum(x-\bar{x})^2 = 10$ $\sum(y-\bar{y})^2 = 179.2$ $\sum(x-\bar{x})(y-\bar{y}) = 40$

Next, calculate the respective standard deviations, s_x and s_y.

$$s_x = \sqrt{\frac{\Sigma(x-\bar{x})^2}{n-1}} = \sqrt{\frac{10}{5-1}} = \sqrt{2.5} \approx 1.58113883 \text{ and}$$

$$s_y = \sqrt{\frac{\Sigma(y-\bar{y})^2}{n-1}} = \sqrt{\frac{179.2}{5-1}} = \sqrt{44.8} \approx 6.693280212$$

Put these values all together in the formula for the correlation coefficient r:

$$r = \frac{\Sigma(x-\bar{x})(y-\bar{y})}{(n-1)s_x s_y} = \frac{40}{(5-1)(1.58113883)(6.693280212)} \approx 0.9449111826$$

$$b_1 = r \cdot \frac{s_y}{s_x} = (0.9449111826) \cdot \left(\frac{6.693280212}{1.58113883}\right) = 4,\ b_0 = \bar{y} - (b_1 \cdot \bar{x})$$

$$= 19.6 - (4 \cdot (-3)) = 31.6$$

Therefore, $\hat{y} = 4x + 31.6$.

(b) For each increase of 1 unit, the estimated value of y increases by 4 units. When $x = 0$ the estimated value of y is 31.6.

17. **(a)** First, we calculate the respective means, $\bar{x}$ and $\bar{y}$.

$$\bar{x} = \frac{\Sigma x}{n} = \frac{5+10+15+20+25+30}{6} = \frac{105}{6} = 17.5 \text{ and}$$

$$\bar{y} = \frac{\Sigma y}{n} = \frac{7+8+8+8+7+8}{6} = \frac{46}{6} \approx 7.666666667 \approx 7.6667$$

x	y	$(x-\bar{x})$	$(x-\bar{x})^2$	$(y-\bar{y})$	$(y-\bar{y})^2$	$(x-\bar{x})(y-\bar{y})$
5	7	−12.5	156.25	−0.6667	0.44448889	8.33375
10	8	−7.5	56.25	0.3333	0.11108889	−2.49975
15	8	−2.5	6.25	0.3333	0.11108889	−0.83325
20	8	2.5	6.25	0.3333	0.11108889	0.83325
25	7	7.5	56.25	−0.6667	0.44448889	−5.00025
30	8	12.5	156.25	0.3333	0.11108889	4.16625

$\sum(x-\bar{x})^2 = 437.5$ $\sum(y-\bar{y})^2 = 1.33333334$ $\sum(x-\bar{x})(y-\bar{y}) = 5$

Next, calculate the respective standard deviations, s_x and s_y.

$$s_x = \sqrt{\frac{\Sigma(x-\bar{x})^2}{n-1}} = \sqrt{\frac{437.5}{6-1}} = \sqrt{87.5} \approx 9.354143467 \text{ and}$$

$$s_y = \sqrt{\frac{\Sigma(y-\bar{y})^2}{n-1}} = \sqrt{\frac{1.33333334}{6-1}} = \sqrt{0.266666668} \approx 0.5163977808$$

Put these values all together in the formula for the correlation coefficient r:

$r = \frac{\sum(x-\bar{x})(y-\bar{y})}{(n-1)s_x s_y} = \frac{5}{(6-1)(9.354143467)(0.5163977808)} \approx 0.2070196673$

$b_1 = r \cdot \frac{s_y}{s_x} = (0.2070196673) \cdot \left(\frac{0.5163977808}{9.354143467}\right) \approx 0.0114285714 \approx 0.0114, b_0 = \bar{y} - (b_1 \cdot \bar{x})$ $\approx 7.666666667 - (0.0114285714 \cdot 17.5) \approx 7.4667$ Therefore, $\hat{y} = 0.0114x + 7.4667$

(b) For each increase of 1 unit, the estimated value of y increases by 0.0114 unit. When $x = 0$ the estimated value of y is 7.4667.

19. **(a)** First, we calculate the respective means, $\bar{x}$ and $\bar{y}$.

$\bar{x} = \frac{\sum x}{n} = \frac{(-70)+(-60)+(-50)+(-40)+(-30)+(-20)+(-10)+0}{8} = \frac{-280}{8} = -35$ and

$\bar{y} = \frac{\sum y}{n} = \frac{5+10+15+20+20+15+10+5}{8} = \frac{100}{8} = 12.5$

x	y	$(x-\bar{x})$	$(x-\bar{x})^2$	$(y-\bar{y})$	$(y-\bar{y})^2$	$(x-\bar{x})(y-\bar{y})$
−70	5	−5	1225	−7.5	56.25	262.5
−60	10	−25	625	−2.5	6.25	62.5
−50	15	−15	225	2.5	6.25	−37.5
−40	20	−5	25	7.5	56.25	−37.5
−30	20	5	25	7.5	56.25	37.5
−20	15	15	225	2.5	6.25	37.5
−10	10	25	625	−2.5	6.25	−62.5
0	5	35	1225	−7.5	56.25	−262.5
			$\sum(x-\bar{x})^2 = 4200$		$\sum(y-\bar{y})^2 = 250$	$\sum(x-\bar{x})(y-\bar{y}) = 0$

Next, calculate the respective standard deviations, s_x and s_y.

$s_x = \sqrt{\frac{\sum(x-\bar{x})^2}{n-1}} = \sqrt{\frac{4200}{8-1}} = \sqrt{600} \approx 24.49489743$ and

$s_y = \sqrt{\frac{\sum(y-\bar{y})^2}{n-1}} = \sqrt{\frac{250}{8-1}} = \sqrt{35.71428571} \approx 5.976143047$

Put these values all together in the formula for the correlation coefficient r:

$r = \frac{\sum(x-\bar{x})(y-\bar{y})}{(n-1)s_x s_y} = \frac{0}{(8-1)(24.49489743)(5.976143047)} = 0$

$b_1 = r \cdot \frac{s_y}{s_x} = (0) \cdot \left(\frac{5.976143047}{24.49489743}\right) = 0, b_0 = \bar{y} - (b_1 \cdot \bar{x}) = 12.5 - (0 \cdot (-35)) = 12.5.$

Therefore, $\hat{y} = 0x + 12.5 = 12.5$.

(b) For each increase of 1 unit, the estimated value of y remains 12.5

When $x = 0$ the estimated value of y is 12.5.

21. **(a)** From Exercise 17 in Section 4.1, $\bar{x} = 67, \bar{y} = 134.2, s_x \approx 1.58113883$, $s_y \approx 12.27599283$, and $r = 0.9273546941$.

$b_1 = r \cdot \frac{s_y}{s_x} = (0.9273546941) \cdot \left(\frac{12.27599283}{1.58113883}\right) = 7.2, b_0 = \bar{y} - (b_1 \cdot \bar{x}) = 134.2 - (7.2 \cdot 67) =$ -348.2. Therefore, $\hat{y} = 7.2x - 348.2$.

(b) For each increase of 1 inch, the estimated weight increases by 7.2 pounds. When $x = 0$ inches, the estimated weight is –348.2 pounds.

23. **(a)** From Exercise 19 in Section 4.1, $\bar{x} = 5.6, \bar{y} = 252$, $s_x \approx 3.361547263, s_y \approx 110.770032$, and $r \approx 0.9896371485$.

$b_1 = r \cdot \frac{s_y}{s_x} = (0.9896371485) \cdot \left(\frac{110.770032}{3.361547263,}\right) \approx 32.61, b_0 = \bar{y} - (b_1 \cdot \bar{x}) = 252 - (32.61 \cdot 5.6) \approx 69.38, \hat{y} = 32.61x + 69.38$

(b) For each increase of 1 hour of labor, the estimated cost of the repairs increases by \$32.61. When $x = 0$ hours of labor, the estimated cost of the repairs is \$69.38.

25. **(a)** From Exercise 13, $\hat{y} = 0.34x - 1.5$. When $x = 30, \hat{y} = 0.34x - 1.5$.
$= 0.34(30) - 1.5 = 8.7$.

(b) The prediction error is $y - \hat{y} = 9 - 8.7 = 0.3$. Since the prediction error is positive, the data point lies above the regression line. Thus, the actual y-value of 9 is greater than the predicted y-value of 8.7.

(c) Since $x = 30$ is the x-value of one of the data points, this estimate does not represent extrapolation.

27. **(a)** From Exercise 15, $\hat{y} = 4x + 31.6$. When $x = -5, \hat{y} = 4x + 31.6 = 4(-5) + 31.6 = 11.6$.

(b) The prediction error is $y - \hat{y} = 10 - 11.6 = -1.6$. Since the prediction error is negative, the data point lies below the regression line, Thus, the actual y-value of 10 is less than the predicted y-value of 11.6.

(c) Since $x = -5$ is the x-value of one of the data points, this estimate does not represent extrapolation.

29. **(a)** From Exercise 17, $\hat{y} = 0.0114x + 7.4667$. When $x = 0, \hat{y} = 0.0114x + 7.4667 = 0.0114(0) + 7.4667 = 7.4667$.

(b) There are no points in the data set that have an x-value of $x = 0$, so the prediction error $y - \hat{y}$ can't be found.

(c) Since $x = 0$ is outside the range $5 \leq x \leq 30$ of the data set, this estimate represents extrapolation.

31. **(a)** From Exercise 19, $\hat{y} = 12.5$. When $x = 0, \hat{y} = 12.5$,

(b) The prediction error is $y - \hat{y} = 5 - 12.5 = -7.5$. Since the prediction error is negative, the data point lies below the regression line. Thus, the actual y-value of 5 is less than the predicted y-value of 12.5.

(c) Since $x = 0$ is the x-value of one of the data points, this estimate does not represent extrapolation.

33. **(a)** From Exercise 21, $\hat{y} = 7.2x - 348.2$. When $x = 68, \hat{y} = 7.2x - 348.2 = 7.2(68) - 348.2 = 141.4$ pounds

(b) The prediction error is $y - \hat{y} = 138 - 141.4 = -3.4$ pounds. Since the prediction error is negative, the data point lies below the regression line. Thus, the actual weight of 138 pounds is less than the predicted weight of 141.4 pounds.

(c) Since $x = 68$ is the x-value of one of the data points, this estimate does not represent extrapolation.

35. **(a)** From Exercise 23, $\hat{y} = 32.61x + 69.38$. When $x = 400$, $\hat{y} = 32.61x + 69.38 = 32.61(400) + 69.38 = \$13{,}113.38$.

(b) There are no points in the data set that have an x-value of $x = 400$, so the prediction error $y - \hat{y}$ can't be found.

(c) Since $x = 400$ is outside the range $2 \le x \le 10$ of the data set, this estimate represents extrapolation.

37. (a) From Exercise 35 in Section 4.1, $\bar{x} = 30.8$, $\bar{y} = 0.92$ $s_x \approx 27.88727308$,

$s_y \approx 0.5718391382$, and $r \approx 0.7402630387$.

Then $b_1 = r \cdot \frac{s_y}{s_x} = (0.7402630387) \cdot \left(\frac{0.5718391382}{27.88727308}\right) \approx 0.0151793751 \approx 0.015$ and $b_0 = \bar{y} - (b_1 \cdot \bar{x}) \approx 0.92 - (0.0151793751 \cdot 30.8) \approx 0.45$. Therefore, $\hat{y} = 0.015x + 0.45$.

(b) The predicted total sales in millions of game units of a video game is 0.015 times the number of weeks on the top 30 list plus 0.45 million game units.

(c) For each increase of 1 week on the top 30 list the predicted number of game units of that game sold increases by 0.015 million units.

(d) The predicted number of game units for a game that has been on the top 30 list for $x = 0$ weeks is 0.45 million units.

39. **(a)** From Exercise 37 in Section 4.1, $\bar{x} = 3.575$, $\bar{y} = 9.45$, $s_x \approx 10.73495625$, $s_y \approx 22.86325811$, and. $r \approx 0.6639494316$.

Then $b_1 = r \cdot \frac{s_y}{s_x} = (0.6639494316) \cdot \left(\frac{22.86325811}{10.73495625}\right) \approx 1.413076301 \approx 1.41$ and $b_0 = \bar{y} - (b_1 \cdot \bar{x}) \approx 9.45 - (1.413076301 \cdot 3.575) \approx 4.39$. Therefore, $\hat{y} = 1.41x + 4.39$.

(b) The predicted gain or loss in one day by the portfolio selected by the darts is 1.41 times the gain or loss of the DJIA plus \$4.39.

(c) For each increase of \$1 in the DJIA the predicted value of the portfolio selected by the darts increases by \$1.41.

(d) The predicted loss or gain in one day by the portfolio selected by the darts for a day when the gain or loss in the DJIA is $x = \$0$ is \$4.39.

41. **(a)** From Exercise 39 in Section 4.1, $\bar{x} = 17$, $\bar{y} = 2.2$

$s_x \approx 3.771236166$, $s_y \approx 0.9189365835$, .and $r \approx 0.064123647$.

Then $b_1 = r \cdot \frac{s_y}{s_x} = (0.064123647) \cdot \left(\frac{0.9189365835}{3.771236166}\right) \approx 0.015625 \approx 0.016$ and $b_0 = \bar{y} - (b_1 \cdot \bar{x}) \approx 2.2 - (0.015625 \cdot 17) \approx 1.93$. Therefore, $\hat{y} = 0.016x + 1.93$.

(b) The predicted number of shots that a person will get is 0.02 times the person's age in years plus 1.93.

(c) For each increase of 1 year of age the predicted number of shots that a person will get increases by 0.02.

(d) The predicted number of shots that a person who is $x = 0$ years old will get is 1.93.

43. **(a)** From Exercise 41 in Section 4.1, $\bar{x} = 192.2$, $\bar{y} = 2.12$,

$s_x \approx 116.2762993$, $s_y \approx 2.418815137$, and $r \approx 0.4482198436$.

Then $b_1 = r \cdot \frac{s_y}{s_x} = (0.4482198436) \cdot \left(\frac{2.418815137}{116.2762993}\right) \approx 0.0093240063 \approx 0.0093$ and $b_0 = \bar{y} - (b_1 \cdot \bar{x}) \approx 2.12 - (0.0093240063 \cdot 192.2) \approx 0.33$. Therefore, $\hat{y} = 0.0093x + 0.33$.

(b) The predicted number of grams of saturated fat in a food item is 0.01 times the number of calories in the food item plus 0.33 gram.

(c) For each increase of 1 calorie in a food item the predicted number of grams of saturated fat increases by 0.01 gram.

(d) The predicted number of grams of saturated fat in a food item with $x = 0$ calories is 0.33 gram.

45. **(a)** From Exercise 43 in section 4.1, $\bar{x} = 32.44$, $\bar{y} = 38.12$,

$s_x \approx 5.919121556$, $s_y \approx 4.138652491$, and $r \approx 0.9195704278$.

Then $b_1 = r \cdot \frac{s_y}{s_x} = (0.9195704278) \cdot \left(\frac{4.138652491}{5.919121556}\right) \approx 0.6429640625 \approx 0.64$ and $b_0 = \bar{y} - (b_1 \cdot \bar{x}) \approx 38.12 - (0.6429640625 \cdot 32.44) \approx 17.26$. Therefore, $\hat{y} = 0.64x + 17.26$.

(b) The predicted percentage of college students in a state who have completed their college degrees is 0.46 times the percentage of people who have attended college plus 17.26%.

(c) For each increase of 1% in the percent of people in a state who have attended college the predicted percentage of college students who have completed their degree increases by 0.64%.

(d) The predicted percent of college students who will complete their degree in a state with $x = 0\%$ of its residents attending college is 17.26%.

47. **(a)** From Exercise 45 in Section 4.1, $\bar{x} = 61.7$, $\bar{y} = 25.7$,

$s_x \approx 3.665281124$, $s_y \approx 7.40077216$, and $r \approx 0.9058813765$.

Then $b_1 = r \cdot \frac{s_y}{s_x} = (0.9058813765) \cdot \left(\frac{7.40077216}{3.665281124}\right) \approx 1.82911527 \approx 1.83$ and $b_0 = \bar{y} - (b_1 \cdot \bar{x}) \approx 25.7 - (1.82911527 \cdot 61.7) \approx -87.2$. Therefore, $\hat{y} = 1.83x - 87.2$.

(b) The predicted teenage birth rate of a state is 1.83 times the overall birth rate of the state minus 87.2 live births per 1000 women aged 15–19.

(c) For each increase of 1 live birth per 1000 women the predicted teenage birth rate increases by 1.83 live births per 1000 women aged 15–19.

(d) The predicted teenage birth rate for a state with an overall birth rate of $x = 0$ live births per 1000 women is –87.2 live births per 1000 women aged 15–19.

49. **(a)** From Exercise 47 in Section 4.1, $\bar{x} \approx 75.6167$, $\bar{y} = 86.55$,
$s_x \approx 4.237888268$, $s_y \approx 6.32462719$, and $r \approx 0.4295641969$.

Then $b_1 = r \cdot \frac{s_y}{s_x} = (0.4295641969) \cdot \left(\frac{6.32462719}{4.237888258}\right) \approx 0.6410818866 \approx 0.641$ and $b_0 = \bar{y} - (b_1 \cdot \bar{x}) \approx 86.55 - (0.6410818866 \cdot 75.6167) \approx 38.1$. Therefore, $\hat{y} = 0.641x + 38.1$.

(b) The predicted consumer sentiment for incomes \$75,000 or higher is 0.641 times the consumer sentiment for incomes under \$75,000 plus 38.1.

(c) For each increase of 1 in the consumer sentiment for incomes under \$75,000 the predicted consumer sentiment for incomes \$75,000 or higher increases by 0.641.

(d) The predicted consumer sentiment for incomes \$75,000 or higher in a month when consumer incomes for under \$75,000 is $x = 0$ is 38.1.

51. **(a)** From Exercise 49 in Section 4.1, $\bar{x} = 0.3165$, $\bar{y} = 84.5$,
$s_x \approx 0.0156080464$, $s_y \approx 7.891627068$, and $r \approx 0.2331862492$.

Then $b_1 = r \cdot \frac{s_y}{s_x} = (0.2331862492) \cdot \left(\frac{7.891627068}{0.0156080464}\right) \approx 117.9019378 \approx 118$ and $b_0 = \bar{y} - (b_1 \cdot \bar{x}) \approx 84.5 - (117.9019378 \cdot 0.3165) \approx 47.2$. Therefore, $\hat{y} = 118x + 47.2$

(b) The predicted number of runs scored by a player is 118 times the player's batting average plus 47.2 runs.

(c) For each increase of 1 in a player's batting average the predicted number of runs scored increases by 118 runs.

(d) The predicted number of runs scored by a player with a batting average of $x = 0$ is 47.2.

53. **(a)** When $x = 10$ years, $\hat{y} \approx -1.24x + 26.19 = -1.24(10) + 26.19 = -12.4 + 26.19 = 13.79$.

(b) When $x = 15$ years, $\hat{y} \approx -1.24x + 26.19 = -1.24(15) + 26.19 = -18.6 + 26.19 = 7.59$.

(c) The range of x = years of education is $5 \leq x \leq 16$, so $x = 20$ years is outside of the range of the data set.

(d) The result in part **(a)** is the predicted unemployment rate for individuals with 10 years of education while 20.6 is the actual unemployment rate for individuals with 10 years of education.

(e) $y - \hat{y} = 20.6 - 13.79 = 6.81$, above the regression line. The observed unemployment rate of 20.6 is greater than the predicted unemployment rate of 13.79 for 10 years of education.

55. **(a)** From Exercise 40, $\hat{y} = -0.11x + 67.58$. When $x = 40$ years old, $\hat{y} = -0.11x + 67.58 = -0.11(40) + 67.58 = 63.18$ inches.

(b) No. A newborn baby will not be 67.58 inches tall.

(c) It is misleading to use the regression equation to predict the height of a 50-year-old person. An age of $x = 50$ years old is outside the range of the given x-values which is from 19–40 years old.

(d) The height in (a) is the predicted height of a 40–year-old while the first height in the table is the actual height of a 40-year-old.

(e) The prediction error for the prediction in (a) is $y - \hat{y} = 63.5 - 63.18 = 0.32$ inch. The actual height of the 40-year-old of $y = 63.5$ inches is greater than the predicted height of $\hat{y} = 63.18$ inches.

57. **(a)** Since the engine displacements are the x-values, all of the x-values would be decreased by the same amount.. Since all of the x-values are decreased, $\sum x$ will also decrease. Therefore, $\bar{x} = \frac{\sum x}{n}$ will decrease.

(b) No change. The y-values aren't changed.

(c) Since $\bar{y}$ and b_1 stay the same, $\bar{x}$ decreases, and b_1 is negative, $b_0 = \bar{y} - (b_1 \cdot \bar{x})$ will decrease. Subtracting a negative number is the same as adding a positive number. Therefore, $b_0 = \bar{y} - (b_1 \cdot \bar{x}) = \bar{y} + \lfloor b_1 \rfloor \bar{x}$. Thus, a smaller value of $\bar{x}$ means a smaller number is added to $\bar{y}$ to find b_0

Another way to think about it is that the slope of the regression line is negative and decreasing all of the x-values by the same amount shifts the regression line to the left. Therefore the regression line will cross the y-axis at a lower point.

(d) No change. Since r, s_x, and s_y all stay the same, $b_1 = r \cdot \frac{s_y}{s_x}$ stays the same.

(e) No change. The correlation coefficient is $r = \frac{\sum(x-\bar{x})(y-\bar{y})}{(n-1)s_x s_y}$ where $s_x = \sqrt{\frac{\sum(x-\bar{x})^2}{n-1}}$ and $s_y = \sqrt{\frac{\sum(y-\bar{y})^2}{n-1}}$. Since the y-values aren't changed, s_y and $(y - \bar{y})$ aren't changed. Since x and $\bar{x}$ are decreased by the same amount, s_x and $(x - \bar{x})$ aren't changed. Since the number of data values hasn't changed, n hasn't changed. Therefore r does not c

59. **(a)** 10.5. In a state with 0 households 10.5% of the households are headed by women. This does not make sense. Since all states have households, the value $x = 0$ would not occur. Since any percent of 0 is 0 there is no reason why the percent of 0 households that are headed by woman would have to be 10.5%.

(b) This estimate would be considered extrapolation because the value of $x = 0$ is outside the range of x-values in the data set.

(c) 0.000000282. For each increase of one household the percentage of households headed by women increases by 0.000000282.

(d) Percentage of households headed by women = 10.5 + 0.000000282 (total number of households). The estimated percentage of households headed by women equals 10.5 plus 0.000000282 times the total number of households. **(e)** Positive, because the slope is positive.

61. **(a)** When $x = 7{,}000{,}000$ households, then the estimated percent of households headed by women is 10.5 + 0.000000282 (7,000,000) = 12.474%.

(b) Since $x = 100{,}000$ is not in the range of the x-values in the data set, it is not appropriate to use the regression equation to estimate the percentage of households headed by women.

63. Curved.

65. **(a)** Darts is the y variable and DJIA is the x variable.

(b) −2.49

(c) When the Dow Jones Industrial Average changes 0%, the portfolios chosen by the darts decreases by 2.49%. Since the Dow Jones Industrial Average is based on a different set of stocks than the stock portfolio selected by the darts, this situation makes sense.

(d) The estimate in **(c)** would not be considered extrapolation because a value of $x = 0$ is inside of the range of the x-values of our data set.

67. **(a)** The percent change in the stocks selected by the darts for Contest A is 1.032 (10) = 10.32% more than the percent change in the stocks selected by the darts for Contest B.

(b) The percent change in the stocks selected by the darts for Contest C is 1.032 (5) = 5.16% less than the percent change in the stocks selected by the darts for Contest D.

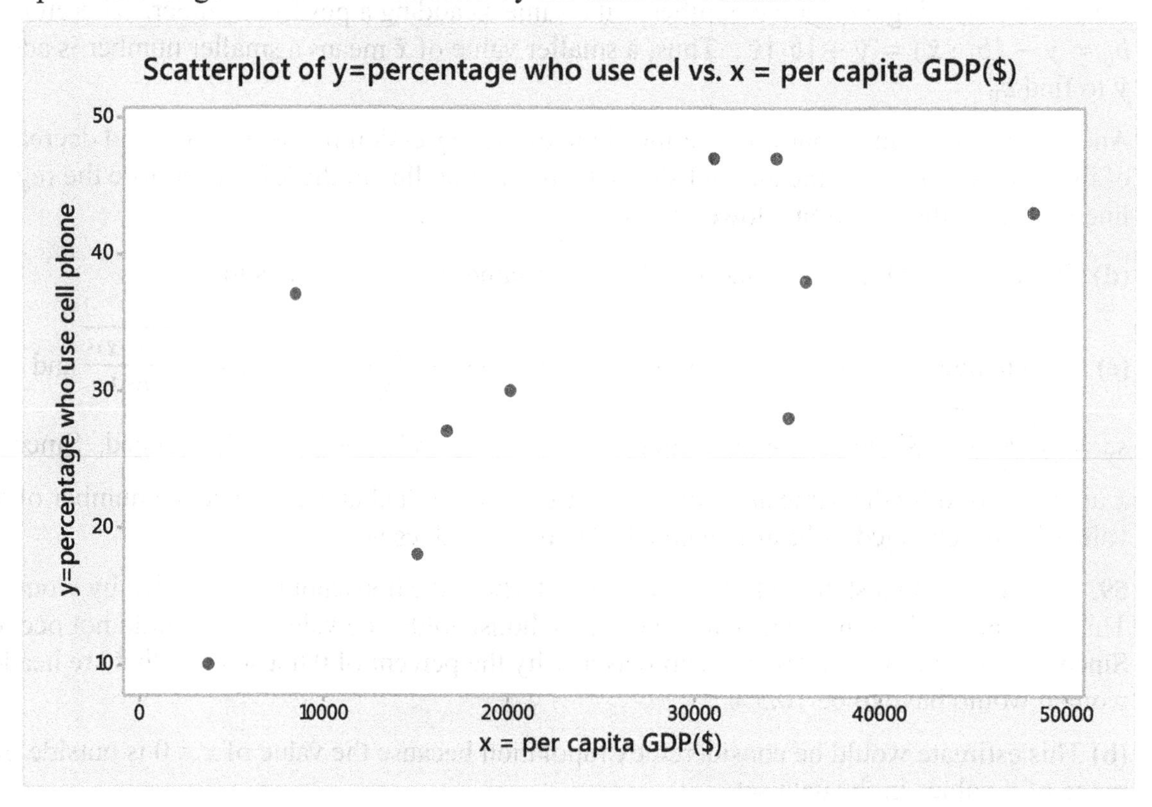

69.

Per capita GDP ($) and percentage who use cell phone to browse Internet have a positive linear relationship.

71.

x	y	xy
48,147	43	2,070,321
35,974	38	1,367,012
35,048	28	981,344
16,687	27	450,549
20,136	30	604,080
31,004	47	1,437,188
8,394	37	310,578
34,362	47	1,615,014
3,703	10	37,030
15,121	18	272,178

$\sum x = 248{,}576$ $\sum y = 325$ $\sum xy = 9{,}165{,}294$

From the TI-83/84: $s_x \approx 14{,}143.50036$ and $s_y \approx 12.26784415$. Then

73. For each increase of $1 in the country's per capita GDP the estimated percentage of cell phone users who use their cell phones to browse the Internet increases by 0.000604 percent. The estimated percentage of cell phone users who use their cell phones to browse the Internet for a country with a per capita GDP of $x =$ \$0 is 17.50%. Since $x =$ \$0 does not lie within the range of x values, this estimate represents extrapolation.

75. United States. The prediction error is $y - \hat{y} = 43 - 46.58 =$ -3.58. The actual percentage of cell phone users who use their cell phones to browse the Internet for the United States of 43% lies below the estimated percentage of cell phone users who use their cell phones to browse the Internet for the United States of 46.58%.

77. **(a)** 23,000,000; **(b)** 900

79. 3500.

81. **(a)** Negative. **(b)** −0.9, the dots lie in a pattern that is close to a straight line. **(c)** Negative, the dots lie in a pattern that is close to a straight line with a negative slope.

83. **(a)** $b_1 = \frac{\sum xy - (\sum x \sum y)/n}{\sum x^2 - (\sum x)^2/n} = \frac{609.3-((28.8)(226))/10}{90.52-28.8^2/10} = \frac{609.3-6508.8/10}{90.52-829.44/10} = \frac{609.3-650.88}{90.52-82.944} = \frac{-41.58}{7.576} \approx -5.488384372 \approx -5.49$

Yes, this agrees with the prediction from Exercise 81(c).

(b) $\bar{x} = \frac{\sum x}{n} = \frac{28.8}{10} = 2.88, \bar{y} = \frac{\sum y}{n} = \frac{226}{10} = 22.6$

$b_0 = \bar{y} - (b_1 \cdot \bar{x}) \approx 22.6 - ((-5.488384372) \cdot (2.88)) \approx 38.41.$

(c) The slope of $b_1 \approx -5.49$ means that the combined mpg will decrease by 5.49 mpg for each 1 liter increase in engine size.

The y-intercept of $b_0 \approx 38.41$ is the predicted combined miles per gallon for an engine size of 0 liters.

85. Answers will vary.

87. **(a)** Answers will vary. One possible answer:

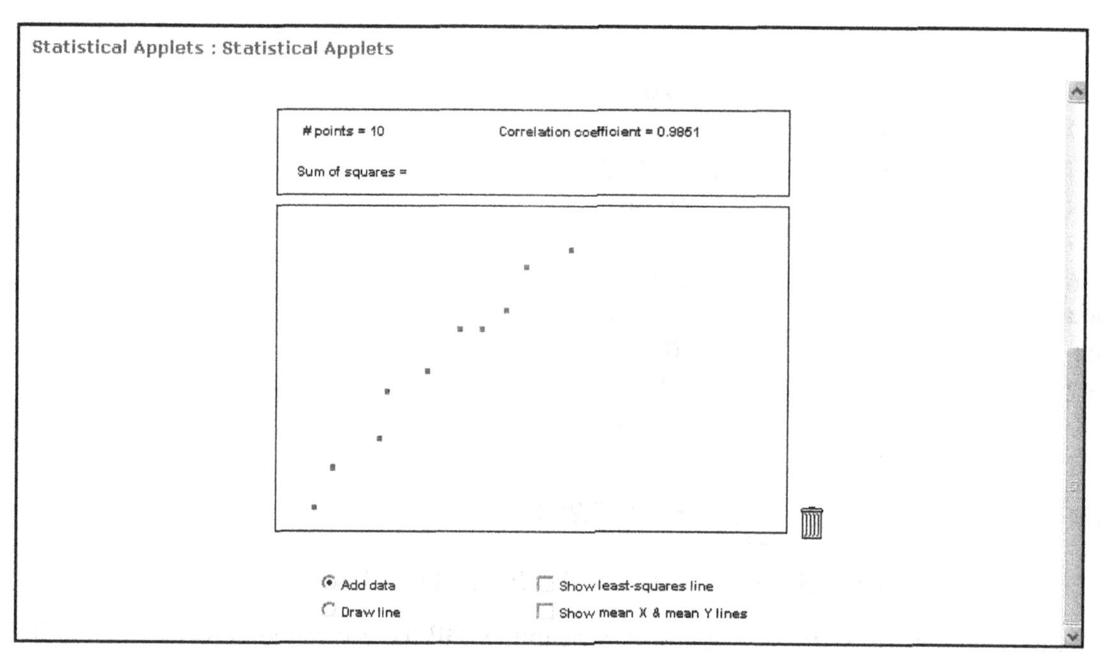

(b) Answers will vary. One possible answer:

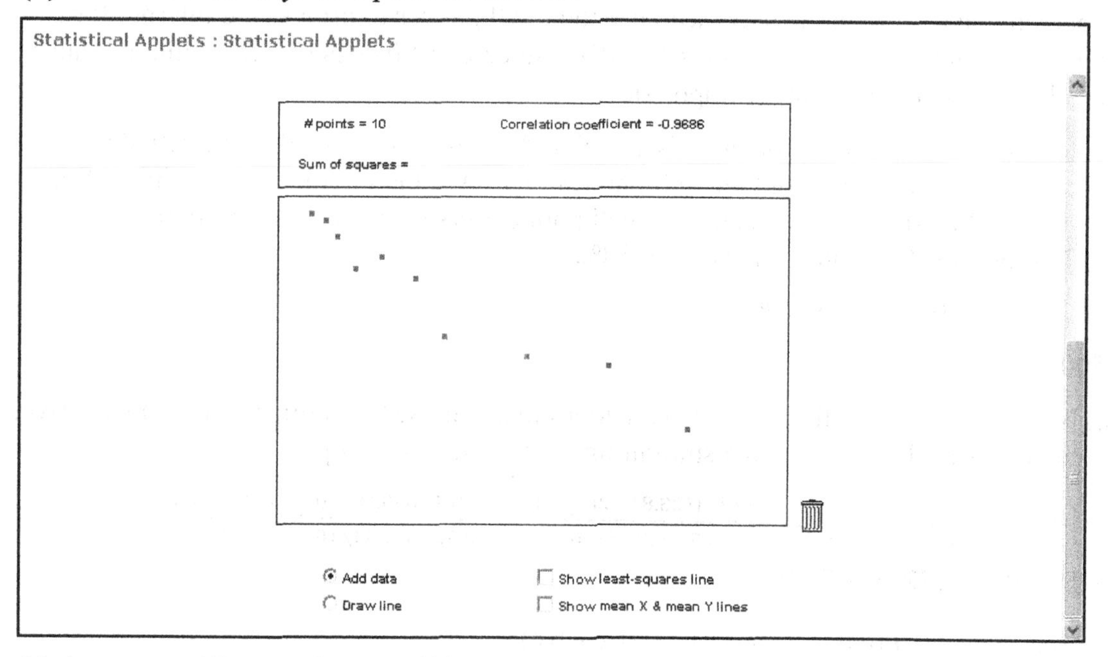

(c) Answers will vary. One possible answer:

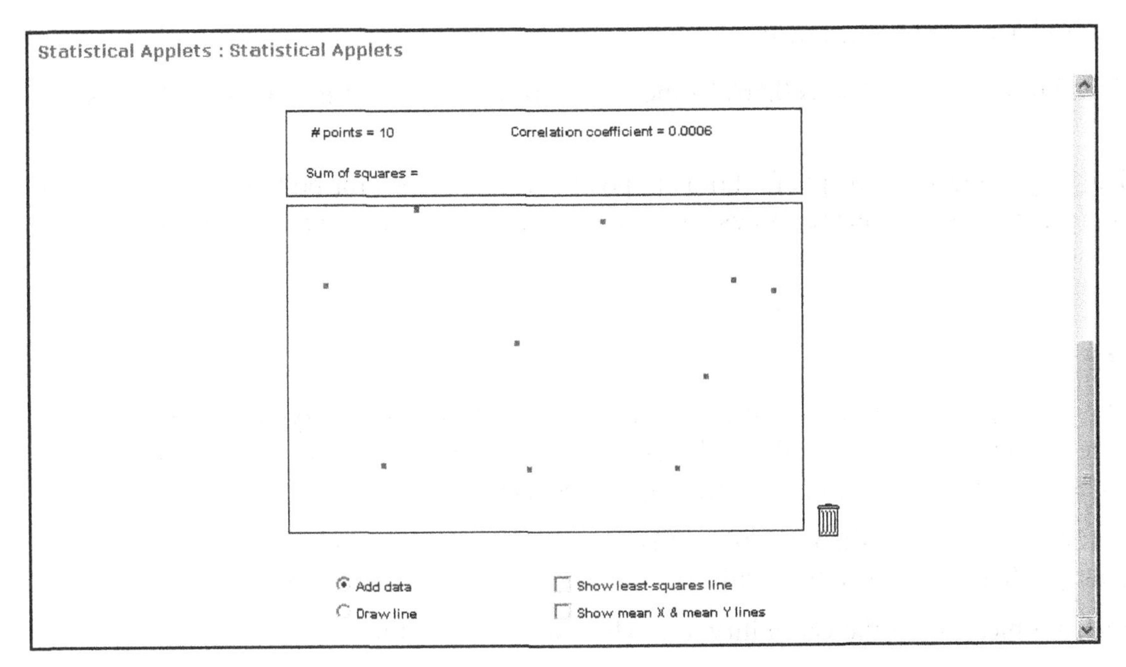

89. Minitab output

```
Regression Equation

Weight = -96.2 + 3.541 Height
```

Weight = −96.2 + 3.541 Height

$\hat{y} = 3.541x - 96.2$

91. The predicted weight for a woman who is $x = 63.5$ inches tall is

$\hat{y} = 3.541x - 96.2 = 3.541(63.5) - 96.2 =$128.6535 pounds.

93. (a) The predicted weight for a woman who is $x = 65$ inches tall is
$\hat{y} = 3.541x - 96.2 = 3.541(65) - 96.2 =$133.965 pounds

(b) The predicted weight for a woman who is $x = 70$ inches tall is
$\hat{y} = 3.541x - 96.2 = 3.541(70) - 96.2 = 151.67$ pounds

95. For each increase of 1 inch in height the estimated weight of a man increases by 4.386 pounds. The estimated weight of a man who is $x = 0$ inches tall is -134.7 pounds.

97. The prediction error is $y - \hat{y} = 144.6 - 165.741 = -21.141$ pounds

99. Minitab output

```
Regression Equation

Hip girth = 43.80 + 0.7428 Waist girth
```

Hip girth = 43.80 + 0.7428 waist girth

The predicted hip girth of a woman is 0.7428 times her waist girth (in centimeters) plus 43.80 centimeters.

101. Minitab output

```
Regression Equation

Hip girth = 49.82 + 0.5671 Waist girth
```

Hip girth = 49.82 + 0.5671 waist girth

The predicted hip girth of a man is 0.5671 times his waist girth (in centimeters) plus 49.82 centimeters.

103. The slopes and y-intercepts for both are positive. The slopes for both are less than 1 and the y-intercepts for both are in the forties. The slope is larger for women and the y-intercept is larger for men.

Section 4.3

1. The standard error of the estimate s is a measure of the size of the typical difference between the predicted value of y and the observed value of y. We would want s to be small. If the prediction error is large, the regression line may not be useful.

3. SSE measures the prediction errors. It is the sum of the squared prediction errors. Since we want our prediction errors to be small, we want SSE to be as small as possible.

5. SST is a measure of the variability in y. The variance s^2 of the y's.

7. No.

9. 64% of the variability in the variable y is accounted for by the linear relationship between x and y.

11. **(a)** From Exercise 13 in Section 4.2, $\hat{y} = 0.34x - 1.5$.

x	y	$\hat{y} = 0.34x - 1.5$	Residual $(y - \hat{y})$	$(\text{Residual})^2$ $(y - \hat{y})^2$
10	2	1.9	0.1	0.01
20	5	5.3	−0.3	0.09
30	9	8.7	0.3	0.09
40	12	12.1	−0.1	0.01

$$\sum (y - \hat{y})^2 = 0.2$$

So SSE $= \sum(y - \hat{y})^2 = 0.2$.

(b) $s = \sqrt{\frac{SSE}{n-2}} = \sqrt{\frac{0.2}{4-2}} \approx 0.316228$. The typical difference between the predicted value of y and the actual observed value of y is 0.316228.

(c) $\bar{y} = \frac{\sum y}{n} = \frac{2+5+9+12}{4} = \frac{28}{4} = 7$

y	$(y - \bar{y})$	$(y - \bar{y})^2$
2	−5	25
5	−2	4
9	2	4
12	5	25

$$\sum (y - \bar{y})^2 = 58$$

So SST$= \sum(y - \bar{y})^2 = 58$.

(d) SSR = SST - SSE = 58 – 0.2 = 57.8

(e) $r^2 = \frac{\text{SSR}}{\text{SST}} = \frac{57.8}{58} \approx 0.9965517241 \approx 0.9966$. Therefore, 99.66% of the variability in y is accounted for by the linear relationship between y and x.

(f) The correlation coefficient r is positive since $b_1 = 0.34$ is positive. Therefore, $r = \sqrt{r^2} \approx \sqrt{0.9965517241} \approx 0.9983$.

13. **(a)** From Exercise 15 in Section 4.2, $\hat{y} = 4x + 31.6$.

x	y	$\hat{y} = 4x + 31.6$	**Residual** $(y - \hat{y})$	**(Residual)^2** $(y - \hat{y})^2$
−5	10	11.6	−1.6	2.56
−4	18	15.6	2.4	5.76
−3	18	19.6	−1.6	2.56
−2	26	23.6	2.4	5.76
−1	26	27.6	−1.6	2.56

$$\sum (y - \hat{y})^2 = 19.20$$

So SSE = $\Sigma(y - \hat{y})^2 = 19.20$.

(b) $s = \sqrt{\frac{SSE}{n-2}} = \sqrt{\frac{19.20}{5-2}} \approx 2.52982$. The typical difference between the predicted value of y and the actual observed value of y is 2.52982.

(c) $\bar{y} = \frac{\Sigma y}{n} = \frac{10+18+18+26+26}{5} = \frac{98}{5} = 19.6$

y	$(y - \bar{y})$	$(y - \bar{y})^2$
10	−9.6	92.16
18	−1.6	2.56
18	−1.6	2.56
26	6.4	40.96
26	6.4	40.96

$$\sum (y - \bar{y})^2 = 179.2$$

So SST= $\Sigma(y - \bar{y})^2$ =179.20

(d) SSR = SST– SSE = 179.20 – 19.20 = 160

(e) $r^2 = \frac{\text{SSR}}{\text{SST}} = \frac{160}{179.20} \approx 0.8928571429 \approx 0.8929$. Therefore, 89.29% of the variability in y is accounted for by the linear relationship between y and x.

(f) The correlation coefficient r is positive since $b_1 = 4$ is positive. Therefore, $r = \sqrt{r^2} \approx \sqrt{0.8928571429} \approx 0.9449$.

15. **(a)** From Exercise 17, $\hat{y} = 0.0114x + 7.4667$.

x	y	$\hat{y} = 0.0114x + 7.4667$	Residual $(y - \hat{y})$	$(\text{Residual})^2$ $(y - \hat{y})^2$
5	7	7.5237	−0.5237	0.27426169
10	8	7.5807	0.4193	0.17581249
15	8	7.6377	0.3623	0.13126129
20	8	7.6947	0.3053	0.09320809
25	7	7.7517	−0.7517	0.56505289
30	8	7.8087	0.1913	0.03659569
				$\sum(y - \hat{y})^2 = 1.27619214$

So $\text{SSE} = \sum(y - \hat{y})^2 = 1.27619214 \approx 1.27619$

(b) $s = \sqrt{\frac{SSE}{n-2}} \approx \sqrt{\frac{1.27619214}{6-2}} \approx 0.564843$. The typical difference between the predicted value of y and the actual observed value of y is 0.564843.

(c) $\bar{y} = \frac{\sum y}{n} = \frac{7+8+8+8+7+8}{6} = \frac{46}{6} \approx 7.666666667 \approx 7.6667$

y	$(y - \bar{y})$	$(y - \bar{y})^2$
7	− 0.6667	0.44448889
8	0.3333	0.11108889
8	0.3333	0.11108889
8	0.3333	0.11108889
7	−0.6667	0.44448889
8	0.3333	0.11108889
		$\sum(y - \bar{y})^2 = 1.33333334$

So $\text{SST} = \sum(y - \bar{y})^2 \approx 1.33333334 \approx 1.33333$.

(d) $\text{SSR} = \text{SST} - \text{SSE} \approx 1.33333334 - 1.27619214 \approx 0.05714$

(e) $r^2 = \frac{\text{SSR}}{\text{SST}} \approx \frac{0.05714}{1.33333} \approx 0.0429$. Therefore, 4.29% of the variability in y is accounted for by the linear relationship between y and x.

(f) The correlation coefficient r is positive since $b_1 \approx 0.0114$ is positive. Therefore, $r = \sqrt{r^2} \approx \sqrt{0.0429} \approx 0.2071$.

17. **(a)** From Exercise 19 in Section 4.2, $\hat{y} = 12.5$.

x	y	$\hat{y} = 12.5$	Residual $(y - \hat{y})$	$(\text{Residual})^2$ $(y - \hat{y})^2$
−70	5	12.5	−7.5	56.25
−60	10	12.5	−2.5	6.25
−50	15	12.5	2.5	6.25
−40	20	12.5	7.5	56.25
−30	20	12.5	7.5	56.25
−20	15	12.5	2.5	6.25
−10	10	12.5	−2.5	6.25
0	5	12.5	−7.5	56.25
				$\sum(y - \hat{y})^2 = 250$

So $\text{SSE} = \sum(y - \hat{y})^2 = 250$

(b) $s = \sqrt{\frac{SSE}{n-2}} = \sqrt{\frac{230}{8-2}} \approx 6.45497$. The typical difference between the predicted value of y and the actual observed value of y is 6.45497.

(c) $\bar{y} = \frac{\sum y}{n} = \frac{5+10+15+20+20+15+10+5}{8} = \frac{100}{8} = 12.5$

y	$(y-\bar{y})$	$(y-\bar{y})^2$
5	−7.5	56.25
10	−2.5	6.25
15	2.5	6.25
20	7.5	56.25
20	7.5	56.25
15	2.5	6.25
10	−2.5	6.25
5	−7.5	56.25
		$\sum(y-\bar{y})^2 = 250$

So SST= $\sum(y-\bar{y})^2 = 250$

(d) SSR = SST−SSE = 250 − 250 = 0

(e) $r^2 = \frac{\text{SSR}}{\text{SST}} = \frac{0}{250} = 0$. Therefore, 0% of the variability in y is accounted for by the linear relationship between y and x.

(f) $r = \sqrt{r^2} = \sqrt{0} = 0$

19. **(a)** From Exercise 21 in Section 4.3, $\hat{y} = 7.2x - 348.2$.

x	y	$\hat{y} = 7.2x - 348.2$	**Residual** $(y-\hat{y})$	**(Residual)^2** $(y-\hat{y})^2$
66	122	127	−5	25
67	133	134.2	−1.2	1.44
69	153	148.6	4.4	19.36
68	138	141.4	−3.4	11.56
65	125	119.8	5.2	27.04
				$\sum(y-\hat{y})^2 = 84.4$

So SSE = $\sum(y-\hat{y})^2 = 84.40$

(b) $s = \sqrt{\frac{SSE}{n-2}} = \sqrt{\frac{84.40}{5-2}} \approx 5.30409$ pounds. The typical difference between the predicted value of y = weight and the actual observed value of y = weight is 5.30409 pounds.

(c) $\bar{y} = \frac{\sum y}{n} = \frac{122+133+153+138+125}{5} = \frac{671}{5} = 134.2$

y	$(y-\bar{y})$	$(y-\bar{y})^2$
122	−12.2	148.84
133	−1.2	1.44
153	18.8	353.44
138	3.8	14.44
125	−9.2	84.64

$$\Sigma(y-\bar{y})^2 = 602.8$$

So SST$= \Sigma(y-\bar{y})^2 = 602.80$.

(d) SSR = SST−SSE = 602.80 – 84.40 = 518.40

(e) $r^2 = \frac{\text{SSR}}{\text{SST}} = \frac{518.40}{602.80} \approx 0.8599867286 \approx 0.8600$. Therefore, 86.00% of the variability in y = weight is accounted for by the linear relationship between y = weight and x = height.

(f) The correlation coefficient r is positive since $b_1 = 7.2$ is positive. Therefore, $r = \sqrt{r^2} \approx \sqrt{0.8599867286} \approx 0.9274$.

21. **(a)** From Exercise 23 in Section 4.2, $\hat{y} = 32.61x + 69.3$

x	y	$\hat{y} = 32.61x + 69.38$	Residual $(y-\hat{y})$	(Residual)^2 $(y-\hat{y})^2$
2	120	134.6	−14.6	213.16
3	180	167.21	12.79	163.5841
5	230	232.43	−2.43	5.9049
8	350	330.26	19.74	389.6676
10	380	395.48	−15.48	239.6304

$$\Sigma(y-\hat{y})^2 = 1011.947 \approx 1012$$

So SSE $= \Sigma(y-\hat{y})^2 \approx 1012$.

(b) $s = \sqrt{\frac{SSE}{n-2}} \approx \sqrt{\frac{1011.947}{5-2}} \approx 18.3662$. The typical difference between the predicted value of y = cost and the actual observed value of y = cost is \$18.3662.

(c) $\bar{y} = \frac{\Sigma y}{n} = \frac{120+180+230+350+390}{5} = \frac{1260}{5} = 252$

y	$(y-\bar{y})$	$(y-\bar{y})^2$
120	-132	17,424
180	-72	5,184
230	-22	484
350	98	9,604
380	128	16,384

$$\Sigma(y-\bar{7})^2 = 49{,}080$$

So SST$= \Sigma(y-\bar{y})^2 = 49{,}080$.

(d) SSR = SST−SSE = 49,080 – 1012 = 48,068

(e) $r^2 = \frac{\text{SSR}}{\text{SST}} \approx \frac{48.069}{49{,}080} \approx 0.9793806031 \approx 0.9794$. Therefore, 97.94% of the variability in y = cost is accounted for by the linear relationship between y = cost and x = hours.

(f) The correlation coefficient r is positive since $b_1 \approx 32.61$ is positive. Therefore, $r = \sqrt{r^2} \approx \sqrt{0.9793806031} \approx 0.9896$.

23. Minitab output

Regression Analysis: Total sales versus Weeks

```
Analysis of Variance
Source          DF   Adj SS    Adj MS  F-Value  P-Value
Regression       1  0.71677  0.71677      3.64    0.153
  Weeks          1  0.71677  0.71677      3.64    0.153
Error            3  0.59123  0.19708
  Lack-of-Fit    2  0.09123  0.04561      0.09    0.920
  Pure Error     1  0.50000  0.50000
Total            4  1.30800
Model Summary
       S    R-sq  R-sq(adj)  R-sq(pred)
0.443933  54.80%     39.73%       0.00%
Coefficients
Term          Coef  SE Coef  T-Value  P-Value   VIF
Constant     0.452    0.315     1.43    0.247
Weeks      0.01518  0.00796     1.91    0.153  1.00
Regression Equation
Total sales = 0.452 + 0.01518 Weeks
```

(a) From Exercise 37 in Section 4.2, $\hat{y} = 0.015x + 0.45$.

x	y	$\hat{y} = 0.015x + 0.45$	**Residual** $(y - \hat{y})$	**(Residual)^2** $(y - \hat{y})^2$
78	1.7	1.62	0.08	0.0064
27	0.6	0.855	−0.255	0.065025
29	0.9	0.885	0.015	0.000225
10	1.2	0.6	0.6	0.36
10	0.2	0.6	−0.4	0.16
				$\sum(y - \hat{y})^2 \approx 0.59165$

So $SSE = \sum(y - \hat{y})^2 \approx 0.59165$.

From Minitab: SSE = 0.59123

(b) $s = \sqrt{\frac{SSE}{n-2}} \approx \sqrt{\frac{0.59165}{5-2}} \approx 0.444091$

From Minitab: $s = 0.443933$. The typical difference between the predicted number of total sales in millions of units and the actual observed number of total sales in millions of units is 0.443933 million units.

(c) $\bar{y} = \frac{\sum y}{n} = \frac{1.7+0.6++0.9+1.2+0.2}{5} = \frac{4.6}{5} = 0.92$

y	$(y-\bar{y})$	$(y-\bar{y})^2$
1.7	0.78	0.6084
0.6	−0.32	0.1024
0.9	−0.02	0.0004
1.2	0.28	0.0784
0.2	−0.72	0.5184
		$\sum(y-\bar{y})^2 = 1.308$

$\text{SST} = \sum(y-\bar{y})^2 = 1.308$

(d) SSR = SST − SSE = 1.308− 0.59165 = 0.71635

From Minitab: SSR = 0.71677

(e) $r^2 = \frac{\text{SSR}}{\text{SST}} \approx \frac{071635}{1.308} \approx 0.5476681957 \approx 0.5477$

From Minitab: $r^2 \approx 0.5480$. Therefore, 54.80% of the variability in total sales in millions of units is accounted for by the linear relationship between total sales in millions of units and weeks on top 30 list.

(f) Since $b_1 \approx 0.015$ is positive, r is positive.

$r = \sqrt{r^2} \approx \sqrt{0.5476681957} \approx 0.7400$

Correlation: Weeks, Total sales

```
Pearson correlation of Weeks and Total sales = 0.740
```

From Minitab: $r \approx 0.7400$. This value of r is close to the maximum value $r = 1$. We would therefore say that the total sales in millions of units and weeks in the top 30 are positively correlated. As the number of weeks in the top 30 increases, the total sales in millions of units also tends to increase.

25. **(a)** From Exercise 39 in Section 4.2, $\hat{y} = 1.41x + 4.39$.

x	y	$\hat{y} = 1.41x + 4.39$	**Residual** $(y-\hat{y})$	**(Residual)^2** $(y-\hat{y})^2$
−12.8	−27.4	−13.658	−13.742	188.842564
9.3	18.7	17.503	1.197	1.432809
8.0	42.2	15.67	26.53	703.8409
−8.5	−16.3	−7.595	−8.705	75.777025
15.8	11.2	26.668	−15.468	239.259024
10.6	28.5	19.336	9.164	83.978896
11.5	1.8	20.605	−18.805	353.628025
−5.3	16.9	−3.083	19.983	399.320289
				$\sum(y-\hat{y})^2 \approx 2046.079532 \approx 2046$

So $\text{SSE} = \sum(y-\hat{y})^2 \approx 2046$

(b) $s = \sqrt{\frac{SSE}{n-2}} \approx \sqrt{\frac{2046.07953}{8-2}} \approx 18.4665$. The typical difference between the predicted value of the change in the stocks in the portfolio of stocks selected by the darts and the actual observed value of the change in the stocks in the portfolio of stocks selected by the darts is 18.4665.

(c) $\bar{y} = \frac{\sum y}{n} = \frac{(-27.4)+18.7+42.2+(-16.3)+11.2+28.5+1.8+16.9}{8} = \frac{75.6}{8} = 9.45$

y	$(y - \bar{y})$	$(y - \bar{y})^2$
−27.4	−36.85	1357.9225
18.7	9.25	85.5625
42.2	32.75	1072.5625
−16.3	−25.75	663.0625
11.2	1.75	3.0625
28.5	19.05	362.9025
1.8	−7.65	58.5225
16.9	7.45	55.5025
		$\sum(y - \bar{y})^2 = 3659.1$

$\text{SST} = \sum(y - \bar{y})^2 = 3659.1 \approx 3659$

(d) $\text{SSR} = \text{SST} - \text{SSE} = 3659 - 2046 = 1613$

(e) $r^2 = \frac{\text{SSR}}{\text{SST}} \approx \frac{1613}{3659} \approx 0.4408308281 \approx 0.4408$

Therefore, 44.08% of the variability in the change in the stocks in the portfolio of stocks selected by the darts is accounted for by the linear relationship between the change in the stocks in the portfolio of stocks selected by the darts and the change in the stocks in the DJIA.

(f) Since $b_1 \approx 1.41$ is positive, r is positive.

$r = \sqrt{r^2} \approx \sqrt{0.4408} \approx 0.6639$. This value of r is positive. We would therefore say that the change in the stocks in the portfolio of stocks selected by the darts and the change in the stocks in the DJIA are positively correlated. As the change in the stocks in the DJIA increases, the change in the stocks in the portfolio of stocks selected by the darts also tends to increase.

27. **Minitab output:**

Regression Analysis: Shots versus Age

Analysis of Variance

Source	DF	Adj SS	Adj MS	F-Value	P-Value
Regression	1	0.03125	0.03125	0.03	0.860
Age	1	0.03125	0.03125	0.03	0.860
Error	8	7.56875	0.94609		
Lack-of-Fit	6	5.06875	0.84479	0.68	0.700
Pure Error	2	2.50000	1.25000		
Total	9	7.60000			

Model Summary

S	R-sq	R-sq(adj)	R-sq(pred)
0.972674	0.41%	0.00%	0.00%

Coefficients

Term	Coef	SE Coef	T-Value	P-Value	VIF
Constant	1.93	1.49	1.30	0.231	
Age	0.0156	0.0860	0.18	0.860	1.00

Regression Equation

```
Shots = 1.93 + 0.0156 Age
```

(a) From Exercise 41 in Section 4.2, $\hat{y} = 0.016x + 1.93$

x	y	$\hat{y} = 0.016 + 1.93$	Residual $(y - \hat{y})$	$(\text{Residual})^2$ $(y - \hat{y})^2$
13	3	2.138	0.862	0.743044
21	3	2.266	0.734	0.538756
16	3	2.186	0.814	0.662596
17	2	2.202	−0.202	0.040804
17	3	2.202	0.798	0.636804
18	1	2.218	−1.218	1.483524
25	2	2.33	−0.33	0.1089
15	3	2.17	0.83	0.6889
12	1	2.122	−1.122	1.258884
16	1	2.186	−1.186	1.406596
				$\Sigma(y - \hat{y})^2 \approx 7.568808$

So SSE $= \Sigma(y - \hat{y})^2 \approx 7.568808$

Minitab: SSE $\approx$ 7.56875

(b) $s = \sqrt{\frac{\text{SSE}}{n-2}} \approx \sqrt{\frac{7.568808}{10-2}} \approx 0.972677$ shot

Minitab: 0.972674 shot. The typical difference between the predicted number of shots and the actual observed number of shots is 0.972674.

(c) $\bar{y} = \frac{\Sigma y}{n} = \frac{63.5+63+64.4+63+63.8+68+61.8+69}{10} = \frac{22}{10} = 2.2$

y	$(y - \bar{y})$	$(y - \bar{y})^2$
3	0.8	0.64
3	0.8	0.64
3	0.8	0.64
2	− 0.2	0.04
3	0.8	0.64
1	−1.2	1.44
2	−0.2	0.04
3	0.8	0.64
1	−1.2	1.44
1	−1.2	1.44
		$\Sigma(y - \bar{y})^2 = 7.6$

SST $= \Sigma(y - \bar{y})^2 = 7.6$

(d) SSR = SST – SSE = 7.6 - 7.56875 = 0.03125

(e) $r^2 = \frac{\text{SSR}}{\text{SST}} \approx \frac{0.03125}{7.6} \approx 0.0041$. Therefore, 0.41% of the variability in the number of shots is accounted for by the linear relationship between the number of shots and age.

(f) Since $b_1 \approx 0.016$ is positive, r is positive.

$r = \sqrt{r^2} \approx \sqrt{0.0041} \approx 0.0640$. This value of r is close to $r = 0$. We would therefore say that there is no linear relationship between the number of shots a child gets and the child's age.

29. **(a)** From Exercise 43 in Section 4.2, $\hat{y} = 0.0093x + 0.33$.

x	y	$\hat{y} = 0.0093x + 0.33$	**Residual** $(y - \hat{y})$	**(Residual)^2** $(y - \hat{y})^2$
216	7	2.3388	4.6612	21.72678544
364	5.6	3.7152	1.8848	3.55247104
149	1.9	1.7157	0.1843	0.03396649
75	0.6	1.0275	−0.4275	0.18275625
276	3.2	2.8968	0.3032	0.09193024
375	0.3	3.8175	−3.5175	12.37280625
146	1.6	1.6878	−0.0878	0.00770884
59	0.5	0.8787	−0.3787	0.14341369
67	0.2	0.9531	−0.7531	0.56715961
195	0.3	2.1435	−1.8435	3.39849225
				$\Sigma(y - \hat{y})^2 \approx 42.0774901$

So $\text{SSE} = \Sigma(y - \hat{y})^2 \approx 42.0774901 \approx 42.08$

(b) $s = \sqrt{\frac{\text{SSE}}{n-2}} \approx \sqrt{\frac{42.0774901}{10-2}} \approx 2.29340$ grams.

The typical difference between the predicted amount of saturated fat and the actual observed amount of saturated fat is 2.29340 grams.

(c) $\bar{y} = \frac{\Sigma y}{n} = \frac{7+5.6+1.9+0.6+3.2+0.3+1.6+0.5+0.2+03}{10} = \frac{21.2}{10} = 2.12$

y	$(y - \bar{y})$	$(y - \bar{y})^2$
7	4.88	23.8144
5.6	3.48	12.1104
1.9	−0.22	0.484
0.6	−1.52	2.3104
3.2	1.08	1.1664
0.3	−1.82	3.3124
1.6	−0.52	0.2704
0.5	−1.62	2.6244
0.2	−1.92	3.6864
0.3	−1.82	3.3124
		$\Sigma(y - \bar{y})^2 = 52.656$

$\text{SST} = \Sigma(y - \bar{y})^2 = 52.656 \approx 52.66$

(d) $\text{SSR} = \text{SST} - \text{SSE} = 52.66 - 42.08 = 10.58$

(e) $r^2 = \frac{\text{SSR}}{\text{SST}} \approx \frac{10.58}{52.66} \approx 0.2009$. Therefore, 20.09% of the variability in the amount of saturated fat is accounted for by the linear relationship between the amount of saturated fat and the calories per serving.

(f) Since $b_1 \approx 0.0093$ is positive, r is positive.

$r = \sqrt{r^2} \approx \sqrt{0.6936} \approx 0.4482$. This value of r is positive. We would therefore say that the number of calories per serving and the number of grams of saturated fat per serving are positively correlated. As the number of calories in a serving of food increases, the number of grams of saturated fat also tends to increase.

31. Minitab output:

Regression Analysis: Completes college versus Attends college

```
Analysis of Variance

Source              DF   Adj SS    Adj MS   F-Value  P-Value
Regression           1  130.356   130.356     43.82    0.000
  Attends college    1  130.356   130.356     43.82    0.000
Error                8   23.800     2.975
  Lack-of-Fit        7   22.820     3.260      3.33    0.399
  Pure Error         1    0.980     0.980
Total                9  154.156

Model Summary

      S    R-sq  R-sq(adj)  R-sq(pred)
1.72483  84.56%     82.63%      78.26%

Coefficients

Term                Coef  SE Coef  T-Value  P-Value   VIF
Constant           17.26     3.20     5.40    0.001
Attends college   0.6430   0.0971     6.62    0.000  1.00

Regression Equation

Completes college = 17.26 + 0.6430 Attends college
```

(a) From Exercise 45 in Section 4.2, $\hat{y} = 0.64x + 17.26$.

x	y	$\hat{y} = 0.64x + 17.26$	**Residual** $(y - \hat{y})$	**(Residual)^2** $(y - \hat{y})^2$
30.9	38.8	37.036	1.764	3.111696
26.6	35.5	34.284	1.216	1.478656
30.1	34.1	36.524	−2.424	5.875776
35.7	39.1	40.108	−1.008	1.016064
45.2	45.9	46.188	−0.288	0.082944
38.7	42.8	42.028	0.772	0.595984
30.1	35.5	36.524	−1.024	1.048576
28.4	37.1	35.436	1.664	2.768896
32.5	40.2	38.06	2.14	4.5796
26.2	32.2	34.028	−1.828	3.341584
				$\Sigma(y - \hat{y})^2 \approx 23.899776$

So $\text{SSE} = \Sigma(y - \hat{y})^2 \approx 23.899776 \approx 23.900$

Minitab: $\text{SSE} \approx 23.800$

(b) $s = \sqrt{\frac{\text{SSE}}{n-2}} \approx \sqrt{\frac{23.899776}{10-2}} \approx 1.72483\%$. The typical difference between the predicted percent of a state's population that completes college and the actual observed percent of the state's population that completes college is 1.74283%.

(c) $\bar{y} = \frac{\Sigma y}{n} = \frac{38.8+35.5+34.1++39.1+45.9+42.8+35.5+37.1+40.2+32.2}{10} = \frac{381.2}{10} = 38.12$

y	$(y-\bar{y})$	$(y-\bar{y})^2$
38.8	0.68	0.4624
35.5	−2.62	6.8644
34.1	−4.02	16.1604
39.1	0.98	0.9604
45.9	7.78	60.5284
42.8	4.68	21.9024
35.5	−2.62	6.8644
37.1	−1.02	1.0404
40.2	2.08	4.3264
32.2	−5.92	35.0464
		$\Sigma(y-\bar{y})^2 = 154.156$

SST $= \Sigma(y-\bar{y})^2 = 154.156$

(d) SSR = SST − SSE = 154.156 − 23.800 = 130.356

(e) $r^2 = \frac{\text{SSR}}{\text{SST}} \approx \frac{130.356}{154.156} \approx 0.8456$. Therefore, 84.56% of the variability in the percent of a state's population that completes college is accounted for by the linear relationship between the percent of a state's population that completes college and the percent of a state's population that attends college.

(f) Since $b_1 \approx 0.64$ is positive, r is positive.

$r = \sqrt{r^2} \approx \sqrt{0.8456} \approx 0.9196$. This value of r is very close to the maximum value $r = 1$. We would therefore say that the percent of a state's population that completes college and the percent of a state's population that attends college are positively correlated. As the percent of a state's population that attends college increases, the percent of a state's population that completes college tends to increase.

y	$(y-\bar{y})$	$(y-\bar{y})^2$
0.9	−0.02	0.0004
0.8	−0.12	0.0144
0.9	−0.02	0.0004
0.5	−0.42	0.1764
0.3	−0.62	0.3844
0.4	−0.52	0.2704
0.4	−0.52	0.2704
2.1	1.18	1.3924
0.4	− 0.52	0.2704
2.5	1.58	2.4964
		$\Sigma(y-\bar{y})^2 = 5.276$

SST $= \Sigma(y-\bar{y})^2 = 5.27600$

(d) SSR = SST − SSE = 5.27600 − 4.97525 = 0.30075

(e) $r^2 = \frac{\text{SSR}}{\text{SST}} \approx \frac{0.30075}{5.27600} \approx 0.0570$. Therefore, 5.70% of the variability in the percent of a city's population that bikes to work is accounted for by the linear relationship between the percent of a city's population that bikes to work and the percent of a city's population that walks to work.

(f) Since $b_1 \approx 0.093$ is positive, r is positive.

$r = \sqrt{r^2} \approx \sqrt{0.0570} \approx 0.2387$. This value of r is close to $r = 0$. We would therefore say that the percent of a city's population that bikes to work and the percent of a city's population that walks to work have no apparent linear relationship.

33. Minitab output:

Regression Analysis: Teen birth rate versus Overall birth rate

```
Analysis of Variance

Source                DF   Adj SS   Adj MS  F-Value  P-Value

Regression             1    222.2    222.2     0.02    0.894

  Overall birth rate   1    222.2    222.2     0.02    0.894

Error                  6  68727.8  11454.6

Total                  7  68950.1

Model Summary

      S    R-sq  R-sq(adj)  R-sq(pred)

107.026   0.32%      0.00%       0.00%

Coefficients

Term                 Coef  SE Coef  T-Value  P-Value   VIF

Constant              -35      682    -0.05    0.961

Overall birth rate    1.5     11.0     0.14    0.894  1.00

Regression Equation

Teen birth rate = -35 + 1.5 Overall birth rate
```

(a) From Exercise 47 in Section 4.2, $\hat{y} = 1.83x - 87.2$

x	y	$\hat{y} = 1.83x - 87.2$	Residual $(y - \hat{y})$	(Residual)^2 $(y - \hat{y})^2$
62	23.6	26.26	−2.66	7.0756
59.3	24.6	21.319	3.281	10.764961
61.6	30.5	25.528	4.972	24.720784
58.8	17.7	20.404	−2.704	7.311616
62.7	27.2	27.541	−0.341	0.116281
58.4	20.9	19.672	1.228	1.507984
69.9	41.0	40.717	0.283	0.080089
60.9	20.1	24.247	4.147	17.197609
				$\Sigma(y - \hat{y})^2 \approx 68.774924$

So $\text{SSE} = \Sigma(y - \hat{y})^2 \approx 68.774924 \approx 68.77$

(b) $s = \sqrt{\frac{\text{SSE}}{n-2}} \approx \sqrt{\frac{68.774924}{8-2}} \approx 3.38563$

Minitab: $s \approx 3.38560$. The typical difference between the predicted teen birth rate and the actual observed teen birth rate is 3.38560.

(c) $\bar{y} = \frac{\Sigma y}{n} = \frac{23.6+24.6+30.5+17.7+27.2+20.9+41+20.1}{8} = \frac{205.6}{8} = 25.7$

y	$(y - \bar{y})$	$(y - \bar{y})^2$

23.6	−2.1	4.41
24.6	−1.1	1.21
30.5	4.8	23.04
17.7	−8.0	64.00
27.2	1.5	2.25
20.9	−4.8	23.04
41.0	15.3	234.09
20.1	−5.6	31.36
		$\sum(y-\bar{y})^2 = 383.4$

$\text{SST} = \sum(y-\bar{y})^2 = 383.40$

(d) SSR = SST − SSE = 383.40 − 68.77 = 314.63

(e) $r^2 = \frac{\text{SSR}}{\text{SST}} \approx \frac{314.63}{383.40} \approx 0.8206$. Therefore, 82.06% of the variability in the teen birth rate is accounted for by the linear relationship between the teen birth rate and the overall birth rate.

(f) Since $b_1 \approx 1.83$ is positive, r is positive.

$r = \sqrt{r^2} \approx \sqrt{0.8206} \approx 0.9059$. This value of r is very close to the maximum value $r = 1$. We would therefore say that the teen birth rate and the overall birth rate are positively correlated. As the overall birth rate increases, the teen birth rate also tends to increase.

35. Minitab output:

Regression Analysis: Higher Income versus Lower Income

```
Analysis of Variance

Source          DF  Adj SS  Adj MS  F-Value  P-Value
Regression       1   81.19   81.19     2.26    0.163
  Lower Income   1   81.19   81.19     2.26    0.163
Error           10  358.82   35.88
Total           11  440.01

Model Summary

      S    R-sq  R-sq(adj)  R-sq(pred)
5.99013  18.45%     10.30%       0.00%

Coefficients

Term           Coef  SE Coef  T-Value  P-Value   VIF
Constant       38.1     32.3     1.18    0.265
Lower Income  0.641    0.426     1.50    0.163  1.00

Regression Equation

Higher Income = 38.1 + 0.641 Lower Income
```

(a) From Exercise 49 in Section 4.2, $\hat{y} = 0.641x + 38.1$

x	y	$\hat{y} = 0.641x + 38.1$	**Residual** $(y-\hat{y})$	**(Residual)^2** $(y-\hat{y})^2$
71.6	80.2	83.9956	−3.7956	14.40657936
75.7	82.4	86.6237	−4.2237	17.83964169
78.3	83.7	88.2903	−4.5903	21.07085409

74.5	79.8	85.8545	−6.0545	36.65697025
80.3	94.1	89.5723	4.5277	20.50006729
76.1	98.9	86.8801	12.0199	144.47799601
82.4	90.0	90.9184	−0.9184	0.84345856
78.0	89.6	88.098	1.502	2.256004
72.3	86.2	84.4443	1.7557	3.08248249
71.4	77.0	83.8674	−6.8674	47.16118276
67.9	88.7	81.6239	7.0761	50.07119121
78.9	88.0	88.6749	−0.6749	0.45549001
				$\sum(y-\hat{y})^2 \approx 358.8219177$

So $\text{SSE} = \sum(y-\hat{y})^2 \approx 358.8219177 \approx 358.82$

(b) $s = \sqrt{\frac{\text{SSE}}{n-2}} \approx \sqrt{\frac{358.8219177}{12-2}} \approx 5.99017$

Minitab: $s \approx 5.99013$. The typical difference between the predicted consumer sentiment for incomes \$75,000 or higher and the actual observed consumer sentiment for incomes \$75,000 or higher is 5.99013.

(c) $\bar{y} = \frac{\sum y}{n} = \frac{80.2+82.4+83.7+79.8+84.1+98.9+90.0+89.6+86.2+77.0+88.7+99.0}{12} = \frac{1038.6}{12} = 86.55$

y	$(y-\bar{y})$	$(y-\bar{y})^2$
80.2	25 −6.35	40.3225
82.4	−4.15	17.2225
83.7	−2.85	8.1225
79.8	−6.75	45.5625
94.1	7.55	57.0025
98.9	12.35	152.5225
90.0	3.45	11.9025
89.6	3.05	9.3025
86.2	−0.35	0.1225
77.0	−9.55	91.2025
88.7	2.15	4.6225
88.0	1.45	2.1025
		$\sum(y-\bar{y})^2 = 440.01$

$\text{SST} = \sum(y-\bar{y})^2 = 440.01$

(d) $\text{SSR} = \text{SST} - \text{SSE} = 440.01 - 358.82 = 81.19$

(e) $r^2 = \frac{\text{SSR}}{\text{SST}} \approx \frac{81.89}{440.01} \approx 0.1845$. Therefore, 18.45% of the variability in the consumer sentiment for incomes \$75,000 or higher is accounted for by the linear relationship between the consumer sentiment for incomes \$75,000 or higher and the consumer sentiment for incomes under \$75,000.

(f) Since $b_1 \approx 0.641$ is positive, r is positive.

$r = \sqrt{r^2} \approx \sqrt{0.1845} \approx 0.4295$. This value of r is positive. We would therefore say that the consumer sentiment for incomes \$75,000 or higher and the consumer sentiment for incomes under \$75,000 are positively correlated. As the consumer sentiment for incomes under \$75,000 increases, the consumer sentiment for incomes \$75,000 or higher also tends to increase.

37. **Minitab output:**

Regression Analysis: Runs scored versus Batting average

```
Analysis of Variance
Source               DF   Adj SS   Adj MS   F-Value   P-Value
Regression            1    30.48    30.48      0.46     0.517
  Batting average     1    30.48    30.48      0.46     0.517
Error                 8   530.02    66.25
Total                 9   560.50
Model Summary
      S    R-sq   R-sq(adj)   R-sq(pred)
8.13958   5.44%       0.00%        0.00%
Coefficients
Term               Coef   SE Coef   T-Value   P-Value    VIF
Constant           47.2      55.1      0.86     0.417
Batting average     118       174      0.68     0.517   1.00
Regression Equation
Runs scored = 47.2 + 118 Batting average
```

(a) From Exercise 51 in Section 4.2, $\hat{y} = 118x + 47.2$

x	y	$\hat{y} = 118x + 47.2$	Residual $(y - \hat{y})$	$(\text{Residual})^2$ $(y - \hat{y})^2$
0.341	85	87.438	−2.438	5.943844
0.335	87	86.73	0.27	0.0729
0.327	94	85.786	8.214	67.469796
0.324	79	85.432	−6.432	41.370624
0.317	80	84.606	−4.606	21.215236
0.314	77	84.252	−7.252	52.591504
0.313	101	84.134	16.866	284.461956
0.301	81	82.718	−1.718	2.951524
0.300	76	82.6	−6.6	43.56
0.293	85	81.774	3.226	10.407076
				$\Sigma(y - \hat{y})^2 \approx 530.04446$

SSE$= \Sigma(y - \hat{y})^2 \approx 530.04446 \approx 530.04$

Minitab: SSE ≈ 530.02

(b) $s = \sqrt{\frac{\text{SSE}}{n-2}} \approx \sqrt{\frac{530.04446}{10-2}} \approx 8.13975$

Minitab: $s \approx 8.13958$. The typical difference between the predicted number of runs scored by a player and the actual observed runs scored by a player is 8.13958.

(c) $\bar{y} = \frac{\Sigma y}{n} = \frac{85+87+94+79+80+77+101+81+76+85}{10} = \frac{845}{10} = 84.5$

y	$(y - \bar{y})$	$(y - \bar{y})^2$
85 5	0.5	0.25
87	2.5	6.25
94	9.5	90.25

79	-5.5	30.25
80	-4.5	20.25
77	-7.5	56.25
101	16.5	272.25
81	-3.5	12.25
76	-8.5	72.25
85	0.5	0.25
		$\Sigma(y-\bar{y})^2 = 560.5$

$\text{SST} = \Sigma(y-\bar{y})^2 = 560.50$

(d) SSR – SST – SSE = 560.50 – 530.02 = 30.48

(e) $r^2 = \frac{\text{SSR}}{\text{SST}} \approx \frac{30.48}{560.50} \approx 0.0544$. Therefore, 5.44% of the variability in the runs scored is accounted for by the linear relationship between the batting averages and the runs scored.

(f) Since $b_1 \approx 118$ is positive, r is positive.

$r = \sqrt{r^2} \approx \sqrt{0.0544} \approx 0.2332$. This value of r is close to the value $r = 0$. We would therefore say that the batting averages and the runs scored have no linear relationship.

39. First, we calculate the respective means, $\bar{x}$ and $\bar{y}$.

$$\bar{x} = \frac{\Sigma x}{n} = \frac{39.6+41.7+5.7+27.9+36.0}{5} = \frac{150.9}{5} = 30.18 \text{ and}$$

$$\bar{y} = \frac{\Sigma y}{n} = \frac{1.82+2.79+1.28+1.88+10.62}{5} = \frac{18.39}{5} = 3.678$$

x	y	$(x-\bar{x})$	$(x-\bar{x})^2$	$(y-\bar{y})$	$(y-\bar{y})^2$	$(x-\bar{x})(y-\bar{y})$
39.6	1.82	25 9.42	88.7364	−1.858	3.452164	−17.50236
41.7	2.79	11.52	132.7104	−0.888	0.788544	−10.22976
5.7	1.28	−24.48	599.2704	−2.398	5.750404	58.70304
27.9	1.88	−2.28	5.1984	−1.798	3.232804	4.09944
36.0	10.62	5.82	33.8724	6.942	48.191364	40.40244
			$\Sigma(x-\bar{x})^2 =$ 859.788		$\Sigma(y-\bar{y})^2 =$ 61.41528	$\Sigma(x-\bar{x})(y-\bar{y}) =$ 75.4728

Next, calculate the respective standard deviations, s_x and s_y.

$$s_x = \sqrt{\frac{\Sigma(x-\bar{x})^2}{n-1}} = \sqrt{\frac{859.788}{5-1}} = \sqrt{214.947} \approx 14.6610709 \text{ and}$$

$$s_y = \sqrt{\frac{\Sigma(y-\bar{y})^2}{n-1}} = \sqrt{\frac{61.41528}{5-1}} = \sqrt{15.35382} \approx 3.91839508$$

Put these values all together in the formula for the correlation coefficient r:

$$r = \frac{\Sigma(x-\bar{x})(y-\bar{y})}{(n-1)s_x s_y} = \frac{75.4728}{(5-1)(14.6610709)(3.91839508)} \approx 0.3283961299 \approx 0.3284$$

Then $b_1 = r \cdot \frac{s_y}{s_x} = (0.3283961299) \cdot \left(\frac{3.91839508}{14.6610709}\right) \approx 0.0877686132 \approx 0.088$ and $b_0 = \bar{y} - (b_1 \cdot \bar{x}) \approx 3.678 - (0.0877686132 \cdot 30.18) \approx 1.03$.

Therefore, $\hat{y} = 0.088x + 1.03$.

$b_1 = 0.088$ means that for each increase of $1 billion in the net worth of a company the current ratio increases by 0.088.

$b_0 = 1.03$ means that the predicted current ratio of a company with a net worth of $x = \$0$ is 1.03.

41. **(a)**

x	y	$\hat{y} = 0.088x + 1.03$	**Residual** $(y - \hat{y})$	**(Residual)^2** $(y - \hat{y})^2$
39.6	1.82	4.5148	−2.6948	7.26194704
41.7	2.79	4.6996	−1.9096	3.64657216
5.7	1.28	1.5316	−0.2516	0.06330256
27.9	1.88	3.4852	−1.6052	2.57666704
36.0	10.62	4.198	6.422	41.242084
				$\Sigma(y - \hat{y})^2 \approx 54.7905728$

$\text{SSE} = \Sigma(y - \hat{y})^2 \approx 54.7905728$

y	$(y - \bar{y})$	$(y - \bar{y})^2$
1.82	25 −1.858	3.452164
2.79	−0.888	0.788544
1.28	−2.398	5.750404
1.88	−1.798	3.232804
10.62	6.942	48.191364
		$\Sigma(y - \bar{y})^2 = 61.41528$

$\text{SST} = \Sigma(y - \bar{y})^2 = 61.41528$

$\text{SSR} = \text{SST} - \text{SSE} = 61.41528 - 54.7905728 = 6.6247072$

$r^2 = \frac{\text{SSR}}{\text{SST}} \approx \frac{6.6247072}{61.41528} \approx 0.1079$ or 10.79%.

(b)

x	y	$\hat{y} = 0.175x + 0.69$	Residual $(y - \hat{y})$	(Residual)^2 $(y - \hat{y})^2$
12.51	1.82	2.87925	−1.05925	1.1220105625
18.44	2.79	3.917	−1.127	1.270129
10.95	1.28	2.60625	−1.32625	1.7589390625
24.57	1.88	4.98975	−3.10975	9.6705450625
18.87	10.62	3.99225	6.62775	43.9270700625
				$\Sigma(y - \hat{y})^2 \approx 57.74869375$

$\text{SSE} = \Sigma(y - \hat{y})^2 \approx 57.74869375$

y	$(y - \bar{y})$	$(y - \bar{y})^2$
1.82	−1.858	3.452164
2.79	−0.888	0.788544
1.28	−2.398	5.750404
1.88	−1.798	3.232804
10.62	6.942	48.191364
		$\Sigma(y - \bar{y})^2 = 61.41528$

$\text{SST} = \Sigma(y - \bar{y})^2 = 61.41528$

$\text{SSR} = \text{SST} - \text{SSE} = 61.41528 - 57.74869375 = 3.66658625$

$r^2 = \frac{\text{SSR}}{\text{SST}} \approx \frac{3.66658625}{61.41528} \approx 0.0597$ or 5.97%

43. Net worth, since it has the largest value of r^2 and the smallest value of s.

45. **(a)** First, we calculate the respective means, $\bar{x}$ and $\bar{y}$.

$$\bar{x} = \frac{\Sigma x}{n} = \frac{13.4+10.5+11.3+11.6+9.0+11.2+11.2+12.9+11.7+10.1}{10} = \frac{112.9}{10} = 11.29 \text{ and}$$

$$\bar{y} = \frac{\Sigma y}{n} = \frac{100.0+88.7+84.0+80.7+78.7+78.7+77.0+75.5+75.2+75.2}{10} = \frac{813.7}{10} = 81.37$$

x	y	$(x - \bar{x})$	$(x - \bar{x})^2$	$(y - \bar{y})$	$(y - \bar{y})^2$	$(x - \bar{x})(y - \bar{y})$
13.4	100.0	2.11	4.4521	18.63	347.0769	39.3093
10.5	88.7	−0.79	0.6241	7.33	53.7289	−5.7907
11.3	84.0	0.01	0.0001	2.63	6.9169	0.0263
11.6	80.7	0.31	0.0961	−0.67	0.4489	−0.2077
9.0	78.7	−2.29	5.2441	−2.67	7.1289	6.1143
11.2	78.7	−0.09	0.0081	−2.67	7.1289	0.2403
11.2	77.0	−0.09	0.0081	−4.37	19.0969	0.3933
12.9	75.5	1.61	2.5921	−5.87	34.4569	−9.4507
11.7	75.2	0.41	0.1681	−6.17	38.0689	−2.5297
10.1	75.2	−1.19	1.4161	−6.17	38.0689	7.3423
			$\Sigma(x - \bar{x})^2 = 14.609$		$\Sigma(y - \bar{y})^2 = 552.121$	$\Sigma(x - \bar{x})(y - \bar{y}) = 35.447$

Next, calculate the respective standard deviations, s_x and s_y.

$$s_x = \sqrt{\frac{\Sigma(x-\bar{x})^2}{n-1}} = \sqrt{\frac{14.609}{10-1}} = \sqrt{1.623222222} \approx 1.274057386 \text{ and}$$

$$s_y = \sqrt{\frac{\Sigma(y-\bar{y})^2}{n-1}} = \sqrt{\frac{552.121}{10-1}} = \sqrt{61.34677778} \approx 7.832418386$$

Put these values all together in the formula for the correlation coefficient r:

$$r = \frac{\sum(x-\bar{x})(y-\bar{y})}{(n-1)s_x s_y} = \frac{35.447}{(10-1)(1.274057386)(7.832418386)} \approx 0.3946863508 \approx 0.3947$$

Then $b_1 = r \cdot \frac{s_y}{s_x} = (0.3946863508) \cdot \left(\frac{7.832418386}{1.274057386}\right) \approx 2.426380997 \approx 2.43$ and $b_0 = \bar{y} - (b_1 \cdot \bar{x}) \approx 81.37 - (2.426380997 \cdot 11.29) \approx 54.0$.

Therefore, $\hat{y} = 2.43x + 54.0$.

y = overall dating score = 54.0 + 2.43 percentage 18–24 years old

(b) First, we calculate the respective means, $\bar{x}$ and $\bar{y}$.

$$\bar{x} = \frac{\sum x}{n} = \frac{81.2+74.2+79.4+82.9+83.9+82.7+75.6+90.3+89.8+82.3}{10} = \frac{822.3}{10} = 82.23 \text{ and}$$

$$\bar{y} = \frac{\sum y}{n} = \frac{100.0+88.7+84.0+80.7+78.7+78.7+77.0+75.5+75.2+75.2}{10} = \frac{813.7}{10} = 81.37$$

x	y	$(x - \bar{x})$	$(x - \bar{x})^2$	$(y - \bar{y})$	$(y - \bar{y})^2$	$(x - \bar{x})(y - \bar{y})$
81.2	100.0	−1.03	1.0609	18.63	347.0769	−19.1889
74.2	88.7	−8.03	64.4809	7.33	53.7289	−58.8599
79.4	84.0	−2.83	8.0089	2.63	6.9169	−7.4429
82.9	80.7	0.67	0.4489	−0.67	0.4489	−0.4489
83.9	78.7	1.67	2.7889	−2.67	7.1289	−4.4589
82.7	78.7	0.47	0.2209	−2.67	7.1289	−1.2549
75.6	77.0	−6.63	43.9569	−4.37	19.0969	28.9731
90.3	75.5	8.07	65.1249	−5.87	34.4569	−47.3709
89.8	75.2	7.57	57.3049	−6.17	38.0689	−46.7069
82.3	75.2	0.07	0.0049	−6.17	38.0689	−0.4319
			$\sum(x-\bar{x})^2 =$ 243.401		$\sum(y-\bar{y})^2 =$ 552.121	$\sum(x-\bar{x})(y-\bar{y}) =$ −157.191

Next, calculate the respective standard deviations, s_x and s_y.

$$s_x = \sqrt{\frac{\sum(x-\bar{x})^2}{n-1}} = \sqrt{\frac{243.401}{10-1}} = \sqrt{27.04455556} \approx 5.200438016 \text{ and}$$

$$s_y = \sqrt{\frac{\sum(y-\bar{y})^2}{n-1}} = \sqrt{\frac{552.121}{10-1}} = \sqrt{61.34677778} \approx 7.832418386 .$$

Put these values all together in the formula for the correlation coefficient r:

$$r = \frac{\sum(x-\bar{x})(y-\bar{y})}{(n-1)s_x s_y} = \frac{-157.191}{(10-1)(5.200438016)(7.832418386)} \approx -0.4287946566 \approx -0.4288$$

Then $b_1 = r \cdot \frac{s_y}{s_x} = (-0.4287946566) \cdot \left(\frac{7.832418386}{5.200438016}\right) \approx -0.6458108224 \approx -0.646$ and $b_0 = \bar{y} - (b_1 \cdot \bar{x}) \approx 81.37 - \left((-0.6458108224) \cdot 82.23\right) \approx 134.5$.

Therefore, $\hat{y} = -0.646x + 134.5$.

y = overall dating score = 134.5–0.646 Percentage 18–24 who are single

(c) First, we calculate the respective means, $\bar{x}$ and $\bar{y}$.

$$\bar{x} = \frac{\sum x}{n} = \frac{77.8+88.9+77.4+79.2+100.0+66.9+82.9+51.1+63.5+50.2}{10} = \frac{737.9}{10} = 73.79 \text{ and}$$

$$\bar{y} = \frac{\sum y}{n} = \frac{100.0+88.7+84.0+80.7+78.7+78.7+77.0+75.5+75.2+75.2}{10} = \frac{813.7}{10} = 81.37$$

x	y	$(x - \bar{x})$	$(x - \bar{x})^2$	$(y - \bar{y})$	$(y - \bar{y})^2$	$(x - \bar{x})(y - \bar{y})$
77.8	100.0	4.01	16.0801	18.63	347.0769	74.7063
88.9	88.7	15.11	228.3121	7.33	53.7289	110.7563
77.4	84.0	3.61	13.0321	2.63	6.9169	9.4943
79.2	80.7	5.41	29.2681	−0.67	0.4489	−3.6247
100.0	78.7	26.21	686.9641	−2.67	7.1289	−69.9807
66.9	78.7	−6.89	47.4721	−2.67	7.1289	18.3963
82.9	77.0	9.11	82.9921	−4.37	19.0969	−39.8107
51.1	75.5	−22.69	514.8361	−5.87	34.4569	133.1903
63.5	75.2	−10.29	105.8841	−6.17	38.0689	63.4893
50.2	75.2	−23.59	556.4881	−6.17	38.0689	145.5503
			$\sum(x - \bar{x})^2 =$ 2281.329		$\sum(y - \bar{y})^2 =$ 552.121	$\sum(x - \bar{x})(y - \bar{y}) =$ 442.167

Next, calculate the respective standard deviations, s_x and s_y.

$$s_x = \sqrt{\frac{\sum(x-\bar{x})^2}{n-1}} = \sqrt{\frac{2281.329}{10-1}} = \sqrt{253.481} \approx 15.92108665 \text{ and}$$

$$s_y = \sqrt{\frac{\sum(y-\bar{y})^2}{n-1}} = \sqrt{\frac{552.121}{10-1}} = \sqrt{61.34677778} \approx 7.832418386\,.$$

Put these values all together in the formula for the correlation coefficient r:

$$r = \frac{\sum(x-\bar{x})(y-\bar{y})}{(n-1)s_x s_y} = \frac{442.167}{(10-1)(15.92108665)(7.832418386)} \approx 0.3939809593 \approx 0.3940$$

Then $b_1 = r \cdot \frac{s_y}{s_x} = (0.3939809593) \cdot \left(\frac{7.832418386}{15.92108665}\right) \approx 0.193819918 \approx 0.194$ and $b_0 = \bar{y} - (b_1 \cdot \bar{x}) \approx 81.37 - (0.193819918 \cdot 73.79) \approx 67.1$.

Therefore, $\hat{y} = 0.194x + 67.1$.

y = overall dating score = 67.1 + 0.194 online dating score

47. **(a)** $s = \sqrt{\frac{\text{SSE}}{n-2}} \approx \sqrt{\frac{466.155125}{10-2}} \approx 7.633$

(b) $s = \sqrt{\frac{\text{SSE}}{n-2}} \approx \sqrt{\frac{450.6062471}{10-2}} \approx 7.505$

(c) $s = \sqrt{\frac{\text{SSE}}{n-2}} \approx \sqrt{\frac{466.4407869}{10-2}} \approx 7.636$

49. First, we calculate the respective means, $\bar{x}$ and $\bar{y}$.

$$\bar{x} = \frac{\sum x}{n} = \frac{34.9+34.9+30.9+35.5+40.1+36.6+39.4+30.5+31.5+34.5+31.7+41.2+40.1+39.7+40.1+36.4+35.8+39.6+40.7}{19} = \frac{694.1}{19} \approx 36.5316$$

and

$\bar{y} = \frac{\sum y}{n} = \frac{4055+4055+5559+4103+3368+3970+3535+5333+5031+4354+4925+3179+3368+3334+3368+3919+4284+3467+3336}{19} = \frac{76,543}{19} \approx 4028.5789$

x	y	$(x - \bar{x})$	$(x - \bar{x})^2$	$(y - \bar{y})$	$(y - \bar{y})^2$	$(x - \bar{x})(y - \bar{y})$
34.9	4055	−1.6316	2.66211856	26.4211	698.07452521	-43.10866676
34.9	4055	−1.6316	2.66211856	26.4211	698.07452521	−43.10866676
30.9	5559	−5.6316	31.71491856	1530.4211	2,342,188.7433252	−8618.71946676
35.5	4103	−1.0316	1.06419856	74.4211	5,538.50012521	−76.77290676
40.1	3368	3.5684	12.73347856	−660.5789	436,364.48312521	−2357.20974676
36.6	3970	0.0684	0.00467856	−58.5789	3,431.48752521	−4.00679676
39.4	3535	2.8684	8.22771856	−493.5789	243,620.13052521	−1415.78171676
30.5	5333	−6.0316	36.38019856	1304.4211	1,701,514.4061252	−7867.74630676
31.5	5031	−5.0316	25.31699856	1002.4211	1,004,848.0617252	−5043.78200676
34.5	4354	−2.0316	4.12739856	325.4211	105,898.89232521	−661.12550676
31.7	4925	−4.8316	23.34435856	896.4211	803,570.78852521	−4331.14818676
41.2	3179	4.6684	21.79395856	−849.5789	721,784.30732521	−3966.17413676
40.1	3368	3.5684	12.73347856	−660.5789	436,364.48312521	−2357.20974676
39.7	3334	3.1684	10.03875856	−694.5789	482,439.84832521	−2200.70378676
40.1	3368	3.5684	12.73347856	−660.5789	436,364.48312521	−2357.20974676
36.4	3919	−0.1316	0.01731856	−109.5789	12,007.53532521	14.42058324
35.8	4284	−0.7316	0.53523856	225.4211	65,239.93832521	−186.86607676
39.6	3467	3.0684	9.41507856	−561.5789	315,370.86092521	−1723.14869676
40.7	3336	4.1684	17.37555856	−692.5789	479,665.53272521	−2886.94588676
			$\sum(x - \bar{x})^2 =$ 232.8810526		$\sum(y - \bar{y})^2 =$ 9,597,608.632	$\sum(x - \bar{x})(y - \bar{y}) =$ −46,126.34737

Next, calculate the respective standard deviations, s_x and s_y.

$s_x = \sqrt{\frac{\sum(x-\bar{x})^2}{n-1}} = \sqrt{\frac{232.8810526}{19-1}} = \sqrt{12.93783626} \approx 3.596920386$ and

$s_y = \sqrt{\frac{\sum(y-\bar{y})^2}{n-1}} = \sqrt{\frac{9,597,608.632}{19-1}} = \sqrt{533,200.4795} \approx 730.2057789$

Put these values all together in the formula for the correlation coefficient r:

$r = \frac{\sum(x-\bar{x})(y-\bar{y})}{(n-1)s_x s_y} = \frac{-46,126.34737}{(19-1)(3.596920386)(730.2057789)} \approx -0.9756446502$

Then $b_1 = r \cdot \frac{s_y}{s_x} = (-0.9756446502) \cdot \left(\frac{730.2057789}{3.596920386}\right) \approx -198.0642564 \approx -198.1$ and $b_0 = \bar{y} - (b_1 \cdot \bar{x}) \approx 4028.5789 - ((-198.0642564) \cdot 36.5316) \approx 11,264.$

Therefore, $\hat{y} = -198.1x + 11,264.$

Heating degree-days = 11264−198.1 average January temperatures

51.

x	y	$\hat{y} = -198.1x + 11,264.$	**Residual** $(y - \hat{y})$	**(Residual)^2** $(y - \hat{y})^2$
34.9	4055	4350.31	–295.31	87,207.9961
34.9	4055	4350.31	–295.31	87.207.9961
30.9	5559	5142.71	416.29	173,297.3541
35.5	4103	4231.45	–128.45	16,499.4025
40.1	3368	3320.19	47.81	2285.7961
36.6	3970	4013.54	–43.54	1895.7316
39.4	3535	3458.86	76.14	5797.2996
30.5	5333	5221.95	111.05	12,332.1025
31.5	5031	5023.85	7.15	51.1225
34.5	4354	4429.55	–75.55	5,707.8025
31.7	4925	4984.23	–59.23	3,508.1929
41.2	3179	3102.28	76.72	5,885.9584
40.1	3368	3320.19	47.81	2,285.7961
39.7	3334	3399.43	–65.43	4,281.0849
40.1	3368	3320.19	47.81	2,285.7961
36.4	3919	4053.16	–134.16	17,998.9056
35.8	4284	4172.02	111.98	12,539.5204
39.6	3467	3419.24	47.76	2,281.0176
40.7	3336	3201.33	134.67	18,136.0089
				$\Sigma(y - \hat{y})^2 \approx 461{,}484.8945$

$\text{SSE} = \Sigma(y - \hat{y})^2 \approx 461{,}484.8945$

$$s = \sqrt{\frac{\text{SSE}}{n-2}} \approx \sqrt{\frac{461{,}484.8945}{19-2}} \approx 164.761$$

Minitab: $s \approx 164.753$

53. First, we calculate the respective means, $\bar{x}$ and $\bar{y}$.

$$\bar{x} = \frac{\Sigma x}{n} = \frac{79.2+79.2+71.1+76.9+79.1+78.8+78.5+73.5+75.2+75.1+75.7+80.3+79.1+79.6+79.1+77.9+76.2+78.5+78.8}{19} = \frac{1471.8}{19} \approx 77.4632 \text{ and}$$

$$\bar{y} = \frac{\Sigma y}{n} = \frac{1531+1531+533+1212+1612+1418+1432+758+911+1075+1075+1682+1612+1619+1612+1435+1134+1427+1482}{19} = \frac{25{,}091}{19} \approx 1320.5789$$

x	y	$(x-\bar{x})$	$(x-\bar{x})^2$	$(y-\bar{y})$	$(y-\bar{y})^2$	$(x-\bar{x})(y-\bar{y})$
79.2	1531	1.736	3.01647424	210.4211	44,277.03932521	365.45936648
79.2	1531	1.7368	3.01647424	210.4211	44,277.03932521	365.45936648
71.1	533	−6.3632	40.49031424	−787.5789	620,280.52372521	5,011.52205648
76.9	1212	−0.5632	0.31719424	−108.5789	11,789.37752521	61.15163648
79.1	1612	1.6368	2.67911424	291.4211	84,926.25752521	476.99805648
78.8	1418	1.3368	1.78703424	97.4211	9,490.87072521	130.23252648
78.5	1432	1.0368	1.07495424	111.4211	12,414.66152521	115.52139648
73.5	758	−3.9632	15.70695424	−562.5789	316,495.01872521	2,229.61269648
75.2	911	−2.2632	5.12207424	−409.5789	167,754.87532521	926.95896648
75.1	1075	−2.3632	5.58471424	−245.5789	60,308.99612521	580.35205648
75.7	1075	−1.7632	3.10887424	−45.5789	60,308,99612521	433.00471648
80.3	1682	2.8368	8.04743424	361.4211	130,625.21152521	1,025.27937648
79.1	1612	1.6368	2.67911424	291.4211	84,926.25752521	476.99805648
79.6	1619	2.1368	4.56591424	298.4211	89,055.15292521	637.66620648
79.1	1612	1.6368	2.67911424	291.4211	84,926.25752521	476.99805648
77.9	1435	0.4368	0.19079424	114.4211	13,092.18812521	49.97913648
76.2	1134	−1.2632	1.59567424	−186.5789	34,811.68592521	235.68646648
78.5	1427	1.0368	1.07495424	106.4211	11,325.45052521	110.3373948
78.8	1482	1.3368	1.78703424	161.4211	26,056.77152521	215.78772636
			$\sum(x-\bar{x})^2$ = 104.5242106		$\sum(y-\bar{y})^2$ = 1,907,142.632	$\sum(x-\bar{x})(y-\bar{y})$ = 13,925.00526

Next, calculate the respective standard deviations, s_x and s_y.

$$s_x = \sqrt{\frac{\Sigma(x-\bar{x})^2}{n-1}} = \sqrt{\frac{104.5242106}{19-1}} = \sqrt{5.806900587} \approx 2.409751146 \text{ and}$$

$$s_y = \sqrt{\frac{\Sigma(y-\bar{y})^2}{n-1}} = \sqrt{\frac{1{,}907{,}142.632}{19-1}} = \sqrt{105{,}952.3684} \approx 325.5032541$$

Put these values all together in the formula for the correlation coefficient r:

$$r = \frac{\Sigma(x-\bar{x})(y-\bar{y})}{(n-1)s_x s_y} = \frac{13{,}925.00526}{(19-1)(2.409751146)(325.5032541)} \approx 0.9862688865$$

Then $b_1 = r \cdot \frac{s_y}{s_x} = (0.9862688865) \cdot \left(\frac{325.5032541}{2.409751146}\right) \approx 133.2227739 \approx 133.22$ and $b_0 = \bar{y} - (b_1 \cdot \bar{x}) \approx 1320.5789 - (133.2227739 \cdot 77.4632) \approx -8999.$

Therefore, $\hat{y} = 133.22x - 8999.$

Cooling degree-days = −8999 + 133.22 average July temperatures

55.

x	y	$\hat{y} = 133.22x - 8999$	Residual $(y - \hat{y})$	$(\text{Residual})^2$ $(y - \hat{y})^2$
79.2	1531	1552.024	−21.024	442.008576
79.2	1531	1552.024	−21.024	442.008576
71.1	533	472.942	60.058	3606.963364
76.9	1212	1245.618	−33.618	1130.169924
79.1	1612	1538.702	73.298	5372.596804
78.8	1418	1498.736	−80.736	6518.301696
78.5	1432	1458.77	−26.77	716.6329
73.5	758	792.67	−4.67	1202.0089
75.2	911	1019.144	−108.144	11,695.124736
75.1	1075	1005.822	69.178	4785.595684
75.7	1075	1085.754	−10.754	115.648516
80.3	1682	1698.566	−16.566	274.432356
79.1	1612	1538.702	73.298	5372.596804
79.6	1619	1605.312	13.688	187.361344
79.1	1612	1538.702	73.298	5372.596804
77.9	1435	1378.838	56.162	3154.170244
76.2	1134	1152.364	−18.364	337.236496
78.5	1427	1458.77	−31.77	1009.3329
78.8	1482	1498.736	−16.736	280.093696
				$\Sigma(y - \hat{y})^2 \approx 52{,}014.88032$

$\text{SSE} = \Sigma(y - \hat{y})^2 \approx 52{,}014.88032$

$$s = \sqrt{\frac{\text{SSE}}{n-2}} \approx \sqrt{\frac{52{,}014.88032}{19-2}} \approx 55.3145$$

57. **(a)** $x = 10$ *years of education*, $y = 20.6 =$ *unemployment rate*

It doesn't follow the trend of the higher the number of years of education, the lower the unemployment rate.

(b) Since $r^2 \approx 0.6824$, 68.24% of the variability in the variable $y =$ *unemployment rate* is accounted for by the linear relationship between $x =$ *years of education* and $y =$ *unemployment rate.* Hence, the statement is not true.

(c)

x = years of education	y = unemployment rate	$\hat{y} = -1.24x + 26.19$	$(y - \hat{y})$	$\lvert y - \hat{y} \rvert$
5	16.8	19.99	−3.19	3.19
7.5	17.1	16.89	0.21	0.21
8	15.3	16.27	−0.97	0.97
10	20.6	13.79	6.81	6.81
12	11.7	11.31	0.39	0.39
14	8.1	8.83	−0.73	0.73
16	3.8	6.35	−2.55	2.55

Since the absolute values of the residuals for 5, 10, and 16 years of education are more than

1%, this claim is not always true.

(d) Since $b_1 = -1.24$, we can say that each additional year of education drops the predicted unemployment rate by 1.24%.

59. SST = 228, SSR = 216

61. **(a)** $s = \sqrt{\frac{\text{SSE}}{n-2}} \approx \sqrt{\frac{20.193012}{10-2}} \approx 1.58874998 \approx 1.5887$

TI-83/84: $s \approx 1.588748636 \approx 1.5887$

(b) If we know the car's engine size (x), then our estimate of the combined mpg will typically differ from the actual mpg by 1.5887 mpg.

63. **(a)** $b_1 \approx -5.49$, so r is negative. Therefore,

$r = -\sqrt{r^2} \approx -\sqrt{0.9187076812} \approx -0.9585$.

(b) This value of r is very close to the minimum value $r = -1$. We would therefore say that the gas mileage and the engine displacement are negatively correlated. As the engine displacement of a car increases, the gas mileage of the car tends to decrease.

65. Since the variables are negatively correlated, the slope of the regression line is negative. An increase in the new engine size would result in an increase in the x-coordinate of the point $(\bar{x}, \bar{y})$. Since the point $(\bar{x}, \bar{y})$ is on the regression line, an increase in the x-coordinate would stretch the regression line to the right. This would result in the regression line becoming less steep, so the slope would become less negative and the left side of the line would not intersect the y-axis as high. Therefore, the slope would increase and the y-intercept would decrease.

Scatterplot with regression line with point $(\bar{x}, \bar{y})$ = (2.88, 21.70) added to data set and scatterplot with regression line with point $(\bar{x} + d, \bar{y}) = (2.88 + d, 21.70)$ where d is some unknown positive number added to data set

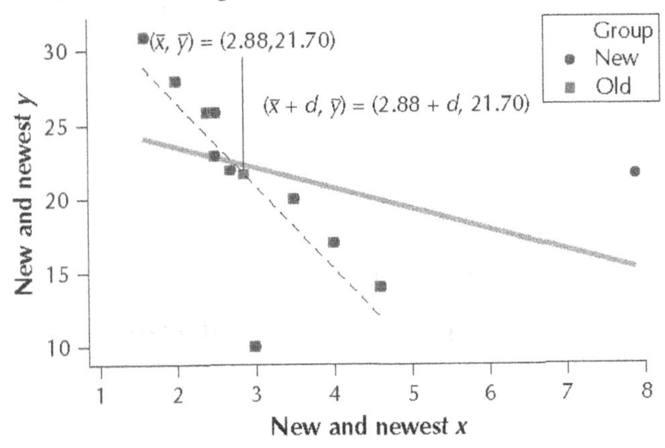

67. Since $b_1 = 0$, the regression equation is $\hat{y} = \bar{y} = 25.029$. Thus, $\hat{y} - \bar{y} = 0$ for all of the $\hat{y} - \bar{y}$'s. Hence SSR = 0, so SSR would decrease. Since SST = SSR + SSE and SSR = 0, SSE = SST = 358.5622634, so both SSE and SST increase. Since the regression line doesn't include any information from the x-values, SSR = 0, $r^2 = 0$, and $r = 0$. Since SSR = 0, $r^2 = \frac{\text{SSR}}{\text{SST}}$ and $r = \sqrt{r^2}$, $r^2 = 0$, and $r = 0$. Since $s = \sqrt{\frac{\text{SSE}}{n-2}}$ and SSE increases, s increases.

69. **(a)**

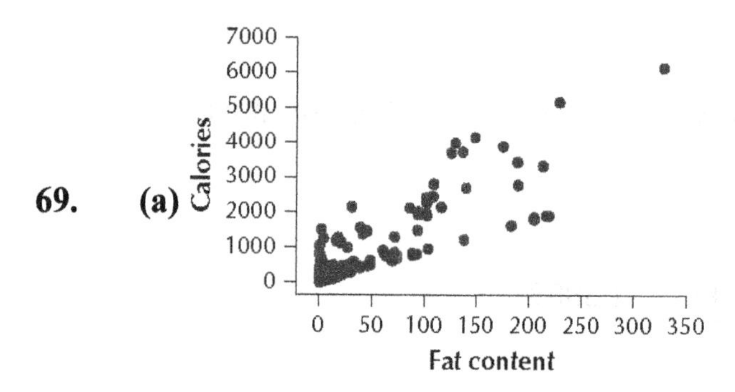

(b) $\hat{y} = 8.12x + 1.28$. The estimated calories per gram equals 8.12 times the amount of fat per gram, plus 1.28. **(c)** $r^2 = 0.736$, so 73.6% of the variability in the number of calories per gram is accounted for by the linear relationship between calories per gram and fat per gram. **(d)** $s = 0.9944$. The typical difference between the predicted number of calories per gram and the actual calories per gram is 0.9944. **(e)** $r = \sqrt{r^2} = \sqrt{0.736} \approx 0.8579$.

71–75. Answers will vary.

77. **(a)** Answers will vary. One possible answer:

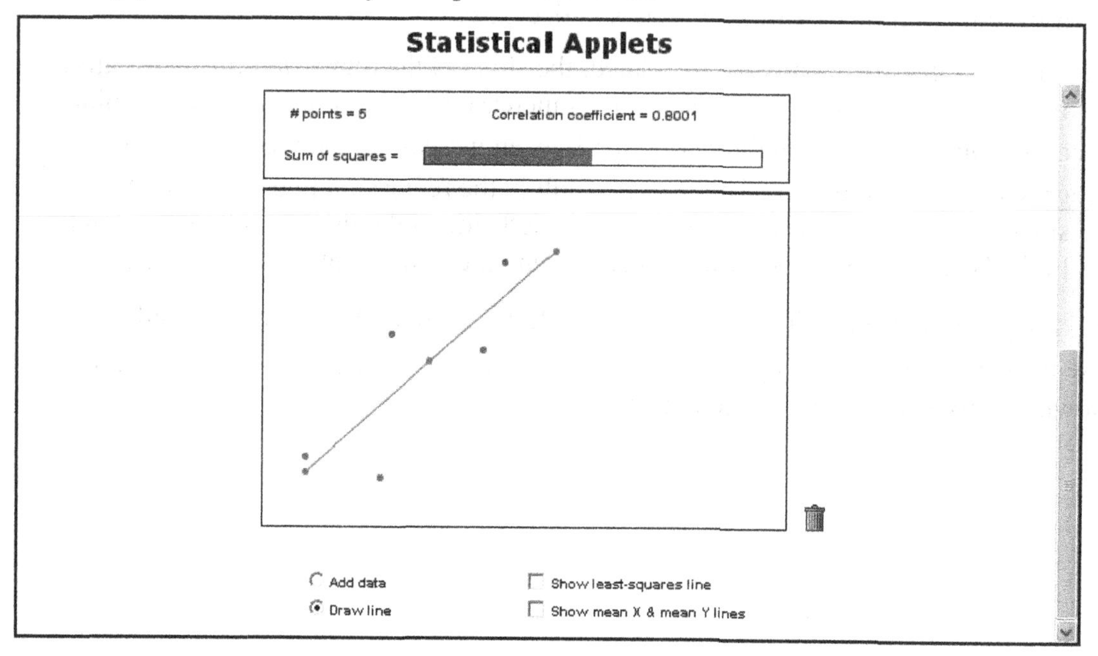

(b) That our adjusted line matches the least squares regression line. **(c)** Yes.

79. $s \approx 19.1723$ pounds. The typical difference between the predicted weight of a woman and the actual weight of a woman is 19.1723 pounds.

81. Since $b_1 \approx 3.541$ is positive, r is positive.

$r = \sqrt{r^2} \approx \sqrt{0.1852} \approx 0.4303$

83. $r^2 \approx 0.2863$. Therefore, 28.63% of the variability in men's weights is accounted for by the linear relationship between men's weights and men's heights.

85. **(a)** The values of s for men and women are both between 19 and 20 but are not the same.

(b) The values of r^2 for men and women are both low but the value of r^2 for men is higher than the value of r^2 for women.

87. $s \approx 1.79822$ centimeters. The typical difference between the predicted bicep girth of a woman and the actual bicep girth of a woman is 1.79822 centimeters.

89. Bicep girth = 8.12 + 0.4651 hip girth

$\hat{y} = 0.4651x + 8.12$

The predicted bicep girth of a man is 0.4651 times his thigh girth plus 8.12.

$b_1 = 0.4651$ means that for each increase of 1 centimeter in a man's thigh girth the predicted bicep girth increases by 0.4651 centimeter.

$b_0 = 8.12$ means that the predicted bicep girth of a man with a thigh girth of $x = 0$ centimeters is 8.12 centimeters.

91. $r^2 \approx 0.4388$. Therefore, 43.88% of the variability in men's bicep girths is accounted for by the linear relationship between men's bicep girths and men's thigh girths.

Chapter 4 Review Exercises

1.

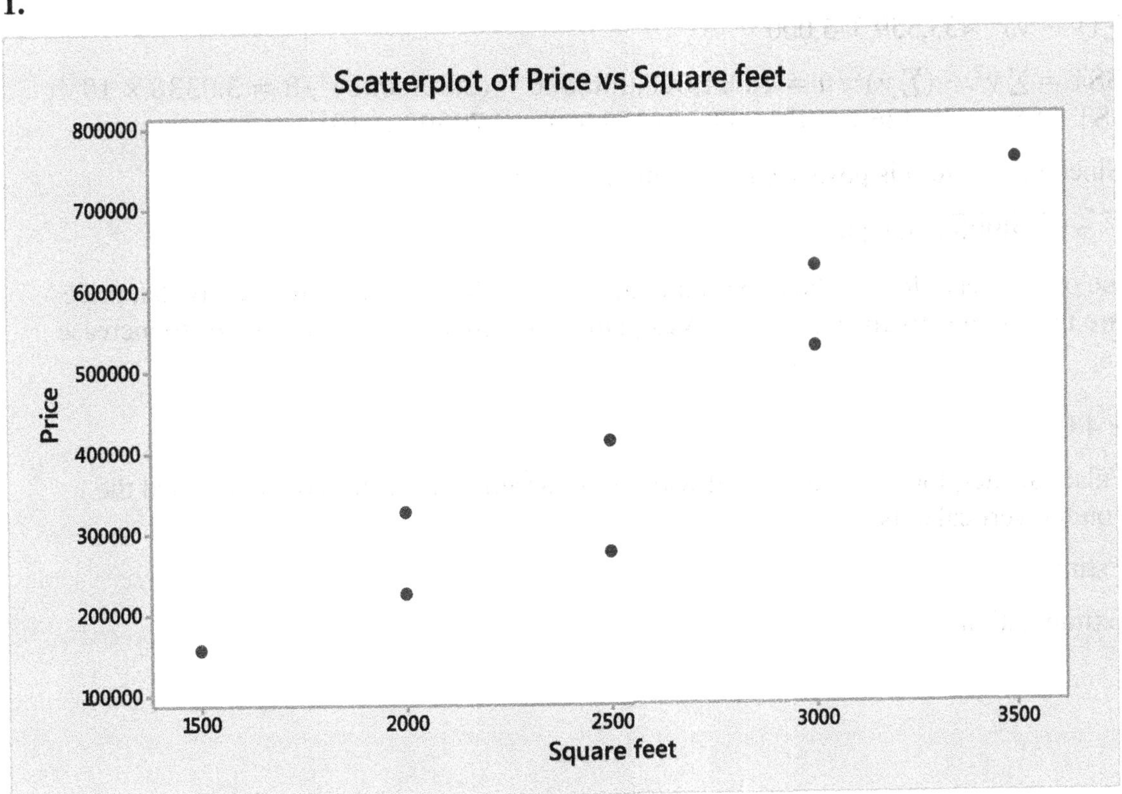

3. As the number of square feet increases, the price also tends to increase.

5. This value of r is very close to the maximum value $r = 1$. We would therefore say that price and square feet are positively correlated. As square feet increases, price also tends to increase.

7. The predicted price of a house is 300.0 times the number of square feet minus \$337,500.

9. $b_0 = -337{,}500$ means that the predicted price for a home with $x = 0$ square feet is $-\$337{,}500$.

11. The prediction error is $y - \hat{y} = \$756{,}250 - \$712{,}500 = \$43{,}750$. The prediction error is positive. Therefore, the actual observed price of house with 3500 square feet lies above the predicted price of a house with 3500 square feet.

13.

x	y	$\hat{y} = 300.0x - 337{,}500$	**Residual** $(y - \hat{y})$	**(Residual)^2** $(y - \hat{y})^2$
1500	156,250	112,500	43,750	1,914,062,500
2000	225,000	262,500	−37,500	1,406,250,000
2000	325,000	262,500	62,500	3,906,250,000
2500	412,500	412,500	0	0
2500	275,000	412,500	−137,500	18,906,250,000
3000	525,000	562,500	−37,500	1,406,250,000
3000	625,000	562,500	62,500	3,906,250,000
3500	756,250	712,500	43,750	1,914,062,500
				$\Sigma(y - \hat{y})^2 \approx 33{,}359{,}375{,}000$

$\text{SSE} = \Sigma(y - \hat{y})^2 \approx 33{,}359{,}375{,}000$

15. $\text{SST} = \Sigma y^2 - (\Sigma y)^2/n = 1{,}664{,}610{,}000{,}000 - (3{,}300{,}000)^2/8 = 3.0336 \times 10^{11}$,
$\text{SSR} = \text{SST} - \text{SSE} = 3.0336 \times 10^{11} - 33{,}359{,}375{,}000 \approx 2.70000 \times 10^{11}$

17. Since $b_1 = 300.0$ is positive, r is positive.

$r = \sqrt{r^2} \approx \sqrt{0.8900} \approx 0.9434$

This value of r is very close to the maximum value $r = 1$. We would therefore say that price and square feet are positively correlated. As square feet increases, price also tends to increase.

Chapter 4 Quiz

1. False. Scatterplots are constructed with the x variable on the horizontal axis and the y variable on the vertical axis.

3. Estimate.

5. Extrapolation.

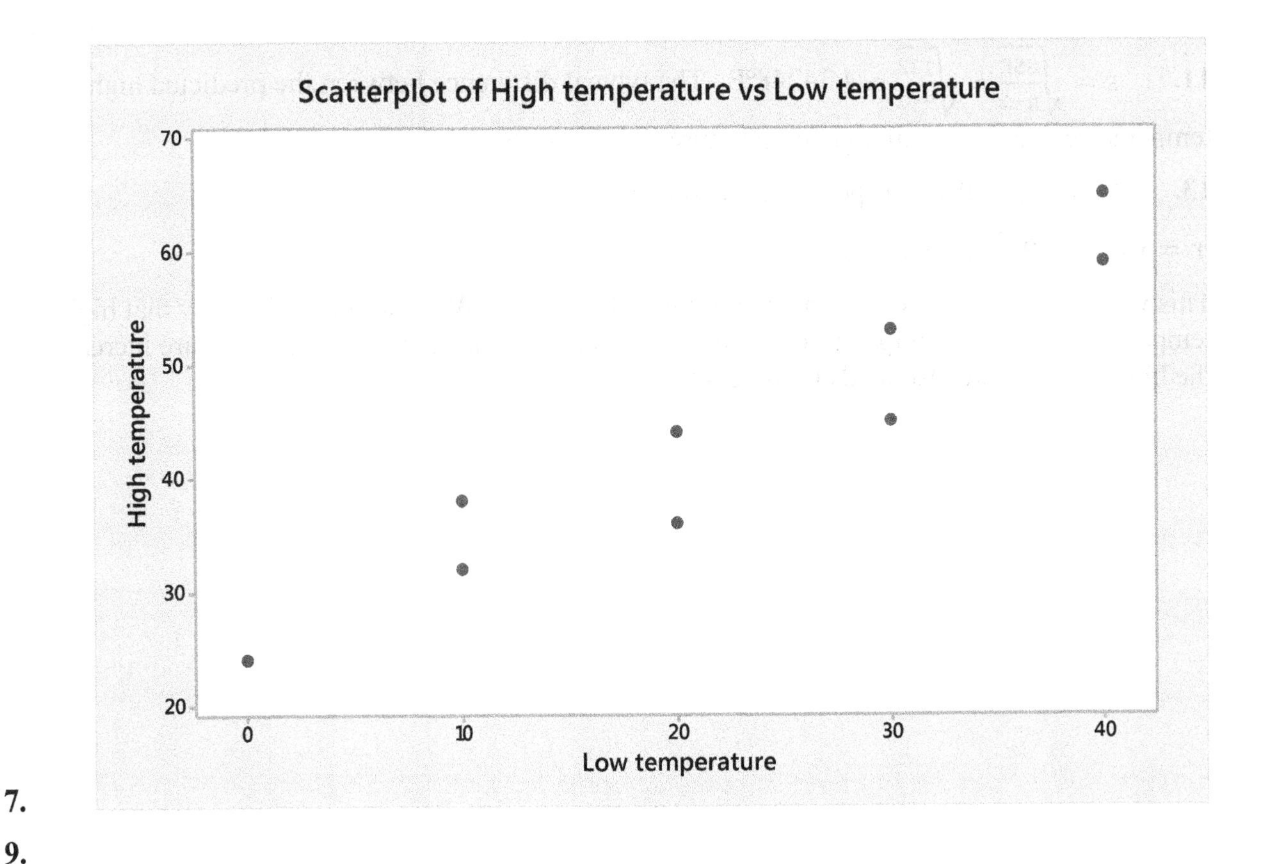

7.

9.

x	y	x^2	y^2	xy
0	24	0	576	0
10	32	100	1024	320
10	38	100	1444	380
20	36	400	1296	720
20	44	400	1936	880
30	45	900	2025	1350
30	53	900	2809	1590
40	59	1600	3481	2360
40	65	1600	4225	2600

$\sum x = 200 \quad \sum y = 396 \quad \sum x^2 = 6000 \quad \sum y^2 = 18{,}816 \quad \sum xy = 10{,}200$

$$b_1 = \frac{\sum xy - (\sum x \sum y)/n}{\sum x^2 - (\sum x)^2/n} = \frac{10{,}200 - \big((200)(396)\big)/9}{6000 - (200)^2/9} = 0.900$$

$$\bar{x} = \frac{\sum x}{n} = \frac{200}{9} \approx 22.22222222$$

$$\bar{y} = \frac{\sum y}{n} = \frac{396}{9} = 44$$

$$b_0 = \bar{y} - (b_1 \cdot \bar{x}) = 44 - \big((0.900)(22.22222222)\big) = 24$$

Thus,

$$\hat{y} = 0.900x + 24.00$$

11. $s = \sqrt{\frac{\text{SSE}}{n-2}} \approx \sqrt{\frac{132}{9-2}} \approx 4.34248$°F. The typical difference between the predicted high temperature and the actual high temperature is 4.34248°F.

13. Since $b_1 = 0.900$ is positive, r is positive.

$r = \sqrt{r^2} \approx \sqrt{0.9052} \approx 0.9514$

This value of r is very close to the maximum value $r = 1$. We would therefore say that high temperatures and low temperatures are positively correlated. As the low temperature increases, the high temperature also tends to increase.

Chapter 5: Probability

Section 5.1

1. Answers will vary; chance, likelihood.

3. **(a), (b), and (c)** Answers will vary.

5. The experiment has equally likely outcomes.

7. We consider all available information, tempered by our experience and intuition, and then assign a probability value that expresses our estimate of the likelihood that the outcome will occur.

9. First find out how many students are at your college and find out how many of them like Hip-Hop music. Then calculate the relative frequency of students who like Hip-Hop music. Use the relative frequency method.

11. Not a probability model. The probability for males is negative and the probabilities don't add up to 1.

13. Not a probability model. The probabilities don't add up to 1.

15. Probability model. All probabilities are between 0 and 1 and the sum of the probabilities is 1.

17. The chance of recovering from the surgery is near 100% which means that the probability of recovering from the surgery is near 1. Therefore it is almost certain that the person will recover from the surgery.

19. The chances are high that the blue chip stock will gain in value this year which means that the probability that the blue chip stock will gain in value this year is near 1. Therefore it is almost certain that the blue chip stock will gain in value this year.

For Problems 21 through 24, there are 52 cards in a deck of cards. Therefore, $N(S) = 52$.

21. There are 4 jacks in a deck of cards: {J♣, J♦, J♥, J♠}. Therefore, *P*(a jack) $= \frac{4}{52} = \frac{1}{13}$.

23. There is only 1 jack of clubs in the deck: { J♣}. Therefore, *P*(The jack of clubs)$= \frac{1}{52}$.

For Exercises 25 through 30, the sample space is $\{1, 2, 3, 4, 5, 6\}$. There are 6 outcomes in the sample space. Therefore, $N(S) = 6$.

25. There is only one 2 in the sample space: $\{2\}$. Therefore, *P*(roll a 2) $= \frac{1}{6}$.

27. There are 4 numbers in the sample space that are greater than 2: $\{3, 4, 5, 6\}$. Therefore, *P*(roll a number greater than 2) $= \frac{4}{6} = \frac{2}{3}$.

29. There are two numbers in the sample space that are either 2 or 3: {2, 3}

Therefore, *P*(roll a 2 or a 3} $= \frac{2}{6} = \frac{1}{3}$

31.

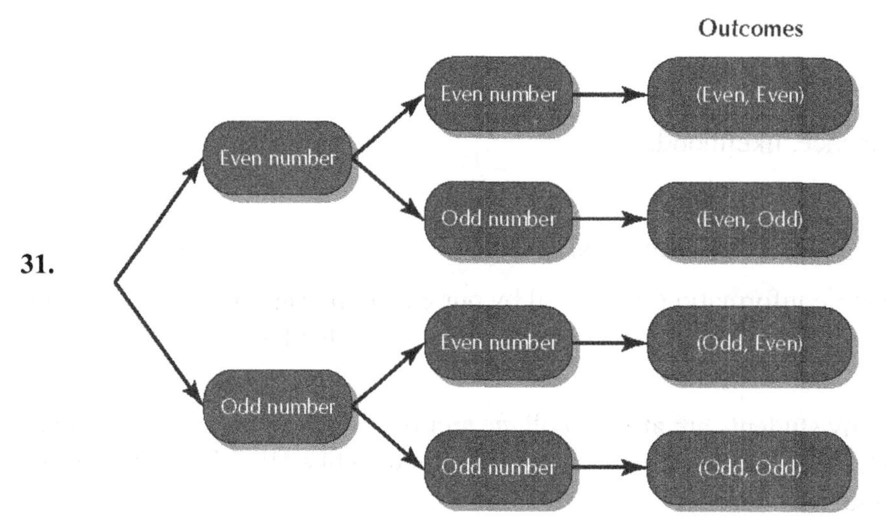

33. Let L = tossing a number less than 4 and G = tossing a number greater than or equal to 4.

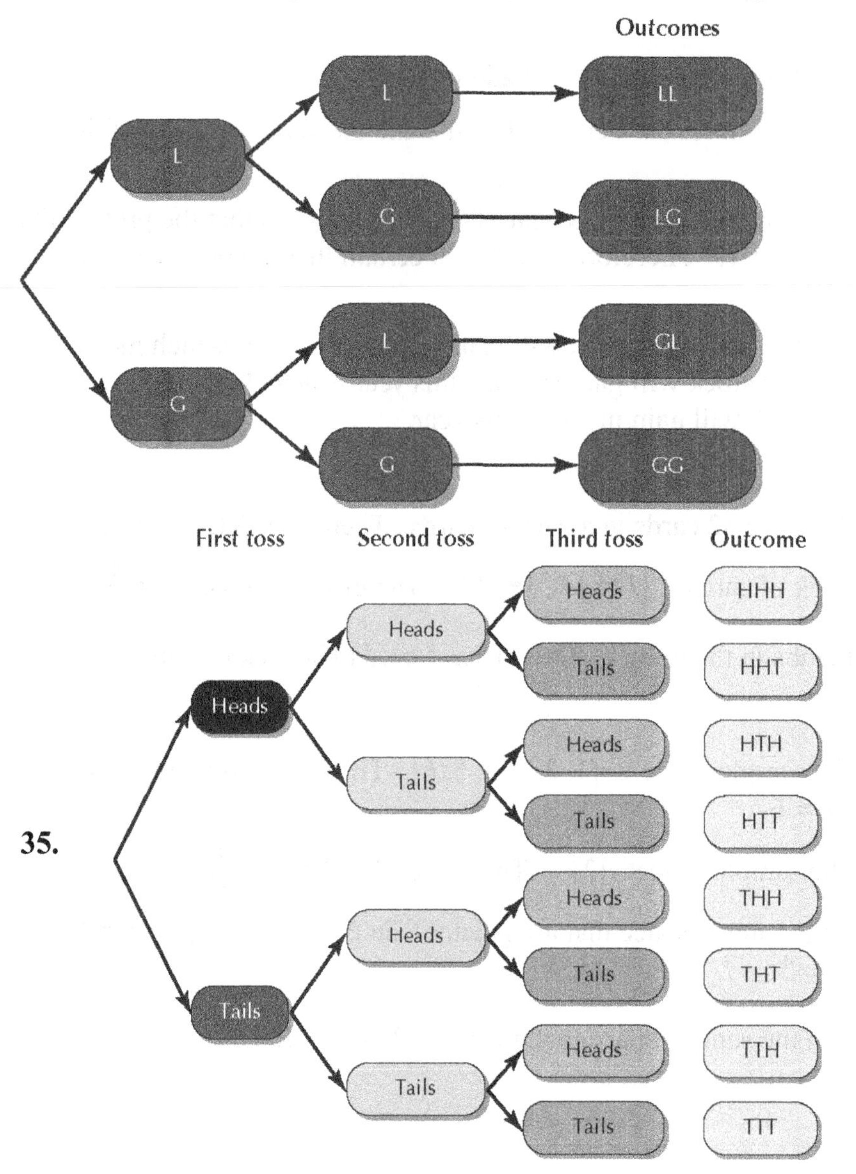

37. The tree diagram helps us see all of the possible outcomes for each toss of the coin. We can then follow the branches to get all possible outcomes.

For Problems 39 through 42, the sample space is {HH, HT, TH, TT}. Therefore, $N(S) = 4$.

39. There is 1 outcome in the sample space that consists of 0 heads: {TT}. Therefore, *P*(0 heads) $= \frac{1}{4}$.

41. There is 1 outcome in the sample space that consists of 2 heads: (HH} Therefore, *P*(2 heads) $= \frac{1}{4}$.

For Problems 43 through 50, the sample space is given in Figure 3 on page 247. There are 36 outcomes in the sample space, so $N(S) = 36$.

43. There are 4 outcomes in the sample space that have a sum of 9:

{(3, 6), (4, 5), (5, 4), (6, 3)}. Therefore, *P*(sum is 9) $= \frac{4}{36} = \frac{1}{9}$.

45. There are 2 outcomes in the sample space that have a sum of 11: {(5, 6), (6, 5)}. Therefore, *P*(sum of 11) $= \frac{2}{36} = \frac{1}{18}$.

47. There are 0 outcomes in the sample space that have a sum of 1. Therefore,

P(sum of 1) $= \frac{0}{36} = 0$.

49. Sum of 7

51. $\frac{35}{100} = \frac{7}{20} = 0.35$

53 $\frac{20}{100} = \frac{1}{5} = 0.2$

55. Relative frequency method was used since the probabilities were calculated using actual data.

57. $\frac{80}{200} = \frac{2}{5} = 0.4$

59. $\frac{40}{200} = \frac{1}{5} = 0.2$

61.

Favorite color	Probability
Red	$\frac{25}{100}=\frac{1}{4}=0.25$
Blue	$\frac{25}{100}=\frac{1}{4}=0.25$
Green	$\frac{20}{100}=\frac{1}{5}=0.20$
Black	$\frac{10}{100}=\frac{1}{10}=0.10$
Violet	$\frac{10}{100}=\frac{1}{10}=0.10$
Yellow	$\frac{10}{100}=\frac{1}{10}=0.10$

63. **(a)** Let STA denote Super Tech Adopter, SPR denote Smart Phone Reliant, MT denote Mature Technophile, and TAO denote Tech-Averse Older.

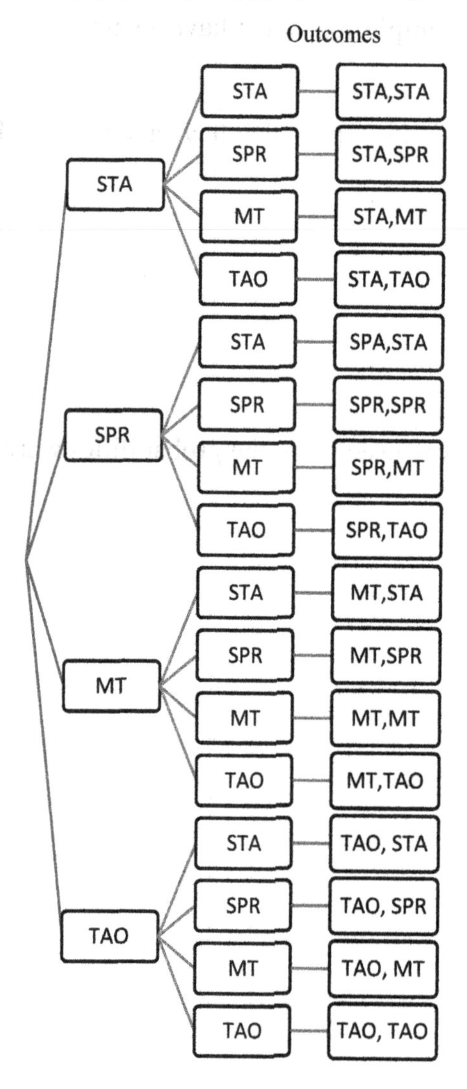

(b) {{Super Tech Adopter, Super Tech Adopter}, {Super Tech Adopter, Smart Phone Reliant}, {Super Tech Adopter, Mature Technophile}, {Super Tech Adopter, Tech-Averse Older}, {Smart Phone Reliant, Super Tech Reliant}, {Smart Phone Reliant, Smart Phone Reliant}, {Smart Phone Reliant, Mature Technophile}, {Smart Phone Reliant, Tech-Averse Older}, {Mature Technophile, Super Tech Adopter}, {Mature Technophile, Smart Phone Reliant}, {Mature Technophile, Mature Technophile}, {Mature Technophile, Tech-Averse Older}, {Tech-Averse Older, Super Tech Adopter}, {Tech-Averse Older, Smart Phone Reliant}, {Tech-Averse Older, Mature Technophile}. {Tech-Averse Older, Tech-Averse Older}}

65. **(a)** $\frac{14}{144} = \frac{7}{72}$

(b) $1 - \frac{7}{72} = \frac{65}{72}$

(c) Relative frequency method was used since the probabilities were calculated using actual data.

67. **(a)**

	Frequency	Relative frequency
Girls	18	$18/44 = 0.4091$
Boys	26	$26/44 = 0.5909$
Total	44	$44/44 = 1.0000$

(b)

Outcome	Probability
Girl	$18/44 = 0.4091$
Boy	$26/44 = 0.5909$

69. **(a)** There are 36 outcomes in the sample space. There are 6 outcomes in the sample space that have a sum of at least 10: {(4, 6), (5, 5), (6, 4), (5, 6), (6, 5), (6, 6)}.

Therefore, the probability of rolling a sum of at least 10 and winning the game is $\frac{6}{36} = \frac{1}{6}$.

(b) The probability of not winning the game is $1 - \frac{1}{6} = \frac{5}{6}$.

(c) The probability of winning the game is $\frac{1}{6}$. This means that in the long run someone wins the game about $\frac{1}{6}$ of the time. So in the long run, someone wins 1 out of every 6 games. Therefore, the average winnings will be about $\frac{\$10}{6} \approx \$1.67$ per game. Therefore, a person should pay \$1.67 to play to make the game fair.

71. There is 1 outcome with zero heads: TTT. Therefore, the probability of zero heads is 1/8 = 0.125.

73. There are 3 outcomes with exactly 2 heads: HHT, HTH, and THH. Therefore, the probability of 2 heads is 3/8 = 0.375.

75.

Number of heads	**Probability**
0	1/8 = 0.125
1	3/8 = 0.375
2	3/8 = 0.375
3	1/8 = 0.125

Tree diagram for Problems 77–80.

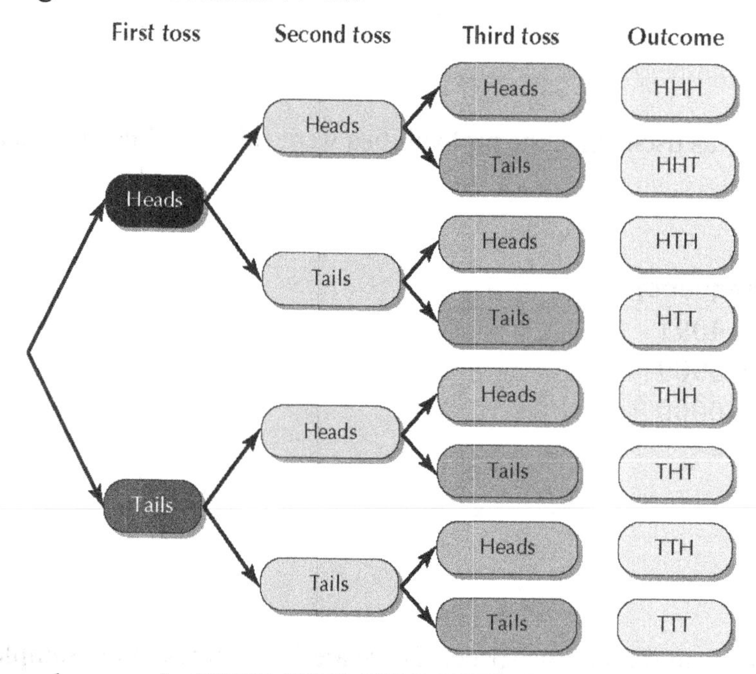

The sample space is {HHH, HHT, HTH, HTT, THH, THT, TTH, TTT}. There are 8 outcomes in the sample space, so $N(S) = 8$.

77. There is 1 outcome in the sample space that consists of 3 heads: {HHH}.

Therefore, *P*(tossing 3 heads) $= \frac{1}{8}$.

79. There are 3 outcomes in the sample space that consists of tossing 2 tails:

{HTT, THT, TTH}. Therefore, *P*(tossing 2 tails) $= \frac{3}{8}$.

81. **(a)** Answers will vary. One possible answer:

Tally for Discrete Variables: C5

C5	Count	Percent
1	1	10.00
2	3	30.00
3	2	20.00
4	2	20.00
5	1	10.00
6	1	10.00
N=	10	

Tally for Discrete Variables: C5

C5	Count	Percent
1	8	8.00
2	18	18.00
3	12	12.00
4	20	20.00
5	15	15.00
6	27	27.00
N=	100	

Tally for Discrete Variables: C5

C5	Count	Percent
1	91	9.10
2	176	17.60
3	167	16.70
4	163	16.30
5	201	20.10
6	202	20.20
N=	1000	

Tally for Discrete Variables: C5

C5	Count	Percent
1	867	8.67
2	1642	16.42
3	1713	17.13
4	1607	16.07
5	1965	19.65
6	2206	22.06
N=	10000	

Here, Under $25,000 = 1, $25,000 to $49,999 = 2, $50,000 to $74,999 = 3, $75,000 to $99,999 = 4, $100,000 to $149,999 = 5, $150,000 or more = 6.

(b) As the sample size increases the relative frequencies approach the probabilities.

83. **(a)**

Type of music	Probability
Hip-Hop/rap	0.27
Pop	0.23
Rock/punk	0.17
Alternative	0.07
Christian/gospel	0.06
R&B	0.06
Country	0.05
Techno/house	0.04
Jazz	0.01
Other	0.04

(b) Yes. **(c)** Answers will vary. One possible answer:

Tally for Discrete Variables: C4

C4	Count	Percent
1	4	40.00
2	2	20.00
5	2	20.00
6	2	20.00
N=	10	

Tally for Discrete Variables: C4

C4	Count	Percent
1	26	26.00
2	23	23.00
3	19	19.00
4	8	8.00
5	7	7.00
6	3	3.00
7	5	5.00
8	6	6.00
9	2	2.00
10	1	1.00
N=	100	

Tally for Discrete Variables: C4

C4	Count	Percent
1	270	27.00
2	222	22.20
3	173	17.30
4	60	6.00
5	58	5.80
6	60	6.00
7	51	5.10
8	44	4.40
9	13	1.30
10	49	4.90
N=	1000	

Tally for Discrete Variables: C4

C4	Count	Percent
1	2601	26.01
2	2329	23.29
3	1695	16.95
4	729	7.29
5	637	6.37
6	595	5.95
7	512	5.12
8	386	3.86
9	113	1.13
10	403	4.03
N=	10000	

Here Hip-Hop/rap = 1, Pop = 2, Rock/punk = 3, Alternative = 4, Christian/gospel = 5, R&B = 6, Country = 7, Techno/house = 8, Jazz = 9, Other = 10. **(d)** As the sample size increases the relative frequencies approach the probabilities.

85. Let p be the probability that Paul correctly predicts the winner of each World Cup Soccer match. **(a)** Since $p > 1/2$, $p^8 > (1/2)^8$, so the probability of Paul predicting all 8 matches correctly would be greater than the probability calculated in the previous exercise. **(b)** Since $p > 1/2$, $p^9 > (1/2)^9$, so the probability of Paul predicting all 9 matches correctly would be greater than the probability calculated in the previous exercise.

87. $\begin{Bmatrix} (1,1), (1,2), (1,3), (1,4), (1,5), (1,6), (2,1), (2,2), (2,3), (2,4), (2,5), (2,6) \\ (3,1), (3,2), (3,3), (3,4), (3,5), (3,6), (4,1), (4,2), (4,3), (4,4), (4,5), (4,6) \\ (5,1), (5,2), (5,3), (5,4), (5,5), (5,6), (6,1), (6,2), (6,3), (6,4), (6,5), (6,6) \end{Bmatrix}$

Example 7: Experiment is tossing 2 fair dice.

89. 1/36. Classical probability method, have the sample space but no actual data and can assume outcomes are equally likely.

91. From Exercise 87, $N(S) = 36$.

(a) Let E be the event 2 high die results. Then event E consists of the outcomes {(5, 5), (5, 6), (6, 5), (6, 6)}, so $N(E) = 4$. Therefore, $P(E) = \frac{N(E)}{N(S)} = \frac{4}{36} = \frac{1}{9}$.

(b) Let E be the event 1 medium die result. Then event E consists of the outcomes {(3, 1), (3, 2), (3, 5), (3, 6), (4, 1), (4, 2), (4, 5), (4, 6), (1, 3), (2, 3), (5, 3), (6, 3), (1, 4), (2, 4), (5, 4), (6, 4)}, so $N(E) = 16$. Therefore, $P(E) = \frac{N(E)}{N(S)} = \frac{16}{36} = \frac{4}{9}$.

(c) Let E be the event no low die results. Then event E consists of the outcomes {(3, 3), (3, 4), (3, 5), (3, 6), (4, 3), (4, 4), (4, 5), (4, 6), (5, 3), (5, 4), (5, 5), (5, 6), (6, 3), (6, 4), (6, 5), (6, 6)}, so $N(E) = 16$. Therefore, $P(E) = \frac{N(E)}{N(S)} = \frac{16}{36} = \frac{4}{9}$

(d) Let E be the event at least 1 high die result. Then event E consists of the outcomes {(1, 5), (1, 6), (2, 5), (2, 6), (3, 5), (3, 6), (4, 5), (4, 6), (5, 1), (5, 2), (5, 3), (5, 4), (5, 5), (5, 6), (6, 1), (6, 2), (6, 3), (6, 4), (6, 5), (6, 6)}, so $N(E) = 20$. Therefore, $P(E) = \frac{N(E)}{N(S)} = \frac{20}{36} = \frac{5}{9}$.

(e) Let E be the event at most 1 medium die result. Then event E consists of the outcomes {(3, 1), (3, 2), (3, 5), (3, 6), (4, 1), (4, 2), (4, 5), (4, 6), (1, 3), (2, 3), (5, 3), (6, 3), (1, 4), (2, 4), (5, 4), (6, 4), (1, 1), (1, 2), (1, 5), (1, 6), (2, 1), (2, 2), (2, 5), (2, 6), (5, 1), (5, 2), (5, 5), (5, 6), (6, 1), (6, 2), (6, 5), (6, 6)} , so $N(E) = 32$. Therefore,

$P(E) = \frac{N(E)}{N(S)} = \frac{32}{36} = \frac{8}{9}$.

93. $\frac{3}{20}$

95. All probabilities are between 0 and 1 and the sum of the probabilities is

$$\frac{3}{20}+\frac{2}{20}+\frac{4}{20}+\frac{3}{20}+\frac{4}{20}+\frac{1}{20}+\frac{3}{20}=\frac{20}{20}=1$$

97.

Type	Relative frequency
Adventure	$\frac{2}{20}=\frac{1}{10}$
Platform	$\frac{2}{20}=\frac{1}{10}$
Sports	$\frac{1}{20}$
Shooter	$\frac{8}{20}=\frac{2}{5}$
Action	$\frac{5}{20}=\frac{1}{4}$
Role-playing	$\frac{1}{20}$

99. **(a)** Answers will vary. One possible answer:

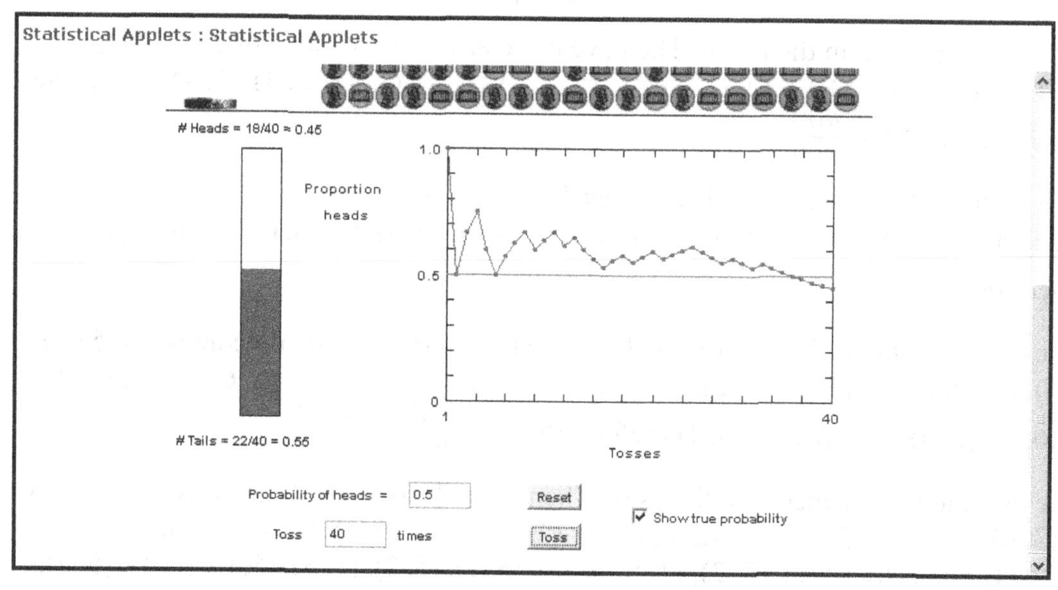

(b) Answers will vary. One possible answer:

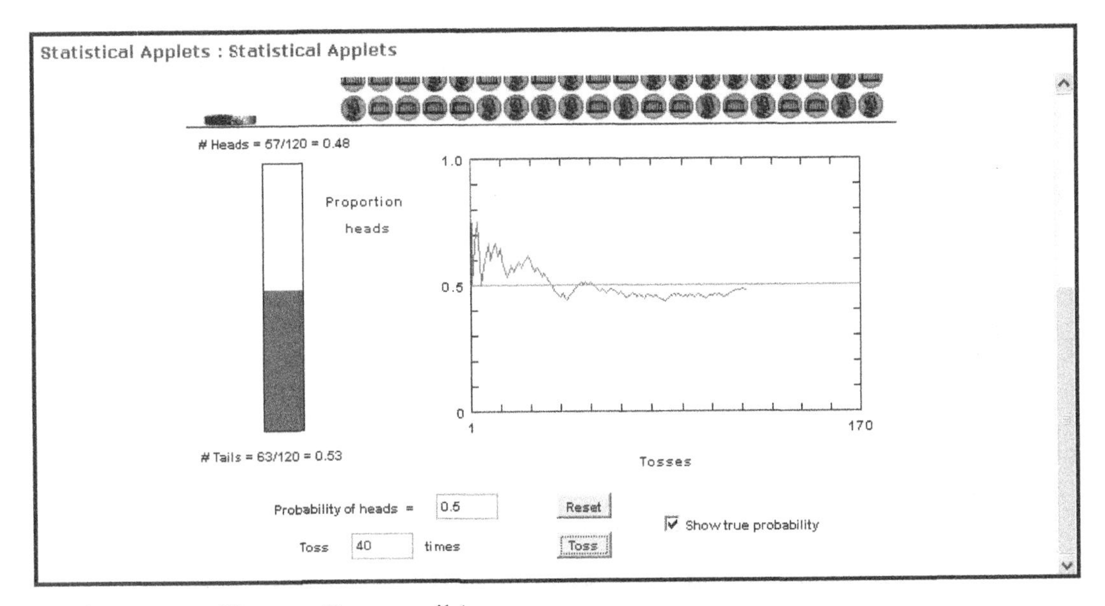

(c) Answers will vary. One possible answer:

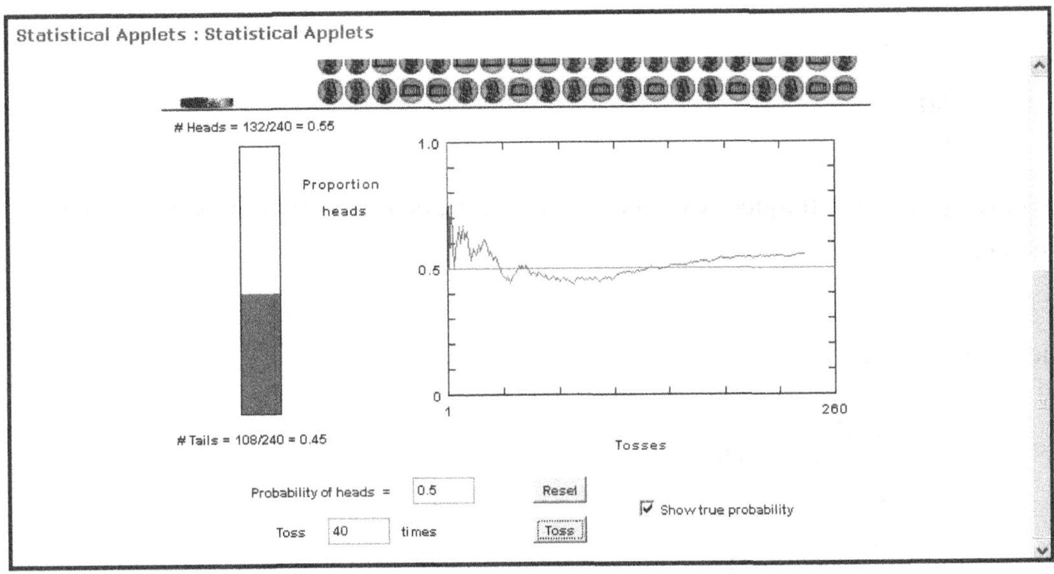

(d) Answers will vary. One possible answer:

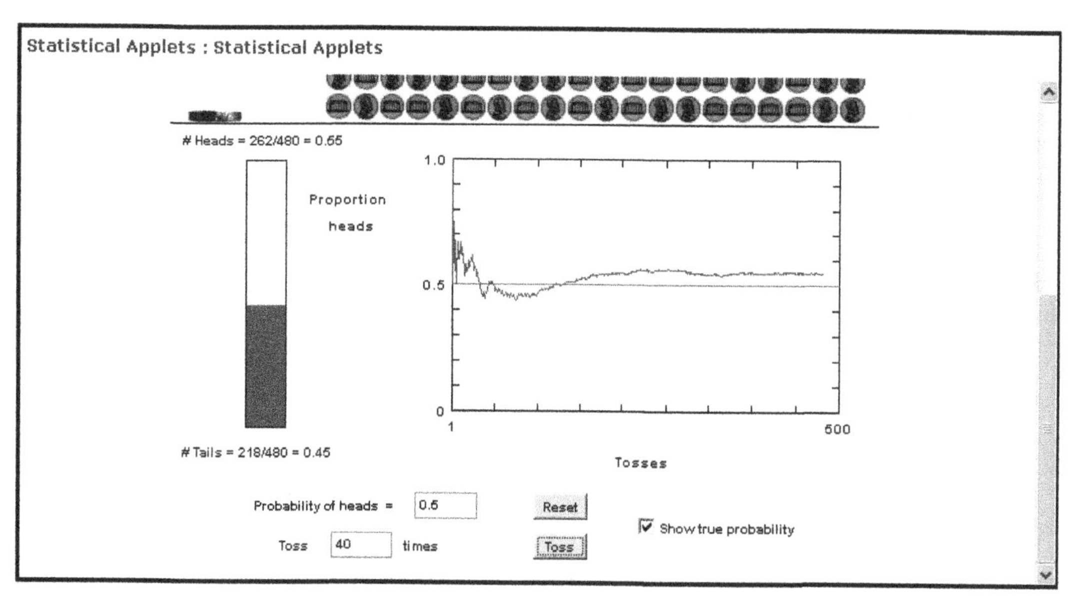

101.

Clinic location	Frequency
Suburban	963
Urban	450
Total	1413

103. $\frac{450}{1413} \approx 0.3185$. Relative frequency method was used since the probabilities were calculated using actual data.

105.

Insurance Type	Relative frequency ≈ Probability
Hospital based	$\frac{84}{1413} \approx 0.0594$
Med assistance	$\frac{275}{1413} \approx 0.1946$
Military	$\frac{331}{1413} \approx 0.2343$
Private payer	$\frac{723}{1413} \approx 0.5117$

Section 5.2

1. Two events are mutually exclusive if they have no outcomes in common.

3. It is all of the outcomes in each of the events. There are no outcomes in both.

5. You are more likely to select a male than a male football player. All male football players are males, but most males are not football players. Therefore, there are many more males than male football players at any college or university.

7. There are 5 outcomes in the sample space that are not a 6: $\{1, 2, 3, 4, 5\}$.

Therefore, P(roll a number not a 6) $= \frac{5}{6}$. Or:

There is 1 number in the sample space that is a 6: {6}. Then P(roll a 6} $= \frac{1}{6}$. Therefore, P(roll a number not a 6) = 1 – P(roll a 6) = $1 - \frac{1}{6} = \frac{5}{6}$.

9. The event E has 3 outcomes in it. Therefore, $P(E) = \frac{3}{6} = \frac{1}{2}$. Then $(E^C) = 1 - P(E) = 1 - \frac{1}{2} = \frac{1}{2}$.

11. The event E has 3 outcomes in it. Therefore, $P(E) = \frac{3}{6} = \frac{1}{2}$. Then $(E^C) = 1 - P(E) = 1 - \frac{1}{2} = \frac{1}{2}$.

13. {J♣, J♠}

15 {2♠, 3♠, 4♠, 5♠, 6♠, 7♠, 8♠, 9♠, 10♠, J♠, Q♠, K♠, A♠}

17. {J♦, J♥, J♠, J♣, 2♠, 3♠, 4♠, 5♠, 6♠, 7♠, 8♠, 9♠, 10♠, Q♠, K♠, A♠}

For Exercises 19–24, $N(S) = 52$.

19. From Exercise 13, the event $J \cap B$ has 2 outcomes. Therefore, $P(J \cap B) = \frac{2}{52} = \frac{1}{26}$.

21. From Exercise 15, the event $B \cap S$ has 13 outcomes. Therefore, $P(B \cap S) = \frac{13}{52} = \frac{1}{4}$.

23. From Exercise 17, the event $J \cup S$ has 16 outcomes. Therefore, $(J \cup S) = \frac{16}{52} = \frac{4}{13}$.

25. The event Child or Adult is highlighted in the chart below.

	Child	**Adult**	**Total**
Did not survive	52	1438	1490
Survived	57	654	711
Total	109	2092	2201

Then *P*(Child or Adult) = *P*(Child) + *P*(Adult) $= \frac{109}{2201} + \frac{2092}{2201} = \frac{2201}{2201} = 1$

27. The event Child and Survived is highlighted in the chart below.

	Child	**Adult**	**Total**
Did not survive	52	1438	1490
Survived	57	654	711
Total	109	2092	2201

Therefore, *P*(Child and Survived) $= \frac{57}{2201}$.

29. The event Child and Did not survive is highlighted below.

	Child	**Adult**	**Total**
Did not survive	52	1438	1490
Survived	57	654	711
Total	109	2092	2201

Therefore, *P*(Child and Did not survive) $= \frac{52}{2201}$.

31. There was no one who both survived and did not survive.

	Child	Adult	Total
Did not survive	52	1438	1490
Survived	57	654	711
Total	109	2092	2201

Therefore, $P(\text{Survived and Did not survive}) = \frac{0}{2201} = 0$.

33. The event Adult or Survived is highlighted below.

	Child	Adult	Total
Did not survive	52	1438	1490
Survived	57	654	711
Total	109	2092	2201

Therefore, $P(\text{Adult or Survived}) = P(\text{Adult}) + P(\text{Survived}) - P(\text{Adult and Survived}) = \frac{2092}{2201} + \frac{711}{2201} - \frac{654}{2201} = \frac{2149}{2201}$.

35. The event Adult or Did not survive is highlighted below.

	Child	Adult	Total
Did not survive	52	1438	1490
Survived	57	654	711
Total	109	2092	2201

Therefore, $P(\text{Adult or Did not Survive}) = P(\text{Adult}) + P(\text{Did not survive}) - P(\text{Adult and Did not Survive}) = \frac{2092}{2201} + \frac{1490}{2201} - \frac{1438}{2201} = \frac{2144}{2201}$.

37. The companies from China are highlighted below.

		Assets		
		Over \$1 trillion	**Under \$1 trillion**	**Total**
Country	**China**	4	1	5
	USA	2	3	5
	Total	6	4	10

Therefore, $P(\text{China}) = \frac{5}{10} = \frac{1}{2}$.

39. The companies with over \$1 trillion in assets are highlighted below.

		Assets		
		Over \$1 trillion	**Under \$1 trillion**	**Total**
Country	**China**	4	1	5
	USA	2	3	5
	Total	6	4	10

Therefore, $P(\text{Over \$1 trillion}) = \frac{6}{10} = \frac{3}{5}$.

41. The companies that are from China and have Over \$1 trillion in assets are highlighted below.

		Assets		
		Over \$1 trillion	**Under \$1 trillion**	**Total**
Country	**China**	4	1	5

	USA	2	3	5
	Total	6	4	10

Therefore, P(China and Over \$1 trillion) $= \frac{4}{10} = \frac{2}{5}$.

43. The companies that are from the United States or have Over \$1 trillion in assets are highlighted below

		Assets		
		Over \$1 trillion	**Under \$1 trillion**	**Total**
Country	**China**	4	1	5
	USA	2	3	5
	Total	6	4	10

Therefore, P(USA or Over \$1 trillion) = P(USA) + P(Over \$1 trillion) – P(USA and Over \$1 trillion) $= \frac{5}{10} + \frac{6}{10} - \frac{2}{10} = \frac{9}{10}$.

45. P(2 nomination or 3 nominations) = P(2 nominations) + P(3 nominations)

$\frac{5}{12} + \frac{3}{12} = \frac{8}{12} = \frac{2}{3}$.

47. P(3 nomination or 4 nominations) = P(3 nominations) + P(4 nominations)

$= \frac{3}{12} + \frac{0}{12} = \frac{3}{12} = \frac{1}{4}$.

49. **(a)** P(Super Tech Adopter or Smart Phone Reliant) = P(Super Tech Adopter) + P(Smart Phone Reliant) $= 0.31 + 0.19 = 0.50$.

(b) (Super Tech Adopter or Mature Technophile) = P(Super Tech Adopter) + P(Mature Technophile) $= 0.31 + 0.22 = 0.53$.

51. **(a)** P(Super Tech Adopter or Smart Phone Reliant or Mature Technophile) = P(Super Tech Adopter) + P(Smart Phone Reliant) + P(Mature Technophile) $= 0.31 + 0.19 + 0.22 = 0.72$.

(b) P(Neither Super Tech Adopter nor Smart Phone Reliant nor Mature Technophile) $= 1 -$ P(Super Tech Adopter or Smart Phone Reliant or Mature Technophile) $= 1 - 0.72 = 0.28$.

53. From Example 7 in Section 5.1, $N(S) = 36$.

There are 6 outcomes in the sample space that have a sum of 7:

$\{(1, 6), (2, 5), (3, 4), (4, 3), (5, 2), (6, 1)\}$. Therefore, P(sum of 7) $= \frac{6}{36} = \frac{1}{6}$.

There are 2 outcomes in the sample space that have a sum of 11: $\{(5, 6), (6, 5)\}$. Therefore, P(sum of 11) $= \frac{2}{36} = \frac{1}{18}$.

Then P(sum of 7 or sum of 11) = P(sum of 7) + P(sum of 11) $= \frac{6}{36} + \frac{2}{36} = \frac{8}{36} = \frac{2}{9}$.

55. The probability of winning the game is $\frac{4}{13}$. This means that in the long run someone wins the game about $\frac{4}{13}$ of the time. So in the long run, someone wins 4 out of every 13 games.

Therefore, the average winnings will be about $\left(\frac{4}{13}\right) \cdot \$3 = \frac{\$12}{13} \approx \$0.92$ per game. Therefore, a person should pay $0.92 to play to make the game fair.

57. The probability of winning the game is $\frac{7}{13}$. This means that in the long run someone wins the game about $\frac{7}{13}$ of the time. So in the long run, someone wins 7 out of every 13 games. Therefore, the average winnings will be about $\left(\frac{7}{13}\right) \cdot \$3 = \frac{\$21}{13} \approx \$1.62$ per game. Therefore, a person should pay $1.62 to play to make the game fair.

59. **(a)** The females are highlighted in the table below.

	Physical Appearance				
Gender	**Very attractive**	**Attractive**	**Average**	**Prefer not to answer**	**Total**
Female	3,113	16,181	6,093	3,478	28,865
Male	1,415	12,454	7,274	2,809	23,952
Total	4,528	28,635	13,367	6,287	52,817

Therefore, $P(\text{Female}) = \frac{28{,}865}{52{,}817} \approx 0.5465$.

(b) The people who self-reported as attractive are highlighted in the table below.

	Physical Appearance				
Gender	**Very attractive**	**Attractive**	**Average**	**Prefer not to answer**	**Total**
Female	3,113	16,181	6,093	3,478	28,865
Male	1,415	12,454	7,274	2,809	23,952
Total	4,528	28,635	13,367	6,287	52,817

Therefore, $P(\text{Self-reported as attractive}) = \frac{28{,}635}{52{,}817} \approx 0.5422$.

(c) The females who self-reported as attractive are highlighted in the table below.

	Physical Appearance				
Gender	**Very attractive**	**Attractive**	**Average**	**Prefer not to answer**	**Total**
Female	3,113	16,181	6,093	3,478	28,865
Male	1,415	12,454	7,274	2,809	23,952
Total	4,528	28,635	13,367	6,287	52,817

Therefore, $P(\text{Female who self-reported as attractive}) = \frac{16{,}181}{52{,}817} \approx 0.3064$.

(d) The people who are Females or who self-reported as attractive are highlighted in the table below.

	Physical Appearance				
Gender	**Very attractive**	**Attractive**	**Average**	**Prefer not to answer**	**Total**
Female	3,113	16,181	6,093	3,478	28,865
Male	1,415	12,454	7,274	2,809	23,952
Total	4,528	28,635	13,367	6,287	52,817

Therefore, $P(\text{Female or Self-reported as attractive}) = P(\text{Female}) + P(\text{Self-reported as attractive}) - P(\text{Female who self-reported as attractive}) = \frac{28{,}865}{52{,}817} + \frac{28{,}635}{52{,}817} - \frac{16{,}181}{52{,}817}$

$= \frac{41{,}319}{52{,}817} \approx 0.7823.$

(e) The people who self-reported as attractive or who self-reported as average are highlighted in the table below.

	Physical Appearance				
Gender	**Very attractive**	**Attractive**	**Average**	**Prefer not to answer**	**Total**
Female	3,113	16,181	6,093	3,478	28,865
Male	1,415	12,454	7,274	2,809	23,952
Total	4,528	28,635	13,367	6,287	52,817

Therefore, *P*(Self-reported as attractive or Self-reported as average) = *P*(Self-reported as attractive) + *P*(Self-reported as average) $= \frac{28{,}635}{52{,}817} + \frac{13{,}367}{52{,}817} = \frac{42{,}002}{52{,}817} \approx 0.7952$

61. **(a)** The compact cars are highlighted in the table below.

	Regular	Premium	Total
Compact	2	1	3
Midsize	2	2	4
Large	1	2	3
Total	5	5	10

Therefore, *P*(Compact car) $= \frac{3}{10} = 0.3.$

(b)

	Regular	Premium	Total
Compact	2	1	3
Midsize	2	2	4
Large	1	2	3
Total	5	5	10

P(Not a Compact car) = 1 - *P*(Compact car) $= 1 - \frac{3}{10} = \frac{7}{10} = 0.7.$

(c) The midsize cars are highlighted below.

	Regular	Premium	Total
Compact	2	1	3
Midsize	2	2	4
Large	1	2	3
Total	5	5	10

Therefore, *P*(Midsize car) $= \frac{4}{10} = \frac{2}{5} = 0.4.$

(d)

	Regular	Premium	Total
Compact	2	1	3
Midsize	2	2	4
Large	1	2	3
Total	5	5	10

P(Not a Midsize car) = 1 – *P*(Midsize car) = $1 - \frac{2}{5} = \frac{3}{5} = 0.6$.

63. **(a)** The midsize cars that use premium gasoline are highlighted in the table below.

	Regular	Premium	Total
Compact	2	1	3
Midsize	2	2	4
Large	1	2	3
Total	5	5	10

Therefore, *P*(Midsize car that uses premium gasoline) = $\frac{2}{10} = \frac{1}{5} = 0.2$.

(b) The midsize cars that use regular gasoline are highlighted in the table below.

	Regular	Premium	Total
Compact	2	1	3
Midsize	2	2	4
Large	1	2	3
Total	5	5	10

Therefore, *P*(Midsize car that uses regular gasoline) = $\frac{2}{10} = \frac{1}{5} = 0.2$.

(c) The large cars that use premium gasoline are highlighted in the table below.

	Regular	Premium	Total
Compact	2	1	3
Midsize	2	2	4
Large	1	2	3
Total	5	5	10

Therefore, *P*(Large car that uses premium gasoline) = $\frac{2}{10} = \frac{1}{5} = 0.2$.

(d) The large car that uses regular gasoline is highlighted in the table below.

	Regular	Premium	Total
Compact	2	1	3
Midsize	2	2	4
Large	1	2	3
Total	5	5	10

Therefore, *P*(Large car that uses regular gasoline) = $\frac{1}{10} = 0.1$.

65. $P(\text{Fox and grossed over \$500 million}) = \frac{1}{10} = 0.1$

67. $P(\text{Fox}) = \frac{3}{10} = 0.3$

$P(\text{Grossed over \$500 million}) = \frac{4}{10} = \frac{2}{5} = 0.4$

$P(\text{Fox or grossed over \$500 million}) = P(\text{Fox}) + P(\text{Grossed over \$500 million})$

$-P(\text{Fox and grossed over \$500 million}) = \frac{3}{10} + \frac{4}{10} - \frac{1}{10} = \frac{6}{10} = \frac{3}{5} = 0.6$

69. $P(\text{Warner and grossed over \$500 million}) = \frac{1}{10} = 0.1$

71. $P(\text{Warner and produced in 21st century}) = \frac{2}{10} = \frac{1}{5} = 0.2$

73. $P(\text{Not Warner and produced in 21st century}) = \frac{4}{10} = \frac{2}{5} = 0.4$

75. $P(E) = 0.13$

77. $P(\text{Consonant}) = 1 - P(\text{Vowel}) = 1 - 0.378 = 0.622$

79. The 8 letters with the highest frequencies are E (130), T (93), N (78), R (77), I (74), O (74), A (73), and S (63). Of these, 4 are vowels (E, I, O, A).

81. If you had no money to buy a vowel it would be best to guess the 5 consonants with the highest frequencies: T (93), N (78), R (77), S (63), and D (44).

83. The males are highlighted in the chart below.

	GPA < 2.5	GPA 2.5 to < 3.0	GPA 3.0 to < 3.5	GPA 3.5 or Higher	Total
Females	2,698	2,439	3,314	3,868	12,319
Males	2,535	1,975	2,324	2,351	9,185
Total	5,233	4,414	5,638	6,219	21,504

Therefore, $P(\text{Male}) = \frac{9{,}185}{21{,}504} \approx 0.4271$

85. The students that are female and have a GPA < 2.5 are highlighted in the table below.

	GPA < 2.5	GPA 2.5 to < 3.0	GPA 3.0 to < 3.5	GPA 3.5 or Higher	Total
Females	2,698	2,439	3,314	3,868	12,319
Males	2,535	1,975	2,324	2,351	9,185
Total	5,233	4,414	5,638	6,219	21,504

Therefore, $P(\text{Female and GPA} < 2.5) = \frac{2{,}698}{21{,}504} \approx 0.1255.$

87. The students that are female or have a GPA < 2.5 are highlighted in the table below.

	GPA < 2.5	GPA 2.5 to < 3.0	GPA 3.0 to < 3.5	GPA 3.5 or Higher	Total
Females	2,698	2,439	3,314	3,868	12,319
Males	2,535	1,975	2,324	2,351	9,185
Total	5,233	4,414	5,638	6,219	21,504

Therefore, $P(\text{Female or GPA} < 2.5) = P(\text{Female}) + P(\text{GPA} < 2.5) - P(\text{Female and GPA} < 2.5)$

$= \frac{12{,}319}{21{,}504} + \frac{5{,}233}{21{,}504} - \frac{2{,}698}{21{,}504} = \frac{14{,}854}{21{,}504} \approx 0.6908$

89.

	GPA < 2.5	GPA 2.5 to < 3.0	GPA 3.0 to < 3.5	GPA 3.5 or Higher	Total
Females	2,698	2,439	3,314	3,868	12,319
Males	2,535	1,975	2,324	2,351	9,185
Total	5,233	4,414	5,638	6,219	21,504

$P(\text{Does not have a GPA} < 2.5) = 1 - P(\text{GPA} < 2.5) = 1 - \frac{5{,}233}{21{,}504} = \frac{16{,}271}{21{,}504} \approx 0.7566$

91. The students that are male and don't have a GPA < 2.5 are highlighted in the table below.

	GPA < 2.5	GPA 2.5 to < 3.0	GPA 3.0 to < 3.5	GPA 3.5 or Higher	Total
Females	2,698	2,439	3,314	3,868	12,319
Males	2,535	1,975	2,324	2,351	9,185
Total	5,233	4,414	5,638	6,219	21,504

Therefore, *P*(Male and doesn't have a GPA < 2.5) = *P*(Male and has a GPA 2.5 to 3.0 or Male and has a GPA 3.0 to 3.5 or Male and has a GPA 3.5 or higher)

= *P*(Male and has a GPA 2.5 to 3.0) + *P*(Male and has a GPA 3.0 to 3.5) + *P*(Male and has a GPA 3.5 or higher) $= \frac{1{,}975}{21{,}504} + \frac{2{,}324}{21{,}504} + \frac{2{,}351}{21{,}504} = \frac{6{,}650}{21{,}504} \approx 0.3092$

93. The students that are male or don't have a GPA < 2.5 are highlighted in the table below

	GPA < 2.5	GPA 2.5 to < 3.0	GPA 3.0 to < 3.5	GPA 3.5 or Higher	Total
Females	2,698	2,439	3,314	3,868	12,319
Males	2,535	1,975	2,324	2,351	9,185
Total	5,233	4,414	5,638	6,219	21,504

Therefore, *P*(Male or doesn't have a GPA < 2.5) = *P*(Male) + *P*(Doesn't have a GPA < 2.5) – *P*(Male and doesn't have a GPA < 2.5) $= \frac{9{,}185}{21{,}504} + \frac{16{,}271}{21{,}504} - \frac{6{,}650}{21{,}504}$

$$= \frac{18{,}806}{21{,}504} \approx 0.8745$$

95.

Studio	Probability
Sony	$\frac{3}{20}$
MS	$\frac{2}{20}=\frac{1}{10}$
Nintendo	$\frac{4}{20}=\frac{1}{5}$
Electronic Arts	$\frac{3}{20}$
Activision	$\frac{4}{20}=\frac{1}{5}$
Focus	$\frac{1}{20}$
Take-Two	$\frac{3}{20}$

97. **(a)** P(Not made by Sony, made by some other studio) = 1 − P(Made by Sony) $= 1 - \frac{3}{20} = \frac{17}{20}$

(b) P(Not made by Nintendo, made by some other studio) = 1 − P(Made by Nintendo) $= 1 - \frac{1}{5} = \frac{4}{5}$

99. **(a)** inFamous: Second Son for PS4

(b) Minecraft for PS3, MLB 14 The Show for PS4, inFamous: Second Son for PS4, Super Mario Brothers U for WiiU, Grand Theft Auto V for PS3, Grand Theft Auto V for Xbox 360, Bound by Flame for PS4

(c) Kirby: Triple Deluxe for 3DS, Super Luigi U for WiiU

(d) Kirby: Triple Deluxe for 3DS, Pokeman X/Y for 3DS, Super Luigi U for WiiU, Super Mario Brothers U for WiiU

(e) Super Mario Brothers U for WiiU

(f) Kirby: Triple Deluxe for 3DS, Pokeman X/Y for 3DS, Super Luigi U for WiiU, Super Mario Brothers U for WiiU, inFamous: Second Son for PS4, Grand Theft Auto V for PS3, Grand Theft Auto V for Xbox 360, Bound by Flame for PS4

101. **(a)** None

(b) Minecraft for PS3, MLB 14 The Show for PS4, inFamous: Second Son for PS4, Kirby: Triple Deluxe for 3DS, Pokeman X/Y for 3DS, Super Luigi U for WiiU, Super Mario Brothers U for WiiU

(c) None

(d) inFamous: Second Son for PS4, Grand Theft Auto V for PS3, Grand Theft Auto V for Xbox 360, Bound by Flame for PS4, Kirby: Triple Deluxe for 3DS, Super Luigi U for WiiU, Super Mario Brothers U for WiiU

103.

Had Medical Assistance	Frequency	Relative Frequency
No	1138	$\frac{1138}{1413} \approx 0.8054$
Yes	275	$\frac{275}{1413} \approx 0.1946$
Total	1413	$\frac{1413}{1413} = 1.0000$

Completed Medical Treatment	Frequency	Relative Frequency
No	944	$\frac{944}{1413} \approx 0.6681$
Yes	469	$\frac{469}{1413} \approx 0.3319$
Total	1413	$\frac{1413}{1413} = 1.0000$

(a) $P(\text{Had medical assistance}) \approx \frac{275}{1413} \approx 0.1946$

(b) $P(\text{Completed medical treatment}) \approx \frac{469}{1413} \approx 0.3319$

105. **(a)** The patients that had medical assistance and completed the treatment are highlighted in the table below.

	Did not complete treatment	Completed treatment	Total
Did not have medical assistance	724	414	1138
Had medical assistance	220	55	275
Total	944	469	1413

Therefore, $P(\text{Had medical assistance and completed the treatment}) \approx \frac{55}{1413} \approx 0.0389$.

(b) The patients that had medical assistance and completed the treatment are highlighted in the table below.

	Did not complete treatment	Completed treatment	Total
Did not have medical assistance	724	414	1138
Had medical assistance	220	55	275
Total	944	469	1413

Therefore, $P(\text{Had medical assistance and did not complete the treatment}) \approx \frac{220}{1413} \approx 0.1557$.

(c) The patients that did not have medical assistance and completed the treatment are highlighted in the table below.

	Did not complete treatment	Completed treatment	Total
Did not have medical assistance	724	414	1138
Had medical assistance	220	55	275
Total	944	469	1413

Therefore, P(Did not have medical assistance and completed the treatment) $\approx \frac{414}{1413} \approx 0.2930$.

(d) The patients that did not have medical assistance and did not complete the treatment are highlighted in the table below.

	Did not complete treatment	Completed treatment	Total
Did not have medical assistance	724	414	1138
Had medical assistance	220	55	275
Total	944	469	1413

Therefore, P(Did not have medical assistance and did not complete the treatment) $\approx \frac{724}{1413} \approx 0.5124$.

107. **(a)** The patients that had medical assistance or completed the treatment are highlighted in the table below.

	Did not complete treatment	Completed treatment	Total
Did not have medical assistance	724	414	1138
Had medical assistance	220	55	275
Total	944	469	1413

Therefore, P(Had medical assistance or completed the treatment) = P(Had medical assistance) + P(completed the treatment) – P(Had medical assistance and completed the treatment) $\approx \frac{275}{1415} + \frac{469}{1413} - \frac{55}{1413} = \frac{689}{1413} \approx 0.4876$

(b) The patients that had medical assistance or did not have medical assistance are highlighted in the table below.

	Did not complete treatment	Completed treatment	Total
Did not have medical assistance	724	414	1138
Had medical assistance	220	55	275
Total	944	469	1413

Therefore, P(Had medical assistance or did not have medical assistance) = P(Had medical assistance) + P(Did not have medical assistance) $\approx \frac{275}{1415} + \frac{1138}{1413} = \frac{1413}{1413} = 1$

Section 5.3

1. **(a)** Yes. **(b)** The probability of winning the football game depends on whether the star quarterback can play in the game.

3. For P(A | B), we assume that the event B has occurred, and now need to find the probability of event A, given event B. On the other hand, for P(A ∩ B), we do not assume that event B has occurred, and instead need to determine the probability that both events occurred.

5. The Gambler's Fallacy is the mistaken belief that each time an event doesn't happen, the probability that it will happen increases. For example, if a coin is flipped several times and lands on tails every time, then one might think that the coin is bound to land on heads soon. But if the coin is fair, then the probability of it landing on heads is 0.5 on each flip no matter what happened on the previous flips.

7. **(a)** Independent; sampling with replacement. **(b)** Dependent; sampling without replacement.

9. The mammals in E are highlighted in the table below and the mammals that are in A and E are in bold in the table below.

	Africa (A)	Australia (B)	North America (C)	Total
Threatened (T)	1	1	2	4
Endangered (E)	**3**	1	2	6
Total	4	2	4	10

Then $P(\text{A given E}) = \frac{N(A \text{ and E})}{N(\text{E})} = \frac{3}{6} = \frac{1}{2}$.

11. The mammals in E are highlighted in the table below and the mammals that are in C and E are in bold in the table below.

	Africa (A)	Australia (B)	North America (C)	Total
Threatened (T)	1	1	2	4
Endangered (E)	3	1	**2**	6
Total	4	2	4	10

Then $P(\text{C given E}) = \frac{N(\text{C and E})}{N(\text{E})} = \frac{2}{6} = \frac{1}{3}$.

13. The mammals in T are highlighted in the table below and the mammals that are in B and T are in bold in the table below.

	Africa (A)	Australia (B)	North America (C)	Total
Threatened (T)	1	**1**	2	4
Endangered (E)	3	1	2	6
Total	4	2	4	10

Then $P(\text{B given T}) = \frac{N(\text{B and T})}{N(\text{T})} = \frac{1}{4}$

15. The mammals in A are highlighted in the table below and the mammals that are in E and A are in bold in the table below.

	Africa (A)	Australia (B)	North America (C)	Total
Threatened (T)	1	1	2	4
Endangered (E)	**3**	1	2	6
Total	4	2	4	10

Then $P(\text{E given A}) = \frac{N(\text{E and A})}{N(\text{A})} = \frac{3}{4}$.

17. The mammals in C are highlighted in the table below and the mammals that are in E and C are in bold in the table below.

	Africa (A)	Australia (B)	North America (C)	Total
Threatened (T)	1	1	2	4
Endangered (E)	3	1	**2**	6
Total	4	2	4	10

Then $P(\text{E given C}) = \frac{N(\text{E and C})}{N(\text{C})} = \frac{2}{4} = \frac{1}{2}$.

19. The mammals in B are highlighted in the table below and the mammals that are in T and B are in bold in the table below.

	Africa (A)	Australia (B)	North America (C)	Total
Threatened (T)	1	**1**	2	4
Endangered (E)	3	1	2	6
Total	4	2	4	10

Then $P(\text{T given B}) = \frac{N(\text{T and B})}{N(\text{B})} = \frac{1}{2}$.

21. $P(\text{E given A}) = \frac{3}{4}$ is greater than $P(\text{T given A}) = \frac{1}{4}$, so a higher proportion of mammals from Africa are endangered.

23. The students in U are highlighted in the table below and the students that are in D and U are in bold in the table below.

	Desktops (D)	Smartphones (S)	Total
Undergrads (U)	**2**	4	6
Graduate students (G)	3	6	9
Total	5	10	15

Then $P(\text{D given U}) = \frac{N(\text{D and U})}{N(\text{U})} = \frac{2}{6} = \frac{1}{3}$.

25. The students in U are highlighted in the table below and the students that are in S and U are in bold in the table below.

	Desktops (D)	Smartphones (S)	Total
Undergrads (U)	2	**4**	6
Graduate students (G)	3	6	9
Total	5	10	15

Then $P(\text{S given U}) = \frac{N(\text{S and U})}{N(\text{U})} = \frac{4}{6} = \frac{2}{3}$.

27. The students in U are highlighted in the table below and the students that are in D and U are in bold in the table below.

	Desktops (D)	Smartphones (S)	Total
Undergrads (U)	**2**	4	6
Graduate students (G)	3	6	9
Parents (P)	5	0	5
Total	10	10	20

Then $P(\text{D given U}) = \frac{N(\text{D and U})}{N(\text{U})} = \frac{2}{6} = \frac{1}{3}$.

29. The people in P are highlighted in the table below and the students that are in D and P are in bold in the table below.

	Desktops (D)	Smartphones (S)	Total
Undergrads (U)	2	4	6
Graduate students (G)	3	6	9
Parents (P)	**5**	0	5
Total	10	10	20

Then $P(\text{D given P}) = \frac{N(\text{D and P})}{N(\text{P})} = \frac{5}{5} = 1$.

31. The students in G are highlighted in the table below and the students that are in S and G are in bold in the table below.

	Desktops (D)	Smartphones (S)	Total
Undergrads (U)	2	4	6
Graduate students (G)	3	**6**	9
Parents (P)	5	0	5
Total	10	10	20

Then $P(\text{S given G}) = \frac{N(\text{S and G})}{N(\text{G})} = \frac{6}{9} = \frac{2}{3}$.

33. From Exercise 9, $P(\text{A given E}) = \frac{3}{6} = \frac{1}{2}$.

The mammals in A are highlighted in the table below.

	Africa (A)	Australia (B)	North America (C)	Total
Threatened (T)	1	1	2	4
Endangered (E)	3	1	2	6
Total	4	2	4	10

Then $P(\text{A}) = \frac{4}{10} = \frac{2}{5}$.

Then $P(\text{A given E}) = \frac{3}{6} = \frac{1}{2} \neq P(\text{A}) = \frac{4}{10} = \frac{2}{5}$, so A and E are not independent.

Or:

From Exercise 15, $P(\text{E given A}) = \frac{3}{4}$.

The mammals in E are highlighted in the table below.

	Africa (A)	Australia (B)	North America (C)	Total
Threatened (T)	1	1	2	4
Endangered (E)	3	1	2	6
Total	4	2	4	10

Then $P(\text{E}) = \frac{6}{10} = \frac{3}{5}$.

$P(\text{E given A}) = \frac{3}{4} \neq P(\text{E}) = \frac{6}{10} = \frac{3}{5}$, so A and E are not independent.

35. From Exercise 11, $P(\text{C given E}) = \frac{2}{6} = \frac{1}{3}$.

The mammals in C are highlighted in the table below.

	Africa (A)	Australia (B)	North America (C)	Total
Threatened (T)	1	1	2	4
Endangered (E)	3	1	2	6
Total	4	2	4	10

Then $P(\text{C}) = \frac{4}{10} = \frac{2}{5}$.

Then $P(\text{C given E}) = \frac{2}{6} = \frac{1}{3} \neq P(\text{C}) = \frac{4}{10} = \frac{2}{5}$, so C and E are not independent.

Or:

From Exercise 17, $P(\text{E given C}) = \frac{2}{4} = \frac{1}{2}$.

The mammals in E are highlighted in the table below.

	Africa (A)	Australia (B)	North America (C)	Total
Threatened (T)	1	1	2	4
Endangered (E)	3	1	2	6
Total	4	2	4	10

Then $P(\text{E}) = \frac{6}{10} = \frac{3}{5}$.

$P(\text{E given C}) = \frac{2}{4} = \frac{1}{2} \neq P(\text{E}) = \frac{6}{10} = \frac{3}{5}$, so C and E are not independent.

37. From Exercise 13, $P(\text{B given T}) = \frac{1}{4}$.

The mammals in B are highlighted in the table below.

	Africa (A)	Australia (B)	North America (C)	Total
Threatened (T)	1	1	2	4
Endangered (E)	3	1	2	6
Total	4	2	4	10

Then $P(\text{B}) = \frac{2}{10} = \frac{1}{5}$.

Then $P(\text{B given T}) = \frac{1}{4} \neq P(\text{B}) = \frac{2}{10} = \frac{1}{5}$, so B and T are not independent.

Or:

From Exercise 19, $P(\text{T given B}) = \frac{1}{2}$.

The mammals in T are highlighted in the table below.

	Africa (A)	Australia (B)	North America (C)	Total
Threatened (T)	1	1	2	4
Endangered (E)	3	1	2	6
Total	4	2	4	10

Then $P(\text{T}) = \frac{4}{10} = \frac{2}{5}$.

$P(\text{T given A}) = \frac{1}{2} \neq P(\text{T}) = \frac{4}{10} = \frac{2}{5}$, so A and T are not independent.

39. From Exercise 23, $P(\text{D given U}) = \frac{2}{6} = \frac{1}{3}$. The students in D are highlighted in the table below.

	Desktops (D)	Smartphones (S)	Total
Undergrads (U)	2	4	6
Graduate students (G)	3	6	9
Total	5	10	15

Then $P(\text{D}) = \frac{5}{15} = \frac{1}{3}$.

$P(\text{D given U}) = \frac{2}{6} = \frac{1}{3} = P(\text{D}) = \frac{5}{15} = \frac{1}{3}$. Therefore, D and U are independent.

41. From Exercise 25, $P(\text{S given U}) = \frac{4}{6} = \frac{2}{3}$. The students in S are highlighted in the table below.

	Desktops (D)	Smartphones (S)	Total
Undergrads (U)	2	4	6
Graduate students (G)	3	6	9
Total	5	10	15

Then $P(\text{S}) = \frac{10}{15} = \frac{2}{3}$.

$P(\text{S given U}) = \frac{4}{6} = \frac{2}{3} = P(\text{S}) = \frac{10}{15} = \frac{2}{3}$. Therefore, S and U are independent.

43. From Exercise 27, $P(\text{D given U}) = \frac{2}{6} = \frac{1}{3}$. The students in D are highlighted in the table below.

	Desktops (D)	Smartphones (S)	Total
Undergrads (U)	2	4	6
Graduate students (G)	3	6	9
Parents (P)	5	0	5
Total	10	10	20

Then $P(\text{D}) = \frac{10}{20} = \frac{1}{2}$.

$P(\text{D given U}) = \frac{2}{6} = \frac{1}{3} \neq P(\text{D}) = \frac{10}{20} = \frac{1}{2}$. Therefore, D and U are not independent.

45. From Exercise 29, $P(\text{D given P}) = \frac{5}{5} = 1$. The students in D are highlighted in the table below

	Desktops (D)	Smartphones (S)	Total
Undergrads (U)	2	4	6
Graduate students (G)	3	6	9
Parents (P)	5	0	5
Total	10	10	20

Then $P(\text{D}) = \frac{10}{20} = \frac{1}{2}$.

$P(\text{D given P}) = \frac{5}{5} = 1 \neq P(\text{D}) = \frac{10}{20} = \frac{1}{2}$. Therefore, D and P are not independent.

47. From Exercise 31, $P(\text{S given G}) = \frac{6}{9} = \frac{2}{3}$. The students in S are highlighted in the table below.

	Desktops (D)	Smartphones (S)	Total
Undergrads (U)	2	4	6
Graduate students (G)	3	6	9
Parents (P)	5	0	5
Total	10	10	20

Then $P(\text{S}) = \frac{10}{20} = \frac{1}{2}$.

$P(\text{S given G}) = \frac{6}{9} = \frac{2}{3} \neq P(\text{G}) = \frac{10}{20} = \frac{1}{2}$. Therefore, S and G are not independent.

49. It is not true. The probabilities in Problems 33–38 indicate that the continent a mammal lives on and whether the mammal is endangered or threatened are not independent.

51. It is not true. The probabilities in Problems 39–42 indicate that whether a student prefers a desktop or a laptop is independent of whether the student is an undergraduate or a graduate.

53. P(male who applied for federal financial aid)
= P(applied for federal financial aid) · P(male given applied for federal financial aid)
$= (0.71)(0.475) = 0.33725$

55. P(take statistics and pass statistics)
= P(take statistics) · P(pass statistics given take statistics)
$= (0.3)(0.9) = 0.27$

57. P(male and taking a biology course this semester)
= P(male) · P(taking a biology course this semester given male)
$= (0.25)(0.50) = 0.125$

59. $P(\text{F1 and F2}) = P(\text{F1}) \cdot P(\text{F2}) = (0.36)(0.36) = 0.1296$

61. $P(\text{M1 and F2}) = P(\text{M1}) \cdot P(\text{F2}) = (0.38)(0.36) = 0.1368$

63. $P(\text{F2 given F1}) = \frac{P(\text{F1 and F2})}{P(\text{F1})} = \frac{0.1296}{0.36} = 0.36$

65. $P(\text{F1 and F2}) = (0.36)(0.36) = 0.1296$ whereas $P(\text{F2 given F1}) = 0.36$. Less than 1% of population was sampled, so the 1% Guideline applies. Therefore, F1 and F2 can be considered independent events, so $(\text{F1 and F2}) = (0.36)(0.36) = 0.1296$ and $P(\text{F2 given F1}) = \frac{P(\text{F1 and F2})}{P(\text{F1})} = \frac{P(\text{F1})P(\text{F2})}{P(\text{F1})} = P(\text{F2}) = 0.36.$

For Problems 67–70, we are sampling with replacement, so we have independent events. Therefore, we can use the Multiplication Rule for Independent Events.

$P(\text{R1}) = \frac{26}{52} = \frac{1}{2}$

$P(\text{R2}) = \frac{26}{52} = \frac{1}{2}$

$P(\text{H1}) = \frac{13}{52} = \frac{1}{4}$

$P(\text{H2}) = \frac{13}{52} = \frac{1}{4}$

67. $P(\text{R1 and R2}) = P(\text{R1}) \cdot P(\text{R2}) = \left(\frac{1}{2}\right)\left(\frac{1}{2}\right) = \frac{1}{4} = 0.25$

69. $P(\text{R1 and H2}) = P(\text{R1}) \cdot P(\text{H2}) = \left(\frac{1}{2}\right)\left(\frac{1}{4}\right) = \frac{1}{8} = 0.125$

For Exercises 71–74, we are sampling without replacement. Therefore we have dependent events. A sample of size $n = 2$ is $\left(\frac{2}{52}\right) \cdot 100\% \approx 3.85\%$ of the population and a sample of size $n = 3$ is $\left(\frac{3}{52}\right) \cdot 100\% \approx 5.77\%$ of the population. Therefore, the 1% Guideline does not apply.

$P(\text{R1}) = \frac{26}{52} = \frac{1}{2}$

$P(\text{R2 given R1}) = \frac{25}{51}$

$P(\text{R3 given R1 and R2}) = \frac{24}{50} = \frac{12}{25}$

$P(\text{H1}) = \frac{13}{52} = \frac{1}{4}$

$P(\text{H2 given H1}) = \frac{12}{51} = \frac{4}{17}$

$P(\text{H3 given H1 and H2}) = \frac{11}{50}$

$P(\text{R2 given H1}) = \frac{12}{51} = \frac{4}{17}$

$P(\text{R3 given H1 and R2}) = \frac{11}{50}$

$P(\text{R3 given H1 and H2}) = \frac{24}{50} = \frac{12}{25}$

71. $P(\text{H1 and R2 and R3}) = P(\text{H1}) \cdot P(\text{R2 given H1}) \cdot P(\text{R3 given H1 and R2}) = \left(\frac{1}{2}\right)\left(\frac{25}{51}\right)\left(\frac{11}{50}\right) = \frac{11}{102} \approx 0.1078$

73. $P(\text{R1 and R2 and R3}) = P(\text{R1}) \cdot P(\text{R2 given R1}) \cdot P(\text{R3 given R1 and R2}) = \left(\frac{1}{2}\right)\left(\frac{25}{51}\right)\left(\frac{12}{25}\right) = \frac{2}{17} \approx 0.1176$

75. **(a)** A sample of size $n = 2$ is $\frac{2}{1{,}500{,}000} \cdot 100\% \approx 0.00013\%$ of the population, which is less than 1%. Therefore the 1% Guideline applies.

(b) A sample of size $n = 3$ is $\frac{3}{1,500,000} \cdot 100\% = 0.0002\%$ of the population, which is less than 1%. Therefore the 1% Guideline applies.

(c) A sample of size $n = 5$ is $\frac{5}{1,500,000} \cdot 100\% \approx 0.00033\%$ of the population, which is less than 1%. Therefore the 1% Guideline applies.

For Exercises 76–78 let S_i denote the event that the ith student who took the Natural Sciences subject exam in 2014 had taken four years of science in high school. From Exercise 75, S_1, S_2, S_3, S_4, and S_5 are all independent. Then
$P(S_1) = P(S_2) = P(S_3) = P(S_4) = P(S_5) = 0.48.$

77. $P(S_1 \text{ and } S_2 \text{ and } S_3) = P(S_1) \cdot P(S_2) \cdot P(S_3) = (0.48)(0.48)(0.48) = 0.110592 \approx 0.1106$

79. The manga that are in V and A are highlighted in the table below.

	VIZ Media (V)	Kodansha (K)	Seven Seas	Total
1 week (A)	3	1	0	4
2 weeks (B)	3	0	1	4
>2 weeks (C)	0	2	0	2
Total	6	3	1	10

Then $P(\text{V and A}) = \frac{3}{10} = 0.3$.

The manga that are in V are highlighted below.

	VIZ Media (V)	Kodansha (K)	Seven Seas	Total
1 week (A)	3	1	0	4
2 weeks (B)	3	0	1	4
>2 weeks (C)	0	2	0	2
Total	6	3	1	10

Therefore, $P(\text{V}) = \frac{6}{10} = 0.6$.

The manga that are in A are highlighted below.

	VIZ Media (V)	Kodansha (K)	Seven Seas	Total
1 week (A)	3	1	0	4
2 weeks (B)	3	0	1	4
>2 weeks (C)	0	2	0	2
Total	6	3	1	10

Therefore, $P(\text{A}) = \frac{4}{10} = 0.4$.

Then $P(\text{V}) \cdot P(\text{A}) = (0.6)(0.4) = 0.24 \neq 0.3 = P(\text{V and A})$.

Therefore, V and A are not independent.

81. The manga that are in V and B are highlighted in the table below.

	VIZ Media (V)	Kodansha (K)	Seven Seas	Total
1 week (A)	3	1	0	4
2 weeks (B)	3	0	1	4
>2 weeks (C)	0	2	0	2
Total	6	3	1	10

Then $P(\text{V and B}) = \frac{3}{10} = 0.3$.

The manga that are in V are highlighted below.

	VIZ Media (V)	Kodansha (K)	Seven Seas	Total
1 week (A)	3	1	0	4
2 weeks (B)	3	0	1	4
>2 weeks (C)	0	2	0	2
Total	6	3	1	10

Therefore, $P(\text{V}) = \frac{6}{10} = 0.6$.

The manga that are in B are highlighted below.

	VIZ Media (V)	Kodansha (K)	Seven Seas	Total
1 week (A)	3	1	0	4
2 weeks (B)	3	0	1	4
>2 weeks (C)	0	2	0	2
Total	6	3	1	10

Therefore, $P(\text{B}) = \frac{4}{10} = 0.4$.

Then $P(\text{V}) \cdot P(\text{B}) = (0.6)(0.4) = 0.24 \neq 0.3 = P(\text{V and B})$.

Therefore, V and B are not independent.

83. The manga that are in V and C are highlighted in the table below.

	VIZ Media (V)	Kodansha (K)	Seven Seas	Total
1 week (A)	3	1	0	4
2 weeks (B)	3	0	1	4
>2 weeks (C)	0	2	0	2
Total	6	3	1	10

Then $P(\text{V and C}) = \frac{0}{10} = 0$.

The manga that are in V are highlighted below.

	VIZ Media (V)	Kodansha (K)	Seven Seas	Total
1 week (A)	3	1	0	4
2 weeks (B)	3	0	1	4
>2 weeks (C)	0	2	0	2
Total	6	3	1	10

Therefore, $P(\text{V}) = \frac{6}{10} = 0.6$.

The manga that are in C are highlighted below.

	VIZ Media (V)	Kodansha (K)	Seven Seas	Total
1 week (A)	3	1	0	4
2 weeks (B)	3	0	1	4
>2 weeks (C)	0	2	0	2
Total	6	3	1	10

Therefore, $P(\text{C}) = \frac{2}{10} = 0.2$.

Then $P(\text{V}) \cdot P(\text{C}) = (0.6)(0.2) = 0.12 \neq 0 = P(\text{V and C})$.

Therefore, V and C are not independent.

85.

	VIZ Media (V)	Kodansha (K)	Seven Seas	Total
1 week (A)	3	1	0	4
2 weeks (B)	3	0	1	4
>2 weeks (C)	0	2	0	2
Total	6	3	1	10

V and K are mutually exclusive. Thus $P(\text{V and K}) = \frac{0}{10} = 0$.

The manga that are in V are highlighted below.

	VIZ Media (V)	Kodansha (K)	Seven Seas	Total
1 week (A)	3	1	0	4
2 weeks (B)	3	0	1	4
>2 weeks (C)	0	2	0	2
Total	6	3	1	10

Therefore, $P(\text{V}) = \frac{6}{10} = 0.6$.

The manga that are in K are highlighted below.

	VIZ Media (V)	Kodansha (K)	Seven Seas	Total
1 week (A)	3	1	0	4
2 weeks (B)	3	0	1	4
>2 weeks (C)	0	2	0	2
Total	6	3	1	10

Therefore, $P(\text{K}) = \frac{3}{10} = 0.3$.

Then $P(\text{V}) \cdot P(\text{K}) = (0.6)(0.3) = 0.18 \neq 0 = P(\text{V and K})$.

Therefore, V and K are not independent.

87.

	VIZ Media (V)	Kodansha (K)	Seven Seas	Total
1 week (A)	3	1	0	4
2 weeks (B)	3	0	1	4
>2 weeks (C)	0	2	0	2
Total	6	3	1	10

A and C are mutually exclusive. Thus $P(\text{A and C}) = \frac{0}{10} = 0$.

The manga that are in A are highlighted below.

	VIZ Media (V)	Kodansha (K)	Seven Seas	Total
1 week (A)	3	1	0	4
2 weeks (B)	3	0	1	4
>2 weeks (C)	0	2	0	2
Total	6	3	1	10

Therefore, $P(\text{A}) = \frac{4}{10} = 0.4$.

The manga that are in C are highlighted below.

	VIZ Media (V)	Kodansha (K)	Seven Seas	Total
1 week (A)	3	1	0	4
2 weeks (B)	3	0	1	4
>2 weeks (C)	0	2	0	2
Total	6	3	1	10

Therefore, $P(\text{C}) = \frac{2}{10} = 0.2$.

Then $P(\text{A}) \cdot P(\text{C}) = (0.4)(0.2) = 0.08 \neq 0 = P(\text{A and C})$.

Therefore, A and C are not independent.

89. If $P(X) \neq 0$ and $P(Y) \neq 0$, then $P(X) \cdot P(Y) \neq 0 = P(X \text{ and } Y)$ so X and Y are not independent. If either $P(X) = 0$ or $P(Y) = 0$, then $P(X) \cdot P(Y) = 0 = P(X \text{ and } Y)$ and X and Y are independent.

91. If the intersection of events W and Z is empty, then $P(W \text{ and } Z) = 0$. If $P(W) \neq 0$ and $P(Z) \neq 0$, then $P(W) \cdot P(Z) \neq 0 = P(W \text{ and } Z)$ so W and Z are not independent. If either $P(W) = 0$ or $P(Z) = 0$, then $P(W) \cdot P(Z) = 0 = P(W \text{ and } Z)$ and W and Z are independent.

For 93–96, the sample space is $\{1, 2, 3, 4, 5, 6\}$. There are 3 outcomes in the sample space that are even: $\{2, 4, 6\}$. Therefore, $P(E) = \frac{3}{6} = \frac{1}{2}$.

Let E_i be the event roll an even number on the ith toss. Then $P(E_i) = \frac{1}{2}$.

93. $P(E_1 \text{ and } E_2 \text{ and } E_3) = P(E_1) \cdot P(E_2) \cdot P(E_3) = \left(\frac{1}{2}\right)\left(\frac{1}{2}\right)\left(\frac{1}{2}\right) = \frac{1}{8}$.

95. $P(E_1 \text{ and } E_2 \text{ and } E_3 \text{ and } E_4 \text{ and } E_5) = P(E_1) \cdot P(E_2) \cdot P(E_3) \cdot P(E_4) \cdot P(E_5) =$ $\left(\frac{1}{2}\right)\left(\frac{1}{2}\right)\left(\frac{1}{2}\right)\left(\frac{1}{2}\right)\left(\frac{1}{2}\right) = \frac{1}{32}$

For 97–100, the sample space is $\{1, 2, 3, 4, 5, 6\}$. There are 3 outcomes in the sample space that are greater than 3: $\{4, 5, 6\}$. Therefore, $P(H) = \frac{3}{6} = \frac{1}{2}$.

Let H_i be the event roll a number greater than 3 on the ith toss. Then $P(H_i) = \frac{1}{2}$ and $P(H_i^C) = 1 - P(H_i) = 1 - \frac{1}{2} = \frac{1}{2}$.

97. $P(H$ occurs at least once$) = 1 - P(H$ does not occur$) = 1 - P(H_1^C$ and H_2^C and $H_3^C)$
$= 1 - [P(H_1^C) \cdot P(H_2^C) \cdot P(H_3^C)] = 1 - \left(\frac{1}{2}\right)\left(\frac{1}{2}\right)\left(\frac{1}{2}\right) = 1 - \frac{1}{8} = \frac{7}{8}$

99. $P(H$ occurs at least once$) = 1 - P(H$ does not occur$) = 1 -$
$P(H_1^C$ and H_2^C and H_3^C and H_4^C and $H_5^C) = 1 - [P(H_1^C) \cdot P(H_2^C) \cdot P(H_3^C) \cdot P(H_4^C) \cdot P(H_5^C)] = 1 -$
$\left(\frac{1}{2}\right)\left(\frac{1}{2}\right)\left(\frac{1}{2}\right)\left(\frac{1}{2}\right)\left(\frac{1}{2}\right) = 1 - \frac{1}{32} = \frac{31}{32}$

For Exercises 101–104, let E_i be the event the ith email is encrypted. Then $P(E_i) = 0.5$ and $P(E_i^C) = 1 - P(E_i) = 1 - 0.5 = 0.5$.

101. P(at least one email is encrypted) = 1 – P(no emails are encrypted) = 1 – $P(E_1^C$ and $E_2^C)$
$= 1 - [P(E_1^C) \cdot P(E_2^C)] = 1 - (0.5)(0.5) = 1 - 0.25 = 0.75$

103. P(at least one email is encrypted) = 1 – P(no emails are encrypted) = 1 –
$P(E_1^C$ and E_2^C and E_3^C and $E_4^C) = 1 -$
$[P(E_1^C) \cdot P(E_2^C) \cdot P(E_3^C) \cdot P(E_4^C)] = 1 - (0.5)(0.5)(0.5)(0.5) = 1 - 0.0625 = 0.9375$

105. $P(A \text{ given } B) = \frac{P(A) \cdot P(B|A)}{P(A) \cdot P(B|A) + P(A^C) \cdot P(B|A^C)} = \frac{(0.5)(0.3)}{(0.5)(0.3)+(0.5)(0.5)} = \frac{0.15}{0.15+0.25} = \frac{0.15}{0.40} = 0.375$

107. $P(A \text{ given } B) = \frac{P(A) \cdot P(B|A)}{P(A) \cdot P(B|A) + P(A^C) \cdot P(B|A^C)} = \frac{(0.4)(0.25)}{(0.4)(0.25)+(0.6)(0.4)} = \frac{0.1}{0.1+0.24} = \frac{0.1}{0.34} \simeq 0.2941$

109. Let M_i be the event that the ith YouTube user accesses YouTube with a mobile device. Then $P(M_i) = 0.4$ and $P(M_i^C) = 1 - P(M_i) = 1 - 0.4 = 0.6$.

(a) $P(M_1 \text{ and } M_2) = P(M_1) \cdot P(M_2) = (0.4)(0.4) = 0.16$

(b) $P(M_1 \text{ and } M_2 \text{ and } M_3) = P(M_1) \cdot P(M_2) \cdot P(M_3) = (0.4)(0.4)(0.4) = 0.064$

(c) $P(M_1 \text{ and } M_2 \text{ and } M_3 \text{ and } M_4 \text{ and } M_5) = P(M_1) \cdot P(M_2) \cdot P(M_3) \cdot P(M_4) \cdot P(M_5) = (0.4)(0.4)(0.4)(0.4)(0.4) = 0.01024$

(d) P(at least one YouTube user accesses YouTube with a mobile device) =

1 – P(none of the YouTube users access YouTube with a mobile evice)

$= 1 - P(M_1^C \text{ and } M_2^C \text{ and } M_3^C \text{ and } M_4^C) = 1 - \text{P}(M_1^C) \cdot \text{P}(M_2^C) \cdot \text{P}(M_3^C) \cdot \text{P}(M_4^C)$

$1 - (0.6)(0.6)(0.6)(0.6) = 1 - 0.1296 = 0.8704.$

111. Let V_i be the event that the ith social media user plays video games. 60% of 10 is $(0.6)(10) = 6$ of the 10 social media users play video games.

(a) $P(V_1) = \frac{6}{10} = \frac{3}{5}$

(b) $P(V_2|V_1) = \frac{5}{9}$

(c) $P(V_3|V_1 \text{ and } V_2) = \frac{4}{8} = \frac{1}{2}$

113. **(a)** The females are highlighted in the table below and the females that are in Associate's degree programs are in bold.

	Associates	**Bachelor's**	**Total**
Female	**5,276**	5.491	10,767
Male	3,819	4,483	8,302
Total	9,095	9,974	19,069

Therefore, $P(\text{Associate's degree} \mid \text{Female}) = \frac{N(\text{Associate's degree and Female})}{N(\text{Female})} = \frac{5{,}276}{10{,}767} \approx 0.4900.$

(b) The males are highlighted in the table below and the males that are in Associate's degree programs are in bold.

	Associates	**Bachelor's**	**Total**
Female	5,276	5,491	10,767
Male	**3,819**	4,483	8,302
Total	9,095	9,974	19,069

Therefore, $P(\text{Associate's degree} \mid \text{Male}) = \frac{N(\text{Associate's degree and Male})}{N(\text{Male})} = \frac{3{,}819}{8{,}302} \approx 0.4600.$

(c) The females are highlighted in the table below and the females that are in Bachelor's degree programs are in bold.

	Associates	**Bachelor's**	**Total**
Female	5,276	**5,491**	10,767
Male	3,819	4,483	8,302
Total	9,095	9,974	19,069

Therefore, $P(\text{Bachelor's degree} \mid \text{Female}) = \frac{N(\text{Bachelor's degree and Female})}{N(\text{Female})} = \frac{5{,}491}{10{,}767} \approx 0.5100.$

(d) The males are highlighted in the table below and the males that are in Bachelor's degree programs are in bold.

	Associates	**Bachelor's**	**Total**
Female	5,276	5,491	10,767
Male	3,819	**4,483**	8,302
Total	9,095	9,974	19,069

Therefore, $P(\text{Bachelor's degree} \mid \text{Male}) = \frac{N(\text{Bachelor's degree and Male})}{N(\text{Male})} = \frac{4{,}483}{8{,}302} \approx 0.5400.$

115. **(a)** The people in B are highlighted in the table below.

	Completed Bachelor's degree or more (B)	**Completed less than a Bachelor's degree (B^C)**	**Total**
Millennials: 2013 (M)	6,686	11,662	18,348
Gen X'ers: 1995 (M^C)	4,458	10,897	15,355
Total	11,144	22,559	33,703

Therefore, $P(B) = \frac{11{,}144}{33{,}703} \approx 0.3307.$

(b) The people in M are highlighted in the table below.

	Completed Bachelor's de-	**Completed less than a Bache-**	**Total**

	gree or more (*B*)	lor's degree (B^C)	
Millennials: 2013 (*M*)	6,686	11,662	18,348
Gen X'ers: 1995 (M^C)	4,458	10,897	15,355
Total	11,144	22,559	33,703

Therefore, $P(M) = \frac{18,348}{33,703} \approx 0.5444$.

117. **(a)** The people in *M* are highlighted in the table below and the people that are in *M* and *B* are in bold in the table below.

	Completed Bachelor's degree or more (*B*)	Completed less than a Bachelor's degree (B^C)	Total
Millennials: 2013 (*M*)	**6,686**	11,662	18,348
Gen X'ers: 1995 (M^C)	4,458	10,897	15,355
Total	11,144	22,559	33,703

Therefore, $P(B\,|M) = \frac{N(B \text{ and } M)}{N(M)} = \frac{6,686}{18,348} \approx 0.3644$

(b) The people in *M* are highlighted in the table below and the people that are in *M* and B^C are in bold in the table below.

	Completed Bachelor's degree or more (*B*)	Completed less than a Bachelor's degree (B^C)	Total
Millennials: 2013 (*M*)	6,686	**11,662**	18,348
Gen X'ers: 1995 (M^C)	4,458	10,897	15,355
Total	11,144	22,559	33,703

Therefore, $P(B^C\,|M) = \frac{N(B^C \text{ and } M)}{N(M)} = \frac{11,662}{18,348} \approx 0.6356$

(c) The people in M^C are highlighted in the table below and the people that are in M^C and *B* are in bold in the table below.

	Completed Bachelor's degree or more (*B*)	Completed less than a Bachelor's degree (B^C)	Total
Millennials: 2013 (*M*)	6,686	11,662	18,348
Gen X'ers: 1995 (M^C)	**4,458**	10,897	15,355
Total	11,144	22,559	33,703

Therefore, $P(B\,|M^C) = \frac{N(B \text{ and } M^C)}{N(M^C)} = \frac{4,458}{15,355} \approx 0.2903$

(d) The people in M^C are highlighted in the table below and the people that are in M^C and B^C are in bold in the table below.

	Completed Bachelor's degree or more (B)	Completed less than a Bachelor's degree (B^C)	Total
Millennials: 2013 (M)	6,686	11,662	18,348
Gen X'ers: 1995 (M^C)	4,458	**10,897**	15,355
Total	11,144	22,559	33,703

Therefore, $P(B^C \mid M^C) = \frac{N(B^C \text{ and } M^C)}{N(M^C)} = \frac{10{,}897}{15{,}355} \approx 0.7097$

119. **(a)** The females are highlighted in the table below.

		Happiness of Marriage			
		Very happy	**Pretty happy**	**Not too happy**	**Total**
Gender	**Male**	242	115	9	366
	Female	257	149	17	423
	Total	499	264	26	789

Therefore, $P(\text{Female}) = \frac{423}{789} \approx 0.5361$

(b) The males are highlighted in the table below.

		Happiness of Marriage			
		Very happy	**Pretty happy**	**Not too happy**	**Total**
Gender	**Male**	242	115	9	366
	Female	257	149	17	423
	Total	499	264	26	789

Therefore, $P(\text{Male}) = \frac{366}{789} \approx 0.4639$

(c) The people who said that they were not too happy in their marriage are highlighted in the table below.

		Happiness of Marriage			
		Very happy	**Pretty happy**	**Not too happy**	**Total**
Gender	**Male**	242	115	9	366
	Female	257	149	17	423
	Total	499	264	26	789

Therefore, $P(\text{Not}) = \frac{26}{789} \approx 0.0330$

121. Not independent. $P(F) \cdot P(Not) = (0.5361) \cdot (0.0330) = 0.0176913 \neq 0.0215 = P(F \text{ and } Not)$ and $P(M) \cdot P(Not) = (0.4639) \cdot (0.0330) = 0.0153087 \neq 0.0114 = P(M \text{ and } Not)$

123. Either reject the batch if at least one computer is defective or increase the sample size.

125. **(a)** The patients in F and C are highlighted in the table below.

		Practice Type			
		Pediatric	**Family (*F*)**	**OB/GYN (*G*)**	**Total**
	No	353	259	332	944
Completed	**Yes (*C*)**	162	106	201	469
	Total	515	365	533	1413

Therefore, $P(F \text{ and } C) = \frac{106}{1413} \approx 0.0750$

(b) The patients in G and C are highlighted in the table below.

		Practice Type			
		Pediatric	Family (*F*)	OB/GYN (*G*)	Total
	No	353	259	332	944
Completed	Yes (*C*)	162	106	201	469
	Total	515	365	533	1413

Therefore, $P(G \text{ and } C) = \frac{201}{1413} \approx 0.1423$

127. **(a)** $P(F) \cdot P(C) = \left(\frac{365}{1413}\right)\left(\frac{469}{1413}\right) = \frac{171{,}185}{1{,}996{,}569} \approx 0.0857 \neq 0.0750 \approx \frac{106}{1413} = P(F \text{ and } C)$. Therefore F and C are not independent.

Or:

$P(C\,|F) = \frac{106}{365} \approx 0.2904 \neq 0.3319 \approx \frac{469}{1413} = P(C)$. Therefore F and C are not independent.

(b) $P(G) \cdot P(C) = \left(\frac{533}{1413}\right)\left(\frac{469}{1413}\right) = \frac{249{,}977}{1{,}996{,}569} \approx 0.1252 \neq 0.1423 \approx \frac{201}{1413} = P(G \text{ and } C)$. Therefore C and C are not independent.

Or:

$P(C\,|G) = \frac{201}{533} \approx 0.3771 \neq 0.3319 \approx \frac{469}{1413} = P(C)$. Therefore G and C are not independent.

129. **(a)** The patients in C are highlighted in the table below.

	Did not complete treatment	Completed treatment (C)	Total
11–17 years old (A)	454	247	701
18–26 years old (B)	490	222	712
Total	944	469	1413

Therefore, $P(C) = \frac{469}{1413} \approx 0.3319$

(b) The patients in A are highlighted in the table below.

	Did not complete treatment	Completed treatment (C)	Total
11–17 years old (A)	454	247	701
18–26 years old (B)	490	222	712
Total	944	469	1413

Therefore, $P(A) = \frac{701}{1413} \approx 0.4961$

(c) The patients in B are highlighted in the table below.

	Did not complete treatment	Completed treatment (C)	Total
11–17 years old (A)	454	247	701
18–26 years old (B)	490	222	712
Total	944	469	1413

Therefore, $P(B) = \frac{712}{1413} \approx 0.5039$

(d) The patients in C and A are highlighted in the table below.

	Did not complete treatment	Completed treatment (C)	Total
11–17 years old (A)	454	247	701
18–26 years old (B)	490	222	712
Total	944	469	1413

Therefore, $P(C \text{ and } A) = \frac{247}{1413} \approx 0.1748$

(e) The patients in C and B are highlighted in the table below.

	Did not complete treatment	Completed treatment (C)	Total
11–17 years old (A)	454	247	701
18–26 years old (B)	490	222	712
Total	944	469	1413

Therefore, $P(C \text{ and } B) = \frac{222}{1413} \approx 0.1571$

(f) The patients in A are highlighted in the table below and the patients in C and A are in bold in the table below.

	Did not complete treatment	Completed treatment (C)	Total
11–17 years old (A)	454	**247**	701
18–26 years old (B)	490	222	712
Total	944	469	1413

Therefore, $P(C \mid A) = \frac{N(C \text{ and } A)}{N(A)} = \frac{247}{701} \approx 0.3524$

(g) The patients in B are highlighted in the table below and the patients in C and B are in bold in the table below.

	Did not complete treatment	Completed treatment (C)	Total
11–17 years old (A)	454	247	701
18–26 years old (B)	490	**222**	712
Total	944	469	1413

Therefore, $P(C \mid B) = \frac{N(C \text{ and } B)}{N(B)} = \frac{222}{712} \approx 0.3118$

131. The age group that had a higher proportion finishing the treatment is the 11–17 years old. This represents true variation. People 11–17 years old are more likely to have their parents making sure they get their vaccines.

Section 5.4

1. Tree diagram.

3. In a permutation, order is important. In a combination, order is not important.

5. Acceptance sampling refers to the process of (1) selecting a random sample from a batch of items, (2) evaluating the sample for defectives, and (3) either accepting or rejecting the entire batch based on the evaluation of the sample.

7.

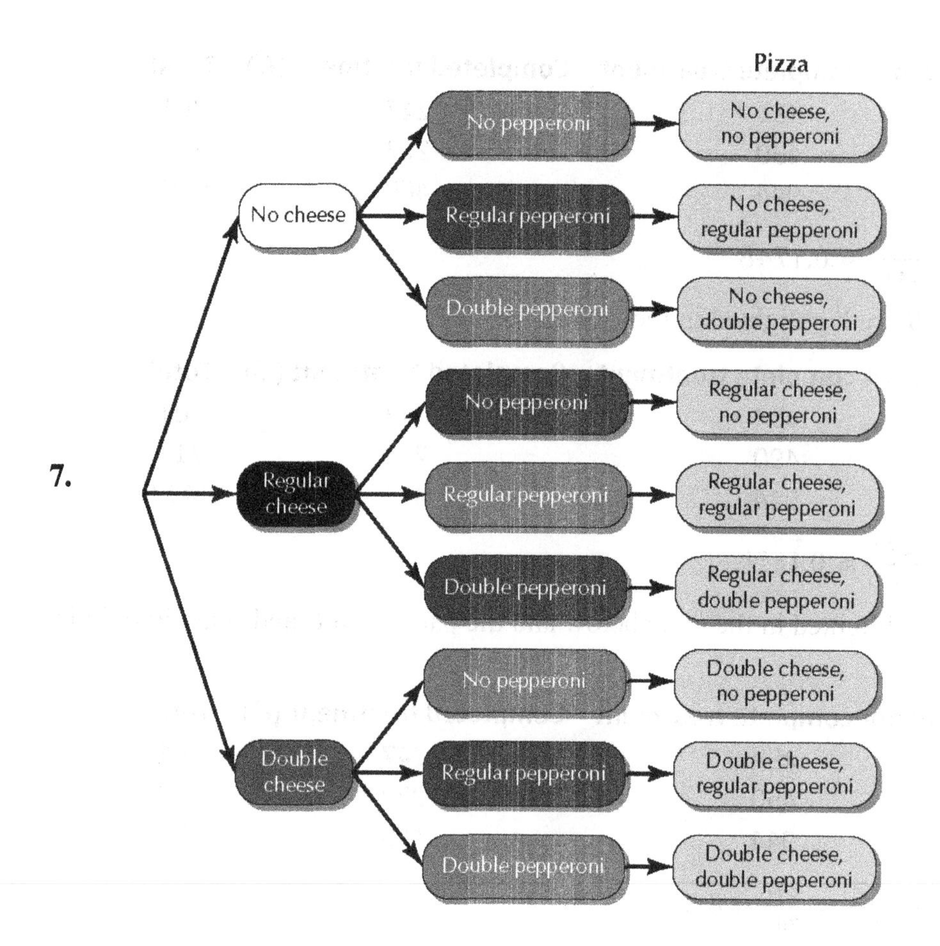
Pizza
No cheese
Regular cheese
Double cheese
No pepperoni
Regular pepperoni
Double pepperoni
No pepperoni
Regular pepperoni
Double pepperoni
No pepperoni
Regular pepperoni
Double pepperoni
No cheese, no pepperoni
No cheese, regular pepperoni
No cheese, double pepperoni
Regular cheese, no pepperoni
Regular cheese, regular pepperoni
Regular cheese, double pepperoni
Double cheese, no pepperoni
Double cheese, regular pepperoni
Double cheese, double pepperoni

9.

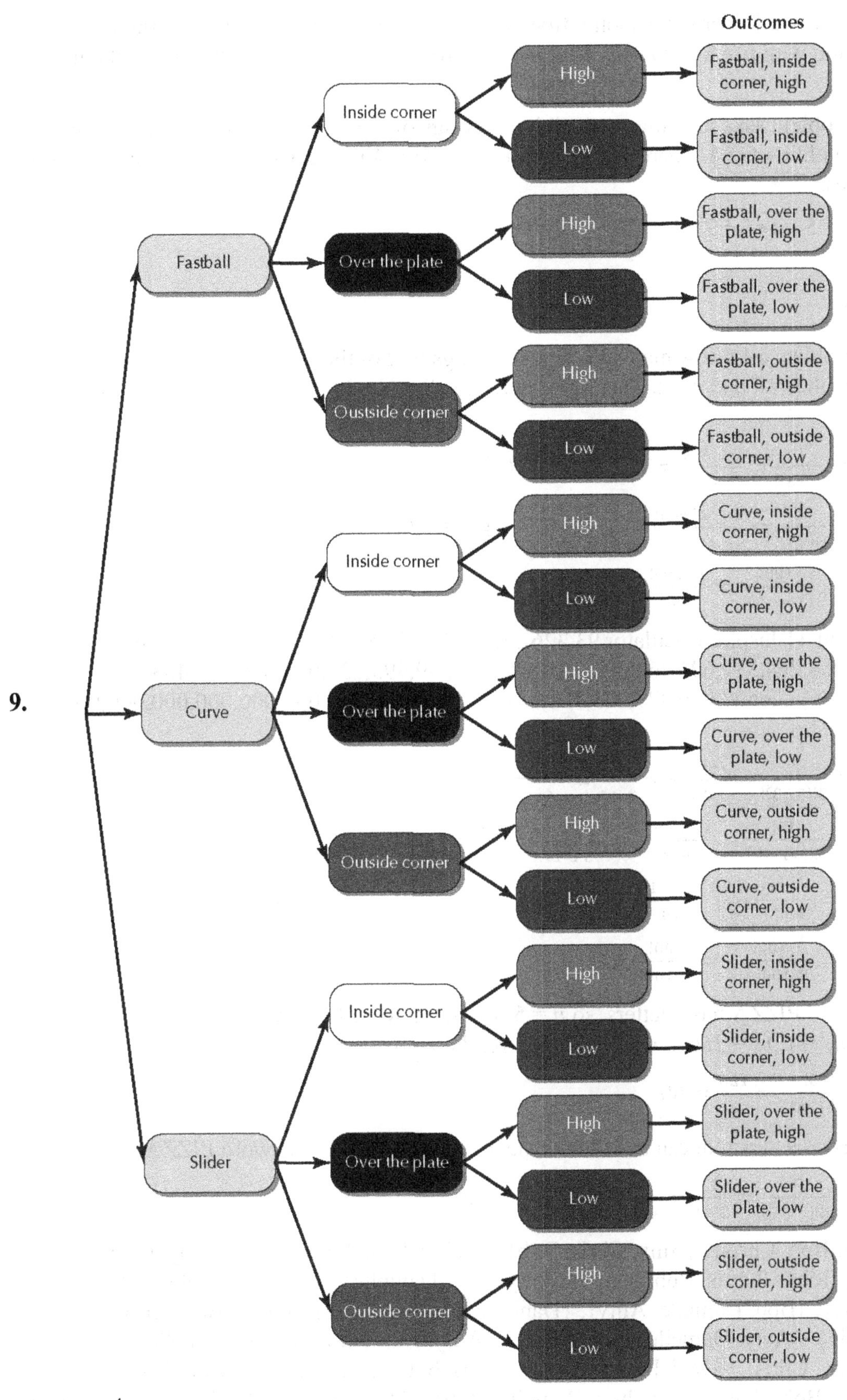
Outcomes
Fastball
Curve
Slider
Inside corner
Over the plate
Outside corner
Inside corner
Over the plate
Outside corner
Inside corner
Over the plate
Outside corner
High
Low
Fastball, inside corner, high
Fastball, inside corner, low
Fastball, over the plate, high
Fastball, over the plate, low
Fastball, outside corner, high
Fastball, outside corner, low
Curve, inside corner, high
Curve, inside corner, low
Curve, over the plate, high
Curve, over the plate, low
Curve, outside corner, high
Curve, outside corner, low
Slider, inside corner, high
Slider, inside corner, low
Slider, over the plate, high
Slider, over the plate, low
Slider, outside corner, high
Slider, outside corner, low

11. 26^4

13. The survey asks for the person's first and second favorite flavors out of 5 flavors, so this is a permutation with $n = 5$ and $r = 2$. Therefore, there are ${}_5P_2 = 5 \cdot 4 = 20$ possible sets of favorites.

15. The problem asks how many possible orderings of the 4 sororities are there, so this is a permutation with $n = r = 4$. Therefore, there are ${}_4P_4 = 4! = 4 \cdot 3 \cdot 2 \cdot 1 = 24$ possible orderings of the 4 sororities.

17. $6! = 6 \cdot 5 \cdot 4 \cdot 3 \cdot 2 \cdot 1 = 720$

19. $0! = 1$

21. $1! = 1$

23. The problem asks how many possible orderings of 2 of the 4 sororities there are, so this is a permutation with $n = 4$ and $r = 2$. Therefore, there are ${}_4P_2 = 4 \cdot 3 = 12$ possible orderings of 2 of the 4 sororities.

25. ${}_7P_3 = \frac{7!}{(7-3)!} = \frac{7 \cdot 6 \cdot 5 \cdot 4!}{4!} = 7 \cdot 6 \cdot 5 = 210$

27. ${}_8P_5 = \frac{8!}{(8-5)!} = \frac{8 \cdot 7 \cdot 6 \cdot 5 \cdot 4 \cdot 3!}{3!} = 8 \cdot 7 \cdot 6 \cdot 5 \cdot 4 = 6720$

29. ${}_{100}P_1 = \frac{100!}{(100-1)!} = \frac{100 \cdot 99!}{99!} = 100$

31. Done on TI-Inspire calculator 93,326,215,443,944,152,681,699,238,856,266, 700,490,715,968,264,381,621,468,592,963,895,217,599,993,229,915,608,941,463, 976,156,518,286,253,697,920,827,223,758,251,185,210,916,864,000,000,000,000,000,000,000,0 00

33. ${}_7C_3 = \frac{7!}{3!(7-3)!} = \frac{7 \cdot 6 \cdot 5 \cdot 4!}{3 \cdot 2 \cdot 1 \cdot 4!} = \frac{7 \cdot 6 \cdot 5}{3 \cdot 2 \cdot 1} = \frac{210}{6} = 35$

35. ${}_{11}C_8 = \frac{11!}{8!(11-8)!} = \frac{11 \cdot 10 \cdot 9 \cdot 8!}{8! \cdot 3!} = \frac{11 \cdot 10 \cdot 9}{3 \cdot 2 \cdot 1} = \frac{990}{6} = 165$

37. ${}_{11}C_{10} = \frac{11!}{10!(11-10)!} = \frac{11 \cdot 10!}{10! \cdot 1!} = \frac{11}{1} = 11$

39. ${}_{100}C_0 = \frac{100!}{0!(100-0)!} = \frac{100!}{0! \cdot 100!} = 1$

41. The word PIZZA has 5 letters, so $n = 5$. Also, the word PIZZA has 1 P, 1 I, 2 Z's, and 1 A, so $n_1 = 1$, $n_2 = 1$, $n_3 = 2$, and $n_4 = 1$. Therefore, there are

$$\frac{n!}{n_1!n_2!n_3!n_4!} = \frac{5!}{1!1!2!1!} = \frac{120}{2} = 60$$

distinct strings of letters that can be made using all of the letters in the word PIZZA.

43. ${}_7C_3 = \frac{7!}{3!(7-3)!} = \frac{7!}{3! \cdot 4!} = \frac{7!}{4! \cdot 3!} = \frac{7!}{4!(7-4)!} = {}_7C_4$

45. {Amy, Bob, Chris}, {Amy, Chris, Bob}, {Bob, Chris, Amy}, {Bob, Amy, Chris}, {Chris, Amy, Bob}, {Chris, Bob, Amy}, {Amy, Bob, Danielle}, {Amy, Danielle, Bob}, {Bob, Amy, Danielle}, {Bob, Danielle, Amy}, {Danielle, Amy, Bob}, {Danielle, Bob, Amy}, {Amy, Chris, Danielle}, {Amy, Danielle, Chris},{Chris, Amy, Danielle}, {Chris, Danielle, Amy}, {Danielle, Amy, Chris}, {Danielle, Chris, Amy}, {Bob, Chris, Danielle}, {Bob, Danielle, Chris}, {Chris, Bob, Danielle}, {Chris, Danielle, Bob}, {Danielle, Bob, Chris}, {Danielle, Chris,

Bob} $_4P_3 = 24$.

47. Order is important in a permutation, but not in a combination. Thus {Amy, Bob, Chris}, {Amy, Chris, Bob}, {Chris, Amy, Bob}, {Chris, Bob, Amy}, {Bob, Amy, Chris}, and {Bob, Chris, Amy} are all different permutations but the same combination.

49. r!

51 **(a)**

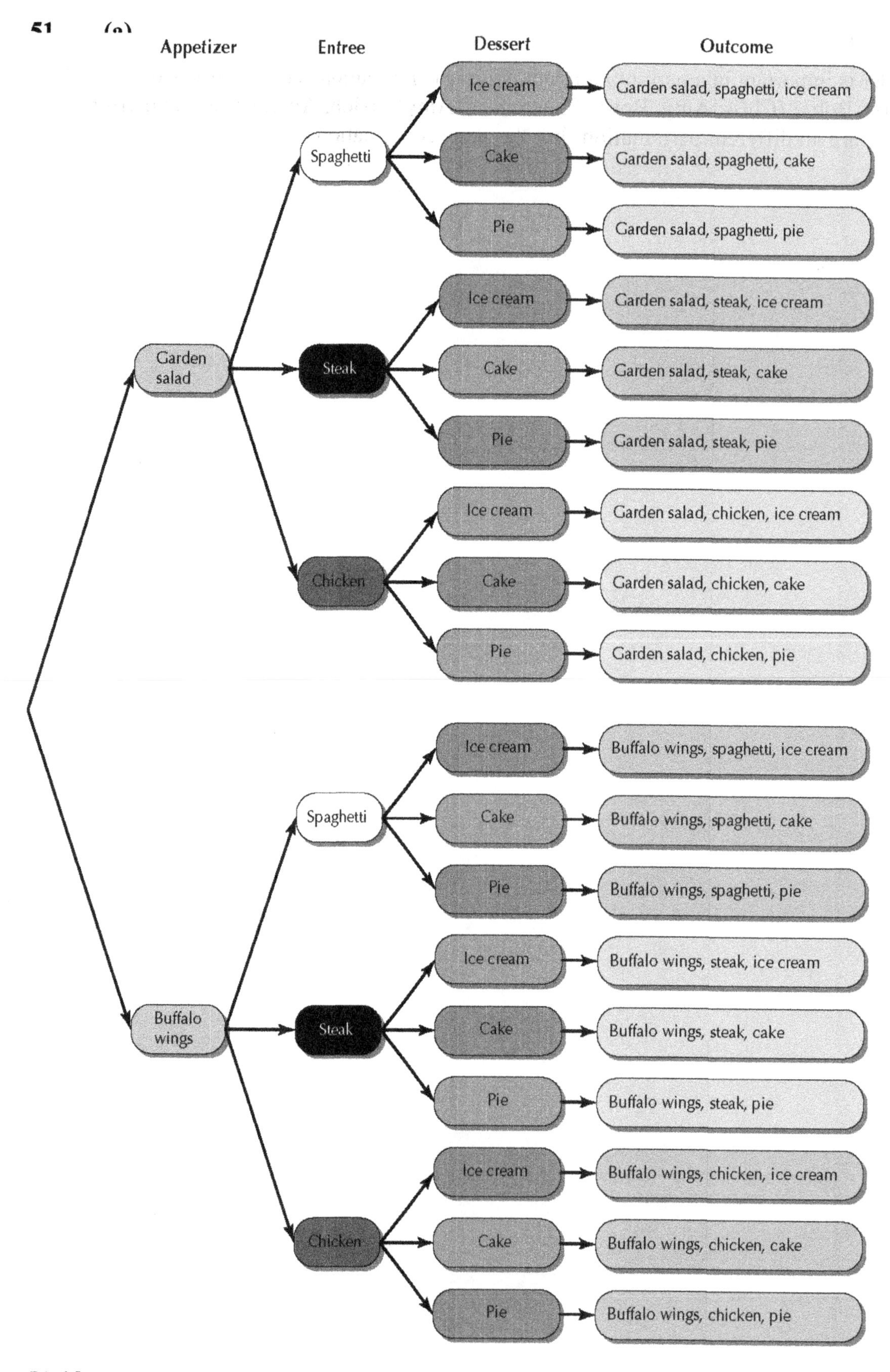
Appetizer
Entree
Dessert
Outcome
Garden salad
Buffalo wings
Spaghetti
Steak
Chicken
Ice cream
Cake
Pie
Garden salad, spaghetti, ice cream
Garden salad, spaghetti, cake
Garden salad, spaghetti, pie
Garden salad, steak, ice cream
Garden salad, steak, cake
Garden salad, steak, pie
Garden salad, chicken, ice cream
Garden salad, chicken, cake
Garden salad, chicken, pie
Buffalo wings, spaghetti, ice cream
Buffalo wings, spaghetti, cake
Buffalo wings, spaghetti, pie
Buffalo wings, steak, ice cream
Buffalo wings, steak, cake
Buffalo wings, steak, pie
Buffalo wings, chicken, ice cream
Buffalo wings, chicken, cake
Buffalo wings, chicken, pie

(b) 18

53. The key word here is "arrange." This means that order is important, so of all $n = r = 10$ friends. Therefore, there are ${}_{10}P_{10} = 10! = 3{,}628{,}800$ ways that the student can arrange her 10 friends top to bottom.

55. What makes the routes different is the order in which the different countries are visited, so this is a permutation of all $n = r = 6$ countries. Therefore, there are ${}_6P_6 = 6! = 720$ ways to visit the 6 countries.

57. Since one child is throwing the ball and one child is catching the ball, order is important. Thus, this is a permutation with $n = 5$ and $r = 2$. Therefore, there are

$${}_5P_2 = \frac{5!}{(5-2)!} = \frac{5 \cdot 4 \cdot 3!}{3!} = 5 \cdot 4 = 20$$

ways that one child can throw a ball to another child once.

59. Since both people that are shaking hands with each other are doing the same thing, order is unimportant. Therefore this is a combination. Since it takes2 people to shake hands, this is a combination of $n = 25$ people taken $r = 2$ at a time. Therefore, there are a total of

$${}_{25}C_2 = \frac{25!}{2!(25-2)!} = \frac{25 \cdot 24 \cdot 23!}{2! \cdot \cdot 23!} = \frac{25 \cdot 24}{2!} = \frac{600}{2} = 300 \text{ handshakes}$$

61. Since what matters in a random sample is what people or objects are selected and not the order in which they are selected, this is a combination with $n = 20$ and $r = 1$. Therefore, there are

$${}_{20}C_1 = \frac{20!}{1!(20-1)!} = \frac{20 \cdot 19!}{1! \cdot \cdot 19!} = \frac{20}{1} = 20$$

random samples of size 1 that can be chosen from a population of size 20.

63. Since what matters in a random sample is what people or objects are selected and not the order in which they are selected, this is a combination with $n = 20$ and $r = 10$. Therefore, there are

$${}_{20}C_{10} = \frac{20!}{10!(20-10)!} = \frac{20 \cdot 19 \cdot 18 \cdot 17 \cdot 16 \cdot 15 \cdot 14 \cdot 13 \cdot 12 \cdot 11 \cdot 10!}{10 \cdot 9 \cdot 8 \cdot 7 \cdot 6 \cdot 5 \cdot 4 \cdot 3 \cdot 2 \cdot 1 \cdot 10!} = \frac{670{,}442{,}572{,}800}{3{,}628{,}800} = 184{,}756$$

random samples of size 10 that can be chosen from a population of size 20.

65. We are seeking the number of permutations of $n = 8$ items, of which $n_1 = 3$ letter S's, $n_2 = 1$ letter B, $n_3 = 1$ letter U, $n_4 = 1$ letter I, $n_5 = 1$ letter N, and $n_6 = 1$ letter E. Using the formula for the number of permutations of nondistinct items,

$$\frac{n!}{n_1!n_2!n_3!n_4!n_5!n_6!} = \frac{8!}{3!1!1!1!1!1!} = \frac{40{,}320}{6 \cdot 1 \cdot 1 \cdot 1 \cdot 1 \cdot 1} = 6720.$$

There are 6720 distinct strings of letters that can be made using all of the letters in the word BUSINESS.

Chapter 5 Review Exercises

For Exercises 1–5, the sample space is {HHH, HHT, HTH, THH, TTH, THT, HTT, TTT} and $N(S) = 8$.

1. Let E be the event 2 heads. Then event E consists of the outcomes: {HHT, HTH, THH}, so $N(E) = 3$. Therefore, $P(E) = \frac{N(E)}{N(S)} = \frac{3}{8}$.

3. Let E be the event 4 heads. Since the coin is only tossed 3 times, there can't be 4 heads. Therefore there are no outcomes in event E, so $N(E) = 0$. Therefore, $P(E) = \frac{N(E)}{N(S)} = \frac{0}{8} = 0$.

5. Let E be the event at most 1 tail. Then event E consists of the outcomes with 0 or 1 tail: {HHH, HHT, HTH, THH}, so $N(E) = 4$. Therefore, $P(E) = \frac{N(E)}{N(S)} = \frac{4}{8} = \frac{1}{2}$.

7. **(a)** Let E be the event that a noncitizen farm worker is a high school graduate and let F be the event that a noncitizen farm worker has some college. Then since the educational levels attained are mutually exclusive,

$$P(E \cup F) = P(E) + P(F) = \frac{59{,}784}{376{,}000} + \frac{20{,}304}{376{,}000} = \frac{80{,}088}{376{,}000} \approx 0.213.$$

(b) Let E be the event that a citizen farm worker is a high school graduate and let F be the event that a citizen farm worker has some college. Then, because the educational levels attained are mutually exclusive,

$$P(E \cup F) = P(E) + P(F) = \frac{222{,}144}{624{,}000} + \frac{187{,}200}{624{,}000} = \frac{409{,}344}{624{,}000} \approx 0.656.$$

(c) Since the education levels are mutually exclusive, the probability that a noncitizen farm worker has less than a ninth-grade education and some college is 0.

9. Since there is only one research study that did not favor the drug, the probability that the second study does not favor the drug given that the first study does not favor the drug is 0.

11. **(a)** $P(F \cap D) = \frac{N(F \cap D)}{N(S)} = \frac{50}{300} = \frac{1}{6}$.

(b) $P(M \cap D) = \frac{N(M \cap D)}{N(S)} = \frac{50}{300} = \frac{1}{6}$

13. The men's TV channel, since $P(\text{Dog}|\text{Male}) = 5/12 > 5/18 = P(\text{Dog}|\text{Female})$.

15. $5 \cdot 4 \cdot 3 = 60$

Chapter 5 Quiz

1. False. An *event* is a collection of a series of events from the sample space of an experiment. An *outcome* is the result of a single trial of an experiment.

3. 0, 1

5. 0.5

7. with replacement

9. **(a)** Let E be the event sum of the 2 dice equals 5. Then event E consists of the outcomes: {(1,4), (2,3), (3,2), (4,1)}, so $N(E) = 4$. Therefore,

$$P(E) = \frac{N(E)}{N(S)} = \frac{4}{36} = \frac{1}{9}.$$

(b) Let E be the event sum of the 2 dice equals 5. Then the event the sum of the 2 dice does not equal 5 is the complement of the event the sum of the 2 dice equals 5, so $P(E^C) = 1 - P(E) = 1 - \frac{1}{9} = \frac{8}{9}$.

(c) Let F be the event 1 of the 2 dice shows 2. Then event F consists of the outcomes: {(1,2), (3,2), (4,2), (5,2), (6,2), (2,1), (2,3), (2,4), (2,5), (2,6)}, so $N(F) = 10$. Therefore, $P(E) = \frac{N(E)}{N(S)} = \frac{10}{36} = \frac{5}{18}$.

(d) Let E be the event sum of the 2 dice equals 5 and let F be the event 1 of the 2 dice shows 2. Then event $E \cap F$ consists of the outcomes: {(2,3), (3,2)}, so $N(E \cap F) = 2$. Therefore, $P(E \cap F) = \frac{N(E \cap F)}{N(S)} = \frac{2}{36} = \frac{1}{18}$.

(e) Let E be the event sum of the 2 dice equals 5 and let F be the event 1 of the2 dice shows 2. Then $P(E \cup F) = P(E) + P(F) - P(E \cap F) = \frac{4}{36} + \frac{10}{36} - \frac{2}{36} = \frac{12}{36} = \frac{1}{3}$.

11. $P(A \cap B) = P(B) \bullet P(A|B) = 0.85 \cdot 0.25 = 0.2125$.

For Exercises 13–18, there are 36 outcomes in the sample space. Therefore, $N(S) = 36$.

There are 4 outcomes in event X: $\{(3, 6), (4, 5), (5, 4), (6, 3)\}$. Therefore, $P(X) = \frac{4}{36} = \frac{1}{9}$.

There are 3 outcomes in event Y: $\{(4, 6), (5, 5), (6, 4)\}$. Therefore, $P(Y) = \frac{3}{36} = \frac{1}{12}$.

There are 6 outcomes in event Z: $\{(1, 1), (2, 2), (3, 3), (4, 4), (5, 5), (6, 6)\}$. Therefore, $P(Z) = \frac{6}{36} = \frac{1}{6}$.

There are 6 outcomes in event W: $\{(1, 6), (2, 6), (3, 6), (4, 6), (5, 6), (6, 6)\}$. Therefore, $P(W) = \frac{6}{36} = \frac{1}{6}$.

13. There are 0 outcomes in X and Z Therefore, $P(X \text{ and } Z) = \frac{0}{6} = 0 \neq \frac{1}{54} = \left(\frac{1}{9}\right)\left(\frac{1}{6}\right) = P(X) \cdot P(Z)$. Thus, X and Z are not independent.

15. There is 1 outcome in X and W: $\{(3, 6)\}$ Therefore, $P(X \text{ and } W) = \frac{1}{36} \neq \frac{1}{54} = \left(\frac{1}{9}\right)\left(\frac{1}{6}\right) = P(X) \cdot P(W)$. Thus, X and W are not independent.

17. There are 0 outcomes in X and Y Therefore, $P(X \text{ and } Y) = \frac{0}{6} = 0 \neq \frac{1}{108} = \left(\frac{1}{9}\right)\left(\frac{1}{12}\right) = P(X) \cdot P(Y)$. Thus, X and Y are not independent.

19. Since we are only interested in which three teams make the playoffs and not the order in which they are selected, this is a combination with $n = 4$ and $r = 3$. There are

${}_4C_3 = \frac{4!}{3!(4-3)!} = \frac{4 \cdot 3!}{3! \cdot (4-3)!} = \frac{4}{1} = 4$ different sets of teams making the playoffs.

Chapter 6: Probability Distributions

Note to instructors and students: Some answers may differ by 1 or 2 in the last digit depending on whether you round for intermediate steps or wait until you get the final answer to round. Also, different software and different forms of technology may give slightly different answers.

Section 6.1

1. A random variable is a variable whose values are determined by chance. Answers will vary.

3. A discrete random variable can take either a finite or a countable number of values. Each value can be graphed as a separate point on the number line with space between each point. A continuous random variable can take infinitely many values. The values of a continuous random variable form an interval on the number line.

5. □$P(X) = 1$ and $0 \le P(X) \le 1$

7. Since the number of siblings a person has is finite and may be written as a list of numbers, it represents a discrete random variable.

9. Volume is something that must be measured, not counted. Volume can take infinitely many different possible values, with these values forming an interval on the number line. Thus, how much coffee there is in your next cup of coffee is a continuous random variable.

11. Since the number of correct answers a person has on a multiple-choice quiZ is finite and may be written as a list of numbers, it represents a discrete random variable.

13. {0, 1, 2, 3, 4, 5, 6, 7, 8, 9, 10, 11, 12, 13, 14, 15}

15. {0, 1, 2, 3, 4}

17. **(a)**

X = Number of CDs	0	1	2	3	4
$P(X)$	0.06	0.24	0.38	0.22	0.10

(b)

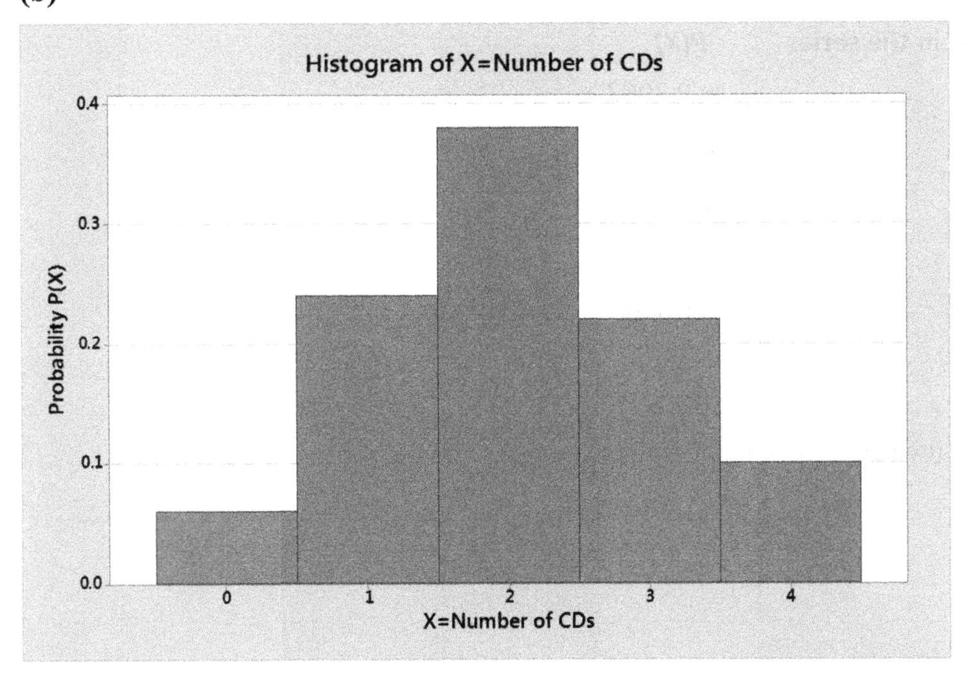

19. (a)

X = Money gained	−\$10,000	\$10,000	\$50,000
$P(X)$	1/3	1/2	1/6

(b)

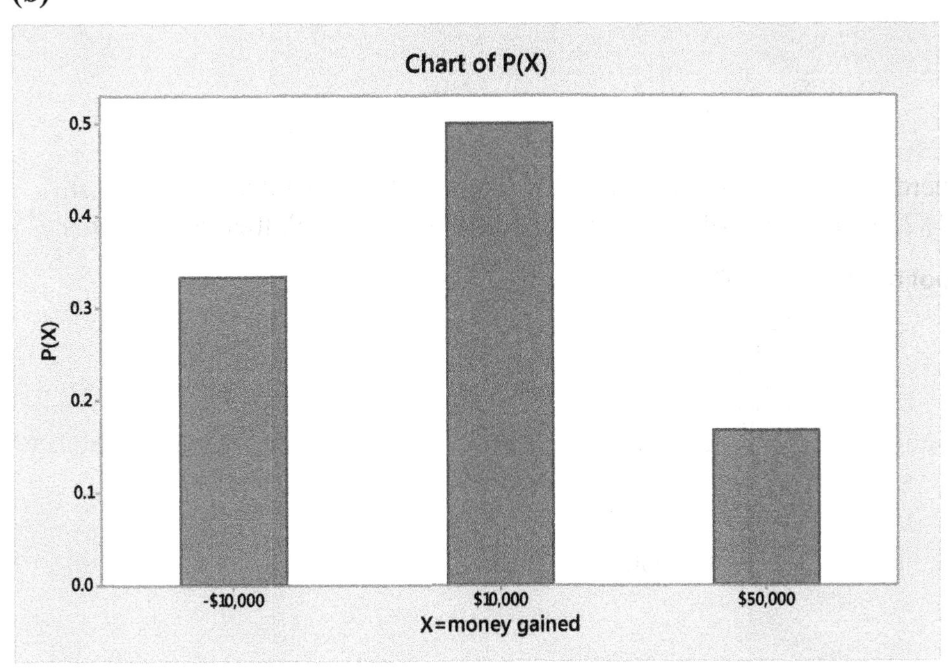

21. (a)

X **= Number of games in the series**	***P(X)***
4	$\frac{5}{24} \approx 0.2083$
5	$\frac{5}{24} \approx 0.2083$
6	$\frac{7}{24} \approx 0.2917$
7	$\frac{7}{24} \approx 0.2917$
Total	$\frac{24}{24} = 1.0000$

(b)

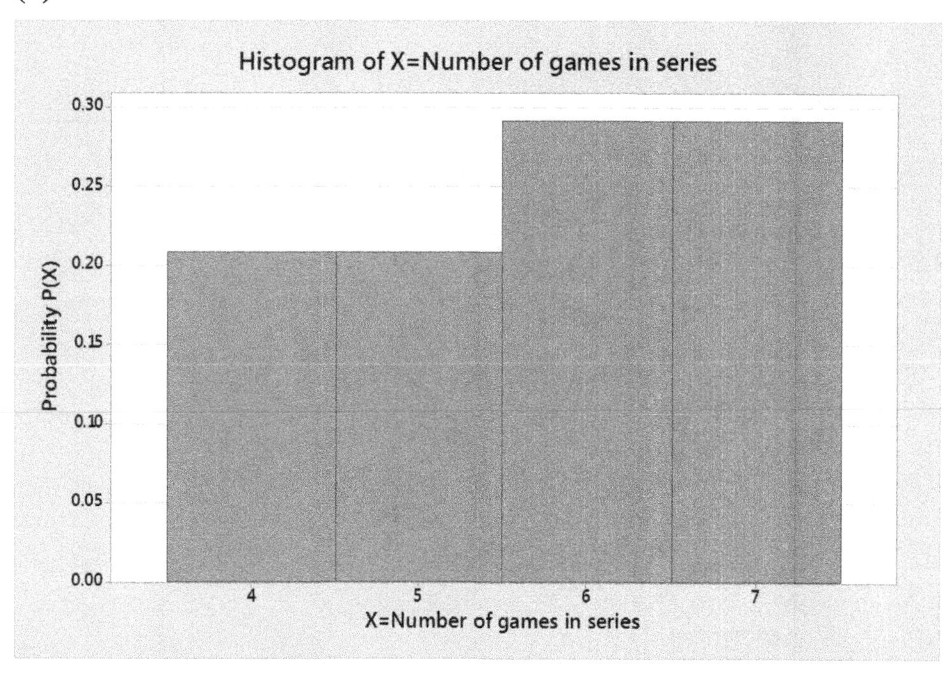

23. (a) In 2014, there were a total of 742,000 + 209,000 + 30,000 + 13,000 = 994,000 students who had either 1, 2, 3, or 4 years of high school math when they took their SAT exams.

X **= Years of high school math**	***P(X)***
4	$\frac{742,000}{994,000} \approx 0.7465$
3	$\frac{209,000}{994,000} \approx 0.2103$
2	$\frac{30,000}{994,000} \approx 0.0302$
1	$\frac{13,000}{994,000} \approx 0.0131$
Total	$\frac{994,000}{994,000} \approx 1.0001$

(b)

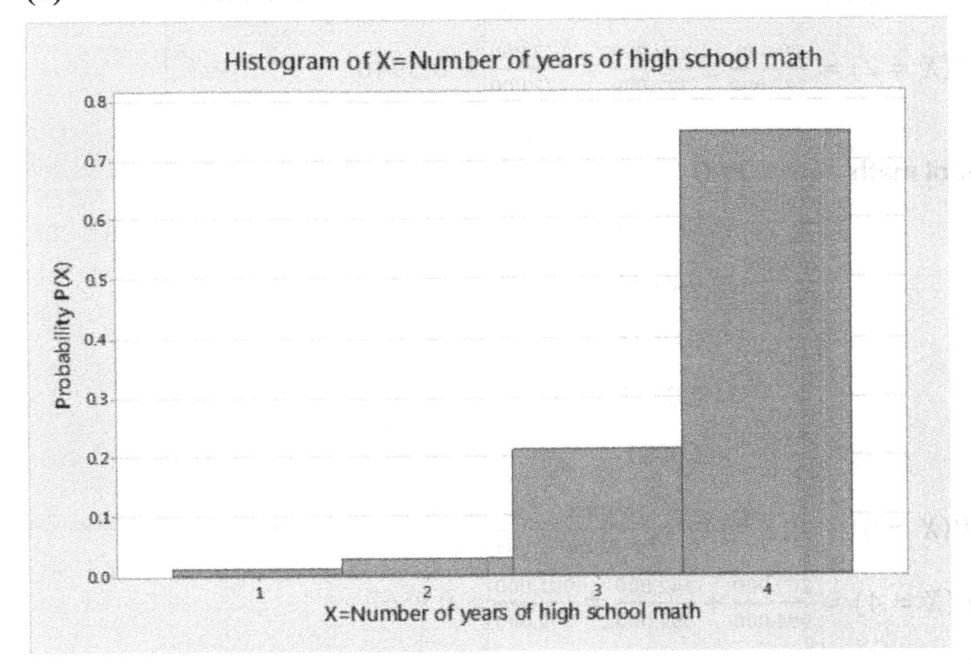

25. Not a valid probability distribution. The probabilities don't add up to 1.

27. Not a valid probability distribution. $P(X = 1)$ is negative.

For Exercise 29–32:

***X* = Number of games in the series**	***P(X)***
4	$\frac{5}{24} \approx 0.2083$
5	$\frac{5}{24} \approx 0.2083$
6	$\frac{7}{24} \approx 0.2917$
7	$\frac{7}{24} \approx 0.2917$
Total	$\frac{24}{24} = 1.0000$

29. $P(X \geq 6) = P(X = 6) + P(X = 7) = \frac{7}{24} + \frac{7}{24} = \frac{14}{24} \approx 0.5833$

31. $P(5 \leq X \leq 7) = P(X = 5) + P(X = 6) + P(X = 7) = \frac{5}{24} + \frac{7}{24} + \frac{7}{24} = \frac{19}{24} \approx 0.7917$

For Exercises 33–36:

***X* = Number of SAT subject tests**	***P(X)***
1	$\frac{29,000}{222,000} \approx 0.1306$
2	$\frac{103,000}{222,000} \approx 0.4640$
3	$\frac{90,000}{222,000} \approx 0.4054$
Total	$\frac{222,000}{222,000} = 1.0000$

33. $P(X \geq 2) = P(X = 2) + P(X = 3) = \frac{103{,}000}{222{,}000} + \frac{90{,}000}{222{,}000} = \frac{193{,}000}{222{,}000} \approx 0.8694$

35. $P(X = 1) + P(X = 2) = \frac{29{,}000}{222{,}000} + \frac{103{,}000}{222{,}000} = \frac{132{,}000}{222{,}000} \approx 0.5946$

For Exercises 37–40:

X **= Years of high school math**	***P(X)***
4	$\frac{742{,}000}{994{,}000} \approx 0.7465$
3	$\frac{209{,}000}{994{,}000} \approx 0.2103$
2	$\frac{30{,}000}{994{,}000} \approx 0.0302$
1	$\frac{13{,}000}{994{,}000} \approx 0.0131$
Total	$\frac{994{,}000}{994{,}000} \approx 1.0001$

37. $P(X \geq 3) = P(X = 3) + P(X = 4) = \frac{209{,}000}{994{,}000} + \frac{742{,}000}{994{,}000} = \frac{951{,}000}{994{,}000} \approx 0.9567$

39. $P(X = 3) + P(X = 4) = \frac{209{,}000}{994{,}000} + \frac{742{,}000}{994{,}000} = \frac{951{,}000}{994{,}000} \approx 0.9567$

For Exercises 41–44:

X **= Number of major hurricanes**	***P(X)***
0	$\frac{2}{25} = 0.08$
1	$\frac{4}{25} = 0.16$
2	$\frac{7}{25} = 0.28$
3	$\frac{3}{25} = 0.12$
4	$\frac{2}{25} = 0.08$
5	$\frac{4}{25} = 0.16$
6	$\frac{2}{25} = 0.08$
7	$\frac{1}{25} = 0.04$
Total	$\frac{25}{25} = 1.00$

41. $P(X \leq 1) = P(X = 0) + P(X = 1) = \frac{2}{25} + \frac{4}{25} = \frac{6}{25} = 0.24$

43. $P(X = 10) = 0$

45. **(a)**

X **= Number of CDs**	***P*(*X*)**	***X* · *P*(*X*)**
0	0.06	$0(0.06) = 0$
1	0.24	$1(0.24) = 0.24$
2	0.38	$2(0.38) = 0.76$
3	0.22	$3(0.22) = 0.66$
4	0.10	$4(0.10) = 0.40$
Total	1.00	$\mu = \sum[X \cdot P(X)] = 2.06$

$\mu = 2.06$ CDs

(b) $X = 2$ CDs has the highest probability of occurring, so 2 CDs is the most likely value of X.

(c) $E(X) = \mu = 2.06$ CDs

47. **(a)**

X = Money gained	$P(X)$	$X \cdot P(X)$
– \$10,000	1/3	$-\$10{,}000(1/3) = \frac{-\$10{,}000}{3}$
\$10,000	1/2	$\$10{,}000(1/2) = \$5{,}000$
\$50,000	1/6	$\$50{,}000(1/6) = \frac{\$25{,}000}{3}$
Total	1.00	$\mu = \sum[X \cdot P(X)] = \$10{,}000$

$\mu = \$10{,}000$

(b) $X = \$10{,}000$ has the highest probability of occurring, so \$10,000 is the most likely value of X.

(c) $E(X) = \mu = \$10{,}000$

49. **(a)**

X = Number of games in the series	$P(X)$	$X \cdot P(X)$
4	$\frac{5}{24}$	$4\left(\frac{5}{24}\right) = \frac{20}{24} = \frac{5}{6}$
5	$\frac{5}{24}$	$5\left(\frac{5}{24}\right) = \frac{25}{24}$
6	$\frac{7}{24}$	$6\left(\frac{7}{24}\right) = \frac{42}{24} = \frac{7}{4}$
7	$\frac{7}{24}$	$7\left(\frac{7}{24}\right) = \frac{49}{24}$
Total	$\frac{24}{24} = 1.0000$	$\mu = \sum[X \cdot P(X)] = \frac{136}{24} \approx 5.6667$

$\mu \approx 5.6667$ games

(b) $X = 6$ games and $X = 7$ games have the highest probability of occurring, so 6 and 7 games are the most likely values of X.

(c) $E(X) = \mu \approx 5.6667$ games

51. **(a)**

X = Years of high school math	$P(X)$	$X \cdot P(X)$
4	$\frac{742{,}000}{994{,}000}$	$4\left(\frac{742{,}000}{994{,}000}\right) = \frac{2{,}968{,}000}{994{,}000}$
3	$\frac{209{,}000}{994{,}000}$	$3\left(\frac{209{,}000}{994{,}000}\right) = \frac{627{,}000}{994{,}000}$
2	$\frac{30{,}000}{994{,}000}$	$2\left(\frac{30{,}000}{994{,}000}\right) = \frac{60{,}000}{994{,}000}$
1	$\frac{13{,}000}{994{,}000}$	$1\left(\frac{13{,}000}{994{,}000}\right) = \frac{13{,}000}{994{,}000}$
Total	$\frac{994{,}000}{994{,}000}$	$\mu = \sum[X \cdot P(X)] = \frac{3{,}668{,}000}{994{,}000} \approx 3.6901$

$\mu \approx 3.6901$ years

(b) $X = 4$ years has the highest probability of occurring, so 4 years is the most likely value of X.

(c) $E(X) = \mu \approx 3.6901$ years

53. **(a)** From Exercise 45 (a), $\mu = 2.06$ CDs.

X = Number of CDs	$P(X)$	$(x-\mu)^2 \cdot P(X)$
0	0.06	$(0-2.06)^2(0.06) = 0.254616$
1	0.24	$(1-2.06)^2(0.24) = 0.269664$
2	0.38	$(2-2.06)^2(0.38) = 0.001368$
3	0.22	$(3-2.06)^2(0.22) = 0.194392$
4	0.10	$(4-2.06)^2(0.10) = 0.37636$
Total	1.00	$\sigma^2 = \sum(X-\mu)^2 \cdot P(X) = 1.0964$

$\sigma^2 = \sum(X-\mu)^2 \cdot P(X) = 1.0964$ CDs squared

(b) $\sigma = \sqrt{\sigma^2} = \sqrt{1.0964} \approx 1.0471$ CDs

(c) For $X = 0$ CDs, $Z = \frac{X-\mu}{\sigma} = \frac{0-2.06}{1.0471} \approx -1.97$.

For $X = 1$ CD, $Z = \frac{X-\mu}{\sigma} = \frac{1-2.06}{1.0471} \approx -1.01$.

For $X = 2$ CDs, $Z = \frac{X-\mu}{\sigma} = \frac{2-2.06}{1.0471} \approx -0.06$.

For $X = 3$ CDs, $Z = \frac{X-\mu}{\sigma} = \frac{3-2.06}{1.0471} \approx 0.90$.

For $X = 4$ CDs, $Z = \frac{X-\mu}{\sigma} = \frac{4-2.06}{1.0471} \approx 1.85$.

All of the Z-scores are between –2 and 2. Therefore, there are no outliers or unusual data values.

55. **(a)** From Exercise 47 (a), $\mu = \$10,000$.

X = Money gained	$P(X)$	$(x-\mu)^2 \cdot P(X)$
– \$10,000	1/3	$(-10,000-10,000)^2(1/3) = \frac{400,000,000}{3}$
\$10,000	1/2	$(10,000-10,000)^2(1/2) = 0$
\$50,000	1/6	$(50,000-10,000)^2(1/6) = \frac{800,000,000}{3}$
Total	1.00	$\sigma^2 = \sum(X-\mu)^2 \cdot P(X) = 400,000,000$

$\sigma^2 = \sum(X-\mu)^2 \cdot P(X)$ =400,000,000 dollars squared

(b) $\sigma = \sqrt{\sigma^2} = \sqrt{400,000,000} = \$20,000$

(c) For $X = -\$10,000$, $Z = \frac{X-\mu}{\sigma} = \frac{-\$10,000-\$10,000}{\$20,000} = -1$.

For $X = \$10,000$, $Z = \frac{X-\mu}{\sigma} = \frac{\$10,000-\$10,000}{\$20,000} = 0$.

For $X = \$50,000$, $Z = \frac{X-\mu}{\sigma} = \frac{\$50,000-\$10,000}{\$20,000} = 2$.

For $X = \$50{,}000$, $2 \le Z\text{-score} < 3$, so $X = \$50{,}000$ is moderately unusual.

The Z-scores for $X = -\$10{,}000$ and $X = \$10{,}000$ are between –2 and 2, so neither value is moderately unusual or an outlier.

57. **(a)** From Exercise 49 (a), $\mu \approx 5.6667$ games.

X = Number of games in the series	P(X)	$(x-\mu)^2 \cdot P(X)$
4	$\frac{5}{24}$	$(4-5.6667)^2\left(\frac{5}{24}\right) \approx 0.5787268521$
5	$\frac{5}{24}$	$(5-5.6667)^2\left(\frac{5}{24}\right) \approx 0.0926018521$
6	$\frac{7}{24}$	$(6-5.6667)^2\left(\frac{7}{24}\right) \approx 0.0324009263$
7	$\frac{7}{24}$	$(7-5.6667)^2\left(\frac{7}{24}\right) \approx 0.5184925929$
Total	$\frac{24}{24}=1.0000$	$\sigma^2 = \sum(X-\mu)^2 \cdot P(X) \approx 1.222222223$

$\sigma^2 = \sum(X-\mu)^2 \cdot P(X) \approx 1.222222223$ games squared

(b) $\sigma = \sqrt{\sigma^2} = \sqrt{1.222222223} \approx 1.1055$ games

(c) For $X = 4$ games, $Z = \frac{X-\mu}{\sigma} = \frac{4-5.6667}{1.1055} \approx -1.51$.

For $X = 5$ game, $Z = \frac{X-\mu}{\sigma} = \frac{5-5.6667}{1.1055} \approx -0.60$.

For $X = 6$ games, $Z = \frac{X-\mu}{\sigma} = \frac{6-5.6667}{1.1055} \approx 0.30$.

For $X = 7$ games, $Z = \frac{X-\mu}{\sigma} = \frac{7-5.6667}{1.1055} \approx 1.21$.

All of the Z-scores are between –2 and 2. Therefore, there are no outliers or unusual data values

59. **(a)** From Exercise 51 (a), $\mu \approx 3.6901$ years.

X = Years of high school math	P(X)	$(x-\mu)^2 \cdot P(X)$
4	$\frac{742{,}000}{994{,}000}$	$(4-3.6901)^2\left(\frac{742{,}000}{994{,}000}\right) \approx 0.0716903455$
3	$\frac{209{,}000}{994{,}000}$	$(3-3.6901)^2\left(\frac{209{,}000}{994{,}000}\right) \approx 0.1001345514$
2	$\frac{30{,}000}{994{,}000}$	$(2-3.6901)^2\left(\frac{30{,}000}{994{,}000}\right) \approx 0.0862104027$
1	$\frac{13{,}000}{994{,}000}$	$(1-3.6901)^2\left(\frac{13{,}000}{994{,}000}\right) \approx 0.0946441591$
Total	$\frac{994{,}000}{994{,}000}$	$\sigma^2 = \sum(X-\mu)^2 \cdot P(X) \approx 0.3526794587$

$\sigma^2 = \sum(X-\mu)^2 \cdot P(X) \approx 0.3526794587$ year squared

(b) $\sigma = \sqrt{\sigma^2} = \sqrt{0.3526794587} \approx 0.5939$ year

(c) For $X = 4$ years, $Z = \frac{X-\mu}{\sigma} = \frac{4-3.6901}{0.5939} \approx 0.52$.

For $X = 3$ years, $Z = \frac{X-\mu}{\sigma} = \frac{3-3.6901}{0.5939} \approx -1.16$.

For $X = 2$ years, $Z = \frac{X-\mu}{\sigma} = \frac{2-3.6901}{0.5939} \approx -2.85$.

For $X = 1$ year, $Z = \frac{X-\mu}{\sigma} = \frac{1-3.6901}{0.5939} \approx -4.53$.

The Z-score for $X = 1$ year is ≤ -3, so $X = 1$ year of high school math is considered an outlier.

For $X = 2$ years, $-3 < Z\text{-score} < -2$, so $X = 2$ years of high school math is considered moderately unusual.

The Z-scores for $X = 3$ years of high school math and $X = 4$ years of high school math are between –2 and 2, so 3 or 4 years of high school math is not moderately unusual or an outlier.

61. **(a)** A faculty member from all degree-granting institutions in the United States is randomly selected.

(b) The number of classes taught by a faculty member is counted.

(c)

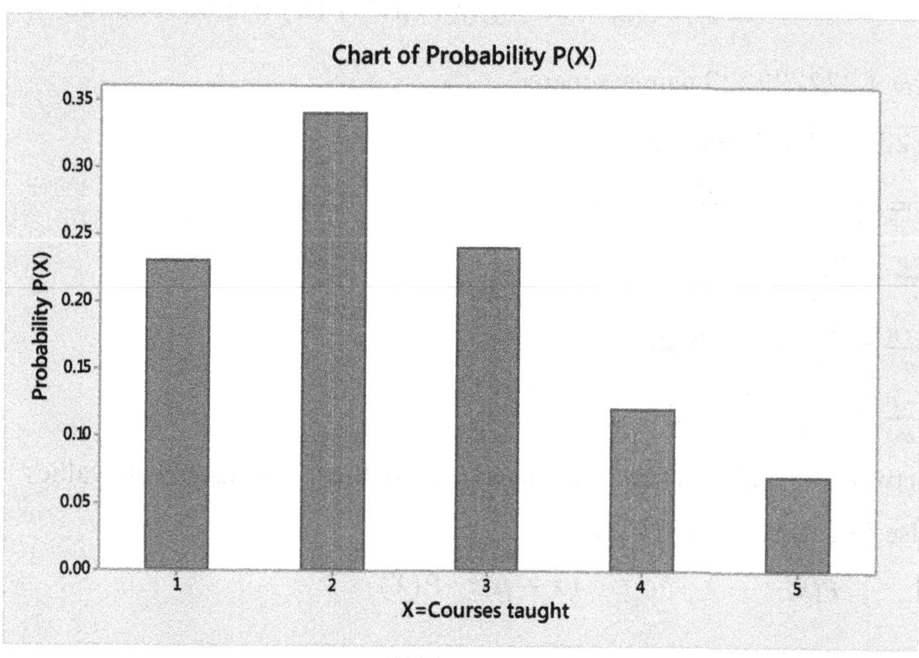

(d) $P(X \geq 3) = P(X = 3) + P(X = 4) + P(X = 5) = 0.24 + 0.12 + 0.07 = 0.43$

(e) $P(X > 3) = P(X = 4) + P(X = 5) = 0.12 + 0.07 = 0.19$

(f) $X = 2$ classes has the highest probability of occurring, so 2 classes is the most likely value of X.

63. **(a)** $n = 288 + 3151 + 109 + 12 + 4 + 1 = 3565$

X = Number of vehicles involved	$P(X)$
1	$\frac{288}{3565} \approx 0.0808$
2	$\frac{3151}{3565} \approx 0.8839$
3	$\frac{109}{3565} \approx 0.0306$
4	$\frac{12}{3565} \approx 0.0034$
5	$\frac{4}{3565} \approx 0.0011$
8	$\frac{1}{3565} \approx 0.0003$
Total	$\frac{3565}{3565} \approx 1.0001$

(b) $P(X \geq 3) = P(X = 3) + P(X = 4) + P(X = 5) + P(X = 8) = \frac{109}{3565} + \frac{12}{3565} + \frac{4}{3565} + \frac{1}{3565} = \frac{126}{3565} \approx 0.0353$

(c) $P(X = 6) = \frac{0}{3565} = 0$

(d) $P(1 \leq X \leq 3) = P(X = 1) + P(X = 2) + P(X = 3) = \frac{288}{3565} + \frac{3151}{3565} + \frac{109}{3565} = \frac{3548}{3565} \approx 0.9952$

(e) $P(1 < X < 3) = P(X = 2) = \frac{3151}{3565} \approx 0.8839$

65. **(a)**

X	$P(X)$	$X \cdot P(X)$
1	0.23	$(1)(0.23) = 0.23$
2	0.34	$(2)(0.34) = 0.68$
3	0.24	$(3)(0.24) = 0.72$
4	0.12	$(4)(0.12) = 0.48$
5	0.07	$(5)(0.07) = 0.35$
Total	1.00	$E(X) = \mu = \sum[X \cdot P(X)] = 2.46$

The expected number of classes taught is 2.46.

(b)

X	$P(X)$	$(x-\mu)^2 \cdot P(X)$
1	0.23	$(1 - 2.46)^2(0.23) = 0.490268$
2	0.34	$(2 - 2.46)^2(0.34) = 0.071944$
3	0.24	$(3 - 2.46)^2(0.24) = 0.069984$
4	0.12	$(4 - 2.46)^2(0.12) = 0.284592$
5	0.07	$(5 - 2.46)^2(0.07) = 0.451612$
Total	1.00	$\sigma^2 = \sum(X - \mu)^2 \cdot P(X) = 1.3684$

$\sigma^2 = 1.3684$ classes squared

$\sigma = \sqrt{\sigma^2} = \sqrt{1.3684} \approx 1.1698$ classes.

(c) For $X = 5$ classes, $Z = \frac{X-\mu}{\sigma} = \frac{5-2.46}{1.1698} \approx 2.17$.

For $X = 5$ classes, $2 \leq Z$-score < 3. Therefore $X = 5$ classes is moderately unusual.

Thus, it is unusual to teach 5 classes.

67. **(a)**

X = Number of vehicles involved	*P(X)*	$X \cdot P(X)$
1	$\frac{288}{3565}$	$1\left(\frac{288}{3565}\right) = \frac{288}{3565}$
2	$\frac{3151}{3565}$	$2\left(\frac{3151}{3565}\right) = \frac{6302}{3565}$
3	$\frac{109}{3565}$	$3\left(\frac{109}{3565}\right) = \frac{327}{3565}$
4	$\frac{12}{3565}$	$4\left(\frac{12}{3565}\right) = \frac{48}{3565}$
5	$\frac{4}{3565}$	$5\left(\frac{4}{3565}\right) = \frac{20}{3565}$
8	$\frac{1}{3565}$	$8\left(\frac{1}{3565}\right) = \frac{8}{3565}$
Total	$\frac{3565}{3565} \approx 1.0001$	$E(X) = \mu = \sum[X \cdot P(X)] = \frac{6993}{3565} \approx 1.9616$

The expected number of vehicles involved in a crash is 1.9616 vehicles.

(b)

X = Number of vehicles involved	*P(X)*	$(x-\mu)^2 \cdot P(X)$
1	$\frac{288}{3565}$	$\left(1 - \frac{6993}{3565}\right)^2 \left(\frac{288}{3565}\right) \approx 0.0746956845$
2	$\frac{3151}{3565}$	$\left(2 - \frac{6993}{3565}\right)^2 \left(\frac{3151}{3565}\right) \approx 0.0013053018$
3	$\frac{109}{3565}$	$\left(3 - \frac{6993}{3565}\right)^2 \left(\frac{109}{3565}\right) \approx 0.0329701349$
4	$\frac{12}{3565}$	$\left(4 - \frac{6993}{3565}\right)^2 \left(\frac{12}{3565}\right) \approx 0.0139866261$
5	$\frac{4}{3565}$	$\left(5 - \frac{6993}{3565}\right)^2 \left(\frac{4}{3565}\right) \approx 0.0103585434$
8	$\frac{1}{3565}$	$\left(8 - \frac{6993}{3565}\right)^2 \left(\frac{1}{3565}\right) \approx 0.0102279458$
Total	$\frac{3565}{3565} \approx 1.0001$	$\sigma^2 = \sum(X-\mu)^2 \cdot P(X) \approx 0.1435442365$

$\sigma^2 = 0.1438442365$ vehicle squared

$\sigma = \sqrt{\sigma^2} = \sqrt{0.1438442365} \approx 0.3787$ vehicle

(c) For $X = 1$ vehicle, $Z = \frac{X-\mu}{\sigma} = \frac{1-1.9619}{0.3787} \approx -2.54$.

For $X = 2$ vehicles, $Z = \frac{X-\mu}{\sigma} = \frac{2-1.9619}{0.3787} \approx 0.10$.

For $X = 3$ vehicles, $Z = \frac{X-\mu}{\sigma} = \frac{3-1.9619}{0.3787} \approx 2.74$.

For $X = 4$ vehicles, $Z = \frac{X-\mu}{\sigma} = \frac{4-1.9619}{0.3787} \approx 5.38$.

For $X = 5$ vehicles, $Z = \frac{X-\mu}{\sigma} = \frac{5-1.9619}{0.3787} \approx 8.02$.

For $X = 8$ vehicles, $Z = \frac{X-\mu}{\sigma} = \frac{8-1.9619}{0.3787} \approx 15.94$.

For $X = 1$ vehicle, $-3 < Z$-score ≤ -2. Therefore, $X = 1$ vehicle is considered moderately unusual.

For $X = 3$ vehicles, $2 \leq Z$-score < 3. Therefore, $X = 3$ vehicles is considered moderately unusual.

For $X = 4, 5$, or 8 vehicles, Z-score ≥ 3. Therefore, 4, 5, and 8 vehicles are outliers.

69. Since there are 36 outcomes in the sample space, $N(S) = 36$. The only outcome that gives a sum of 2 is $\{(1, 1)\}$, so

$P(X = 2) = \frac{1}{36}$.

The outcomes that give a sum of 3 are $\{(1,2), (2,1)\}$, so

$P(X = 3) = \frac{2}{36} = \frac{1}{18}$.

The outcomes that give a sum of 4 are $\{(1,3), (2,2), (3,1)\}$, so

$P(X = 4) = \frac{3}{36} = \frac{1}{12}$.

The outcomes that give a sum of 5 are $\{(1,4), (2,3), (3,2), (4,1)\}$, so

$P(X = 5) = \frac{4}{36} = \frac{1}{9}$.

The outcomes that give a sum of 6 are $\{(1,5), (2,4), (3,3), (4,2), (5,1)\}$, so

$P(X = 6) = \frac{5}{36}$.

The outcomes that give a sum of 7 are $\{(1,6), (2,5), (3,4), (4,3), (5,2), (6,1)\}$, so

$P(X = 7) = \frac{6}{36} = \frac{1}{6}$.

The outcomes that give a sum of 8 are $\{(2,6), (3,5), (4,4), (5,3), (6,2)\}$, so

$P(X = 8) = \frac{5}{36}$.

The outcomes that give a sum of 9 are $\{(3,6), (4,5), (5,4), (6,3)\}$, so

$P(X = 9) = \frac{4}{36} = \frac{1}{9}$.

The outcomes that give a sum of 10 are $\{(4,6), (5,5), (6,4)\}$, so

$P(X = 10) = \frac{3}{36} = \frac{1}{12}$.

The outcomes that give a sum of 11 are $\{(5,6), (6,5)\}$, so

$P(X = 11) = \frac{2}{36} = \frac{1}{18}$.

The only outcome that gives a sum of 12 is $\{(6,6)\}$, so

$P(X = 12) = \frac{1}{36}$.

X = **Sum of dice**	2	3	4	5	6	7	8	9	10	11	12

$P(X)$	1/36	1/18	1/12	1/9	5/36	1/6	5/36	1/9	1/12	1/18	1/36

71. $\mu = \sum[X \cdot P(X)] = 2\left(\frac{1}{36}\right) + 3\left(\frac{1}{18}\right) + 4\left(\frac{1}{12}\right) + 5\left(\frac{1}{9}\right) + 6\left(\frac{5}{36}\right) + 7\left(\frac{1}{6}\right) + 8\left(\frac{5}{36}\right) + 9\left(\frac{1}{9}\right) + 10\left(\frac{1}{12}\right) + 11\left(\frac{1}{18}\right) + 12\left(\frac{1}{36}\right) = 7.$

The estimate is equal to the actual value. If we were to consider tossing 2 dice an infinite number of times, the mean sum of the dice would be 7.

73. Since the value $X = 2$ is $\left|\frac{X-\mu}{\sigma}\right| \approx \left|\frac{2-7}{2.4152}\right| \approx 2.07$ standard deviations from the mean and because $2 \leq 2.07 < 3$, snake eyes is considered moderately unusual. By symmetry, so is 12.

75. Symmetric, one mode.

77. **(a)** An NFL team is selected at random.

(b) The number of games a team has won is counted.

(c)

X = Number of games won	*P(X)*
2	$\frac{1}{32} = 0.03125$
3	$\frac{1}{32} = 0.03125$
4	$\frac{5}{32} = 0.15625$
5	$\frac{1}{32} = 0.03125$
6	$\frac{1}{32} = 0.03125$
7	$\frac{4}{32} = 0.12500$
8	$\frac{7}{32} = 0.21875$
9	$\frac{1}{32} = 0.03125$
10	$\frac{2}{32} = 0.06250$
11	$\frac{4}{32} = 0.12500$
12	$\frac{3}{32} = 0.09375$
13	$\frac{2}{32} = 0.06250$
Total	$\frac{32}{32} = 1.00000$

(d)

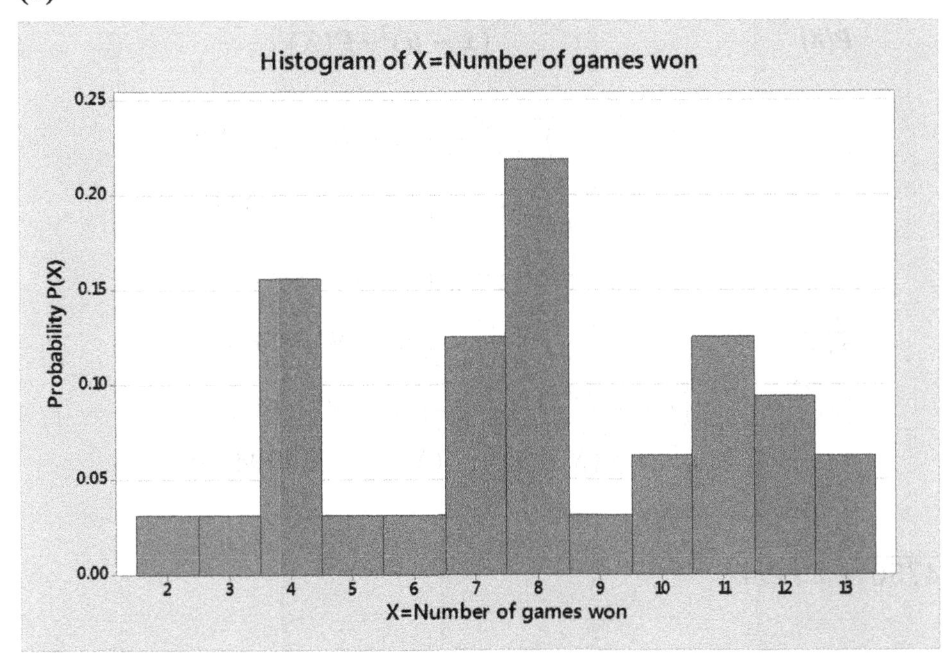

(e) $P(X \leq 5) = P(X = 2) + P(X = 3) + P(X = 4) + P(X = 5) = \frac{1}{32} + \frac{1}{32} + \frac{5}{32} + \frac{1}{32} = \frac{8}{32} = \frac{1}{4} = 0.25$

(f) Since $X = 8$ games won has the highest probability of occurring, 8 games won is the most likely value of X.

79. There are finitely many possible exam scores.

81. All probabilities are between 0 and 1 and the probabilities add up to 1.

83. **(a)** $P(X \geq 2) = P(X = 2) + P(X = 3) + P(X = 4) + P(X = 5) = \frac{4{,}937}{25{,}900} + \frac{6{,}333}{25{,}900} + \frac{5{,}108}{25{,}900} + \frac{3{,}464}{25{,}900} = \frac{19{,}842}{25{,}900} \approx 0.7661$

(b) $P(X > 1) = P(X = 2) + P(X = 3) + P(X = 4) + P(X = 5) = \frac{4{,}937}{25{,}900} + \frac{6{,}333}{25{,}900} + \frac{5{,}108}{25{,}900} + \frac{3{,}464}{25{,}900} = \frac{19{,}842}{25{,}900} \approx 0.7661$

(c) $P(X \leq 2) = P(X = 1) + P(X = 2) = \frac{6{,}058}{25{,}900} + \frac{4{,}937}{25{,}900} = \frac{10{,}995}{25{,}900} \approx 0.4245$

(d) $P(X < 3) = P(X = 1) + P(X = 2) = \frac{6{,}058}{25{,}900} + \frac{4{,}937}{25{,}900} = \frac{10{,}995}{25{,}900} \approx 0.4245$

(e) $P(X \geq 2) = P(X > 1)$. $P(X \leq 2) = P(X < 3)$. X is a discrete random variable.

85. $X = 3$ has the highest probability of occurring, so 3 is the most likely value of X.

87.

X = AP Statistics Exam Score	**P(X)**	$(x-\mu)^2 \cdot P(X)$
1	$\frac{6{,}058}{25{,}900}$	$\left(1-\frac{72{,}683}{25{,}900}\right)^2\left(\frac{6{,}058}{25{,}900}\right) \approx 0.7631433295$
2	$\frac{4{,}937}{25{,}900}$	$\left(2-\frac{72{,}683}{25{,}900}\right)^2\left(\frac{4{,}937}{25{,}900}\right) \approx 0.1239223418$
3	$\frac{6{,}333}{25{,}900}$	$\left(3-\frac{72{,}683}{25{,}900}\right)^2\left(\frac{6{,}333}{25{,}900}\right) \approx 0.0091748379$
4	$\frac{5{,}108}{25{,}900}$	$\left(4-\frac{72{,}683}{25{,}900}\right)^2\left(\frac{5{,}108}{25{,}900}\right) \approx 0.2810258618$
5	$\frac{3{,}464}{25{,}900}$	$\left(5-\frac{72{,}683}{25{,}900}\right)^2\left(\frac{3{,}464}{25{,}900}\right) \approx 0.6436283846$
Total		$\sigma^2 = \sum(X-\mu)^2 \cdot P(X) \approx 1.820894756$

$\sigma^2 = 1.820894756$

$\sigma = \sqrt{\sigma^2} = \sqrt{1.820894756} \approx 1.3494.$

89.

X = Number of cylinders	**Relative frequency**
4	0.4075
6	0.3637
8	0.2287
Total	0.9999

91.

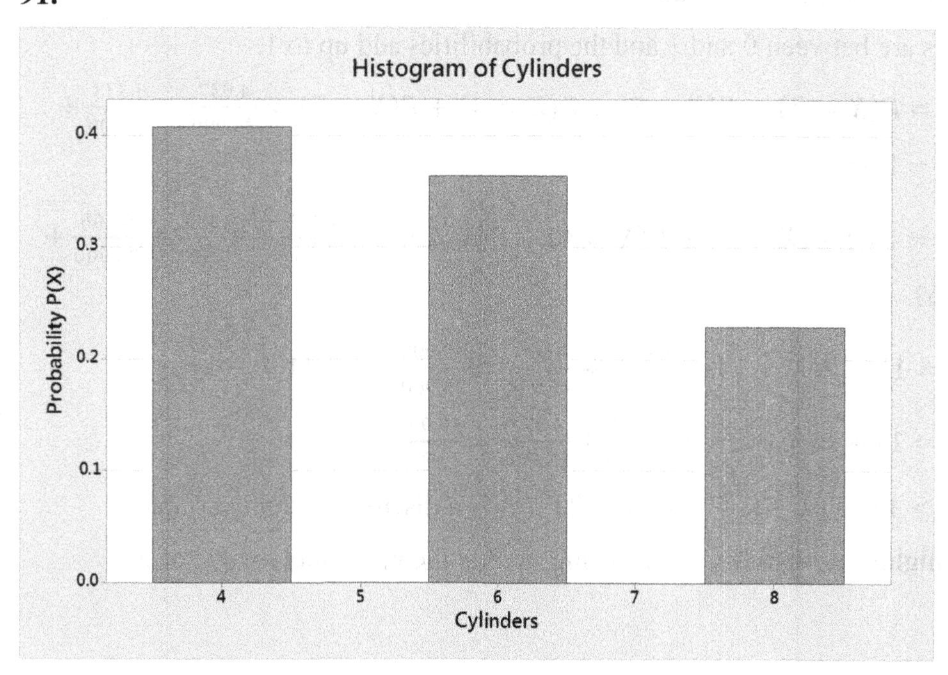

93. $X = 4$ cylinders has the highest probability of occurring, so 4 cylinders is the most likely value of X.

95.

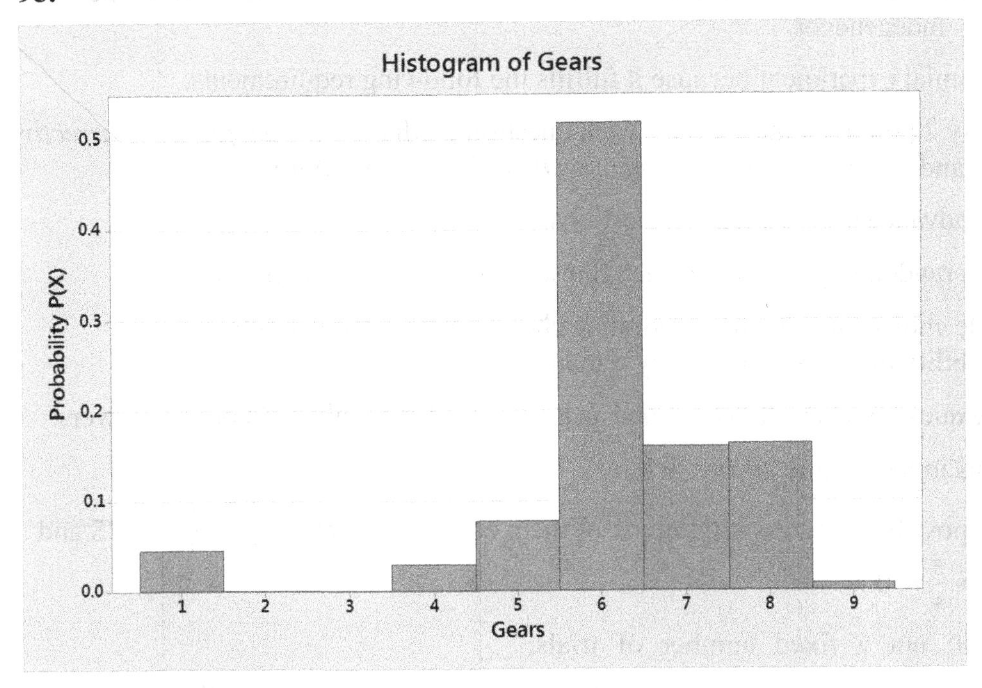

97.

X = Number of gears	***P(X)***	$(x-\mu)^2 \cdot P(X)$
1	0.0447	$(1-6.1493)^2(0.0447) = 1.185233484903$
4	0.0289	$(4-6.1493)^2(0.0289) = 0.133503275161$
5	0.0771	$(5-6.1493)^2(0.0771) = 0.101840656779$
6	0.5188	$(6-6.1493)^2(0.5188) = 0.011564306212$
7	0.1604	$(7-6.1493)^2(0.1604) = 0.116079954596$
8	0.1621	$(8-6.1493)^2(0.1621) = 0.555207168429$
9	0.0079	$(9-6.1493)^2(0.0079) = 0.064199274871$
Total	0.9999	$\sigma^2 = \sum(X-\mu)^2 \cdot P(X) \approx 2.167628121$

$\sigma^2 = 2.167628121$ gears squared

$\sigma = \sqrt{\sigma^2} = \sqrt{2.167628121} \approx 1.4723$ gears.

Section 6.2

1. Each trial of the experiment has only 2 possible mutually exclusive outcomes (or is defined in such a way that the number of outcomes is reduced to 2). One outcome is denoted a success and the other a failure.

(i) There is a fixed number of trials, known in advance of the experiment.

(ii) The experimental outcomes are independent of each other.

(iii) The probability of observing a success remains the same from trial to trial.

3. If you perform an experiment *n* times, you can't have more than *n* successes. For example, if you flip a coin 10 times, you can't get 11 heads.

5. Not binomial; the events "Person A comes to party" and "Person B comes to party" may not be independent.

7. This is a binomial experiment because it fulfills the following requirements:

(i) There are only 2 possible outcomes for each question, with *answer the question correctly* defined as a success and *answer the question incorrectly* defined as a failure.

(ii) We know in advance that there are 8 questions.

(iii) The person is randomly guessing on each question, so the trials are independent.

(iv) Each multiple choice question has 4 possible choices, with only 1 of them correct. Therefore, the probability of guessing correctly remains the same for each question.

Since answering the question correctly is defined as a success, X = number of correct answers.

There are 8 questions in the sample, so $n = 8$.

Each question has 4 possible choices, with only 1 of them correct. Therefore, $p = \frac{1}{4} = 0.25$ and $q = 1 - p = 1 - \frac{1}{4} = \frac{3}{4} = 0.75$.

9. Not binomial; not a fixed number of trials.

11. This is a not binomial experiment because it does not fulfill requirement (iii).The 4 cards are selected without replacement, and 4 is $\frac{1}{13} \cdot 100\% \approx 7.69\%$ of the population, which is more than 1% of the population. Therefore, the trials are not independent.

13. Binomial; $n = 2$, X = number of games won, $p = 0.25$, $1 - p = 0.75$.

15. $P(X = 1) = ({}_5C_1)(0.25)^1(1 - 0.25)^{5-1} \approx 0.3955078125 \approx 0.3955$

17. $P(X = 7) = ({}_{10}C_7)(0.5)^7(1 - 0.5)^{10-7} = 0.1171875 \approx 0.1172$

19. $P(X = 10) = ({}_{12}C_{10})(0.9)^{10}(1 - 0.9)^{12-10} \approx 0.2301277705 \approx 0.2301$

21. $P(X \leq 1) = P(X = 0) + P(X = 1) = ({}_5C_0)(0.25)^0(1 - 0.25)^{5-0} + ({}_5C_1)(0.25)^1(1 - 0.25)^{5-1} \approx 0.2373046875 + 0.3955078125 = 0.6328125 \approx 0.6328$

23. $P(X = 7 \text{ or } X = 8) = P(X = 7) + P(X = 8) = ({}_{10}C_7)(0.5)^7(1 - 0.5)^{10-7} + ({}_{10}C_8)(0.5)^8(1 - 0.5)^{10-8} \approx 0.1171875 + 0.0439453125 = 0.1611328125 \approx 0.1611$

25. $P(X \geq 10) = P(X = 10) + P(X = 11) + P(X = 12) = ({}_{12}C_{10})(0.9)^{10}(1 - 0.9)^{12-10} + ({}_{12}C_{11})(0.9)^{11}(1 - 0.9)^{12-11} + ({}_{12}C_{12})(0.9)^{12}(1 - 0.9)^{12-12} \approx 0.2301277705 + 0.3765727153 + 0.2824295365 = 0.8891300223 \approx 0.8891$

27. $P(9 \leq X \leq 12) = P(X = 9) + P(X = 10) + P(X = 11) + P(X = 12) = ({}_{12}C_9)(0.9)^9(1 - 0.9)^{12-9} + ({}_{12}C_{10})(0.9)^{10}(1 - 0.9)^{12-10} + ({}_{12}C_{11})(0.9)^{11}(1 - 0.9)^{12-11} + ({}_{12}C_{12})(0.9)^{12}(1 - 0.9)^{12-12} \approx 0.0852325076 + 0.2301277705 + 0.3765727153 + 0.2824295365 = 0.9743625299 \approx 0.9744$

29. $p = 0.25, q = 1 - p = 1 - 0.25 = 0.75, n = 3, X = 0.$

$P(X = 0) = ({}_3C_0)(0.25)^0(0.75)^{3-0} = (1)(1)(0.421875) = 0.421875$

31. $p = 0.25, q = 1 - p = 1 - 0.25 = 0.75, n = 3, X = 2.$

$P(X = 2) = ({}_3C_2)(0.25)^2(0.75)^{3-2} = (3)(0.0625)(0.75) = 0.140625$

33. $P(X \geq 1) = 1 - P(X < 1) = 1 - P(X = 0) = 1 - 0.421875 = 0.578125$

35. $p = 0.075, q = 1 - p = 1 - 0.075 = 0.925, n = 4, X = 3.$

$P(X = 3) = ({}_4C_3)(0.075)^3(0.925)^{4-3} = (4)(0.000421875)(0.925) = 0.0015609375$

37. $p = 0.075, q = 1 - p = 1 - 0.075 = 0.925, n = 4, X = 0.$

$P(X = 0) = ({}_4C_0)(0.075)^0(0.925)^{4-0} = (1)(1)(0.7320941406) = 0.7320941406$

39. $P(1 \leq X \leq 4) = P(X = 1) + P(X = 2) + P(X = 3) + P(X = 4) =$

$({}_4C_1)(0.075)^1(0.925)^{4-1} + ({}_4C_2)(0.075)^2(0.925)^{4-2} + ({}_4C_3)(0.075)^3(0.925)^{4-3} +$
$({}_4C_4)(0.075)^4(0.925)^{4-4} = (4)(0.075)(0.791453125) + (6)(0.005625)(0.855625) +$
$(4)(0.000421875)(0.925) + (1)(0.000031640625)(1) =$
$0.2374359375 + 0.0288773438 + 0.0015609375 + 0.000031640625 = 0.2679058594$

41. $p = 0.10, q = 1 - p = 1 - 0.10 = 0.90, n = 5, X = 0.$

$P(X = 0) = ({}_5C_0)(0.10)^0(0.90)^{5-0} = (1)(1)(0.59049) = 0.59049$

43. $P(X \leq 2) = P(X = 0) + P(X = 1) + P(X = 2) = ({}_5C_0)(0.10)^0(0.90)^{5-0} +$
$({}_5C_1)(0.10)^1(0.90)^{5-1} + ({}_5C_2)(0.10)^2(0.90)^{5-2} = (1)(1)(0.59049) + (5)(0.1)(0.6561) +$
$(10)(0.01)(0.729) = 0.59049 + 0.32805 + 0.0729 = 0.99144$

45. $p = 040, q = 1 - p = 1 - 0.40 = 0.60, n = 6, X = 6.$

$P(X = 6) = ({}_6C_6)(0.40)^6(0.60)^{6-6} = (1)(0.004096)(1) = 0.004096$

47. $P(3 \leq X \leq 5) = P(X = 3) + P(X = 4) + P(X = 5) = ({}_6C_3)(0.40)^3(0.60)^{6-3} +$
$({}_6C_4)(0.40)^4(0.60)^{6-4} + ({}_6C_5)(0.40)^5(0.60)^{6-5} =$
$(20)(0.064)(0.216) + (15)(0.0256)(0.36) + (6)(0.01024)(0.6) = 0.27648 + 0.13824 +$
$0.036864 = 0.451584$

49. **(a)** $p = 0.25, q = 1 - p = 1 - 0.25 = 0.75, n = 3,$

$\mu = np = (3)(0.25) = 0.75$ business major. If we take an infinite number of random samples of siZe 3 from all of the Masters degrees granted in 2012 and calculated the mean number of business majors in the samples it would be 0.75 business major.

(b) $\sigma^2 = npq = (3)(0.25)(0.75) = 0.5625$ business major squared

(c) $\sigma = \sqrt{\sigma^2} = \sqrt{0.5625} = 0.75$ business major

51. **(a)** $p = 0.10, q = 1 - p = 1 - 0.10 = 0.90, n = 5,$

$\mu = np = (5)(0.10) = 0.5$ American 25–29 years old. If we take an infinite number of random samples of siZe 5 of Americans aged 25–29 and calculated the mean number of them living alone it would be 0.5 American.

(b) $\sigma^2 = npq = (5)(0.10)(0.90) = 0.45$ American squared

(c) $\sigma = \sqrt{\sigma^2} = \sqrt{0.45} \approx 0.6708$ American

53. **(a)**

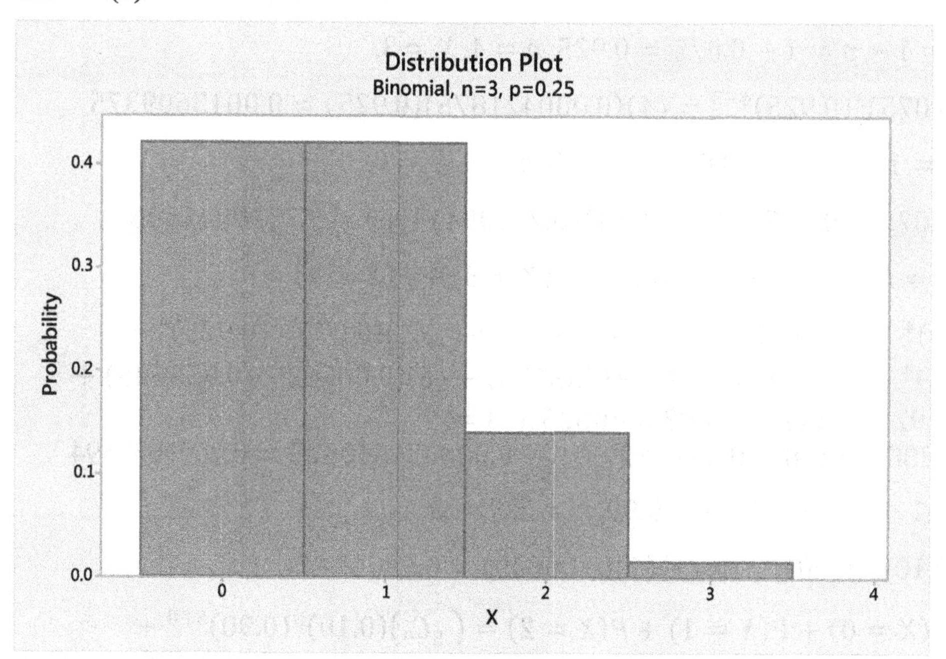

(b) The modes are the values of X with the tallest rectangles. Therefore, 0 business majors and 1 business major are the modes.

55. **(a)**

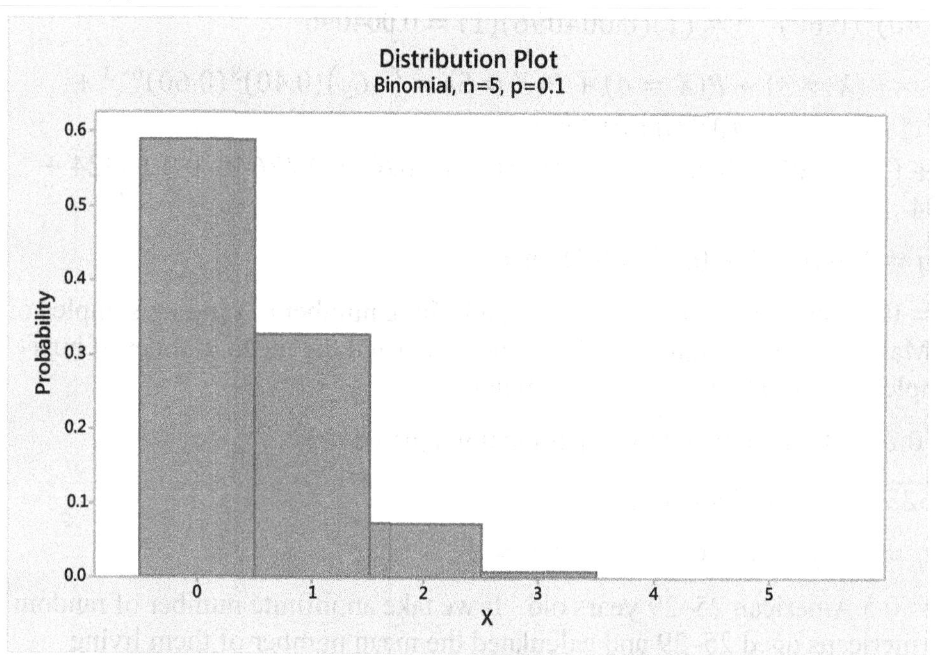

(b) The mode is the value of X with the tallest rectangle. Therefore, the mode is 0 25–29-year-olds

57. **(a)** It fulfills the following requirements:

(i) There are only two possible outcomes for each trial: correct answer or incorrect answer.

(ii) We know in advance that the quiZ will have 5 questions.

(iii) Since you are randomly guessing the answer to each question, the trials are independent.

(iv) Since each question has 4 responses, the probability of guessing correctly remains the same from question to question.

(b) $n = 5, p = 1/4 = 0.25$.

(c) $P(X \geq 3) = P(X = 3) + P(X = 4) + P(X = 5) = ({}_5C_3)(0.25)^3(1 - 0.25)^{5-3} + ({}_5C_4)(0.25)^4(1 - 0.25)^{5-4} + ({}_5C_5)(0.25)^5(1 - 0.25)^{5-5} \approx 0.087890625 + 0.0146484375 + 0.0009765625 = 0.103515625 \approx 0.1035$

(d) $P(X < 3) = 1 - P(X \geq 3) \approx 1 - 0.1035 = 0.8965$

59. **(a)** $p = 041, q = 1 - p = 1 - 0.41 = 0.59, n = 12$

$P(X = 4) = ({}_{12}C_4)(0.41)^4(0.59)^{12-4} = (495)(0.02825761)(0.0146830438) = 0.205379324 \approx 0.2054$

(b) $P(X \leq 4) = P(X = 0) + P(X = 1) + P(X = 2) + P(X = 3) + P(X = 4)$

$= ({}_{12}C_0)(0.41)^0(0.59)^{12-0} + ({}_{12}C_1)(0.41)^1(0.59)^{12-1} + ({}_{12}C_2)(0.41)^2(0.59)^{12-2} + ({}_{12}C_3)(0.41)^3(0.59)^{12-3} + ({}_{12}C_4)(0.41)^4(0.59)^{12-4} =$
$(1)(1)(0.0017791974) + (12)(0.41)(0.0030155888) + (66)(0.1681)(0.0051111675) + (220)(0.068921)(0.0086629958) + (495)(0.02825761)(0.0146830438) =$
$0.0017791974 + 0.0148366969 + 0.0567063589 + 0.1313537134 + 0.205379324 =$
$0.4100552906 \approx 0.4101$

(c) $P(4 \leq X \leq 6) = P(X = 4) + P(X = 5) + P(X = 6) = ({}_{12}C_4)(0.41)^4(0.59)^{12-4}$

$+({}_{12}C_5)(0.41)^5(0.59)^{12-5} + ({}_{12}C_6)(0.41)^6(0.59)^{12-6} =$
$(495)(0.02825761)(0.0146830438) + (792)(0.0115856201)(0.0248865148) + (924)(0.0047501042)(0.042180536) =$
$0.205379324 + 0.2283538592 + 0.1851344281 = 0.6188677063 \approx 0.6189$

61. **(a)** $\mu = np = (5)(0.25) =$ 1.25 correct answers. If we repeat this experiment an infinite number of times, record the number of correct answers for each quiz taken, and take the mean of all of the quizzes, the mean number of correct answers will equal $\mu = 1.25$.

$\sigma^2 = npq = (5)(0.25)(0.75) = 0.9375$ correct answer squared.

$\sigma = \sqrt{\sigma^2} = \sqrt{0.9375}$ correct answer.

(b) For $X = 0, Z = \frac{X-\mu}{\sigma} \approx \frac{0-1.25}{0.9682458366} \approx -1.2910.$

For $X = 1, Z = \frac{X-\mu}{\sigma} \approx \frac{1-1.25}{0.9682458366} \approx -0.2582.$

For $X = 2, Z = \frac{X-\mu}{\sigma} \approx \frac{2-1.25}{0.9682458366} \approx 0.7746..$

For $X = 3, Z = \frac{X-\mu}{\sigma} \approx \frac{3-1.25}{0.9682458366} \approx 1.8074.$

For $X = 4, Z = \frac{X-\mu}{\sigma} \approx \frac{4-1.25}{0.9682458366} \approx 2.8402.$

For $X = 5, Z = \frac{X-\mu}{\sigma} \approx \frac{5-1.25}{0.9682458366} \approx 2.8730.$

Since the Z-scores for 0, 1, 2, and 3 correct answers are all between −2 and 2, none of these values are outliers or moderately unusual. Since $2 \le Z\text{-score} < 3$ for 4 correct answers, 4 correct answers is moderately unusual. Since $Z\text{-score} \ge 3$ for 5 correct answers, 5 correct answers is an outlier.

(c)

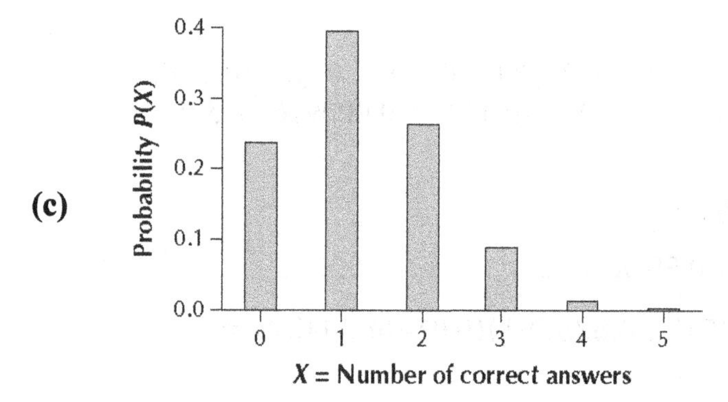

Mode is 1 correct answer.

63. **(a)** $p = 0.41, q = 1 - p = 1 - 0.41 = 0.59, n = 12$

$\mu = np = (12)(0.41) = 4.92$ telephone users who have abandoned their landline; $\sigma^2 = npq = (12)(0.41)(0.59) = 2.9028$ telephone users who have abandoned their landline squared; $\sigma = \sqrt{\sigma^2} = \sqrt{2.9028} \approx 1.7038$ telephone users who have abandoned their landline. If we take infinitely many random samples of telephone users of size 12 and calculate the mean number of telephone users who have abandoned their landline it would be 4.92.

(b) For $X = 0$ telephone users who have abandoned their landline, $Z = \frac{X-\mu}{\sigma} = \frac{0-4.92}{1.7038} \approx -2.89$. Since $-3 < Z - \text{score} \le -2$, 0 telephone users who have abandoned their landline is considered moderately unusual.

65. **(a)** $\mu = np = (5)(0.12) = 0.6$ woman. If we repeat this experiment an infinite number of times, record the number of women affected by a depressive disorder for each sample, and take the mean of all of the samples, the mean number of women living with a depressive disorder will equal $\mu = 0.6$.

(b) Not possible. The expected value of X is not an integer.

(c)

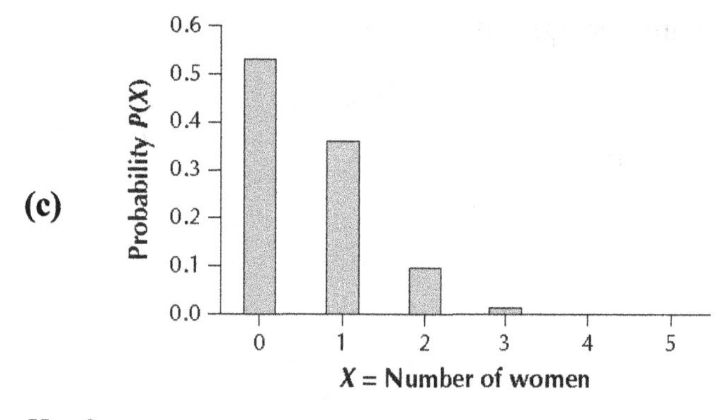

$X = 0$ women.

(d) $P(X = 0) = ({}_5C_0)(0.12)^0(1 - 0.12)^{5-0} \approx 0.5277319168 \approx 0.5277$.

67. **(a)** Since the mean $\mu = 90$ students is a whole number, the median = the mean = 90 students.

(b) Since the mean $\mu = 90$ students is a whole number, the mode = the mean = 90 students.

(c) Since the mode is the value with the highest probability, the mode is the most likely value of X. Therefore, 90 students is the most likely value of X.

69. **(a)**

$$P(X = 5) = \frac{({}_{N_1}C_X)({}_{N_2}C_{n-X})}{({}_{N}C_n)} = \frac{({}_{13}C_5)({}_{39}C_{5-5})}{({}_{52}C_5)} = \frac{(1287)(1)}{2{,}598{,}960} = \frac{1287}{2{,}598{,}960} \approx 0.0005$$

(b) $$P(X = 4) = \frac{({}_{N_1}C_X)({}_{N_2}C_{n-X})}{({}_{N}C_n)} = \frac{({}_{13}C_4)({}_{39}C_{5-4})}{({}_{52}C_5)} = \frac{(715)(39)}{2{,}598{,}960} = \frac{27{,}885}{2{,}598{,}960} \approx 0.0107$$

(c) $$P(X \geq 4) = P(X = 4) + P(X = 5) = \frac{27{,}885}{2{,}598{,}960} + \frac{1287}{2{,}598{,}960} = \frac{29{,}172}{2{,}598{,}960} \approx 0.0112$$

(d) $$P(X = 3) = \frac{({}_{N_1}C_X)({}_{N_2}C_{n-X})}{({}_{N}C_n)} = \frac{({}_{13}C_3)({}_{39}C_{5-3})}{({}_{52}C_5)} = \frac{(286)(741)}{2{,}598{,}960} = \frac{211{,}926}{2{,}598{,}960} \approx 0.0815$$

(e) $$P(X \leq 2) = P(X = 0) + P(X = 1) + P(X = 2) = \frac{({}_{13}C_0)({}_{39}C_{5-0})}{({}_{52}C_5)} + \frac{({}_{13}C_1)({}_{39}C_{5-1})}{({}_{52}C_5)} + \frac{({}_{13}C_2)({}_{39}C_{5-2})}{({}_{52}C_5)} = \frac{(1)(575{,}757)}{2{,}598{,}960} + \frac{(13)(82{,}251)}{2{,}598{,}960} + \frac{(78)(9139)}{2{,}598{,}960} = \frac{575{,}757}{2{,}598{,}960} + \frac{1{,}069{,}263}{2{,}598{,}960} + \frac{712{,}842}{2{,}598.960} = \frac{2{,}357{,}862}{2{,}598{,}960} \approx 0.9072.$$

71. **(i)** Either a new job is provided by a small company or it is not provided by a small company. These are the only 2 possible outcomes and they are mutually exclusive.

(ii) It is known in advance than exactly 10 new jobs will be selected.

(iii) The sample is random, so the outcomes are independent.

(iv) The sample is quite small compared to the siZe of the population, so that the probability that a new job was provided by a small business remains the same from job to job.

73. **(a)** $P(X > 8) = P(X = 9) + P(X = 10) = 0.1877 + 0.0563 = 0.2440$

(b) $P(X \geq 9) = P(X = 9) + P(X = 10) = 0.1877 + 0.0563 = 0.2440$

(c)
$P(X \leq 4) = P(X = 0) + P(X = 1) + P(X = 2) + P(X = 3) + P(X = 4) =$
$0.0000009536743164 + 0.00002861022949 + 0.0003862380981 + 0.0030899048 +$
$0.0163330001 = 0.0197277069 \approx 0.0197$

(d)
$P(X < 5) = P(X = 0) + P(X = 1) + P(X = 2) + P(X = 3) + P(X = 4) =$
$0.0000009536743164 + 0.00002861022949 + 0.0003862380981 + 0.0030899048 +$
$0.0163330001 = 0.0197277069 \approx 0.0197$

X is a discrete variable.

75. Not possible because $\mu = 7.5$ jobs is not a whole number.

77. The mode is the value with the highest probability, so the mode is 8 jobs.

79. For $X = 0$ jobs, $Z = \frac{X-\mu}{\sigma} = \frac{0-7.5}{1.3693} \approx -5.48$

For $X = 1$ job, $Z = \frac{X-\mu}{\sigma} = \frac{1-7.5}{1.3693} \approx -4.75$

For $X = 2$ jobs, $Z = \frac{X-\mu}{\sigma} = \frac{2-7.5}{1.3693} \approx -4.02$

For $X = 3$ jobs, $Z = \frac{X-\mu}{\sigma} = \frac{3-7.5}{1.3693} \approx -3.29$

For $X = 4$ jobs, $Z = \frac{X-\mu}{\sigma} = \frac{4-7.5}{1.3693} \approx -2.56$

For $X = 5$ jobs, $Z = \frac{X-\mu}{\sigma} = \frac{5-7.5}{1.3693} \approx -1.83$

For $X = 6$ jobs, $Z = \frac{X-\mu}{\sigma} = \frac{6-7.5}{1.3693} \approx -1.10$

For $X = 7$ jobs, $Z = \frac{X-\mu}{\sigma} = \frac{7-7.5}{1.3693} \approx -0.37$

For $X = 8$ jobs, $Z = \frac{X-\mu}{\sigma} = \frac{8-7.5}{1.3693} \approx 0.37$

For $X = 9$ jobs, $Z = \frac{X-\mu}{\sigma} = \frac{9-7.5}{1.3693} \approx 1.10$

For $X = 10$ jobs, $Z = \frac{X-\mu}{\sigma} = \frac{10-7.5}{1.3693} \approx 1.83$

For $X = 0, 1, 2, 3$ jobs, Z-score ≤ -3, so 0, 1, 2, and 3 jobs are outliers. For $X = 4$ jobs, $-3 < Z$-score ≤ -2, 4 jobs is moderately unusual.

81. $\mu = np = (100)(0.1639) = 16.39$ cars

83. Answers will vary

Section 6.3

1. The Poisson probability distribution is a discrete probability distribution that is used when observing the number of occurrences of an event within a fixed interval of space and time.

3. The requirements for the Poisson probability distribution are (1) the occurrences must be random, (2) each occurrence must be independent, and (3) the occurrences must be uniformly distributed over the given interval.

5. *X* does not follow a Poisson distribution. Sheep are more likely to be in the pasture where the food (grass) is than in the forest where there is very little food for them. This violates the requirement that the occurrences be uniformly distributed throughout the interval.

7. *X* does follow a Poisson distribution. The cars are random, independent, and occur uniformly over the 1-hour period.

9. $P(X = 8) = \frac{\mu^X e^{-\mu}}{X!} = \frac{10^8 e^{-10}}{8!} \approx 0.1126.$

11. The phrase "at most 3" means $X = 0$ or $X = 1$ or $X = 2$ or $X = 3$.

$P(X = 0) = \frac{\mu^X e^{-\mu}}{X!} = \frac{3^0 e^{-3}}{0!} \approx 0.0498.$

$P(X = 1) = \frac{\mu^X e^{-\mu}}{X!} = \frac{3^1 e^{-3}}{1!!} \approx 0.1494.$

$P(X = 2) = \frac{\mu^X e^{-\mu}}{X!} = \frac{3^2 e^{-3}}{2!} \approx 0.2240.$

$P(X = 3) = \frac{\mu^X e^{-\mu}}{X!} = \frac{3^3 e^{-3}}{3!} \approx 0.2240.$

$P(X \leq 3) = P(0) + P(1) + P(2) + P(3) = 0.0498 + 0.1494 + 0.2240 + 0.2240 = 0.6472$

13. The phrase "at most 2" means $X = 0$ or $X = 1$ or $X = 2$.

$P(X = 0) = \frac{\mu^X e^{-\mu}}{X!} = \frac{4^0 e^{-4}}{0!} \approx 0.0183.$

$P(X = 1) = \frac{\mu^X e^{-\mu}}{X!} = \frac{4^1 e^{-4}}{1!!} \approx 0.0733.$

$P(X = 2) = \frac{\mu^X e^{-\mu}}{X!} = \frac{4^2 e^{-4}}{2!} \approx 0.1465.$

$P(X \leq 2) = P(0) + P(1) + P(2) = 0.0183 + 0.0733 + 0.1465 = 0.2381$

15. **(a)** $\sigma = \sqrt{\mu} = \sqrt{20} \approx 4.4721.$

(b) $\mu - 2\sigma = 20 - 2(4.4721) = 11.0558$

$\mu + 2\sigma = 20 + 2(4.4721) = 28.9442$

Thus any values of X that are less than or equal to 11 or greater than or equal to 29 will be considered moderately unusual.

17. **(a)** $\sigma = \sqrt{\mu} = \sqrt{10} \approx 3.1623.$

(b) $\mu - 2\sigma = 10 - 2(3.1623) = 3.6754$

$\mu + 2\sigma = 10 + 2(3.1623) = 16.3246$

Thus any values of X that are less than or equal to 3 or greater than or equal to 17 will be considered moderately unusual.

19. $n = 200, p = 0.05$

(a) $n = 200 \geq 100$ and $np = 200 \cdot 0.05 = 10 \leq 10$

(b) $\mu = np = 200 \cdot 0.05 = 10, X = 12$

$P(X = 12) = \frac{\mu^X e^{-\mu}}{X!} = \frac{10^{12} e^{-10}}{12!} \approx 0.0948.$

21. $n = 200, p = 0.04$

(a) $n = 200 \geq 100$ and $np = 200 \cdot 0.04 = 8 \leq 10$

(b) $\mu = np = 200 \cdot 0.04 = 8, X = 10$

$P(X = 2) = \frac{\mu^X e^{-\mu}}{X!} = \frac{8^{10} e^{-8}}{10!} \approx 0.0993.$

23. $\mu = 100$

(a) $P(X = 100) = \frac{\mu^X e^{-\mu}}{X!} = \frac{100^{100} e^{-100}}{100!} \approx 0.0399.$

(b) $P(X = 101) = \frac{\mu^X e^{-\mu}}{X!} = \frac{100^{101}e^{-100}}{101!} \approx 0.0395.$

(c) $P(X = 95) = \frac{\mu^X e^{-\mu}}{X!} = \frac{100^{95}e^{-100}}{95!} \approx 0.0360.$

25. $n = 247{,}000{,}000$, $p = 0.0000000214$

(a) $n = 247{,}000{,}000 \geq 100$ and $np = 247{,}000{,}000 \cdot 0.0000000214 = 5.2858 \leq 10.$

(b) $\mu = np = 247{,}000{,}000 \cdot 0.0000000214 = 5.2858$

(c) $P(X = 0) = \frac{\mu^X e^{-\mu}}{X!} = \frac{5.2858^0 e^{-5.2858}}{0!} \approx 0.0051$

(d) $P(X \geq 1) = 1 - P(X < 1) = 1 - P(0) = 1 - 0.0051 = 0.9949$

(e) $P(X = 0) = \frac{\mu^X e^{-\mu}}{X!} = \frac{5.2858^0 e^{-5.2858}}{0!} \approx 0.0051$

$P(X = 1) = \frac{\mu^X e^{-\mu}}{X!} = \frac{5.2858^1 e^{-5.2858}}{1!} \approx 0.0268$

$P(X = 2) = \frac{\mu^X e^{-\mu}}{X!} = \frac{5.2858^2 e^{-5.2858}}{2!} \approx 0.0707$

$P(X \leq 2) = P(X = 0) + P(X = 1) + P(X = 2) = 0.0051 + 0.0268 + 0.0707 = 0.1026$

27. $\mu = 5.19$

(a) $X = 5,\ P(X = 5) = \frac{\mu^X e^{-\mu}}{X!} = \frac{5.19^5 e^{-5.19}}{5!} \approx 0.1749$

(b) $X = 4,\ P(X = 4) = \frac{\mu^X e^{-\mu}}{X!} = \frac{5.19^4 e^{-5.19}}{4!} \approx 0.1684$

(c) $X = 0,\ P(X = 0) = \frac{\mu^X e^{-\mu}}{X!} = \frac{5.19^0 e^{-5.19}}{0!} \approx 0.0056$

(d) Since $\mu = 5.19$ is not a whole number $P(X = \mu) = 0$.

29. $\mu = 3.8$

(a) $P(X = 3) = \frac{\mu^X e^{-\mu}}{X!} = \frac{3.8^3 e^{-3.8}}{3!} \approx 0.2046$

(b) $P(X = 2) = \frac{\mu^X e^{-\mu}}{X!} = \frac{3.8^2 e^{-3.8}}{2!} \approx 0.1615$

(c) $P(X = 0) = \frac{\mu^X e^{-\mu}}{X!} = \frac{3.8^0 e^{-3.8}}{0!} \approx 0.0224$

(d) $P(X = 1) = \frac{\mu^X e^{-\mu}}{X!} = \frac{3.8^1 e^{-3.8}}{1!} \approx 0.0850$

$P(X \leq 3) = P(0) + P(1) + P(2) + P(3) = 0.0224 + 0.0850 + 0.1615 + 0.2046 = 0.4735$

$P(X > 3) = 1 - P(X \leq 3) = 1 - 0.4735 = 0.5265$

31. **(a)** $n = 100 \geq 100$ and $np = (100)(0.10) = 10 \leq 10$

(b) $\mu = np = (100)(0.10) = 10$

(c) $P(X = 10) = \frac{10^{10} \cdot e^{-10}}{10!} \approx 0.1251$

(d) $P(X=10)+P(X=11)+P(X=12)=\frac{10^{10}\cdot e^{-10}}{10!}+\frac{10^{11}\cdot e^{-10}}{11!}+\frac{10^{12}\cdot e^{-10}}{12!}$

$=0.1251100357+0.1137363961+0.0947803301=0.3336267619\approx 0.3336$

Section 6.4

1. For a continuous random variable *X*, the probability that *X* equals some particular value is always Zero.

3. Area under the normal distribution curve above an interval.

5. False. It is always shaped like a rectangle.

7. $\mu=0$

9. True.

11. $P(50<X<100)=\frac{100-50}{100-0}=\frac{50}{100}=0.5$

13. $P(25<X<90)=\frac{90-25}{100-0}=\frac{65}{100}=0.65$

15. $P(24<X<25)=\frac{25-24}{100-0}=\frac{1}{100}=0.01$

17. $P(0\le X\le 5)=\frac{5-0}{5-(-5)}=\frac{5}{10}=0.5$

19. $P(-5\le X\le -4)=\frac{(-4)-(-5)}{5-(-5)}=\frac{1}{10}=0.1$

21. *A* has mean 10; *B* has mean 25. The peak of a normal curve is at the mean; from the graphs we see that the mean of *A* is less than the mean of *B*.

23. $\mu=0, \sigma=1$

25. $\mu=-10, \sigma=10$

27. $\mu=100$, 125 is 2 standard deviations from the mean of $\mu=100$, so $\sigma=\frac{1}{2}(125-100)=12.5$.

29. $\mu=0$. 6 is 3 standard deviations above the mean of $\mu=0$, so $\sigma=\frac{1}{3}(6-0)=2$.

Use the following graph for Exercises 31–36.

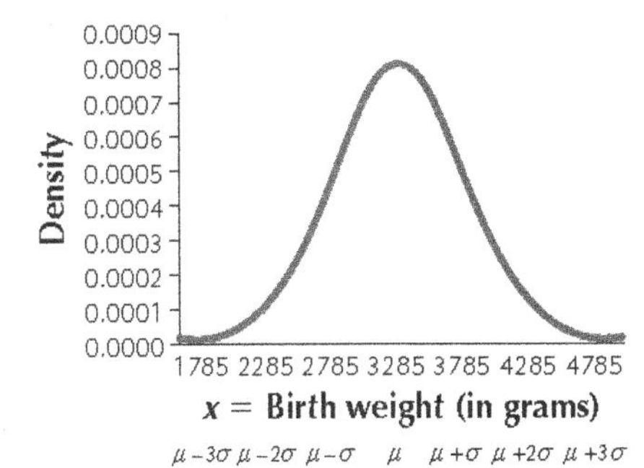

31. $P(X = 3285) = 0$

33. $P(X \geq 3285) = 0.5$

35. Less than 0.5. Since $X = 4285$ is greater than the mean of 3285 and the area to the right of $\mu = 3285$ is 0.5, the area to the right of $X = 4285$ is less than the area to the right of $X = 3285$.

37. **(a)**

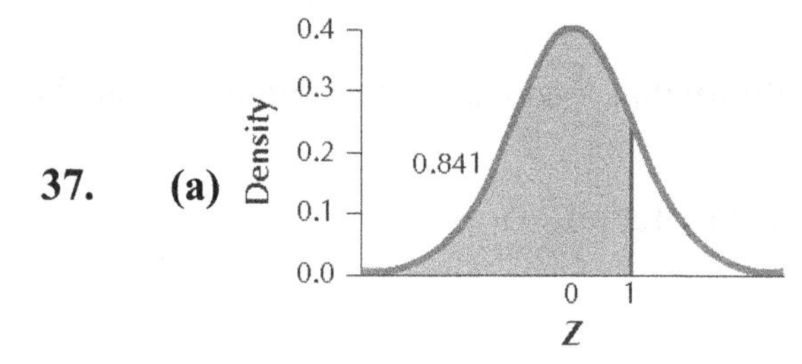

(b) From Table C, the area to the left of $Z = 1.00$ is 0.8413.

39. **(a)**

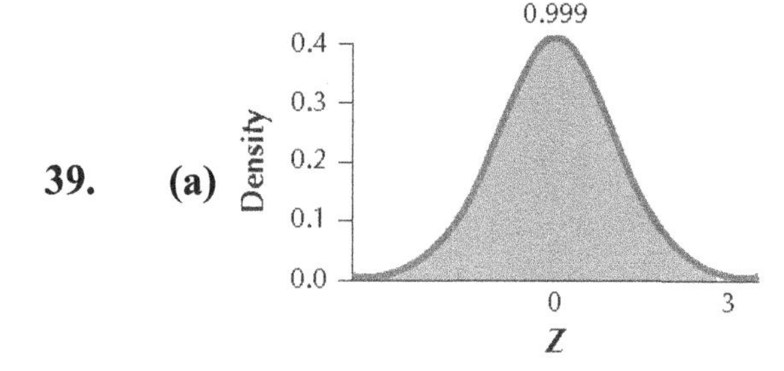

(b) From Table C, the area to the left of $Z = 3.00$ is 0.9987.

41. **(a)**

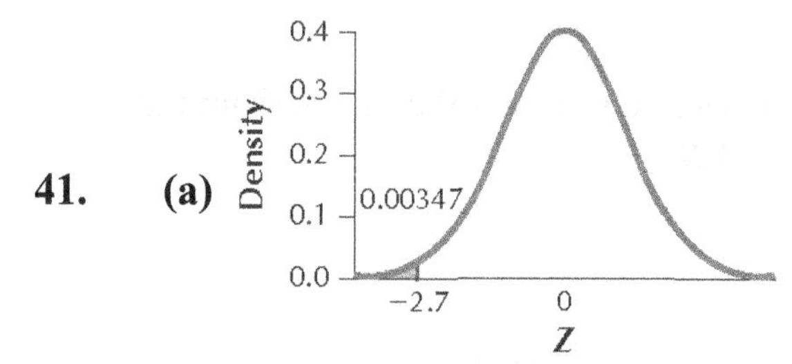

(b) From Table C, the area to the left of $Z = -2.70$ is 0.0035.

43. **(a)**

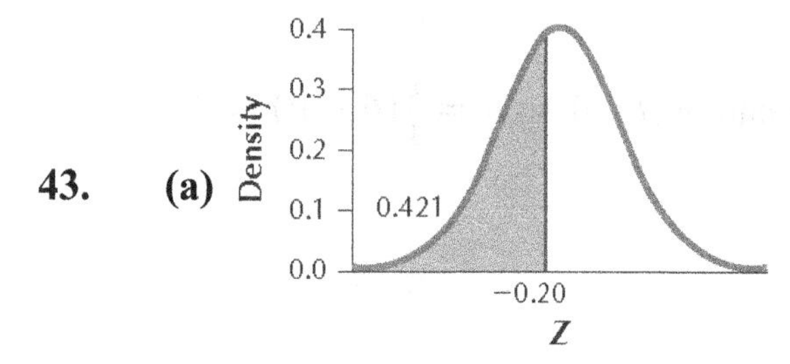

(b) From Table C, the area to the left of $Z = -0.20$ is 0.4207.

45. **(a)**

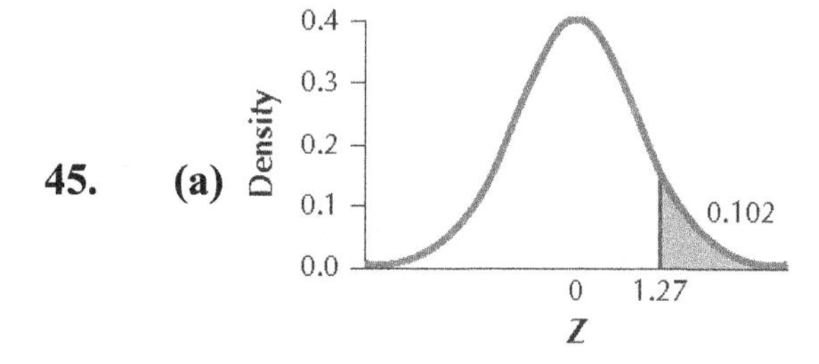

(b) From Table C, the area to the right of $Z = 1.27$ is $1 - 0.8980 = 0.1020$.

47. **(a)**

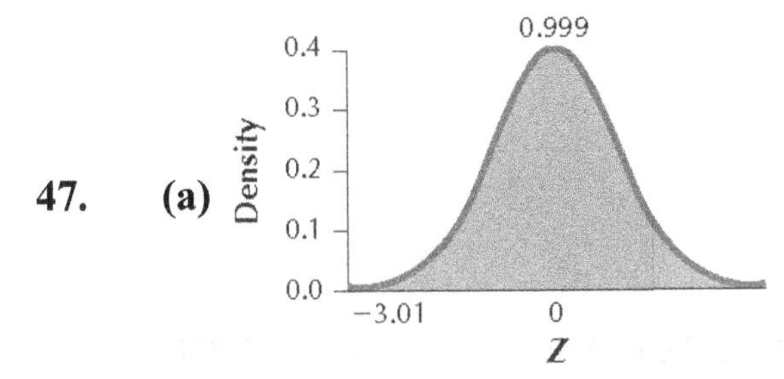

(b) From Table C, the area to the right of $Z = -3.01$ is $1 - 0.0013 = 0.9987$.

49. **(a)**

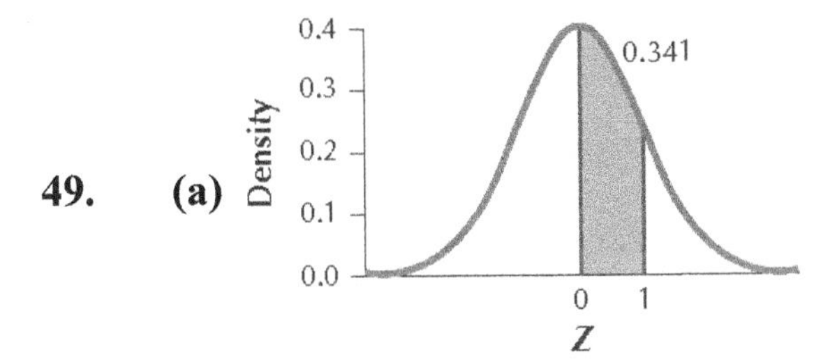

(b) From Table C, the area between $Z = 0$ and $Z = 1.00$ is $0.8413 - 0.5 = 0.3413$.

51. **(a)**

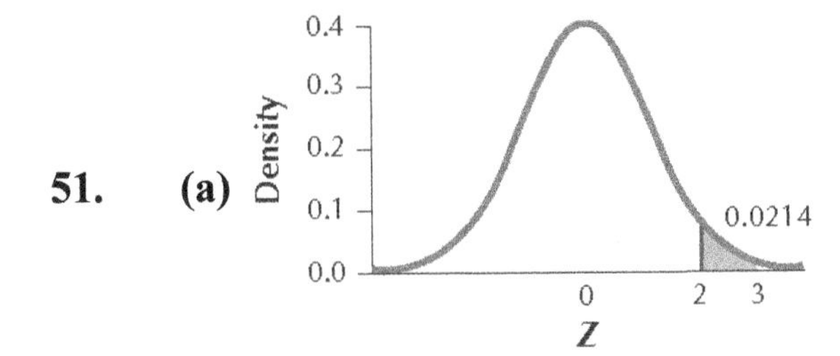

(b) From Table C, the area between $Z = 2.00$ and $Z = 3.00$ is $0.9987 - 0.9772 = 0.0215$.

53. **(a)**

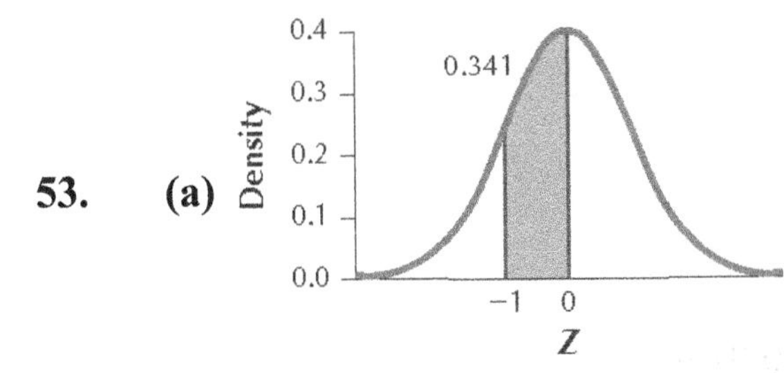

(b) From Table C, the area between $Z = -1.00$ and $Z = 0$ is $0.5000 - 0.1587 = 0.3413$.

55. **(a)**

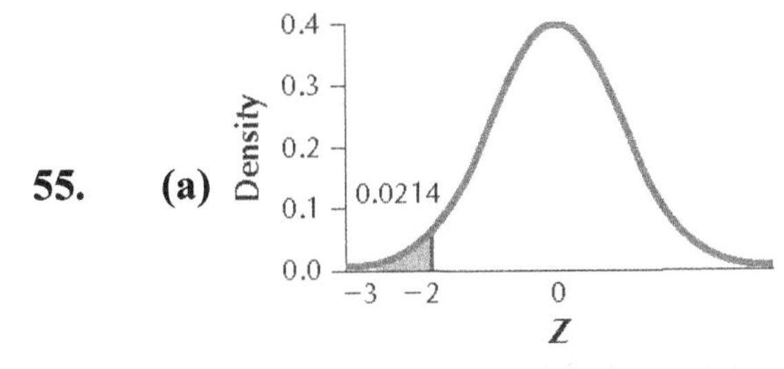

(b) From Table C, the area between $Z = -3.00$ and $Z = -2.00$ is $0.0228 - 0.0013 = 0.0215$. From Minitab, the area is 0.0214.

57. **(a)**

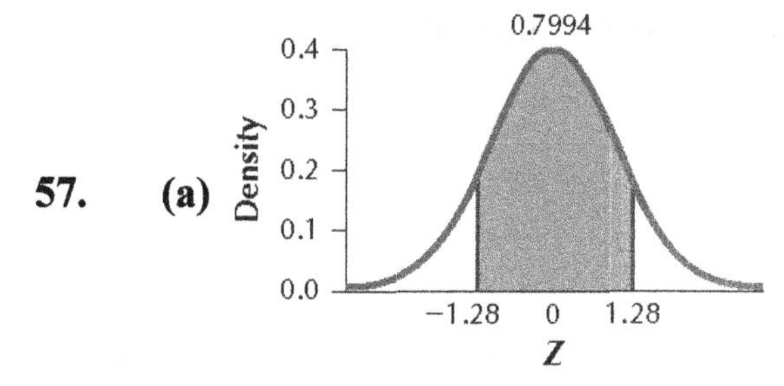

(b) From Table C, the area between $Z = -1.28$ and $Z = 1.28$ is $0.8997 - 0.1003 = 0.7994$.

59. **(a)**

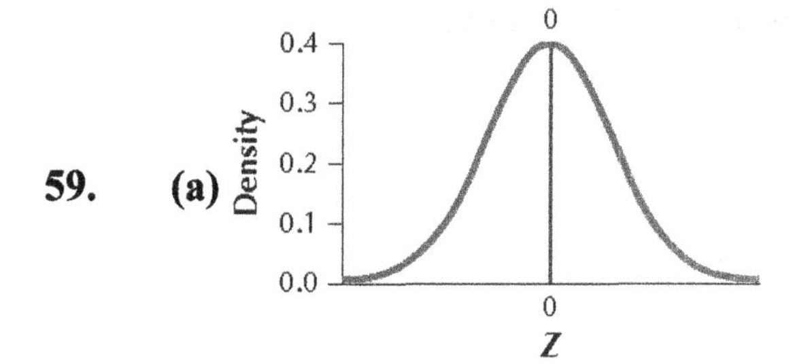

(b) 0

61. **(a)**

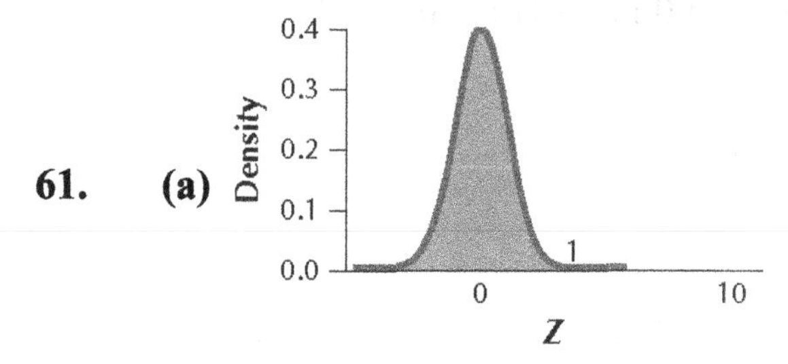

(b) 1

63. **(a)**

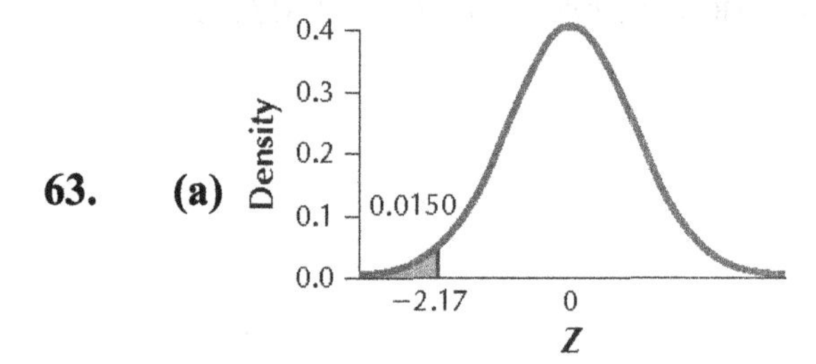

(b) From Table C, the area to the left of $Z = -2.17$ is 0.0150.

65. **(a)**

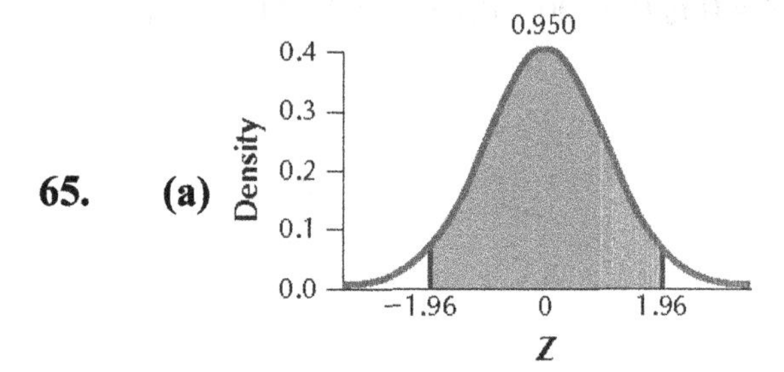

(b) From Table C, the area between $Z = -1.96$ and $Z = 1.96$ is $0.9750 - 0.0250 = 0.9500$.

67. **(a)**

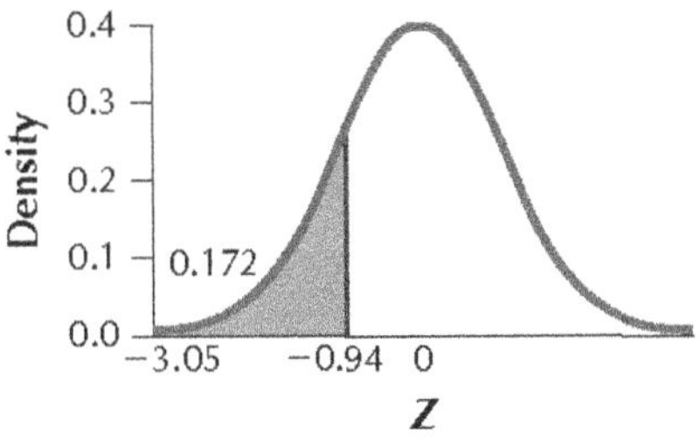

(b) From Table C, the area between $Z = -3.05$ and $Z = -0.94$ is $0.1736 - 0.0011 = 0.1725$.

69. **(a)**

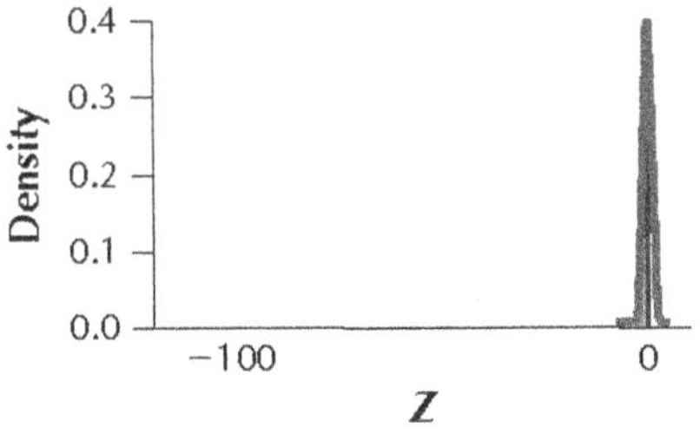

(b) 0.5000

71.

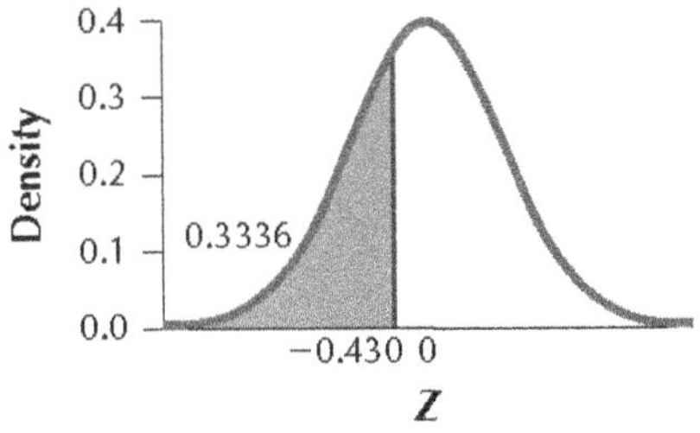

$Z = -0.43$.

73.

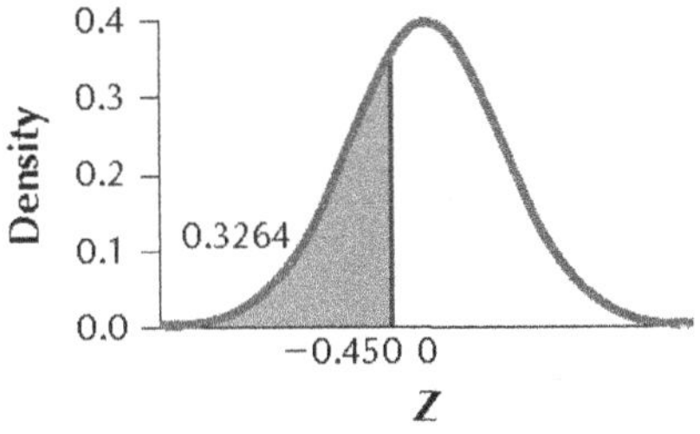

$Z = -0.45$.

75.

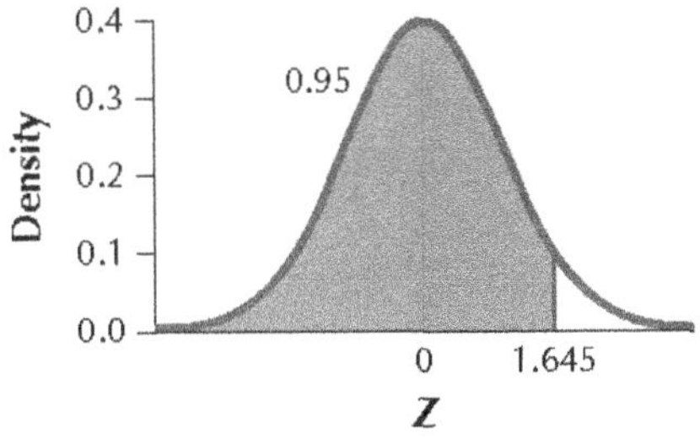

$Z = 1.65$, TI-83/84: $Z = 1.645$.

77.

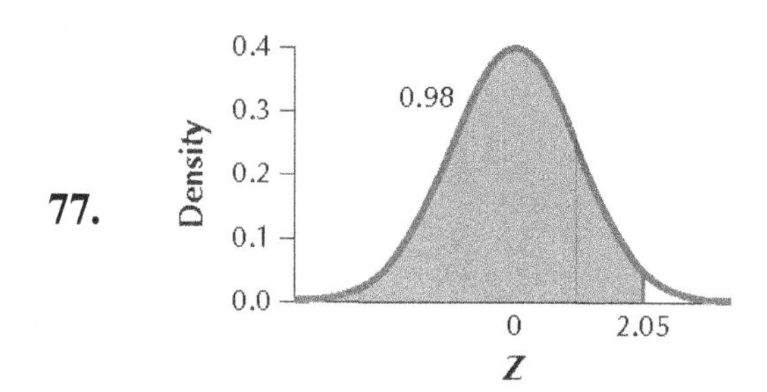

$Z = 2.05$.

79.

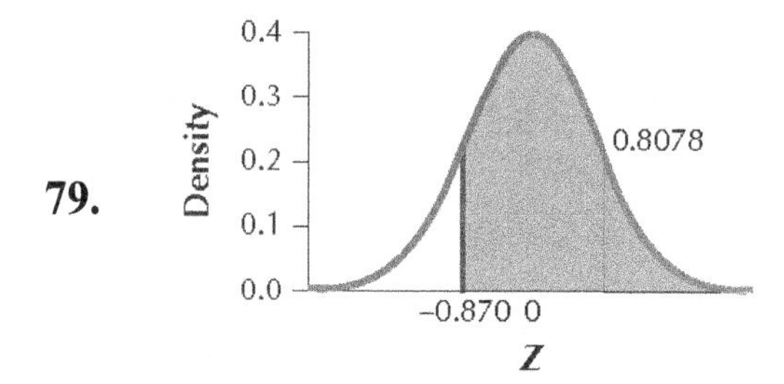

$Z = -0.87$.

81.

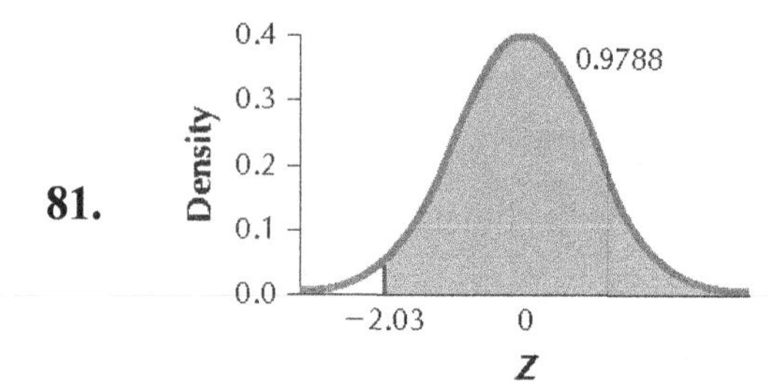

$Z = -2.03$.

83.

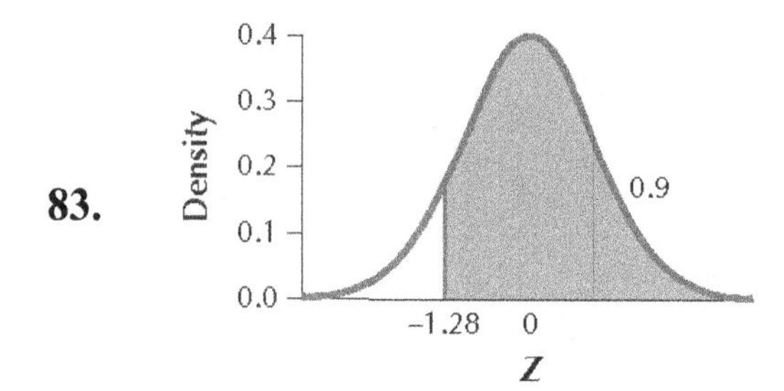

$Z = -1.28$.

85.

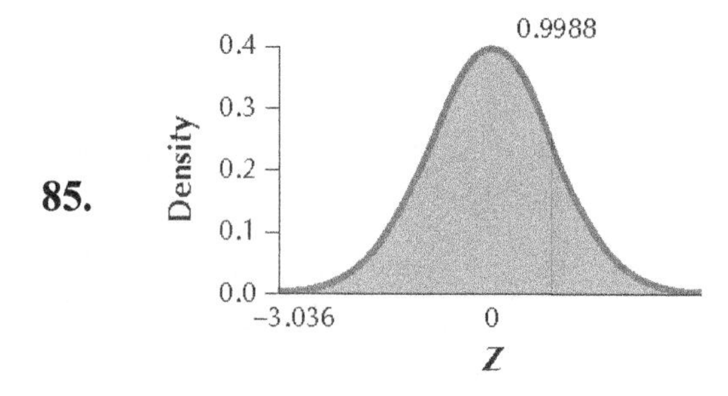

$Z = -3.04$.

87.

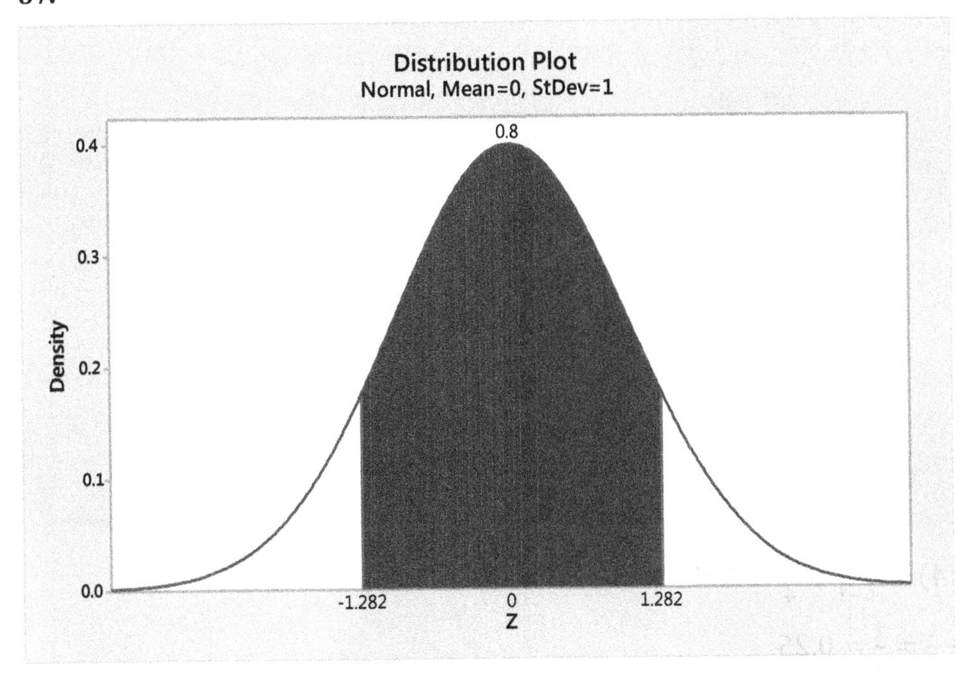

–1.28 and 1.28

89.

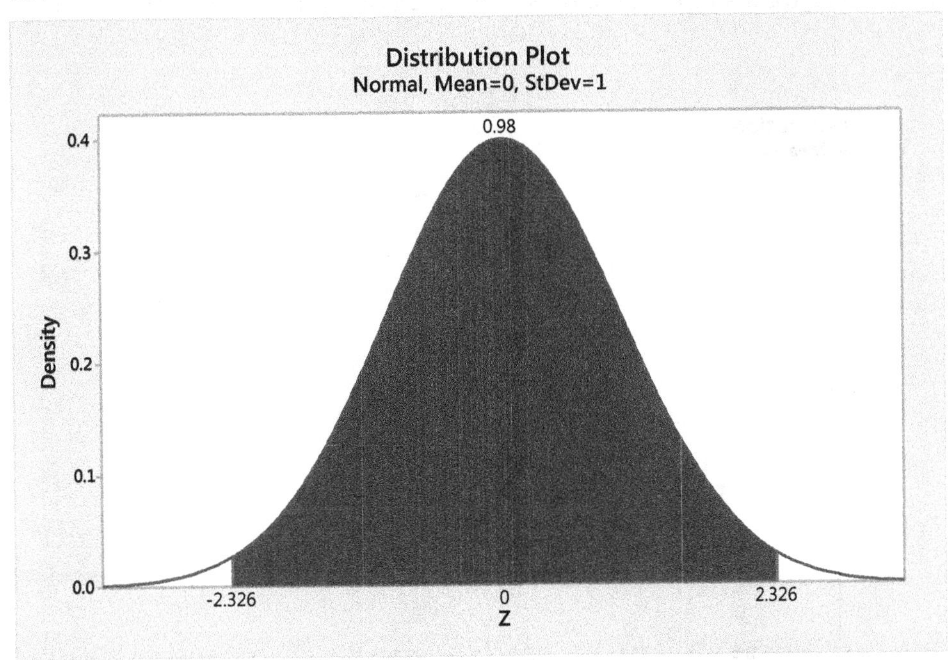

–2.33 and 2.33

91. $Z = 0$

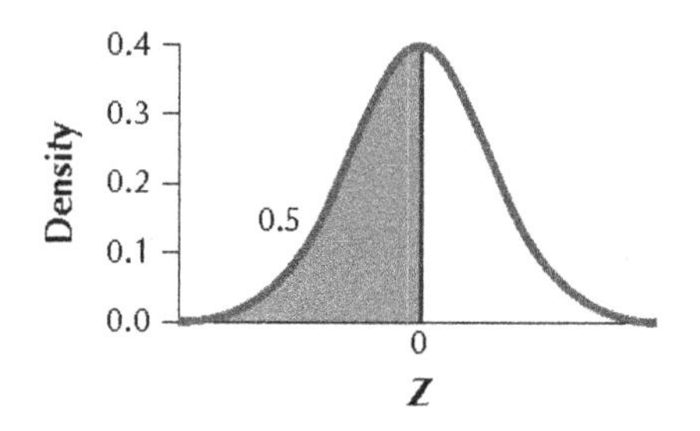

93. $Z = 2.58$

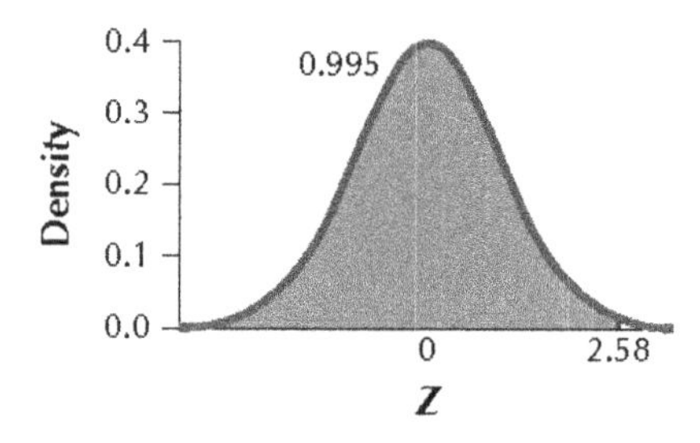

95. **(a)** $P(3 \le X \le 4) = \frac{4-3}{5-1} = \frac{1}{4} = 0.25.$

(b) $P(1 \le X \le 2) = \frac{2-1}{5-1} = \frac{1}{4} = 0.25.$

(c) Since $P(X) = 0$ for X outside of the interval $1 \le X \le 5$, $P(X < 1) = 0$.

(d) 0; Area underneath the curve for a single value of X is the area of a line which is 0.

97.

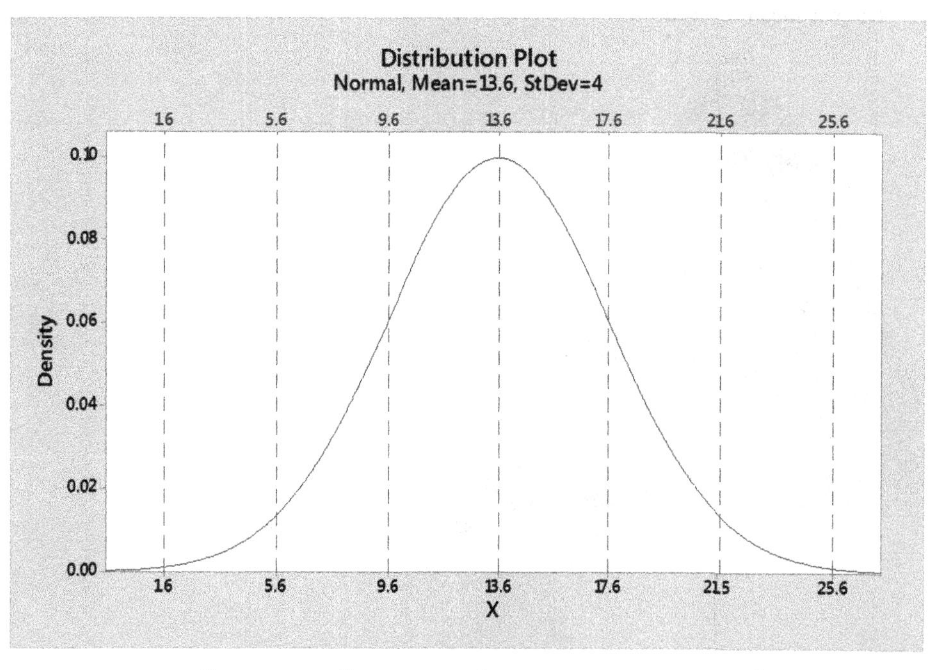

$\mu - 3\sigma$ $\mu - 2\sigma$ $\mu - 1\sigma$ μ $\mu + 1\sigma$ $\mu + 2\sigma$ $\mu + 3\sigma$

99.

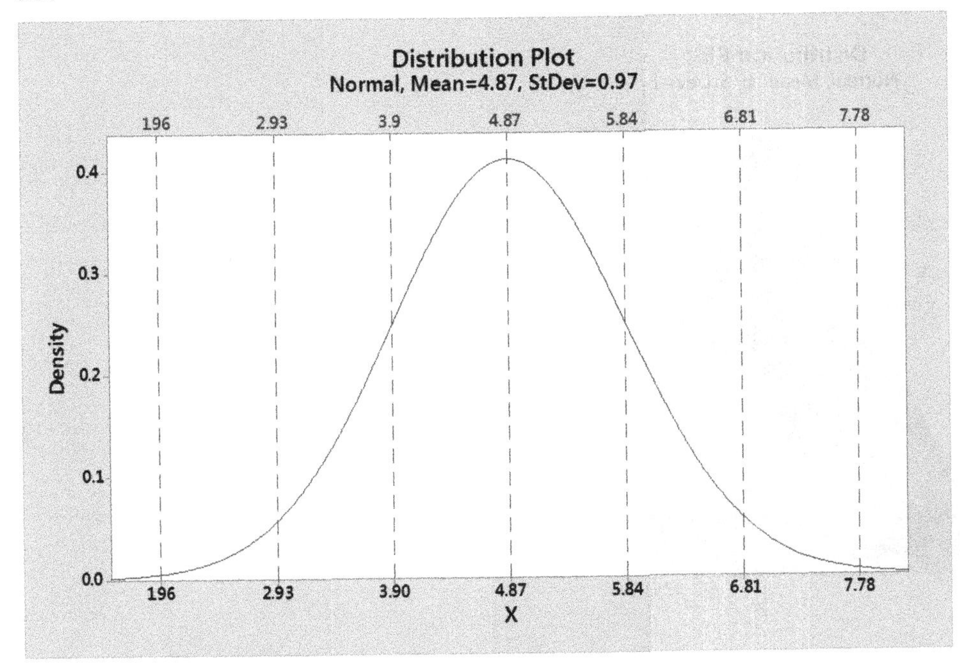

$\mu - 3\sigma$ $\quad$ $\mu - 2\sigma$ $\quad$ $\mu - 1\sigma$ $\quad$ μ $\quad$ $\mu + 1\sigma$ $\quad$ $\mu + 2\sigma$ $\quad$ $\mu + 3\sigma$

101. **(a)** $P(0 \leq X \leq X_1) = \frac{X_1 - 0}{10 - 0} = \frac{X_1}{10} = 0.95$, so $X_1 = 10(0.95) = 9.5$ minutes.

(b) $P(0 \leq X \leq X_1) = \frac{X_1 - 0}{10 - 0} = \frac{X_1}{10} = 0.9$, so $X_1 = 10(0.9) = 9$ minutes.

(c) $P(0 \leq X \leq X_1) = \frac{X_1 - 0}{10 - 0} = \frac{X_1}{10} = 0.975$, so $X_1 = 10(0.975) = 9.75$ minutes.

(d) $P(0 \leq X \leq X_1) = \frac{X_1 - 0}{10 - 0} = \frac{X_1}{10} = 0.05$, so $X_1 = 10(0.05) = 0.5$ minute.

(e) $P(0 \leq X \leq X_1) = \frac{X_1 - 0}{10 - 0} = \frac{X_1}{10} = 0.1$, so $X_1 = 10(0.1) = 1$ minute.

(f) $P(0 \leq X \leq X_1) = \frac{X_1 - 0}{10 - 0} = \frac{X_1}{10} = 0.025$, so $X_1 = 10(0.025) = 0.25$ minute.

103. From Table C, the area to the left of $Z = 1.96$ is 0.9750.

105. From Table C, the area between $Z = 0$ and $Z = 2.10$ is $0.9821 - 0.5 = 0.4821$.

107. From Table C, the area to the right of $Z = 2.10$ is $1 - 0.9821 = 0.0179$.

109. From Table C, the area between $Z = -2.20$ and $Z = 0.90$ is $0.8159 - 0.0139 = 0.8020$.

111. From Table C, the area between $Z = 0.80$ and $Z = 1.90$ is $0.9713 - 0.7881 = 0.1832$.

113. From Table C, the area to the left of $Z = 1.80$ is 0.9641.

115.

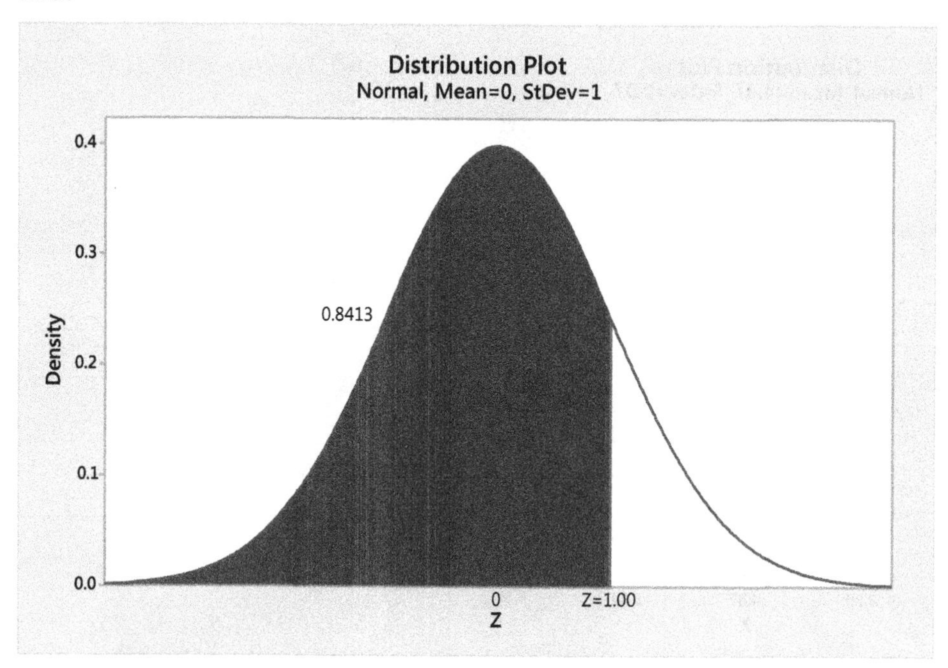

$P(Z < 1.00) = 0.8413$

Therefore, Nicholas scored higher than 84.13% of test takers.

117.

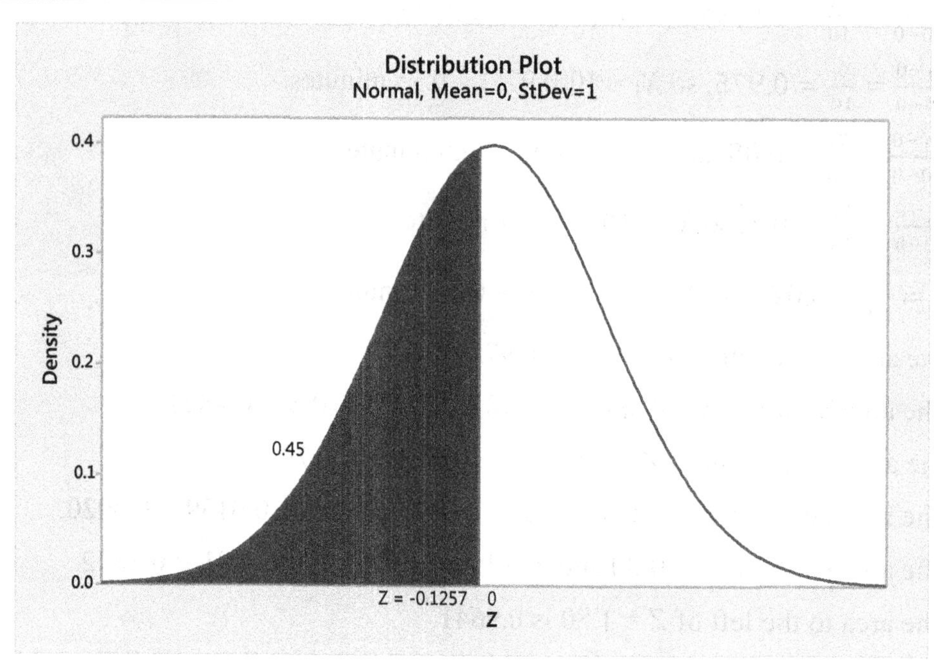

$Z = -0.13$

119.

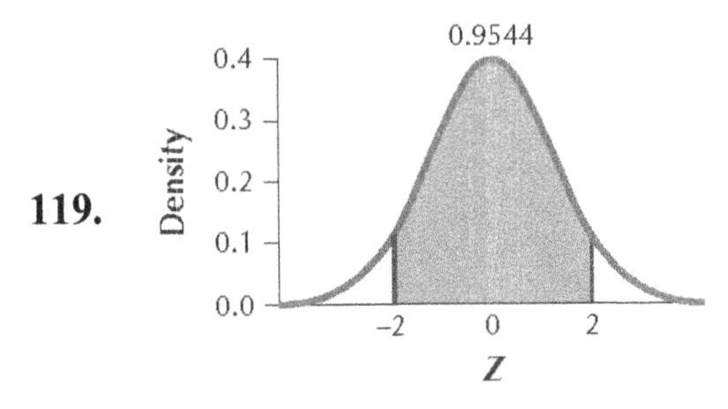

The area between $Z = -2$ and $Z = 2$ is 0.9544.

From Table C, the area between $Z = -2.00$ and $Z = 2.00$ is $0.9772 - 0.0228 = 0.9544$. By the Empirical Rule, the area between $Z = -2$ and $Z = 2$ is about 0.95.

121. **(a)** By symmetry, the area to the right of $Z = 1.5$ equals the area to the left of $Z = -1.5$. Thus, the area to the right of $Z = 1.5$ is 0.0668.

(b) The area to the right of $Z = -1.5$ is 1 − the area to the left of $Z = -1.5 = 1 - 0.0668 = 0.9332$.

(c) By symmetry, the area to the left of $Z = 1.5$ equals the area to the right of $Z = -1.5$. Thus, from (b), the area to the left of $Z = 1.5$ equals 0.9332. Therefore, the area between $Z = -1.5$ and $Z = 1.5$ is the area to the left of $Z = 1.5$ – the area to the left of $Z = -1.5 = 0.9332 - 0.0668 = 0.8664$.

123. $Z = -2.58$ and $Z = 2.58$

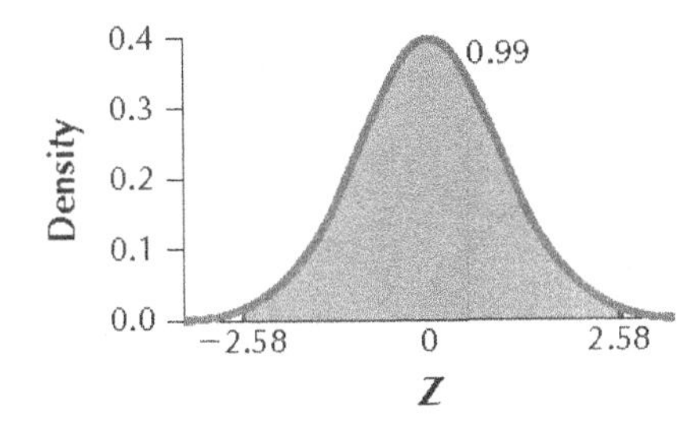

125. For the 25th percentile:

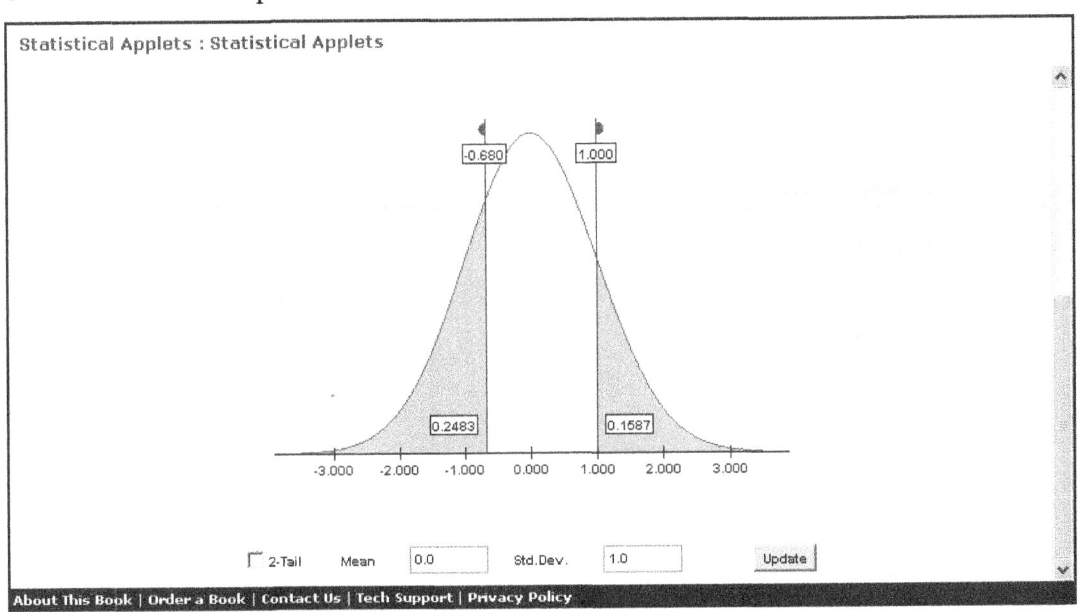

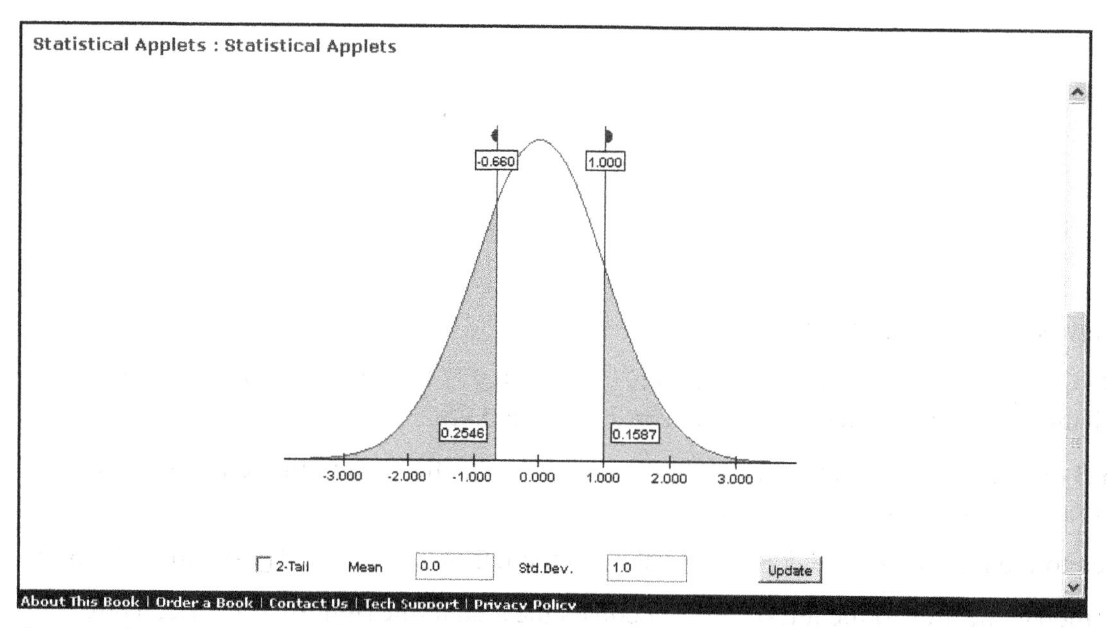

So the 25th percentile is about $Z = -0.67$.

For the 50th percentile:

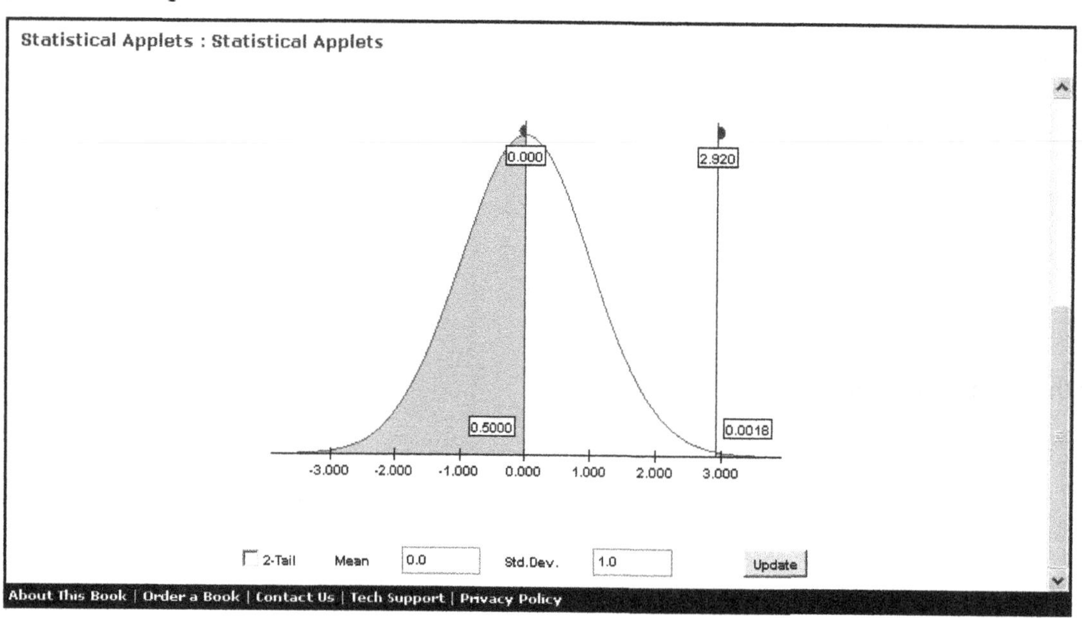

So the 50th percentile is $Z = 0$.

For the 75th percentile:

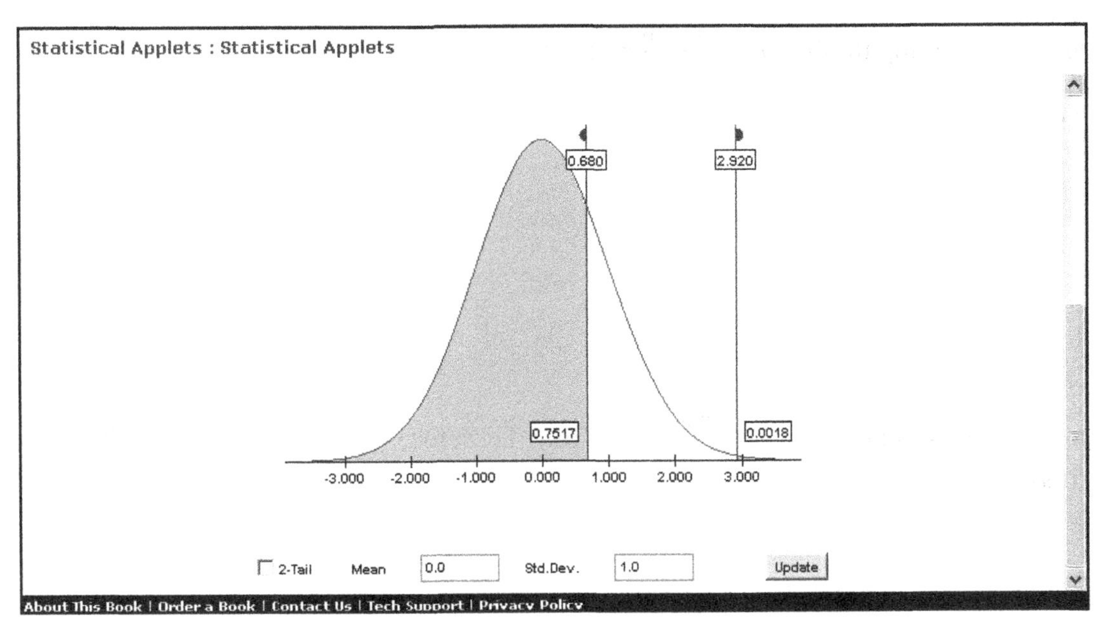

So the 75th percentile is about $Z = 0.67$.

127. The peak would still be at $\mu = 496$ but since the standard deviation is larger the curve would be flatter and more spread out.

129. They are both equal to 0.5. The property that the mean equals the median.

131. $P(X > 611) = P(Z > 1) = 0.1587$ and $P(X > 726) = P(Z > 2) = 0.0228$; yes.

Section 6.5

1. To standardize things means to make them all the same, uniform, or equivalent. To standardize a normal random variable *X*, we transform *X* into the standard normal random variable *Z* by the formula $Z = \frac{X-\mu}{\sigma}$. We do this so that we can use the standard normal table to find the probabilities for *X*.

3.

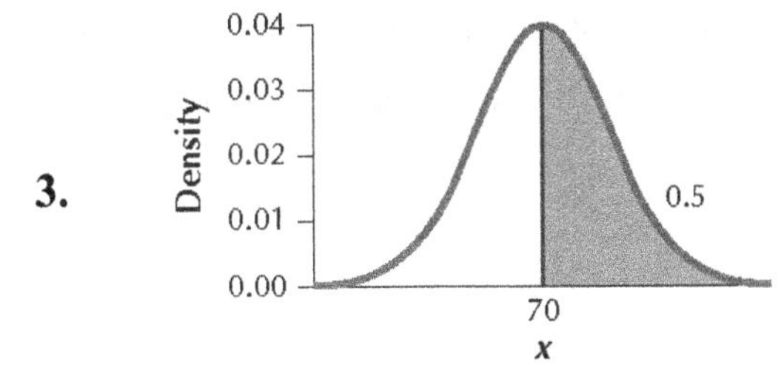

The *Z*-value corresponding to 70 is $Z = \frac{70-70}{10} = 0$. Therefore, $P(X > 70) = P(Z > 0) = 0.5$.

5.

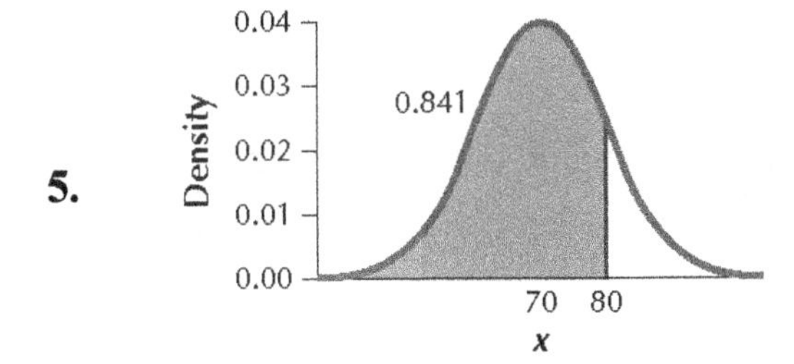

The Z-value corresponding to 80 is $Z = \frac{80-70}{10} = 1.00$. Therefore,

$P(X < 80) = P(Z < 1.00) = 0.8413$.

7. 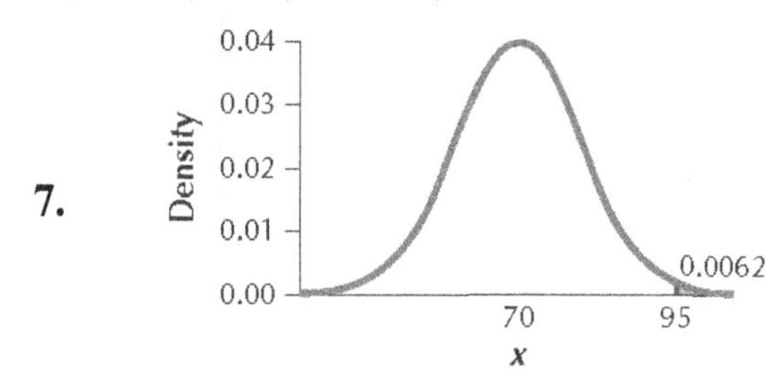

The Z-value corresponding to 95 is $Z = \frac{95-70}{10} = 2.50$. Therefore, $P(X \geq 95) = P(Z \geq 2.50) = 1 - 0.9938 = 0.0062$.

9. 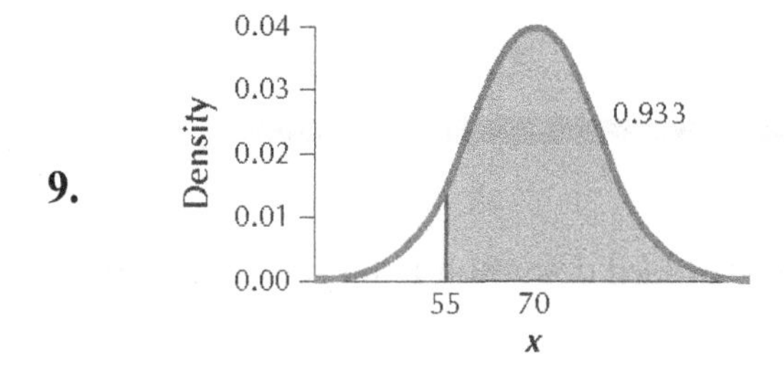

The Z-value corresponding to 55 is $Z = \frac{55-70}{10} = -1.50$. Therefore, $P(X \geq 55) = P(Z \geq -1.50) = 1 - 0.0668 = 0.9332$.

11. 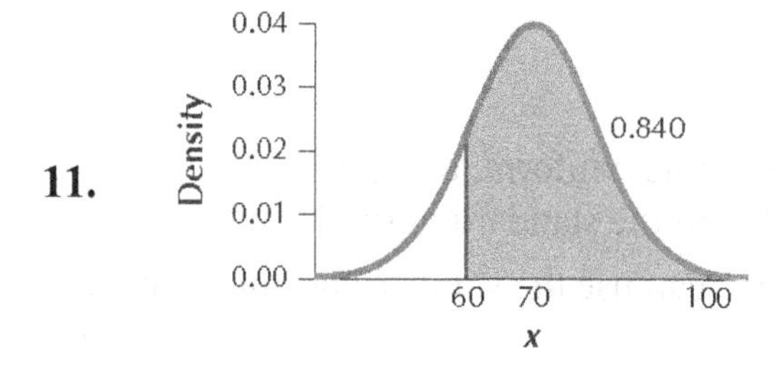

The Z-value corresponding to 60 is $Z = \frac{60-70}{10} = -1.00$. and the Z-value corresponding to 100 is $Z = \frac{100-70}{10} = 3.00$. Therefore, $P(60 \leq X \leq 100) = P(-1.00 \leq Z \leq 3.00) = 0.9987 - 0.1587 = 0.8400$.

13. 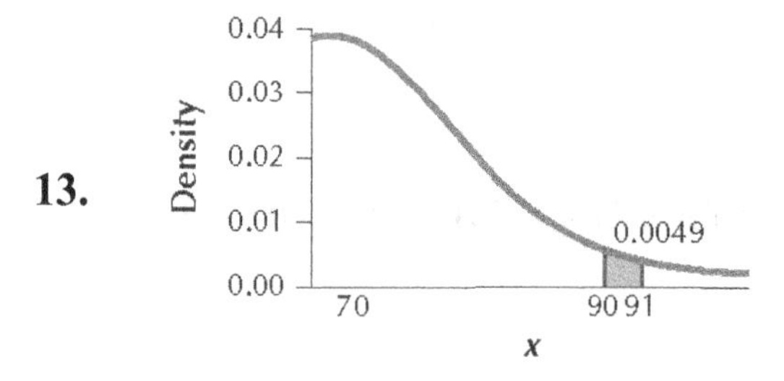

The Z-value corresponding to 90 is $Z = \frac{90-70}{10} = 2.00$ and the Z-value corresponding to 91 is $Z = \frac{91-70}{10} = 2.10$. Therefore, $P(90 \leq X \leq 91) = P(2.00 \leq Z \leq 2.10) = 0.9821 - 0.9772 = 0.004$

15.

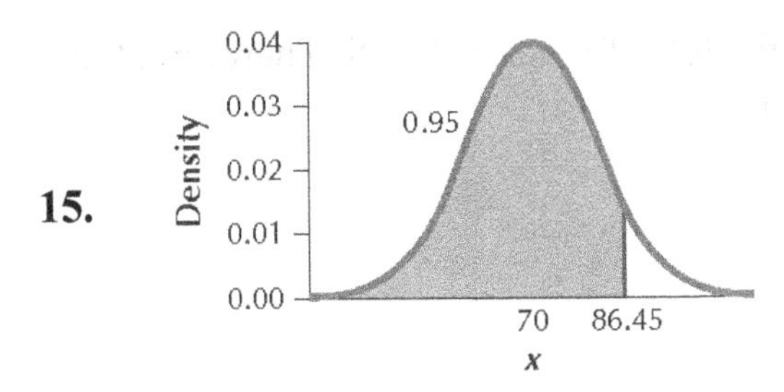

From Table C, the value of Z with an area of 0.9500 to the left of it is $Z = 1.645$. Therefore, the 95th percentile is $X = Z\sigma + \mu = (1.645)(10) + 70 = 86.45$.

17. $X_l = Z\sigma + \mu = (1.96)(10) + 70 = 89.6$

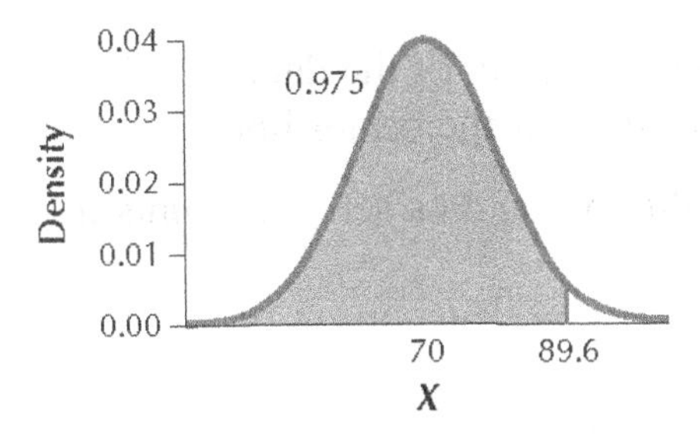

19. $X_l = Z\sigma + \mu = (-2.33)(10) + 70 = 46.7$

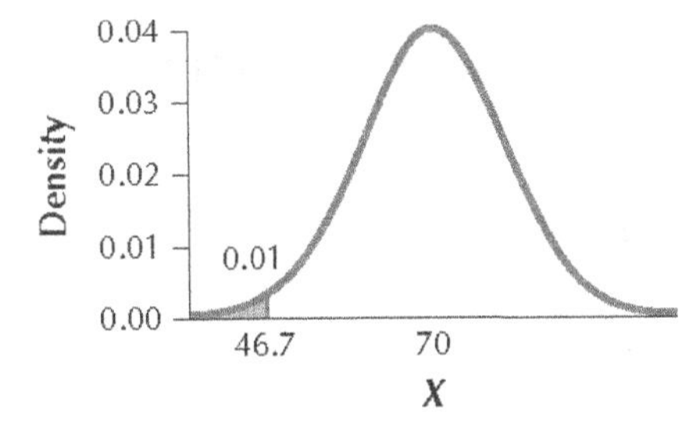

21. $X_l = Z\sigma + \mu = (-2.58)(10) + 70 = 44.2$

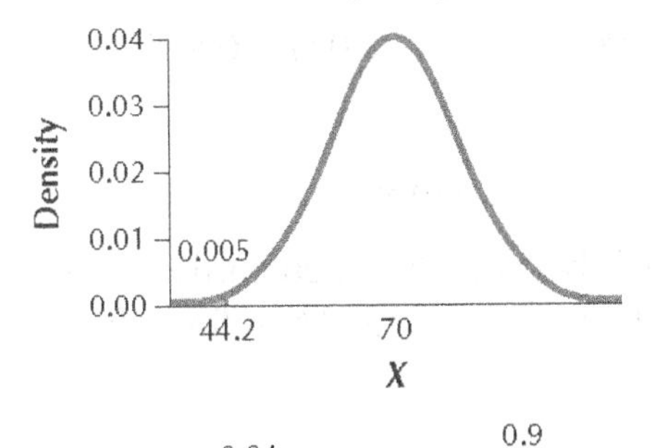

23.

0.9

0.04 0.03 0.02 0.01 0.00

Density

53.55 70 86.45

x

The two symmetric values of X that contain the central 90% of X between them are the 5th percentile and the 95th percentile. From Table C, the value of Z with an area of 0.0500 to the left of it is $Z = -1.645$. Therefore, the 5th percentile is $X = Z\sigma + \mu = (-1.645)(10) + 70 = 53.55$.

From Table C, the value of Z with an area of 0.9500 to the left of it is $Z = 1.645$. Therefore, the 95th percentile is $X = Z\sigma + \mu = (1.645)(10) + 70 = 86.45$.

25. 46.7 and 93.3

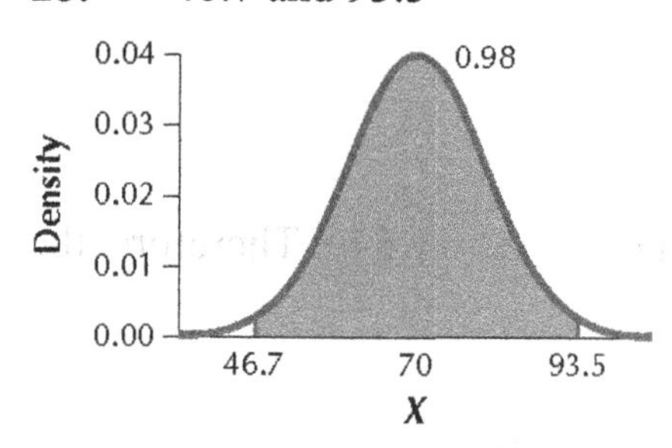

27. The normal probability plot does not indicate acceptable normality of the data set. Several points lie outside of the curved lines and most points do not lie near the center line

29. The normal probability plot indicates acceptable normality of the data set. All points lie between the curved lines and all points lie near the center line.

31. **(a)** $P(X < 15) = P\left(Z < \frac{15-15}{2}\right) = P(Z < 0) = 0.5.$

(b) $P(X > 17) = P\left(Z > \frac{17-15}{2}\right) = P(Z > 1) = 1 - 0.8413 = 0.1587.$ $1 - 0.8413 = 0.1587.$

(c), $P(17 \leq X \leq 19) = P\left(\frac{17-15}{2} \leq \frac{X-15}{2} \leq \frac{19-15}{2}\right) = P(1.00 \leq Z \leq 2.00) = 0.9772 - 0.8413 =$ 0.1359.

33. **(a)** $P(X < 7.2) = P\left(Z < \frac{7.2-13.6}{6}\right) = P(Z < -1.07) = 0.1423.$

(b) $P(X > 20) = P\left(Z > \frac{20-13.6}{6}\right) = P(Z > 1.07) = 1 - 0.8577 = 0.1423.$

So 0.1423 of the days in July have a wind speed greater than 20 mph.

(c) Therefore, $P(15 \leq X \leq 20) = P\left(\frac{15-13.6}{6} < Z < \frac{20-13.6}{6}\right) \approx P(0.23 \leq Z \leq 1.07) = 0.8577 - 0.5910 = 0.2667.$

Therefore, 26.67% of days in July have wind speeds between 15 and 20 mph.

(d) The value of X higher than 99% of all other wind speeds in July is the 99th percentile. From Table C, the value of Z with an area to the left of it equal to 0.9900 is $Z = 2.33$. Therefore, the 99th percentile is $X = Z\sigma + \mu = (2.33)(6) + 13.6 = 27.6$ mph.

(e) The Z-score for $X = 0$ mph is $Z = \frac{X-\mu}{\sigma} = \frac{0-13.6}{6} \approx -2.27$. Since $-3 < -2.27 \leq -2$, a day in July with no wind at all should be considered moderately unusual

35. **(a)** $X_2 = Z\sigma + \mu = (1.645)(2) + 15 = 18.29$ ounces.

(b) $X_1 = Z\sigma + \mu = (-1.645)(2) + 15 = 11.71$ ounces.

(c) 11.71 ounces and 18.29 ounces.

37. **(a)** The value of X higher than 90% of all other wind speeds in July is the 90th percentile. From Table C, the value of Z with an area to the left of it equal to 0.9000 is $Z = 1.28$. Therefore, the 99th percentile is $X = Z\sigma + \mu = (1.28)(6) + 13.6 = 21.28$ mph.

(b) $X_1 = Z\sigma + \mu = (-1.28)(6) + 13.6 = 5.92$ mph. **(c)** 5.92 mph and 21.28 mph.

39. **(a)**

$P(X > 7) = P\left(\frac{X-\mu}{\sigma} > \frac{7-\mu}{\sigma}\right) = P\left(\frac{X-4.9}{1} > \frac{7-4.9}{1}\right) = P(Z > 2.10) = 1 - 0.9821$
$= 0.0179$

(b) $P(3 \leq X \leq 5) = P\left(\frac{3-\mu}{\sigma} \leq \frac{X-\mu}{\sigma} \leq \frac{5-\mu}{\sigma}\right) = P\left(\frac{3-4.9}{1} \leq \frac{X-4.9}{1} \leq \frac{5-4.9}{1}\right)$

$P(-1.90 \leq Z \leq 0.10) = 0.5398 - 0.0287 = 0.5111$

(c) 4.9 days; mean = median

(d) For $X = 8$ days, $Z = \frac{X-\mu}{\sigma} = \frac{8-4.9}{1} = 3.1$. Since Z-score = 3.1 ≥ 3, 8 days is considered an outlier.

41. **(a)** $P(X < \$100) = P\left(\frac{X-\mu}{\sigma} < \frac{\$100-\mu}{\sigma}\right) = P\left(\frac{X-\$105}{\$8} < \frac{\$100-\$105}{\$8}\right) = P(Z <$
$-0.63)$ =0.2643

(b) $P(\$90 < X < \$115) = P\left(\frac{\$90-\mu}{\sigma} < \frac{X-\mu}{\sigma} < \frac{\$115-\mu}{\sigma}\right) = P\left(\frac{\$90-\$105}{\$8} < \frac{X-\$105}{\$8} < \frac{\$115-\$105}{\$8}\right) =$
$P(-1.88 < Z < 1.25) = 0.8944 - 0.0301 = 0.8643$

(c)

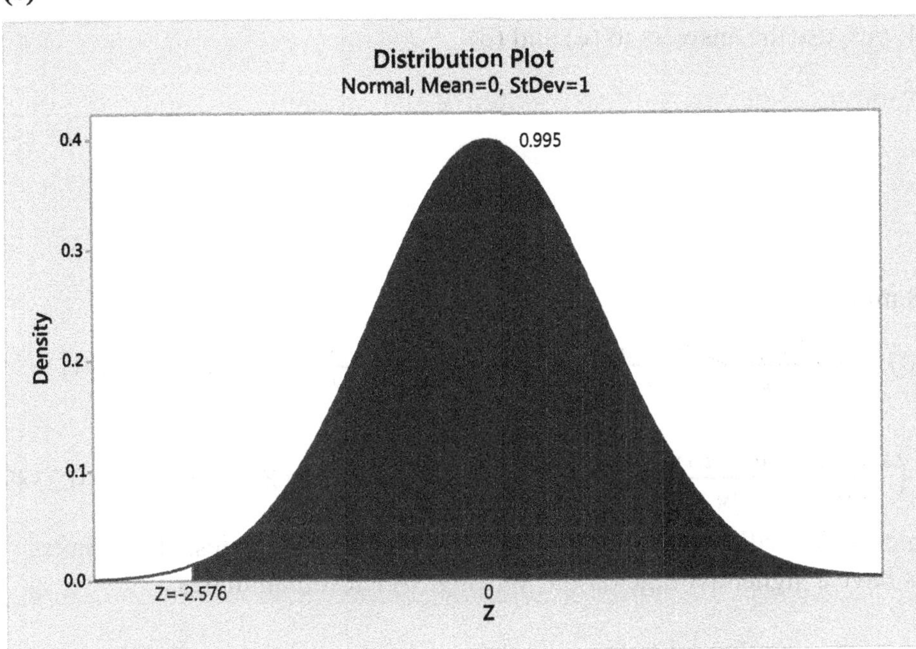

From the table, $Z = -2.58$ is the value of Z corresponding to the 0.5th percentile Therefore, $X = Z\sigma + \mu = (-2.58)(\$8) + \$105 = \$84.36$ is the 0.5th percentile

(d)

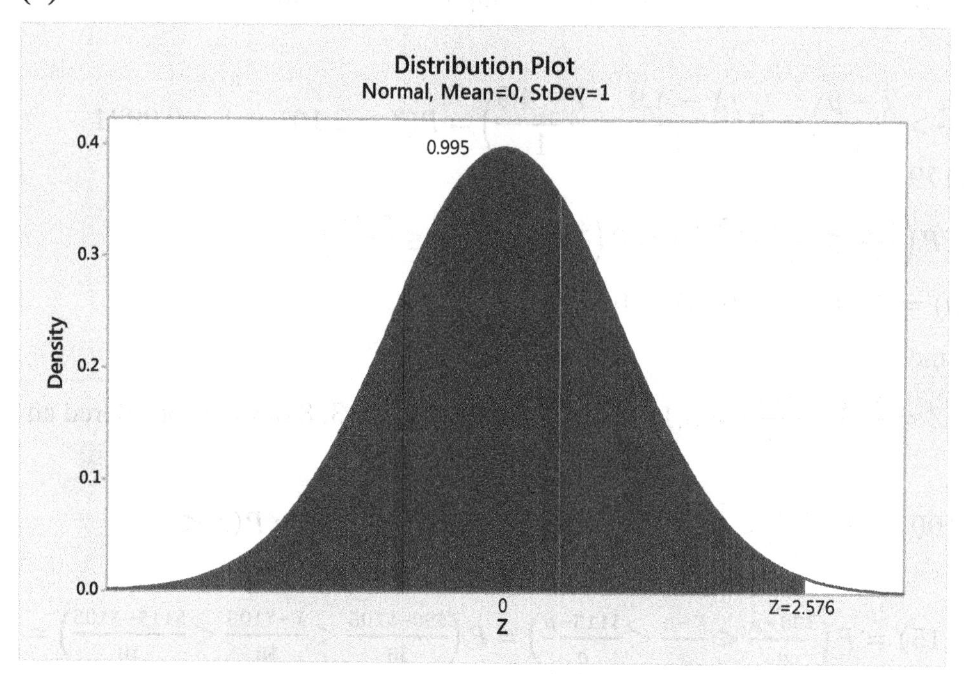

From the table, $Z = 2.58$ is the value of Z corresponding to the 99.5th percentile Therefore, $X = Z\sigma + \mu = (2.58)(\$8) + \$105 = \$125.64$ is the 99.5th percentile

(e) \$84.36 and \$125.64; yes, use the answers to (c) and (d)

43. **(a)** 18.29 ounces

(b) 11.71 ounces

(c) \$171.40

(d) 21.28 mph

(e) 5.92 mph and 21.28 mph

45. **(a)** $P(X_F < 400) = P\left(\frac{X_F - \mu_F}{\sigma_F} < \frac{400 - \mu_F}{\sigma_F}\right) = P\left(\frac{X_F - 493}{112} < \frac{400 - 493}{112}\right) = P(Z < -0.83) =$ 0.2033

(b) $P(X_M < 400) = P\left(\frac{X_M - \mu_M}{\sigma_M} < \frac{400 - \mu_M}{\sigma_M}\right) = P\left(\frac{X_M - 482}{115} < \frac{400 - 482}{115}\right) = P(Z < -0.71) = 0.2389$

(c) A higher proportion of males will be identified as at-risk writers than females. This makes sense given that females have a higher average on the Writing SAT test than males.

47.

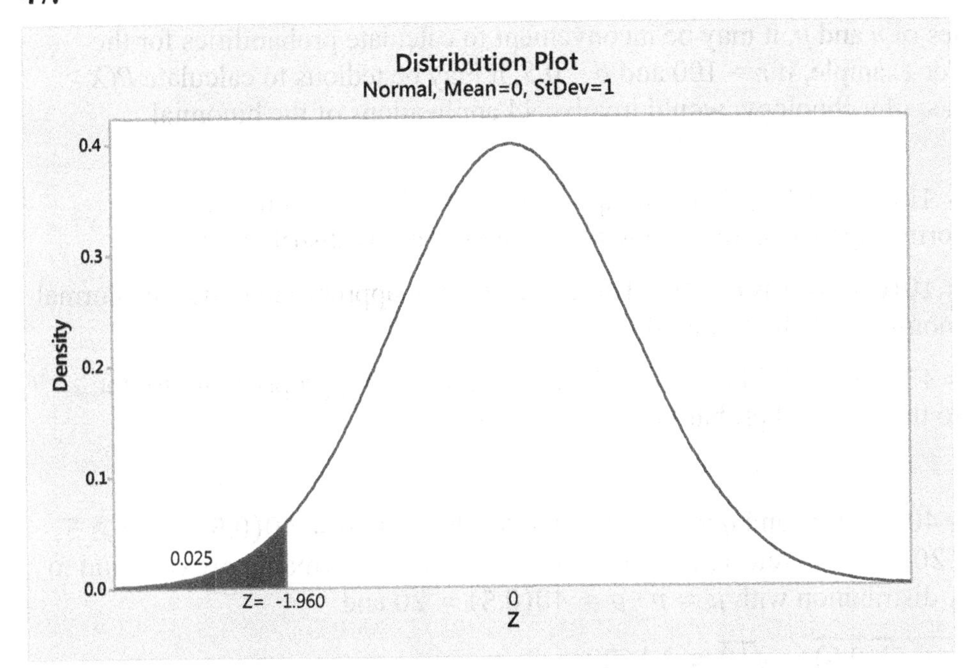

From the table, $Z = -1.96$ is the value of Z corresponding to the 2.5th percentile Therefore, $X_1 = Z\sigma_M + \mu_M = (-1.96)(115) + 482 = 256.6$ is the 2.5th percentile

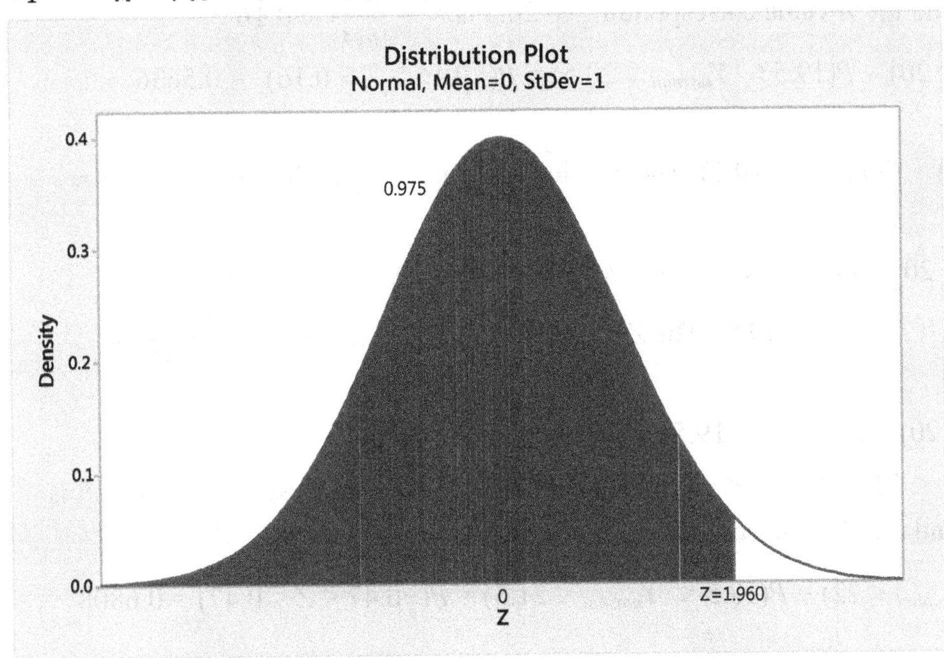

From the table, $Z = 1.96$ is the value of Z corresponding to the 97.5th percentile Therefore, $X_2 = Z\sigma_M + \mu_M = (1.96)(115) + 482 = 707.4$ is the 97.5th percentile

Therefore, 256.6 and 707.4 are the two symmetric male Writing SAT scores that contain 95% of the X-values between them.

Section 6.6

1. For certain values of n and p, it may be inconvenient to calculate probabilities for the binomial distribution. For example, if $n = 100$ and $p = 0.5$, it may be tedious to calculate $P(X \geq 57)$, which, in the absence of technology, would involve 44 applications of the binomial probability formula.

3. We have $n \cdot p = 10(0.5) = 5 \geq 5$ and. $n \cdot q = 10(0.5) = 5 \geq 5$ Therefore, it is appropriate to use the normal approximation to the binomial probability distribution.

5. We have $n \cdot p = 10(0.4) = 4$ is not ≥ 5 Therefore, it is not appropriate to use the normal approximation to the binomial probability distribution.

7. We have $n \cdot p = 45(0.1) = 4.5$ is not ≥ 5. Therefore, it is not appropriate to use the normal approximation to the binomial probability distribution.

For Exercises 9–16, $n = 40$, $p = 0.5$, and $q = 1 - 0.5 = 0.5$. We have $n \cdot p = 40(0.5) = 20 \geq 5$ and. $n \cdot q = 40(0.5) = 20 \geq 5$. Therefore, it is appropriate to use the normal approximation to the binomial probability distribution with $\mu = n \cdot p = 40(0.5) = 20$ and

$\sigma = \sqrt{n \cdot p \cdot q} = \sqrt{(40)(0.5)(0.5)} = \sqrt{10} \approx 3.1623$.

9. $P(X_{binomial} = 20) \approx P(19.5 \leq Y_{normal} \leq 20.5)$. The Z-value corresponding to 19.5 is $Z \approx \frac{19.5-20}{3.1623} \approx -0.16$ and the Z-value corresponding to 20.5 is $\approx \frac{20.5-20}{3.1623} \approx 0.16$.

Therefore, $P(X_{binomial} = 20) \approx P(19.5 \leq Y_{normal} \leq 20.5) = P(-0.16 \leq Z \leq 0.16) = 0.5636 - 0.4364 = 0.1272$.

11. $P(X_{binomial} > 20) \approx P(Y_{normal} > 20.5)$. The Z-value corresponding to 20.5 is $\approx \frac{20.5-20}{3.1623} \approx 0.16$.

Therefore, $P(X_{binomial} > 20) \approx P(Y_{normal} > 20.5) = P(Z > 0.16) = 1 - 0.5636 = 0.4364$.

13. $P(X_{binomial} < 20) \approx P(Y_{normal} < 19.5)$. The Z-value corresponding to 19.5 is $\approx \frac{19.5-20}{3.1623} \approx -0.16$.

Therefore, $P(X_{binomial} < 20) \approx P(Y_{normal} < 19.5) = P(Z < -0.16) = 0.4364$.

15. $P(18 < X_{binomial} < 22) \approx P(18.5 < Y_{normal} < 21.5)$. The Z-value corresponding to 18.5 is $Z \approx \frac{18.5-20}{3.1623} \approx -0.47$ and the Z-value corresponding to 21.5 is $Z \approx \frac{21.5-20}{3.1623} \approx 0.47$..

Therefore, $P(18 < X_{binomial} < 22) \approx P(18.5 < Y_{normal} < 21.5) = P(-0.47 < Z < 0.47) = 0.6808 - 0.3192 = 0.3616$.

For Exercises 17–24, $n = 120$, $p = 0.1$, and $q = 1 - 0.1 = 0.9$. We have $n \cdot p = 120(0.1) = 12 \geq 5$ and. $n \cdot q = 120(0.9) = 108 \geq 5$. Therefore, it is appropriate to use the normal approximation to the binomial probability distribution with $\mu = n \cdot p = 120(0.1) = 12$ and

$\sigma = \sqrt{n \cdot p \cdot q} = \sqrt{(120)(0.1)(0.9)} = \sqrt{10.8} \approx 3.2863$.

17. $P(X_{binomial} \geq 10) \approx P(Y_{normal} \geq 9.5)$. The Z-value corresponding to 9.5 is

$Z \approx \frac{9.5-12}{3.2863} \approx -0.76$.

Therefore, $P(X_{binomial} \geq 10) \approx P(Y_{normal} \geq 9.5) = P(Z \geq -0.76) = 1 - 0.2236 = 0.7764$.

19. $P(X_{binomial} > 10) \approx P(Y_{normal} > 10.5)$. The Z-value corresponding to 10.5 is $Z \approx \frac{10.5-12}{3.2863} \approx -0.46$.

Therefore, $P(X_{binomial} > 10) \approx P(Y_{normal} > 10.5) = P(Z > -0.46) = 1 - 0.3228 = 0.6772$.

21. $P(X_{binomial} < 8) \approx P(Y_{normal} < 7.5)$. The Z-value corresponding to 7.5 is

$Z \approx \frac{7.5-12}{3.2863} \approx -1.37$.

Therefore, $P(X_{binomial} < 8) \approx P(Y_{normal} < 7.5) = P(Z < -1.37) = 0.0853$.

23. $P(9 < X_{binomial} < 11) \approx P(9.5 < Y_{normal} < 10.5)$. The Z-value corresponding to 9.5 is $Z \approx \frac{9.5-12}{3.2863} \approx -0.76$, and the Z-value corresponding to 10.5 is $Z \approx \frac{10.5-12}{3.2863} \approx -0.46$,

Therefore, $P(9 < X_{binomial} < 11) \approx P(9.5 < Y_{normal} < 10.5) = P(-0.76 < Z < -0.46) = 0.3228 - 0.2236 = 0.0992$.

25. $n = 1000, p = 0.065, q = 1 - p = 1 - 0.065 = 0.935$

We have $n \cdot p = 1000 \cdot 0.065 = 65 \geq 5$ and $n \cdot q = 1000 \cdot 0.935 = 935 \geq 5$. Therefore, it is appropriate to use the normal approximation to the binomial distribution. $\mu = n \cdot p = 1000 \cdot 0.065 = 65$ and $\sigma = \sqrt{npq} = \sqrt{(1000)(0.065)(0.935)} \approx 7.7958$.

(a) $P(X_{\text{binomial}} > 65) \approx P(Y_{\text{normal}} \geq 65.5)$. The Z-value corresponding to 65.5 is

$$Z = \frac{65.5 - 65}{7.7958} \approx 0.06$$

Therefore, $P(X_{\text{binomial}} > 65) \approx P(Y_{\text{normal}} \geq 65.5) \approx P(Z \geq 0.06) = 1 - 0.5239 = 0.4761$.

(b) $P(X_{\text{binomial}} \geq 65) \approx P(Y_{\text{normal}} \geq 64.5)$. The Z-value corresponding to 64.5 is

$$Z = \frac{64.5 - 65}{7.7958} \approx -0.06$$

Therefore, $P(X_{\text{binomial}} \geq 65) \approx P(Y_{\text{normal}} \geq 64.5) \approx P(Z \geq -0.06) = 1 - 0.4761 = 0.5239$.

(c) $P(X_{\text{binomial}} < 65) \approx P(Y_{\text{normal}} \leq 64.5)$. The Z-value corresponding to 64.5 is

$$Z = \frac{64.5 - 65}{7.7958} \approx -0.06$$

Therefore, $P(X_{\text{binomial}} < 65) \approx P(Y_{\text{normal}} \leq 64.5) \approx P(Z \leq -0.06) = 0.4761$.

(d) $P(60 \leq X_{\text{binomial}} \leq 70) \approx P(59.5 \leq Y_{\text{normal}} \leq 70.5)$. The Z-value corresponding to 59.5 is $Z = \frac{59.5-65}{7.7958} \approx -0.71$ and the Z-value corresponding to 70.5 is $Z = \frac{70.5-65}{7.7958} \approx 0.71$.

Therefore, $P(60 \leq X_{\text{binomial}} \leq 70) \approx P(59.5 \leq Y_{\text{normal}} \leq 70.5) \approx P(-0.71 \leq Z \leq 0.71) = 0.7611 - 0.2389 = 0.5222$.

27. For people living in New Orleans, $n = 100$, $p = 0.19$, and $q = 1 - p = 1 - 0.19 = 0.81$. We have $n \cdot p = 100 \cdot 0.19 = 19 \geq 5$ and $n \cdot q = 100\ (0.81) = 81 \geq 5$. Therefore, it is appropriate to use the normal approximation to the binomial probability distribution with $\mu = n \cdot p = 100 \cdot 0.19 = 19$ and $\sigma = \sqrt{n \cdot p \cdot q} = \sqrt{(100)(0.19)(0.81)} = \sqrt{15.39} \approx 3.9230$.

For people living in other areas, $n = 100$, $p = 0.57$, and $q = 1 - p = 1 - 0.57 = 0.43$. We have $n \cdot p = 100 \cdot 0.57 = 57 \geq 5$ and $n \cdot q = 100\,(0.43) = 43 \geq 5$. Therefore, it is appropriate to use the normal approximation to the binomial probability distribution with $\mu = n \cdot p = 100 \cdot 0.57 = 57$ and $\sigma = \sqrt{n \cdot p \cdot q} = \sqrt{(100)(0.57)(0.43)} = \sqrt{24.51} \approx 4.9508$.

(a) $P(X_{\text{binomial}} \geq 30) \approx P(Y_{\text{normal}} \geq 29.5)$. The Z-value corresponding to 29.5 is $Z \approx \frac{29.5-19}{3.9230} \approx 2.68$,

Therefore, $P(X_{\text{binomial}} \geq 30) \approx P(Y_{\text{normal}} \geq 29.5) = P(Z \geq 2.68) = 1 - 0.9963 = 0.0037$.

(b) $P(X_{\text{binomial}} \geq 30) \approx P(Y_{\text{normal}} \geq 29.5)$. The Z-value corresponding to 29.5 is $Z \approx \frac{29.5-57}{4.9508} \approx -5.55$,

Therefore, $P(X_{\text{binomial}} \geq 30) \approx P(Y_{\text{normal}} \geq 29.5) = P(Z \geq -5.55) = 1 - 0 = 1$.

(c) $P(X_{\text{binomial}} < 20) \approx P(Y_{\text{normal}} < 19.5)$. The Z-value corresponding to 19.5 is $Z \approx \frac{19.5-19}{3.9230} \approx 0.13$

Therefore, $P(X_{\text{binomial}} < 20) \approx P(Y_{\text{normal}} < 19.5) = P(Z < 0.13) = 0.5517$.

(d) $P(X_{\text{binomial}} < 20) \approx P(Y_{\text{normal}} < 19.5)$. The Z-value corresponding to 19.5 is $Z \approx \frac{19.5-57-12}{4.9508} \approx -7.57$,

Therefore, $P(X_{\text{binomial}} < 20) \approx P(Y_{\text{normal}} < 19.5) = P(Z < -7.57) = 0$.

29.

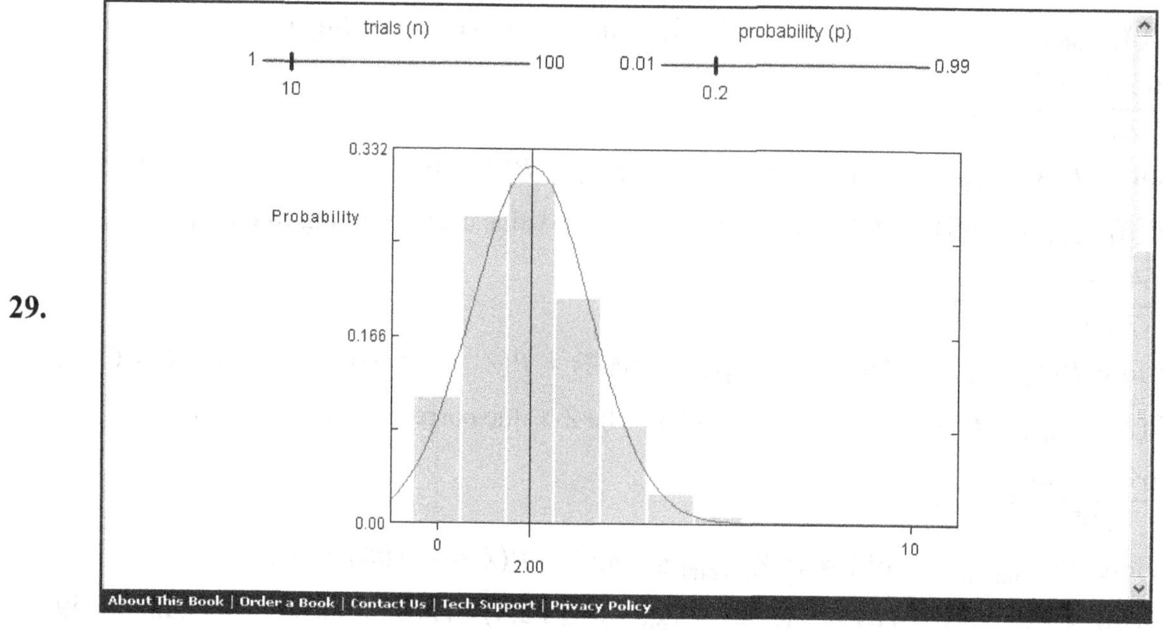

(a) No.

(b) The normal distribution is not a good approximation to the binomial distribution.

(c) We have $n \cdot p = 10 \cdot 0.2 = 2 < 5$. Therefore, it is not appropriate to use the normal approximation to the binomial probability distribution.

Chapter 6 Review Exercises

1. We will define our random variable X to be X = the number of friends that show up for lunch. The probability distribution for X is:

X = Number of friends	0	1	2	3	More than 3
$P(X)$	–	0.25	0.35	0.20	0.05

(a) $P(X \geq 1) = P(X = 1) + P(X = 2) + P(X = 3) + P(X > 3) = 0.25 + 0.35 + 0.20 + 0.05 = 0.85$. Therefore, $P(X = 0) = 1 - P(X \geq 1) = 1 - 0.85 = 0.15$.

(b) $P(X > 1) = P(X = 2) + P(X = 3) + P(X > 3) = 0.35 + 0.20 + 0.05 = 0.60$.

3. $n = 20$ and $p = P(\text{gestational diabetes}) = 0.08$. Since $p = 0.08$ is not in Table B in the Appendix, we need to use the formula or technology.

Minitab generated the following table for $n = 20$ and $p = 0.08$.

X	**Probability**	X	**Probability**	X	**Probability**
0	0.1887	7	0.0005	14	0.0000
1	0.3282	8	0.0001	15	0.0000
2	0.2711	9	0.0000	16	0.0000
3	0.1414	10	0.0000	17	0.0000
4	0.0523	11	0.0000	18	0.0000
5	0.0145	12	0.0000	19	0.0000
6	0.0032	13	0.0000	20	0.0000

(a) From the above table, $P(X = 0) = 0.1887$. Or you can use the formula with $n = 20$, $p = 0.08$, $1 - p = 1 - 0.08 = 0.92$, and $X = 0$. Therefore,

$P(X = 0) = ({}_nC_X)\, p^X (1 - p)^{n-X} = ({}_{20}C_0)\, 0.08^0 (0.92)^{20-0} = (1)(1)(0.92)^{20} = 0.1887$.

(b) $P(X \geq 1) = 1 - P(X = 0) = 1 - 0.1887 = 0.8113$

(c) From the above table,

$P(X \leq 2) = P(X = 0) + P(X = 1) + P(X = 2) = 0.1887 + 0.3282 + 0.2711 = 0.7880$.

5. $P(X = 5) = \frac{\mu^X \cdot e^{-\mu}}{X!} = \frac{5^5 \cdot e^{-5}}{5!} \approx 0.1754673698 \approx 0.1755$.

7. $P(X < 3) = P(X = 0) + P(X = 1) + P(X = 2) = \frac{5^0 \cdot e^{-5}}{0!} + \frac{5^1 \cdot e^{-5}}{1!} + \frac{5^2 \cdot e^{-5}}{2!} \approx 0.006737947 + 0.033689735 + 0.0842243375 = 0.1246520195 \approx 0.1247$

9. **(a)** $P(X = 1) = \frac{\mu^X \cdot e^{-\mu}}{X!} = \frac{1.65^1 \cdot e^{-1.65}}{1!} \approx 0.3168823492 \approx 0.3169$.

(b) $(X = 0) = \frac{\mu^X \cdot e^{-\mu}}{X!} = \frac{1.65^0 \cdot e^{-1.65}}{0!} \approx 0.1920499086 \approx 0.1920$.

(c) $P(X \leq 2) = P(X = 0) + P(X = 1) + P(X = 2) = \frac{1.65^0 \cdot e^{-1.65}}{0!} + \frac{1.65^1 \cdot e^{-1.65}}{1!} + \frac{1.65^2 \cdot e^{-1.65}}{2!} \approx 0.1920499086 + 0.3168823492 + 0.2614279381 = 0.7703601959 \approx 0.7704$.

(d) $P(X > 2) = 1 - P(X \leq 2) \approx 1 - 0.7704 = 0.2296$.

11. Since 106 mm Hg is the mean, $P(X > 106) = 0.5$.

13. From the graph, the area between 98 mm Hg and 114 mm Hg is the area between $\mu - \sigma$ and $\mu + \sigma$. Courtesy of the Empirical Rule, the area between $\mu - \sigma$ and $\mu + \sigma$ is about 0.68. Therefore, the probability that a randomly chosen systolic blood pressure is between 98 and 114 mm Hg is about 0.68.

15. **(a)**

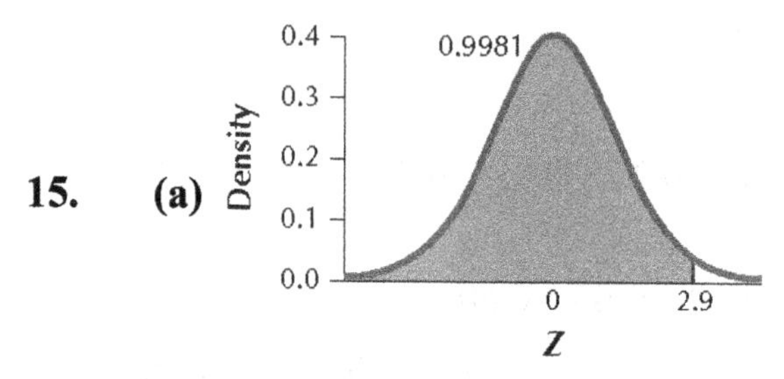

(b) From Table C, $P(Z < 2.9) = 0.9981$.

17. **(a)**

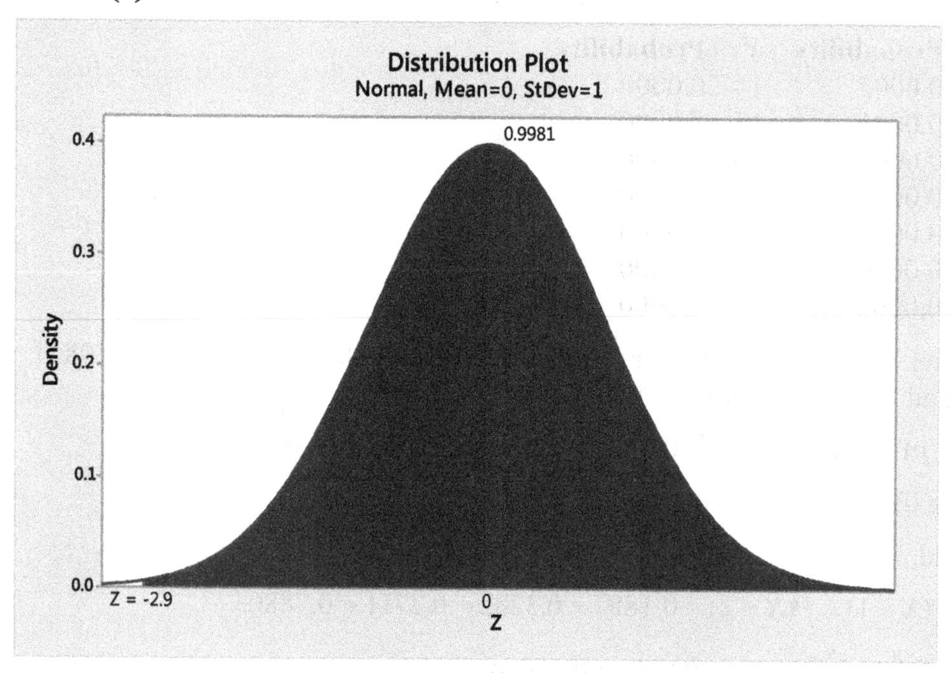

(b) 0.9981

19. **(a)**

0.7853
Density
0.4
0.3
0.2
0.1
0.0
−1.04 0 1.51
Z

(b) From Table C, $P(-1.04 < Z < 1.51) = P(Z < 1.51) - P(Z < -1.04) = 0.9345 - 0.1492 = 0.7853$.

21. **(a)** $P(X > 178) = P\left(\frac{X-\mu}{\sigma} > \frac{178-\mu}{\sigma}\right) = P\left(\frac{X-153}{10} > \frac{178-153}{10}\right) = P(Z > 2.50) = 1 - 0.9938 = 0.0062$

(b) $P(153 \le X \le 178) = P\left(\frac{153-\mu}{\sigma} \le \frac{X-\mu}{\sigma} \le \frac{178-\mu}{\sigma}\right) = P\left(\frac{153-153}{10} \le \frac{X-153}{10} \le \frac{178-153}{10}\right)$

$P(0.00 \leq Z \leq 2.50) = 0.9938 - 0.5000 = 0.4938$

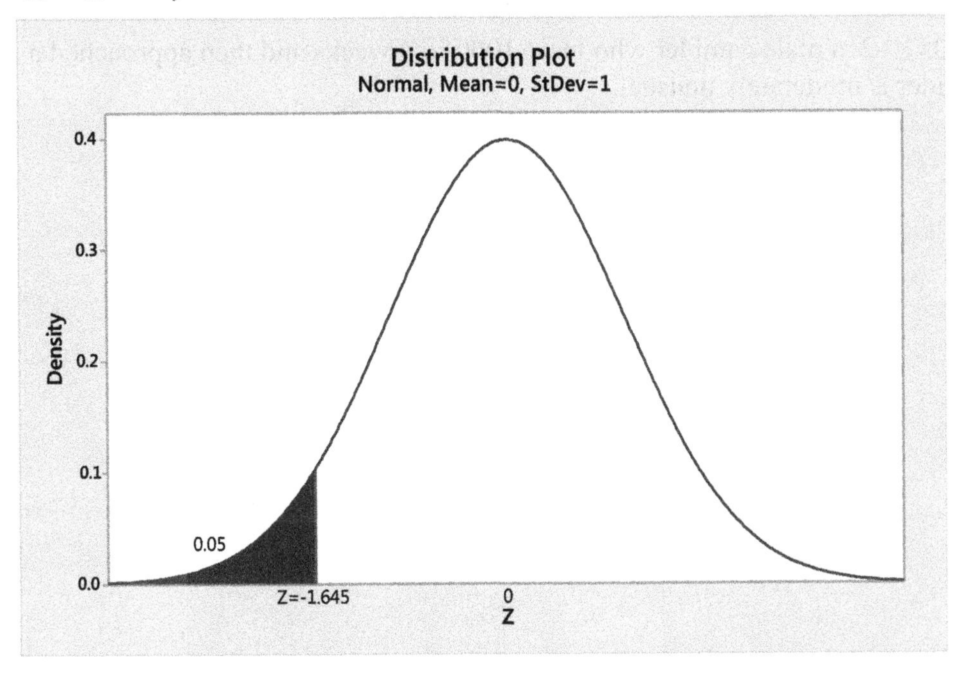

(c)

From the table, $Z = -1.645$ is the value of Z corresponding to the 5th percentile Therefore, $X = Z\sigma + \mu = (-1.645)(10) + 153 = 136.55$ is the 5th percentile

(d) For $X = 188$, $Z = \frac{X-\mu}{\sigma} = \frac{188-153}{10} = 3.5$. Since Z-score = $3.5 \geq 3$, a score of 188 is considered an outlier.

Chapter 6 Quiz

1. True.

3. False.

5. 0

7. Discrete.

9. Mean $\mu = 0$, standard deviation $\sigma = 1$.

11. **(a)** From Table C,

$P(X > 4000) = P\left(Z > \frac{4000-2849}{900}\right) = P(Z > 1.28) = 1 - P(Z \leq 1.28) = 1 - 0.8997 = 0.1003.$

(b) From Table C,

$P(3000 \leq X \leq 4000) = P\left(\frac{3000-2849}{900} < Z < \frac{4000-2849}{900}\right) \approx P(0.17 \leq Z \leq 1.28) = 0.8997 - 0.5675 =$ 0.3322.

Thus, 33.22% of males lost between \$3000 and \$4000.

(c) From Table C, the value of Z with an area of 0.9500 to the left of it is $Z = 1.645$. Therefore, the 95th percentile is $X = Z\sigma + \mu = (1.645)(\$900) + \$2849 = \$4329.50$.

(d) The Z-score for $X = \$1000$ is $Z = \frac{X-\mu}{\sigma} = \frac{1000-2849}{900} \approx -2.05$.

Since $-3 < -2.05 \leq -2$, a male gambler who lost \$1000 in 4 weeks and then approached a treatment provider is moderately unusual.

Chapter 7: Sampling Distributions

Note to instructors and students: Some answers may differ by 1 or 2 in the last digit depending on whether you round for intermediate steps or wait until you get the final answer to round. Also, different software and different forms of technology may give slightly different answers.

Section 7.1

1. The *sampling distribution of the sample mean* $\bar{x}$ *for a given size n* consists of the sample means of all possible samples of size *n* from the population. Sampling distributions can tell us about the expected location and variability of a statistic. In general, the sampling distribution of any particular statistic for a given sample size *n* consists of the collection of that sample statistic across all possible samples of size *n*.

3. True.

5. $n = 30$

7. 4 times as large.

9. **(a)** $\mu_{\bar{x}} = \mu = 10$

(b) $\sigma_{\bar{x}} = \frac{\sigma}{\sqrt{n}} = \frac{2}{\sqrt{25}} = 0.4$

(c) and (d)

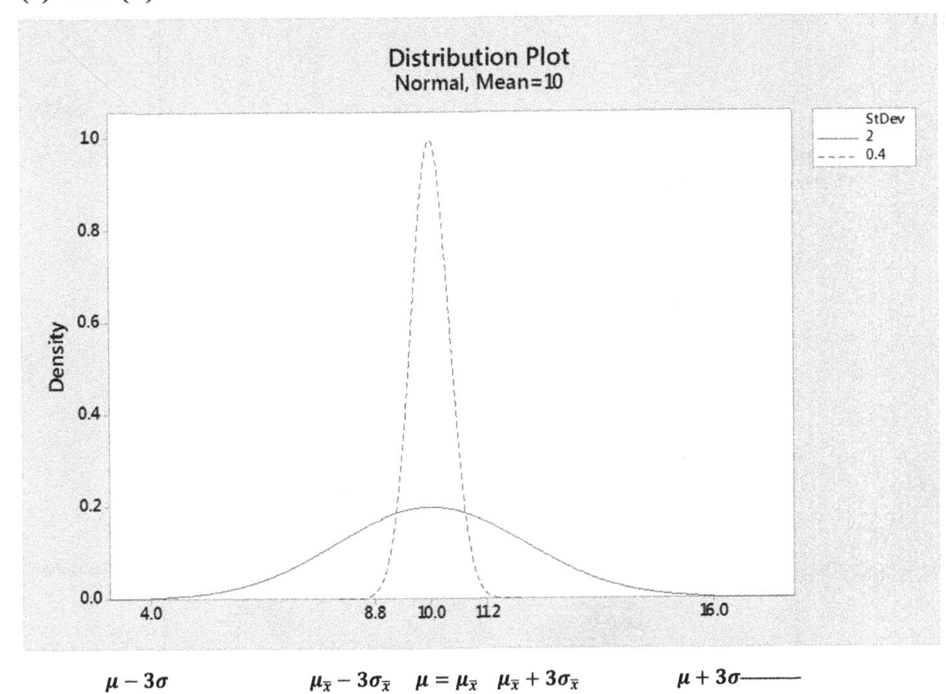

11. **(a)** $\mu_{\bar{x}} = \mu = 0$

(b) $\sigma_{\bar{x}} = \frac{\sigma}{\sqrt{n}} = \frac{1}{\sqrt{9}} \approx 0.3333$

(c) and (d)

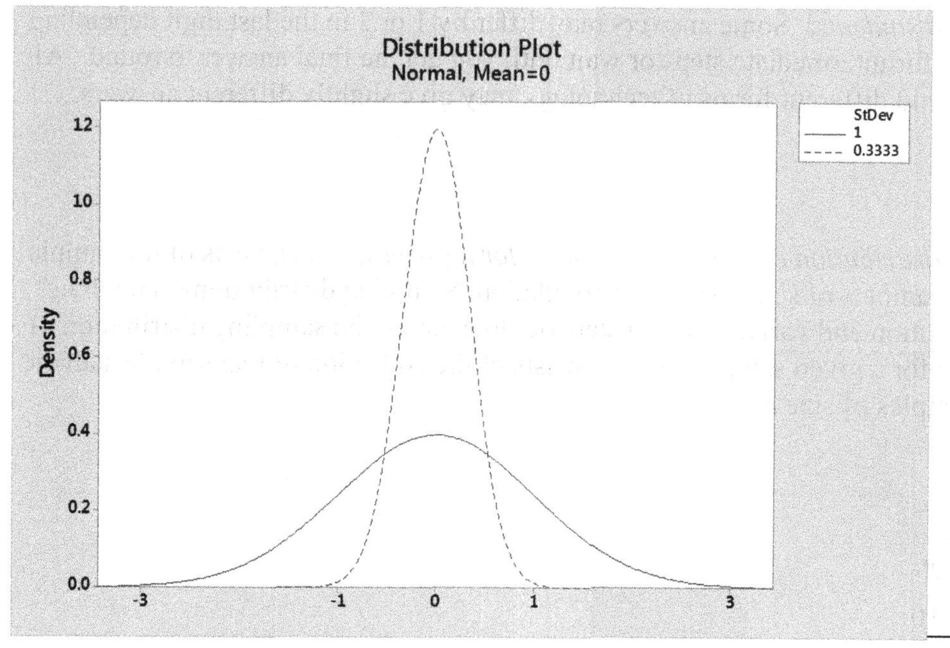

$\mu - 3\sigma$ $\quad \mu_{\bar{x}} - 3\sigma_{\bar{x}}$ $\quad \mu = \mu_{\bar{x}}$ $\quad \mu_{\bar{x}} + 3\sigma_{\bar{x}}$ $\quad \mu + 3\sigma$——

13. **(a)** $\mu_{\bar{x}} = \mu = 100$

(b) $\sigma_{\bar{x}} = \frac{\sigma}{\sqrt{n}} = \frac{10}{\sqrt{100}} = 1$

(c) and (d)

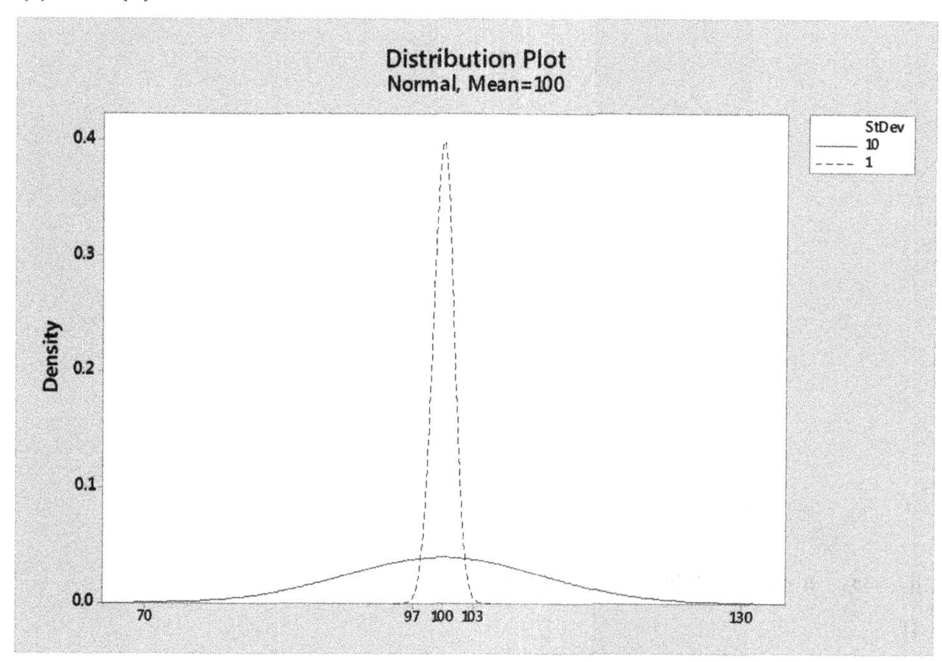

$\mu - 3\sigma$ $\quad \mu_{\bar{x}} - 3\sigma_{\bar{x}}$ $\quad \mu = \mu_{\bar{x}}$ $\quad \mu_{\bar{x}} + 3\sigma_{\bar{x}}$ $\quad \mu + 3\sigma$——

15. **(a)** $\mu_{\bar{x}} = \mu = 75$

(b) $\sigma_{\bar{x}} = \frac{\sigma}{\sqrt{n}} = \frac{15}{\sqrt{36}} = 2.5$

(c) and (d)

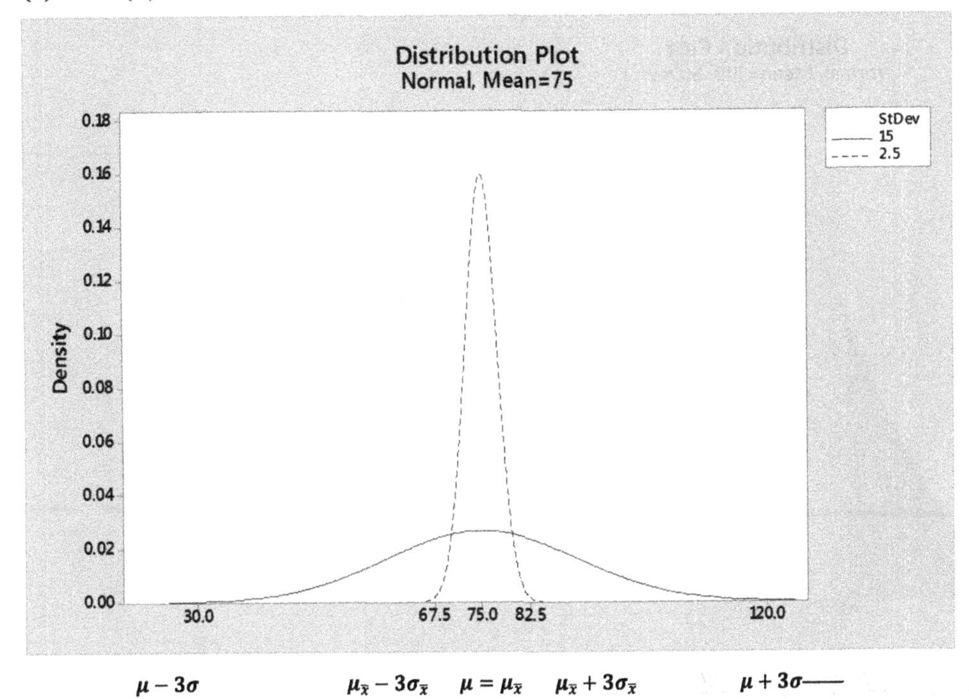

$\mu - 3\sigma$ $\quad$ $\mu_{\bar{x}} - 3\sigma_{\bar{x}}$ $\quad$ $\mu = \mu_{\bar{x}}$ $\quad$ $\mu_{\bar{x}} + 3\sigma_{\bar{x}}$ $\quad$ $\mu + 3\sigma$——

17. Systolic blood pressure readings are not normally distributed, but the sample size of $n = 49$ is ≥ 30. Thus, the Central Limit Theorem applies, and the distribution of $\bar{x}$ is approximately normal.

19. The gas mileage for a 2014 Toyota Prius hybrid vehicle is not normally distributed and the sample size of $n = 16$ is not ≥ 30. Therefore, we have insufficient information to conclude that the sampling distribution of the sample mean $\bar{x}$ is either normal or approximately normal.

21. The pollen count distribution for Los Angeles in September is not normally distributed and the sample size of $n = 16$ is not ≥ 30. Therefore, we have insufficient information to conclude that the sampling distribution of the sample mean $\bar{x}$ is either normal or approximately normal.

23. The distribution of X (prices for boned trout) is normally distributed, so the distribution of $\bar{x}$ is normally distributed, regardless of the sample size.

25. Accountant incomes are not normally distributed and the sample size of $n = 10$ is not ≥ 30. Therefore, we have insufficient information to conclude that the sampling distribution of the sample mean $\bar{x}$ is either normal or approximately normal.

27. $\mu_{\bar{x}} = \mu = 100, \sigma_{\bar{x}} = \frac{\sigma}{\sqrt{n}} = \frac{15}{\sqrt{25}} = 3$

29.

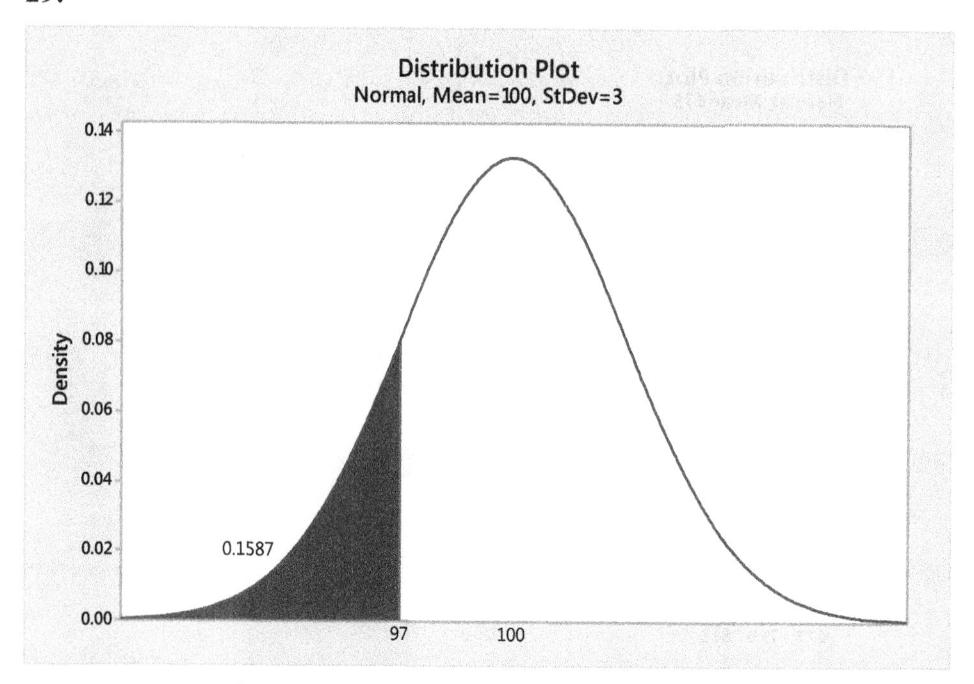

$$P(\bar{x} < 97) = P\left(\frac{\bar{x}-\mu_{\bar{x}}}{\sigma_{\bar{x}}} < \frac{97-\mu_{\bar{x}}}{\sigma_{\bar{x}}}\right) = P\left(\frac{\bar{x}-\mu}{\sigma/\sqrt{n}} < \frac{97-\mu}{\sigma/\sqrt{n}}\right)$$

$$= P\left(\frac{\bar{x}-100}{15/\sqrt{25}} < \frac{97-100}{15/\sqrt{25}}\right) = P\left(\frac{\bar{x}-100}{3} < \frac{97-100}{3}\right) = P(Z < -1) = 0.1587$$

31.

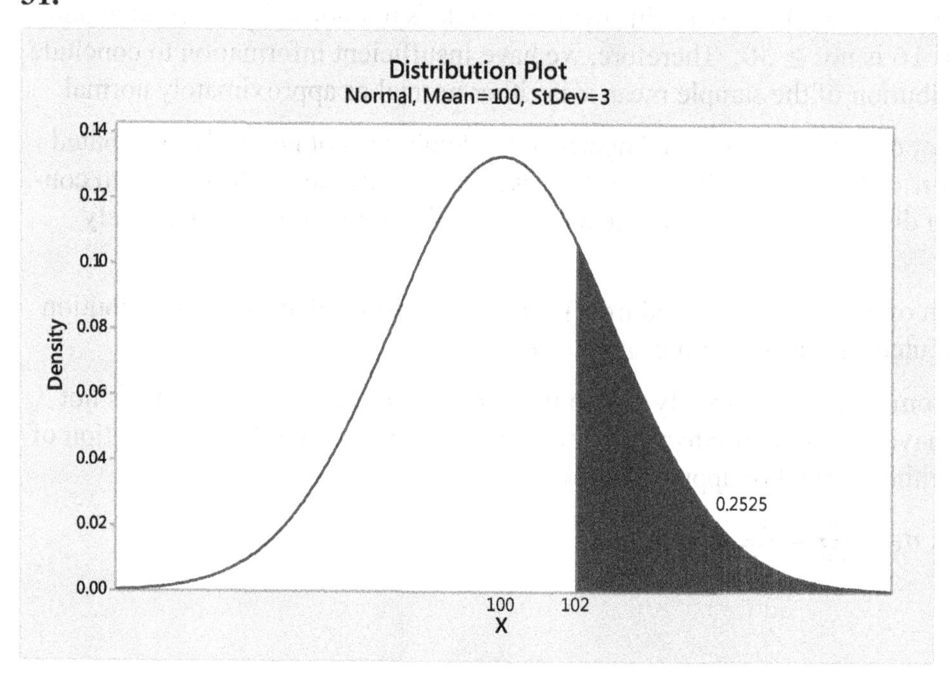

In Exercise 30, 98 is 2 units below the mean. In this exercise, 102 is 2 units above the mean. Therefore, using the symmetry of the normal distribution, $P(\bar{x} > 102) = P(\bar{x} < 98) = 0.2525$.

33. $\mu_{\bar{x}} = \mu = 0, \sigma_{\bar{x}} = \frac{\sigma}{\sqrt{n}} = \frac{1}{\sqrt{100}} = 0.1$

35.

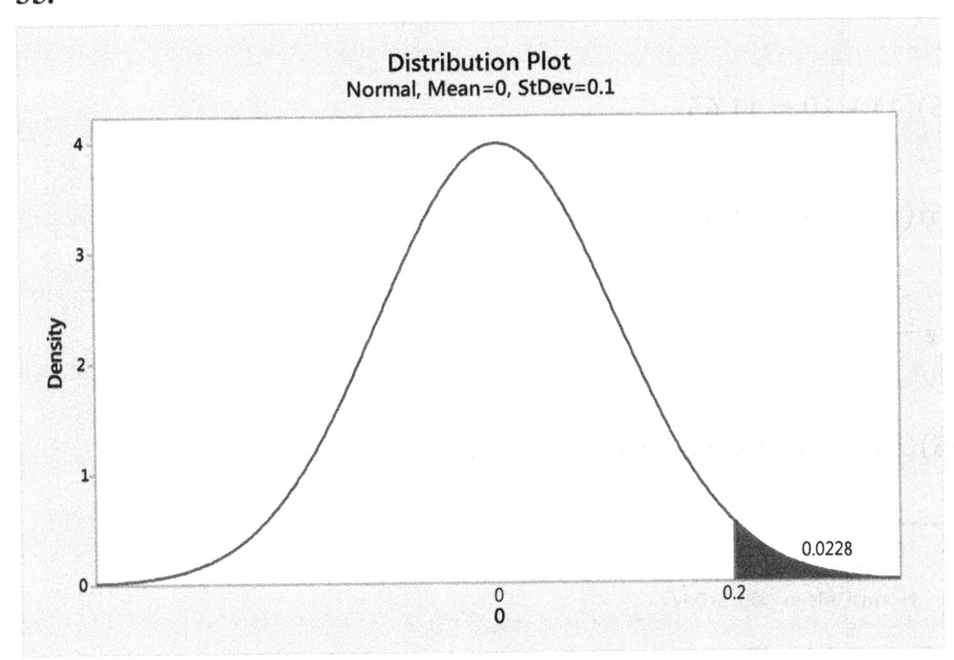

$P(\bar{x} > 0.2) = P\left(\frac{\bar{x}-\mu_{\bar{x}}}{\sigma_{\bar{x}}} > \frac{0.2-\mu_{\bar{x}}}{\sigma_{\bar{x}}}\right) = P\left(\frac{\bar{x}-\mu}{\sigma/\sqrt{n}} > \frac{0.2-\mu}{\sigma/\sqrt{n}}\right) = P\left(\frac{\bar{x}-0}{1/\sqrt{100}} > \frac{0.2-0}{1/\sqrt{100}}\right) = P\left(\frac{\bar{x}-0}{0.1} > \frac{0.2-0}{0.1}\right) = P(Z > 2.00) = 1 - 0.9772 = 0.0228$

37.

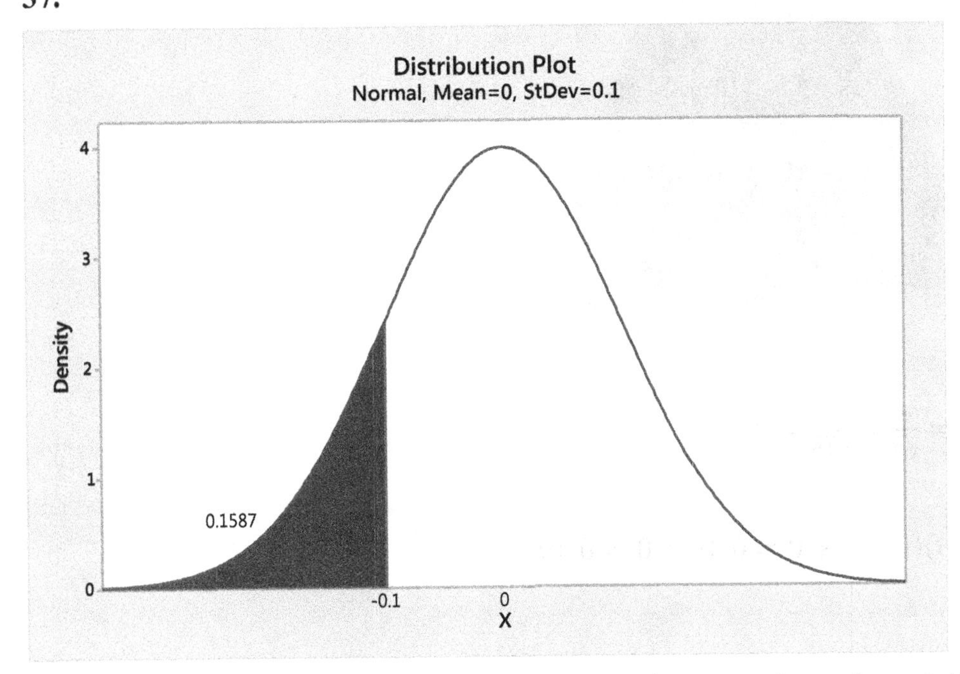

Using the symmetry of the normal distribution, $P(\bar{x} < -0.1) = P(\bar{x} > 0.1) = 0.1587$.

39. $\mu_{\bar{x}} = \mu = 10, \sigma_{\bar{x}} = \frac{\sigma}{\sqrt{n}} = \frac{3}{\sqrt{9}} = 1$

41. $Z = 1.65$

$\bar{x} = Z \cdot \sigma_{\bar{x}} + \mu_{\bar{x}} = (1.65)(1) + 10 = 11.65$

43. $Z = 1.96$

$\bar{x} = Z \cdot \sigma_{\bar{x}} + \mu_{\bar{x}} = (1.96)(1) + 10 = 11.96$

45. 8.35 and 11.65

47. $\mu_{\bar{x}} = \mu = 100, \sigma_{\bar{x}} = \frac{\sigma}{\sqrt{n}} = \frac{15}{\sqrt{25}} = 3$

49. $Z = 1.65$

$\bar{x} = Z \cdot \sigma_{\bar{x}} + \mu_{\bar{x}} = (1.65)(3) + 100 = 4.95 + 100 = 104.95$

51.

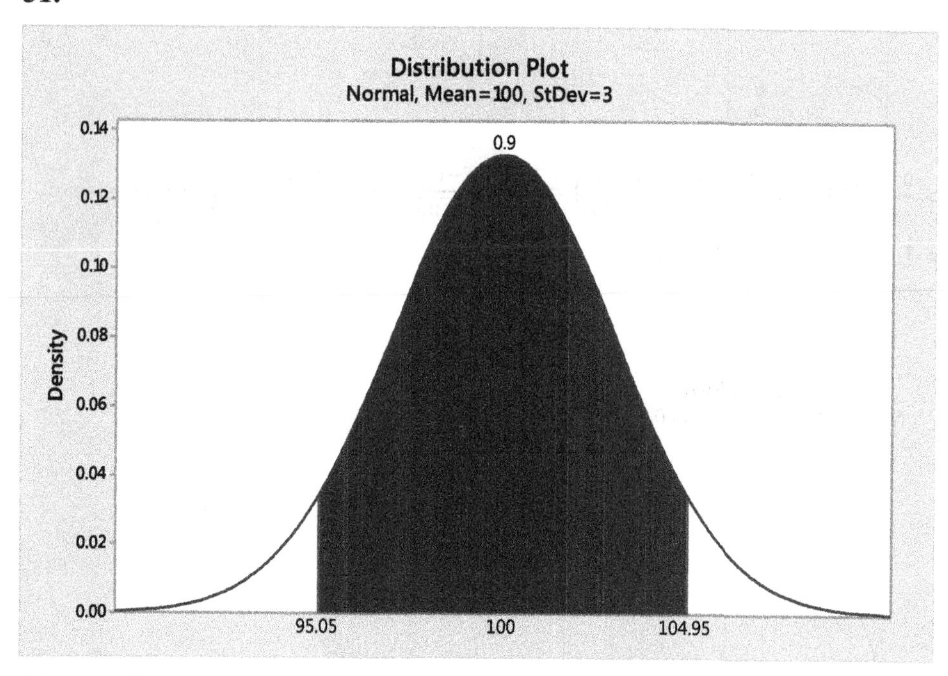

95.05 and 104.95

53. $\mu_{\bar{x}} = \mu = 0, \sigma_{\bar{x}} = \frac{\sigma}{\sqrt{n}} = \frac{1}{\sqrt{16}} = 0.25$

55. $Z = 1.96$

$\bar{x} = Z \cdot \sigma_{\bar{x}} + \mu_{\bar{x}} = (1.96)(0.25) + 0 = 0.49 + 0 = 0.49$

57.

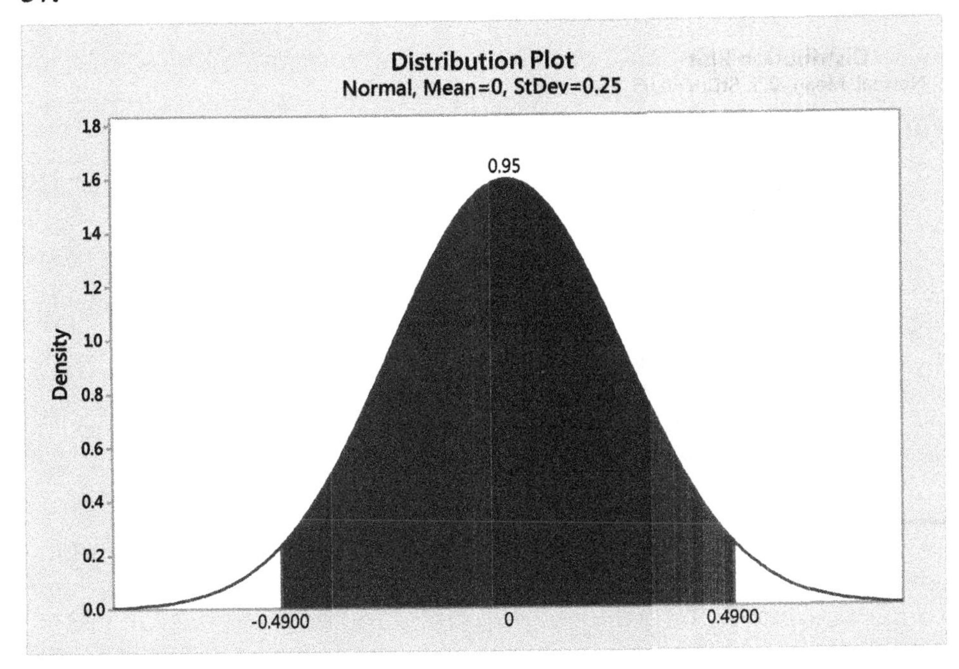

-0.49 and 0.49

59. $\mu_{\bar{x}} = \mu = 2.5,\ \sigma_{\bar{x}} = \frac{\sigma}{\sqrt{n}} = \frac{0.5}{\sqrt{100}} = 0.05$

61.

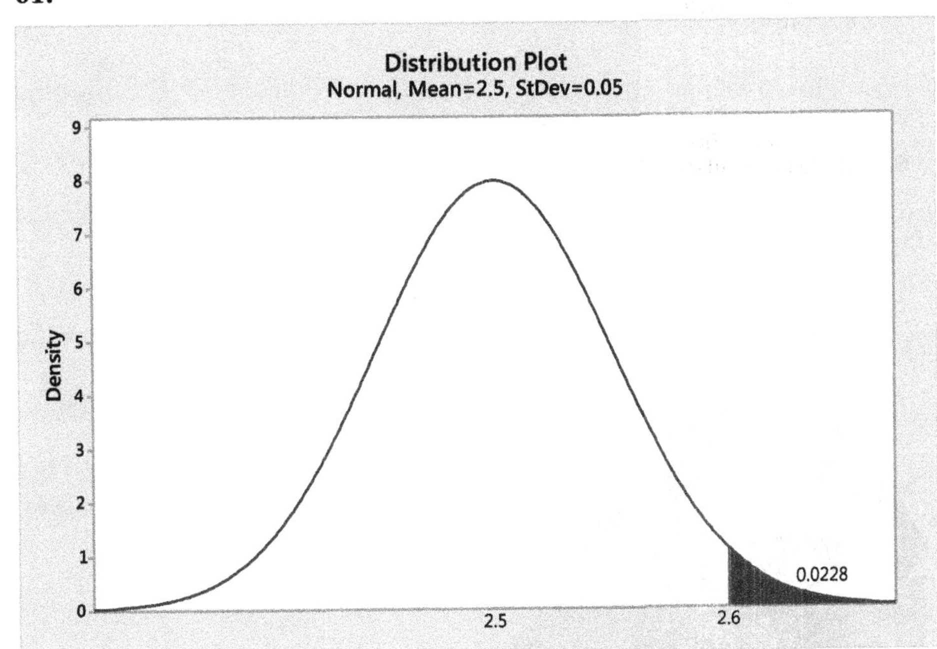

$P(\bar{x} > 2.6) = P\left(\frac{\bar{x}-\mu_{\bar{x}}}{\sigma_{\bar{x}}} > \frac{2.6-\mu_{\bar{x}}}{\sigma_{\bar{x}}}\right) = P\left(\frac{\bar{x}-\mu}{\sigma/\sqrt{n}} > \frac{2.6-\mu}{\sigma/\sqrt{n}}\right) = P\left(\frac{\bar{x}-2.5}{0.5/\sqrt{100}} > \frac{2.6-2.5}{0.5/\sqrt{100}}\right) = P\left(\frac{\bar{x}-2.5}{0.05} > \frac{2.6-2.5}{0.05}\right) = P(Z > 2.00) = 1 - 0.9772 = 0.0228$

63.

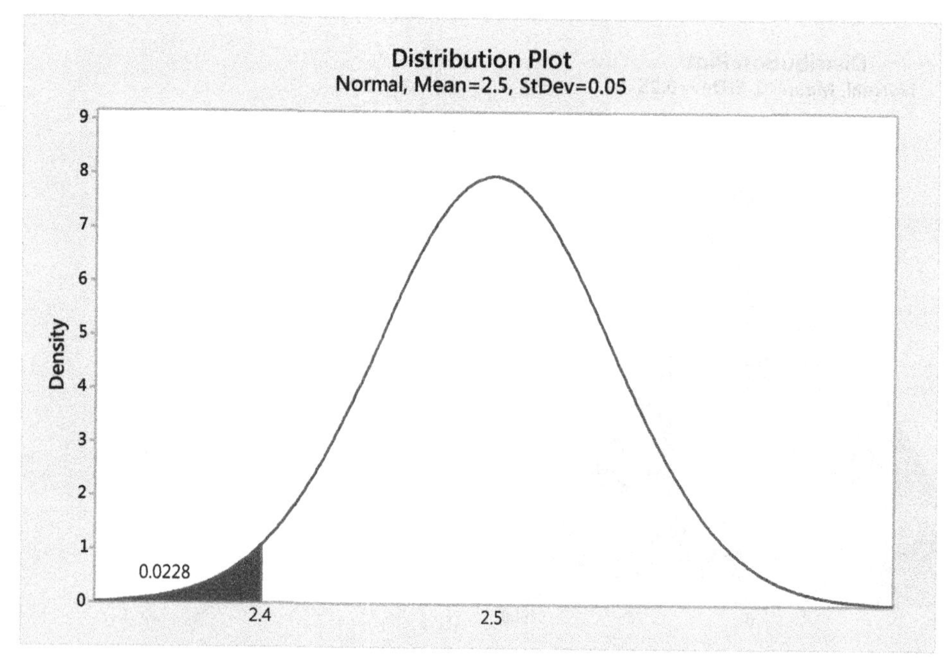

$P(\bar{x} < 2.4) = P\left(\frac{\bar{x}-\mu_{\bar{x}}}{\sigma_{\bar{x}}} < \frac{2.4-\mu_{\bar{x}}}{\sigma_{\bar{x}}}\right) = P\left(\frac{\bar{x}-\mu}{\sigma/\sqrt{n}} < \frac{2.4-\mu}{\sigma/\sqrt{n}}\right) = P\left(\frac{\bar{x}-2.5}{0.5/\sqrt{100}} < \frac{2.4-2.5}{0.5/\sqrt{100}}\right) = P\left(\frac{\bar{x}-2.5}{0.05} < \frac{2.4-2.5}{0.05}\right) = P(Z < -2.00) = 0.0228$

65. $\mu_{\bar{x}} = \mu = -5\,,\ \sigma_{\bar{x}} = \frac{\sigma}{\sqrt{n}} = \frac{4}{\sqrt{64}} = 0.5$

67.

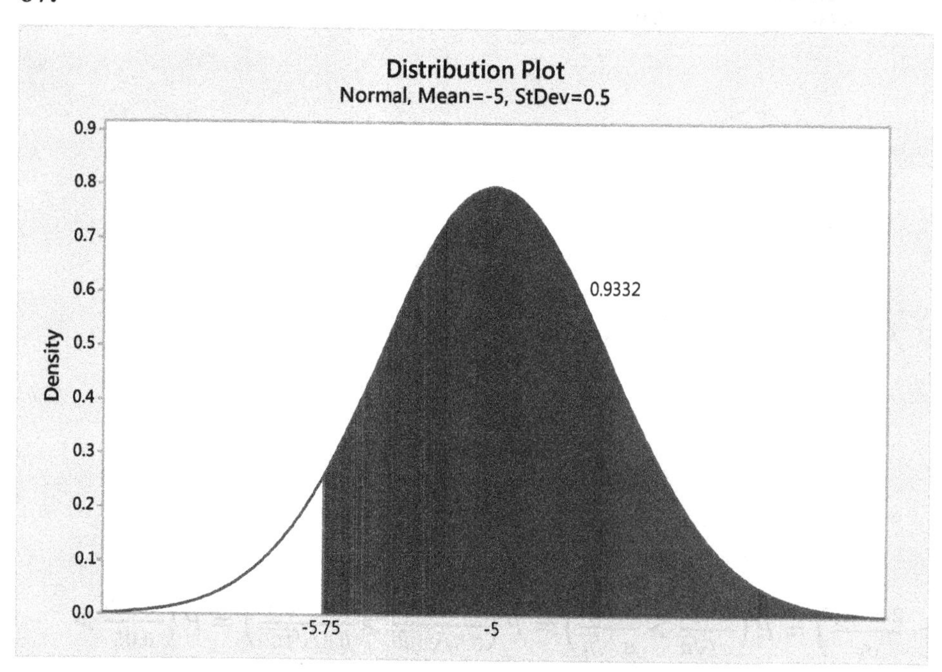

$P(\bar{x} > -5.75) = P\left(\frac{\bar{x}-\mu_{\bar{x}}}{\sigma_{\bar{x}}} > \frac{-5.75-\mu_{\bar{x}}}{\sigma_{\bar{x}}}\right) = P\left(\frac{\bar{x}-\mu}{\sigma/\sqrt{n}} > \frac{-5.75-\mu}{\sigma/\sqrt{n}}\right)$

$= P\left(\frac{\bar{x}-(-5)}{4/\sqrt{64}} > \frac{-5.75-(-5)}{4/\sqrt{64}}\right) = P\left(\frac{\bar{x}-(-5)}{0.5} > \frac{-5.75-(-5)}{0.5}\right) = P(Z > -1.50)$

$= 1 - 0.0668 = 0.9332$

69.

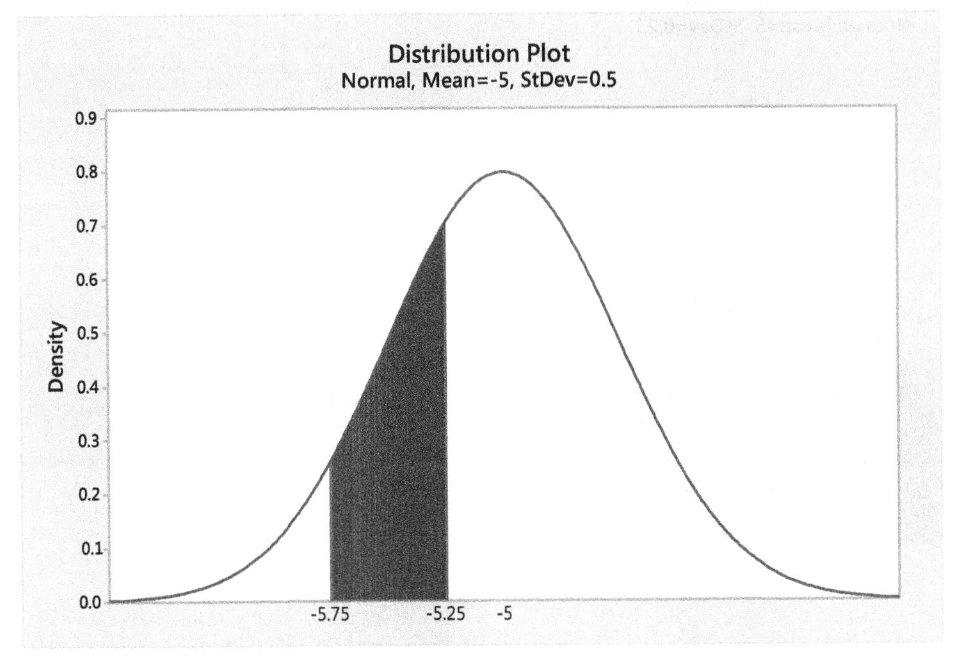

71. $\mu_{\bar{x}} = \mu = 80, \sigma_{\bar{x}} = \frac{\sigma}{\sqrt{n}} = \frac{6}{\sqrt{36}} = 1$

73. $Z = 2.58$

$\bar{x} = Z \cdot \sigma_{\bar{x}} + \mu_{\bar{x}} = (2.58)(1) + 80 = 2.58 + 80 = 82.58$

75.

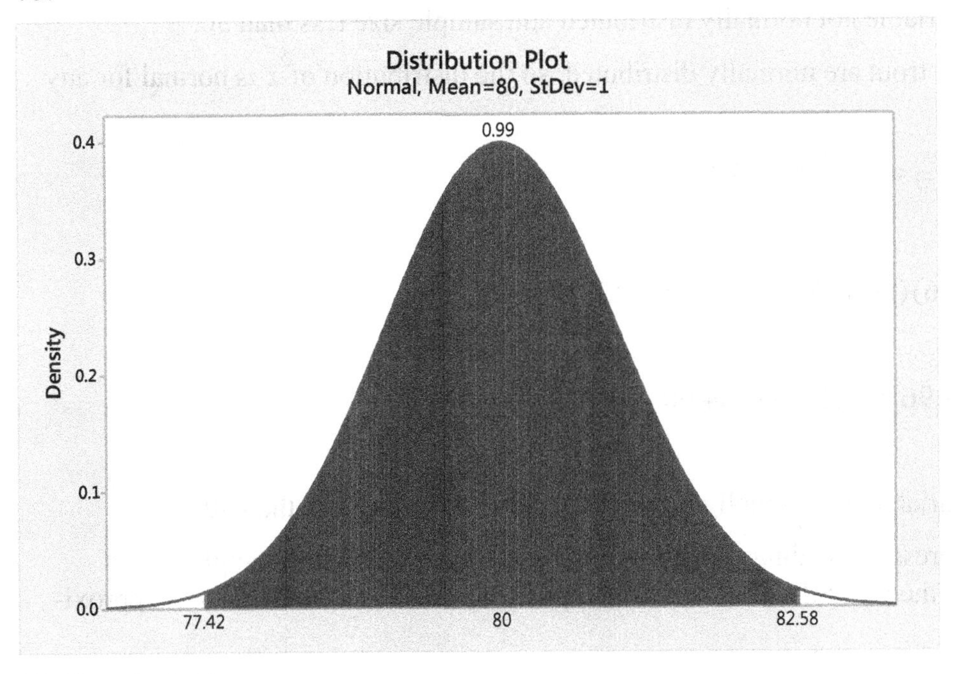

77.42 and 82.58

77. $\mu_{\bar{x}} = \mu = 5, \sigma_{\bar{x}} = \frac{\sigma}{\sqrt{n}} = \frac{2}{\sqrt{64}} = 0.25$

79.

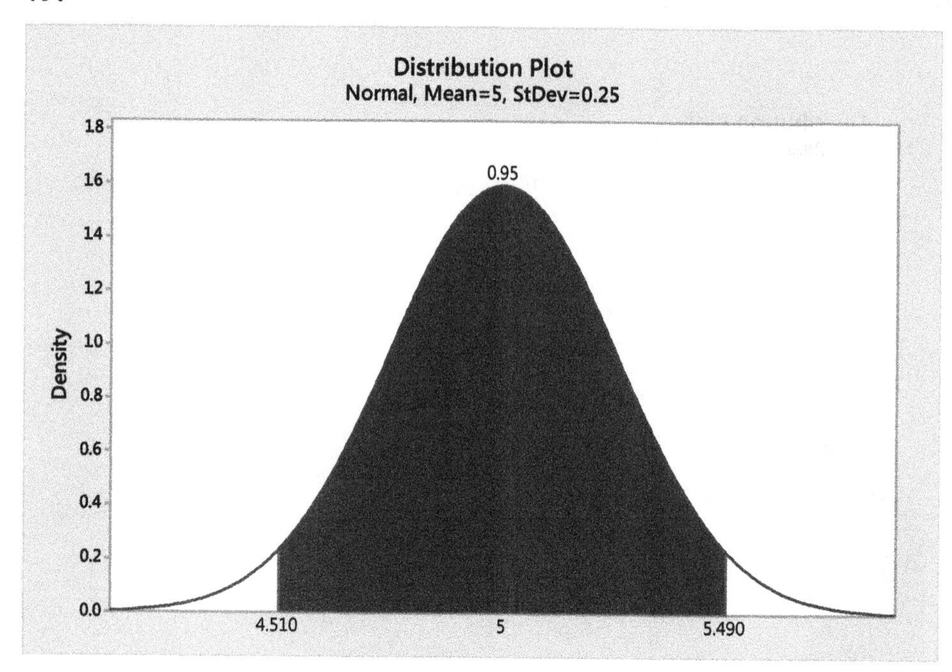

$Z = 1.96$

$\bar{x} = Z \cdot \sigma_{\bar{x}} + \mu_{\bar{x}} = (1.96)(0.25) + 5 = 5.49$

$Z = -1.96$

$\bar{x} = Z \cdot \sigma_{\bar{x}} + \mu_{\bar{x}} = (-1.96)(0.25) + 5 = 4.51$

4.51 and 5.49

81. Not possible; variable not normally distributed and sample size less than 30.

83. Prices for boned trout are normally distributed, so the distribution of $\bar{x}$ is normal for any sample size n.

$\mu_{\bar{x}} = \mu = \$4.00, \sigma_{\bar{x}} = \frac{\sigma}{\sqrt{n}} = \frac{\$0.50}{\sqrt{16}} = \$0.125$

$Z = 1.96$

$\bar{x} = Z \cdot \sigma_{\bar{x}} + \mu_{\bar{x}} = (1.96)(\$0.125) + \$4.00 = \$4.245$

$Z = -1.96$

$\bar{x} = Z \cdot \sigma_{\bar{x}} + \mu_{\bar{x}} = (-1.96)(\$0.125) + \$4.00 = \$3.755$

\$3.755 and \$4.245

85. Not possible; variable not normally distributed and sample size less than 30.

87. Systolic blood pressure readings are not normally distributed, but the sample size of $n = 49$ is ≥ 30. Thus, the Central Limit Theorem applies, and the distribution of $\bar{x}$ is approximately normal.

$\mu_{\bar{x}} = \mu = 78, \sigma_{\bar{x}} = \frac{\sigma}{\sqrt{n}} = \frac{7}{\sqrt{49}} = 1$

$Z = 1.96$

$\bar{x} = Z \cdot \sigma_{\bar{x}} + \mu_{\bar{x}} = (1.96)(1) + 78 = 79.96$

$Z = -1.96$

$\bar{x} = Z \cdot \sigma_{\bar{x}} + \mu_{\bar{x}} = (-1.96)(1) + 78 = 76.04$

76.04 and 79.96

89. Not possible; variable not normally distributed and sample size less than 30.

91. **(a)** $\mu_{\bar{x}} = \mu = 31.4$ micrograms, $\sigma_{\bar{x}} = \frac{\sigma}{\sqrt{n}} = \frac{14.2}{\sqrt{16}} = 3.55$ micrograms

(b) $P(\bar{x} < 27.85) = P\left(\frac{\bar{x}-\mu_{\bar{x}}}{\sigma_{\bar{x}}} < \frac{27.85-\mu_{\bar{x}}}{\sigma_{\bar{x}}}\right) = P\left(\frac{\bar{x}-\mu}{\sigma/\sqrt{n}} < \frac{27.85-\mu}{\sigma/\sqrt{n}}\right)$

$= P\left(\frac{\bar{x}-31.4}{14.2/\sqrt{16}} < \frac{27.85-31.4}{14.2/\sqrt{16}}\right) = P\left(\frac{\bar{x}-31.4}{3.55} < \frac{27.85-31.4}{3.55}\right) = P(Z < -1.00) = 0.1587$

(c) $P(\bar{x} > 38.5) = P\left(\frac{\bar{x}-\mu_{\bar{x}}}{\sigma_{\bar{x}}} > \frac{38.5-\mu_{\bar{x}}}{\sigma_{\bar{x}}}\right) = P\left(\frac{\bar{x}-\mu}{\sigma/\sqrt{n}} > \frac{38.5-\mu}{\sigma/\sqrt{n}}\right)$

$= P\left(\frac{\bar{x}-31.4}{14.2/\sqrt{16}} > \frac{38.5-31.4}{14.2/\sqrt{16}}\right) = P\left(\frac{\bar{x}-31.4}{3.55} > \frac{38.5-31.4}{3.55}\right) = P(Z > 2.00) = 1 - 0.9772 = 0.0228$

93. $\mu_{\bar{x}} = \mu = 66.2, \sigma_{\bar{x}} = \frac{\sigma}{\sqrt{n}} = \frac{10}{\sqrt{25}} = 2$

(a) $P(\bar{x} > 76.2) = P\left(\frac{\bar{x}-\mu_{\bar{x}}}{\sigma_{\bar{x}}} > \frac{76.2-\mu_{\bar{x}}}{\sigma_{\bar{x}}}\right) = P\left(\frac{\bar{x}-\mu}{\sigma/\sqrt{n}} > \frac{76.2-\mu}{\sigma/\sqrt{n}}\right)$

$= P\left(\frac{\bar{x}-66.2}{10/\sqrt{25}} > \frac{76.2-66.2}{10/\sqrt{25}}\right) = P\left(\frac{\bar{x}-66.2}{2} > \frac{76.2-66.2}{2}\right) = P(Z > 5.00) \approx 1 - 1 = 0$

(b) $P(56.2 < \bar{x} < 76.2) = P\left(\frac{56.2-\mu_{\bar{x}}}{\sigma_{\bar{x}}} < \frac{\bar{x}-\mu_{\bar{x}}}{\sigma_{\bar{x}}} < \frac{76.2-\mu_{\bar{x}}}{\sigma_{\bar{x}}}\right) = P\left(\frac{56.2-\mu}{\sigma/\sqrt{n}} < \frac{\bar{x}-\mu}{\sigma/\sqrt{n}} < \frac{76.2-\mu}{\sigma/\sqrt{n}}\right) =$
$P\left(\frac{56.2-66.2}{10/\sqrt{25}} < \frac{\bar{x}-66.2}{10/\sqrt{25}} < \frac{76.2-66.2}{10/\sqrt{25}}\right) = P\left(\frac{56.2-66.2}{2} < \frac{\bar{x}-66.2}{2} < \frac{76.2-66.2}{2}\right) = P(-5.00 < Z <$
$5.00) \approx 1 - 0 = 1$

(c) $P(\bar{x} < 56.2) = P\left(\frac{\bar{x}-\mu_{\bar{x}}}{\sigma_{\bar{x}}} < \frac{56.2-\mu_{\bar{x}}}{\sigma_{\bar{x}}}\right) = P\left(\frac{\bar{x}-\mu}{\sigma/\sqrt{n}} < \frac{56.2-\mu}{\sigma/\sqrt{n}}\right)$

$= P\left(\frac{\bar{x}-66.2}{10/\sqrt{25}} < \frac{56.2-66.2}{10/\sqrt{25}}\right) = P\left(\frac{\bar{x}-66.2}{2} < \frac{56.2-66.2}{2}\right) = P(Z < -5.00) \approx 0$

95. **(a)** $Z = 1.65$

$\bar{x} = Z \cdot \sigma_{\bar{x}} + \mu_{\bar{x}} = (1.65)(3.55) + 31.4 = 37.2575$ micrograms

(b) $Z = -1.65$

$\bar{x} = Z \cdot \sigma_{\bar{x}} + \mu_{\bar{x}} = (-1.65)(3.55) + 31.4 = 25.5425$ micrograms

(c) 25.5425 micrograms and 37.2575 micrograms

97. **(a)** $Z = 2.58$

$\bar{x} = Z \cdot \sigma_{\bar{x}} + \mu_{\bar{x}} = (2.58)(2) + 66.2 = 71.36$

$Z = -2.58$

$\bar{x} = Z \cdot \sigma_{\bar{x}} + \mu_{\bar{x}} = (-2.58)(2) + 66.2 = 61.04$

61.05 and 71.35

(b)

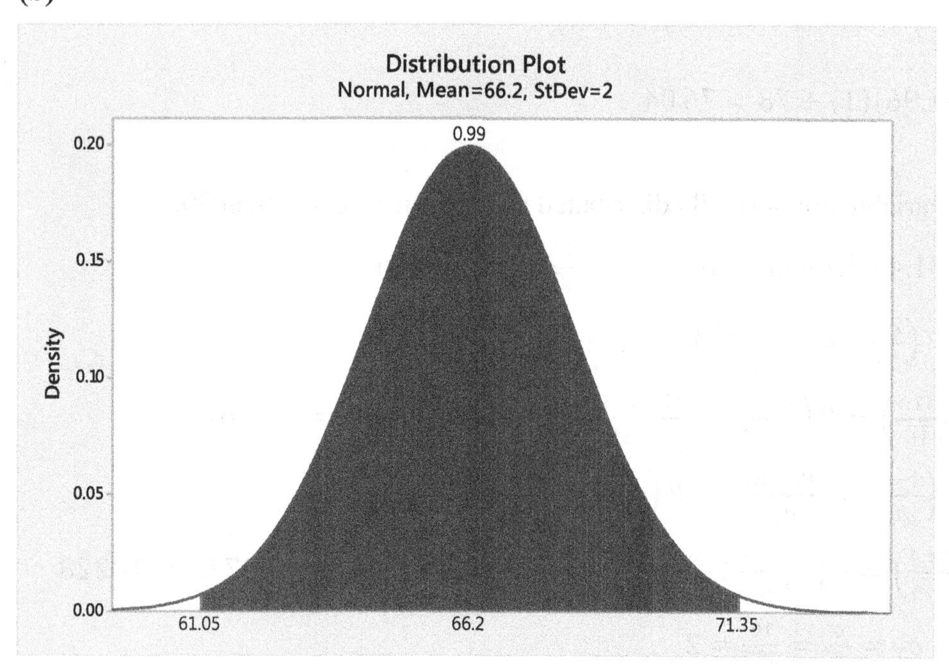

99. **(a)** Since the sample size of $n = 36$ is large ($n \geq 30$), the sampling distribution of $\bar{x}$ is approximately normal. Thus, $P(\bar{X} > 212) = P\left(Z > \frac{212-202}{45/\sqrt{36}}\right) \approx P(Z > 1.33) = 1 - 0.9082 = 0.0918$.

TI-83/84: 0.0912.

(b)

$$P(192 < \bar{x} < 212) = P\left(\frac{192-202}{45/\sqrt{36}} < \frac{\bar{x}-202}{45/\sqrt{36}} < \frac{212-202}{45/\sqrt{36}}\right)$$

$$= P\left(\frac{192-202}{7.5} < \frac{\bar{x}-202}{7.5} < \frac{212-202}{7.5}\right) \approx P(-1.33 < Z < 1.33)$$

$$= 0.9082 - 0.0918 = 0.8164; \text{ TI-83/84: } 0.8176.$$

101. (a) $P(\bar{x} < 110) = P\left(\frac{\bar{x}-124}{50/\sqrt{100}} < \frac{110-124}{50/\sqrt{100}}\right) = P\left(\frac{\bar{x}-124}{5} < \frac{110-124}{5}\right) = P(Z < -2.8) = 0.0026.$

(b)

$$P(110 < \bar{x} < 124) = P\left(\frac{110-124}{50/\sqrt{100}} < \frac{\bar{x}-124}{50/\sqrt{100}} < \frac{124-124}{50/\sqrt{100}}\right) = P\left(\frac{110-124}{5} < \frac{\bar{x}-124}{5} < \frac{124-124}{5}\right)$$

$$= P(-2.8 < Z < 0) = 0.5000 - 0.0026 = 0.4974.$$

103. **(a)** $\bar{x} = Z \cdot \frac{\sigma}{\sqrt{n}} + \mu = (1.96)\left(\frac{10}{\sqrt{36}}\right) + 38.6 \approx 41.87°$

(b) $\bar{x} = Z \cdot \frac{\sigma}{\sqrt{n}} + \mu = (-1.96)\left(\frac{10}{\sqrt{36}}\right) + 38.6 \approx 35.33°$

(c)

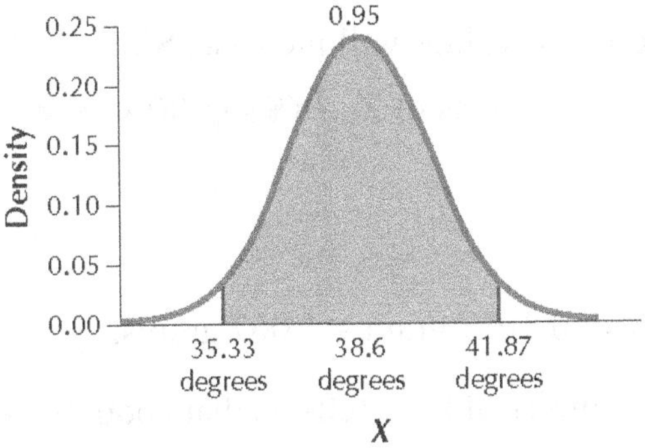

0.95

0.25 0.20 0.15 0.10 0.05 0.00

Density

35.33 degrees 38.6 degrees 41.87 degrees

X

105. $\mu_{\bar{x}} = \mu = 397$ ppm, $\sigma_{\bar{x}} = \frac{\sigma}{\sqrt{n}} = \frac{25}{\sqrt{25}} = 5$ ppm

(a) $P(387 < \bar{x} < 407) = P\left(\frac{387-\mu_{\bar{x}}}{\sigma_{\bar{x}}} < \frac{\bar{x}-\mu_{\bar{x}}}{\sigma_{\bar{x}}} < \frac{407-\mu_{\bar{x}}}{\sigma_{\bar{x}}}\right) = P\left(\frac{387-\mu}{\sigma/\sqrt{n}} < \frac{\bar{x}-\mu}{\sigma/\sqrt{n}} < \frac{407-\mu}{\sigma/\sqrt{n}}\right)$

$= P\left(\frac{387-397}{25/\sqrt{25}} < \frac{\bar{x}-397}{25/\sqrt{25}} < \frac{407-397}{25/\sqrt{25}}\right) = P\left(\frac{387-397}{5} < \frac{\bar{x}-397}{5} < \frac{407-397}{5}\right)$

$= P(-2.00 < Z < 2.00) = 0.9772 - 0.0228 = 0.9544$

(b) 397 ppm; in a normal distribution the median equals the mean.

(c) $Z = 2.58$

$\bar{x} = Z \cdot \sigma_{\bar{x}} + \mu_{\bar{x}} = (2.58)(5) + 397 = 409.9$ ppm

$Z = -2.58$

$\bar{x} = Z \cdot \sigma_{\bar{x}} + \mu_{\bar{x}} = (-2.58)(5) + 397 = 384.1$ ppm

384.1 ppm and 409.9 ppm

107. In favor of normality. All of the points are between the curved lines and most of the points are close to the center line.

109. $P(\bar{x} > 86) = P\left(\frac{\bar{x}-80}{10/\sqrt{36}} > \frac{86-80}{10/\sqrt{36}}\right) = P(Z > 3.6) \approx 1 - 1 = 0$

TI-83/84: 0.0001591

111. Not possible. The variable is not normally distributed and the sample size is less than 30.

113. **(a)** Remain unchanged. From Fact 1 in Section 7.1. $\mu_{\bar{x}} = \mu$. Thus, $\mu_{\bar{x}}$ does not depend on the sample size *n*.

(b) Decrease. Since $\sigma_{\bar{x}} = \frac{\sigma}{\sqrt{n}}$, an increase in the sample size *n* results in a decrease in $\sigma_{\bar{x}}$.

(c) Insufficient information to tell. If $\bar{x} > 50,000$, then $\bar{x} - 50,000 > 0$ Since $\sigma_{\bar{x}}$ decreases and is positive, $Z = \frac{\bar{x}-50,000}{\sigma_{\bar{x}}}$ will increase. If $\bar{x} = 50,000$, then $\bar{x} - 50,000 > 0$. Thus, $Z = \frac{\bar{x}-50,000}{\sigma_{\bar{x}}}$ will remain 0. If $\bar{x} < 50,000$ then $\bar{x} - 50,000 > 0$. Since $\sigma_{\bar{x}}$ decreases and is positive, $Z = \frac{\bar{x}-50,000}{\sigma_{\bar{x}}}$ will decrease.

(d) Increase. From part (c), $Z = \frac{\bar{x}-50,000}{\sigma_{\bar{x}}} = \frac{60,000-50,000}{\sigma_{\bar{x}}}$ will increase and $Z = \frac{\bar{x}-50,000}{\sigma_{\bar{x}}} =$

$\frac{40{,}000-50{,}000}{\sigma_{\bar{x}}}$ will decrease. Thus, the area between these two values will increase. Since $P(\$40{,}000 < \bar{x} < \$60{,}000)$ is the area between these two values of Z, $P(\$40{,}000 < \bar{x} < \$60{,}000)$ will increase.

115. **(a)** $3\sigma_{\bar{X}} = 0.32$, so $\sigma_{\bar{X}} = \frac{0.32}{3} \approx 0.1067$ gram.

(b) Since $\sigma_{\bar{X}} = \frac{\sigma}{\sqrt{n}} = \frac{\sigma}{\sqrt{100}} = \frac{\sigma}{10} \approx 0.1067$ gram, $\sigma = 10 \cdot 0.1067$ gram $= 1.067$ grams.

(c) Since $P(127.68 < \bar{x} < 128.32) = P(-3 < Z < 3)$, the Empirical Rule tells us that about 0.997 of the mean weights lie between 127.68 and 128.32 grams.

117. **(a)** $P(127.68 < \bar{x} < 128.32) \approx P\left(\frac{127.68-127.3}{0.1067} < \frac{\bar{x}-127.3}{0.1067} < \frac{128.32-127.3}{0.1067}\right)$

$\approx P(3.56 < Z < 9.56) = 1 - 0.9998 = 0.0002.$

(b) 0.0002, $1 - 0.0002 = 0.9998$

(c) The probability that he will get away with it found in part (b) is smaller than the probability that he will get away with it found in the original Case Study in the text, so the value found in the original Case Study in the text favors the Master of the Mint.

119. $n = 2$

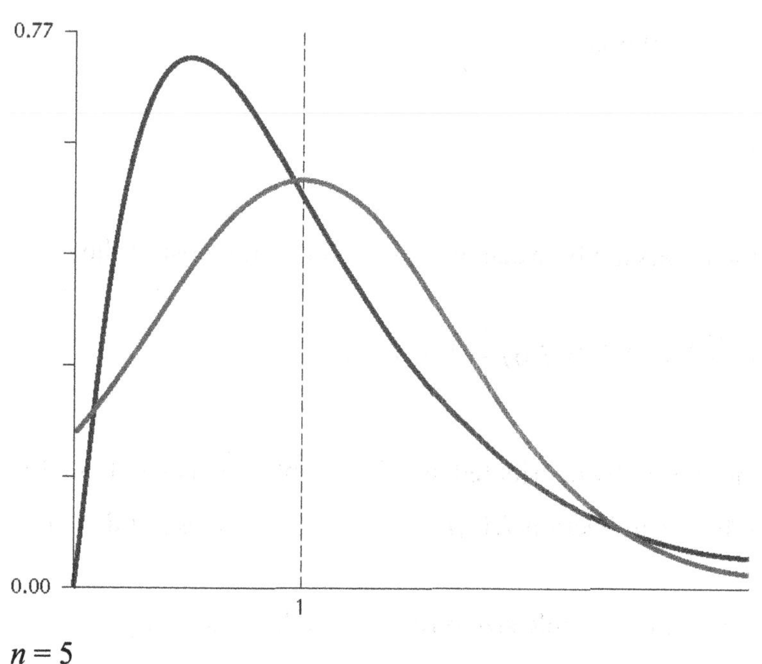

$n = 5$

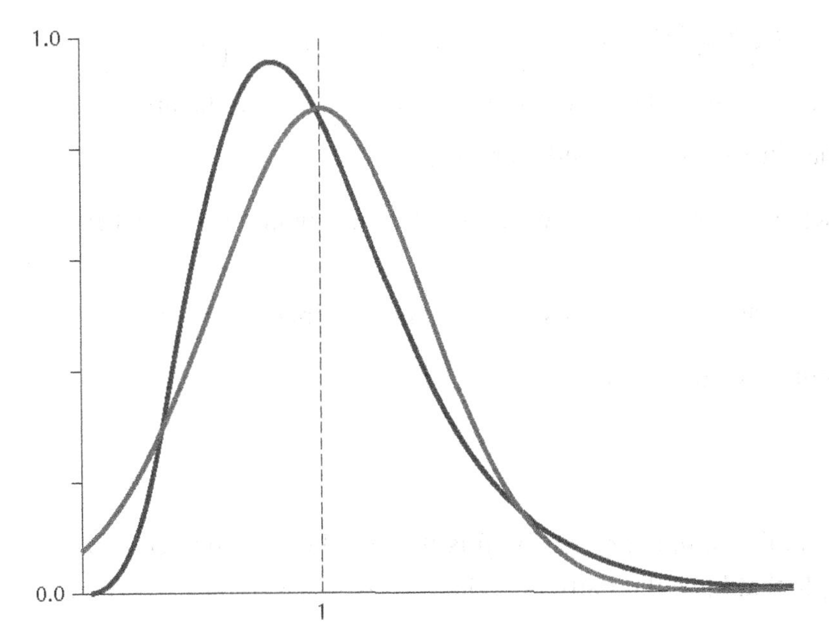

$n = 30$

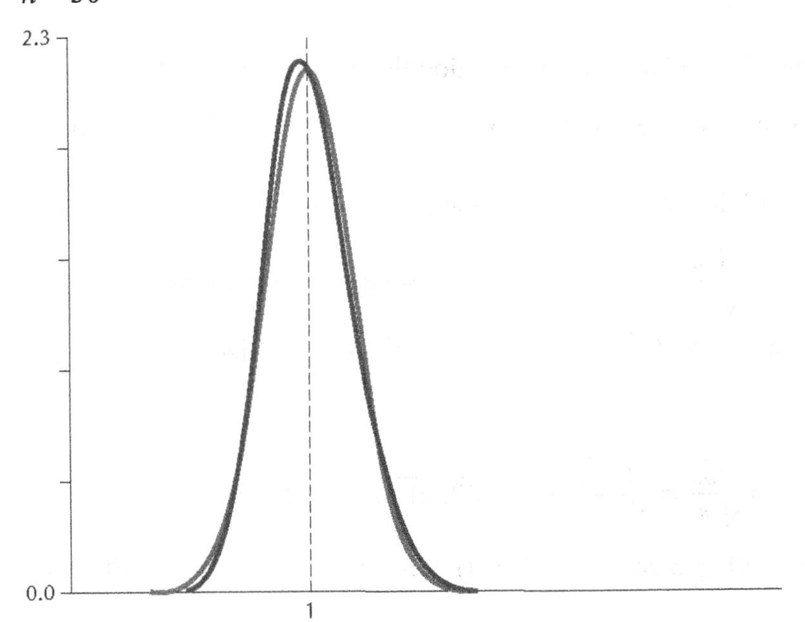

121. $P(x < 500) = P\left(\frac{x-\mu}{\sigma} < \frac{500-\mu}{\sigma}\right) = P\left(\frac{x-514}{116} < \frac{500-514}{116}\right)$

$\approx P(Z < -0.12) = 0.4522$

123. **(a)** $P(\bar{x} < 500) = P\left(\frac{\bar{x}-\mu_{\bar{x}}}{\sigma_{\bar{x}}} < \frac{500-\mu_{\bar{x}}}{\sigma_{\bar{x}}}\right) = P\left(\frac{\bar{x}-\mu}{\sigma/\sqrt{n}} < \frac{500-\mu}{\sigma/\sqrt{n}}\right)$

$= P\left(\frac{\bar{x}-514}{116/\sqrt{16}} < \frac{500-514}{116/\sqrt{16}}\right) = P\left(\frac{\bar{x}-514}{29} < \frac{500-514}{29}\right) \approx P(Z < -0.48) = 0.3156$

(b) Means are less variable than individual values.

125. **(a)** Since $P(x < 500) = P\left(\frac{x-\mu}{\sigma} < \frac{500-\mu}{\sigma}\right) = P\left(\frac{x-514}{\sigma} < \frac{500-514}{\sigma}\right) = P\left(Z < \frac{-14}{\sigma}\right)$, an increase σ would result in an increase in $\frac{-14}{\sigma}$. This would result in an increase in the area to the left of $\frac{-14}{\sigma}$, which would mean that $P(x < 500)$ would increase.

(b) Since $\mu_{\bar{x}} = \mu$, $\mu_{\bar{x}}$ does not depend on σ. Therefore, an increase in σ would not affect $\mu_{\bar{x}}$. Thus, $\mu_{\bar{x}}$ would remain the same.

(c) Since $\sigma_{\bar{x}} = \frac{\sigma}{\sqrt{n}}$ and $\sqrt{n}$ is positive, an increase in σ would result in an increase in $\sigma_{\bar{x}}$.

(d) Still normal with $\mu_{\bar{x}} = 514$ but $\sigma_{\bar{x}}$ would increase.

Section 7.2

1. If we take a sample of size n, the sample proportion $\hat{p}$ is $\hat{p} = \frac{x}{n}$, where x represents the number of individuals in the sample that have the particular characteristic. Examples will vary.

3. $\sigma_{\hat{p}} = \sqrt{\frac{pq}{n}}$, where $q = 1 - p$.

5. Let n_1 be the original sample size and $n_2 = 2n_1$ be the doubled sample size. Let $\sigma_{\hat{p},1} = \sqrt{\frac{pq}{n_1}}$ be the population standard deviation of the sample proportion for samples of size n_1 and $\sigma_{\hat{p},2} = \sqrt{\frac{pq}{n_2}}$ be the population standard deviation of the sample proportion for samples of size $n_2 = 2n_1$. Then $\sigma_{\hat{p},2} = \sqrt{\frac{pq}{n_2}} = \sqrt{\frac{pq}{2n_1}} = \frac{1}{\sqrt{2}} \cdot \sqrt{\frac{pq}{n_1}} = \frac{1}{\sqrt{2}} \cdot \sigma_{\hat{p},1}$, so the population standard deviation of the sample proportion decreases by a factor of $\frac{1}{\sqrt{2}} \approx 0.7071$ when the sample size is doubled.

7. **(a)** $\mu_{\hat{p}} = p = 0.5$

(b) $q = 1 - p = 1 - 0.5 = 0.5$, $\sigma_{\hat{p}} = \sqrt{\frac{pq}{n}} = \sqrt{\frac{(0.5)(0.5)}{400}} = \sqrt{0.000625} = 0.025$

(c) Since both $np = (400)(0.5) = 200 \geq 5$ and $nq = (400)(0.5) = 200 \geq 5$, the distribution of $\hat{p}$ is approximately normal.

9. **(a)** $\mu_{\hat{p}} = p = 0.05$

(b) $q = 1 - p = 1 - 0.05 = 0.95$, $\sigma_{\hat{p}} = \sqrt{\frac{pq}{n}} = \sqrt{\frac{(0.05)(0.95)}{200}} = \sqrt{0.0002375} \approx 0.01541$

(c) Since both $np = (200)(0.05) = 10 \geq 5$ and $nq = (200)(0.95) = 190 \geq 5$, the distribution of $\hat{p}$ is approximately normal

11. **(a)** $\mu_{\hat{p}} = p = 0.3$

(b) $q = 1 - p = 1 - 0.3 = 0.7$, $\sigma_{\hat{p}} = \sqrt{\frac{pq}{n}} = \sqrt{\frac{(0.3)(0.7)}{12}} = \sqrt{0.0175} \approx 0.1323$

(c) Since $np = (12)(0.3) = 3.6$ is not ≥ 5, the distribution of $\hat{p}$ is unknown.

13. **(a)** $\mu_{\hat{p}} = p = 0.002$

(b) $q = 1 - p = 1 - 0.002 = 0.998, \sigma_{\hat{p}} = \sqrt{\frac{pq}{n}} = \sqrt{\frac{(0.002)(0.998)}{2000}} = \sqrt{0.000000998} \approx$ 0.000999

(c) Since $np = (2000)(0.002) = 4$ is not ≥ 5, the distribution of $\hat{p}$ is unknown.

15. **(a)** $\mu_{\hat{p}} = p = 0.02$

(b) $q = 1 - p = 1 - 0.02 = 0.98, \sigma_{\hat{p}} = \sqrt{\frac{pq}{n}} = \sqrt{\frac{(0.02)(0.98)}{250}} = \sqrt{0.0000784} \approx 0.008854$

(c) Since both $np = (250)(0.02) = 5 \geq 5$ and $nq = (250)(0.98) = 245 \geq 5$, the distribution of $\hat{p}$ is approximately normal.

17. **(a)** $\mu_{\hat{p}} = p = 0.01$

(b) $q = 1 - p = 1 - 0.01 = 0.99, \sigma_{\hat{p}} = \sqrt{\frac{pq}{n}} = \sqrt{\frac{(0.01)(0.99)}{500}} = \sqrt{0.0000198} \approx 0.004450$

(c) Since both $np = (500)(0.01) = 5 \geq 5$ and $nq = (500)(0.99) = 495 \geq 5$, the distribution of $\hat{p}$ is approximately normal.

19. $n_1 = \frac{5}{p} = \frac{5}{0.8} = 6.25$ and $n_2 = \frac{5}{q} = \frac{5}{1-p} = \frac{5}{1-0.8} = \frac{5}{0.2} = 25$. The larger of $n_1 = 6.25$ and $n_2 = 25$ is $n_2 = 25$. Therefore, $n = 25$ is the minimum sample size that produces a sampling distribution of $\hat{p}$ that is approximately normal.

21. $n_1 = \frac{5}{p} = \frac{5}{0.9} = 5.555555556$ and $n_2 = \frac{5}{q} = \frac{5}{1-p} = \frac{5}{1-0.9} = \frac{5}{0.1} = 50$. The larger of $n_1 = 5.555555556$ and $n_2 = 50$ is $n_2 = 50$. Therefore, $n = 50$ is the minimum sample size that produces a sampling distribution of $\hat{p}$ that is approximately normal.

23. $n_1 = \frac{5}{p} = \frac{5}{0.99} = 5.050505051$ and $n_2 = \frac{5}{q} = \frac{5}{1-p} = \frac{5}{1-0.99} = \frac{5}{0.01} = 500$. The larger of $n_1 = 5.050505051$ and $n_2 = 500$ is $n_2 = 500$. Therefore, $n = 500$ is the minimum sample size that produces a sampling distribution of $\hat{p}$ that is approximately normal.

25. $q = 1 - p = 1 - 0.5 = 0.5$. Since both $np = (225)(0.5) = 112.5 \geq 5$ and $nq = (225)(0.5) = 112.5 \geq 5$, the distribution of $\hat{p}$ is approximately normal.

$\mu_{\hat{p}} = p = 0.5$ and $\sigma_{\hat{p}} = \sqrt{\frac{pq}{n}} = \sqrt{\frac{(0.5)(0.5)}{225}} = \sqrt{0.0011111111} \approx 0.03333$

$$P(\hat{p} > 0.55) = P\left(\frac{\hat{p}-\mu_{\hat{p}}}{\sigma_{\hat{p}}} > \frac{0.55-\mu_{\hat{p}}}{\sigma_{\hat{p}}}\right) = P\left(\frac{\hat{p}-0.5}{0.03333} > \frac{0.55-0.5}{0.03333}\right)$$

$$\approx P(Z > 1.50) = 1 - 0.9332 = 0.0668$$

27. Not possible since $n \cdot p = (36)(0.03) = 1.08 < 5$.

29. $q = 1 - p = 1 - 0.8 = 0.2$. Since both $np = (50)(0.8) = 40 \geq 5$ and $nq = (50)(0.2) = 10 \geq 5$, the distribution of $\hat{p}$ is approximately normal.

$\mu_{\hat{p}} = p = 0.8$ and $\sigma_{\hat{p}} = \sqrt{\frac{pq}{n}} = \sqrt{\frac{(0.8)(0.2)}{50}} = \sqrt{0.0032} \approx 0.05657$

$P(0.88 < \hat{p} < 0.91) = P\left(\frac{0.88-\mu_{\hat{p}}}{\sigma_{\hat{p}}} < \frac{\hat{p}-\mu_{\hat{p}}}{\sigma_{\hat{p}}} < \frac{0.91-\mu_{\hat{p}}}{\sigma_{\hat{p}}}\right)$

$= P\left(\frac{0.88-0.8}{0.05657} < \frac{\hat{p}-0.8}{0.05657} < \frac{0.91-0.8}{0.05657}\right) \approx P(1.41 < Z < 1.94) = 0.9738 - 0.9207 = 0.0531$

31. Not possible since $n \cdot q = (225)(0.02) = 4.50 < 5$.

33. $q = 1 - p = 1 - 0.6 = 0.4$. Since both $np = (100)(0.6) = 60 \geq 5$ and $nq = (100)(0.4) = 40 \geq 5$, the distribution of $\hat{p}$ is approximately normal.

$\mu_{\hat{p}} = p = 0.6$ and $\sigma_{\hat{p}} = \sqrt{\frac{pq}{n}} = \sqrt{\frac{(0.6)(0.4)}{100}} = \sqrt{0.0024} \approx 0.04899$

$Z = 1.28$

$\hat{p} = Z \cdot \sigma_{\hat{p}} + \mu_{\hat{p}} = (1.28)(0.04899) + 0.6 \approx 0.6627$

35. Not possible since $n \cdot q = (225)(0.01) = 2.25 < 5$.

37. $q = 1 - p = 1 - 0.2 = 0.8$. Since both $np = (225)(0.2) = 45 \geq 5$ and $nq = (225)(0.8) = 180 \geq 5$, the distribution of $\hat{p}$ is approximately normal.

$\mu_{\hat{p}} = p = 0.2$ and $\sigma_{\hat{p}} = \sqrt{\frac{pq}{n}} = \sqrt{\frac{(0.2)(0.8)}{225}} = \sqrt{0.000711111111111} \approx 0.02667$

$Z = -1.96$

$\hat{p} = Z \cdot \sigma_{\hat{p}} + \mu_{\hat{p}} = (-1.96)(0.02667) + 0.2 \approx 0.1477$

39. **(a)** $\mu_{\hat{p}} = p = 0.25$, $\sigma_{\hat{p}} = \sqrt{\frac{pq}{n}} = \sqrt{\frac{(0.25)(0.75)}{36}} \approx 0.0721687836 \approx 0.0722$.

(b) Since both $np = (36)(0.25) = 9 \geq 5$ and $n(1-p) = 36(1-0.25) = 27 \geq 5$, the sampling distribution of $\hat{p}$ is approximately normal (0.25, 0.0722).

(c) $P(\hat{p} > 0.26) = P\left(\frac{\hat{p}-0.25}{\sqrt{\frac{0.25(0.75)}{36}}} > \frac{0.26-0.25}{\sqrt{\frac{0.25(0.75)}{36}}}\right) \approx P(Z > 0.14) = 1 - 0.5557 = 0.4443$. TI-83/84: 0.4449.

41. **(a)** $\mu_{\hat{p}} = p = 0.75$

$q = 1 - p = 1 - 0.75 = 0.25$, $\sigma_{\hat{p}} = \sqrt{\frac{pq}{n}} = \sqrt{\frac{(0.75)(0.25)}{20}} \approx 0.0968245837 \approx 0.0968$.

Since both $np = (20)(0.75) = 15 \geq 5$ and $nq = (20)(0.25) = 5 \geq 5$, the sampling distribution of $\hat{p}$ is approximately normal (0.75, 0.0968).

(b) $P(\hat{p} > 0.69) = P\left(\frac{\hat{p}-0.75}{\sqrt{\frac{0.75(0.25)}{20}}} > \frac{0.69-0.75}{\sqrt{\frac{0.75(0.25)}{20}}}\right) \approx P(Z > -0.62)$

$= 1 - 0.2676 = 0.7324$. TI-83/84: 0.7323.

(c) $P(0.775 < \hat{p} < 0.8) = P\left(\frac{0.8-0.75}{\sqrt{\frac{0.75(0.25)}{20}}} < \frac{\hat{p}-0.75}{\sqrt{\frac{0.75(0.25)}{20}}} < \frac{0.8-0.75}{\sqrt{\frac{0.75(0.25)}{20}}}\right) \approx P(0.26 < Z < 0.8) =$

$0.6985 - 0.6026 = 0.0959.$ TI-83/84: 0.0954

43. **(a)** $\hat{p} = Z \cdot \sqrt{\frac{pq}{n}} + p = (-1.645) \cdot \sqrt{\frac{(0.25)(0.75)}{36}} + 0.25 \approx 0.1313.$

$\hat{p} = Z \cdot \sqrt{\frac{pq}{n}} + p = (1.645) \cdot \sqrt{\frac{(0.25)(0.75)}{36}} + 0.25 \approx 0.3687$

(b)

(c) For $\hat{p} = \frac{x}{n} = \frac{2}{36} = 0.0555555556 \approx 0.0556$, $Z = \frac{\hat{p}-p}{\sqrt{\frac{pq}{n}}} = \frac{0.0555555556-0.25}{\sqrt{\frac{(0.25)(0.75)}{36}}}$

$\approx -2.69.$ Since $-3 <$ *Z*-score ≤ -2, $\hat{p} = \frac{2}{36}$ is considered moderately unusual.

(d) Sample proportions with *Z*-scores ≤ -3 or ≥ 3 are considered outliers. Sample proportions with *Z*-scores ≤ -3 are the sample proportions that are both ≥ 0 and

$\leq \hat{p} = Z \cdot \sqrt{\frac{pq}{n}} + p = (-3) \cdot \sqrt{\frac{(0.25)(0.75)}{36}} + 0.25 \approx 0.0335.$

Sample proportions with *Z*-scores ≥ 3 are sample proportions that are both ≤ 1 and

$\geq \hat{p} = Z \cdot \sqrt{\frac{pq}{n}} + p = (3) \cdot \sqrt{\frac{(0.25)(0.75)}{36}} + 0.25 \approx 0.4665.$

$0.25 \approx 0.4665$. Therefore, sample proportions between 0 and 0.0335 inclusive and between 0.4665 and 1 inclusive would be considered outliers.

45. **(a)**

$\hat{p} = Z \cdot \sqrt{\frac{p(1-p)}{n}} + p = (-2.58) \cdot \sqrt{\frac{0.75(1-0.75)}{20}} + 0.75 \approx 0.5002$

$\hat{p} = Z \cdot \sqrt{\frac{p(1-p)}{n}} + p = (2.58) \cdot \sqrt{\frac{0.75(1-0.75)}{20}} + 0.75 \approx 0.9998$; TI-83/84: 0.5007, 0.9993.

(b)

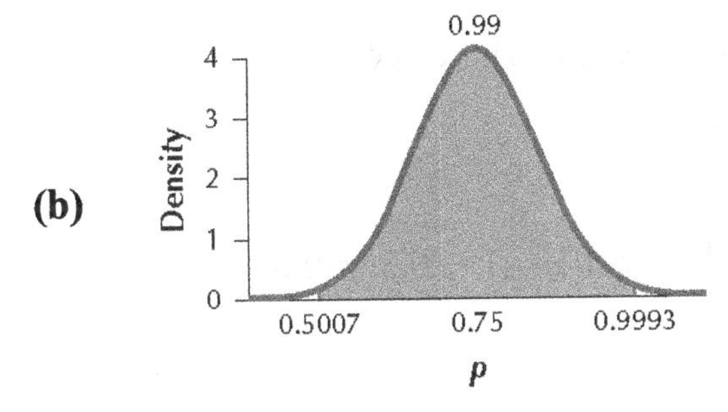

(c) For $\hat{p} = \frac{x}{n} = \frac{14}{20} = 0.7$, $Z = \frac{\hat{p}-p}{\sqrt{\frac{pq}{n}}} = \frac{0.7-0.75}{\sqrt{\frac{(0.75)(0.25)}{20}}} \approx -0.5164$.

≈ -2.69. Since $-2 < Z\text{-score} < 2$, $\hat{p} = 0.7$ is neither moderately unusual nor an outlier.

47. **(a)** Remains the same since $\mu_{\hat{p}} = p$ does not depend on n.

(b) Decrease. Since the sample size n is in the denominator of

$\sigma_{\hat{p}} = \sqrt{\frac{p \cdot q}{n}}$, $\sigma_{\hat{p}}$ decreases as the sample size n increases.

(c) Decrease. Standardizing we get $Z = \frac{0.86-\mu_{\hat{p}}}{\sigma_{\hat{p}}} = \frac{0.86-0.840}{\sigma_{\hat{p}}} = \frac{0.02}{\sigma_{\hat{p}}}$. From (b), $\sigma_{\hat{p}}$ decreases as the sample size n increases. Therefore $Z = \frac{0.02}{\sigma_{\hat{p}}}$ increases as the sample size n increases. Therefore $P(\hat{p} > 0.86) = P\left(Z > \frac{0.02}{\sigma_{\hat{p}}}\right)$ decreases.

(d) Increase. Standardizing we get $Z = \frac{0.82-\mu_{\hat{p}}}{\sigma_{\hat{p}}} = \frac{0.82-0.840}{\sigma_{\hat{p}}} = \frac{-0.02}{\sigma_{\hat{p}}}$ and $Z = \frac{0.86-\mu_{\hat{p}}}{\sigma_{\hat{p}}} = \frac{0.86-0.840}{\sigma_{\hat{p}}} = \frac{0.02}{\sigma_{\hat{p}}}$. From (b), $\sigma_{\hat{p}}$ decreases as the sample size n increases. Therefore $Z = \frac{-0.02}{\sigma_{\hat{p}}}$ decreases and $Z = \frac{0.02}{\sigma_{\hat{p}}}$ increases as the sample size n increases. Thus $P(0.82 < \hat{p} < 0.86) = P\left(\frac{-0.02}{\sigma_{\hat{p}}} < Z < \frac{0.02}{\sigma_{\hat{p}}}\right)$ increases as the sample size n increases.

(e) Decrease. Standardizing we get $Z = \frac{0.82-\mu_{\hat{p}}}{\sigma_{\hat{p}}} = \frac{0.82-0.840}{\sigma_{\hat{p}}} = \frac{-0.02}{\sigma_{\hat{p}}}$. From (b), $\sigma_{\hat{p}}$ decreases as the sample size n increases. Therefore $Z = \frac{-0.02}{\sigma_{\hat{p}}}$ decreases as the sample size n increases. Thus $P(\hat{p} < 0.82) = P\left(Z < \frac{-0.02}{\sigma_{\hat{p}}}\right)$ decreases as the sample size n increases.

(f) Increase. The 2.5th percentile is found by the formula

$\hat{p}_1 = (-1.96)\sigma_{\hat{p}} + \mu_{\hat{p}}$. From (a) $\mu_{\hat{p}}$ remains the same as the sample size n increases and from (b) $\sigma_{\hat{p}}$ decreases as the sample size n increases. Therefore

$\hat{p}_1 = (-1.96)\sigma_{\hat{p}} + \mu_{\hat{p}}$ increases as the sample size n increases.

(g) Decrease. The 97.5th percentile is found by the formula

$\hat{p}_2 = (1.96)\sigma_{\hat{p}} + \mu_{\hat{p}}$. From (a) $\mu_{\hat{p}}$ remains the same as the sample size n increases and from (b) $\sigma_{\hat{p}}$ decreases as the sample size n increases. Therefore

$\hat{p}_2 = (1.96)\sigma_{\hat{p}} + \mu_{\hat{p}}$ decreases as the sample size n increases.

49. **(a)** $P(\hat{p} > 0.65) = P\left(\frac{\hat{p}-0.65}{\sqrt{\frac{0.65(0.35)}{100}}} > \frac{0.65-0.65}{\sqrt{\frac{0.65(0.35)}{100}}}\right) = P(Z > 0) = 1 - 0.5 = 0.5$.

(b) $P(\hat{p} > 0.65) = P\left(\frac{\hat{p}-0.41}{\sqrt{\frac{0.41(0.59)}{100}}} > \frac{0.65-0.41}{\sqrt{\frac{0.41(0.59)}{100}}}\right) \approx P(Z > 4.88) \approx 1 - 1 = 0.$

(c) $P(\hat{p} < 0.41) = P\left(\frac{\hat{p}-0.65}{\sqrt{\frac{0.65(0.35)}{100}}} < \frac{0.41-0.65}{\sqrt{\frac{0.65(0.35)}{100}}}\right) \approx P(Z < -5.03) \approx 0.$

(d) $P(\hat{p} < 0.41) = P\left(\frac{\hat{p}-0.41}{\sqrt{\frac{0.41(0.59)}{100}}} < \frac{0.41-0.41}{\sqrt{\frac{0.41(0.59)}{100}}}\right) = P(Z < 0) = 0.5.$

51. The results of Exercises 49 and 50 do not support this claim. The 97.5th percentile for the males is less than the 2.5th percentile for the females. Also, $P(\hat{p} < 0.41)$ and $P(\hat{p} > 0.65)$ are both very different for males and females.

Chapter 7 Review Exercises

1. $\mu_{\bar{x}} = \mu = 100, \sigma_{\bar{x}} = \frac{\sigma}{\sqrt{n}} = \frac{15}{\sqrt{25}} = 3$

3. $\mu_{\bar{x}} = \mu = 100, \sigma_{\bar{x}} = \frac{\sigma}{\sqrt{n}} = \frac{15}{\sqrt{49}} \approx 2.1429$

5. $\mu_{\bar{x}} = \mu = 0, \sigma_{\bar{x}} = \frac{\sigma}{\sqrt{n}} = \frac{1}{\sqrt{16}} = 0.25$

7. Scores on a psychology test are normally distributed, so the distribution of $\bar{x}$ is normal for any sample size n.

$\mu_{\bar{x}} = \mu = 100, \sigma_{\bar{x}} = \frac{\sigma}{\sqrt{n}} = \frac{15}{\sqrt{25}} = 3$

$P(100 < \bar{x} < 105) = P\left(\frac{100-\mu_{\bar{x}}}{\sigma_{\bar{x}}} < \frac{\bar{x}-\mu_{\bar{x}}}{\sigma_{\bar{x}}} < \frac{105-\mu_{\bar{x}}}{\sigma_{\bar{x}}}\right)$

$= P\left(\frac{100-100}{3} < \frac{\bar{x}-100}{3} < \frac{105-100}{3}\right) \approx P(0 < Z < 1.67)$

$= 0.9525 - 0.5000 = 0.4525$

9. Not possible since $n \cdot q = (40)(0.1) = 4 < 5$.

11. $q = 1 - p = 1 - 0.98 = 0.02$. Since both $np = (400)(0.98) = 392 \geq 5$ and $nq = (400)(0.02) = 8 \geq 5$, the distribution of $\hat{p}$ is approximately normal.

$\mu_{\hat{p}} = p = 0.98$ and $\sigma_{\hat{p}} = \sqrt{\frac{pq}{n}} = \sqrt{\frac{(0.98)(0.02)}{400}} = \sqrt{0.000049} = 0.007$

For the 100 – 75 = 25th percentile, $Z = -0.67$. Therefore, the 25th percentile is

$\hat{p} = Z \cdot \sigma_{\hat{p}} + \mu_{\hat{p}} = (-0.67)(0.007) + 0.98 \approx 0.9753.$

13. $q = 1 - p = 1 - 0.12 = 0.88$. Since both $np = (100)(0.12) = 12 \geq 5$ and $nq = (100)(0.88) = 88 \geq 5$, the distribution of $\hat{p}$ is approximately normal.

$\mu_{\hat{p}} = p = 0.12$ and $\sigma_{\hat{p}} = \sqrt{\frac{pq}{n}} = \sqrt{\frac{(0.12)(0.88)}{100}} = \sqrt{0.001056} \approx 0.03250$

(a) $P(\hat{p} > 0.16) = P\left(\frac{\hat{p}-\mu_{\hat{p}}}{\sigma_{\hat{p}}} > \frac{0.16-\mu_{\hat{p}}}{\sigma_{\hat{p}}}\right) = P\left(\frac{\hat{p}-0.12}{0.03250} > \frac{0.16-0.12}{0.03250}\right)$

$\approx P(Z > 1.23) = 1 - 0.8907 = 0.1093$

(b) $P(0.12 < \hat{p} < 0.16) = P\left(\frac{0.12-\mu_{\hat{p}}}{\sigma_{\hat{p}}} < \frac{\hat{p}-\mu_{\hat{p}}}{\sigma_{\hat{p}}} < \frac{0.16-\mu_{\hat{p}}}{\sigma_{\hat{p}}}\right)$

$= P\left(\frac{0.12-0.12}{0.03250} < \frac{\hat{p}-0.12}{0.03250} < \frac{0.16-0.12}{0.03250}\right) \approx P(0.00 < Z < 1.23) = 0.8907 - 0.5000 = 0.3907$

(c) $P(\hat{p} < 0.16) = 1 - P(\hat{p} > 0.16) \approx 1 - 0.1093 = 0.8907$

(d) For the 5th percentile, $Z = -1.645$. Therefore, the 5th percentile is

$\hat{p} = Z \cdot \sigma_{\hat{p}} + \mu_{\hat{p}} \approx (-1.645)(0.03250) + 0.12 \approx 0.0665.$

For the 95th percentile, $Z = 1.645$. Therefore, the 95th percentile is

$\hat{p} = Z \cdot \sigma_{\hat{p}} + \mu_{\hat{p}} \approx (1.645)(0.03250) + 0.12 \approx 0.1735.$

Chapter 7 Quiz

1. True.

3. Sampling error.

5. No.

7. Because the population is normal, the distribution of $\bar{x}$ is normal for any sample size n.

(a) $P(\bar{x} < 38) = P\left(\frac{\bar{x}-40}{20/\sqrt{100}} < \frac{38-40}{20/\sqrt{100}}\right) = P(Z < -1.00) = 0.1587.$

(b) $P(36.08 < \bar{x} < 43.92) = P\left(\frac{36.08-40}{20/\sqrt{100}} < \frac{\bar{x}-40}{20/\sqrt{100}} < \frac{43.92-40}{20/\sqrt{100}}\right)$

$= P(-1.96 < Z < 1.96) = 0.9730 - 0.0250 = 0.9500.$

(c) $P(\bar{x} > 42.5) = P\left(\frac{\bar{x}-40}{20/\sqrt{100}} > \frac{42{,}5-40}{20/\sqrt{100}}\right) = P(Z > 1.25) = 1 - 0.8944 = 0.1056.$

9. Incomes in Texas are normally distributed, so the distribution of $\bar{x}$ is normal for any sample size n. Incomes in Florida are normally distributed, so the distribution of $\bar{x}$ is normal for any sample size n.

For Texas, $\mu_{\bar{x}} = \mu = \$25{,}800$, $\sigma_{\bar{x}} = \frac{\sigma}{\sqrt{n}} = \frac{\$8{,}000}{\sqrt{100}} = \$800$

For Florida, $\mu_{\bar{x}} = \mu = \$26{,}500$, $\sigma_{\bar{x}} = \frac{\sigma}{\sqrt{n}} = \frac{\$8{,}000}{\sqrt{100}} = \$800$

(a) $P(\bar{x} > \$27{,}000) = P\left(\frac{\bar{x}-\mu_{\bar{x}}}{\sigma_{\bar{x}}} > \frac{\$27{,}000-\mu_{\bar{x}}}{\sigma_{\bar{x}}}\right) = P\left(\frac{\bar{x}-\mu}{\sigma/\sqrt{n}} > \frac{\$27{,}000-\mu}{\sigma/\sqrt{n}}\right)$

$= P\left(\frac{\bar{x}-\$25{,}800}{\$8{,}000/\sqrt{100}} > \frac{\$27{,}000-\$25{,}800}{\$8{,}000/\sqrt{100}}\right) = P\left(\frac{\bar{x}-\$25{,}800}{\$800} > \frac{\$27{,}000-\$25{,}800}{\$800}\right)$

$= P(Z > 1.50) = 1 - 0.9332 = 0.0668$

(b) $P(\bar{x} > \$27{,}000) = P\left(\frac{\bar{x}-\mu_{\bar{x}}}{\sigma_{\bar{x}}} > \frac{\$27{,}000-\mu_{\bar{x}}}{\sigma_{\bar{x}}}\right) = P\left(\frac{\bar{x}-\mu}{\sigma/\sqrt{n}} > \frac{\$27{,}000-\mu}{\sigma/\sqrt{n}}\right)$

$= P\left(\frac{\bar{x}-\$26{,}500}{\$8{,}000/\sqrt{100}} > \frac{\$27{,}000-\$26{,}500}{\$8{,}000/\sqrt{100}}\right) = P\left(\frac{\bar{x}-\$26{,}500}{\$800} > \frac{\$27{,}000-\$26{,}500}{\$800}\right)$

$= P(Z > 0.63) = 1 - 0.7357 = 0.2643$

11. $q = 1 - p = 1 - 0.066 = 0.934$. Since both $np = (400)(0.066) = 26.4 \geq 5$ and $nq = (400)(0.934) = 373.6 \geq 5$, the distribution of $\hat{p}$ is approximately normal.

$\mu_{\hat{p}} = p = 0.066$ and $\sigma_{\hat{p}} = \sqrt{\frac{pq}{n}} = \sqrt{\frac{(0.066)(0.934)}{400}} = \sqrt{0.00015411} \approx 0.01241$

(a) $P(\hat{p} < 0.066) = P\left(\frac{\hat{p}-\mu_{\hat{p}}}{\sigma_{\hat{p}}} < \frac{0.066-\mu_{\hat{p}}}{\sigma_{\hat{p}}}\right) = P\left(\frac{\hat{p}-0.066}{0.01241} < \frac{0.066-0.066}{0.01241}\right)$

$= P(Z < 0) = 0.5$

(b) $P(0.05 < \hat{p} < 0.066) = P\left(\frac{0.05-\mu_{\hat{p}}}{\sigma_{\hat{p}}} < \frac{\hat{p}-\mu_{\hat{p}}}{\sigma_{\hat{p}}} < \frac{0.066-\mu_{\hat{p}}}{\sigma_{\hat{p}}}\right) = P\left(\frac{0.05-0.066}{0.01241} < \frac{\hat{p}-0.066}{0.01241} < \frac{0.066-0.066}{0.01241}\right) \approx P(-1.29 < Z < 0) = 0.5 - 0.0985 = 0.4015$

(c) For the 2.5th percentile, $Z = -1.96$. Therefore, the 2.5th percentile is

$\hat{p} = Z \cdot \sigma_{\hat{p}} + \mu_{\hat{p}} \approx (-1.96)(0.01241) + 0.066 \approx 0.0417.$

For the 97.5th percentile, $Z = 1.96$. Therefore, the 97.5th percentile is

$\hat{p} = Z \cdot \sigma_{\hat{p}} + \mu_{\hat{p}} \approx (1.96)(0.01241) + 0.066 \approx 0.0903.$

Chapter 8: Confidence Intervals

Section 8.1

1. A range of values is more likely to contain μ than a point estimate is to be exactly equal to μ. We have no measure of confidence that our point estimate is close to μ. A confidence level for a confidence interval means that if we take sample after sample for a very long time, then in the long run, the percent of intervals that will contain the population mean μ will equal the confidence level.

3. We are 95% confident that the population mean football score lies between 15 and 25.

5. $\bar{x} \pm E$ is shorthand for writing the two values $\bar{x} - E$ and $\bar{x} + E$.

$\pm$ is shorthand notation for writing two numbers.

7. **(a)** As the confidence level increases, $Z_{\alpha/2}$ increases.

(b) Since the confidence level is $(1 - \alpha) \times 100\%$, as the confidence level increases, $1 - \alpha$ increases. Thus, α and $\alpha/2$ will decrease. Since $\alpha/2$ is the area underneath the standard normal curve to the right of $Z_{\alpha/2}$, a decrease in $\alpha/2$ will result in an increase in $Z_{\alpha/2}$. For a 95% confidence interval, $Z_{\alpha/2} = 1.96$.

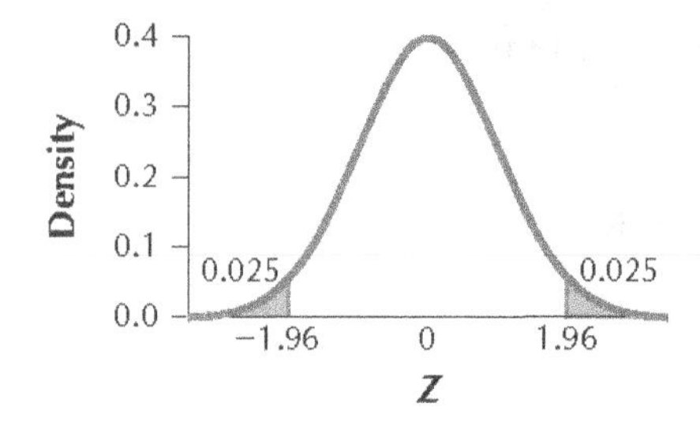

For a 99% confidence interval, $Z_{\alpha/2} = 2.576$.

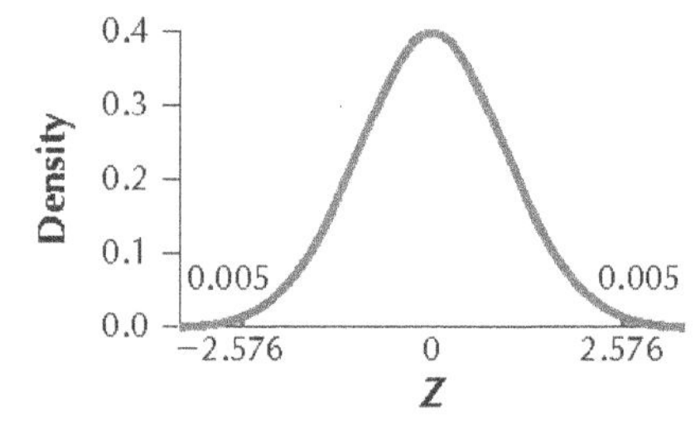

9. The formula for the sample size for estimating the population mean μ within a margin of error E with confidence $100(1 - \alpha)\%$ is given by $n = \left(\frac{(Z_{\alpha/2})\sigma}{E}\right)^2$.

As the confidence level is increased, $Z_{\alpha/2}$ increases, so n increases. As the confidence level is decreased, $Z_{\alpha/2}$ decreases, so n decreases.

11. The point estimate of the population mean μ is the sample mean $\bar{x} = \frac{\sum x}{n} = \frac{4+5+3+5+3}{5} = \frac{20}{5} = 4$.

13. The point estimate of the population mean μ is the sample mean $\bar{x} = \frac{\sum x}{n} = \frac{10+16+13+16+10}{5} = \frac{65}{5} = 13$.

15. Since σ is unknown, we may not use the Z interval.

17. Since the sample size is large ($n \geq 30$) and σ is known, we may use the Z interval.

19. Since the sample size is large ($n \geq 30$) and σ is known, we may use the Z interval.

21. $Z_{\alpha/2} = 2.576$

23. $Z_{\alpha/2} = 1.96$

25. $Z_{\alpha/2} = 1.645$

27. **(a)** $\frac{\sigma}{\sqrt{n}} = \frac{10}{\sqrt{16}} = 2.5$

(b) $Z_{\alpha/2} = 1.96$

(c) $\bar{x} \pm Z_{\alpha/2}\left(\frac{\sigma}{\sqrt{n}}\right) = 75 \pm 1.96\left(\frac{10}{\sqrt{16}}\right) = 75 \pm 4.9 = (70.1, 79.9)$. We are 95% confident that the population mean μ lies between 70.1 and 79.9.

29. **(a)** $\frac{\sigma}{\sqrt{n}} = \frac{8}{\sqrt{9}} \approx 2.6667$

(b) $Z_{\alpha/2} = 1.96$

(c) $\bar{x} \pm Z_{\alpha/2}\left(\frac{\sigma}{\sqrt{n}}\right) = 20 \pm 1.96\left(\frac{8}{\sqrt{9}}\right) \approx 20 \pm 5.2267 = (14.7733, 25.2267)$. We are 95% confident that the population mean μ lies between 14.7733 and 25.2267.

31. **(a)** $\frac{\sigma}{\sqrt{n}} = \frac{7}{\sqrt{49}} = 1$

(b) $Z_{\alpha/2} = 1.96$

(c) $\bar{x} \pm Z_{\alpha/2}\left(\frac{\sigma}{\sqrt{n}}\right) = 20 \pm 1.96\left(\frac{7}{\sqrt{49}}\right) = 20 \pm 1.96 = (18.04, 21.96)$. We are 95% confident that the population mean μ lies between 18.04 and 21.96.

33. **(a)** $\frac{\sigma}{\sqrt{n}} = \frac{2}{\sqrt{400}} = 0.1$

(b) $Z_{\alpha/2} = 1.96$

(c) $\bar{x} \pm Z_{\alpha/2}\left(\frac{\sigma}{\sqrt{n}}\right) = -5 \pm 1.96\left(\frac{2}{\sqrt{400}}\right) = -5 \pm 0.196 = (-5.196, -4.804)$. We are 95% confident that the population mean μ lies between –5.196 and –4.804.

35. **(a)** $E = Z_{\alpha/2}\left(\frac{\sigma}{\sqrt{n}}\right) = 1.96\left(\frac{10}{\sqrt{16}}\right) = 4.9$

(b) We can estimate the population mean μ to within 4.9 with 95% confidence.

37. **(a)** $E = Z_{\alpha/2}\left(\frac{\sigma}{\sqrt{n}}\right) = 1.96\left(\frac{8}{\sqrt{9}}\right) \approx 5.2267$

(b) We can estimate the population mean μ to within 5.2267 with 95% confidence.

39. **(a)** $E = Z_{\alpha/2}\left(\frac{\sigma}{\sqrt{n}}\right) = 1.96\left(\frac{7}{\sqrt{49}}\right) = 1.96$

(b) We can estimate the population mean μ to within 1.96 with 95% confidence.

41. **(a)** $E = Z_{\alpha/2}\left(\frac{\sigma}{\sqrt{n}}\right) = 1.96\left(\frac{2}{\sqrt{400}}\right) = 0.196$

(b) We can estimate the population mean μ to within 0.196 with 95% confidence.

43. **(a)** (95.1, 104.9)

(b) We are 95% confident that the population mean μ lies between 95.1 and 104.9.

(c) The margin of error $E = \frac{104.9-95.1}{2} = \frac{9.8}{2} = 4.9$.

(d) We can estimate the population mean μ to within 4.9 with 95% confidence.

45. **(a)** (2.6807, 2.8193)

(b) We are 95% confident that the population mean μ lies between 2.6807 and 2.8193.

(c) The margin of error $E = \frac{2.8193-2.6807}{2} = \frac{0.1386}{2} = 0.0693$.

(d) We can estimate the population mean μ to within 0.0693 with 95% confidence.

47. $n = \left[\frac{(Z_{\alpha/2})\sigma}{E}\right]^2 = \left[\frac{(1.96)(10)}{32}\right]^2 = 0.37515625$

Round n up to $n = 1$.

49. $n = \left[\frac{(Z_{\alpha/2})\sigma}{E}\right]^2 = \left[\frac{(1.96)(10)}{8}\right]^2 = 6.0025$

Round n up to $n = 7$.

51. $n = \left[\frac{(Z_{\alpha/2})\sigma}{E}\right]^2 = \left[\frac{(1.645)(10)}{10}\right]^2 = 2.706025$

Round n up to $n = 3$.

53. $n = \left[\frac{(Z_{\alpha/2})\sigma}{E}\right]^2 = \left[\frac{(2.576)(10)}{10}\right]^2 = 6.635776$

Round n up to $n = 7$.

55. **(a)** $\bar{x} \pm Z_{\alpha/2}\left(\frac{\sigma}{\sqrt{n}}\right) = 10 \pm 1.645\left(\frac{2}{\sqrt{25}}\right) = 10 \pm 0.658 = (9.342, 10.658)$

(b) $\bar{x} \pm Z_{\alpha/2}\left(\frac{\sigma}{\sqrt{n}}\right) = 10 \pm 1.96\left(\frac{2}{\sqrt{25}}\right) = 10 \pm 0.784 = (9.216, 10.784)$

(c) $\bar{x} \pm Z_{\alpha/2}\left(\frac{\sigma}{\sqrt{n}}\right) = 10 \pm 2.576\left(\frac{2}{\sqrt{25}}\right) = 10 \pm 1.0304 = (8.9696, 11.0304)$

(d) It increases.

57. **(a)** $\bar{x} = 69$ gallons

(b) $\frac{\sigma}{\sqrt{n}} = \frac{20}{\sqrt{30}} \approx 3.65$ gallons

(c) $Z_{\alpha/2} = 1.96$

(d) $n \geq 30$ and the population standard deviation σ is known.

(e) $\bar{x} \pm Z_{\alpha/2}\left(\frac{\sigma}{\sqrt{n}}\right) = 69 \pm 7.16 = (61.84, 76.16)$. We are 95% confident that the true mean amount of soda dispensed μ lies between 61.84 gallons and 76.16 gallons.

59. **(a)** $\bar{x} = 107$ seconds.

(b) $\frac{\sigma}{\sqrt{n}} = \frac{117}{\sqrt{36}}$ = 19.5 seconds

(c) $Z_{\alpha/2}$ = 1.96

(d) $n \geq 30$ and the population standard deviation σ is known.

(e) $\bar{x} \pm Z_{\alpha/2}\left(\frac{\sigma}{\sqrt{n}}\right) = 107 \pm 1.96\left(\frac{117}{\sqrt{36}}\right) = 107 \pm 38.22 = (68.78, 145.22)$.

We are 95% confident that the true mean length of time that boys remain engaged with a science exhibit at a museum μ lies between 68.78 seconds and 145.22 seconds.

61. **(a)** The margin of error is $E = Z_{\alpha/2}\left(\frac{\sigma}{\sqrt{n}}\right) = (1.96)\left(\frac{20}{\sqrt{36}}\right) \approx 7.16$ gallons. We may estimate μ, the mean amount of carbonated beverages consumed by all Americans per year, to within 7.16 gallons with 95% confidence.

(b) $n = \left(\frac{(Z_{\alpha/2})\sigma}{E}\right)^2 = \left(\frac{(1.96)(20)}{25}\right)^2 = 2.458624$. **Round n up to 3.**

(c) $n = \left(\frac{(Z_{\alpha/2})\sigma}{E}\right)^2 = \left(\frac{(1.96)(20)}{5}\right)^2 = 61.4656$. **Round n up to 62.**

63. **(a)** The margin of error is $= Z_{\alpha/2}\left(\frac{\sigma}{\sqrt{n}}\right) = (1.96)\left(\frac{117}{\sqrt{36}}\right) = 38.22$ seconds. We may estimate μ, the mean length of time that boys remain engaged with a science exhibit at a museum, to within 38.22 seconds with 95% confidence.

(b) $n = \left(\frac{(Z_{\alpha/2})\sigma}{E}\right)^2 = \left(\frac{(1.96)(117)}{30}\right)^2 = 58.430736$. **Round n up to 59.**

(c) $n = \left(\frac{(Z_{\alpha/2})\sigma}{E}\right)^2 = \left(\frac{(1.96)(117)}{3}\right)^2 = 5843.0736$. **Round n up to 5844.**

65. **(a)** (2.3040, 2.6960)

(b) We are 95% confident that the population mean μ lies between 2.3040 and 2.6960.

(c) $E = 0.1960$

(d) We can estimate the population mean μ to within 0.1960 with 95% confidence.

67. **(a)** (22.97, 23.22)

(b) We are 95% confident that the population mean μ lies between 22.97 and 23.22.

(c) The margin of error $E = \frac{23.22-22.97}{2} = \frac{0.25}{2} = 0.125$.

(d) We can estimate the population mean μ to within 0.125 with 95% confidence.

69. **(a)**

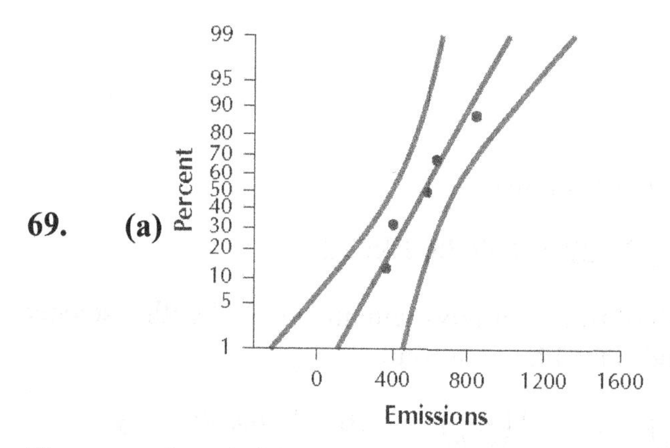

The normal probability plot indicates an acceptable level of normality.

(b) $\bar{x} = \frac{\sum x}{n} = \frac{361+844+398+577+631}{5} = \frac{2811}{5} = 562.2.$

Lower bound = $\bar{x} - Z_{\alpha/2}\left(\frac{\sigma}{\sqrt{n}}\right) = 562.2 - 1.645\left(\frac{200}{\sqrt{5}}\right) \approx 562.2 - 147.133 = 415.067.$

Upper bound = $\bar{x} - Z_{\alpha/2}\left(\frac{\sigma}{\sqrt{n}}\right) = 562.2 + 1.645\left(\frac{200}{\sqrt{5}}\right) \approx 562.2 + 147.133 = 709.333.$

Confidence interval: (415.067, 709.333). TI-83/84: (415.08, 709.32). We are 90% confident that the population mean carbon emissions lies between 415.067 (415.08) million tons and 709.333 (709.32) million tons.

(c) $E = Z_{\alpha/2}\left(\frac{\sigma}{\sqrt{n}}\right) = (1.645)\left(\frac{200}{\sqrt{5}}\right) \approx 147.133$ million tons. We can estimate the population mean emissions level of all nations to within 147.133 million tons with 90% confidence.

(d) $n = \left(\frac{(Z_{\alpha/2})\sigma}{E}\right)^2 = \left(\frac{(1.645)(200)}{50}\right)^2 = 43.2964.$ **Round n up to 44 nations.**

71. **(a)**

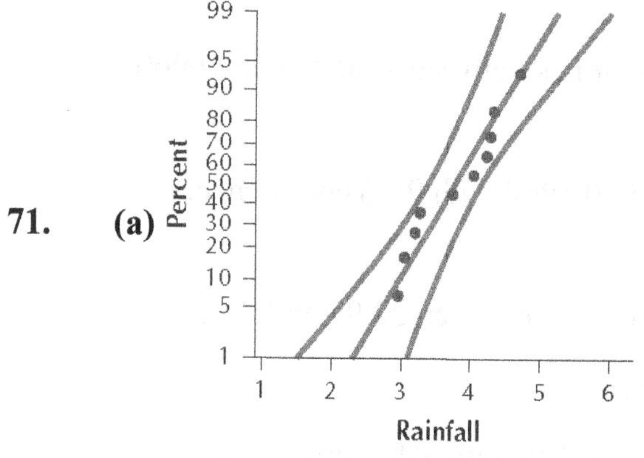

The normal probability plot indicates an acceptable level of normality.

(b) $\bar{x} = \frac{\sum x}{n} = \frac{4.72+4.31+2.96+4.06+3.20+4.25+3.03+4.36+3.75+3.26}{10} = \frac{37.9}{10} = 3.79$

Lower bound = $\bar{x} - Z_{\alpha/2}\left(\frac{\sigma}{\sqrt{n}}\right) = 3.79 - 1.96\left(\frac{0.64}{\sqrt{10}}\right) \approx 3.79 - 0.397 = 3.393.$

Upper bound = $\bar{x} + Z_{\alpha/2}\left(\frac{\sigma}{\sqrt{n}}\right) = 3.79 + 1.96\left(\frac{0.64}{\sqrt{10}}\right) \approx 3.79 + 0.397 = 4.187.$

Confidence interval: (3.393, 4.187). We are 95% confident that the population mean rainfall in

Georgia lies between 3.393 inches and 4.187 inches.

(c) $E = Z_{\alpha/2}\left(\frac{\sigma}{\sqrt{n}}\right) = (1.96)\left(\frac{0.64}{\sqrt{10}}\right) \approx 0.397$ inch. We can estimate the population mean rainfall in Georgia to within 0.397 inch with 95% confidence.

(d) $n = \left(\frac{(Z_{\alpha/2})\sigma}{E}\right)^2 = \left(\frac{(1.96)(0.64)}{0.1}\right)^2 = 157.351936$. **Round *n* up to 158 locations.**

73. **(a)** The margin of error is $E = Z_{\alpha/2}\left(\frac{\sigma}{\sqrt{n}}\right) = (1.96)\left(\frac{3}{\sqrt{30}}\right) \approx 1.07$ miles. We may estimate μ, the mean commuting distance, to within 1.07 miles with 95% confidence.

(b) $\bar{x} = \frac{\sum x}{n} = \frac{298}{30} \approx 9.93$ miles. $\bar{x} \pm Z_{\alpha/2}\left(\frac{\sigma}{\sqrt{n}}\right) \approx 9.93 \pm 1.96\left(\frac{3}{\sqrt{30}}\right) \approx 9.93 \pm 1.07 = (8.86, 11.00)$. We are 95% confident that the true mean commuting distance μ lies between 8.86 miles and 11.00 miles.

75. **(a)** $Z_{\alpha/2} = 1.96$.

(b) $\bar{x}$ is the midpoint of the confidence interval. Thus, $\bar{x} = (2.7152 + 9.1048)/2 = 5.91$.

(c) Since the length of the confidence interval is 2 E, the margin of error is one-half of the length of the confidence interval. Thus, $E = (9.1048 - 2.7152)/2 = 3.1948$. Or, since the confidence interval is $\bar{x} \pm E$, the distance between $\bar{x}$ and either end of the confidence interval is the margin of error. Thus, $E = 9.1048 - 5.91 = 5.91 - 2.7152 = 3.1948$. We may estimate the true mean quality of life score within 3.1948 with 95% confidence.

(d) $E = Z_{\alpha/2}\left(\frac{\sigma}{\sqrt{n}}\right) = (1.96)\left(\frac{\sigma}{\sqrt{36}}\right) = 1.96\left(\frac{\sigma}{6}\right) = 3.1948$, so $\sigma = (3.1948 \times 6)/1.96 = 9.78$.

77. **(a)** $n \geq 30$ and the population standard deviation σ is known.

(b) $\bar{x} \pm Z_{\alpha/2}\left(\frac{\sigma}{\sqrt{n}}\right) = 1 \pm 1.96\left(\frac{0.5}{\sqrt{100}}\right) = 1 \pm 0.098 = (0.902, 1.098)$

(c) We are 95% confident that μ, the population mean lead contamination for all trout on the Spokane River, lies between 0.902 ppm and 1.098 ppm.

79. **(a)** Since σ is a population characteristic, it stays constant and is unaffected by a decrease in confidence level.

(b) The quantity $\frac{\sigma}{\sqrt{n}}$ is unaffected by a decrease in confidence level.

(c) A decrease in the confidence level will result in a decrease in $Z_{\alpha/2}$. The width of the confidence interval is $2E = 2\,Z_{\alpha/2}\left(\frac{\sigma}{\sqrt{n}}\right)$. Thus, a decrease in $Z_{\alpha/2}$ will result in a decrease in the width of the confidence interval.

(d) The quantity $\bar{x}$ only depends on the sample taken, so it will remain unaffected by a decrease in confidence level.

(e) A decrease in the confidence level will result in a decrease in $Z_{\alpha/2}$. Since the margin of error is $E = Z_{\alpha/2}\left(\frac{\sigma}{\sqrt{n}}\right)$, a decrease in $Z_{\alpha/2}$ will result in a decrease in the margin of error.

81. The distribution is approximately normal and σ is known.

83. First calculate the sample mean:

$\bar{x} = \frac{\sum x}{n} = \frac{65+40+74+16+56+36+42+110}{8} = \frac{439}{8} = 54.875$ thousand game units.

$\bar{x} \pm Z_{\alpha/2}\left(\frac{\sigma}{\sqrt{n}}\right) = 54.875 \pm 2.576\left(\frac{30}{\sqrt{8}}\right) = 54.875 \pm 27.32260603 \approx (27.6, 82.2)$. We are 99% confident that μ, the population mean Wii game sales, lies between 27.6 thousand units and 82.2 thousand units.

85. $n = \left[\frac{(Z_{\alpha/2})\sigma}{E}\right]^2 = \left[\frac{(2.576)(30)}{5}\right]^2 = 238.887936.$

Round n up to $n = 239$ games.

87.

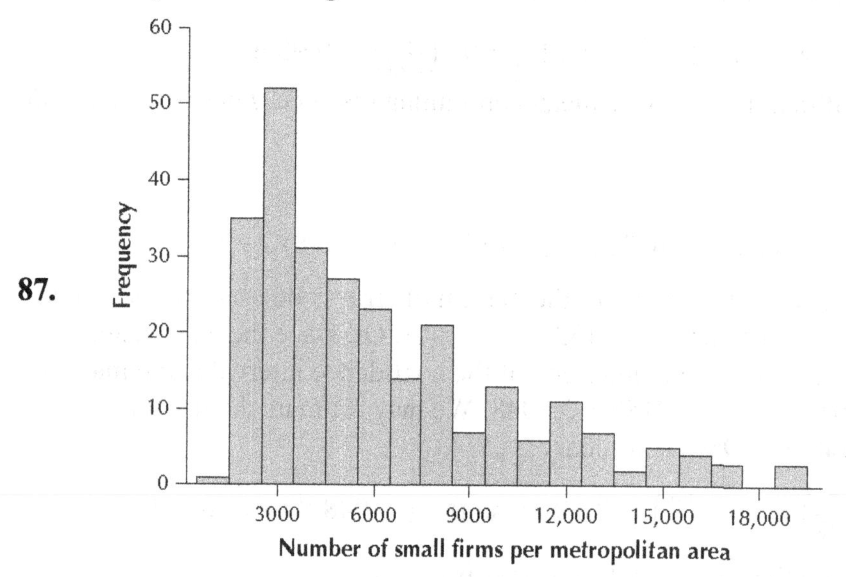

89. $\bar{x} \pm Z_{\alpha/2}\left(\frac{\sigma}{\sqrt{n}}\right) = 6199 \pm 1.96\left(\frac{25{,}000}{\sqrt{265}}\right) = 6199 \pm 3010 \approx (3189, 9209)$. We are 95% confident that the average number of small firms per metropolitan area lies between 3189 and 9209.

91. **(a)** Answers will vary but should be close to 90%. One possible answer:

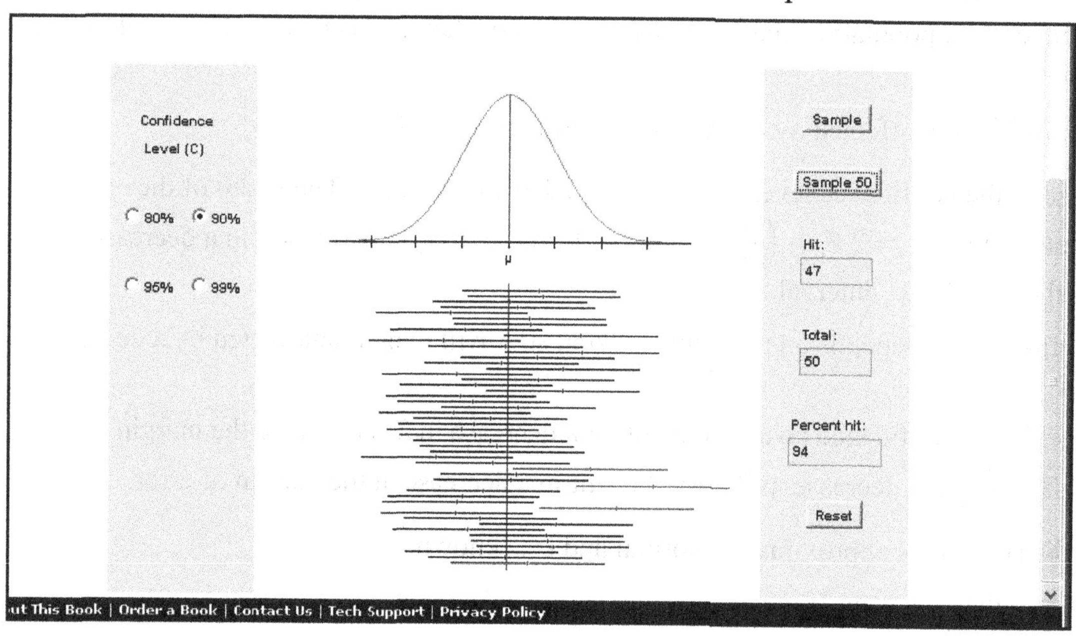

For this particular example, the percent hit is 94%.

(b) Answers will vary but should be close to 90%. One possible answer:

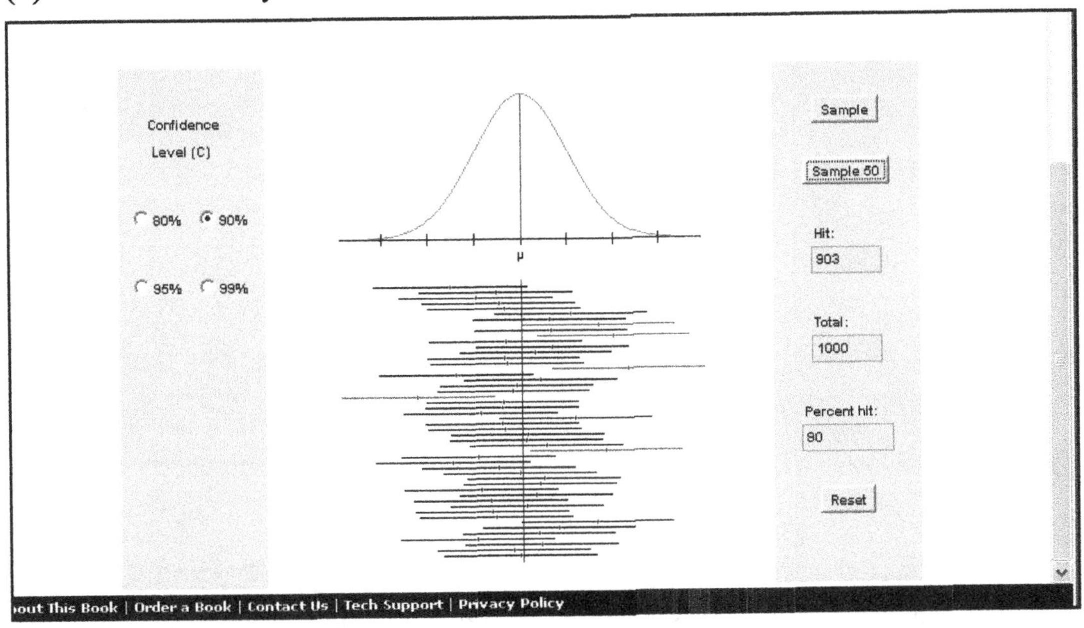

For this particular example, the percent hit is 90%.

(c) If we take sample after sample for a very long time, then in the long run, the proportion of intervals that will contain the parameter μ will equal 90%. One thousand confidence intervals is not enough to qualify for the long run.

93. Answers will vary.

95. Answers will vary.

97. 0.90

Section 8.2

1. In most real-world problems, the population standard deviation σ is unknown, so we may not use the Z interval.

3. As the sample size gets larger and larger, the t curve gets closer and closer to the Z curve.

5. **(a)** 90% confidence level, df $= n - 1 = 21 - 1 = 20,\ t_{\alpha/2} = 1.725$

(b) 95% confidence level, df $= n - 1 = 21 - 1 = 20,\ t_{\alpha/2} = 2.086$

(c) 99% confidence level, df $= n - 1 = 21 - 1 = 20,\ t_{\alpha/2} = 2.845$

7. **(a)** For a given sample size, the value of $t_{\alpha/2}$ increases as the confidence level increases.

(b) As the value of $t_{\alpha/2}$ increases, the confidence interval becomes wider. With a wider confidence interval you can be more confident that your confidence interval contains .

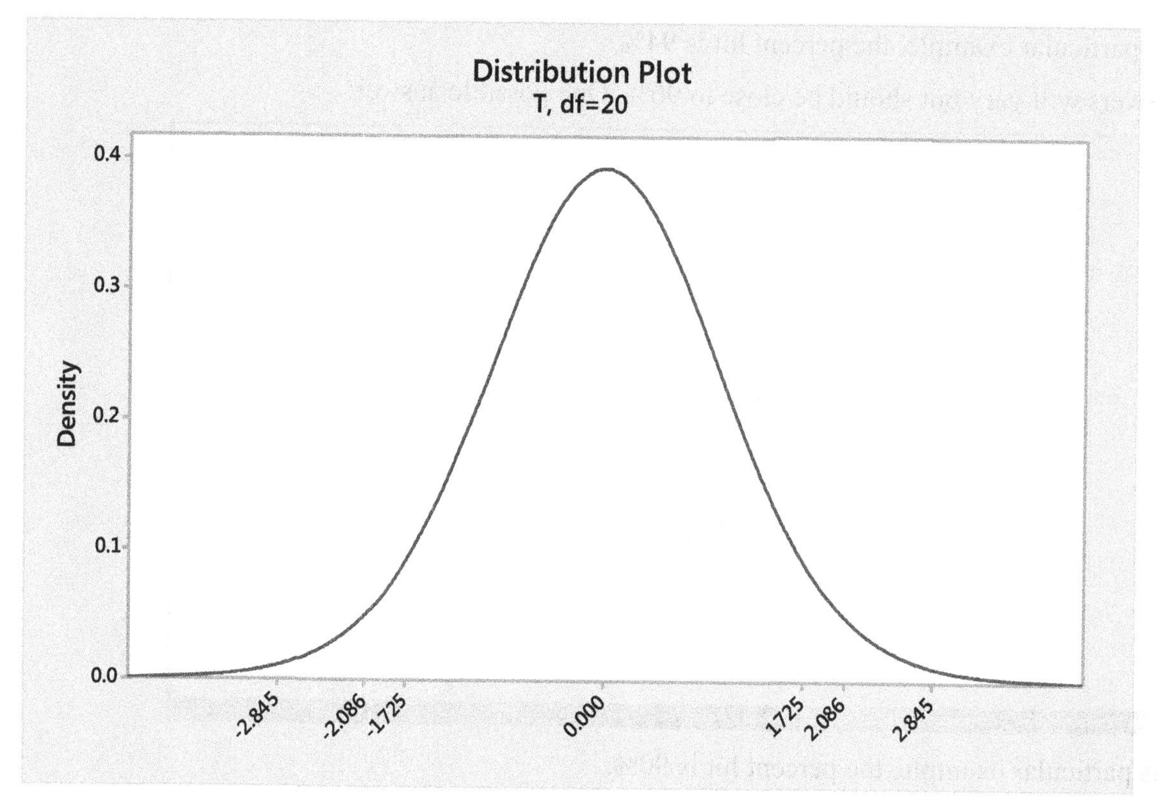

9. The sample size is not large (n is not ≥ 30) and we are not told that the population is normal. Therefore, the conditions are not met for the t interval for μ. It is not okay to construct the t interval.

11. We are not told that the population is normal. However, the sample size is large (n is ≥ 30). Therefore, the conditions are met for the t interval for μ. It is okay to construct the t interval.

13. **(a)** $\bar{x} = \frac{\sum x}{n} = \frac{4+5+3+5+3}{5} = \frac{20}{5} = 4$

$$s = \sqrt{\frac{\sum(x-\bar{x})^2}{n-1}} = \sqrt{\frac{(4-4)^2+(5-4)^2+(3-4)^2+(5-4)^2+(3-4)^2}{5-1}}$$

$$= \sqrt{\frac{(0)^2+(1)^2+(-1)^2+(1)^2+(-1)^2}{4}} = \sqrt{\frac{0+1+1+1+1}{4}} = \sqrt{\frac{4}{4}} = \sqrt{1} = 1$$

(b) 95% confidence interval, $\text{df} = n - 1 = 5 - 1 = 4$, $t_{\alpha/2} = 2.776$

(c) $\bar{x} \pm t_{\alpha/2}\left(\frac{s}{\sqrt{n}}\right) = 4 \pm 2.776\left(\frac{1}{\sqrt{5}}\right) \approx 4 \pm 1.241 = (2.759, 5.241)$. We are 95% confident that the population mean μ lies between 2.759 and 5.241.

15. **(a)** $\bar{x} = \frac{\sum x}{n} = \frac{10+16+13+16+10}{5} = \frac{65}{5} = 13$

$$s = \sqrt{\frac{\sum(x-\bar{x})^2}{n-1}} = \sqrt{\frac{(10-13)^2+(16-13)^2+(13-13)^2+(16-13)^2+(10-13)^2}{5-1}}$$

$$= \sqrt{\frac{(-3)^2+(3)^2+(0)^2+(3)^2+(-3)^2}{4}} = \sqrt{\frac{9+9+0+9+9}{4}} = \sqrt{\frac{36}{4}} = \sqrt{9} = 3$$

(b) 95% confidence interval, df = $n - 1 = 5 - 1 = 4,\ t_{\alpha/2}$=2.776

(c) $\bar{x} \pm t_{\alpha/2}\left(\frac{s}{\sqrt{n}}\right) = 13 \pm 2.776\left(\frac{3}{\sqrt{5}}\right) \approx 13 \pm 3.724$ = (9.276, 16.724). We are 95% confident that the population mean μ lies between 9.276 and 16.724.

17. **(a)** 90% confidence interval, df = $n - 1 = 16 - 1 = 15,\ t_{\alpha/2}$ = 1.753

(b) We are told that the population is normal.

(c) $\bar{x} \pm t_{\alpha/2}\left(\frac{s}{\sqrt{n}}\right) = 10 \pm 1.753\left(\frac{5}{\sqrt{16}}\right) \approx 10 \pm 2.2$ = (7.8, 12.2). We are 90% confident that the population mean μ lies between 7.8 and 12.2.

(d)

7 8 9 10 11 12 13

19. **(a)** 99% confidence interval, df = $n - 1 = 9 - 1 = 8,\ t_{\alpha/2}$= 3.355

(b) We are told that the population is normal.

(c) $\bar{x} \pm t_{\alpha/2}\left(\frac{s}{\sqrt{n}}\right) = 50 \pm 3.355\left(\frac{6}{\sqrt{9}}\right) \approx 50 \pm 6.7$ = (43.3, 56.7). We are 99% confident that the population mean μ lies between 43.3 and 56.7.

(d)

45 50 55

21. **(a)** 95% confidence interval, df = $n - 1 = 400 - 1 = 399$,

df = 399 is not in the table, so we take the next row with smaller df in the table. In this case, we take df = 100. Then $t_{\alpha/2}$=1.984.

(b) We are told that the population is normal.

(c) $\bar{x} \pm t_{\alpha/2}\left(\frac{s}{\sqrt{n}}\right) = -20 \pm 1.984\left(\frac{6}{\sqrt{400}}\right) \approx -20 \pm 0.6$ = (–20.6, –19.4). We are 95% confident that the population mean μ lies between –20.6 and –19.4.

(d)

20.5 20.0 19.5

23. 90% confidence interval, df = $n - 1 = 256 - 1 = 255$, df = 255 is not in the table, so we take the next row with smaller df in the table. In this case, we take df = 100. $t_{\alpha/2}$ =1.660

(b) We are not told that the population is normal. However, the sample size is large (n is ≥ 30).

(c) $\bar{x} \pm t_{\alpha/2}\left(\frac{s}{\sqrt{n}}\right) = 100 \pm 1.660\left(\frac{16}{\sqrt{256}}\right) \approx 100 \pm 1.7$ = (98.3, 101.7). We are 90% confident that the population mean μ lies between 98.3 and 101.7.

(d)

98 99 100 101 102

25. **(a)** 90% confidence interval, df = $n-1=225-1=224$, df = 224 is not in the table, so we take the next row with smaller df in the table. In this case, we take df = 100. $t_{\alpha/2}$=1.660

(b) We are not told that the population is normal. However, the sample size is large (n is $\geq$ 30).

(c) $\bar{x} \pm t_{\alpha/2}\left(\frac{s}{\sqrt{n}}\right) = 35 \pm 1.660\left(\frac{5}{\sqrt{225}}\right) \approx 35 \pm 0.6$ = (34.4, 35.6). We are 90% confident that the population mean μ lies between 34.4 and 35.6.

(d)

34.5 35.0 35.5

27. **(a)** 95% confidence interval, df = $n-1=64-1=63$, df = 63 is not in the table, so we take the next row with smaller df in the table. In this case, we take df = 60. $t_{\alpha/2}$ = 2.000

(b) We are not told that the population is normal. However, the sample size is large (n is $\geq$ 30).

(c) $\bar{x} \pm t_{\alpha/2}\left(\frac{s}{\sqrt{n}}\right) = -20 \pm 2.000\left(\frac{4}{\sqrt{64}}\right) \approx -20 \pm 1.0$ = (–21, –19). We are 95% confident that the population mean μ lies between –21 and –19.

(d)

21.0 20.5 20.0 19.5 19.0

29. $E = t_{\alpha/2}\left(\frac{s}{\sqrt{n}}\right) = 1.753\left(\frac{5}{\sqrt{16}}\right) \approx 2.2$. We can estimate the population mean μ to within 2.2 with 90% confidence.

31. $E = t_{\alpha/2}\left(\frac{s}{\sqrt{n}}\right) = 3.355\left(\frac{6}{\sqrt{9}}\right) \approx 6.7$. We can estimate the population mean μ to within 6.7 with 99% confidence.

33. $E = t_{\alpha/2}\left(\frac{s}{\sqrt{n}}\right) = 1.984\left(\frac{6}{\sqrt{400}}\right) \approx 0.6$. We can estimate the population mean μ to within 0.6 with 95% confidence.

35. $E = t_{\alpha/2}\left(\frac{s}{\sqrt{n}}\right) = 1.660\left(\frac{16}{\sqrt{256}}\right) \approx 1.7$. We can estimate the population mean μ to within 1.7 with 90% confidence.

37. $E = t_{\alpha/2}\left(\frac{s}{\sqrt{n}}\right) = 1.660\left(\frac{5}{\sqrt{225}}\right) \approx 0.6$. We can estimate the population mean μ to within 0.6 with 90% confidence.

39. $E = t_{\alpha/2}\left(\frac{s}{\sqrt{n}}\right) = 2.000\left(\frac{4}{\sqrt{64}}\right) \approx 1$. We can estimate the population mean μ to within 1 with 95% confidence.

41. **(a)** We are not told that the population is normal. However, the sample size is large (n is $\geq$ 30).

(b) 95% confidence interval, df = $n-1=31-1=30$, $t_{\alpha/2}$ = 2.042

(c) $\bar{x} \pm t_{\alpha/2}\left(\frac{s}{\sqrt{n}}\right) = 4.8 \pm 2.042\left(\frac{3}{\sqrt{31}}\right) \approx 4.8 \pm 1.1 = (3.7, 5.9)$; We are 95% confident that μ, the population mean length of stay in hospital for all heart attack victims, lies between 3.7 days and 5.9 days.

43. **(a)** We are not told that the population is normal. However, the sample size is large (n is ≥ 30).

(b) 95% confidence interval, df $= n - 1 = 100 - 1 = 99$, df = 99 is not in the table, so we take the next row with smaller df in the table. In this case, we take df = 90. $t_{\alpha/2} = 1.987$

(c) $\bar{x} \pm t_{\alpha/2}\left(\frac{s}{\sqrt{n}}\right) = 82.5 \pm 1.987\left(\frac{10}{\sqrt{100}}\right) \approx 82.5 \pm 2.0 = (80.5, 84.5)$. We are 95% confident that μ, the population mean consumer sentiment for all consumers, lies between 80.5 and 84.5.

45. **(a)** $E = t_{\alpha/2}\left(\frac{s}{\sqrt{n}}\right) = 2.042\left(\frac{3}{\sqrt{31}}\right) \approx 1.1$ days. We can estimate the population mean length of stay in hospital of all heart attack victims to within 1.1 days with 95% confidence.

(b) The new margin of error is $E = t_{\alpha/2}\left(\frac{s}{\sqrt{n}}\right) = 2.042\left(\frac{3}{\sqrt{400}}\right) \approx 0.31$ day. The increase in the sample size n results in a decrease in the margin of error E because n is in the denominator and $t_{\alpha/2}$, s, and n are all positive.

47. **(a)** $E = t_{\alpha/2}\left(\frac{s}{\sqrt{n}}\right) = 1.987\left(\frac{10}{\sqrt{100}}\right) \approx 2$. We can estimate the population mean consumer sentiment for all consumers to within 2 with 95% confidence.

(b) Decrease the confidence level or increase the sample size; increase the sample size. The only way to have both high confidence and a tight interval is to boost the sample size.

49. **(a)** (72.2, 77.8)

(b) We are 95% confident that μ, the population mean exam score for all exam takers, lies between 72.2 and 77.8.

(c) $E = \frac{77.8 - 72.2}{2} = \frac{5.6}{2} = 2.8$

(d) We can estimate the population mean exam score for all exam takers to within 2.8 with 95% confidence.

51. **(a)** (20.3, 23.3)

(b) We are 95% confident that μ, the population mean number of vegetable farms per county for all counties nationwide, lies between 20.3 vegetable farms per county and 23.3 vegetable farms per county.

(c) $E = \frac{23.3 - 20.3}{2} = \frac{3}{2} = 1.5$ vegetable farms per county

(d) We can estimate the population mean number of vegetable farms per county for all counties nationwide to within 1.5 vegetable farms per county with 95% confidence.

53. **(a)** (22,400, 25,630)

(b) We are 95% confident that μ, the population mean number of small businesses per city for all cities nationwide, lies between 22,400 small businesses per city and 25,630 small businesses per city.

(c) $E = \frac{25{,}630 - 22{,}400}{2} = \frac{3230}{2} = 1615$ small businesses per city

(d) We can estimate the population mean number of small businesses per city for all cities nationwide to within 1615 small businesses per city with 95% confidence.

55. **(a)** (6.08, 7.77)

(b) We are 90% confident that μ, the population mean amount of sugar for all breakfast cereals, lies between 6.08 grams and 7.77 grams.

(c) $E = \frac{7.77 - 6.08}{2} = \frac{1.69}{2} = 0.845$ gram

(d) We can estimate the population amount of sugar for all breakfast cereals to within 0.845 gram with 90% confidence.

57. **(a)** Acceptable normality. All points are between the curved lines and close to the center line.

(b) (320.31, 804.09). We are 95% confident that μ, the population mean carbon emissions for all nations, lies between 320.31 million tons and 804.09 million tons.

(c) $E = 241.89$ million tons. We can estimate the population mean carbon emissions for all nations to within 241.89 million tons with 95% confidence.

(d) Decrease the confidence level or increase the sample size; increase the sample size. The only way to have both high confidence and a tight interval is to boost the sample size.

59 **(a)** Acceptable normality. All points are between the curved lines and close to the center line.

(b) 95% confidence interval, df $= n - 1 = 8 - 1 = 7$, $t_{\alpha/2} = 2.365$

$\bar{x} = \frac{\sum x}{n} = \frac{65+40+74+16+56+36+42+110}{8} = \frac{439}{8} = 54.875$ thousand game units.

x	x^2
65	4,225
40	1,600
74	5,476
16	256
56	3,136
36	1,296
42	1,764
110	12,100
$\sum x = 439$	$\sum x^2 = 29{,}853$

$$s = \sqrt{\frac{\sum x^2 - (\sum x)^2/n}{n-1}} = \sqrt{\frac{29{,}853 - (439)^2/8}{8-1}} = \sqrt{\frac{29{,}853 - 192{,}721/8}{7}} = \sqrt{\frac{29{,}853 - 24{,}090.125}{7}}$$

$$= \sqrt{\frac{5762.875}{7}} = \sqrt{823.2678571} \approx 28.6926 \text{ thousand game units}$$

$\bar{x} \pm t_{\alpha/2}\left(\frac{s}{\sqrt{n}}\right) \approx 54.875 \pm 2.365\left(\frac{28.6926}{\sqrt{8}}\right) \approx 54.875 \pm 24 \approx (31, 79)$. We are 95% confident that μ, the population mean number of Wii game units sold per week for all weeks, lies between 31 thousand units and 79 thousand units.

(c) $E = t_{\alpha/2}\left(\frac{s}{\sqrt{n}}\right) \approx 2.365\left(\frac{28.6926}{\sqrt{8}}\right) \approx 24$ thousand units; We can estimate the population mean number of Wii game units sold per week for all weeks to within 24 thousand units with 95% confidence.

(d) Decrease the confidence level or increase the sample size; increase the sample size. The only way to have both high confidence and a tight interval is to boost the sample size.

61. **(a)**

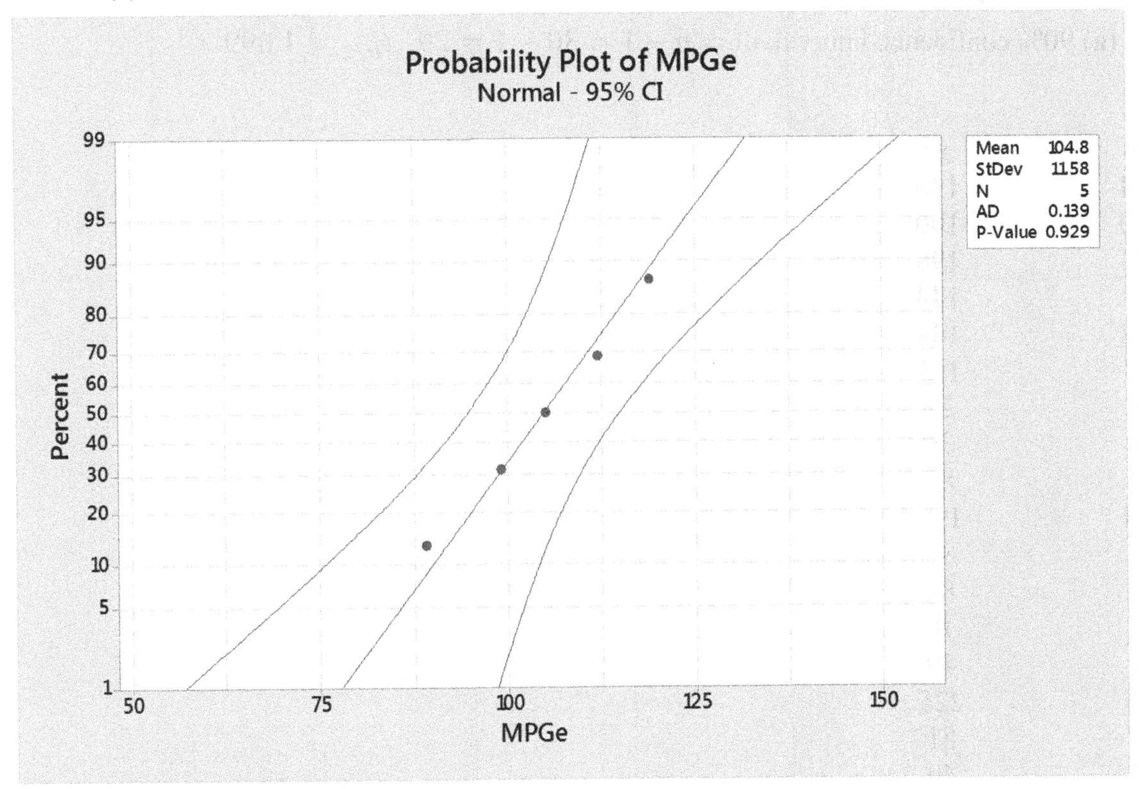

Acceptable normality. All points are between the curved lines and are close to the center line.

(b) 90% confidence interval, df $= n - 1 = 5 - 1 = 4$, $t_{\alpha/2} = 2.132$

(c)

x	x^2
89	7,921
99	9,801
105	11,025
112	12,544
119	14,161

$\sum x = 524 \quad \sum x^2 = 55{,}452$

$$s = \sqrt{\frac{\sum x^2 - (\sum x)^2/n}{n-1}} = \sqrt{\frac{55{,}452-(524)^2/5}{5-1}} = \sqrt{\frac{55{,}452-274{,}576/5}{4}} = \sqrt{\frac{55{,}452-54{,}915.2}{4}}$$

$$= \sqrt{\frac{536.8}{4}} = \sqrt{134.2} \approx 11.5845 \text{ MPGe}$$

$E = t_{\alpha/2}\left(\frac{s}{\sqrt{n}}\right) \approx 2.132\left(\frac{11.5845}{\sqrt{5}}\right) \approx 11.0$ MPGe. We can estimate the population mean mileage to within 11.0 MPGe with 90% confidence.

(d) $\bar{x} = \frac{\sum x}{n} = \frac{89+99+105+112+119}{5} = \frac{524}{5} = 104.8$ MPGe

$\bar{x} \pm t_{\alpha/2}\left(\frac{s}{\sqrt{n}}\right) \approx 104.8 \pm 2.132\left(\frac{11.5845}{\sqrt{5}}\right) \approx 104.8 \pm 11.0 \approx (93.8, 115.8)$. We are 90% confident that μ, the population mean mileage, lies between 93.8 MPGe and 115.8 MPGe.

63. **(a)** 90% confidence interval, df $= n - 1 = 30 - 1 = 29$, $t_{\alpha/2} = 1.699$

(b)

x	x^2
14	196
10	100
14	196
12	144
12	144
11	121
5	25
6	36
9	81
14	196
9	81
9	81
4	16
7	49
15	225
9	81
7	49
7	49
12	144
10	100
15	225
10	100
6	36
11	121
9	81
11	121
10	100
11	121
7	49
12	144

$\sum x = 298 \quad \sum x^2 = 3212$

$s = \sqrt{\frac{\sum x^2 - (\sum x)^2/n}{n-1}} = \sqrt{\frac{3212-(298)^2/30}{30-1}} = \sqrt{\frac{3212-88{,}804/30}{29}} = \sqrt{\frac{3212-2960.133333}{29}}$

$= \sqrt{\frac{251.866667}{29}} = \sqrt{8.685056483} \approx 2.9470$ miles

$E = t_{\alpha/2}\left(\frac{s}{\sqrt{n}}\right) \approx 1.699\left(\frac{2.9470}{\sqrt{30}}\right) \approx 0.9$ mile. We can estimate the population mean commuting distance to within 0.9 mile with 90% confidence.

(c) $\bar{x} = \frac{\sum x}{n} = \frac{298}{30} \approx 9.9$

$\bar{x} \pm t_{\alpha/2}\left(\frac{s}{\sqrt{n}}\right) \approx 9.9 \pm 1.699\left(\frac{2.9470}{\sqrt{30}}\right) \approx 9.9 \pm 0.9 \approx (9.0, 10.8)$. We are 90% confident that μ, the population mean commuting distance, lies between 9.0 miles and 10.8 miles.

65. The graph is symmetric about the middle value with the values with the highest frequency in the middle. This indicates that the normality assumption is valid. Since the normality assumption appears to be valid and σ is unknown, we may use the *t* interval.

Descriptive Statistics: Cigarettes

Variable	N	N*	Mean	SE Mean	StDev	Minimum	Q1	Median	Q3
Cigarettes	8	0	2392.2	97.1	274.6	2010.0	2138.5	2391.5	2654.0

Variable	Maximum
Cigarettes	2791.0

67. 90% confidence interval, df $= n - 1 = 8 - 1 = 7$, $t_{\alpha/2} = 1.895$,

$\bar{x} \pm t_{\alpha/2}\left(\frac{s}{\sqrt{n}}\right) \approx 2392.2 \pm 1.895\left(\frac{274.6}{\sqrt{8}}\right) \approx 2392 \pm 184 \approx (2208, 2576)$. We are 90% confident that μ, the population mean number of cigarettes smoked per capita, lies between 2208 cigarettes and 2576 cigarettes.

69.

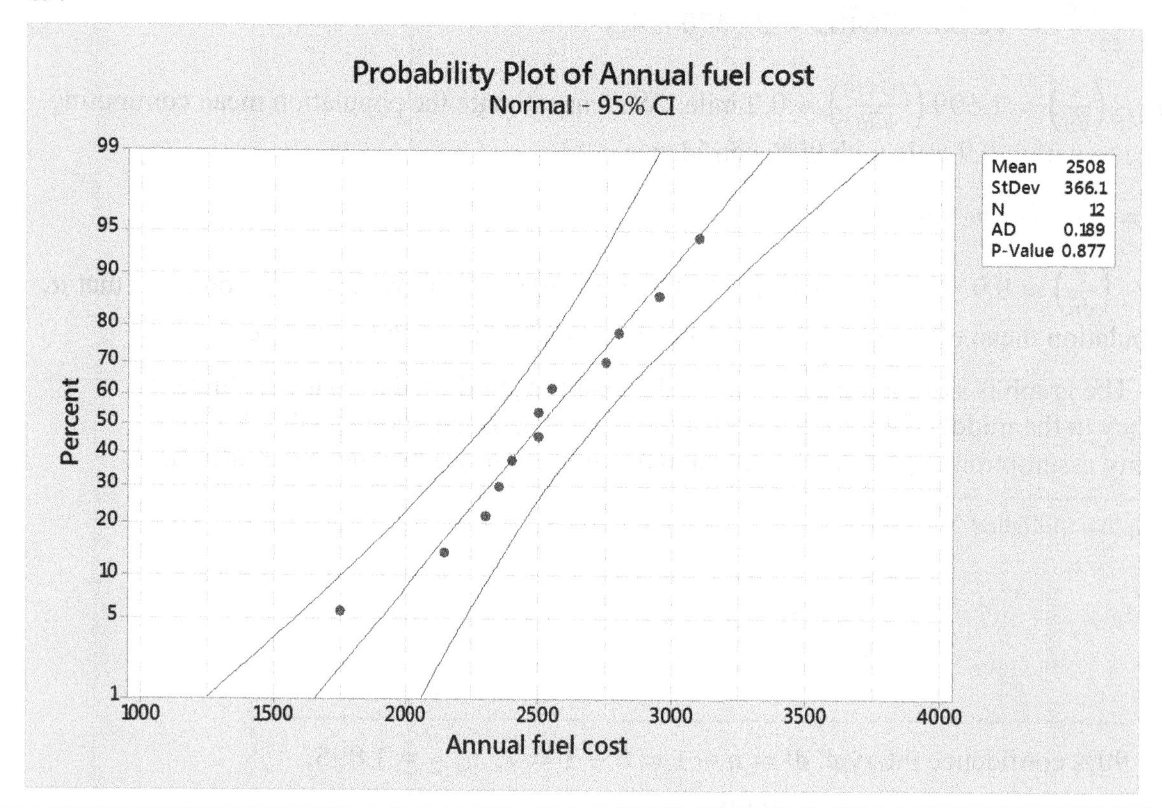

Acceptable normality. All of the points lie between the curved lines and most of the points lie near the center line.

71.

x	x^2
1750	3,062,500
2150	4,622,500
2550	6,502,500
2500	6,250,000
3100	9,610,000
2750	7,562,500
2400	5,760,000
2950	8,702,500
2300	5,290,000
2350	5,522,500
2500	6,250,000
2800	7,840,000

$\sum x = 30{,}100 \quad \sum x^2 = 76{,}975{,}000$

$$s = \sqrt{\frac{\sum x^2 - (\sum x)^2/n}{n-1}} = \sqrt{\frac{76{,}975{,}000 - (30{,}100)^2/12}{12-1}} = \sqrt{\frac{76{,}975{,}000 - 906{,}010{,}000/12}{11}} =$$

$$\sqrt{\frac{76{,}975{,}000 - 75{,}500{,}833.33}{11}} = \sqrt{\frac{1{,}474{,}166.67}{11}} = \sqrt{134{,}015.1518} \approx \$366.1$$

73. $\bar{x} \pm t_{\alpha/2}\left(\frac{s}{\sqrt{n}}\right) \approx 2508 \pm 1.796\left(\frac{366.1}{\sqrt{12}}\right) \approx 2508 \pm 189.8 \approx$ (2318.2, 2697.8). We are 90% confident that μ, the population mean annual fuel cost, lies between \$2318.2 and \$2697.8.

75. Answers will vary.

77. Answers will vary.

79. Answers will vary.

Section 8.3

1. Since the purpose of a confidence interval for p is to estimate p, it would not make sense to construct a confidence interval for p if p is known unless there is some reason to suspect that the value of p has changed.

3. $\hat{p} = \frac{x}{n} = \frac{90}{250} = 0.36$

5. $\hat{p} = \frac{x}{n} = \frac{90}{154} \approx 0.5844$

7. **(a)** $Z_{\alpha/2} = 1.96$

(b) $\hat{p} = \frac{x}{n} = \frac{90}{250} = 0.36, \hat{q} = 1 - \hat{p} = 1 - 0.36 = 0.64$

Both $n \cdot \hat{p} = (250)(0.36) = 90 \geq 5$ and $n \cdot \hat{q} = (250)(0.64) = 160 \geq 5$.

Therefore, the conditions for constructing a confidence interval for p are met.

(c) $\hat{p} \pm Z_{\alpha/2}\sqrt{\frac{\hat{p}\cdot\hat{q}}{n}} = 0.36 \pm 1.96\sqrt{\frac{(0.36)(0.64)}{250}} \approx 0.36 \pm 0.0595 = (0.3005, 0.4195)$

(d)

0.30 0.35 0.40

9. **(a)** $Z_{\alpha/2} = 1.96$

(b) $\hat{q} = \frac{n-x}{n} = \frac{4}{100} = 0.04, \hat{p} = 1 - \hat{q} = 1 - 0.04 = 0.96$

$n \cdot \hat{p} = (100)(0.96) = 96 \geq 5$ but $n \cdot \hat{q} = (100)(0.04) = 4$ is not ≥ 5.

Therefore, the conditions for constructing a confidence interval for p are not met.

(c) Can't do.

(d) Can't do.

11. **(a)** $Z_{\alpha/2} = 1.96$

(b) $\hat{p} = \frac{x}{n} = \frac{4}{154} \approx 0.0260, \hat{q} = 1 - \hat{p} \approx 1 - 0.0260 = 0.9740$

$n \cdot \hat{p} \approx (154)(0.0260) = 4.004$ is not ≥ 5. Therefore, the conditions for constructing a confidence interval for p are not met.

(c) Can't do.

(d) Can't do.

13. **(a)** $Z_{\alpha/2} = 1.96$

(b) $\hat{p} = \frac{x}{n} = \frac{142}{232} \approx 0.6121, \hat{q} = 1 - \hat{p} \approx 1 - 0.6121 = 0.3879$

Both $n \cdot \hat{p} \approx (232)(0.6121) = 142.0072 \geq 5$ and $n \cdot \hat{q} \approx (232)(0.3879) = 89.9928 \geq 5$.

Therefore, the conditions for constructing a confidence interval for p are met.

(c) $\hat{p} \pm Z_{\alpha/2}\sqrt{\frac{\hat{p} \cdot \hat{q}}{n}} \approx 0.6121 \pm 1.96\sqrt{\frac{(0.6121)(0.3879)}{232}} \approx 0.6121 \pm 0.0627 = (0.5494, 0.6748)$

(d) 0.55 0.60 0.65 0.70

15. **(a)** $Z_{\alpha/2} = 1.645$

(b) $\hat{p} = 0.4, \hat{q} = 1 - \hat{p} = 1 - 0.4 = 0.6$

Both $n \cdot \hat{p} = (100)(0.4) = 40 \geq 5$ and $n \cdot \hat{q} = (100)(0.6) = 60 \geq 5$.

Therefore, the conditions for constructing a confidence interval for p are met.

(c) $\hat{p} \pm Z_{\alpha/2}\sqrt{\frac{\hat{p} \cdot \hat{q}}{n}} = 0.4 \pm 1.645\sqrt{\frac{(0.4)(0.6)}{100}} \approx 0.4 \pm 0.0806 = (0.3194, 0.4806)$

(d) 0.30 0.35 0.40 0.45 0.50

17. **(a)** $Z_{\alpha/2} = 1.645$

(b) $\hat{p} = 0.2, \hat{q} = 1 - \hat{p} = 1 - 0.2 = 0.8$

Both $n \cdot \hat{p} = (100)(0.2) = 20 \geq 5$ and $n \cdot \hat{q} = (100)(0.8) = 80 \geq 5$.

Therefore, the conditions for constructing a confidence interval for p are met.

(c) $\hat{p} \pm Z_{\alpha/2}\sqrt{\frac{\hat{p} \cdot \hat{q}}{n}} = 0.2 \pm 1.645\sqrt{\frac{(0.2)(0.8)}{100}} = 0.2 \pm 0.0658 = (0.1342, 0.2658)$

(d) 0.15 0.20 0.25

19. **(a)** $Z_{\alpha/2} = 1.96$

(b) $\hat{p} = 0.8, \hat{q} = 1 - \hat{p} = 1 - 0.8 = 0.2$

Both $n \cdot \hat{p} = (25)(0.8) = 20 \geq 5$ and $n \cdot \hat{q} = (25)(0.2) = 5 \geq 5$.

Therefore, the conditions for constructing a confidence interval for p are met.

(c) $\hat{p} \pm Z_{\alpha/2}\sqrt{\frac{\hat{p} \cdot \hat{q}}{n}} = 0.8 \pm 1.96\sqrt{\frac{(0.8)(0.2)}{25}} = 0.8 \pm 0.1568 = (0.6432, 0.9568)$

(d) 0.6 0.7 0.8 0.9 1.0

21. $E = Z_{\alpha/2}\sqrt{\frac{\hat{p} \cdot \hat{q}}{n}} = 1.645\sqrt{\frac{(0.4)(0.6)}{100}} \approx 0.0806$

23. $E = Z_{\alpha/2}\sqrt{\frac{\hat{p}\cdot\hat{q}}{n}} = 1.645\sqrt{\frac{(0.2)(0.8)}{100}} \approx 0.0658$

25. $E = Z_{\alpha/2}\sqrt{\frac{\hat{p}\cdot\hat{q}}{n}} = 1.645\sqrt{\frac{(0.8)(0.2)}{25}} = 0.1316$

27. $E = Z_{\alpha/2}\sqrt{\frac{\hat{p}\cdot\hat{q}}{n}} = 2.576\sqrt{\frac{(0.8)(0.2)}{25}} \approx 0.2061$

29. **(a)** $Z_{\alpha/2} = 1.96$. $\hat{p} = \frac{x}{n} = \frac{5}{10} = 0.5$. $\hat{q} = 1 - \hat{q} = 1 - 0.5 = 0.5$.

$E = Z_{\alpha/2}\sqrt{\frac{\hat{p}\hat{q}}{n}} = 1.96\sqrt{\frac{(0.5)(0.5)}{10}} \approx 0.3099$.

(b) $Z_{\alpha/2} = 1.96$. $\hat{p} = \frac{x}{n} = \frac{50}{100} = 0.5$. $\hat{q} = 1 - \hat{q} = 1 - 0.5 = 0.5$.

$E = Z_{\alpha/2}\sqrt{\frac{\hat{p}\hat{q}}{n}} = 1.96\sqrt{\frac{(0.5)(0.5)}{100}} \approx 0.098$.

(c) $Z_{\alpha/2} = 1.96$. $\hat{p} = \frac{x}{n} = \frac{500}{1000} = 0.5$. $\hat{q} = 1 - \hat{q} = 1 - 0.5 = 0.5$.

$E = Z_{\alpha/2}\sqrt{\frac{\hat{p}\hat{q}}{n}} = 1.96\sqrt{\frac{(0.5)(0.5)}{1000}} \approx 0.0310$.

(d) $Z_{\alpha/2} = 1.96$. $\hat{p} = \frac{x}{n} = \frac{5{,}000}{10{,}000} = 0.5$. $\hat{q} = 1 - \hat{q} = 1 - 0.5 = 0.5$.

$E = Z_{\alpha/2}\sqrt{\frac{\hat{p}\hat{q}}{n}} = 1.96\sqrt{\frac{(0.5)(0.5)}{10{,}000}} \approx 0.0098$.

31. **(a)** Since the margin of error is $E = Z_{\alpha/2}\sqrt{\frac{\hat{p}\hat{q}}{n}}$, an increase in the sample size while $\hat{p}$ remains constant results in a decrease in the margin of error.

(b) Since the width of the confidence interval is 2 E, an increase in the sample size while $\hat{p}$ remains constant results in a decrease in the width of the confidence interval.

33. $n = \hat{p}\hat{q}\left(\frac{Z_{\alpha/2}}{E}\right)^2 = (0.3)(0.7)\left(\frac{1.96}{0.03}\right)^2 = 896.3733333$. Round n up to 897.

35. $n = \hat{p}\hat{q}\left(\frac{Z_{\alpha/2}}{E}\right)^2 = (0.1)(0.9)\left(\frac{1.96}{0.03}\right)^2 = 384.16$. Round n up to 385.

37. $n = \hat{p}\hat{q}\left(\frac{Z_{\alpha/2}}{E}\right)^2 = (0.001)(0.999)\left(\frac{1.96}{0.03}\right)^2 = 4.264176$. Round n up to 5.

39. $n = \left(\frac{0.5\cdot Z_{\alpha/2}}{E}\right)^2 = \left(\frac{(0.5)(1.645)}{0.03}\right)^2 = 751.6736111$. Round n up to 752.

41. $n = \left(\frac{0.5\cdot Z_{\alpha/2}}{E}\right)^2 = \left(\frac{(0.5)(2.576)}{0.03}\right)^2 = 1843.271111$. Round n up to 1844.

43. $n = \left(\frac{0.5\cdot Z_{\alpha/2}}{E}\right)^2 = \left(\frac{(0.5)(1.96)}{0.0075}\right)^2 = 17073.7778$. Round n up to 17,074.

45. As the confidence level increases, the required sample size increases as well.

47. **(a)** $Z_{\alpha/2} = 1.96$

(b) $\hat{p} = \frac{x}{n} = \frac{183}{830} \approx 0.2205, \hat{q} = 1 - \hat{p} \approx 1 - 0.2205 = 0.7795$

Both $n \cdot \hat{p} \approx (830)(0.2205) = 183.015 \geq 5$ and $n \cdot \hat{q} \approx (830)(0.7795) = 646.985 \geq 5$.

Therefore, the conditions for constructing a confidence interval for p are met.

(c) $\hat{p} \pm Z_{\alpha/2}\sqrt{\frac{\hat{p} \cdot \hat{q}}{n}} \approx 0.2205 \pm 1.96\sqrt{\frac{(0.2205)(0.7795)}{830}} \approx 0.2205 \pm 0.0282 = (0.1923, 0.2487)$.
We are 95% confident that p, the population proportion of millennials who are married, lies between 0.1923 and 0.2487.

0.19 0.20 0.21 0.22 0.23 0.24 0.25 0.26

49. **(a)** $Z_{\alpha/2} = 2.576$

(b) $\hat{p} = 0.83, \hat{q} = 1 - \hat{p} = 1 - 0.83 = 0.17$

Both $n \cdot \hat{p} = (100)(0.83) = 83 \geq 5$ and $n \cdot \hat{q} = (100)(0.17) = 17 \geq 5$.

Therefore, the conditions for constructing a confidence interval for p are met.

(c) $\hat{p} \pm Z_{\alpha/2}\sqrt{\frac{\hat{p} \cdot \hat{q}}{n}} = 0.83 \pm 2.576\sqrt{\frac{(0.83)(0.17)}{100}} \approx 0.83 \pm 0.0968 = (0.7332, 0.9268)$. We are 99% confident that p, the population proportion of college females who agree that heavier drinking occurs on spring break trips than is typically found on campus, lies between 0.7332 and 0.9268.

0.70 0.75 0.80 0.85 0.90 0.95

51. **(a)** $E = Z_{\alpha/2}\sqrt{\frac{\hat{p} \cdot \hat{q}}{n}} \approx 1.96\sqrt{\frac{(0.2205)(0.7795)}{830}} \approx 0.0282$

(b) We can estimate the population proportion of millennials who are married to within 0.0282 with 95% confidence.

53. **(a)** $E = Z_{\alpha/2}\sqrt{\frac{\hat{p} \cdot \hat{q}}{n}} = 2.576\sqrt{\frac{(0.83)(0.17)}{100}} \approx 0.0968$

(b) We can estimate the population proportion of college females who agree that heavier drinking occurs on spring break trips than is typically found on campus to within 0.0968 with 99% confidence.

55. **(a)** (0.61018, 0.78982)

(b) We are 95% confident that p, the population proportion of times the weather forecaster correctly predicted rain, lies between 0.61018 and 0.78982.

(c) The margin of error is $E = \frac{0.78982 - 0.61018}{2} = \frac{0.19764}{2} = 0.08982$.

(d) We can estimate the population proportion of times the weather forecaster correctly predicted rain to within 0.08982 with 95% confidence.

57. **(a)** (0.0383, 0.0497)

(b) We are 95% confident that p, the population proportion of clothing store purchases made online, lies between 0.0383 and 0.0497.

(c) The margin of error is $E = \frac{0.0497-0.0383}{2} = \frac{0.0114}{2} = 0.08982.$

(d) We can estimate the population proportion of clothing store purchases made online to within 0.0057 with 95% confidence.

59. **(a)** $Z_{\alpha/2} = 2.576, \hat{p} = 0.655, \hat{q} = 1 - \hat{p} = 1 - 0.655 = 0.345$. The margin of error is $E = Z_{\alpha/2}\sqrt{\frac{\hat{p}\cdot\hat{q}}{n}} = 2.576\sqrt{\frac{(0.655)(0.345)}{1000}} \approx 0.0387$. We can estimate the population proportion of Hawaii residents that are thriving to within 0.0387 with 99% confidence.

(b) $\hat{p} \pm Z_{\alpha/2}\sqrt{\frac{\hat{p}\cdot\hat{q}}{n}} = 0.655 \pm 2.576\sqrt{\frac{(0.655)(0.345)}{1000}} \approx 0.655 \pm 0.0387 = (0.6163, 0.6937)$. We are 99% confident that p, the population proportion of Hawaii residents that are thriving, lies between 0.6163 and 0.6937.

61. We have $\hat{p} = \frac{x}{n} = \frac{65}{100} = 0.65$, so $n\hat{p} = 100(0.65) = 65 \geq 5$ and $n\hat{q} = 100(0.35) = 35 \geq 5$. Thus, we may use the *Z* interval for *p*.

$$\hat{p} \pm Z_{\alpha/2}\sqrt{\frac{\hat{p}\hat{q}}{n}} = 0.65 \pm 1.96\sqrt{\frac{(0.65)(0.35)}{100}} \approx 0.65 \pm 0.0935 = (0.5565, 0.7435).$$

63. **(a)** Since the margin of error is $E = Z_{\alpha/2}\sqrt{\frac{\hat{p}\cdot\hat{q}}{n}}$ and *n* is in the denominator, an increase in sample size will result in a decrease in the margin of error.

(b) Since $Z_{\alpha/2}$ does not depend on the sample size, an increase in the sample size will leave $Z_{\alpha/2}$ unchanged.

(c) Since the width of the confidence interval is 2*E*, an increase in the sample size will result in a decrease in the margin of error, which will result in a decrease in the width of the confidence interval.

65. $\hat{p} = \frac{X}{n} = \frac{25}{54} \approx 0.4630$

67. $Z_{\alpha/2} = 1.96$

69. 0.4630 ± 0.1330

71. $n = \hat{p}\cdot\hat{q}\left(\frac{Z_{\alpha/2}}{E}\right)^2 \approx (0.4630)(0.5370)\left(\frac{1.96}{0.1330}\right)^2 = 53.9963169$

If *n* is not a whole number, we round *n* up to the next integer. Therefore, round *n* up to 54.

Same margin of error and same confidence level yields the same sample size.

73. We have $\hat{p} = \frac{x}{n} = \frac{39}{40} = 0.975$. Thus $n\hat{p} = 40(0.975) = 39 \geq 5$, but $n\hat{q} = 40(0.025) = 1$, which is not ≥ 5. Thus, we may not use the *Z* interval for *p*.

75. **(a)** A decrease in the confidence level will result in a decrease in $Z_{\alpha/2}$ from 1.96 to 1.645.

(b) Since the margin of error is $E = Z_{\alpha/2}\sqrt{\frac{\hat{p}\cdot\hat{q}}{n}}$, a decrease in the confidence level will result in a decrease in $Z_{\alpha/2}$, which will result in a decrease in the margin of error from 0.0748 to 0.0628.

(c) Since the width of the confidence interval is $2E$, a decrease in the confidence level will result in a decrease in $Z_{\alpha/2}$, which will result in a decrease in the width of the confidence interval from 0.1496 to 0.1256.

77. Answers will vary.

79. Answers will vary.

81. Answers will vary.

83. Answers will vary.

85. 0.1770. Answers will vary.

Section 8.4

1. In order to construct a confidence interval for σ^2 or σ, the population must be normal.

3. We can't just use the "point estimate ± margin of error" method used earlier in this chapter because, in order to use this method, the distribution has to be symmetric and the X^2 curve is not symmetric.

5. False. The X^2 curve is *not* symmetric.

7. True.

9. df = $n - 1 = 50 - 1 = 49$.

df = 49 is not in the χ^2 table, so take the next row with smaller df in the χ^2 table. Therefore, we will use df = 40.

For a 90% confidence interval, $1 - \alpha = 0.90$, $\alpha = 1 - 0.90 = 0.10$, $\alpha/2 = 0.10/2 = 0.05$, and $1 - \alpha/2 = 1 - 0.05 = 0.95$.

$\chi^2_{0.95} = 26.509$ and $\chi^2_{0.05} = 55.758$.

[Using Minitab: $\chi^2_{0.95} = 33.9303$ and $\chi^2_{0.05} = 66.3386$]

11. df = $n - 1 = 50 - 1 = 49$.

df = 49 is not in the χ^2 table, so take the next row with smaller df in the χ^2 table. Therefore, we will use df = 40.

For a 99% confidence interval, $1 - \alpha = 0.99$, $\alpha = 1 - 0.99 = 0.01$, $\alpha/2 = 0.01/2 = 0.005$, and $1 - \alpha/2 = 1 - 0.005 = 0.995$.

$\chi^2_{0.995} = 20.707$ and $\chi^2_{0.005} = 66.766$.

[Using Minitab: $\chi^2_{0.995} = 27.2493$ and $\chi^2_{0.005} = 78.230$]

13. df = $n - 1 = 25 - 1 = 24$.

For a 95% confidence interval, $1 - \alpha = 0.95$, $\alpha = 1 - 0.95 = 0.05$, $\alpha/2 = 0.05/2 = 0.025$, and $1 - \alpha/2 = 1 - 0.025 = 0.975$.

$\chi^2_{0.975} = 12.401$ and $\chi^2_{0.025} = 39.364$.

15. $\chi^2_{1-\alpha/2}$ decreases and $\chi^2_{\alpha/2}$ increases.

17. df = $n - 1 = 100 - 1 = 99.$

df = 99 is not in the χ^2 table, so take the next row with smaller df in the χ^2 table. Therefore, we will use df = 90.

For a 90% confidence interval, $1 - \alpha = 0.90$, $\alpha = 1 - 0.90 = 0.10$, $\alpha/2 = 0.10/2 = 0.05$, and $1 - \alpha/2 = 1 - 0.05 = 0.95$.

$\chi^2_{0.95} = 69.126$ and $\chi^2_{0.05} = 113.145$.

Lower bound $= \frac{(n-1)s^2}{\chi^2_{0.05}} = \frac{(100-1)25}{113.145} \approx 21.87$

Upper bound $= \frac{(n-1)s^2}{\chi^2_{0.95}} = \frac{(100-1)25}{69.126} \approx 35.80$

(21.87, 35.80) [Using Minitab: (20.1, 32.1)]

19. df = $n - 1 = 100 - 1 = 99.$

df = 99 is not in the χ^2 table, so take the next row with smaller df in the χ^2 table. Therefore, we will use df = 90.

For a 99% confidence interval, $1 - \alpha = 0.99$, $\alpha = 1 - 0.99 = 0.01$, $\alpha/2 = 0.01/2 = 0.005$, and $1 - \alpha/2 = 1 - 0.005 = 0.995$.

$\chi^2_{0.995} = 59.196$ and $\chi^2_{0.005} = 128.299$.

Lower bound $= \frac{(n-1)s^2}{\chi^2_{0.005}} = \frac{(100-1)25}{128.299} \approx 19.29$

Upper bound $= \frac{(n-1)s^2}{\chi^2_{0.995}} = \frac{(100-1)25}{59.196} \approx 41.81$

(19.29, 41.81) [Using Minitab: (17.8, 37.2)]

21. df = $n - 1 = 100 - 1 = 99.$

df = 99 is not in the χ^2 table, so take the next row with smaller df in the χ^2 table. Therefore, we will use df = 90.

For a 95% confidence interval, $1 - \alpha = 0.95$, $\alpha = 1 - 0.95 = 0.05$, $\alpha/2 = 0.05/2 = 0.025$, and $1 - \alpha/2 = 1 - 0.025 = 0.975$.

$\chi^2_{0.975} = 65.647$ and $\chi^2_{0.025} = 118.136$.

Lower bound $= \sqrt{\frac{(n-1)s^2}{\chi^2_{0.025}}} = \sqrt{\frac{(100-1)25}{118.136}} \approx 4.58$

Upper bound $= \sqrt{\frac{(n-1)s^2}{\chi^2_{0.95}}} = \sqrt{\frac{(100-1)25}{65.647}} \approx 6.14$

(4.58, 6.14) [Using Minitab: (4.39, 5.81)]

23. Lower bound decreases while the upper bound increases.

25. df = $n - 1 = 30 - 1 = 29.$

For a 95% confidence interval, $1 - \alpha = 0.95$, $\alpha = 1 - 0.95 = 0.05$, $\alpha/2 = 0.05/2 = 0.025$, and $1 - \alpha/2 = 1 - 0.025 = 0.975$.

$\chi^2_{0.975} = 16.047$ and $\chi^2_{0.025} = 45.722$.

Lower bound $= \frac{(n-1)s^2}{\chi^2_{0.025}} = \frac{(30-1)25}{45.722} \approx 15.86$

Upper bound $= \frac{(n-1)s^2}{\chi^2_{0.95}} = \frac{(30-1)25}{16.047} \approx 45.18$

(15.86, 45.18)

27. df $= n - 1 = 50 - 1 = 49$.

df = 49 is not in the χ^2 table, so take the next row with smaller df in the χ^2 table. Therefore, we will use df = 40.

For a 95% confidence interval, $1 - \alpha = 0.95$, $\alpha = 1 - 0.95 = 0.05$, $\alpha/2 = 0.05/2 = 0.025$, and $1 - \alpha/2 = 1 - 0.025 = 0.975$.

$\chi^2_{0.975} = 24.433$ and $\chi^2_{0.025} = 59.342$.

Lower bound $= \frac{(n-1)s^2}{\chi^2_{0.025}} = \frac{(50-1)25}{59.342} \approx 20.64$

Upper bound $= \frac{(n-1)s^2}{\chi^2_{0.95}} = \frac{(50-1)25}{24.433} \approx 50.14$

(20.64, 50.14) [Using Minitab: (17.4, 38.8)]

29. df $= n - 1 = 40 - 1 = 39$.

df = 39 is not in the χ^2 table, so take the next row with smaller df in the χ^2 table. Therefore, we will use df = 30.

For a 95% confidence interval, $1 - \alpha = 0.95$, $\alpha = 1 - 0.95 = 0.05$, $\alpha/2 = 0.05/2 = 0.025$, and $1 - \alpha/2 = 1 - 0.025 = 0.975$.

$\chi^2_{0.975} = 16.791$ and $\chi^2_{0.025} = 46.979$.

Lower bound $= \sqrt{\frac{(n-1)s^2}{\chi^2_{0.025}}} = \sqrt{\frac{(40-1)25}{46.979}} \approx 4.56$

Upper bound $= \sqrt{\frac{(n-1)s^2}{\chi^2_{0.95}}} = \sqrt{\frac{(40-1)25}{16.791}} \approx 7.62$

(4.56, 7.62) [Using Minitab: (4.10, 6.42)]

31. Lower bound increases while the upper bound decreases.

33. **(a)** Acceptable normality. All of the points are between the curved lines and most of the points are close to the center line.

(b) df $= n - 1 = 9 - 1 = 8$.

For a 95% confidence interval, $1 - \alpha = 0.95$, $\alpha = 1 - 0.95 = 0.05$, $\alpha/2 = 0.05/2 = 0.025$, and $1 - \alpha/2 = 1 - 0.025 = 0.975$.

$\chi^2_{0.975} = 2.180$ and $\chi^2_{0.025} = 17.535$

(c) TI-83/84: $s^2 = 700.8452778$

Lower bound $= \frac{(n-1)s^2}{\chi^2_{0.025}} = \frac{(9-1)700.8452778}{17.535} \approx 319.75$

Upper bound $= \frac{(n-1)s^2}{\chi^2_{0.95}} = \frac{(9-1)700.8452778}{2.180} \approx 2571.91$

(319.75, 2571.91). We are 95% confident that the population variance σ^2 lies between 319.75 megawatts squared and 2571.91 megawatts squared.

(d) Lower bound $= \sqrt{\frac{(n-1)s^2}{\chi^2_{0.025}}} = \sqrt{\frac{(9-1)700.8452778}{17.535}} \approx 17.88$

Upper bound $= \sqrt{\frac{(n-1)s^2}{\chi^2_{0.95}}} = \sqrt{\frac{(9-1)700.8452778}{2.180}} \approx 50.71.$

(17.88, 50.71). We are 95% confident that the population standard deviation σ lies between 17.88 megawatts and 50.71 megawatts.

35. **(a)** Megawatts squared

(b) Megawatts

(c) Megawatts

37.

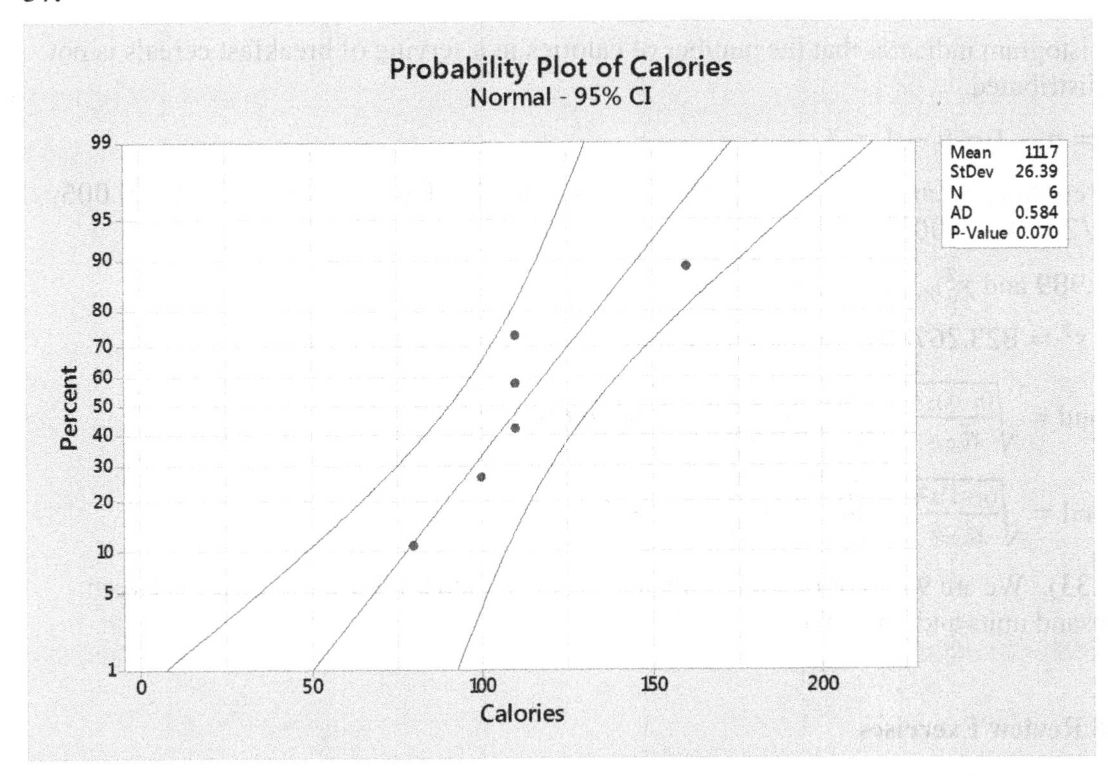

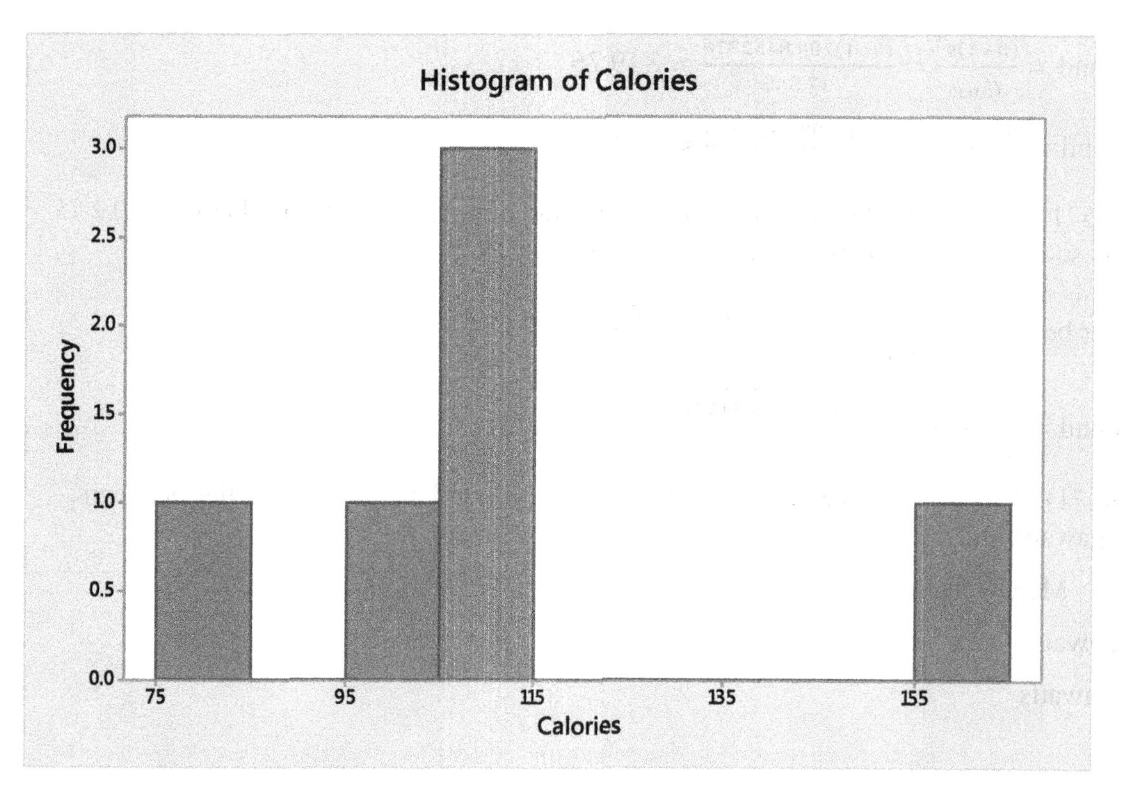

No. The histogram indicates that the number of calories in a serving of breakfast cereals is not normally distributed.

39. df $= n - 1 = 8 - 1 = 7$.

For a 99% confidence interval, $1 - \alpha = 0.99$, $\alpha = 1 - 0.99 = 0.01$, $\alpha/2 = 0.01/2 = 0.005$, and $1 - \alpha/2 = 1 - 0.005 = 0.995$.

$\chi^2_{0.995} = 0.989$ and $\chi^2_{0.005} = 20.278$.

TI-83/84: $s^2 = 823.2678571$ thousand game units

Lower bound $= \sqrt{\frac{(n-1)s^2}{\chi^2_{0.005}}} = \sqrt{\frac{(8-1)823.2678571}{20.278}} \approx 16.86$

Upper bound $= \sqrt{\frac{(n-1)s^2}{\chi^2_{0.995}}} = \sqrt{\frac{(8-1)823.2678571}{0.989}} \approx 76.33$

(16.86, 76.33). We are 99% confident that the population standard deviation σ lies between 16.86 thousand units and 76.33 thousand units.

Chapter 8 Review Exercises

1. **(a)** $\frac{\sigma}{\sqrt{n}} = \frac{20}{\sqrt{16}} = 5$

(b) $Z_{\alpha/2} = 1.96$

(c) $E = Z_{\alpha/2}\left(\frac{\sigma}{\sqrt{n}}\right) = 1.96\left(\frac{20}{\sqrt{16}}\right) = 9.8$. We can estimate the population mean μ to within 9.8 with 95% confidence.

(d) $\bar{x} \pm Z_{\alpha/2}\left(\frac{\sigma}{\sqrt{n}}\right) = 100 \pm 1.96\left(\frac{20}{\sqrt{16}}\right) = 100 \pm 9.8 = (90.2, 109.8)$. We are 95% confident that the population mean μ lies between 90.2 and 109.8.

3. **(a)** $\bar{x} = 7$ points

(b) $\frac{\sigma}{\sqrt{n}} = \frac{2}{\sqrt{45}} \approx 0.2981$ points

(c) $Z_{\alpha/2} = 1.645$

(d) Since the sample size of $n = 45$ is large ($n \geq 30$) and σ is known, we may use the Z interval. The margin of error is $E = Z_{\alpha/2}\left(\frac{\sigma}{\sqrt{n}}\right) = 1.645\left(\frac{2}{\sqrt{45}}\right) \approx 0.4904$ point. We may estimate μ, the mean increase in IQ points for all children after listening to a Mozart piano sonata for about 10 minutes, to within 0.4904 point with 90% confidence.

(e) $\bar{x} \pm Z_{\alpha/2}\left(\frac{\sigma}{\sqrt{n}}\right) = 7 \pm 1.645\left(\frac{2}{\sqrt{45}}\right) = 7 \pm 0.4904 = (6.5096, 7.4904)$.

We are 90% confident that the true mean increase in IQ points for all children after listening to a Mozart piano sonata for about 10 minutes μ lies between 6.5096 points and 7.4904 points.

5. $n = \left[\frac{(Z_{\alpha/2})\sigma}{E}\right]^2 = \left[\frac{(1.96)(30)}{5}\right]^2 = 138.2976.$ Round n up to $n = 139$.

7. $n = \left[\frac{(Z_{\alpha/2})\sigma}{E}\right]^2 = \left[\frac{(1.645)(10)}{2}\right]^2 = 67.650625.$ Round n up to $n = 68$.

9. We are told the population is normal. Therefore, it is appropriate construct the indicated confidence interval.

df $= n - 1 = 25 - 1 = 24$, 90% confidence interval

$t_{\alpha/2}$=1.711, $\bar{x} \pm t_{\alpha/2}\left(\frac{s}{\sqrt{n}}\right) = 42 \pm 1.711\left(\frac{5}{\sqrt{25}}\right) \approx 42 \pm 1.7 = (40.3, 43.7)$

11. $\bar{x} = \frac{\sum x}{n} = \frac{3.7+12+18.5+30+37+44.1+57+68+83}{9} = \frac{353.3}{9} \approx 39.26$ MW.

x	x^2
3.7	13.69
12	144
18.5	342.25
30	900
37	1369
44.1	1944.81
57	3249
68	4624
83	6889

$\sum x = 353.3 \quad \sum x^2 = 19{,}475.75$

$s = \sqrt{\frac{\sum x^2 - (\sum x)^2/n}{n-1}} = \sqrt{\frac{19{,}475.75-(353.3)^2/9}{9-1}} = \sqrt{\frac{19{,}475.75-124{,}820.89/9}{8}} = \sqrt{\frac{19{,}475.75-13{,}868.98778}{8}} =$
$\sqrt{\frac{5{,}606.7622}{8}} = \sqrt{700.845275} \approx 26.47$ MW

95% confidence interval, df $= n - 1 = 9 - 1 = 8$, $t_{\alpha/2} = 2.306$

$\bar{x} \pm t_{\alpha/2}\left(\frac{s}{\sqrt{n}}\right) \approx 39.26 \pm 2.306\left(\frac{26.47}{\sqrt{9}}\right) \approx 39.26 \pm 20.35 \approx (18.91, 59.61)$

We are 95% confident that the population mean μ lies between 18.91 MW and 59.61 MW.

13. **(a)** $Z_{\alpha/2} = 1.96$

(b) $x = n\hat{p} = (500)(0.01) = 5$ is ≥ 5 and $n - x = 500 - 5 = 495$ is ≥ 5.

(c) $\hat{p} = 0.01, \hat{q} = 1 - \hat{p} = 1 - 0.01 = 0.99$ $E = Z_{\alpha/2}\sqrt{\frac{\hat{p} \cdot \hat{q}}{n}} = 1.96\sqrt{\frac{(0.01)(0.99)}{500}} \approx 0.0087$. We can estimate the population proportion to within 0.0087 with 95% confidence.

(d) $\hat{p} \pm Z_{\alpha/2}\sqrt{\frac{\hat{p} \cdot \hat{q}}{n}} = 0.01 \pm 1.96\sqrt{\frac{(0.01)(0.99)}{500}} \approx 0.01 \pm 0.0087 = (0.0013, 0.0187)$. We are 95% confident that the population proportion p lies between 0.0013 and 0.0187.

15. $\hat{p} = 0.1, \hat{q} = 1 - \hat{p} = 1 - 0.1 = 0.9, n = \hat{p} \cdot \hat{q}\left(\frac{Z_{\alpha/2}}{E}\right)^2 = (0.1)(0.9)\left(\frac{2.576}{0.05}\right)^2 =$ 238.887936. Round n up to 239.

17. $\hat{p} = 0.001, \hat{q} = 1 - \hat{p} = 1 - 0.001 = 0.999$,
$n = \hat{p} \cdot \hat{q}\left(\frac{Z_{\alpha/2}}{E}\right)^2 = (0.001)(0.999)\left(\frac{1.96}{0.05}\right)^2 = 1.53510336$. Round n up to 2.

19. $n = \left[\frac{0.5Z_{\alpha/2}}{E}\right]^2 = \left[\frac{0.5(1.96)}{0.05}\right]^2 = 384.16$ Round n up to 385

21. $\text{df} = n - 1 = 100 - 1 = 99$.

df = 99 is not in the χ^2 table, so take the next row with smaller df in the χ^2 table. Therefore, we will use df = 90. For a 90% confidence interval, $1 - \alpha = 0.90, \alpha = 1 - 0.90 = 0.10, \alpha/2 = 0.10/2 = 0.05$, and $1 - \alpha/2 = 1 - 0.05 = 0.95$.

$\chi^2_{0.95} = 69.126$ and $\chi^2_{0.05} = 113.145$.

Lower bound = $\frac{(n-1)s^2}{\chi^2_{0.05}} = \frac{(100-1)256}{113.145} \approx 224.00$

Upper bound = $\frac{(n-1)s^2}{\chi^2_{0.95}} = \frac{(100-1)256}{69.126} \approx 366.63$

(224.00, 366.63)

23. $\text{df} = n - 1 = 100 - 1 = 99$.

df = 99 is not in the χ^2 table, so take the next row with smaller df in the χ^2 table. Therefore, we will use df = 90. For a 90% confidence interval, $1 - \alpha = 0.90, \alpha = 1 - 0.90 = 0.10, \alpha/2 = 0.10/2 = 0.05$, and $1 - \alpha/2 = 1 - 0.05 = 0.95$.

$\chi^2_{0.95} = 69.126$ and $\chi^2_{0.05} = 113.145$.

Lower bound = $\sqrt{\frac{(n-1)s^2}{\chi^2_{0.05}}} = \sqrt{\frac{(100-1)256}{113.145}} \approx 14.97$

Upper bound = $\sqrt{\frac{(n-1)s^2}{\chi^2_{0.95}}} = \sqrt{\frac{(100-1)256}{69.126}} \approx 19.15$

(14.97, 19.15)

25. df $= n - 1 = 7 - 1 = 6$. For a 95% confidence interval, $1 - \alpha = 0.95$, $\alpha = 1 - 0.95 = 0.05$, $\alpha/2 = 0.05/2 = 0.025$, and $1 - \alpha/2 = 1 - 0.025 = 0.975$.

$\chi^2_{0.975} = 1.237$ and $\chi^2_{0.025} = 14.449$.

From Minitab:

```
                    Total
Variable            Count  Variance
Union membership        7    2245.7
```

Thus,

Lower bound $= \sqrt{\frac{(n-1)s^2}{\chi^2_{0.025}}} = \sqrt{\frac{(7-1)2245.7}{14.449}} \approx 30.537$ and

Upper bound $= \sqrt{\frac{(n-1)s^2}{\chi^2_{0.95}}} = \sqrt{\frac{(7-1)2245.7}{1.237}} \approx 104.368$.

We are 95% confident that σ, the population standard deviation of total union membership per state, lies between 30.537 and 104.368.

Chapter 8 Quiz

1. False. If we take sample after sample for a very long time, then *in the long run*, the proportion of intervals that will contain the parameter μ will equal 90%. The 20 samples in Table 2 are not enough to qualify for the phrase *long run.*

3. 4

5. α is a probability.

7. **(a)** $E = Z_{\alpha/2}\left(\frac{\sigma}{\sqrt{n}}\right) = 1.96\left(\frac{\$5{,}000}{\sqrt{225}}\right) \approx \$653.33$. We can estimate μ, the population mean cost of college education per year, to within \$653.33 with 95% confidence.

(b) $\bar{x} \pm Z_{\alpha/2}\left(\frac{\sigma}{\sqrt{n}}\right) = \$30{,}500 \pm 1.96\left(\frac{\$5{,}000}{\sqrt{225}}\right) \approx \$30{,}500 \pm \$653.33 = (\$29{,}846.67,$ $\$31{,}153.33)$. We are 95% confident that μ, the population mean cost of college education per year, lies between \$29,846.67 and \$31,153.33.

9. We have $\hat{p} = \frac{x}{n} = \frac{340}{1000} = 0.34$, so $n\hat{p} = 1000(0.34) = 88.9952 \geq 5$ and

$n\hat{q} = 1000(0.66) = 23.0048 \geq 5$.

(a) The margin of error is $E = Z_{\alpha/2}\sqrt{\frac{\hat{p}\cdot\hat{q}}{n}} = 2.576\sqrt{\frac{(0.34)(0.66)}{1000}} \approx 0.0386$.

We can estimate p, the true proportion of all *Quebecois* who favor independence for the Province of Quebec, within 0.0386 with 99% confidence.

(b) $\hat{p} \pm Z_{\alpha/2}\sqrt{\frac{\hat{p}\hat{q}}{n}} = 0.34 \pm 2.576\sqrt{\frac{(0.34)(0.66)}{1000}} \approx 0.34 \pm 0.0386 = (0.3014, 0.3786)$.

We are 99% confident that the true proportion of all *Quebecois* who favor independence for the Province of Quebec lies between 0.3014 and 0.3786.

11. $n = \left(\frac{0.5 \cdot Z_{\alpha/2}}{E}\right)^2 = \left(\frac{(0.5)(1.645)}{0.03}\right)^2 = 751.6736111$. Round n up to 752.

Chapter 9: Hypothesis Testing

Section 9.1

1. The null hypothesis is assumed true unless the sample evidence indicates that the alternative hypothesis is true instead. It represents what has been tentatively assumed about the value of the parameter. It is the *status quo* hypothesis. The alternative hypothesis represents an alternative claim about the value of the parameter. The researcher concludes that the alternative hypothesis is true only if the evidence provided by the sample data indicates that it is true.

3. There are three forms of the hypotheses in the hypothesis test for the population mean. They are as follows:

Form	Null and alternative hypotheses
Right-tailed test	$H_0: \mu = \mu_0$ versus $H_a: \mu > \mu_0$
Left-tailed test	$H_0: \mu = \mu_0$ versus $H_a: \mu < \mu_0$
Two-tailed test	$H_0: \mu = \mu_0$ versus $H_a: \mu \neq \mu_0$

5. A Type I error occurs when one rejects H_0 when H_0 is true. A Type II error occurs when one does not reject H_0 when H_0 is false.

7. No. It depends on how many standard deviations the sample mean of 90 is below the population mean of 100 and the level of significance of the test.

9. Invalid. Statistics like $\bar{x}$ should not be used in the hypotheses.

So the correct form is: H_0: $\mu = 100$ versus H_a: $\mu < 100$

11. Invalid. The equal sign always goes in H_0. So the correct form is:

H_0: $\mu = 3.14$ versus H_a: $\mu > 3.14$

13. H_0: $\mu = 79$ versus H_a: $\mu > 79$

15. H_0: $\mu = 75$ versus H_a: $\mu \neq 75$

17. H_0: $\mu = 1000$ versus H_a: $\mu \neq 1000$

19. **(a)** H_0: $\mu = 10$ versus H_a: $\mu < 10$

(b) Conclude that the population mean achievement gap has decreased when, in reality, it has remained the same.

(c) Conclude that the population mean achievement gap has remained the same when, in reality, it has decreased.

21. **(a)** H_0: $\mu = 20{,}000$ versus H_a: $\mu < 20{,}000$

(b) Conclude that the population mean cost of the new treatment has decreased from the cost of the previous treatment when, in reality, it has remained the same.

(c) Conclude that the population mean cost of the new treatment has remained the same as the cost of the previous treatment when, in reality, it has decreased.

23. **(a)** H_0: $\mu = 18$ versus H_a: $\mu > 18$

(b) Conclude the population mean number of hours nursing majors study per week is greater than 18 hours when, in reality, it is greater than 18 hours.

Conclude the population mean number of hours nursing majors study per week is equal to 18 hours when, in reality, it is equal to 18 hours.

(c) Conclude the population mean number of hours nursing majors study per week is greater than 18 hours when, in reality, it is equal to 18 hours.

(d) Conclude the population mean number of hours nursing majors study per week is equal to 18 hours when, in reality, it is greater than 18 hours.

25. **(a)** H_0: $\mu = 3$ versus H_a: $\mu < 3$

(b) Conclude that the population mean number of years it takes for owners of hybrid vehicles to recoup their initial increased cost through reduced fuel consumption is less than 3 years when, in reality, it is less than 3 years.

Conclude that the population mean number of years it takes for owners of hybrid vehicles to recoup their initial increased cost through reduced fuel consumption is equal to 3 years when, in reality, it is equal to 3 years.

(c) Conclude that the population mean number of years it takes for owners of hybrid vehicles to recoup their initial increased cost through reduced fuel consumption is less than 3 years when, in reality, it is equal to 3 years.

(d) Conclude that the population mean number of years it takes for owners of hybrid vehicles to recoup their initial increased cost through reduced fuel consumption is equal to 3 years when, in reality, it is less than 3 years.

27. **(a)** ***Step 1.*** **Find the key words.** The key word "increased" is synonymous with "greater than," >.

Step 2. **Determine the form of the hypotheses.** The symbol > indicates that we should use the hypotheses for the right-tailed test in Table 9.1.

$H_0 : \mu = \mu_0$ versus $H_a : \mu > \mu_0$

Step 3. **Find the value for μ_0 and write your hypotheses.** Asking "greater than what?" (or "increased from what?"), we see that $\mu_0 = 3.24$, so the hypotheses are

$H_0 : \mu = 3.24$ versus $H_a : \mu > 3.24$

(b) The two ways that a correct decision could be made are (1) to conclude that the *mean price* of a gallon of milk this year is greater than \$3.24 when the *population mean price* of a gallon of milk this year actually is greater than \$3.24 per gallon and, (2) to conclude that the *mean price* for a gallon of milk this year is equal to \$3.24 when the *population mean price* of a gallon of milk this year actually is equal to \$3.24 per gallon.

(c) A Type I error is concluding that the *mean price* of a gallon of milk this year is greater than \$3.24 when it actually is equal to \$3.24 per gallon.

(d) A Type II error is concluding that the *mean price* for a gallon of milk this year is equal to \$3.24 when it actually is greater than \$3.24.

29. **(a)** H_0: $\mu = 350$ versus H_a: $\mu \neq 350$

(b) Conclude that the population mean amount of caffeine in a 16-ounce Starbucks Park Place brewed coffee has changed from 350 milligrams when, in reality, it has changed from 350 milligrams.

Conclude that the population mean amount of caffeine in a 16-ounce Starbucks Park Place brewed coffee has not changed from 350 milligrams when, in reality, it has not changed from 350 milligrams.

(c) Conclude that the population mean amount of caffeine in a 16-ounce Starbucks Park Place brewed coffee has changed from 350 milligrams when, in reality, it has not changed from 350 milligrams.

(d) Conclude that the population mean amount of caffeine in a 16-ounce Starbucks Park Place brewed coffee has not changed from 350 milligrams when, in reality, it has changed from 350 milligrams.

31. **(a)** H_0: $\mu = 5$ versus H_a: $\mu > 5$

(b) Conclude that the population mean number of visits that customers make to the store in the six-month period is greater than 5 when, in reality, it is greater than 5.

Conclude that the population mean number of visits that customers make to the store in the six-month period is equal to 5 when, in reality, it is equal to 5.

(c) Conclude that the population mean number of visits that customers make to the store in the six-month period is greater than 5 when, in reality, it is equal to 5.

(d) Conclude that the population mean number of visits that customers make to the store in the six-month period is equal to 5 when, in reality, it is greater than 5.

33. **(a)** H_0: $\mu = 20$ versus H_a: $\mu < 20$

(b) Conclude that the population mean number of items that customers are buying is less than 20 when, in reality, it is less than 20.

Conclude that the population mean number of items that customers are buying is equal to 20 when, in reality, it is equal to 20.

(c) Conclude that the population mean number of items that customers are buying is less than 20 when, in reality, it is equal to 20.

(d) Conclude that the population mean number of items that customers are buying is equal to 20 when, in reality, it is less than 20.

35. **(a)** H_0: $\mu = 150$ versus H_a: $\mu < 150$

(b) Concluding that the population mean number of days since purchase is less than 150 when, in reality, it is less than 150.

Concluding that the population mean number of days since purchase is equal to 150 when, in reality, it is equal to 150.

(c) Concluding that the population mean number of days since purchase is less than 150 when, in reality, it is equal to 150.

(d) Concluding that the population mean number of days since purchase is equal to 150 when, in reality, it is less than 150.

Section 9.2

1. When the observed value of $\bar{x}$ is unusual or extreme in the sampling distribution of $\bar{x}$ that assumes H_0 is true, we should reject H_0. Otherwise, we should not reject H_0.

3. A test statistic is a statistic generated from a data set for the purpose of testing a statistical hypothesis.

5. Z_{crit} is the value of Z that separates the critical region from the noncritical region.

7. The critical region for a right-tailed test lies in the right (upper) tail.

9. $Z_{data} = \frac{\bar{x}-\mu_0}{\sigma/\sqrt{n}} = \frac{79-75}{20/\sqrt{25}} = 1$

11. $Z_{data} = \frac{\bar{x}-\mu_0}{\sigma/\sqrt{n}} = \frac{51-50}{10/\sqrt{100}} = 1$

13. $Z_{data} = \frac{\bar{x}-\mu_0}{\sigma/\sqrt{n}} = \frac{98.35-98.6}{1/\sqrt{16}} = -1$

15. $Z_{data} = \frac{\bar{x}-\mu_0}{\sigma/\sqrt{n}} = \frac{19-20}{5/\sqrt{100}} = -2$

17. $Z_{data} = \frac{\bar{x}-\mu_0}{\sigma/\sqrt{n}} = \frac{1005-1000}{100/\sqrt{400}} = 1$

19. $Z_{data} = \frac{\bar{x}-\mu_0}{\sigma/\sqrt{n}} = \frac{2.45-2.5}{0.5/\sqrt{100}} = -1$

21. It increases.

23. **(a)** $Z_{crit} = 1.28$

(b)

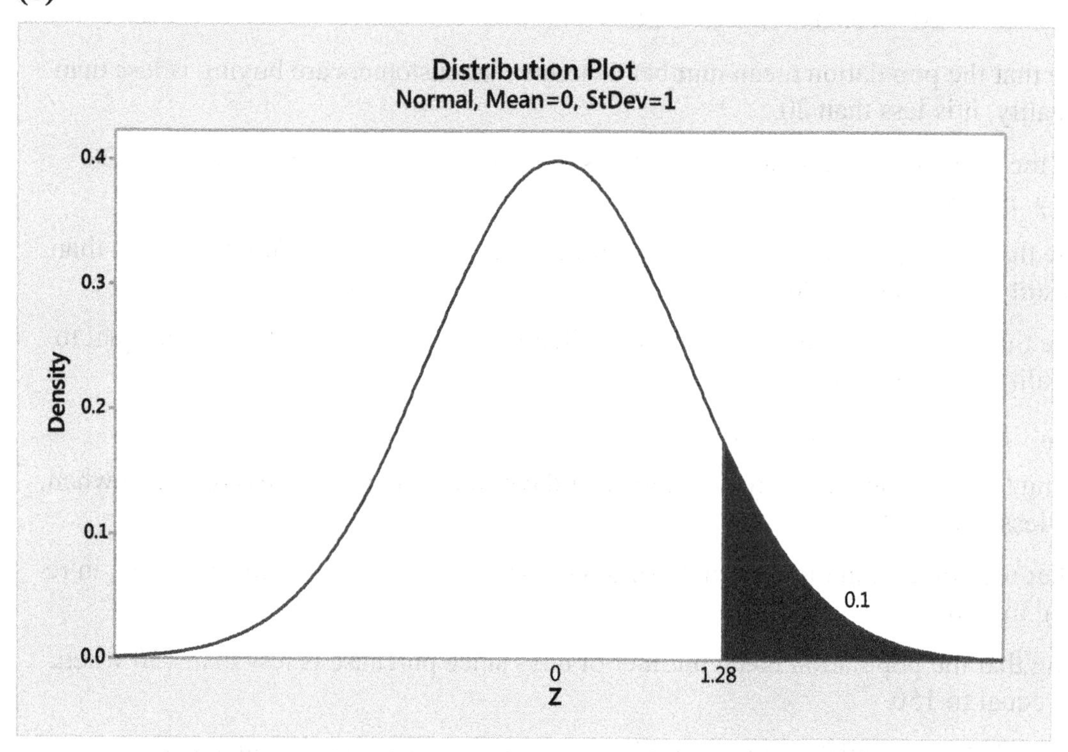

(c) Reject H_0 if $Z_{data} \geq 1.28$.

25. **(a)** $Z_{\text{crit}} = 1.645$

(b)

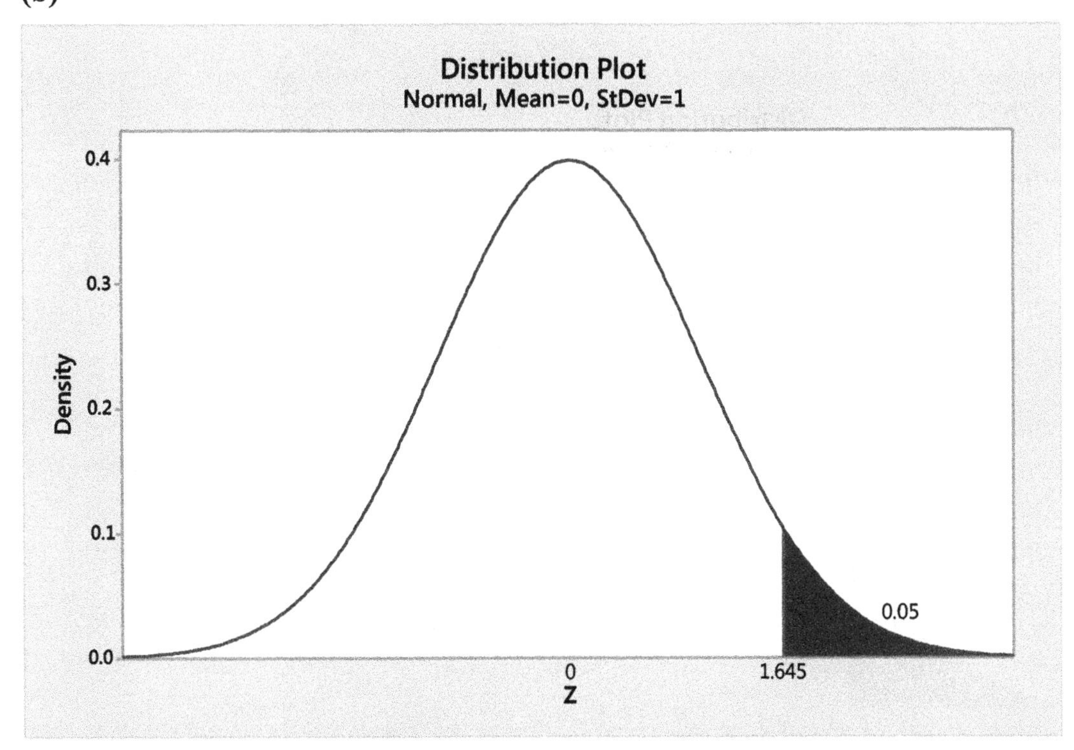

(c) Reject H_0 if $Z_{\text{data}} \geq 1.645$.

27. **(a)** $Z_{\text{crit}} = -1.645$

(b)

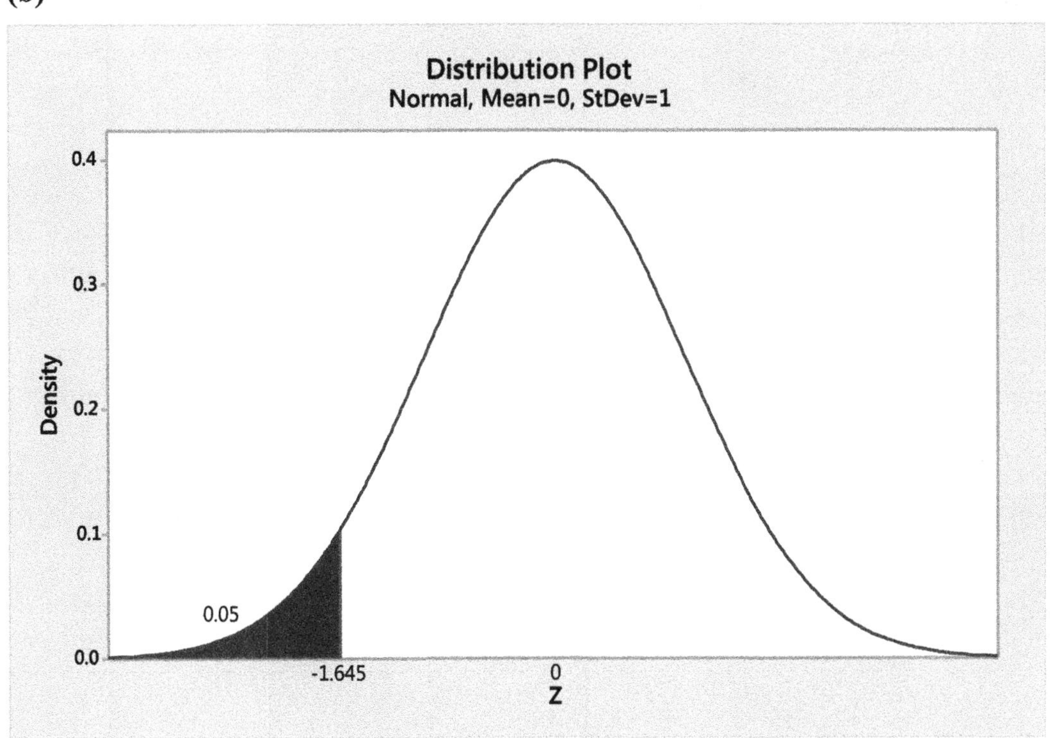

(c) Reject H_0 if $Z_{\text{data}} \leq -1.645$.

29. **(a)** $Z_{\text{crit}} = -1.28$

(b)

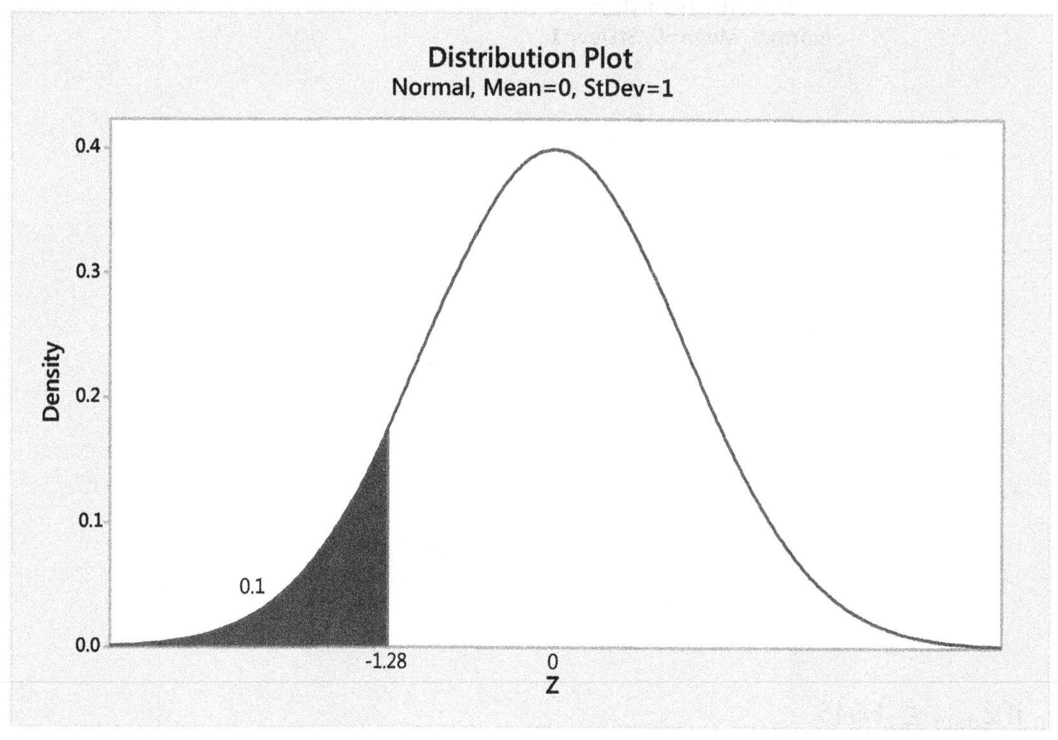

(c) Reject H_0 if $Z_{\text{data}} \leq -1.28$.

31. **(a)** $Z_{\text{crit}} = 1.96$

(b)

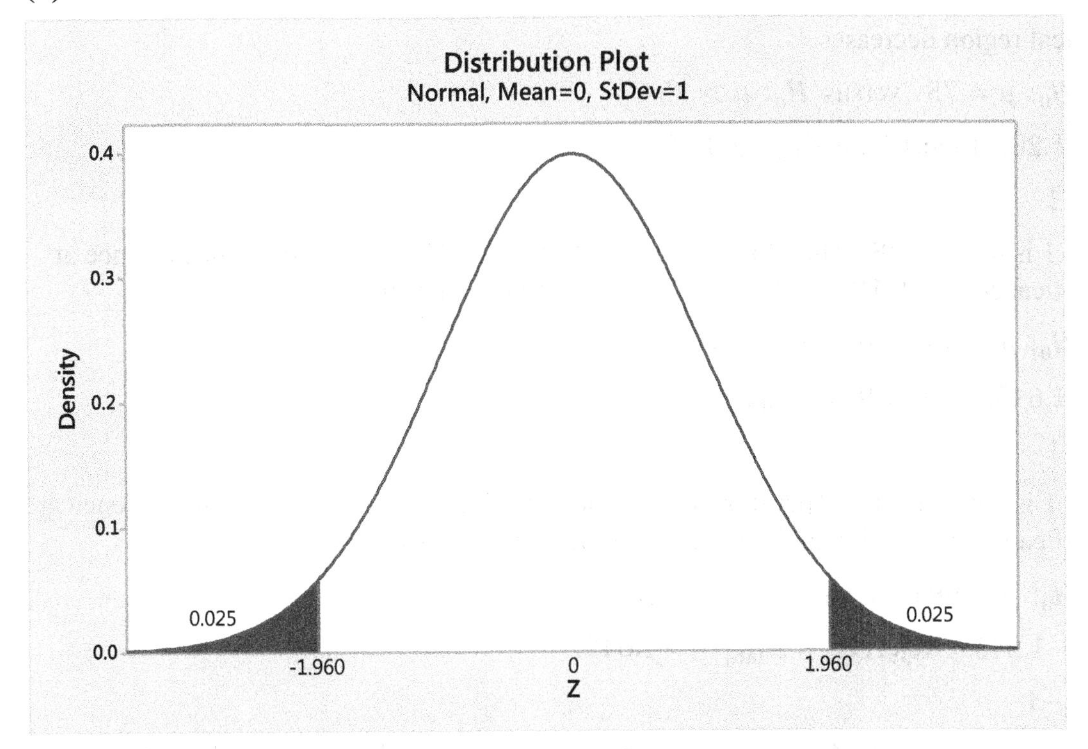

(c) Reject H_0 if $Z_{\text{data}} \leq -1.96$ or $Z_{\text{data}} \geq 1.96$.

33. **(a)** $Z_{\text{crit}} = 1.645$

(b)

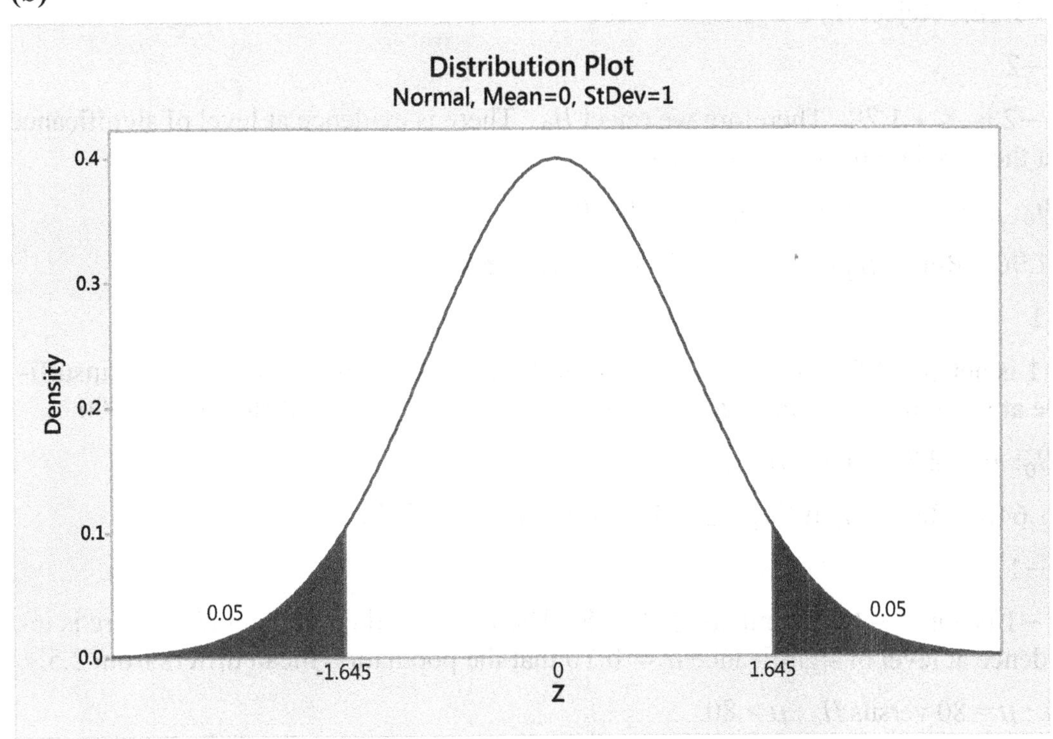

(c) Reject H_0 if $Z_{\text{data}} \leq -1.645$ or $Z_{\text{data}} \geq 1.645$.

35. **(a)** Z_{crit} increases.

(b) The critical region decreases.

37. **(a)** H_0: $\mu = 75$ versus H_a: $\mu > 75$

(b) $Z_{\text{crit}} = 1.28$. Reject H_0 if $Z_{\text{data}} \geq 1.28$.

(c) $Z_{\text{data}} = 1$

(d) $Z_{\text{data}} = 1$ is not ≥ 1.28. Therefore we do not reject H_0. There is insufficient evidence at level of significance $\alpha = 0.10$ that the population mean is greater than 75.

39. **(a)** H_0: $\mu = 50$ versus H_a: $\mu > 50$

(b) $Z_{\text{crit}} = 1.645$. Reject H_0 if $Z_{\text{data}} \geq 1.645$.

(c) $Z_{\text{data}} = 1$

(d) $Z_{\text{data}} = 1$ is not ≥ 1.645. Therefore we do not reject H_0. There is insufficient evidence at level of significance $\alpha = 0.05$ that the population mean is greater than 50.

41. **(a)** H_0: $\mu = 98.6$ versus H_a: $\mu < 98.6$

(b) $Z_{\text{crit}} = -1.645$. Reject H_0 if $Z_{\text{data}} \leq -1.645$.

(c) $Z_{\text{data}} = -1$

(d) $Z_{\text{data}} = -1$ is not ≤ -1.645. Therefore we do not reject H_0. There is insufficient evidence at level of significance $\alpha = 0.05$ that the population mean is less than 98.6.

43. **(a)** H_0: $\mu = 20$ versus H_a: $\mu < 20$

(b) $Z_{\text{crit}} = -1.28$. Reject H_0 if $Z_{\text{data}} \leq -1.28$.

(c) $Z_{\text{data}} = -2$

(d) $Z_{\text{data}} = -2$ is ≤ -1.28. Therefore we reject H_0. There is evidence at level of significance $\alpha = 0.10$ that the population mean is less than 20.

45. **(a)** H_0: $\mu = 1000$ versus H_a: $\mu \neq 1000$

(b) $Z_{\text{crit}} = 1.96$. Reject H_0 if $Z_{\text{data}} \leq -1.96$ or $Z_{\text{data}} \geq 1.96$.

(c) $Z_{\text{data}} = 1$

(d) $Z_{\text{data}} = 1$ is not ≤ -1.96 and not ≥ 1.96. Therefore we do not reject H_0. There is insufficient evidence at level of significance $\alpha = 0.05$ that the population mean differs from 1000.

47. **(a)** H_0: $\mu = 2.5$ versus H_a: $\mu \neq 2.5$

(b) $Z_{\text{crit}} = 1.645$. Reject H_0 if $Z_{\text{data}} \leq -1.645$ or $Z_{\text{data}} \geq 1.645$.

(c) $Z_{\text{data}} = -1$

(d) $Z_{\text{data}} = -1$ is not ≤ -1.645 and not ≥ 1.645. Therefore we do not reject H_0. There is insufficient evidence at level of significance $\alpha = 0.10$ that the population mean differs from 2.5.

49. **(a)** $H_0 : \mu = 80$ versus $H_a : \mu > 80$.

(b) For a right-tailed test with $\alpha = 0.05$, $Z_{\text{crit}} = 1.645$. Reject H_0 if $Z_{\text{data}} \geq 1.645$.

(c) $Z_{\text{data}} = \frac{\bar{x}-\mu_0}{\sigma/\sqrt{n}} = \frac{86-80}{48/\sqrt{64}} = 1.$

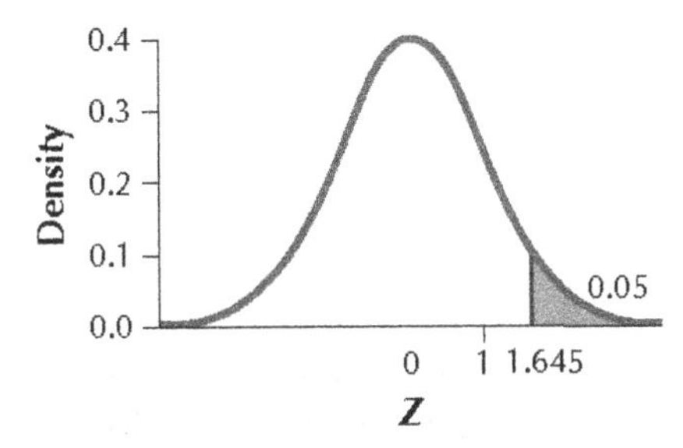

(d) Since $Z_{\text{data}} = 1$ is not ≥ 1.645, the conclusion is do not reject H_0. There is insufficient evidence at the 0.05 level of significance that the population mean number of connections to community pages, groups, and events is greater than 80.

51. **(a)** H_0: $\mu = 60$ versus H_a: $\mu \neq 60$

(b) $Z_{\text{crit}} = 2.58$. Reject H_0 if $Z_{\text{data}} \leq -2.58$ or $Z_{\text{data}} \geq 2.58$.

(c) $Z_{\text{data}} = \frac{\bar{x}-\mu_0}{\sigma/\sqrt{n}} = \frac{69-60}{30/\sqrt{100}} = 3$

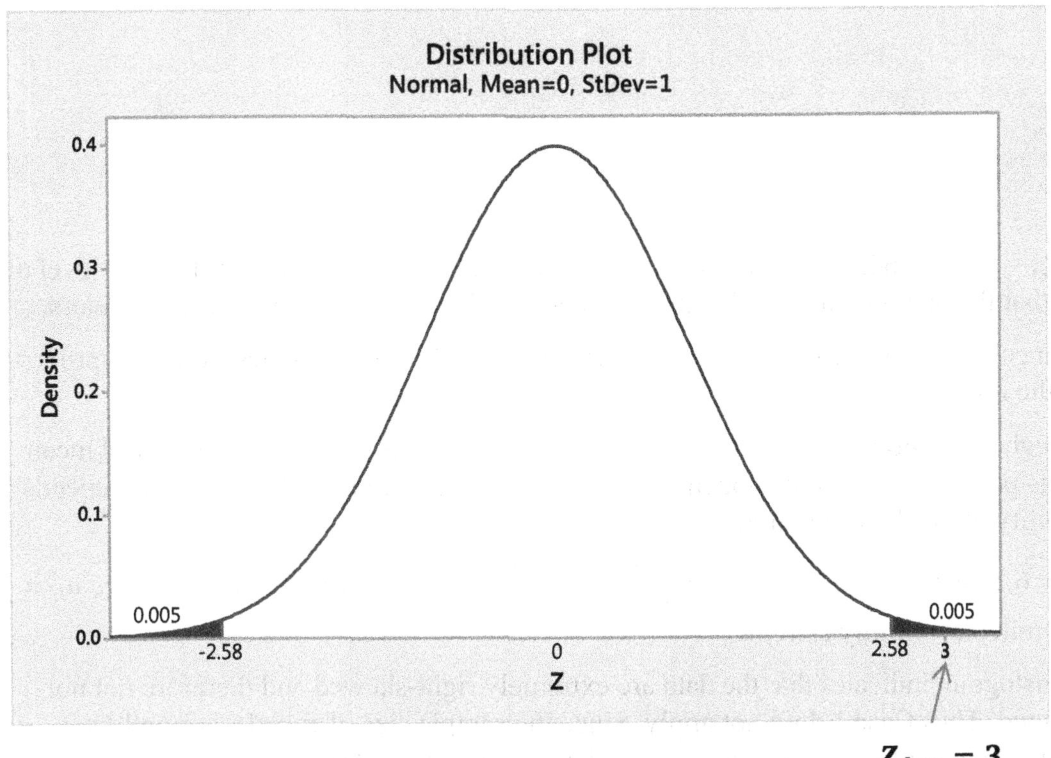

$Z_{\text{data}} = 3$

(d) $Z_{\text{data}} = 3$ is ≥ 2.58. Therefore we reject H_0. There is evidence at level of significance $\alpha = 0.01$ that the population mean number of text messages sent per day for young people ages 12–17 differs from 60.

53. **(a)** $H_0 : \mu = 3.70$ versus $H_a : \mu > 3.70$.

(b) For a right-tailed test with $\alpha = 0.05$, $Z_{\text{crit}} = 1.645$. Reject H_0 if $Z_{\text{data}} \geq 1.645$.

(c) $Z_{\text{data}} = \frac{\bar{x}-\mu_0}{\sigma/\sqrt{n}} = \frac{3.90-3.70}{0.50/\sqrt{25}} = 2.$

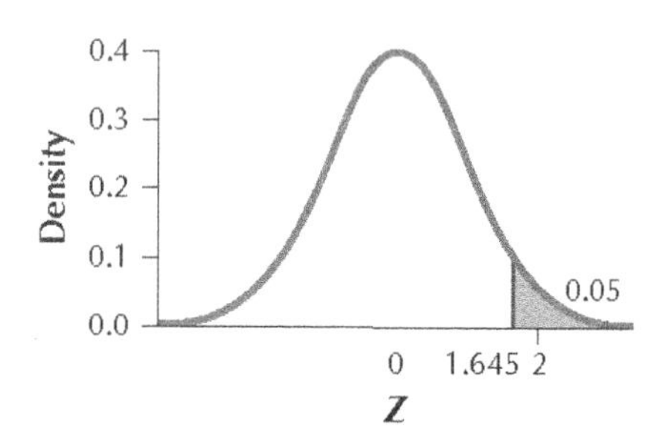

(d) Since $Z_{data} = 2$ is ≥ 1.645, the conclusion is to reject H_0. There is evidence at the 0.05 level of significance that the population mean price of regular gasoline is greater than \$3.70 per gallon. Therefore, we can conclude at the 0.05 level of significance that the population mean price for a gallon of regular gasoline has risen since June 2011.

55. **(a)** $H_0 : \mu = 175$ versus $H_a : \mu \neq 175$.

(b) For a two-tailed test with $\alpha = 0.10$, $Z_{crit} = 1.645$. Reject H_0 if $Z_{data} \leq -1.645$ or if $Z_{data} \geq 1.645$.

(c) $Z_{data} = \frac{\bar{x}-\mu_0}{\sigma/\sqrt{n}} = \frac{176-175}{2.5/\sqrt{400}} = 8.$

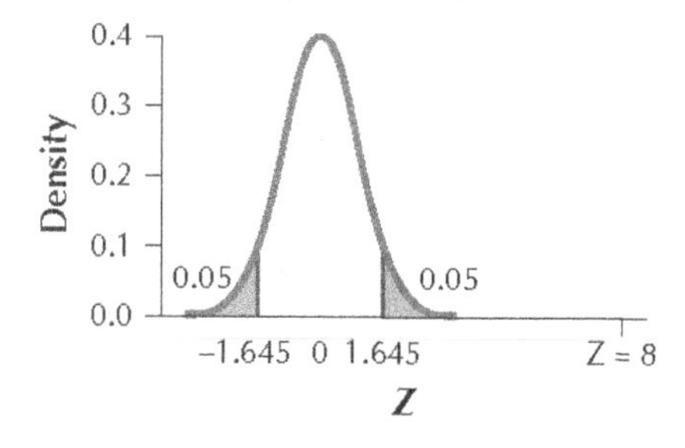

(d) Since $Z_{data} = 8$ is ≥ 1.645, the conclusion is to reject H_0. There is evidence at the 0.10 level of significance that the population mean height of Americans has changed from 175 centimeters.

57. **(a)** Since the sample size of $n = 100$ is large ($n \geq 30$), Case 2 applies, so it is appropriate to apply the Z test.

(b) Even though the sample mean $\bar{x} = 6.2$ cents per mile is greater than the hypothesized mean $\mu_0 = 5.9$ cents per mile, this is not enough by itself to reject the null hypothesis. It also depends on the variability of the data and on α.

(c) Since $\bar{x} = 6.2$ cents per mile is $Z_{data} = \frac{\bar{x}-\mu_0}{\sigma/\sqrt{n}} = \frac{6.2-5.9}{1.5/\sqrt{100}} = 2$ standard deviations above $\mu_0 =$ 5.9 cents per mile, it is mildly extreme.

59. The histogram indicates that the data are extremely right-skewed and therefore not normally distributed. Thus Case 1 does not apply. Since the sample size of $n = 16$ is small ($n <$ 30), Case 2 does not apply. Thus it is not appropriate to apply the Z test.

61. H_0: $\mu = 50$ versus H_a: $\mu > 50$

where μ is the population mean city/highway combined gas mileage for the Toyota Prius hybrid car.

63. $Z_{data} = \frac{\bar{x}-\mu_0}{\sigma/\sqrt{n}} = \frac{50.71-50}{2/\sqrt{16}} = 1.42$

65. **(a)** An increase in the smallest value of x will result in an increase in $\sum x$. Since $\bar{x} = \frac{\sum x}{n}$, this will result in an increase in $\bar{x}$.

(b) Since σ is the population standard deviation, it is not affected by which numbers are selected. Therefore, an increase in the smallest number in the sample will not affect. Thus, σ will remain the same.

(c) An increase in the smallest value of x will not affect the sample size n. There will still be $n = 16$ numbers. Therefore, the sample size n will stay the same.

(d) We have $Z_{\text{data}} = \frac{\bar{x}-\mu_0}{\sigma/\sqrt{n}} = \frac{\bar{x}-50}{2/\sqrt{16}}$. From (a), an increase in the smallest value of x will result in an increase in $\bar{x}$. Since $\bar{x}$ is already greater than $\mu_0 = 50$, $\bar{x} - 50$ will remain positive. Since σ and $\sqrt{n}$ are positive, this will result in an increase in Z_{data}.

(e) Since Z_{crit} does not depend on the sample data, an increase in the smallest value of x will not affect Z_{crit}. Therefore, Z_{crit} will stay the same.

(f) From (e), an increase in the smallest value of x will result in an increase in Z_{data}. Since Z_{data} was already $\geq Z_{\text{crit}} = 1.28$, after an increase in Z_{data} it will still be $\geq Z_{\text{crit}} = 1.28$. Therefore the conclusion will still be reject H_0.

67. Turn to a direct assessment of the strength of evidence against the null hypothesis or obtain more data.

69. Answers will vary.

71. **Minitab output**

Mean of Coupons

```
Mean of Coupons = 0.7724
```

$\mu = 0.7724$ coupon. Answers will vary.

Section 9.3

1. False. The p-value is a probability value and therefore can never be larger than 1.

3. It gives us extra information about whether H_0 was barely rejected or not rejected, or whether it was a no-brainer decision to reject or not to reject H_0.

5. False. For a right-tailed test, the null hypothesis is not rejected whenever $Z_{\text{data}} < Z_{\text{crit}}$. For all hypothesis tests, the null hypothesis is rejected whenever the p-value $\leq \alpha$. Since the null hypothesis is not rejected, the p-value is not $< \alpha$.

7. Right-tailed test.

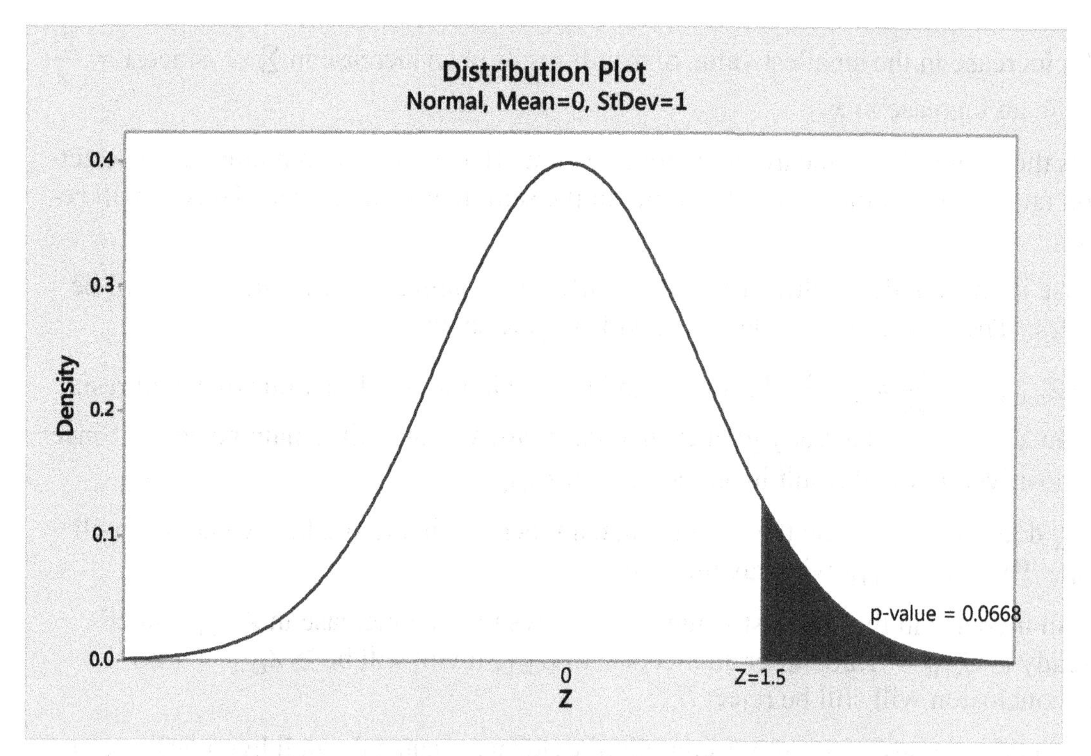

p-value $= P(Z \geq Z_{data}) = P(Z \geq 1.5) = 1 - P(Z < 1.5) = 1 - 0.9332 = 0.0668$

9. Right-tailed test.

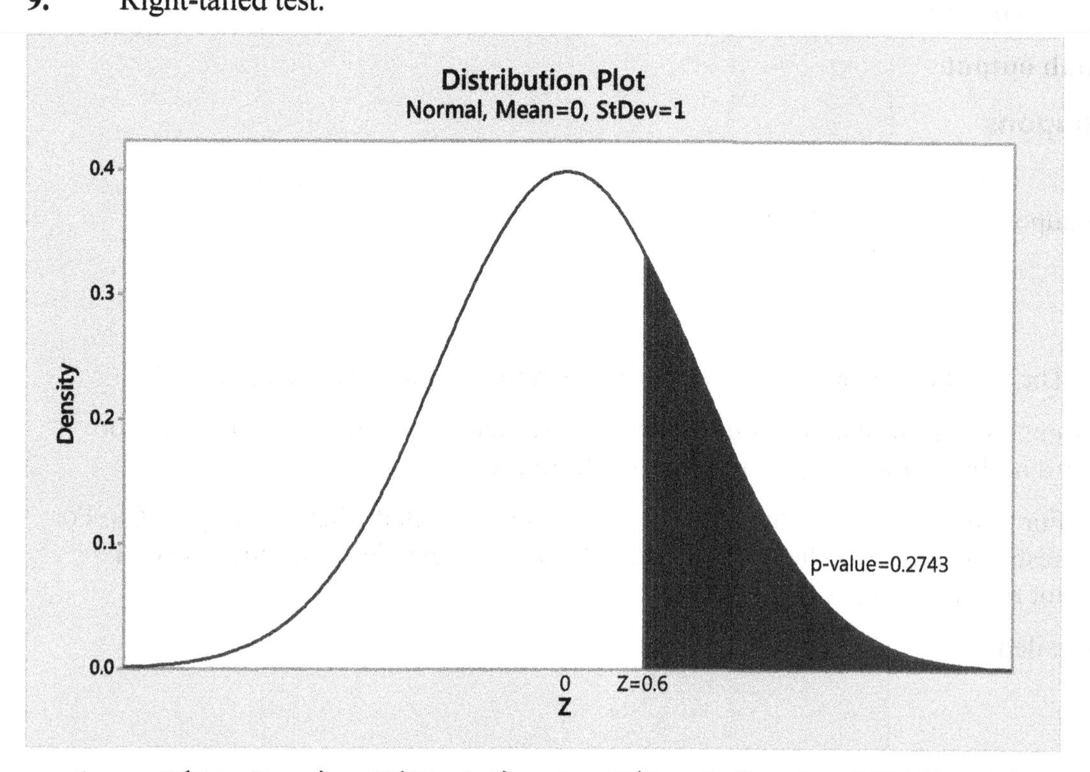

p-value $= P(Z \geq Z_{data}) = P(Z \geq 0.6) = 1 - P(Z < 0.6) = 1 - 0.7257 = 0.2743$

11. Left-tailed test.

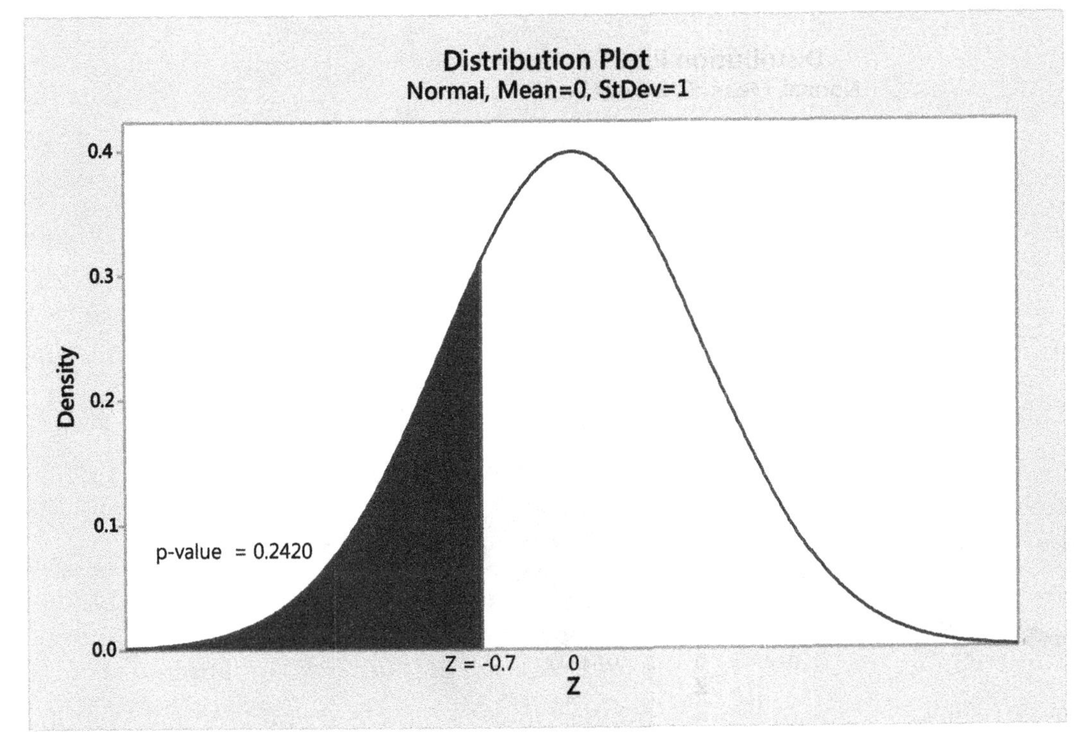

p-value $= P(Z \le Z_{data}) = P(Z \le -0.7) = 0.2420$

13. Left-tailed test.

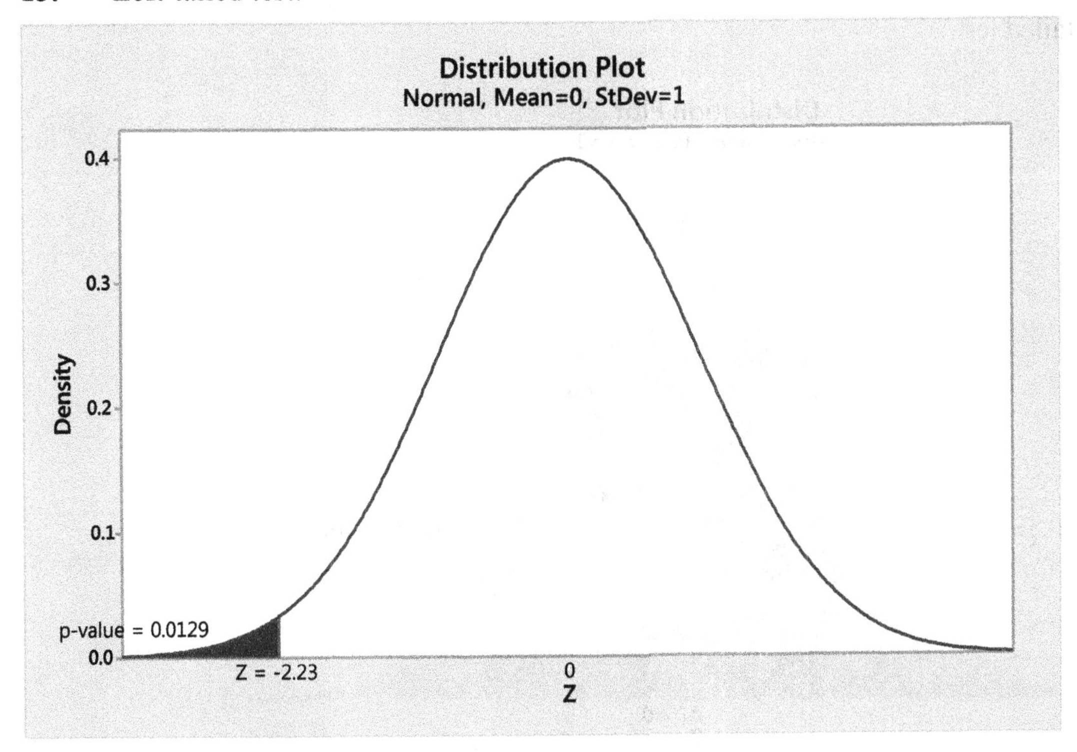

p-value $= P(Z \le Z_{data}) = P(Z \le -2.23) = 0.0129$

15. Two-tailed test.

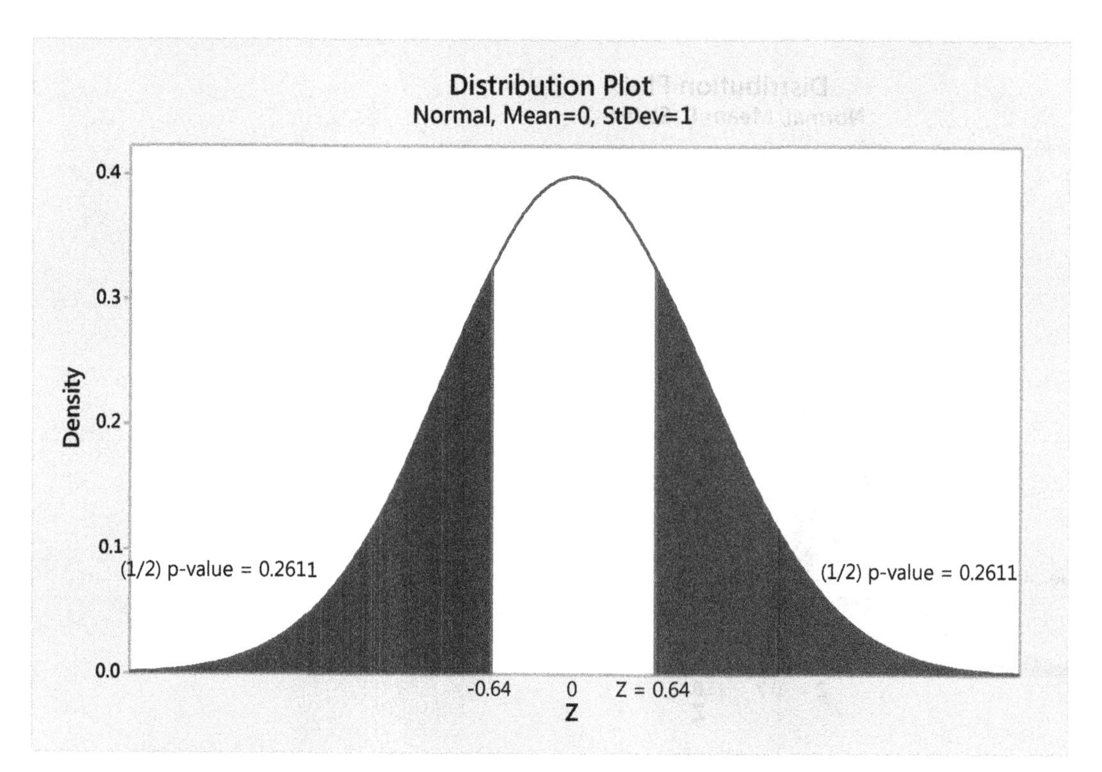

p-value $= 2 \cdot P(Z \geq |Z_{data}|) = 2 \cdot P(Z \geq |0.64|) = 2 \cdot \big(1 - P(Z < 0.64)\big) = 2 \cdot (1 - 0.7389) = 2 \cdot (0.2611) = 0.5222$

17. Two-tailed test.

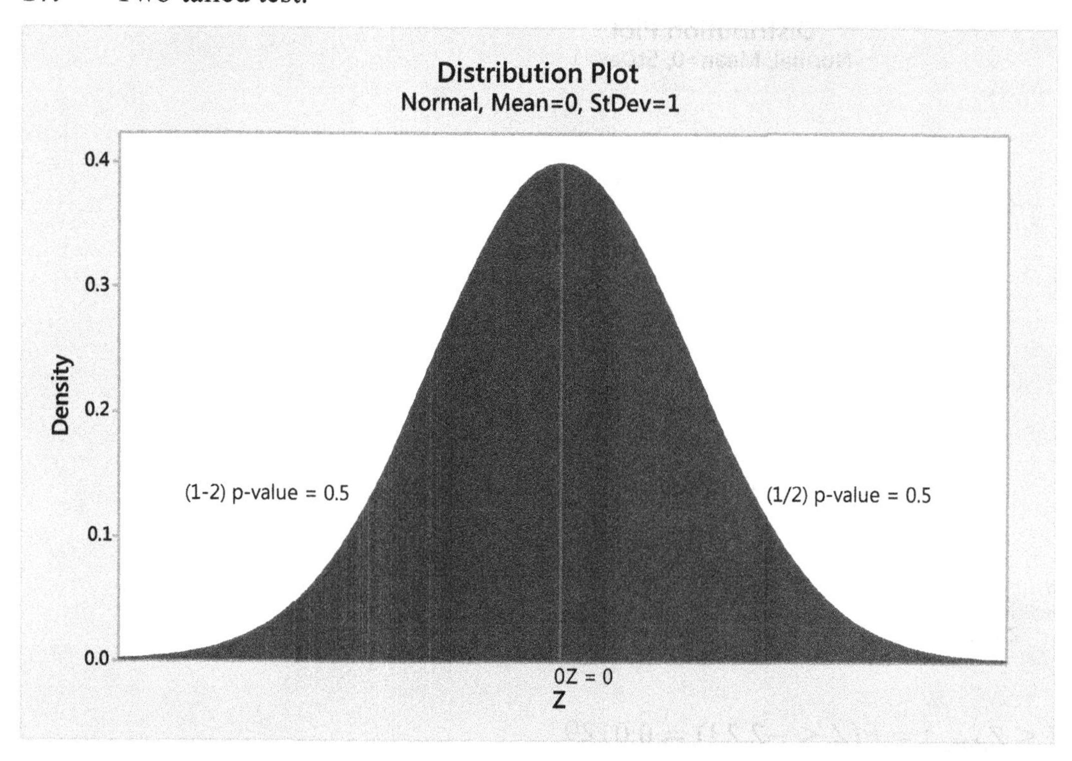

p-value $= 2 \cdot P(Z \geq |Z_{data}|) = 2 \cdot P(Z \geq |0|) = 2 \cdot \big(1 - P(Z < 0)\big) = 2 \cdot (1 - 0.5000) = 2 \cdot (0.5000) = 1$

19. It decreases.

21. **(a)** H_0: $\mu = 3.14$ versus H_a: $\mu > 3.14$

We will reject H_0 if the p-value $\leq \alpha = 0.05$.

(b) $Z_{\text{data}} = \frac{\bar{x}-\mu_0}{\sigma/\sqrt{n}} = \frac{3.2-3.14}{1/\sqrt{100}} = 0.6$

(c) p-value $= P(Z \geq Z_{data}) = P(Z \geq 0.6) = 1 - P(Z < 0.6) = 1 - 0.7257 = 0.2743$

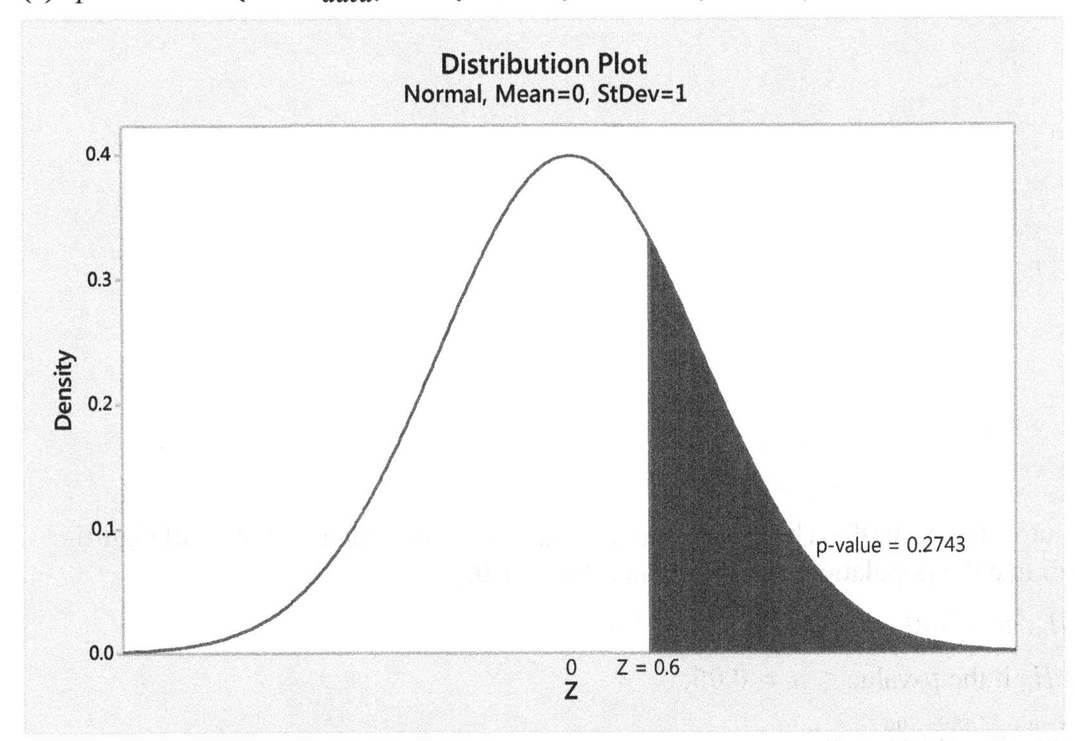

(d) The p-value = 0.2743 is not ≤ 0.05. Therefore we do not reject H_0. There is insufficient evidence at level of significance $\alpha = 0.05$ that the population mean is greater than 3.14.

23. **(a)** H_0: $\mu = -1.0$ versus H_a: $\mu > -1.0$

We will reject H_0 if the p-value $\leq \alpha = 0.05$.

(b) $Z_{\text{data}} = \frac{\bar{x}-\mu_0}{\sigma/\sqrt{n}} = \frac{0-(-1.0)}{1/\sqrt{400}} = 20$

(c) p-value $= P(Z \geq Z_{data}) = P(Z \geq 20) = 1 - P(Z < 20) \approx 1 - 1 = 0$

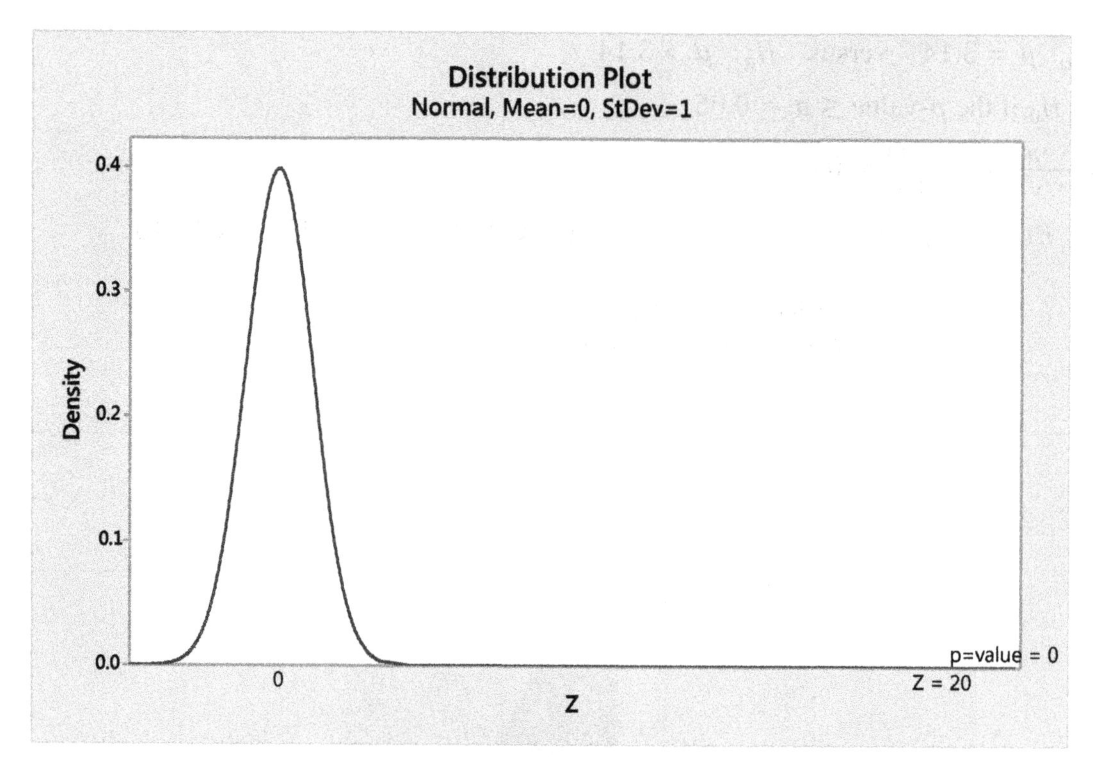

(d) The p-value ≈ 0 is ≤ 0.05. Therefore we reject H_0. There is evidence at level of significance $\alpha = 0.05$ that the population mean is greater than -1.0.

25. **(a)** H_0: $\mu = 500$ versus H_a: $\mu < 500$

We will reject H_0 if the p-value $\leq \alpha = 0.05$.

(b) $Z_{\text{data}} = \frac{\bar{x}-\mu_0}{\sigma/\sqrt{n}} = \frac{450-500}{100/\sqrt{16}} = -2$

(c) p-value $= P(Z \leq Z_{data}) = P(Z \leq -2) = 0.0228$

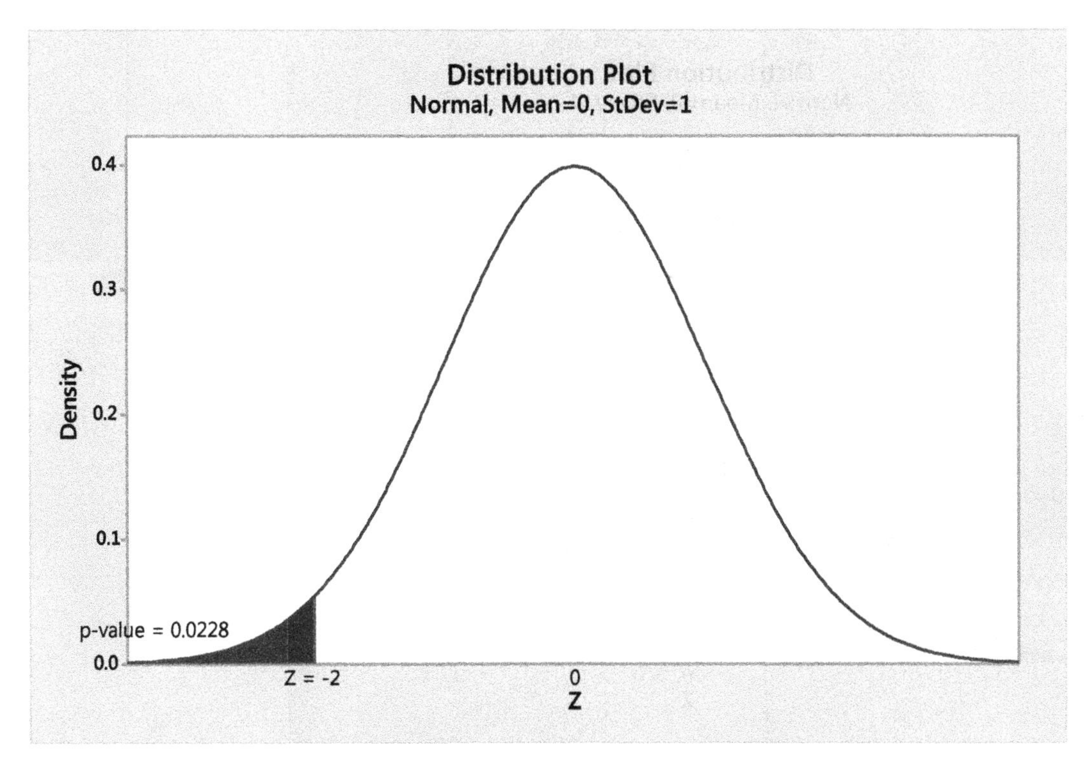

(d) The p-value $= 0.0228$ is ≤ 0.05. Therefore we reject H_0.

There is evidence at level of significance $\alpha = 0.05$ that the population mean is less than 500.

27. **(a)** H_0: $\mu = 10$ versus H_a: $\mu \neq 10$

We will reject H_0 if the p-value $\leq \alpha = 0.05$.

(b) $Z_{\text{data}} = \frac{\bar{x}-\mu_0}{\sigma/\sqrt{n}} = \frac{10-10}{5/\sqrt{100}} = 0$

(c) p-value $= 2 \cdot P(Z \geq |Z_{data}|) = 2 \cdot P(Z \geq |0|) = 2 \cdot (1 - P(Z < 0)) = 2 \cdot (1 - 0.5000) = 2 \cdot (0.5000) = 1$

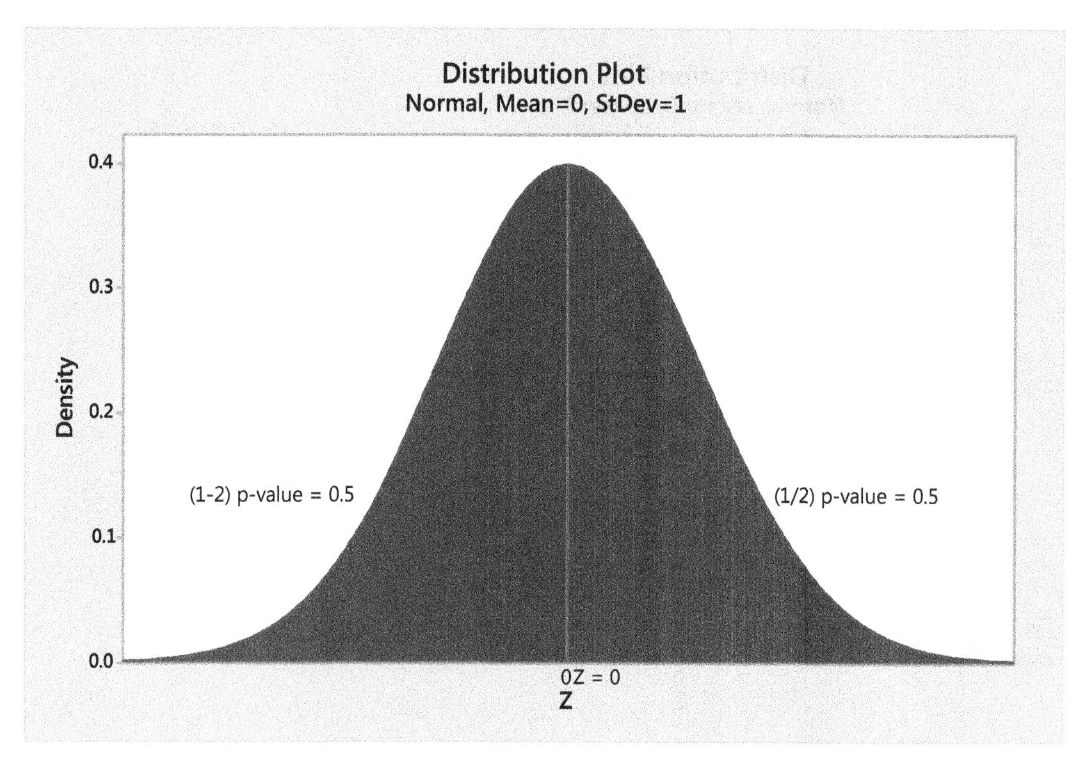

(d) The p-value $= 1$ is not ≤ 0.05. Therefore we do not reject H_0.

There is insufficient evidence at level of significance $\alpha = 0.05$ that the population mean is not equal to 10.

29. **(a)** H_0: $\mu = 0$ versus H_a: $\mu \neq 0$

We will reject H_0 if the p-value $\leq \alpha = 0.05$.

(b) $Z_{data} = \frac{\bar{x}-\mu_0}{\sigma/\sqrt{n}} = \frac{(-0.12)-0}{0.4/\sqrt{81}} = -2.7$

(c) p-value $= 2 \cdot P(Z \geq |Z_{data}|) = 2 \cdot P(Z \geq |-2.7|) = 2 \cdot \left(1 - P(Z < 2.7)\right) = 2 \cdot (1 - 0.9965) = 2 \cdot (0.0035) = 0.0070$

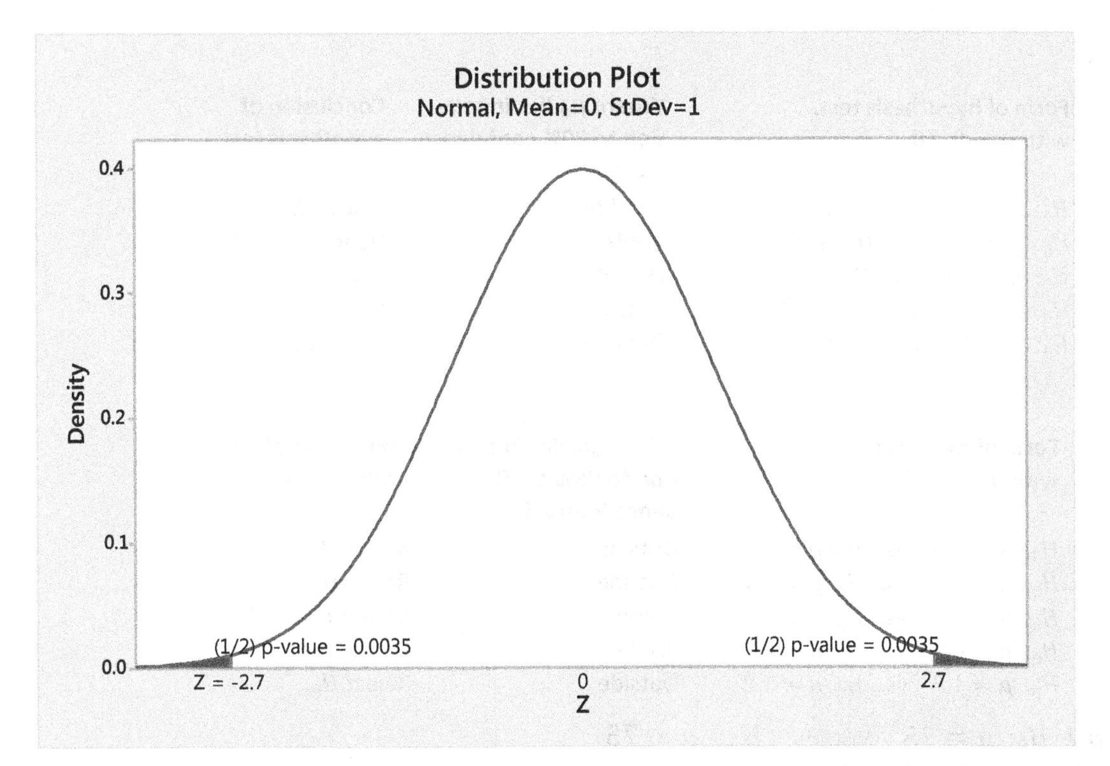

(d) The p-value $= 0.0070$ is ≤ 0.05. Therefore we reject H_0. There is evidence at level of significance $\alpha = 0.05$ that the population mean is not equal to 0.

31. The p-value of 0.2743 implies that there is no evidence against the null hypothesis that the population mean equals 3.14.

33. The p-value of approximately 0 implies that there is extremely strong evidence against the null hypothesis that the population mean equals -1.0.

35. The p-value of 0.0228 implies that there is solid evidence against the null hypothesis that the population mean equals 500.

37. The p-value of 1 implies that there is no evidence against the null hypothesis that the population mean equals 10.

39. The p-value of 0.007 implies that there is very strong evidence against the null hypothesis that the population mean equals 0.

41.

Value of μ_0	Form of hypothesis test, with $\alpha = 0.05$	Where μ_0 lies in relation to 95% confidence interval	Conclusion of hypothesis test
a. −3	H_0: $\mu = -3$ vs. H_a: $\mu \neq -3$	Outside	Reject H_0
b. −2	H_0: $\mu = -2$ vs. H_a: $\mu \neq -2$	Inside	Do not reject H_0
c. 0	H_0: $\mu = 0$ vs. H_a: $\mu \neq 0$	Inside	Do not reject H_0
d. 5	H_0: $\mu = 5$ vs. H_a: $\mu \neq 5$	Inside	Do not reject H_0
e. 7	H_0: $\mu = 7$ vs. H_a: $\mu \neq 7$	Outside	Reject H_0

43.

Value of μ_0	Form of hypothesis test, with $\alpha = 0.10$	Where μ_0 lies in relation to 90% confidence interval	Conclusion of hypothesis test
a. −3	H_0: $\mu = -3$ vs. H_a: $\mu \neq -3$	Outside	Reject H_0
b. −8	H_0: $\mu = -8$ vs. H_a: $\mu \neq -8$	Inside	Do not reject H_0
c. −11	H_0: $\mu = -11$ vs. H_a: $\mu \neq -11$	Outside	Reject H_0
d. 0	H_0: $\mu = 0$ vs. H_a: $\mu \neq 0$	Outside	Reject H_0
e,. 7	H_0: $\mu = 7$ vs. H_a: $\mu \neq 7$	Outside	Reject H_0

45.

Value of μ_0	Form of hypothesis test, with $\alpha = 0.05$	Where μ_0 lies in relation to 95% confidence interval	Conclusion of hypothesis test
a. 1.5	H_0: $\mu = 1.5$ vs. H_a: $\mu \neq 1.5$	Outside	Reject H_0
b. −1	H_0: $\mu = -1$ vs. H_a: $\mu \neq -1$	Outside	Reject H_0
c. 0.5	H_0: $\mu = 0.5$ vs. H_a: $\mu \neq 0.5$	Inside	Do not reject H_0
d. 0.9	H_0: $\mu = 0.9$ vs. H_a: $\mu \neq 0.9$	Inside	Do not reject H_0
e. 1.2	H_0: $\mu = 1.2$ vs. H_a: $\mu \neq 1.2$	Outside	Reject H_0

47. ***Step 1*** H_0: $\mu = 75$ versus H_a: $\mu < 75$

We will reject H_0 if the p-value $\leq \alpha = 0.05$.

Step 2 $Z_{\text{data}} = -1.6$

Step 3 p-value $= 0.0547992894$

Step 4 The p-value $= 0.0547992894$ is not ≤ 0.05. Therefore we do not reject H_0. There is insufficient evidence at level of significance $\alpha = 0.05$ that the population mean is less than 75.

49. ***Step 1*** H_0: $\mu = 70$ versus H_a: $\mu > 70$

We will reject H_0 if the p-value $\leq \alpha = 0.05$.

Step 2 $Z_{\text{data}} = 0.55$

Step 3 p-value $= 0.290$

Step 4 The p-value $= 0.290$ is not ≤ 0.05. Therefore we do not reject H_0.

There is insufficient evidence at level of significance $\alpha = 0.05$ that the population mean is greater than 70.

51. **(a)** H_0: $\mu = 3000$ versus H_a: $\mu > 3000$

We will reject H_0 if the p-value $\leq \alpha = 0.10$.

(b) $Z_{\text{data}} = \frac{\bar{x}-\mu_0}{\sigma/\sqrt{n}} = \frac{3120-3000}{600/\sqrt{9}} = 0.6$

(c) p-value $= P(Z \geq Z_{data}) = P(Z \geq 0.6) = 1 - P(Z < 0.6) = 1 - 0.7257 = 0.2743$

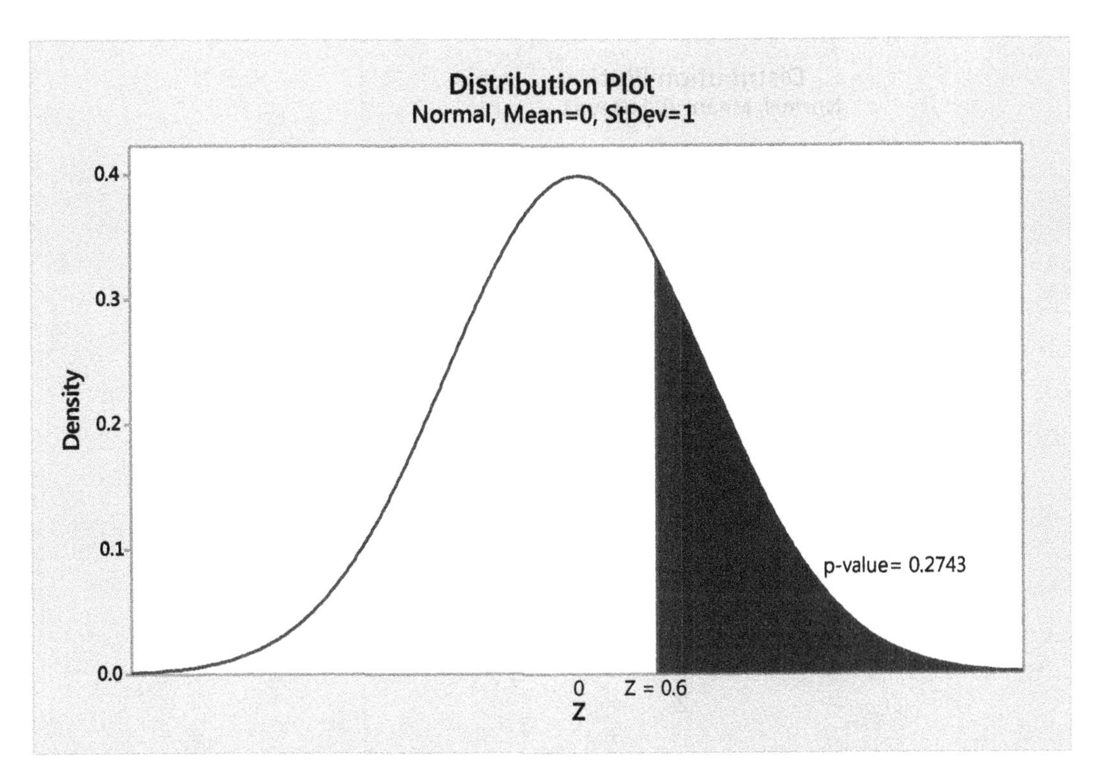

(d) The p-value $= 0.2743$ is not ≤ 0.10. Therefore we do not reject H_0.

There is insufficient evidence at level of significance $\alpha = 0.10$ that the population mean annual car insurance premium for a Porsche Panamera Turbo-S has increased from \$3000.

53. **(a)** $H_0 : \mu = 700$ versus $H_a : \mu < 700$. Reject H_0 if p-value ≤ 0.10.

(b) $Z_{\text{data}} = \frac{\bar{x}-\mu_0}{\sigma/\sqrt{n}} = \frac{650-700}{25/\sqrt{100}} = -20$

(c) p-value $= P(Z < -20) \approx 0$

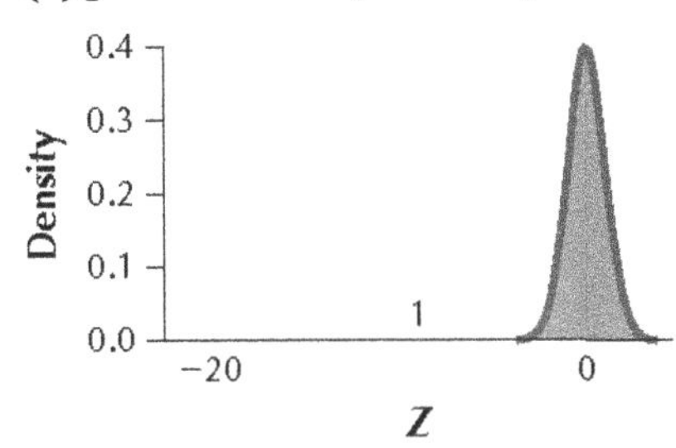

(d) Since p-value ≈ 0 is ≤ 0.10, reject H_0. There is evidence that the population mean number of meals prepared and eaten at home is less than 700.

55. **(a)** H_0: $\mu = 45{,}500$ versus H_a: $\mu > 45{,}500$

We will reject H_0 if the p-value $\leq \alpha = 0.05$.

(b) $Z_{\text{data}} = \frac{\bar{x}-\mu_0}{\sigma/\sqrt{n}} = \frac{50{,}000-45{,}500}{18{,}000/\sqrt{36}} = 1.5$

(c) p-value $= P(Z \geq Z_{data}) = P(Z \geq 1.5) = 1 - P(Z < 1.5) = 1 - 0.9332 = 0.0668$

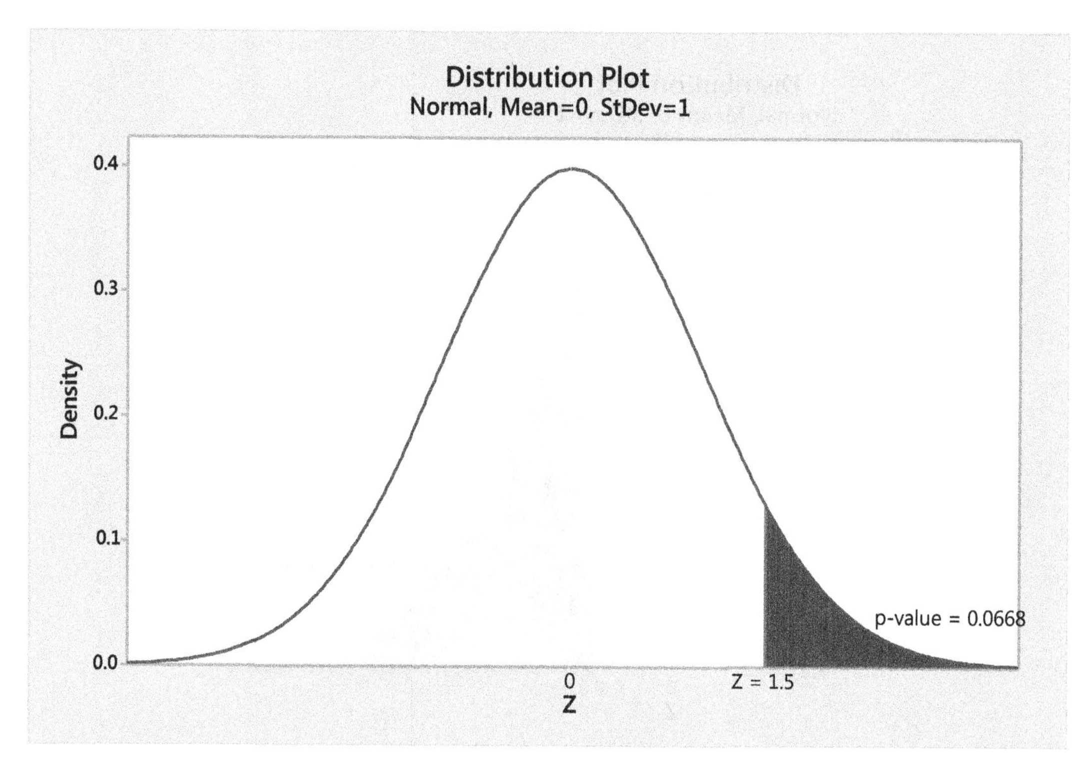

(d) The p-value $= 0.0668$ is not ≤ 0.05. Therefore we do not reject H_0.

There is insufficient evidence at level of significance $\alpha = 0.05$ that the population mean annual income of Millennials with Bachelor's degrees working full time has increased from $45,500.

57. **(a)** $H_0 : \mu = 3$ versus $H_a : \mu < 3$. Reject H_0 if p-value ≤ 0.01.

(b) $Z_{\text{data}} = \frac{\bar{x}-\mu_0}{\sigma/\sqrt{n}} = \frac{2.1-3}{0.2/\sqrt{9}} = -13.5$

(c) p-value $= P(Z < -13.5) \approx 0$

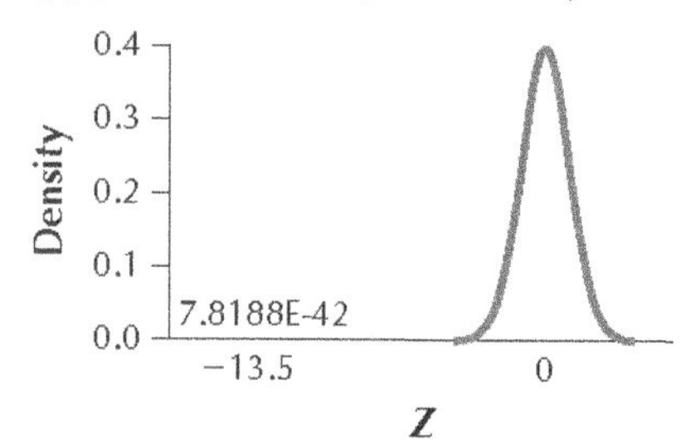

(d) Since p-value ≈ 0 is ≤ 0.01, reject H_0. There is evidence that the population mean time it takes hybrid cars to recoup their initial cost is less than 3 years.

59. The p-value of 0.2743 implies that there is no evidence against the null hypothesis that the population mean equals $3000

61. The p-value of approximately 0 implies that there is extremely strong evidence against the null hypothesis that the population mean equals 700 meals.

63. The p-value of 0.0668 implies that there is moderate evidence against the null hypothesis that the population mean equals $45,500.

65. The p-value of approximately 0 implies that there is extremely strong evidence against the null hypothesis that the population mean equals 3 years.

67. **(a)** $\bar{x} \pm Z_{\alpha/2}\left(\frac{\sigma}{\sqrt{n}}\right) = 2.95 \pm 1.96\left(\frac{0.21}{\sqrt{49}}\right) \approx 2.95 \pm 0.06 = (2.89, 3.01)$

(b)

Value of μ_0	Form of hypothesis test, with $\alpha = 0.05$	Where μ_0 lies in relation to 95% confidence interval	Conclusion of hypothesis test
(i) 2.89	H_0: $\mu = 2.89$ vs. H_a: $\mu \neq 2.89$	Outside	Reject H_0
(ii) 2.90	H_0: $\mu = 2.90$ vs. H_a: $\mu \neq 2.90$	Inside	Do not reject H_0
(iii) 3.00	H_0: $\mu = 3.00$ vs. H_a: $\mu \neq 3.00$	Inside	Do not reject H_0
(iv) 3.01	H_0: $\mu = 3.01$ vs. H_a: $\mu \neq 3.01$	Outside	Reject H_0

69. H_0: $\mu = 16{,}351$ versus H_a: $\mu > 16{,}351$

We will reject H_0 if the p-value $\leq \alpha = 0.05$.

$$Z_{\text{data}} = \frac{\bar{x} - \mu_0}{\sigma/\sqrt{n}} = \frac{17{,}251 - 16{,}351}{5000/\sqrt{100}} = 1.8$$

p-value $= P(Z \geq Z_{data}) = P(Z \geq 1.8) = 1 - P(Z < 1.8) = 1 - 0.9641 = 0.0359$

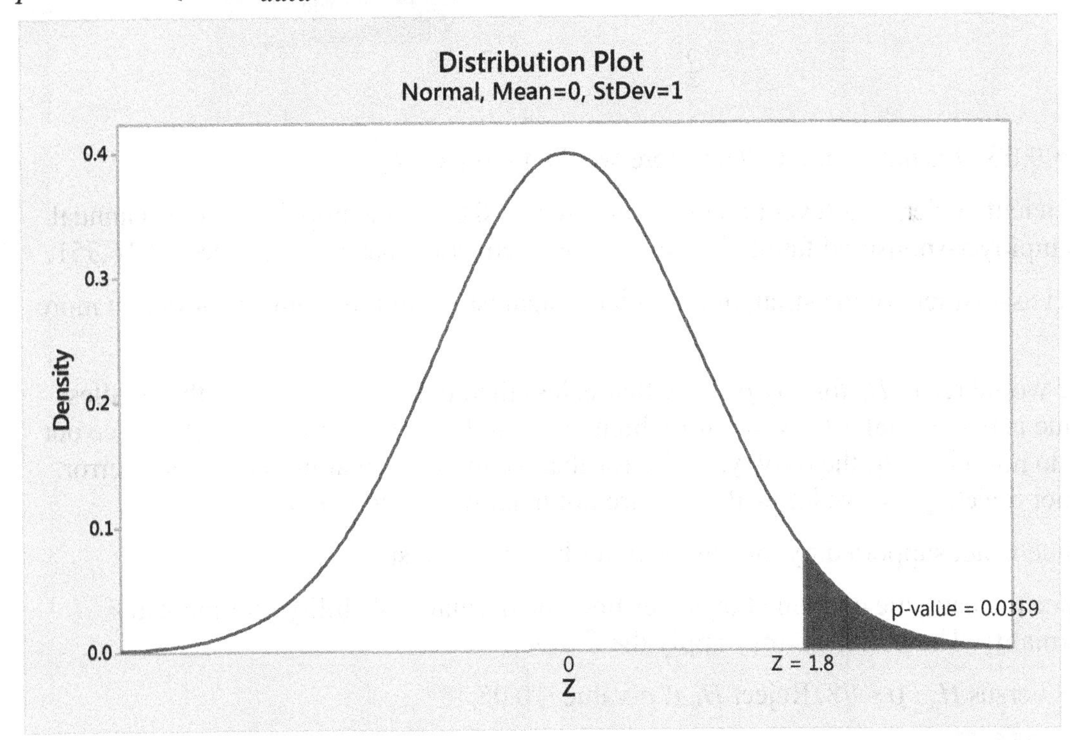

The p-value $= 0.0359$ is ≤ 0.05. Therefore we reject H_0.

There is evidence at level of significance $\alpha = 0.05$ that the population mean annual premium for employer-sponsored family health insurance coverage has increased from \$16,351.

71. H_0: $\mu = 16{,}351$ versus H_a: $\mu > 16{,}351$

We will reject H_0 if the p-value $\leq \alpha = 0.01$.

$Z_{\text{data}} = \frac{\bar{x}-\mu_0}{\sigma/\sqrt{n}} = \frac{17{,}251-16{,}351}{5000/\sqrt{100}} = 1.8$

p-value $= P(Z \geq Z_{data}) = P(Z \geq 1.8) = 1 - P(Z < 1.8) = 1 - 0.9641 = 0.0359$

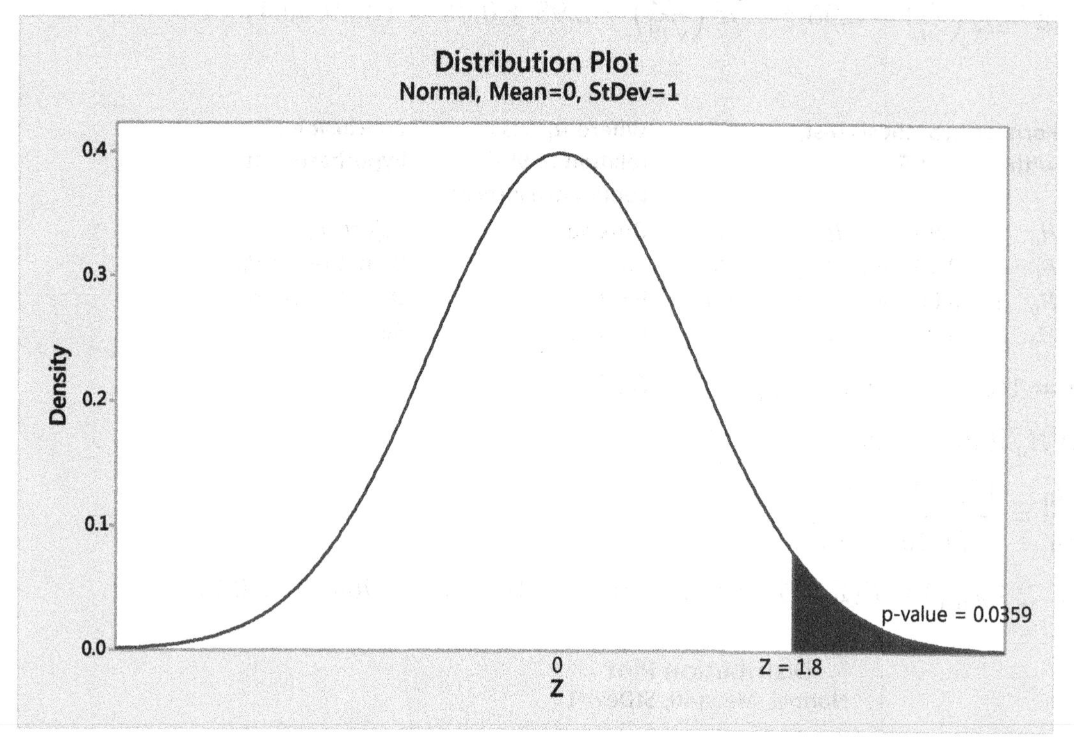

The p-value $= 0.0359$ is not ≤ 0.01. Therefore we do not reject H_0.

There is insufficient evidence at level of significance $\alpha = 0.01$ that the population mean annual premium for employer-sponsored family health insurance coverage has increased from $16,351.

Turn to a direct assessment of the strength of evidence against the null hypothesis or obtain more data.

73. **(a)** We would reject H_0 for any p-value that is less than or equal to α. Since the smallest possible p-value is 0, the smallest p-value for which we would reject H_0 is 0. **(b)** Since our conclusion is do not reject H_0, the only type of error that we might be making is a Type II error. Since we are not rejecting H_0, we know that we are not making a Type I error.

(c) This headline is not supported by the data and our hypothesis test.

75. **(a)** Since the data line up along the center line, the normal probability plot indicates acceptable normality. Therefore, we may apply the Z test.

(b) $H_0 : \mu = 78$ versus $H_a : \mu < 78$. Reject H_0 if p-value ≤ 0.05.

$Z_{\text{data}} = \frac{\bar{x}-\mu_0}{\sigma/\sqrt{n}} = \frac{75.6-78}{9/\sqrt{15}} \approx -1.03$

p-value $= P(Z < -1.03) = 0.1515$

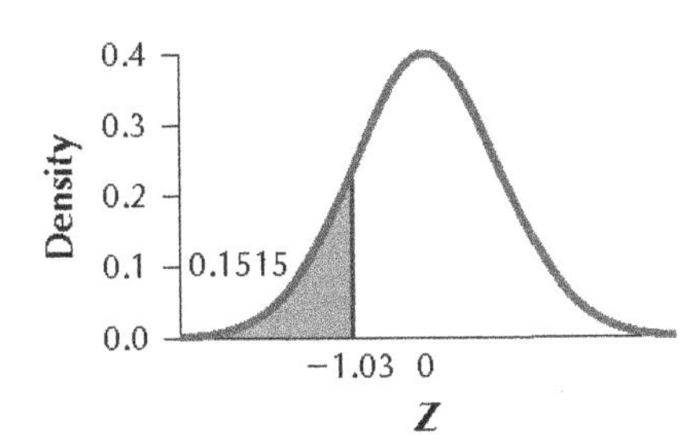

Since p-value = 0.1515 is not ≤ 0.05, do not reject H_0. There is insufficient evidence that the population mean heart rate for all women is less than 78 beats per minute.

(c) $H_0: \mu = 78$ versus $H_a: \mu \neq 78$. Reject H_0 if p-value ≤ 0.05.

$Z_{\text{data}} = \frac{\bar{x}-\mu_0}{\sigma/\sqrt{n}} = \frac{75.6-78}{9/\sqrt{15}} \approx -1.03$

p-value = $2\,P(Z < -1.03) = 2(0.1515) = 0.303$

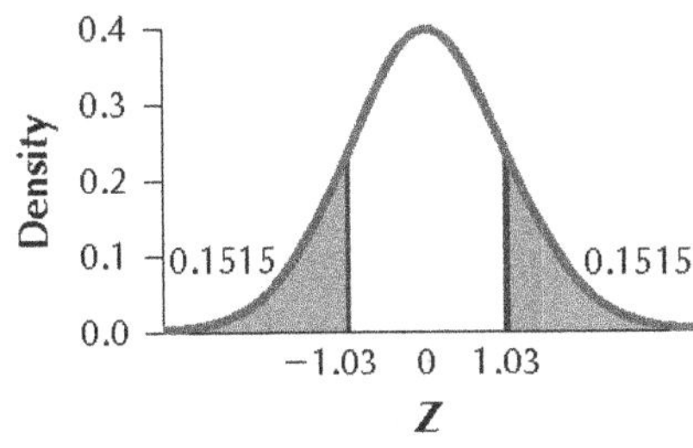

Since p-value = 0.303 is not ≤ 0.05, do not reject H_0. There is insufficient evidence that the population mean heart rate for all women is different than 78 beats per minute.

77. **(a)** Reject H_0.

(b) Do not reject H_0.

(c) Reject H_0.

(d) Do not reject H_0.

79. **(a)** H_0: $\mu = 210$ versus H_a: $\mu < 210$

where μ refers to the population mean sodium content per serving of breakfast cereal

We will reject H_0 if the p-value $\leq \alpha = 0.01$.

(b) $Z_{\text{data}} = \frac{\bar{x}-\mu_0}{\sigma/\sqrt{n}} = \frac{192.39-210}{50/\sqrt{23}} \approx -1.69$

(c) p-value $= P(Z \leq Z_{data}) = P(Z \leq -1.69) = 0.0455$

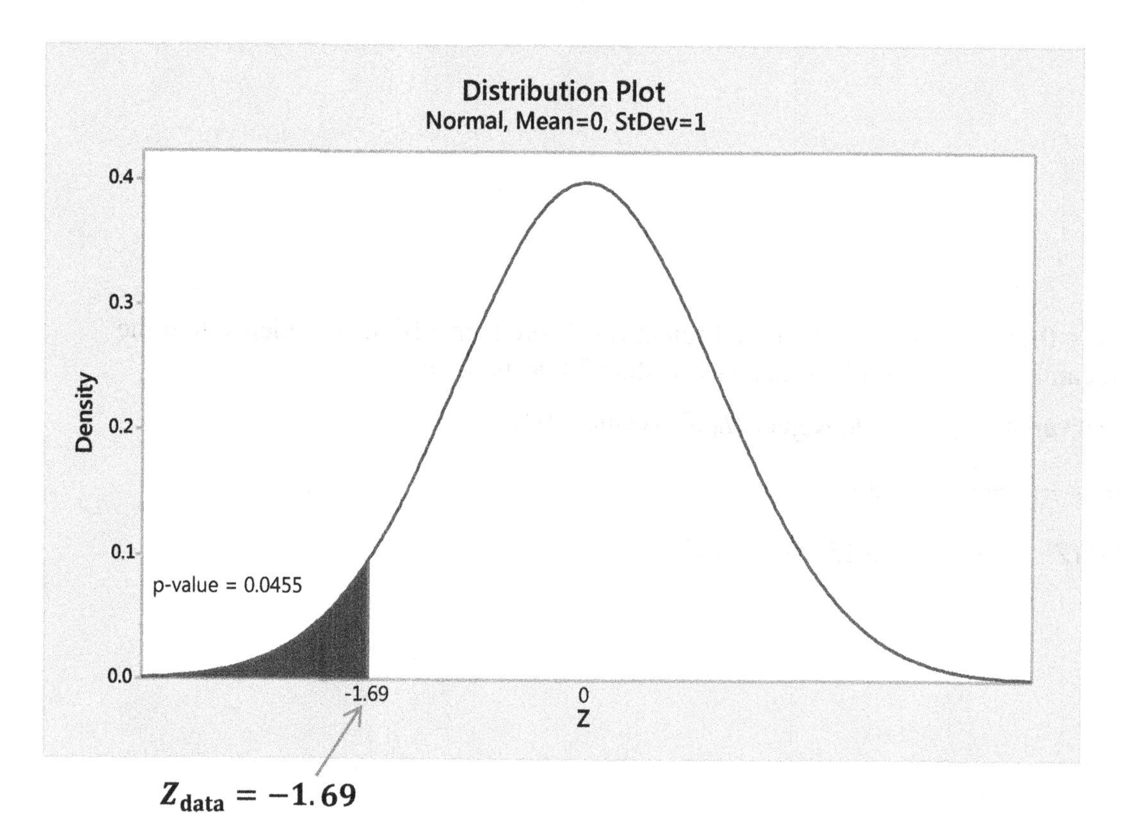

$\boldsymbol{Z_{\text{data}} = -1.69}$

(d) The p-value $= 0.0455$ is not ≤ 0.01. Therefore we do not reject H_0.

There is insufficient evidence at level of significance $\alpha = 0.01$ that the population mean sodium content per serving of breakfast cereal is less than 210 grams.

81. **(a)** $\bar{x} \pm Z_{\alpha/2}\left(\frac{\sigma}{\sqrt{n}}\right) = 192.39 \pm 1.96\left(\frac{50}{\sqrt{23}}\right) \approx 192.39 \pm 20.43 = (171.96, 212.82)$

(b)

Value of μ_0	Form of hypothesis test, with $\alpha = 0.05$	Where μ_0 lies in relation to 95% confidence interval	Conclusion of hypothesis test
(i) 200	H_0: $\mu = 200$ vs. H_a: $\mu \neq 200$	Inside	Do not reject H_0
(ii) 215	H_0: $\mu = 215$ vs. H_a: $\mu \neq 215$	Outside	Reject H_0
(iii) 170	H_0: $\mu = 170$ vs. H_a: $\mu \neq 170$	Outside	Reject H_0
(iv) 160	H_0: $\mu = 160$ vs. H_a: $\mu \neq 160$	Outside	Reject H_0

83. **(a)** $H_0 : \mu = 210$ versus $H_a : \mu < 210$. Reject H_0 if p-value ≤ 0.05.

$$Z_{\text{data}} = \frac{\bar{x} - \mu_0}{\sigma/\sqrt{n}} = \frac{192.39 - 210}{50/\sqrt{23}} \approx -1.69$$

p-value $= P(Z < -1.69) = 0.0455$

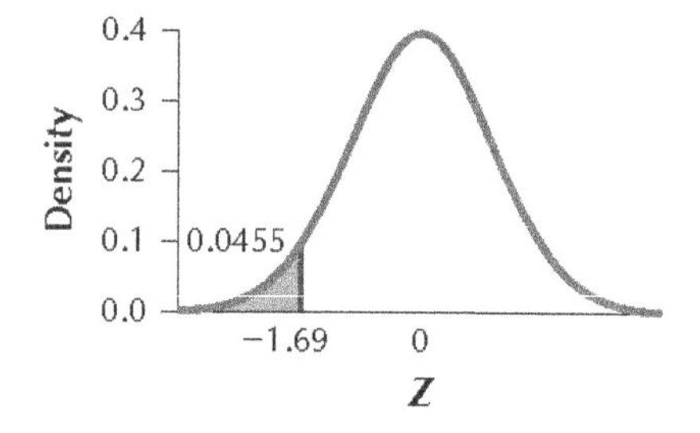

Since p-value = 0.0455 is ≤ 0.05, reject H_0. There is evidence that the population mean sodium content per serving of breakfast cereal is less than 210 grams.

(b) The conclusion is different this time because α was increased to a value greater than or equal to the p-value. The data have not changed.

(c) The first alternative would be to report the p-value and assess the strength of the evidence against the null hypothesis. The second alternative is to obtain more data.

85.

Descriptive Statistics: TOT_OCC

Variable	N	N*	Mean	SE Mean	StDev	Minimum	Q1	Median	Q3	Maximum
TOT_OCC	254	0	23901	5548	88421	42	2347	5767	12406	1026448

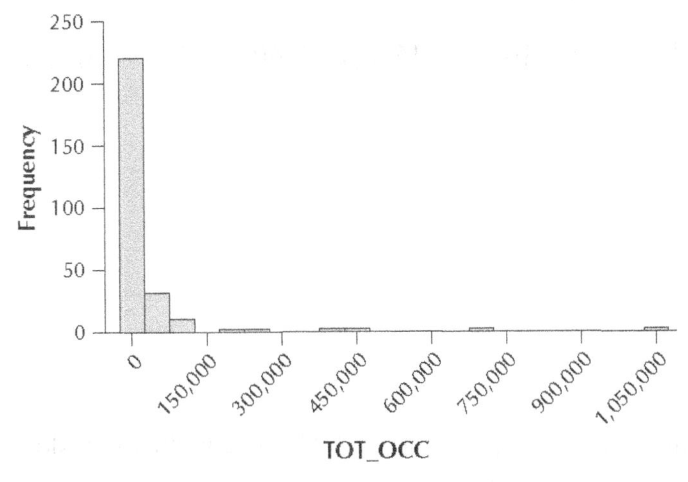

$\bar{x} = 23{,}901$, $s = 88{,}421$: The distribution is right-skewed.

87. Answers will vary.

89. H_0: $\mu = 18$ versus H_a: $\mu < 18$

We will reject H_0 if the p-value $\leq \alpha = 0.05$.

Answers will vary for the rest of the problem.

Section 9.4

1. The population standard deviation σ is known.

3. **(a)** $H_0 : \mu = 22$ versus $H_a : \mu < 22$.

(b) For a left-tailed test with $\alpha = 0.05$ and df $= n - 1 = 31 - 1 = 30$, $t_{crit} = -1.697$. Reject H_0 if $t_{data} \leq -1.697$.

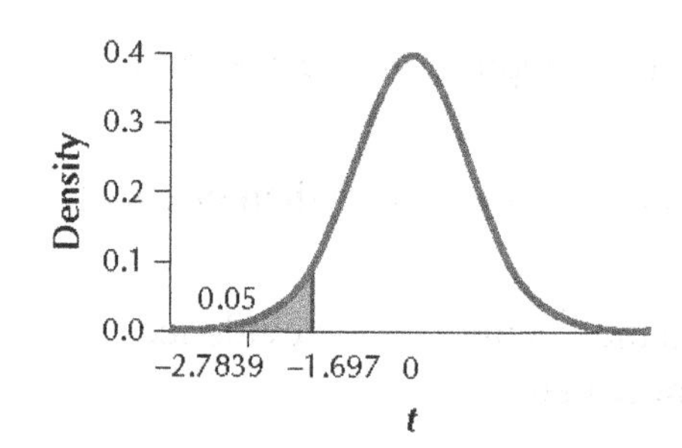

(c) $t_{data} = \frac{\bar{x}-\mu_0}{s/\sqrt{n}} = \frac{20-22}{4/\sqrt{31}} \approx -2.7839$. **(d)** Since $t_{\text{data}} = -2.7839$ is ≤ -1.697, the conclusion is to reject H_0. There is evidence at the 0.05 level of significance that the population mean is less than 22.

5. **(a)** $H_0 : \mu = 11$ versus $H_a : \mu > 11$.

(b) For a right-tailed test with $\alpha = 0.01$ and df $= n - 1 = 16 - 1 = 15$, $t_{\text{crit}} = 2.602$. Reject H_0 if $t_{\text{data}} \geq 2.602$.

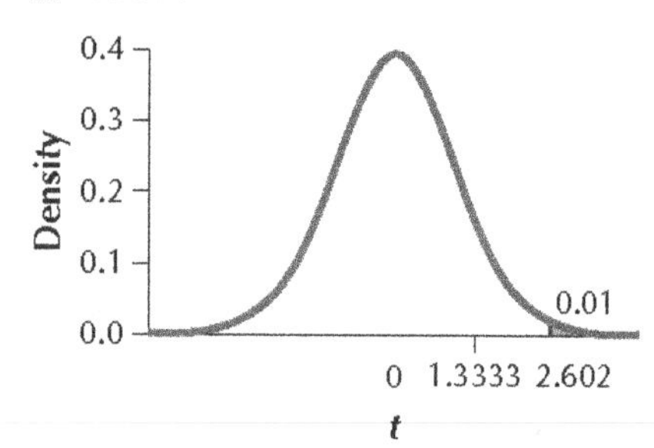

(c) $t_{data} = \frac{\bar{x}-\mu_0}{s/\sqrt{n}} = \frac{12-11}{3/\sqrt{16}} \approx 1.3333$. **(d)** Since $t_{\text{data}} = 1.3333$ is not ≥ 2.602, the conclusion is do not reject H_0. There is insufficient evidence at the 0.01 level of significance that the population mean is greater than 11.

7. **(a)** $H_0 : \mu = 100$ versus $H_a : \mu > 100$.

(b) For a right-tailed test with $\alpha = 0.01$ and df $= n - 1 = 25 - 1 = 24$, $t_{\text{crit}} = 2.492$. Reject H_0 if $t_{\text{data}} \geq 2.492$.

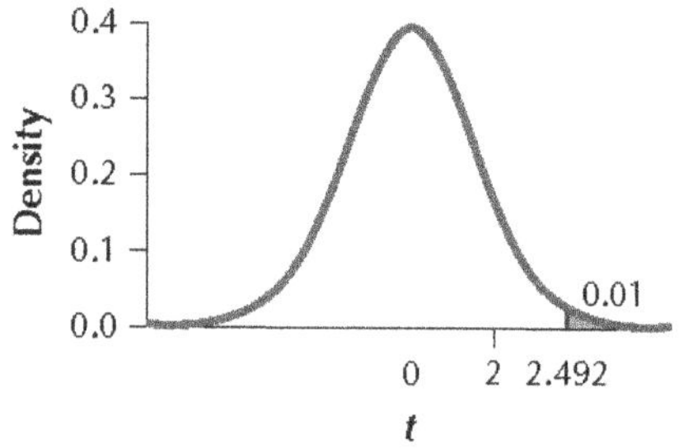

(c) $t_{data} = \frac{\bar{x}-\mu_0}{s/\sqrt{n}} = \frac{104-100}{10/\sqrt{25}} = 2$.

(d) Since $t_{\text{data}} = 2$ is not ≥ 2.492, the conclusion is do not reject H_0. There is insufficient evidence at the 0.01 level of significance that the population mean is greater than 100.

9. **(a)** $H_0 : \mu = 102$ versus $H_a : \mu \neq 102$.

(b) For a two-tailed test with $\alpha = 0.05$ and df $= n - 1 = 81 - 1 = 80$, $t_{\text{crit}} = 1.990$. Reject H_0 if $t_{\text{data}} \leq -1.990$ or if $t_{\text{data}} \geq 1.990$.

0.4
0.3
0.2
0.1
0.0
Density
0.025
0.025
−1.990 0 1.990 3.6
t

(c) $t_{data} = \frac{\bar{x}-\mu_0}{s/\sqrt{n}} = \frac{106-102}{10/\sqrt{81}} = 3.6.$

(d) Since $t_{\text{data}} = 3.6$ is ≥ 1.990, the conclusion is to reject H_0. There is evidence at the 0.05 level of significance that the population mean differs from 102.

11. **(a)** $H_0 : \mu = 1000$ versus $H_a : \mu \neq 1000$.

(b) For a two-tailed test with $\alpha = 0.10$ and df $= n - 1 = 25 - 1 = 24$, $t_{\text{crit}} = 1.711$. Reject H_0 if $t_{\text{data}} \leq -1.711$ or if $t_{\text{data}} \geq 1.711$.

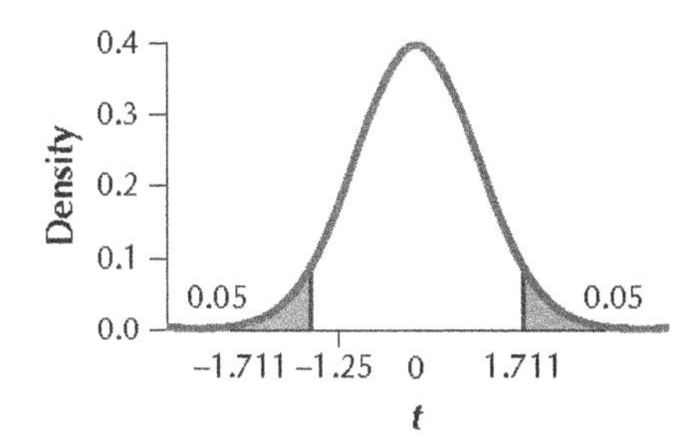

(c) $t_{data} = \frac{\bar{x}-\mu_0}{s/\sqrt{n}} = \frac{975-1000}{100/\sqrt{25}} = -1.25.$

(d) Since $t_{\text{data}} = -1.25$ is not ≤ -1.711 and not ≥ 1.711, the conclusion is do not reject H_0. There is insufficient evidence at the 0.10 level of significance that the population mean differs from 1000.

13. **(a)** $H_0 : \mu = 9$ versus $H_a : \mu \neq 9$.

(b) For a two-tailed test with $\alpha = 0.10$ and df $= n - 1 = 36 - 1 = 35$, $t_{\text{crit}} = 1.690$. Reject H_0 if $t_{\text{data}} \leq -1.690$ or if $t_{\text{data}} \geq 1.690$.

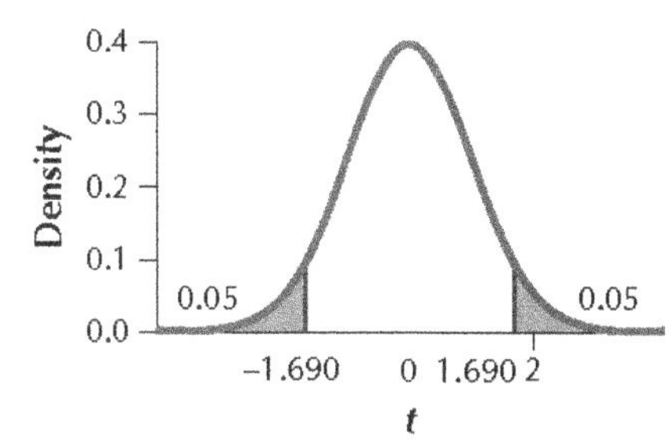

(c) $t_{data} = \frac{\bar{x}-\mu_0}{s/\sqrt{n}} = \frac{10-9}{3/\sqrt{36}} = 2.$

(d) Since $t_{\text{data}} = 2$ is ≥ 1.690, the conclusion is reject H_0. There is evidence at the 0.10 level of significance that the population mean differs from 9.

15. **(a)** $H_0 : \mu = 10$ versus $H_a : \mu < 10$. Reject H_0 if the p-value $\leq \alpha = 0.01$.

(b) $t_{data} = \frac{\bar{x}-\mu_0}{s/\sqrt{n}} = \frac{7-10}{5/\sqrt{81}} = -5.4.$

(c) Left-tailed test; df = $n - 1 = 81 - 1 = 80$.

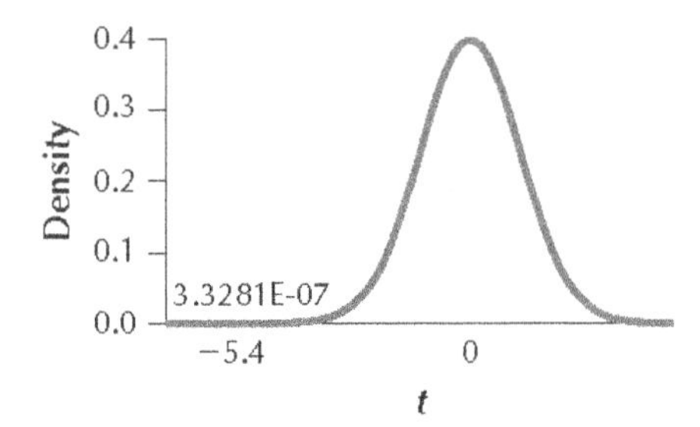

p-value = 0.

(d) Since the p-value = 0 is $\leq \alpha = 0.01$, the conclusion is reject H_0. There is evidence at the 0.01 level of significance that the population mean is less than 10.

17. **(a)** $H_0 : \mu = 100$ versus $H_a : \mu > 100$. Reject H_0 if the p-value $\leq \alpha = 0.10$.

(b) $t_{data} = \frac{\bar{x}-\mu_0}{s/\sqrt{n}} = \frac{120-100}{50/\sqrt{25}} = 2.$

(c) Right-tailed test; df = $n - 1 = 25 - 1 = 24$.

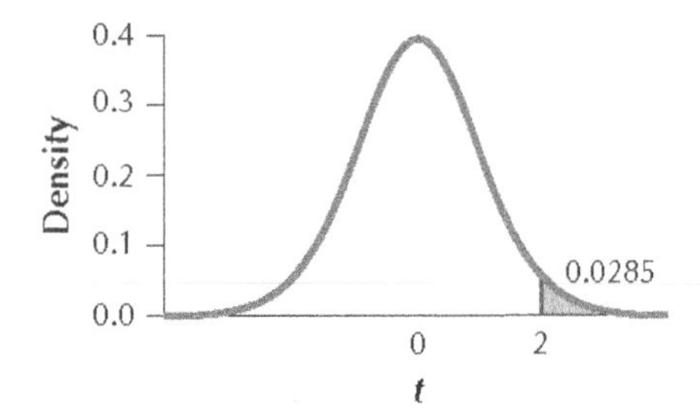

p-value = 0.0285.

(d) Since the p-value = 0.0285 is $\leq \alpha = 0.10$, the conclusion is reject H_0. There is evidence at the 0.10 level of significance that the population mean is greater than 100.

19. **(a)** $H_0 : \mu = 200$ versus $H_a : \mu > 200$. Reject H_0 if the p-value $\leq \alpha = 0.05$.

(b) $t_{data} = \frac{\bar{x}-\mu_0}{s/\sqrt{n}} = \frac{230-200}{5/\sqrt{400}} = 120.$

(c) Right-tailed test; df = $n - 1 = 400 - 1 = 399$.

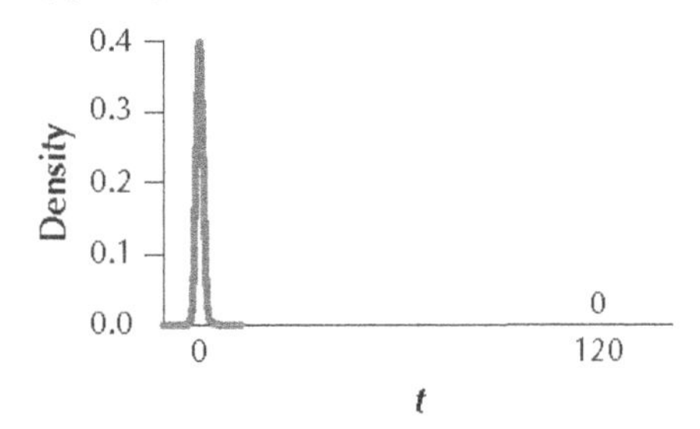

p-value = 0.

(d) Since the p-value = 0 is $\leq \alpha = 0.05$, the conclusion is reject H_0. There is evidence at the 0.05 level of significance that the population mean is greater than 200.

21. **(a)** $H_0 : \mu = 25$ versus $H_a : \mu \neq 25$. Reject H_0 if the p-value $\leq \alpha = 0.01$

(b) $t_{data} = \frac{\bar{x}-\mu_0}{s/\sqrt{n}} = \frac{25-25}{1/\sqrt{31}} = 0.$

(c) Two-tailed test; df = $n - 1 = 31 - 1 = 30$.

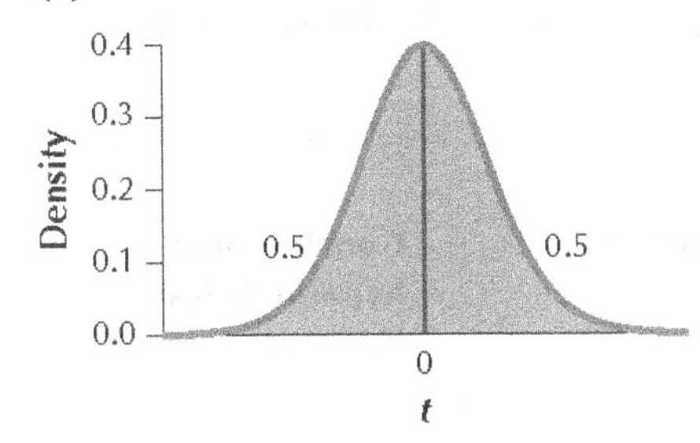

p-value = 2(0.5) = 1.

(d) Since the p-value = 1 is not $\leq \alpha = 0.01$, the conclusion is do not reject H_0. There is insufficient evidence at the 0.01 level of significance that the population mean differs from 25.

23. **(a)** $H_0 : \mu = 3.14$ versus $H_a : \mu \neq 3.14$. Reject H_0 if the p-value $\leq \alpha = 0.10$.

(b) $t_{data} = \frac{\bar{x}-\mu_0}{s/\sqrt{n}} = \frac{3.17-3.14}{0.5/\sqrt{9}} = 0.18.$

(c) Two-tailed test; df = $n - 1 = 9 - 1 = 8$.

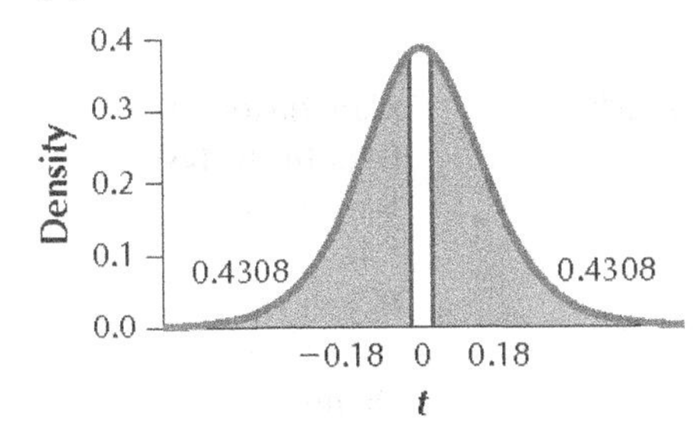

p-value = 2(0.4308) = 0.8616.

(d) Since the p-value = 0.8616 is not $\leq \alpha = 0.10$, the conclusion is do not reject H_0. There is insufficient evidence at the 0.10 level of significance that the population mean differs from 3.14.

25. **(a)** $H_0 : \mu = 0$ versus $H_a : \mu \neq 0$. Reject H_0 if the p-value $\leq \alpha = 0.05$.

(b) $t_{data} = \frac{\bar{x}-\mu_0}{s/\sqrt{n}} = \frac{1-0}{0.5/\sqrt{9}} = 6.$

(c) Two-tailed test; df = $n - 1 = 9 - 1 = 8$.

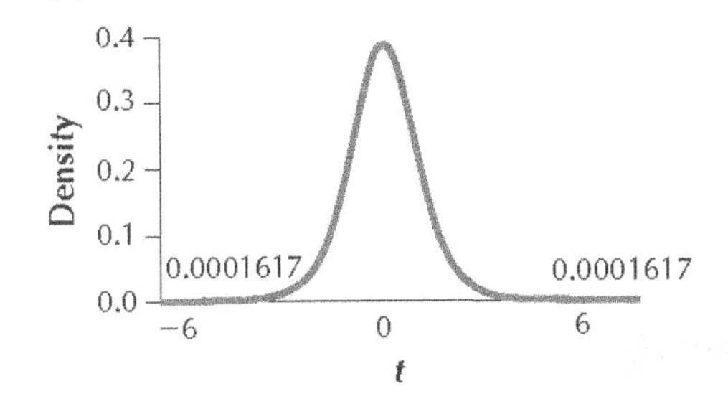

p-value = 2(0.0001617) $\approx$ 0.0003.

(d) Since the p-value $\approx$ 0.0003 is $\leq \alpha = 0.05$, the conclusion is reject H_0. There is evidence at the 0.05 level of significance that the population mean differs from 0.

27. Since this is a left-tailed test with df = $n - 1 = 31 - 1 = 30$, $t_{data} = -2.7839 < -2.750$, so p-value < 0.005.

29. Since this is a two-tailed test df = $n - 1 = 81 - 1 = 80$, $t_{data} = 3.6 > 2.639$, so p-value < 0.01.

31.

	Value of μ_0	**Form of hypothesis test, with $\alpha = 0.05$**	**Where μ_0 lies in relation to 95% confidence interval (1, 4)**	**Conclusion of hypothesis test**
a.	0	$H_0: \mu = 0$ vs. $H_a: \mu \neq 0$	Outside	Reject H_0
b.	2	$H_0: \mu = 2$ vs. $H_a: \mu \neq 2$	Inside	Do not reject H_0
c.	5	$H_0: \mu = 5$ vs. $H_a: \mu \neq 5$	Outside	Reject H_0

33.

	Value of μ_0	**Form of hypothesis test, with $\alpha = 0.10$**	**Where μ_0 lies in relation to 90% confidence interval (–20, –10)**	**Conclusion of hypothesis test**
a.	–21	$H_0: \mu = -21$ vs. $H_a: \mu \neq -21$	Outside	Reject H_0
b.	–5	$H_0: \mu = -5$ vs. $H_a: \mu \neq -5$	Outside	Reject H_0
c.	–12	$H_0: \mu = -12$ vs. $H_a: \mu \neq -12$	Inside	Do not reject H_0

35.

	Value of μ_0	**Form of hypothesis test, with $\alpha = 0.05$**	**Where μ_0 lies in relation to 95% confidence interval (–1, 1)**	**Conclusion of hypothesis test**
a.	1.5	$H_0: \mu = 1.5$ vs. $H_a: \mu \neq 1.5$	Outside	Reject H_0
b.	–1.5	$H_0: \mu = -1.5$ vs. $H_a: \mu \neq -1.5$	Outside	Reject H_0
c.	0	$H_0: \mu = 0$ vs. $H_a: \mu \neq 0$	Inside	Do not reject H_0

37. ***Step 1*** H_0: $\mu = 98.6$ versus H_a: $\mu > 98.6$

We will reject H_0 if the p-value $\leq \alpha = 0.05$.

Step 2 $t_{\text{data}} = 2$

Step 3 p-value = 0.0241198442

Step 4 The p-value = 0.0241198442 is ≤ 0.05. Therefore we reject H_0.

There is evidence at level of significance $\alpha = 0.05$ that the population mean is greater than 98.6.

39. ***Step 1*** H_0: $\mu = 20$ versus H_a: $\mu \neq 20$

We will reject H_0 if the p-value $\leq \alpha = 0.05$.

Step 2 $t_{\text{data}} = 2.308$

Step 3 p-value = 0.021

Step 4 The p-value = 0.021 is ≤ 0.05. Therefore we reject H_0.

There is evidence at level of significance $\alpha = 0.05$ that the population mean is not equal to 20.

41. Critical value method:

$H_0: \mu = 15{,}200$ versus $H_a: \mu > 15{,}200$

This is a right-tailed test with $\alpha = 0.05$ and df $= n - 1 = 400 - 1 = 399$. Since df $= 399$ is not in the table, we use the largest df in the table that is smaller than df $= 399$. Therefore, we use df $= 100$, so $t_{crit} = 1.660$.

Reject H_0 if $t_{data} \geq 1.660$.

$$t_{data} = \frac{\bar{x} - \mu_0}{s/\sqrt{n}} = \frac{16{,}000 - 15{,}200}{5000/\sqrt{400}} = 3.2.$$

Since $t_{data} = 3.2$ is ≥ 1.660, the conclusion is reject H_0. There is evidence at the 0.05 level of significance that the population mean cost of a stay in the hospital for women aged 18–44 is greater than \$15,200. Therefore, we can conclude at level of significance 0.05 that the population mean cost of a stay in the hospital for American women aged 18–24 has increased since 2010.

p-value method:

$H_0 : \mu = 15{,}200$ versus $H_a : \mu > 15{,}200$

Reject H_0 if the p-value $\leq \alpha = 0.05$.

$$t_{data} = \frac{\bar{x} - \mu_0}{s/\sqrt{n}} = \frac{16{,}000 - 15{,}200}{5000/\sqrt{400}} = 3.2.$$

Right-tailed test; df $= n - 1 = 400 - 1 = 399$

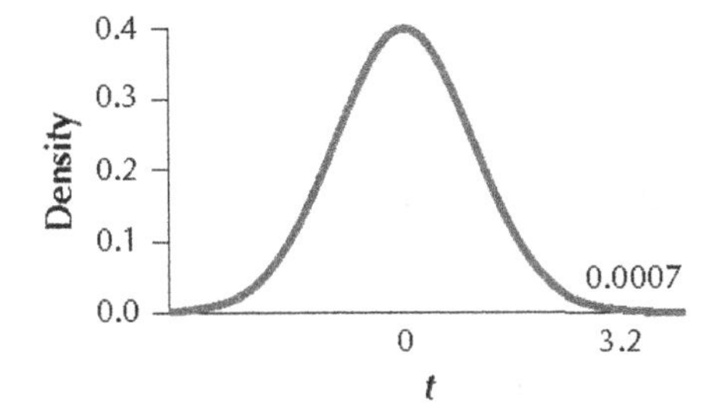

p-value $= 0.0007$.

Since the p-value $= 0.0007$ is $\leq \alpha = 0.05$, the conclusion is reject H_0. There is evidence at the 0.05 level of significance that the population mean cost of a stay in the hospital for women aged 18–44 is greater than \$15,200. Therefore, we can conclude at level of significance 0.05 that the population mean cost of a stay in the hospital for American women aged 18–24 has increased since 2010.

43. Critical value method:

$H_0 : \mu = 130$ versus $H_a : \mu < 130$

This is a left-tailed test with $\alpha = 0.05$ and df $= n - 1 = 100 - 1 = 99$. Since df $= 99$ is not in the table, use the largest df in the table that is less than 99. Use df $= 90$, so $t_{crit} = -1.662$.

Reject H_0 if $t_{data} \leq -1.662$.

$$t_{data} = \frac{\bar{x} - \mu_0}{s/\sqrt{n}} = \frac{110 - 130}{50/\sqrt{100}} = -4.$$

Since $t_{data} = -4$ is ≤ -1.662, the conclusion is reject H_0. There is evidence at the 0.05 level of significance that the population mean number of Facebook friends is less than 130.

p-value method:

$H_0 : \mu = 130$ versus $H_a : \mu < 130$

Reject H_0 if p-value $\leq \alpha = 0.05$.

$t_{data} = \frac{\bar{x}-\mu_0}{s/\sqrt{n}} = \frac{110-130}{50/\sqrt{100}} = -4.$

Left-tailed test; df = $n - 1 = 100 - 1 = 99$

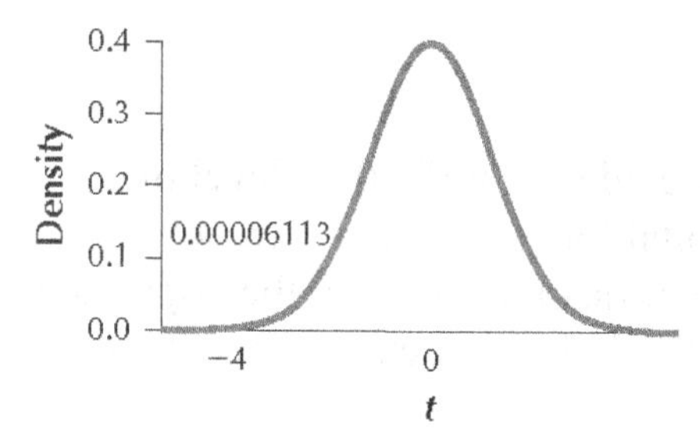

p-value = 0.

Since the p-value = 0 is $\leq \alpha = 0.05$, the conclusion is reject H_0. There is evidence at the 0.05 level of significance that the population mean number of Facebook friends is less than 130.

45. No. The distribution of the variable is not normal and the sample size is less than 30.

47. TI-83/84: $\bar{x} \approx \$0.8416666667$ million, $s \approx \$0.3484489441$ million

Step 1 H_0: $\mu = 500{,}000$ versus H_a: $\mu > 500{,}000$

We will reject H_0 if the p-value $\leq \alpha = 0.10$.

Step 2 TI-83/84: $\bar{x} \approx \$0.8416666667$ million, $s \approx \$0.3484489441$ million

$t_{\text{data}} = \frac{\bar{x}-\mu_0}{s/\sqrt{n}} = \frac{0.8416666667-0.5}{0.3484489441/\sqrt{6}} \approx 2.4018$

Step 3 df $= n - 1 = 6 - 1 = 5$, TI-83/84: p-value $= 0.0307414454$

Step 4 The p-value $= 0.0307414454$ is ≤ 0.10. Therefore we reject H_0.

There is evidence at level of significance $\alpha = 0.10$ that the population mean amount of cleanup money is greater than \$500,000.

49. **(a)** Increasing μ_0 to some value that is still less than $\bar{x}$ will decrease the value of $\bar{x} - \mu_0$. Since s and $\sqrt{n}$ are positive and $t_{\text{data}} = \frac{\bar{x}-\mu_0}{s/\sqrt{n}}$, t_{data} will decrease.

(b) Since t_{crit} does not depend on the value of μ_0, increasing μ_0 will not affect the value of t_{crit}.

(c) This is a right-tailed test, so the p-value $= P(t > t_{data})$. From (a), an increase in μ_0 to some value that is still less than $\bar{x}$ will decrease t_{data}. This will result in an increase in the p-value.

(d) Depends on the new value of μ_0. From (c), the p-value increases. If the increase in μ_0 results in a p-value that is still $\leq \alpha = 0.10$, then the conclusion will still be reject H_0. If the increase in μ_0 results in a p-value that is $> \alpha = 0.10$, then the conclusion will be do not reject H_0.

(e) The p-value in the original problem is greater than $\alpha = 0.01$. From (c), an increase in μ_0 to some value that is still less than $\bar{x}$ will result in an increase in the p-value. Therefore, the p-value will still be greater than $\alpha = 0.01$ so the conclusion will still be do not reject H_0.

(f) Depends on new value of μ_0. From (c), the p-value increases. The new value of μ_0 determines how much the p-value increases. The p-value determines the strength of the evidence against the null hypothesis.

51. ***Step 1*** H_0: $\mu = 4$ versus H_a: $\mu \neq 4$

We will reject H_0 if the p-value $\leq \alpha = 0.10$.

Step 2 TI-83/84: $t_{\text{data}} = -1.044$

Step 3 df $= n - 1 = 10 - 1 = 9$. TI-83/84: p-value $= 0.3237340082$

Step 4 The p-value $= 0.3237340082$ is not ≤ 0.10. Therefore we do not reject H_0. There is insufficient evidence at level of significance $\alpha = 0.10$ that the population mean amount of rainfall differs from 4 inches.

53. Yes. The normal probability plot indicates acceptable normality.

55. ***Step 1*** H_0: $\mu = 3264$ versus H_a: $\mu > 3264$

Step 2 $t_{\text{crit}} = 1.833$ Reject H_0 if $t_{\text{data}} \geq 1.833$.

Step 3 $t_{\text{data}} = \frac{\bar{x}-\mu_0}{s/\sqrt{n}} = \frac{3541-3264}{438/\sqrt{10}} \approx 2.000$

Step 4 $t_{\text{data}} = 2.000$ is ≥ 1.833. Therefore we reject H_0. There is evidence at level of significance $\alpha = 0.05$ that the population mean tuition and fees for community colleges has increased from the previous level of $3264.

57. There is solid evidence against the null hypothesis that the population mean is $3264.

59. The line "Test of $\mu = 3264$ versus $\neq 3264$".

61. Do not reject H_0. Fees have increased. Answers will vary.

63. Since the p-value for the two-tailed test is twice the p-value for the one-tailed test, it is possible to conclude that there is insufficient evidence that the population mean cost has changed but there is evidence that the population mean cost has increased if α is between the two p-values.

65.

One-Sample T: TOT_POP

```
Test of mu = 50000 vs not = 50000

Variable    N   Mean   StDev  SE Mean      95% CI          T
TOT_POP   790  18305  260938     9284  (81, 36529)    -3.41
P
0.001
```

Since p-value $= 0.001$ is ≤ 0.05, we reject H_0. There is evidence that the population mean population of these towns differs from 50,000.

67. H_0: $\mu = 0.4$ versus H_a: $\mu \neq 0.4$

Rest of answer will vary.

69. Answers will vary.

Section 9.5

1. The difference between $\hat{p}$ and p is that $\hat{p}$ is the sample proportion and p is population proportion.

3. When the sample proportion $\hat{p}$ is unusual or extreme in the sampling distribution of $\hat{p}$ in the assumption that H_0 is correct, then we reject H_0. Otherwise, there is insufficient evidence against H_0 and we should not reject H_0.

5. Between 0 and 1 inclusive: $0 \leq p_0 \leq 1$.

7. $\hat{p} = \frac{x}{n} = \frac{30}{50} = 0.6,\ Z_{data} = \frac{\hat{p}-p_0}{\sqrt{\frac{p_0 \cdot q_0}{n}}} = \frac{0.6-0.4}{\sqrt{\frac{(0.4)(0.6)}{50}}} \approx 2.8868$

9. $\hat{p} = \frac{x}{n} = \frac{45}{50} = 0.9,\ Z_{data} = \frac{\hat{p}-p_0}{\sqrt{\frac{p_0 \cdot q_0}{n}}} = \frac{0.9-0.4}{\sqrt{\frac{(0.4)(0.6)}{50}}} \approx 7.2169$

11. $\hat{p} = \frac{x}{n} = \frac{20}{80} = 0.25,\ Z_{data} = \frac{\hat{p}-p_0}{\sqrt{\frac{p_0 \cdot q_0}{n}}} = \frac{0.25-0.5}{\sqrt{\frac{(0.5)(0.5)}{80}}} \approx -4.47$

13. $\hat{p} = \frac{x}{n} = \frac{40}{80} = 0.5,\ Z_{data} = \frac{\hat{p}-p_0}{\sqrt{\frac{p_0 \cdot q_0}{n}}} = \frac{0.5-0.5}{\sqrt{\frac{(0.5)(0.5)}{80}}} = 0$

15. **(a)** We have $np_0 = 225(0.5) = 112.5 \geq 5$ and $n(1-p_0) = 225(1-0.5) = 112.5 \geq 5$, so we may use the Z test for proportions.

(b) $H_0 : p = 0.5$ versus $H_a : p < 0.5$.

(c) For a left-tailed test with $\alpha = 0.05$, $Z_{crit} = -1.645$. Reject H_0 if $Z_{data} \leq -1.645$.

(d) $\hat{p} = \frac{x}{n} = \frac{100}{225} \approx 0.4444,\ Z_{data} = \frac{\hat{p}-p_0}{\sqrt{\frac{p_0 \cdot q_0}{n}}} = \frac{0.4444-0.5}{\sqrt{\frac{(0.5)(0.5)}{225}}} \approx -1.67$

(e) Since $Z_{data} = -1.67$ is ≤ -1.645, we reject H_0. There is evidence that the population proportion is less than 0.5.

17. **(a)** We have $np_0 = 400(0.6) = 240 \geq 5$ and $n(1-p_0) = 400(1-0.6) = 160 \geq 5$, so we may use the Z test for proportions.

(b) $H_0 : p = 0.6$ versus $H_a : p > 0.6$

(c) For a right-tailed test with $\alpha = 0.05$, $Z_{crit} = 1.645$. Reject H_0 if $Z_{data} \geq 1.645$.

(d) $\hat{p} = \frac{x}{n} = \frac{260}{400} = 0.65,\ Z_{data} = \frac{\hat{p}-p_0}{\sqrt{\frac{p_0 \cdot q_0}{n}}} = \frac{0.65-0.6}{\sqrt{\frac{(0.6)(0.4)}{400}}} \approx 2.04$

(e) Since $Z_{data} = 2.04$ is ≥ 1.645, we reject H_0. There is evidence that the population proportion is greater than 0.6.

19. **(a)** We have $np_0 = 100(0.4) = 40 \geq 5$ and $n(1-p_0) = 100(1-0.4) = 60 \geq 5$, so we may use the Z test for proportions.

(b) $H_0 : p = 0.4$ veersus $H_a : p > 0.4$. Reject H_0 if p-value ≤ 0.05.

(c) $\hat{p} = \frac{x}{n} = \frac{44}{100} = 0.44,\ Z_{data} = \frac{\hat{p}-p_0}{\sqrt{\frac{p_0 \cdot q_0}{n}}} = \frac{0.44-0.4}{\sqrt{\frac{(0.4)(0.6)}{100}}} \approx 0.82$

(d) p-value $= P(Z > 0.82) = 1 - 0.7939 = 0.2061$

(e) Since p-value $= 0.2061$ is not ≤ 0.05, we do not reject H_0. There is insufficient evidence that the population proportion is greater than 0.4.

21. **(a)** We have $np_0 = 900(0.5) = 450 \geq 5$ and $n(1 - p_0) = 900(1 - 0.5) = 450 \geq 5$, so we may use the Z test for proportions.

(b) $H_0 : p = 0.5$ versus $H_a : p \neq 0.5$. Reject H_0 if p-value ≤ 0.05.

(c) $\hat{p} = \frac{x}{n} = \frac{475}{900} = 0.5278$, $Z_{data} = \frac{\hat{p} - p_0}{\sqrt{\frac{p_0 \cdot q_0}{n}}} = \frac{0.5278 - 0.5}{\sqrt{\frac{(0.5)(0.5)}{900}}} \approx 1.67$

(d) p-value $= 2P(Z > 1.67 = 2(1 - P(Z < 1.67)) = 2(1 - 0.9525) = 0.095$

(e) Since p-value $= 0.095$ is not ≤ 0.05, we do not reject H_0. There is insufficient evidence that the population proportion is not equal to 0.5.

23.

	Value of p_0	Form of hypothesis test, with $\alpha = 0.05$	Where p_0 lies in relation to 95% confidence interval (0.1, 0.9)	Conclusion of hypothesis test
a.	0	$H_0 : p = 0$ vs. $H_a : p \neq 0$	Outside	Reject H_0
b.	1	$H_0 : p = 1$ vs. $H_a : p \neq 1$	Outside	Reject H_0
c.	0.5	$H_0 : p = 0.5$ vs. $H_a : p \neq 0.5$	Inside	Do not reject H_0

25.

	Value of p_0	Form of hypothesis test, with $\alpha = 0.10$	Where p_0 lies in relation to 90% confidence interval (0.1, 0.2)	Conclusion of hypothesis test
a.	0.09	$H_0 : p = 0.09$ vs. $H_a : p \neq 0.09$	Outside	Reject H_0
b.	0.9	$H_0 : p = 0.9$ vs. $H_a : p \neq 0.9$	Outside	Reject H_0
c.	0.19	$H_0 : p = 0.19$ vs. $H_a : p \neq 0.19$	Inside	Do not reject H_0

27. ***Step 1*** H_0: $p = 0.5$ versus H_a: $p < 0.5$

We reject H_0 if the p-value $\leq \alpha = 0.05$.

Step 2 $Z_{data} = -2.12$

Step 3 p-value $= 0.0169473661$

Step 4 The p-value 0.0169473661 is $\leq \alpha = 0.05$, so we reject H_0. There is evidence at level of significance $\alpha = 0.05$ that the population proportion is less than 0.5.

29. ***Step 1*** H_0: $p = 0.05$ versus H_a: $p \neq 0.05$

We reject H_0 if the p-value $\leq \alpha = 0.05$.

Step 2 $Z_{data} = -1.95$

Step 3 p-value $= 0.052$

Step 4 The p-value 0.052 is not $\leq \alpha = 0.05$, so we do not reject H_0. There is insufficient evidence at level of significance $\alpha = 0.05$ that the population proportion is not equal to 0.05.

31. Both $n \cdot p_0 = 500(0.23) = 115 \geq 5$ and $n \cdot q_0 = 500(1 - 0.23) = 500(0.77) = 385 \geq 5$, so the conditions for using the Z test for proportions are met.

Step 1 H_0: $p = 0.23$ versu H_a: $p < 0.23$

We reject H_0 if the p-value $\leq \alpha = 0.10$.

Step 2 $\hat{p} = \frac{x}{n} = \frac{100}{500} = 0.2$. $Z_{data} = \frac{\hat{p}-p_0}{\sqrt{\frac{p_0 \cdot q_0}{n}}} = \frac{0.2-0.23}{\sqrt{\frac{(0.23)(0.77)}{500}}} \approx -1.59$.

Step 3 p-value $= P(Z \leq Z_{data}) \approx P(Z \leq -1.59) \approx 0.0559$

Step 4 The p-value $= 0.0559$ is $\leq \alpha = 0.10$, so we reject H_0. There is evidence at level of significance $\alpha = 0.10$ that the population proportion of Facebook users ages 18–24 has decreased from 23%.

33. Both $n \cdot p_0 = 900(0.048) = 43.2 \geq 5$ and $n \cdot q_0 = 400(1 - 0.048) = 900(0.952) = 856.8 \geq 5$, so the conditions for using the Z test for proportions are met.

Step 1 H_0: $p = 0.048$ versus H_a: $p > 0.048$

We reject H_0 if the p-value $\leq \alpha = 0.01$.

Step 2 $\hat{p} = \frac{x}{n} = \frac{54}{900} = 0.06$. $Z_{data} = \frac{\hat{p}-p_0}{\sqrt{\frac{p_0 \cdot q_0}{n}}} = \frac{0.06-0.048}{\sqrt{\frac{(0.048)(0.952)}{900}}} \approx 1.68$.

Step 3 p-value $= P(Z > 1.68) = 1 - 0.9535 = 0.0465$

Step 4 The p-value $= 0.0465$ is not $\leq \alpha = 0.01$, so we do not reject H_0. There is insufficient evidence at level of significance $\alpha = 0.01$ that the population proportion of persons ages 12 or older who used a prescription pain reliever nonmedically has increased from 4.8%.

35. Both $n \cdot p_0 = 100(0.58) = 58 \geq 5$ and $n \cdot q_0 = 100(1 - 0.58) = 100(0.42) = 42 \geq 5$, so the conditions for using the Z test for proportions are met.

Step 1 H_0: $p = 0.58$ versus H_a: $p > 0.58$

We reject H_0 if the p-value $\leq \alpha = 0.05$.

Step 2 $\hat{p} = \frac{x}{n} = \frac{67}{100} = 0.67$. $Z_{data} = \frac{\hat{p}-p_0}{\sqrt{\frac{p_0 \cdot q_0}{n}}} = \frac{0.67-0.58}{\sqrt{\frac{(0.58)(0.42)}{100}}} \approx 1.82$.

Step 3 p-value $= P(Z > 1.82) = 1 - 0.9656 = 0.0344$

Step 4 The p-value $= 0.0344$ is $\leq \alpha = 0.05$, so we reject H_0. There is evidence at level of significance $\alpha = 0.05$ that the population proportion of mutual funds that underperform has increased from 58%.

37. **(a)** We have $np_0 = 100(0.153) = 15.3 \geq 5$ and $nq_0 = 100(0.847) = 84.7 \geq 5$, so we may use the Z test for proportions.

(b) **Step 1** $H_0 : p = 0.153$ versus. $H_a : p \neq 0.153$.

Reject H_0 if p-value ≤ 0.01.

Step 2 $\hat{p} = \frac{x}{n} = \frac{23}{100} = 0.23$. $Z_{data} = \frac{\hat{p}-p_0}{\sqrt{\frac{p_0 \cdot q_0}{n}}} = \frac{0.23-0.153}{\sqrt{\frac{(0.153)(0.847)}{100}}} \approx 2.14$.

Step 3 p-value $= 2\,P(Z > 2.14) = 2(1 - P(Z < 2.14)) = 2(1 - 0.9838) = 0.0324$. **Step 4** Since p-value $= 0.0324$ is not ≤ 0.01, we do not reject H_0. There is insufficient evidence that the population proportion of Latino families that had a household income of at least \$75,000 is not equal to 0.153.

39. Both $n \cdot p_0 = 100(0.57) = 57 \geq 5$ and $n \cdot q_0 = 100(1 - 0.57) = 100(0.43) = 43 \geq 5$, so the conditions for using the Z test for proportions are met.

Step 1 H_0: $p = 0.57$ versus H_a: $p \neq 0.57$

We reject H_0 if the p-value $\leq \alpha = 0.10$.

Step 2 $\hat{p} = \frac{x}{n} = \frac{60}{100} = 0.60$. $Z_{data} = \frac{\hat{p}-p_0}{\sqrt{\frac{p_0 \cdot q_0}{n}}} = \frac{0.60-0.57}{\sqrt{\frac{(0.57)(0.43)}{100}}} \approx 0.61$.

Step 3 p-value = $2\,P(Z > 0.61) = 2(1 - P(Z < 0.61)) = 2(1 - 0.7291) = 0.5418$

Step 4 The p-value = 0.5418 is not $\leq \alpha = 0.10$, so we do not reject H_0. There is insufficient evidence at level of significance $\alpha = 0.10$ that the population proportion of young people ages 18–24 who are living with their parents differs from 57%.

41. **(a)** We have $np_0 = 100(0.11) = 11 \geq 5$ and $nq_0 = 100(0.89) = 89 \geq 5$, so we may use the Z test for proportions.

(b) $H_0: p = 0.11$ versus $H_a: p < 0.11$

For a left-tailed test with $\alpha = 0.05$, $Z_{crit} = -1.645$. Reject H_0 if $Z_{data} \leq -1.645$.

$$Z_{data} = \frac{\hat{p}-p_0}{\sqrt{\frac{p_0 \cdot q_0}{n}}} = \frac{0.06-0.11}{\sqrt{\frac{(0.11)(0.89)}{100}}} \approx -1.60.$$

Since $Z_{data} = -1.60$ is not ≤ 1.645, we do not reject H_0. There is insufficient evidence that the population proportion of children age 6 and under exposed to ETS at home on a regular basis is less than 0.11.

43. **(a)** $H_0: p = 0.11$ versus $H_a: p < 0.11$. Reject H_0 if p-value ≤ 0.10.

$$Z_{data} = \frac{\hat{p}-p_0}{\sqrt{\frac{p_0 \cdot q_0}{n}}} = \frac{0.06-0.11}{\sqrt{\frac{(0.11)(0.89)}{100}}} \approx -1.60.$$

p-value = $P(Z < -1.60) = 0.0548$

Since p-value = 0.0548 is ≤ 0.10, we reject H_0. There is evidence that the population proportion of children age 6 and under exposed to environmental tobacco smoke at home on a regular basis is less than 0.11.

(b) The difference in the conclusions is because we changed the value of α and not because we used different methods for the two different hypothesis tests. Since p-value = 0.0548, $0.05 < p$-value < 0.10. Therefore we would reject H_0 for $\alpha = 0.10$ and we would not reject H_0 for $\alpha = 0.05$.

(c) Since $0.05 < p$-value ≤ 0.10, there is mild evidence against the null hypothesis that the population proportion of children age 6 and under exposed to environmental smoke at home on a regular basis is equal to 0.11.

45. Both $n \cdot p_0 = 1000(0.12) = 120 \geq 5$ and $n \cdot q_0 = 1000(1 - 0.12) = 1000(0.88) = 880 \geq 5$, so the conditions for using the Z test for proportions are met.

Step 1 H_0: $p = 0.12$ versus H_a: $p > 0.12$

We reject H_0 if the p-value $\leq \alpha = 0.05$.

Step 2 $\hat{p} = \frac{x}{n} = \frac{134}{1000} = 0.134$. $Z_{data} = \frac{\hat{p}-p_0}{\sqrt{\frac{p_0 \cdot q_0}{n}}} = \frac{0.134-0.12}{\sqrt{\frac{(0.12)(0.88)}{1000}}} \approx 1.36$

Step 3 p-value = $P(Z > 1.36) = 1 - 0.9131 = 0.0869$

Step 4 The p-value = 0.0869 is not $\leq \alpha = 0.05$, so we do not reject H_0. There is insufficient evidence at level of significance $\alpha = 0.05$ that the population proportion of young people ages 18–24 who have had an accident has increased from 12%.

47. **(a)** For values of $p_0 < 0.5$, an increase in p_0 will result in an increase in $p_0 q_0$. Since $\sigma_{\hat{p}} = \sqrt{\frac{p_0 \cdot q_0}{n}}$ and n remains the same, $\sigma_{\hat{p}}$ will increase.

(b) An increase in p_0 such that p_0 is still less than $\hat{p} = 0.134$ will result in a decrease in $\hat{p} - p_0$ but $\hat{p} - p_0$ will still be positive. For values of $p_0 < 0.5$, an increase in p_0 will result in an increase in $p_0 q_0$. Since. $Z_{data} = \frac{\hat{p}-p_0}{\sqrt{\frac{p_0 \cdot q_0}{n}}}$ and n remains the same, Z_{data} will decrease.

(c) From (b), Z_{data} decreases. Since the p-value $= P(Z > Z_{data})$, this will result in an increase in the p-value.

(d) Since α does not depend on p_0 increasing p_0 will not affect α.

(e) The p-value = 0.0869 is not $\leq \alpha = 0.05$, so we do not reject H_0. An increase in p_0 such that p_0 is still less than $\hat{p} = 0.134$, will result in an increase in the p-value from (c). Therefore, the p-value will still not be $\leq \alpha = 0.05$, so we do not reject H_0.

49. Answers will vary.

51 $p = 0.0507$. Answers will vary.

53. Answers will vary.

55. H_0: $p = 0.5$ versus H_a: $p > 0.5$

The rest of the answer will vary.

57. H_0: $p = 0.1$ versus H_a: $p > 0.1$

The rest of the answer will vary.

Section 9.6

1. One instance where an analyst would be interested in performing a hypothesis test about the population standard deviation is the following: A pharmaceutical company that wishes to ensure the safety of a particular new drug would perform statistical tests to make sure that the drug's effect was consistent and did not vary from patient to patient.

3. Since $\sigma = \sqrt{\sigma^2}$ it will never be less than 0. Therefore it does not make sense to test whether $\sigma < 0$.

5. The confidence interval for σ can be used to perform a two-tailed test for σ.

If σ_0 lies in the confidence interval we do not reject H_0.

If σ_0 does not lie in the confidence interval we reject H_0.

7. **(a)** $H_0\colon \sigma = 1$ versus $H_a\colon \sigma > 1$

(b) df $= n - 1 = 21 - 1 = 20$. $\chi_\alpha^2 = \chi_{0.05}^2 = 31.410$. Reject H_0 if $\chi_{\text{data}}^2 \geq 31.410$.

(c) $\chi_{\text{data}}^2 = \frac{(n-1)s^2}{\sigma_0{}^2} = \frac{(21-1)3}{1} = 60$

(d) Since $\chi_{\text{data}}^2 \geq 31.410$, reject H_0. There is evidence that the population standard deviation is greater than 1.

9. **(a)** $H_0\colon \sigma = 3$ versus $H_a\colon \sigma \neq 3$

(b) df $= n - 1 = 16 - 1 = 15$. $\chi_{\alpha/2}^2 = \chi_{0.025}^2 = 27.488$ and $\chi_{1-\alpha/2}^2 = \chi_{0.975}^2 = 6.262$. Reject H_0 if $\chi_{\text{data}}^2 \leq 6.262$ or $\chi_{\text{data}}^2 \geq 27.488$.

(c) $\chi_{\text{data}}^2 = \frac{(n-1)s^2}{\sigma_0{}^2} = \frac{(16-1)2.5^2}{9} \approx 10.417$

(d) Since χ_{data}^2 is not ≤ 6.262 and χ_{data}^2 is not ≥ 27.488, we do not reject H_0. There is insufficient evidence that the population standard deviation is different from 3.

11. **(a)** $H_0\colon \sigma = 20$ versus $H_a\colon \sigma < 20$

(b) df $= n - 1 = 8 - 1 = 7$. $\chi_{1-\alpha}^2 = \chi_{0.90}^2 = 2.833$. Reject H_0 if $\chi_{\text{data}}^2 \leq 2.833$.

(c) $\chi_{\text{data}}^2 = \frac{(n-1)s^2}{\sigma_0{}^2} = \frac{(8-1)350}{400} = 6.125$

(d) Since χ_{data}^2 is not ≤ 2.833, we do not reject H_0. There is insufficient evidence that the population standard deviation is less than 20.

13. **(a)** $H_0\colon \sigma = 1$ versus $H_a\colon \sigma > 1$

Reject H_0 if the p-value $\leq \alpha = 0.05$.

(b) $\chi_{\text{data}}^2 = \frac{(n-1)s^2}{\sigma_0{}^2} = \frac{(21-1)3}{1} = 60$

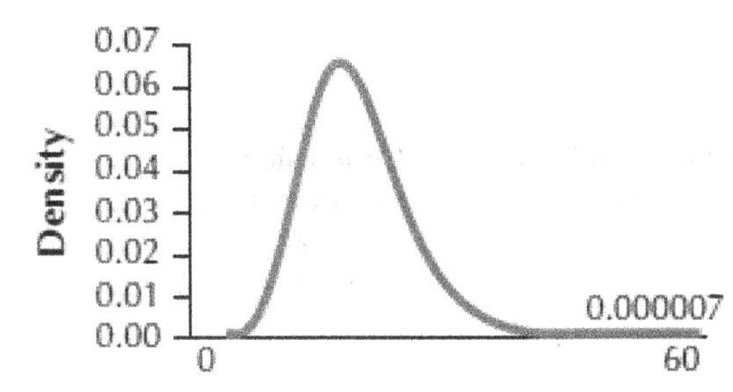

(c) df $= n - 1 = 21 - 1 = 20$. p-value $= P(\chi^2 > \chi_{data}^2) = P(\chi^2 > 60) \approx 7.121750863 \times 10^{-6}$

(d) Since the p-value ≤ 0.05, we reject H_0. There is evidence that the population standard deviation is greater than 1.

15. **(a)** $H_0\colon \sigma = 3$ versus $H_a\colon \sigma \neq 3$

Reject H_0 if the p-value $\leq \alpha = 0.05$.

(b) $\chi_{\text{data}}^2 = \frac{(n-1)s^2}{\sigma_0{}^2} = \frac{(16-1)2.5^2}{9} \approx 10.417$

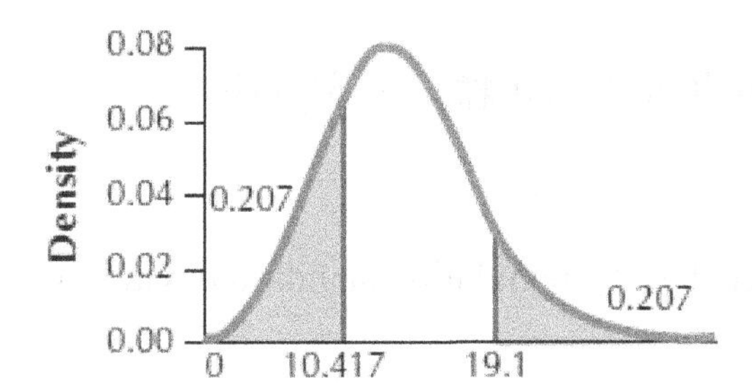

(c) df = $n - 1 = 16 - 1 = 15$. $P(\chi^2 > \chi^2_{data}) = P(\chi^2 > 10.417) \approx 0.7927 > 0.5$, so p-value = $2(\chi^2 < \chi^2_{data}) = 2P(\chi^2 < 10.417) \approx 0.4145552434$

(d) Since the p-value is not ≤ 0.05, we do not reject H_0. There is insufficient evidence that the population standard deviation is different from 3.

17. **(a)** $H_0: \sigma = 20$ versus $H_a: \sigma < 20$

Reject H_0 if the p-value $\le \alpha = 0.05$.

(b) $\chi^2_{\text{data}} = \frac{(n-1)s^2}{\sigma_0{}^2} = \frac{(8-1)350}{400} = 6.125$

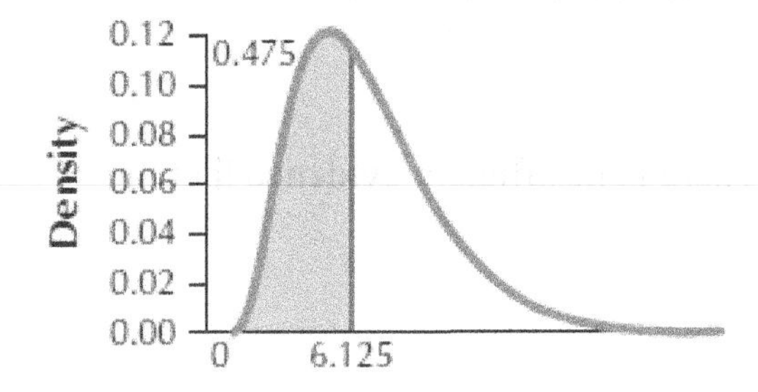

(c) df = $n - 1 = 8 - 1 = 7$. p-value = $P(\chi^2 < \chi^2_{data}) = P(\chi^2 < 6.125) \approx 0.4747679539$

(d) Since the p-value is not ≤ 0.05, we do not reject H_0. There is insufficient evidence that the population standard deviation is less than 20.

19.

	Value of σ_0	**Form of hypothesis test, with $\alpha = 0.05$**	**Where σ_0 lies in relation to 95% confidence interval (1, 4)**	**Conclusion of hypothesis test**
a.	0	$H_0: \sigma = 0$ vs. $H_a: \sigma \neq 0$	Outside	Reject H_0
b.	2	$H_0: \sigma = 2$ vs. $H_a: \sigma \neq 2$	Inside	Do not reject H_0
c.	5	$H_0: \sigma = 5$ vs. $H_a: \sigma \neq 5$	Outside	Reject H_0

21.

	Value of σ_0	**Form of hypothesis test, with $\alpha = 0.10$**	**Where σ_0 lies in relation to 90% confidence interval (100, 200)**	**Conclusion of hypothesis test**
a.	150	$H_0: \sigma = 150$ vs. $H_a: \sigma \neq 150$	Inside	Do not reject H_0
b.	250	$H_0: \sigma = 250$ vs. $H_a: \sigma \neq 250$	Outside	Reject H_0
c.	0	$H_0: \sigma = 0$ vs. $H_a: \sigma \neq 0$	Outside	Reject H_0

23. ***Step 1*** $H_0: \sigma = 25$ versus $H_a: \sigma \neq 25$

Step 2 df = $n - 1 = 9 - 1 = 8$. $\chi^2_{0.975} = 2.180$ and $\chi^2_{0.025} = 17.535$. Therefore we reject H_0 if $\chi^2_{data} \le \chi^2_{0.975} = 2.180$ or $\chi^2_{data} \ge \chi^2_{0.025} = 17.535$.

Step 3 TI-83/84: $s^2 = 700.8452778$ $\chi^2_{\text{data}} = \frac{(n-1)s^2}{\sigma_0{}^2} = \frac{(9-1)700.8452778}{625} \approx 8.97$

Step 4 $\chi^2_{\text{data}} \approx 8.97$ is not $\leq \chi^2_{0.975} = 2.180$ and is not $\geq \chi^2_{0.025} = 17.535$, so we do not reject H_0. There is insufficient evidence at level of significance $\alpha = 0.05$ that the population standard deviation differs from 25 MW.

25. No. In Exercise 37, Section 8.4 it was shown that the population is not normal.

27. ***p*-value method:** H_0: $\sigma = 50$ versus H_a: $\sigma > 50$.

Reject H_0 if p-value≤ 0.05. $\chi^2_{\text{data}} = \frac{(n-1)s^2}{\sigma_0{}^2} = \frac{(101-1)2600}{2500} = 104$. df $= n - 1 = 101 - 1 = 100$. p-value $= P(\chi^2 > \chi^2_{data}) = P(\chi^2 > 104) \approx 0.3721497012$. Since p-value is not ≤ 0.05, we do not reject H_0. There is insufficient evidence that the population standard deviation of test scores for boys is greater than 50 points.

Critical-value method: H_0: $\sigma = 50$ versus H_a: $\sigma > 50$.

df $= n - 1 = 101 - 1 = 100$. $\chi^2_\alpha = \chi^2_{0.05} = 124.342$. Reject H_0 if $\chi^2_{\text{data}} \geq 124.342$

$$\chi^2_{\text{data}} = \frac{(n-1)s^2}{\sigma_0{}^2} = \frac{(101-1)2600}{2500} = 104$$

Since χ^2_{data} is not ≥ 124.342, we do not reject H_0. There is insufficient evidence that the population standard deviation of test scores for boys is greater than 50 points.

Section 9.7

1. A Type II error is not rejecting H_0 when H_0 is false.

3. The probability of rejecting H_0 when H_0 is false.

5. **(a)** We have a right-tailed test $\alpha = 0.10$, so that

$$\bar{x}_{crit} = \mu_0 + Z_{crit}\left(\frac{\sigma}{\sqrt{n}}\right) = 50 + 1.28\left(\frac{4}{\sqrt{25}}\right) = 51.024.$$

(b)

Density
0.5
0.4
0.3
0.2
0.1
0.0
0.5120
51 51.024

(c) $\beta = P(\bar{x} < 51.024) = P\left(Z < \frac{51.024-51}{4/\sqrt{25}}\right) = P(Z < 0.03) = 0.5120.$

7. **(a)** We have a right-tailed test $\alpha = 0.10$, so that

$$\bar{x}_{crit} = \mu_0 + Z_{crit}\left(\frac{\sigma}{\sqrt{n}}\right) = 50 + 1.28\left(\frac{4}{\sqrt{25}}\right) = 51.024$$

(b)

Density
0.5
0.4
0.3
0.2
0.1
0.0
0.006756
51.024
53

(c) $\beta = P(\bar{x} < 51.024) = P\left(Z < \frac{51.024-53}{4/\sqrt{25}}\right) = P(Z < -2.47) = 0.0068$

9. **(a)** We have a right-tailed test $\alpha = 0.10$, so that

$$\bar{x}_{crit} = \mu_0 + Z_{crit}\left(\frac{\sigma}{\sqrt{n}}\right) = 50 + 1.28\left(\frac{4}{\sqrt{25}}\right) = 51.024$$

(b)

Density
0.5
0.4
0.3
0.2
0.1
0.0
3.3476E-07
51.024
55

(c) TI-83/84: $\beta = P(\bar{x} < 51.024) = P\left(Z < \frac{51.024-55}{4/\sqrt{25}}\right) = P(Z < -4.97) = 0.0000003353.$

11. **(a)** We have a left-tailed test with $\alpha = 0.05$, so that

$$\bar{x}_{crit} = \mu_0 - Z_{crit}\left(\frac{\sigma}{\sqrt{n}}\right) = 100 - 1.645\left(\frac{12}{\sqrt{36}}\right) = 96.71.$$

(b)

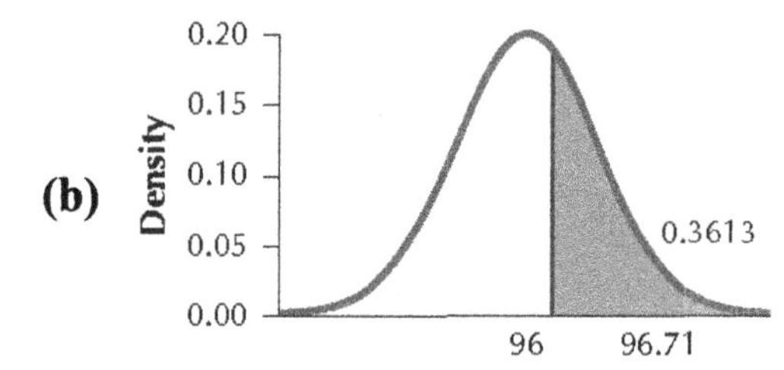

(c) $\beta = P(\bar{x} > 96.71) = P\left(Z > \frac{96.71-96}{12/\sqrt{36}}\right) = P(Z > 0.355) = 0.3613.$

13. **(a)** We have a left-tailed test with $\alpha = 0.05$, so that

$$\bar{x}_{crit} = \mu_0 - Z_{crit}\left(\frac{\sigma}{\sqrt{n}}\right) = 100 - 1.645\left(\frac{12}{\sqrt{36}}\right) = 96.71.$$

(b)

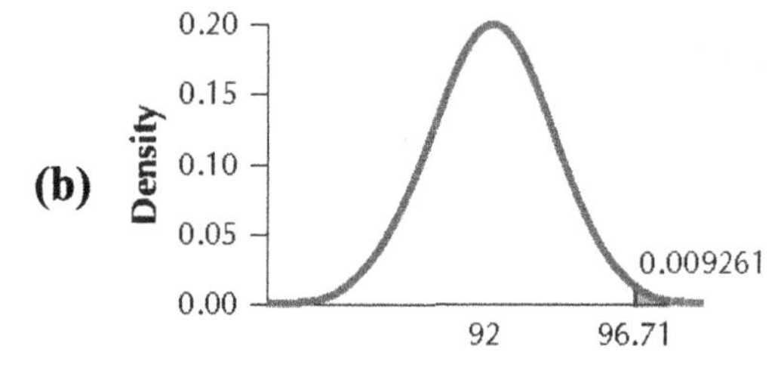

(c) $\beta = P(\bar{x} > 96.71) = P\left(Z > \frac{96.71-92}{12/\sqrt{36}}\right) = P(Z > 2.355) = 0.0093.$

15. **(a)** We have a left-tailed test with $\alpha = 0.05$, so that

$\bar{x}_{crit} = \mu_0 - Z_{crit}\left(\frac{\sigma}{\sqrt{n}}\right) = 100 - 1.645\left(\frac{12}{\sqrt{36}}\right) = 96.71.$

(b)

(c) $\beta = P(\bar{x} > 96.71) = P\left(Z > \frac{96.71-88}{12/\sqrt{36}}\right) = P(Z > 4.355) = 0.000006658.$

17. power = 1 − β = 1 − 0.5120 = 0.4880.

19. power = 1 − β = 1 − 0.0068 = 0.9932.

21. power = 1 − β = 1 − 0.0000003353 = 0.9999996647.

23. power = 1 − β = 1 − 0.3613 = 0.6387.

25. power = 1 − β = 1 − 0.0093 = 0.9907.

27. power = 1 − β = 1 − 0.000006658 = 0.999993342.

29.

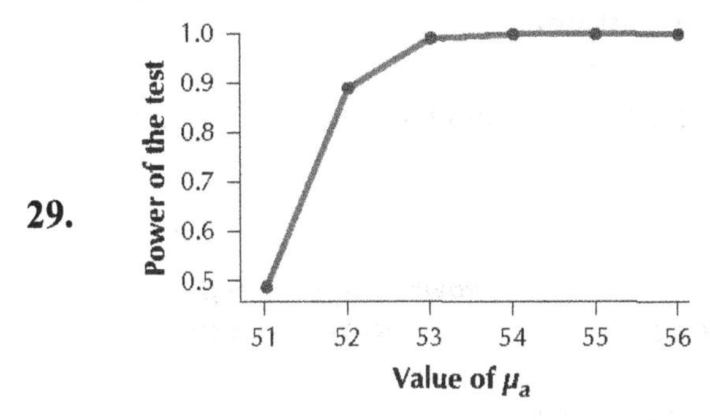

31. **(a)** Concluding that the population mean daily number of shares traded equals 2.9 billion shares when, in reality, it has increased from 2.9 billion shares.

(b) and (c)

μ_a	Probability of Type II error: β	Power of Test: $1-\beta$
(i) 3.0 billion	$P\left(Z < \frac{3.09-3.0}{0.7/\sqrt{36}}\right) = P(Z < 0.77) = 0.7794$	$1 - 0.7794 = 0.2206$
(ii) 3.1 billion	$P\left(Z < \frac{3.09-3.1}{0.7/\sqrt{36}}\right) = P(Z < -0.09) = 0.4641$	$1 - 0.4641 = 0.5359$
(iii) 3.2 billion	$P\left(Z < \frac{3.09-3.2}{0.7/\sqrt{36}}\right) = P(Z < -0.94) = 0.1736$	$1 - 0.1736 = 0.8264$
(iv) 3.3 billion	$P\left(Z < \frac{3.09-3.3}{0.7/\sqrt{36}}\right) = P(Z < -1.8) = 0.0359$	$1 - 0.0359 = 0.9641$

(d)

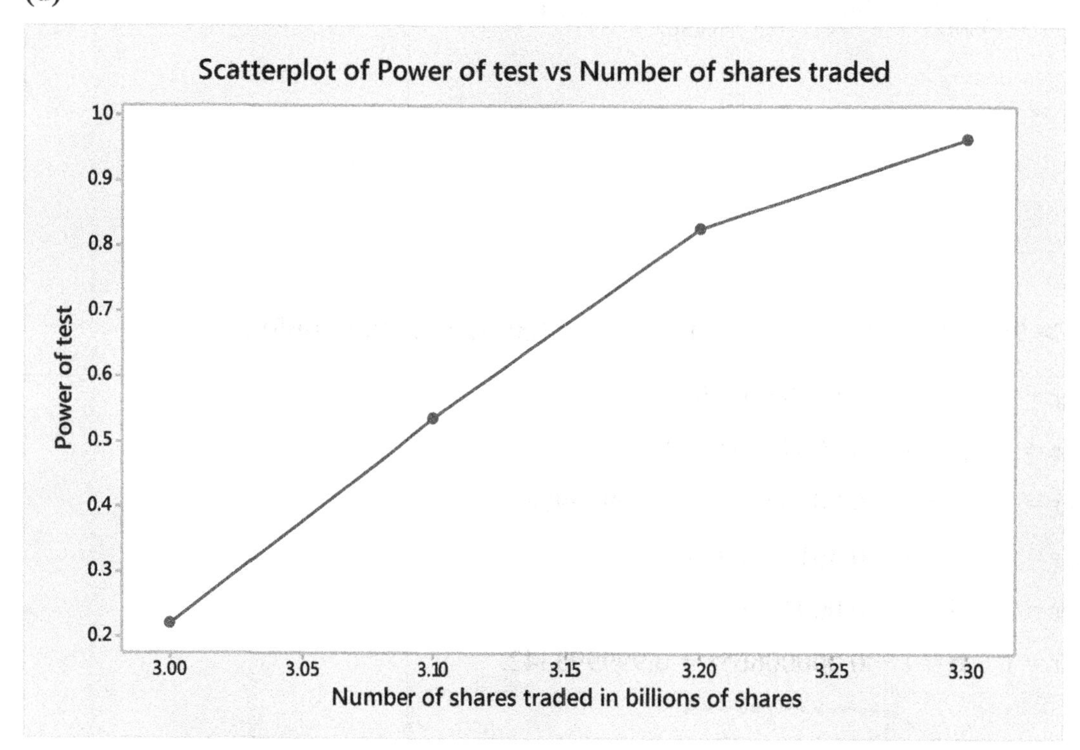

33. **(a)** Concluding that the population mean salary for accountants is equal to $41,560 when, in reality, it has changed from $41,560.

(b) and (c)

μ_a	Probability of Type II error: β	Power of Test: $1-\beta$
(i) $42,000	$P\left(\frac{40,580-42,000}{5000/\sqrt{100}} < Z < \frac{42,540-42,000}{5000/\sqrt{100}}\right)$ $= P(-2.84 < Z < 1.08) = 0.8599 - 0.0023 = 0.8576$	$1 - 0.8576 = 0.1424$
(ii) $43,000	$P\left(\frac{40,580-43,000}{5000/\sqrt{100}} < Z < \frac{42,540-43,000}{5000/\sqrt{100}}\right)$ $= P(-4.84 < Z < -0.92) = 0.1788 - 0 = 0.1788$	$1 - 0.1788 = 0.8212$
(iii) $44,000	$P\left(\frac{40,580-44,000}{5000/\sqrt{100}} < Z < \frac{42,540-44,000}{5000/\sqrt{100}}\right)$ $= P(-6.84 < Z < -2.92) = 0.0018 - 0 = 0.0018$	$1 - 0.0018 = 0.9982$
(iv) $45,000	$P\left(\frac{40,580-45,000}{5000/\sqrt{100}} < Z < \frac{42,540-45,000}{5000/\sqrt{100}}\right)$ $= P(-8.84 < Z < -4.92) = 0 - 0 = 0$	$1 - 0 = 1$

(d)

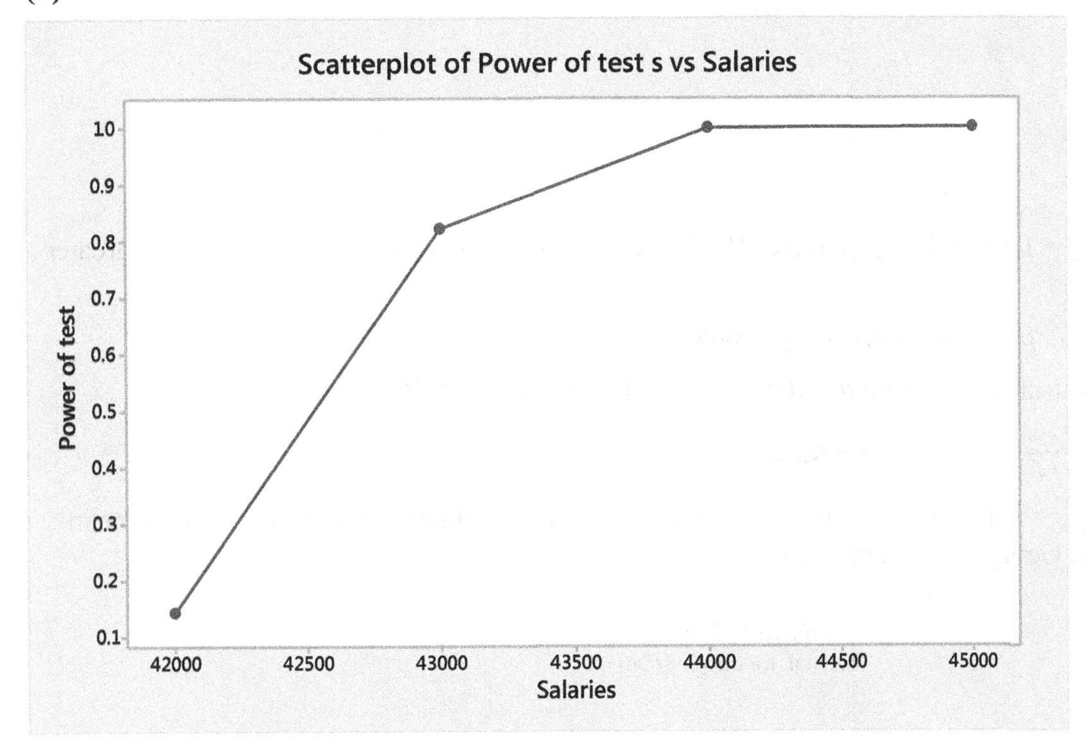

Chapter 9 Review Exercises

1. $H_0 : \mu = 12$ versus $H_a : \mu < 12$

3. $H_0 : \mu = 0$ versus $H_a : \mu < 0$

5. **(a)** $H_0 : \mu = 202.7$ versus $H_a : \mu < 202.7$

(b) The two ways that a correct decision could be made are (1) to conclude that the *population mean number* of speeding-related fatalities is less than 202.7 when the *population mean number* of speeding-related fatalities is actually less than 202.7, and (2) to conclude that the *population mean number* of speeding-related fatalities is equal to 202.7 when the *population mean number* of speeding-related fatalities is actually equal to 202.7.

(c) A Type I error would be concluding that the *population mean number* of speeding-related fatalities is less than 202.7 when it actually is equal to 202.7.

(d) A Type II error would be concluding that the *population mean number* of speeding-related fatalities is equal to 202.7 when it actually is less than 202.7.

7. $Z_{data} = \frac{\bar{x}-\mu_0}{\sigma/\sqrt{n}} = \frac{59-60}{10/\sqrt{100}} = -1.$

9. $Z_{data} = \frac{\bar{x}-\mu_0}{\sigma/\sqrt{n}} = \frac{59-60}{1/\sqrt{100}} = -10.$

11. **(a)** For a right-tailed test with $\alpha = 0.10$, $Z_{\text{crit}} = 1.28$.

(b) Reject H_0 if $Z_{\text{data}} \geq 1.28$.

(c)

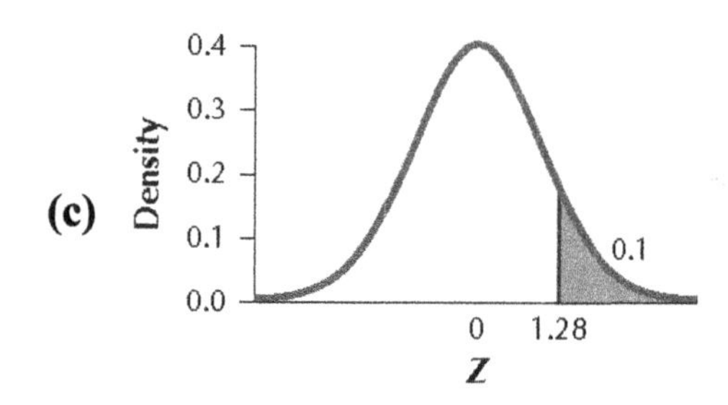

(d) Since $Z_{\text{data}} = 1.5$ is ≥ 1.28, we reject H_0. There is evidence that the population mean is greater than μ_0.

13. **(a)** $H_0 : \mu = 668$ versus $H_a : \mu < 668$

(b) For a right-tailed test with $\alpha = 0.05$, $Z_{\text{crit}} = -1.645$. Reject H_0 if $Z_{\text{data}} \leq -1.645$.

(c) $Z_{data} = \frac{\bar{x} - \mu_0}{\sigma/\sqrt{n}} = \frac{650-668}{50/\sqrt{144}} \approx -4.32.$

(d) Since $Z_{\text{data}} = -4.32$ is ≤ -1.645, we reject H_0. There is evidence that the population mean credit score in Georgia is less than 668.

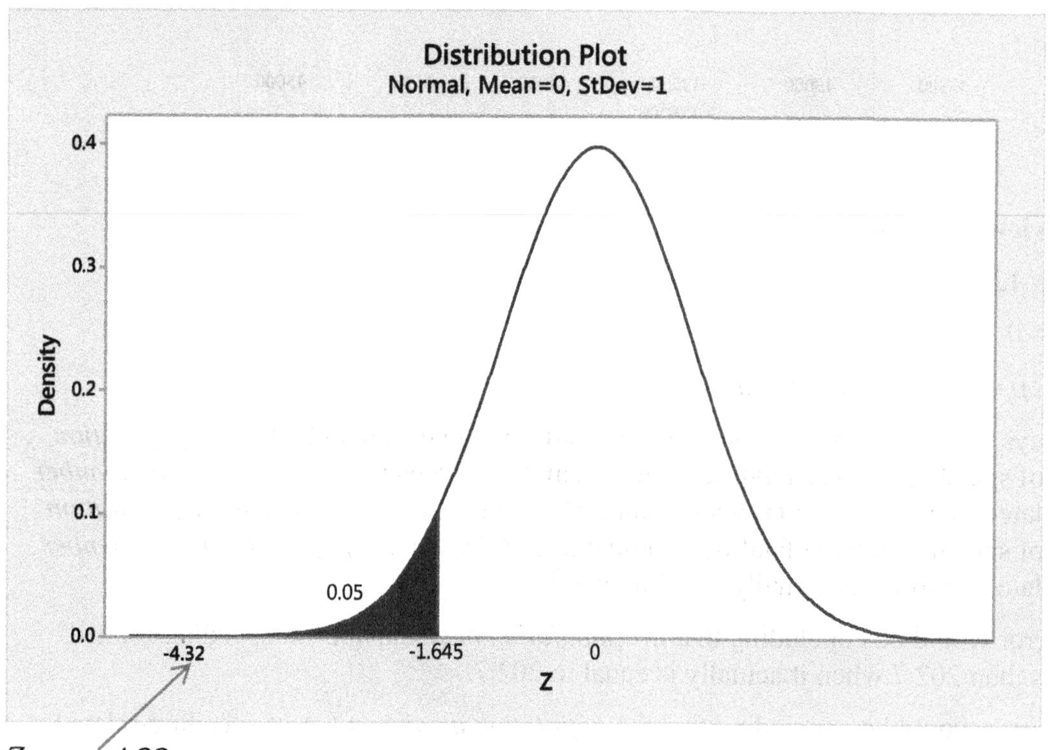

$Z_{\text{data}} = -4.32$

15. ***Step 1*** H_0: $\mu = -10$ versus H_a: $\mu < -10$

We will reject H_0 if the p-value $\leq \alpha = 0.01$.

Step 2 $Z_{data} = \frac{\bar{x} - \mu_0}{\sigma/\sqrt{n}} = \frac{-12-(-10)}{2/\sqrt{25}} = -5.$

Step 3 p-value $= P(Z < Z_{data}) = P(Z < -5) \approx 2.8665 \times 10^{-7}$

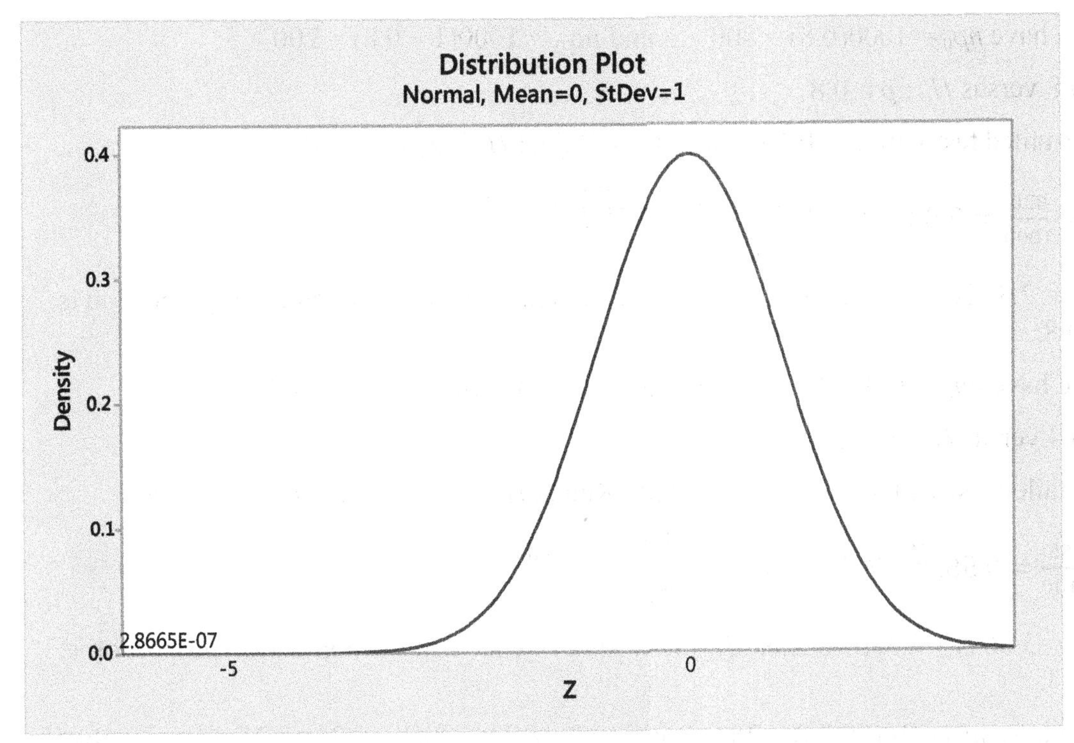

Step 4 The p-value $\approx 2.8665 \times 10^{-7}$ is ≤ 0.01. Therefore we reject H_0.

There is evidence at level of significance $\alpha = 0.01$ that the population mean is less than -10.

17. For a right-tailed test with $\alpha = 0.10$ and $\text{df} = n - 1 = 8 - 1 = 7$, $t_{\text{crit}} = 1.415$.

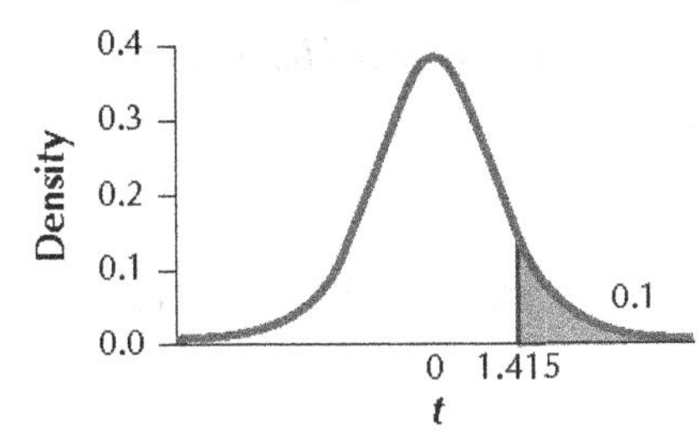

19. For a right-tailed test with $\alpha = 0.01$ and $\text{df} = n - 1 = 8 - 1 = 7$, $t_{\text{crit}} = 2.998$.

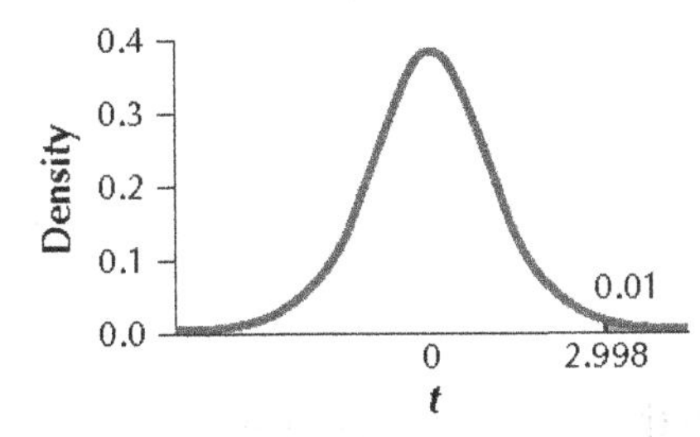

21. $H_0 : \mu = 9$ versus $H_a : \mu \neq 9$. For a two-tailed test with $\alpha = 0.10$ and $\text{df} = n - 1 = 16 - 1 = 15$, $t_{\text{crit}} = 1.753$. Reject H_0 if $t_{\text{data}} \leq -1.753$ or if $t_{\text{data}} \geq 1.753$.

$$t_{data} = \frac{\bar{x} - \mu_0}{s/\sqrt{n}} = \frac{10-9}{3/\sqrt{16}} \approx 1.33.$$

Since $t_{\text{data}} = 1.33$ is not ≤ -1.753 and not ≥ 1.753, we do not reject H_0. There is insufficient evidence that the population mean is different from 9.

23. **(a)** We have $np_0 = 1000(0.8) = 800 \geq 5$ and $nq_0 = 1000(1 - 0.8) = 200 \geq 5$.

(b) $H_0 : p = 0.8$ versus $H_a : p > 0.8$

(c) For a right-tailed test with $\alpha = 0.10$, $Z_{\text{crit}} = 1.28$. Reject H_0 if $Z_{\text{data}} \geq 1.28$.

(d) $\hat{p} = \frac{x}{n} = \frac{830}{1000} = 0.83$, $Z_{data} = \frac{\hat{p} - p_0}{\sqrt{\frac{p_0 \cdot q_0}{n}}} = \frac{0.83 - 0.8}{\sqrt{\frac{(0.8)(0.2)}{1000}}} \approx 2.37$.

(e) Since $Z_{\text{data}} = 2.37$ is ≥ 1.28, we reject H_0. There is evidence that the population proportion is greater than 0.8.

25. **(a)** We have $np_0 = 100(0.4) = 40 \geq 5$ and $n\,(1 - p_0) = 100(1 - 0.4) = 60 \geq 5$.

(b) $H_0 : p = 0.4$ versus $H_a : p \neq 0.4$

(c) For a two-tailed test with $\alpha = 0.01$, $Z_{\text{crit}} = 2.58$. Reject H_0 if $Z_{\text{data}} \leq -2.58$ or $Z_{\text{data}} \geq 2.58$.

(d) $\hat{p} = \frac{x}{n} = \frac{55}{100} = 0.55$, $Z_{data} = \frac{\hat{p} - p_0}{\sqrt{\frac{p_0 \cdot q_0}{n}}} = \frac{0.55 - 0.4}{\sqrt{\frac{(0.4)(0.6)}{100}}} \approx 3.06$.

(e) Since $Z_{\text{data}} = 3.06$ is ≥ 2.58, we reject H_0. There is evidence that the population proportion is not equal to 0.4.

27. **(a)** We have $np_0 = 100(0.25) = 25 \geq 5$ and $n\,(1 - p_0) = 100(1 - 0.25) = 75 \geq 5$. **(b)** $H_0 : p = 0.25$ versus $H_a : p < 0.25$. Reject H_0 if p-value < 0.05.

(c) $\hat{p} = \frac{x}{n} = \frac{25}{100} = 0.25$, $Z_{data} = \frac{\hat{p} - p_0}{\sqrt{\frac{p_0 \cdot q_0}{n}}} = \frac{0.25 - 0.25}{\sqrt{\frac{(0.25)(0.75)}{100}}} = 0$.

(d) p-value $= P\,(Z < 0) = 0.5$ **(e)** Since p-value $= 0.5$ is not ≤ 0.05, we do not reject H_0. There is insufficient evidence that the population proportion is less than 0.25.

29. **(a)** $H_0 : \sigma = 6$ versus $H_a : \sigma > 6$

(b) For a right-tailed test with $\alpha = 0.05$ and df $= n - 1 = 20 - 1 = 19$, $\chi^2_\alpha = \chi^2_{0.05} = 30.144$. Reject H_0 if $\chi^2_{\text{data}} \geq 30.144$.

(c) $\chi^2_{\text{data}} = \frac{(n-1)s^2}{\sigma_0^2} = \frac{(20-1)9^2}{36} \approx 42.75$.

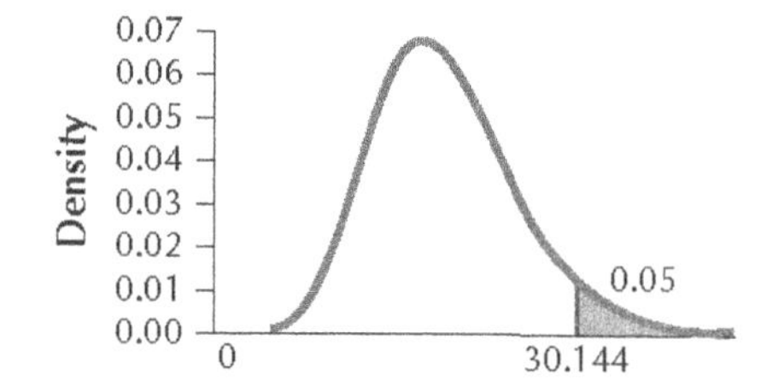

(d) Since $\chi^2_{\text{data}} \approx 42.75$.is ≥ 30.144, we reject H_0. There is evidence that the population standard deviation is greater than 6.

31. **(a)** $H_0 : \sigma = 35$ versus $H_a : \sigma < 35$. Reject H_0 if p-value ≤ 0.05.

(b) $\chi^2_{\text{data}} = \frac{(n-1)s^2}{\sigma_0^2} = \frac{(8-1)1200}{1225} \approx 6.857$.

(c) df $= n - 1 = 8 - 1 = 7$, p-value $= P\,(\chi^2 < 6.857) = 0.5560805474$

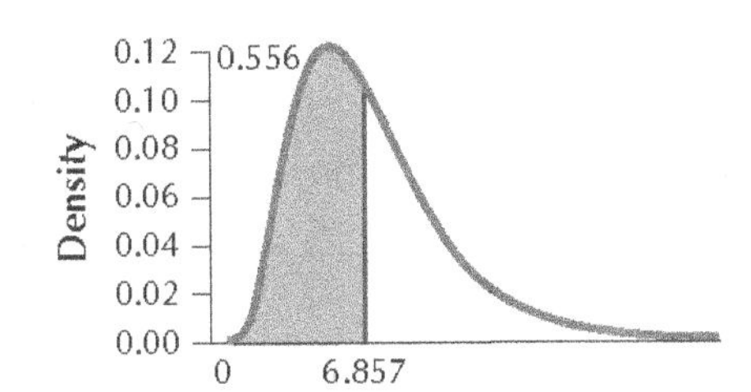

(d) Since p-value = 0.5560805474 is not ≤ 0.05, we do not reject H_0. There is insufficient evidence that the population standard deviation is less than 35.

33. **(a)** We have a two-tailed test with $\alpha = 0.01$, so that

$$\bar{x}_{crit,lower} = \mu_0 - Z_{crit}\left(\frac{\sigma}{\sqrt{n}}\right) = 100 - 2.58\left(\frac{15}{\sqrt{64}}\right) = 95.1625.$$

$$\bar{x}_{crit,upper} = \mu_0 + Z_{crit}\left(\frac{\sigma}{\sqrt{n}}\right) = 100 + 2.58\left(\frac{15}{\sqrt{64}}\right) = 104.8375.$$

(b)

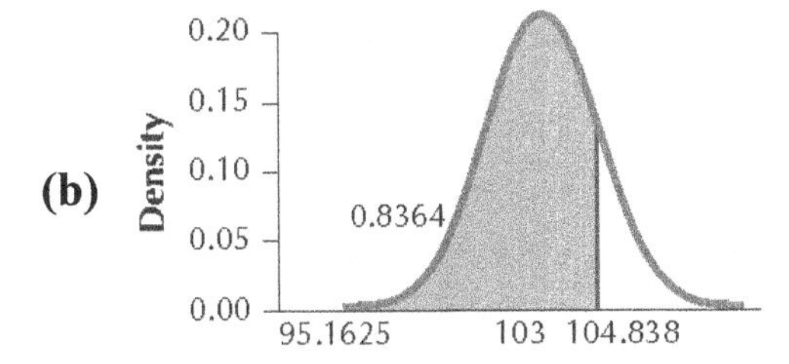

(c) TI-83/84:

$$\beta = P(95.1625 < \bar{x} < 104.8375) = P\left(\frac{95.1625-103}{15/\sqrt{64}} < Z < \frac{104.8375-103}{15/\sqrt{64}}\right)$$

$$\approx P(-4.18 < Z < 0.98) = 0.8364.$$

(d) power = $1 - \beta = 1 - 0.8364 = 0.1636$.

35. **(a)** $\bar{x}_{crit,lower} = \mu_0 - Z_{crit}\left(\frac{\sigma}{\sqrt{n}}\right) = 100 - 2.58\left(\frac{15}{\sqrt{64}}\right) = 95.1625.$

$$\bar{x}_{crit,upper} = \mu_0 + Z_{crit}\left(\frac{\sigma}{\sqrt{n}}\right) = 100 + 2.58\left(\frac{15}{\sqrt{64}}\right) = 104.8375.$$

(b)

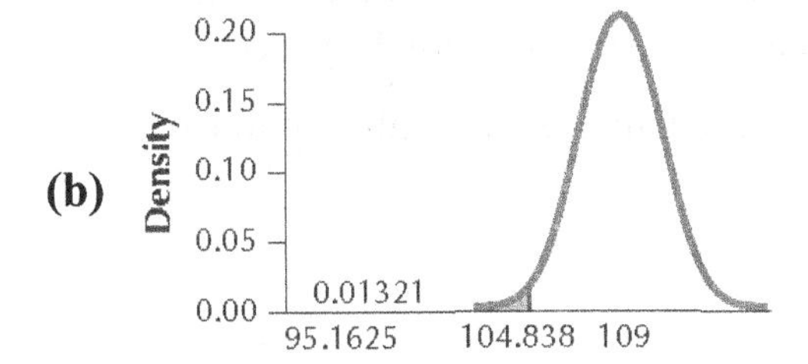

(c) TI-83/84:

$$\beta = P(95.1625 < \bar{x} < 104.8375) = P\left(\frac{95.1625-109}{15/\sqrt{64}} < Z < \frac{104.8375-109}{15/\sqrt{64}}\right)$$

$$\approx P(-7.38 < Z < -2.22) = 0.0132.$$

(d) power = $1 - \beta = 1 - 0.0132 = 0.9868$.

37. **(a)** $\bar{x}_{crit,lower} = \mu_0 - Z_{crit}\left(\frac{\sigma}{\sqrt{n}}\right) = 100 - 2.58\left(\frac{15}{\sqrt{64}}\right) = 95.1625.$

$\bar{x}_{crit,upper} = \mu_0 + Z_{crit}\left(\frac{\sigma}{\sqrt{n}}\right) = 100 + 2.58\left(\frac{15}{\sqrt{64}}\right) = 104.8375.$

(b)

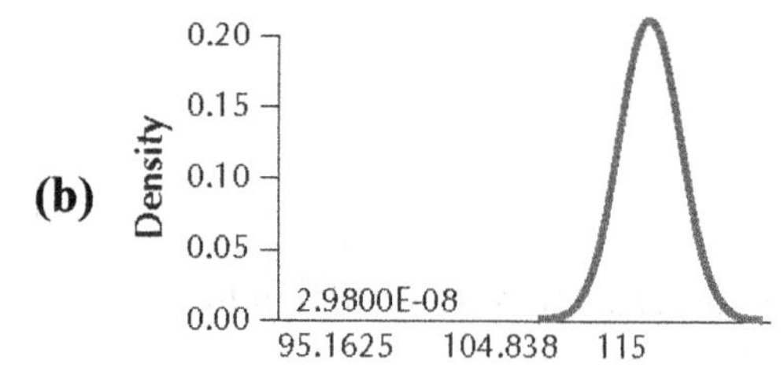

(c) TI-83/84:

$\beta = P(95.1625 < \bar{x} < 104.8375) = P\left(\frac{95.1625-115}{15/\sqrt{64}} < Z < \frac{104.8375-115}{15/\sqrt{64}}\right)$

$\approx P(-10.58 < Z < -5.42) = 0.$

(d) power = $1 - \beta = 1 - 0 = 1$.

Chapter 9 Quiz

1. False. Only one of them is true and the other one is false.

3. True.

5. Small.

7. $np_0 \geq 5$ and $n\,(1 - p_0) \geq 5$

9. No.

11. **(a)** No, because the population standard deviation is not known. $H_0 : \mu = 32$ versus $H_a : \mu \neq 32$. For a two-tailed test with $\alpha = 0.10$ and df $= n - 1 = 36 - 1 = 35$, $t_{crit} = 1.690$. Reject H_0 if $t_{data} \leq -1.690$ or if $t_{data} \geq 1.690$.

$t_{data} = \frac{\bar{x}-\mu_0}{3/\sqrt{n}} = \frac{33.8-32}{6/\sqrt{36}} \approx 1.80.$

Since $t_{data} = 1.80$ is ≥ 1.690, we reject H_0. There is evidence that the population mean years of potential life lost in alcohol-related fatal automobile accidents is different from 32 years.

(b) $0.05 < p\text{-value} = 0.0805 \leq 0.10$, so there is moderate evidence against the null hypothesis.

Chapter 10: Two-Sample Inference

Section 10.1

1. Two samples are *independent* when the subjects selected for the first sample do not determine the subjects in the second sample.

3. Matched-pairs or paired samples

5. Since both samples of games were based on the same players, this is an example of dependent sampling.

7. Since the same students are taking both tests, this is an example of dependent sampling.

9.

Subject	**1**	**2**	**3**	**4**	**5**
Sample 1	3.0	2.5	3.5	3.0	4.0
Sample 2	2.5	2.5	2.0	2.0	1.5
Differences (Sample 1 minus Sample 2)	0.5	0	1.5	1.0	2.5

$$\bar{x}_d = \frac{0.5+0+1.5+1.0+2.5}{5} = \frac{5.5}{5} = 1.1$$

$$s_d = \sqrt{\frac{\sum(x_d-\bar{x}_d)^2}{n-1}} = \sqrt{\frac{(0.5-1.1)^2+(0-1.1)^2+(1.5-1.1)^2+(1.0-1.1)^2+(2.5-1.1)^2}{5-1}} \approx 0.9618$$

11.

Subject	**1**	**2**	**3**	**4**	**5**	**6**	**7**
Sample 1	20	25	15	10	20	30	15
Sample 2	30	30	20	20	25	35	25
Differences (Sample 1 minus Sample 2)	–10	–5	–5	–10	–5	–5	–10

$$\bar{x}_d = \frac{(-10)+(-5)+(-5)+(-10)+(-5)+(-5)+(-10)}{7} = \frac{-50}{7} \approx -7.1429.$$

To find the standard deviation, we first need to find $\sum(x-\bar{x}_d)^2$.

x	$(x-\bar{x}_d)$	$(x-\bar{x}_d)^2$
–10	(–10) – (–7.1429) = –2.8571	8.16302041
–5	(–5) – (–7.1429) = 2.1429	4.59202041
–5	(–5) – (–7.1429) = 2.1429	4.59202041
–10	(–10) – (–7.1429) = –2.8571	8.16302041
–5	(–5) – (–7.1429) = 2.1429	4.59202041
–5	(–5) – (–7.1429) = 2.1429	4.59202041
–10	(–10) – (–7.1429) = –2.8571	8.16302041

$\sum(x-\bar{x}_d)^2 = 42.85714287.$

Thus,

$$s_d = \sqrt{\frac{\sum(x-\bar{x}_d)^2}{n-1}} = \sqrt{\frac{42.85714287}{7-1}} \approx 2.6726.$$

13.

Subject	**1**	**2**	**3**	**4**	**5**	**6**	**7**	**8**
Sample 1	0	0.5	0.75	1.25	1.9	2.5	3.2	3.3
Sample 2	0.25	0.25	0.75	1.5	1.8	2.2	3.3	3.4
Differences	–0.25	0.25	0	–0.25	0.1	0.3	–0.1	–0.1

$$\bar{x}_d = \frac{-0.25 + 0.25 + 0 + (-0.25) + 0.1 + 0.3 + (-0.1) + (-0.1)}{8} = \frac{-0.05}{8} = -0.00625$$

$$s_d = \sqrt{\frac{\sum(x - \bar{x})^2}{n - 1}}$$

$$= \sqrt{\frac{(-0.25 - (-0.00625))^2 + (0.25 - (-0.00625))^2 + (0 - (-0.00625))^2 + (-0.25 - (-0.00625))^2 + (0.1 - (-0.00625))^2 + (0.3 - (-0.00625))^2 + (-0.1 - (-0.00625))^2 + (-0.1 - (-0.00625))^2}{8 - 1}}$$

$$\approx 0.2095$$

15. $H_0 : \mu_d = 0$ versus $H_a : \mu_d > 0$

$\text{df} = n - 1 = 5 - 1 = 4$

This is a right-tailed test with $\alpha = 0.05$.

$t_{\text{crit}} = 2.132$

Reject H_0 if $t_{\text{data}} \geq 2.132$.

$$t_{data} = \frac{\bar{x}_d}{s_d/\sqrt{n}} = \frac{1.1}{0.9618/\sqrt{5}} \approx 2.557.$$

Since $t_{\text{data}} = 2.557$, which is ≥ 2.132, we reject H_0. There is evidence at the $\alpha = 0.05$ level of significance that the population mean difference is greater than 0.

17. $H_0 : \mu_d = 0$ versus $H_a : \mu_d < 0$

$\text{df} = n - 1 = 7 - 1 = 6$

This is a left-tailed test with $\alpha = 0.10$.

$t_{\text{crit}} = -1.440$

Reject H_0 if $t_{\text{data}} \leq -1.440$.

$$t_{\text{data}} = \frac{\bar{x}_d}{s_d/\sqrt{n}} = \frac{-7.1429}{2.6726/\sqrt{7}} \approx -7.071.$$

Since $t_{\text{data}} = -7.071$, which is ≤ -1.440, we reject H_0. There is evidence at the $\alpha = 0.10$ level of significance that the population mean difference is less than 0.

19. $H_0 : \mu_d = 0$ versus $H_a : \mu_d \neq 0$

Reject H_0 if the p-value ≤ 0.05.

$$t_{\text{data}} = \frac{\bar{x}_d}{s_d/\sqrt{n}} = \frac{-0.00625}{0.2095/\sqrt{8}} \approx -0.084.$$

$\text{df} = n - 1 = 8 - 1 = 7$

p-value $= 2P(t < t_{\text{data}}) = 2P(t < -0.084) \approx 0.9351$

Since the p-value $= 0.9351$ is not ≤ 0.05, we do not reject H_0. There is insufficient evidence at the

α = 0.05 level of significance that the population mean difference is not equal to 0.

21. df = $n - 1 = 5 - 1 = 4$,

$\bar{x}_d \pm t_{\alpha/2}(s_d/\sqrt{n}) = 1.1 \pm (2.776)(0.9618/\sqrt{5}) \approx 1.1 \pm 1.1940 = (-0.0940,\ 2.2940)$

23. For a 90% confidence interval with df = $n - 1 = 7 - 1 = 6$, $t_{\alpha/2} = 1.943$.

$\bar{x}_d \pm t_{\alpha/2}(s_d/\sqrt{n}) = -7.1429 \pm (1.943)(2.6726/\sqrt{7}) \approx -7.1429 \pm 1.9627 =$ $(-9.106, -5.180)$

We are 90% confident that the population mean difference lies between −9.106 and −5.180.

25. For a 95% confidence interval with df = $n - 1 = 8 - 1 = 7$, $t_{\alpha/2} = 2.365$.

$\bar{x}_d \pm t_{\alpha/2}(s_d/\sqrt{n}) = -0.00625 \pm (2.365)(0.2095/\sqrt{8}) \approx -0.00625 \pm 0.1752 = (-0.181,$ $0.169)$.

We are 95% confident that the population mean difference lies between −0.181 and 0.169.

27. **(a)** $H_0 : \mu_d = 0$ versus $H_a : \mu_d \neq 0$

$\mu_0 = 0$ lies inside of the interval (−5, 5), so we do not reject H_0 at the α = 0.05 level of significance.

(b) $H_0 : \mu_d = -6$ versus $H_a : \mu_d \neq -6$

$\mu_0 = -6$ lies outside of the interval (−5, 5), so we reject H_0 at the α = 0.05 level of significance.

(c) $H_0 : \mu_d = 4$ versus $H_a : \mu_d \neq 4$

$\mu_0 = 4$ lies inside of the interval (−5, 5), so we do not reject H_0 at the α = 0.05 level of significance.

29. **(a)** $H_0 : \mu_d = -10$ versus $H_a : \mu_d \neq -10$

$\mu_0 = -10$ lies outside of the interval (10, 20), so we reject H_0 at the α = 0.10 level of significance.

(b) $H_0 : \mu_d = 25$ versus $H_a : \mu_d \neq 25$

$\mu_0 = 25$ lies outside of the interval (10, 20), so we reject H_0 at the α = 0.10 level of significance.

(c) $H_0 : \mu_d = 0$ versus $H_a : \mu_d \neq 0$

$\mu_0 = 0$ lies outside of the interval (10, 20), so we reject H_0 at the α = 0.10 level of significance.

31. First find the differences by subtracting the 2013 price from the 2014 price.

Toyota Camry: $20,672 – $20,284 = $388

Honda Civic: $17,069 – $16,499 = $570

Ford F-150: $24,362 – $22,674 = $1,688

Chevy Corvette: $45,684 – $44,021 = $1,663

Tesla Model S: $68,738 – $68,674 = $64

(a) $\bar{x}_d = \frac{\$388+\$570+\$1{,}688+\$1{,}663+\$64}{5} = \frac{4373}{5} = \$874.6.$

$$s_d = \sqrt{\frac{\Sigma(x-\bar{x}_d)^2}{n-1}} = \sqrt{\frac{(388-874.6)^2+(570-874.6)^2+(1{,}688-874.6)^2+(1{,}663-874.6)^2+(64-874.6)^2}{5-1}}$$

$$= \sqrt{\frac{2{,}269{,}827.2}{4}} = \sqrt{567{,}456.8} \approx \$753.2972853$$

(b) $H_0: \mu_d = 0$ versus $H_a: \mu_d > 0$

Reject H_0 if the p-value $\leq \alpha = 0.05$.

$$t_{\text{data}} = \frac{\bar{x}_d}{s_d/\sqrt{n}} = \frac{\$874.6.}{\$753.2972853\ /\sqrt{5}} \approx 2.5961.$$

df = $n - 1 = 5 - 1 = 4$

p-value = $P(t > t_{\text{data}}) \approx P(t > 2.5961) \approx 0.0301463228$

The p-value = 0.0301463228 is $\leq \alpha = 0.05$, so we reject H_0. There is evidence at level of significance $\alpha = 0.05$ that the population mean μ_d of the differences in price of cars is greater than 0. That is, there is evidence at level of significance $\alpha = 0.05$ that the 2014 models of cars are on average more expensive than the 2013 models of cars.

33. First find the differences by subtracting the low temperature from the high temperature.

Day 1: $9.4 - 0.8 = 8.6$

Day 2: $6.1 - (-8.9) = 15$

Day 3: $5.9 - (-1.3) = 7.2$

Day 4: $29.1 - 19.3 = 9.8$

Day 5: $11.9 - 6.7 = 5.2$

Day 6: $30.6 - 21.5 = 9.1$

Day 7: $23.1 - 10.5 = 12.6$

Day 8: $33.1 - 18.7 = 14.4$

Day 9: $14.8 - 7.4 = 7.4$

Day 10: $0.1 - (-9.9) = 10$

(a) $\bar{x}_d = \frac{8.6+15+7.2+9.8+5.2+9.1+12.6+14.4+7.4+10}{10} = \frac{99.3}{10} = 9.93.$

$$s_d = \sqrt{\frac{\Sigma(x-\bar{x}_d)^2}{n-1}} = \sqrt{\frac{(8.6-9.93)^2+(15-9.93)^2+(7.2-9.93)^2+(9.8-9.93)^2+(5.2-9.93)^2 +(9.1-9.93)^2+(12.6-9.93)^2+(14.4-9.93)^2+(7.4-9.93)^2+(10-9.93)^2}{10-1}}$$

$$= \sqrt{\frac{91.521}{9}} = \sqrt{10.169} \approx 3.188886953$$

(b) df = $n - 1 = 10 - 1 = 9$. $t_{\alpha/2} = 2.262$

$\bar{x}_d \pm t_{\alpha/2}\left(\frac{s_d}{\sqrt{n}}\right) = 9.93 \pm 2.262\left(\frac{3.188886953}{\sqrt{10}}\right) = 9.93 \pm 2.281 \approx (7.649, 12.211)$. We are 95% confident that the population mean of the differences between high and low temperatures in May in Waterloo, Ontario, Canada, lies between 7.649 degrees Celsius and 12.211 degrees Celsius.

35. df = $n - 1 = 5 - 1 = 4$. $t_{\alpha/2} = 2.776$

$\bar{x}_d \pm t_{\alpha/2}\left(\frac{s_d}{\sqrt{n}}\right) = \$874.6 \pm 2.776\left(\frac{\$753.2972853}{\sqrt{5}}\right) = \$874.6 \pm \$935.2 \approx (-\$60.6, \$1809.8)$;
[TI-83/84: $(-\$60.74, \$1809.9)$]

We are 95% confident that the population mean of the differences between prices of 2014 model cars and prices of 2013 model cars lies between −\$60.6 and \$1809.8.

37. df = $n - 1 = 10 - 1 = 9$. $t_{\alpha/2} = 3.250$

$\bar{x}_d \pm t_{\alpha/2}\left(\frac{s_d}{\sqrt{n}}\right) = 9.93 \pm 3.250\left(\frac{3.188886953}{\sqrt{10}}\right) = 9.93 \pm 3.277 \approx (6.653, 13.207)$. We are 99% confident that the population mean of the differences between high and low temperatures in May in Waterloo, Ontario, Canada, lies between 6.653 degrees Celsius and 13.207 degrees Celsius.

39. The mean mathematics score in 2007 and 2011 are given for the same country for each of the 15 different countries.

41.

$H_0: \mu_d = 0$ versus $H_a: \mu_d > 0$

Reject H_0 if the p-value $\leq \alpha = 0.05$.

TI-83/84: $t_{data} \approx 1.7741$

TI-83/84: p-value = 0.04889052778

The p-value = 0.04889052778 is $\leq \alpha = 0.05$, so we reject H_0. There is evidence at level of significance $\alpha = 0.05$ that the population mean μ_d of the differences in math scores is greater than 0. That is, there is evidence at level of significance $\alpha = 0.05$ that the mean math score in 2011 is higher than the mean math score in 2007.

42. TI-83/84: $(-1.212, 12.812)$. We are 95% confident that the population mean of the

43. **(a)** H_0: $\mu_d = 15$ versus H_a: $\mu_d \neq 15$

$\mu_0 = 15$ lies outside the interval, so we reject H_0.

(b) H_0: $\mu_d = 5$ versus H_a: $\mu_d \neq 5$

$\mu_0 = 5$ lies inside the interval, so we do not reject H_0.

(c) H_0: $\mu_d = 0$ versus H_a: $\mu_d \neq 0$

$\mu_0 = 0$ lies inside the interval, so we do not reject H_0.

45.

Descriptive Statistics: PHOSPHOR

```
Variable    N  N*    Mean  SE Mean   StDev  Minimum     Q1  Median      Q3
PHOSPHOR  961   0  129.81     6.60  204.63     0.00  24.00   61.00  149.00

Variable  Maximum
PHOSPHOR  2017.00
```

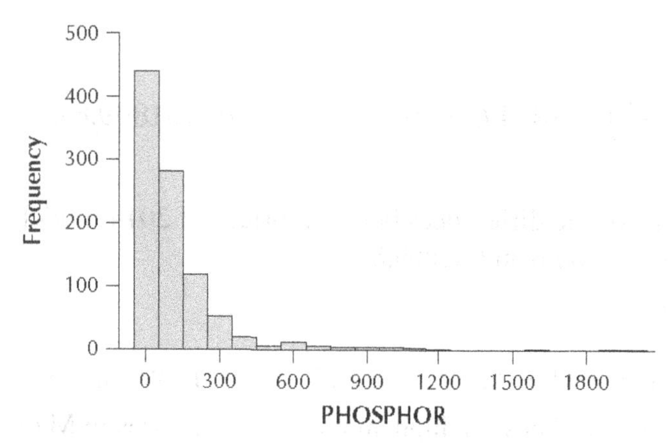

$\bar{x} = 129.81$, $s = 204.63$

47. $H_0: \mu_d = 0$ versus $H_a: \mu_d \neq 0$. Reject H_0 if the p-value ≤ 0.05. $t_{\text{data}} = -15.22$. p-value ≈ 0. Since the p-value ≈ 0 is ≤ 0.05, we reject H_0. There is evidence at the $\alpha = 0.05$ level of significance that the population mean difference is not equal to 0.

49. Answers will vary.

51. $\mu_d = -3.47$ miligrams. Rest of answer will vary.

Section 10.2

1. The two populations are normally distributed. The sample sizes are large (at least 30).

3. **(a)** $H_0 : \mu_1 = \mu_2$ versus $H_a : \mu_1 \neq \mu_2$

(b) $n_1 - 1 = 36 - 1 = 35$ and $n_2 = 36 - 1 = 35$, so df = 35. $\alpha = 0.10$. $t_{crit} = 1.690$. Reject H_0 if $t_{data} \leq -1.690$ or $t_{data} \geq 1.690$.

(c) $t_{\text{data}} = \dfrac{(\bar{x}_1 - \bar{x}_2)}{\sqrt{\frac{s_1^2}{n_1} + \frac{s_2^2}{n_2}}} = \dfrac{(10-8)}{\sqrt{\frac{2^2}{36} + \frac{2^2}{36}}} \approx 4.243.$

(d) Since $t_{data} = 4.243$ is ≥ 1.690, we reject H_0. There is evidence that the population mean for Population 1 is different than the population mean for Population 2.

5. **(a)** $H_0 : \mu_1 = \mu_2$ versus $H_a : \mu_1 < \mu_2$.

(b) The degrees of freedom is the smaller of $n_1 - 1 = 100 - 1 = 99$ and $n_2 - 1 = 50 - 1 = 49$. Thus, df = 49, which is not in the table, so we use the conservative df = 40. This is a left-tailed test with $\alpha = 0.01$, so $t_{crit} = -2.423$. Reject H_0 if $t_{data} \leq -2.423$.

(c) $t_{\text{data}} = \dfrac{(\bar{x}_1 - \bar{x}_2)}{\sqrt{\frac{s_1^2}{n_1} + \frac{s_2^2}{n_2}}} = \dfrac{(70-80)}{\sqrt{\frac{10^2}{100} + \frac{12^2}{30}}} \approx -5.077.$

(d) Since $t_{data} \approx -5.077$, which is ≤ -2.423, we reject H_0. There is evidence at the $\alpha = 0.01$ level of significance that the population mean of Population 1 is less than the population mean of Population 2.

7. **(a)** $H_0 : \mu_1 = \mu_2$ versus $H_a : \mu_1 \neq \mu_2$. Reject H_0 if the p-value ≤ 0.10.

(b) $t_{\text{data}} = \dfrac{(\bar{x}_1 - \bar{x}_2)}{\sqrt{\frac{s_1^2}{n_1} + \frac{s_2^2}{n_2}}} = \dfrac{(0-1)}{\sqrt{\frac{3^2}{64} + \frac{1^2}{49}}} \approx -2.492.$

(c) The degrees of freedom is the smaller of $n_1 - 1 = 64 - 1 = 63$ and $n_2 - 1 = 49 - 1 = 48$. Thus, df = 48. This is a two-tailed test, so p-value $= 2P(t < t_{data}) = 2P(t < -2.492) \approx 0.0162$.

(d) Since the p-value ≈ 0.0162, which is ≤ 0.10, we reject H_0. There is evidence at the $\alpha = 0.10$ level of significance that the population mean of Population 1 is different than the population mean of Population 2.

9. **(a)** $H_0 : \mu_1 = \mu_2$ versus $H_a : \mu_1 < \mu_2$. Reject H_0 if the p-value ≤ 0.05.

(b) $t_{\text{data}} = \dfrac{(\bar{x}_1 - \bar{x}_2)}{\sqrt{\frac{s_1^2}{n_1} + \frac{s_2^2}{n_2}}} = \dfrac{(50-75)}{\sqrt{\frac{10^2}{100} + \frac{15^2}{100}}} \approx -13.868.$

(c) The degrees of freedom is the smaller of $n_1 - 1 = 100 - 1 = 99$ and $n_2 - 1 = 100 - 1 = 99$. Thus, df = 99. This is a right-tailed test, so p-value $= P(t < t_{data}) = P(t < -13.868) \approx 0$.

(d) Since the p-value ≈ 0, which is ≤ 0.05, we reject H_0. There is evidence at the $\alpha = 0.05$ level of significance the population mean of Population 1 is less than the population mean of Population 2.

11. **(a)** $\bar{x}_1 - \bar{x}_2 = 10 - 8 = 2.$

(b) The degrees of freedom is the smaller of $n_1 - 1 = 36 - 1 = 35$ and $n_2 - 1 = 36 - 1 = 35$. Thus, df = 35. For a 90% confidence interval with df = 35, $t_{\alpha/2} = 1.690$.

$$E = t_{\alpha/2}\sqrt{\frac{s_1^2}{n_1} + \frac{s_2^2}{n_2}} = 1.690\sqrt{\frac{2^2}{36} + \frac{2^2}{36}} \approx 0.797.$$

We can estimate the difference in the population means of Population 1 and Population 2 to within 0.797 with 90% confidence.

(c) $(\bar{x}_1 - \bar{x}_2) \pm t_{\alpha/2}\sqrt{\frac{s_1^2}{n_1} + \frac{s_2^2}{n_2}} = (10-8) \pm 1.690\sqrt{\frac{2^2}{36} + \frac{2^2}{36}} \approx 2 \pm 0.797 = (1.203,\ 2.797).$
We are 90% confident that the difference in the population means of Population 1 and Population 2 lies between 1.203 and 2.797.

13. **(a)** $\bar{x}_1 - \bar{x}_2 = 70 - 80 = -10.$

(b) The degrees of freedom is the smaller of $n_1 - 1 = 100 - 1 = 99$ and $n_2 - 1 = 50 - 1 = 49$. Thus, df = 49, which is not in the table, so we use the conservative df = 40. For a 99% confidence interval with df = 40, $t_{\alpha/2} = 2.704$.

$$E = t_{\alpha/2}\sqrt{\frac{s_1^2}{n_1} + \frac{s_2^2}{n_2}} = 2.704\sqrt{\frac{10^2}{100} + \frac{12^2}{50}} \approx 5.326.$$

We can estimate the difference in the population means of Population 1 and Population 2 to within 5.326 with 99% confidence.

(c)

$$(\bar{x}_1 - \bar{x}_2) \pm t_{\alpha/2}\sqrt{\frac{s_1^2}{n_1} + \frac{s_2^2}{n_2}} = (70-80) \pm 2.704\sqrt{\frac{10^2}{100} + \frac{12^2}{50}} \approx -10 \pm 5.326 =$$

$(-15.326, -4.674)$. We are 99% confident that the difference in the population means of Population 1 and Population 2 lies between -15.326 and -4.674.

15. **(a)** $\bar{x}_1 - \bar{x}_2 = 0 - 1 = -1$.

(b) The degrees of freedom is the smaller of $n_1 - 1 = 64 - 1 = 63$ and $n_2 - 1 = 49 - 1 = 48$. Thus, df = 48, which is not in the table, so we use the conservative df = 40. For a 95% confidence interval with df = 40, $t_{\alpha/2} = 2.021$.

$$E = t_{\alpha/2}\sqrt{\frac{s_1^2}{n_1} + \frac{s_2^2}{n_2}} = 2.021\sqrt{\frac{3^2}{64} + \frac{1^2}{49}} \approx 0.811.$$

We can estimate the difference in the population means of Population 1 and Population 2 to within 0.811 with 95% confidence.

(c) $(\bar{x}_1 - \bar{x}_2) \pm t_{\alpha/2}\sqrt{\frac{s_1^2}{n_1} + \frac{s_2^2}{n_2}} = (0 - 1) \pm 2.021\sqrt{\frac{3^2}{64} + \frac{1^2}{49}} \approx -1 \pm 0.811 = (-1.811, -0.189)$.
We are 95% confident that the difference in the population means of Population 1 and Population 2 lies between -1.811 and -0.189.

17. **(a)** $H_0 : \mu_1 - \mu_2 = 0$ versus $H_a : \mu_1 - \mu_2 \neq 0$. $\mu_0 = 0$ lies outside of the interval (10, 15), so we reject H_0 at the $\alpha = 0.05$ level of significance.

(b) $H_0 : \mu_1 - \mu_2 = 12$ versus $H_a : \mu_1 - \mu_2 \neq 12$. $\mu_0 = 12$ lies inside of the interval (10, 15), so we do not reject H_0 at the $\alpha = 0.05$ level of significance.

(c) $H_0 : \mu_1 - \mu_2 = 16$ versus $H_a : \mu_1 - \mu_2 \neq 16$. $\mu_0 = 16$ lies outside of the interval (10, 15), so we reject H_0 at the $\alpha = 0.05$ level of significance.

19. **(a)** $H_0 : \mu_1 - \mu_2 = -10.1$ versus $H_a : \mu_1 - \mu_2 \neq -10.1$. $\mu_0 = -10.1$ lies outside of the interval $(-10, 10)$, so we reject H_0 at the $\alpha = 0.10$ level of significance.

(b) $H_0 : \mu_1 - \mu_2 = -9.9$ versus $H_a : \mu_1 - \mu_2 \neq -9.9$. $\mu_0 = -9.9$ lies inside of the interval $(-10, 10)$, so we do not reject H_0 at the $\alpha = 0.10$ level of significance.

(c) $H_0 : \mu_1 - \mu_2 = 0$ versus $H_a : \mu_1 - \mu_2 \neq 0$. $\mu_0 = 0$ lies inside of the interval $(-10, 10)$, so we do not reject H_0 at the $\alpha = 0.10$ level of significance.

21. $H_0 : \mu_1 = \mu_2$ versus $H_1 : \mu_1 > \mu_2$

df $= n_1 + n_2 - 2 = 36 + 36 - 2 = 70$

This is a right-tailed test with $\alpha = 0.10$.

$t_{\text{crit}} = 1.294$

Reject H_0 if $t_{\text{data}} \geq 1.294$.

$$s_{\text{pooled}}^2 = \frac{(n_1-1)s_1^2+(n_2-1)s_2^2}{n_1+n_2-2} = \frac{(36-1)10^2+(36-1)11^2}{36+36-2} = 110.5$$

$$t_{\text{data}} = \frac{\bar{x}_1-\bar{x}_2}{\sqrt{s_{pooled}^2\left(\frac{1}{n_1}+\frac{1}{n_2}\right)}} = \frac{54-52}{\sqrt{110.5\left(\frac{1}{36}+\frac{1}{36}\right)}} \approx 0.807$$

Since $t_{\text{data}} \approx 0.807$ is not ≥ 1.294, we do not reject H_0. There is insufficient evidence at the $\alpha =$ 0.10 level of significance that the population mean of Population 1 is greater than the population mean of Population 2.

23. For a 95% confidence interval with df $= n_1 + n_2 - 2 = 36 + 36 - 2 = 70$. $t_{\alpha/2} = 1.994$.

$$(\bar{x}_1 - \bar{x}_2) \pm t_{\alpha/2}\sqrt{s^2_{\text{pooled}}\left(\frac{1}{n_1} + \frac{1}{n_2}\right)} = (54 - 52) \pm 1.994\sqrt{110.5\left(\frac{1}{36} + \frac{1}{36}\right)} \approx 2 \pm 4.940 = (-2.940, 6.940).$$

We are 95% confident that the difference in the population means of Population 1 and Population 2 lies between –2.940 and 6.940.

25. $H_0 : \mu_1 = \mu_2$ versus $H_1 : \mu_1 > \mu_2$

$Z_{\text{crit}} = 1.645$

Reject H_0 if $Z_{\text{data}} \geq 1.645$

$$Z_{\text{data}} = \frac{\bar{x}_1 - \bar{x}_2}{\sqrt{\frac{\sigma_1^2}{n_1} + \frac{\sigma_2^2}{n_2}}} = \frac{100-99}{\sqrt{\frac{1^2}{49} + \frac{2^2}{36}}} \approx 2.757$$

Since $Z \approx 2.757$, which is ≥ 1.645, we reject H_0. There is evidence at the $\alpha = 0.05$ level of significance that the population mean of Population 1 is greater than the population mean of Population 2.

27. $(\bar{x}_1 - \bar{x}_2) \pm Z_{\alpha/2}\sqrt{\frac{\sigma_1^2}{n_1} + \frac{\sigma_2^2}{n_2}} = (100 - 99) \pm 1.96\sqrt{\frac{1^2}{49} + \frac{2^2}{36}} \approx 1 \pm 0.711 = (0.289, 1.711)$. We are 95% confident that the difference in the population means of Population 1 and Population 2 lies between 0.289 and 1.711.

29. Since both sample sizes are less than 30 and the distribution of both populations is unknown, it is not appropriate to use Welch's t test.

31. **(a)** $\bar{x}_1 - \bar{x}_2 = \$31{,}987 - \$33{,}179 = -\$1192$.

(b) The degrees of freedom is the smaller of $n_1 - 1 = 36 - 1 = 35$ and $n_2 - 1 = 49 - 1 = 48$. Thus, df = 35. For a 95% confidence interval with df = 35, $t_{\alpha/2} = 2.030$.

$$E = t_{\alpha/2}\sqrt{\frac{s_1^2}{n_1} + \frac{s_2^2}{n_2}} = 2.030\sqrt{\frac{5000^2}{36} + \frac{6000^2}{49}} \approx \$2426.7954.$$

(c) $(\bar{x}_1 - \bar{x}_2) \pm t_{\alpha/2}\sqrt{\frac{s_1^2}{n_1} + \frac{s_2^2}{n_2}} = (\$31{,}987 - \$33{,}179) \pm 2.030\sqrt{\frac{5000^2}{36} + \frac{6000^2}{49}} \approx -\$1192 \pm \$2426.7954 \approx (-3618.7954,\ 1234.7954)$. We are 95% confident that the difference in population mean incomes $\mu_1 - \mu_2$ lies between $-\$3618.7954$ and $\$1234.7954$.

(d) $H_0: \mu_1 = \mu_2$ versus $H_a: \mu_1 < \mu_2$

Reject H_0 if the p-value $\leq \alpha = 0.05$.

$$t_{\text{data}} = \frac{(\bar{x}_1 - \bar{x}_2)}{\sqrt{\frac{s_1^2}{n_1} + \frac{s_2^2}{n_2}}} = \frac{(31{,}987 - 33{,}179)}{\sqrt{\frac{5000^2}{36} + \frac{6000^2}{49}}} \approx -0.9971.$$

This is a left-tailed test, so p-value $= P(t < t_{data}) = P(t < -0.9971) \approx 0.1627799835$ (TI-84: 0.1608312442)

The p-value $= 0.1627799835$ is not $\leq \alpha = 0.05$, so we do not reject H_0. There is insufficient evidence at level of significance $\alpha = 0.05$ that the population mean income of Sacramento County is less than that of Los Angeles County.

(e) No. The hypothesis test in (d) is a one-tailed test and confidence intervals can only be used for two-tailed tests.

33. **(a)** The degrees of freedom is the smaller of $n_1 - 1 = 36 - 1 = 35$ and $n_2 - 1 = 64 - 1 = 63$. Thus, df = 35. For a 99% confidence interval with df = 35, $t_{\alpha/2} = 2.724$.

$$(\bar{x}_1 - \bar{x}_2) \pm t_{\alpha/2}\sqrt{\frac{s_1^2}{n_1} + \frac{s_2^2}{n_2}} = (20.9 - 19.3) \pm 2.724\sqrt{\frac{5^2}{36} + \frac{4^2}{64}} \approx 1.6 \pm 2.647 = (-1.047, 4.247).$$

We are 99% confident that the interval (−1.047, 4.247) captures the difference in the population mean number of children per teacher in the towns of Cupertino, California, and Santa Rosa, California.

(b) $H_0: \mu_1 - \mu_2 = 0$ versus $H_a: \mu_1 - \mu_2 \neq 0$. $\mu_0 = 0$ lies inside of the interval (−1.047, 4.247), so we do not reject H_0.

There is insufficient evidence at the $\alpha = 0.01$ level of significance that the population mean number of children per teacher in the town of Cupertino, California, differs from the population mean number of children per teacher in the town of Santa Rosa, California.

35. $H_0: \mu_1 = \mu_2$ versus $H_a: \mu_1 \neq \mu_2$

Reject H_0 if the p-value $\leq \alpha = 0.10$

$$t_{\text{data}} = \frac{(\bar{x}_1 - \bar{x}_2)}{\sqrt{\frac{s_1^2}{n_1} + \frac{s_2^2}{n_2}}} = \frac{(62{,}719 - 62{,}535)}{\sqrt{\frac{10{,}000^2}{100} + \frac{10{,}000^2}{100}}} \approx 0.1301.$$

The degrees of freedom is the smaller of $n_1 - 1 = 100 - 1 = 99$ and $n_2 - 1 = 100 - 1 = 99$. Thus, df = 99. This is a two-tailed test, so p-value = $2\,P(t > t_{data}) = 2\,P(t > 0.1301) \approx 0.896751292$ (TI-83/84: 0.8966133443)

The p-value = 0.896751292 is not $\leq \alpha = 0.10$, so we do not reject H_0. There is insufficient evidence at level of significance $\alpha = 0.10$ that the population mean salary of engineering graduates has changed from 2013 to 2014.

37. $H_0: \mu_1 = \mu_2$ versus $H_a: \mu_1 > \mu_2$

Reject H_0 if p-value ≤ 0.05.

$$t_{\text{data}} = \frac{(\bar{x}_1 - \bar{x}_2)}{\sqrt{\frac{s_1^2}{n_1} + \frac{s_2^2}{n_2}}} = \frac{(29 - 21)}{\sqrt{\frac{59^2}{100} + \frac{52^2}{100}}} \approx 1.017.$$

$n_1 - 1 = 100 - 1 = 99$ and $n_2 = 100 - 1 = 99$, so df = 99.

p-value = $P(t > t_{data}) = P(t > 1.017) = 0.1558$

Since p-value is not ≤ 0.05, we do not reject H_0. There is insufficient evidence that the population mean coached SAT score improvement is greater than the population mean noncoached SAT score improvement.

Critical value method:

$H_0: \mu_1 = \mu_2$ *versus* $H_a: \mu_1 > \mu_2$

The degrees of freedom is the smaller of $n_1 - 1 = 100 - 1 = 99$ and $n_2 - 1 = 100 - 1 = 99$. Thus, df = 99, which is not in the table, so we use the conservative df = 90.

This is a right-tailed test with $\alpha = 0.05$, so

$t_{crit} = 1.662$

Reject H_0 if $t_{data} \geq 1.662$.

$$t_{\text{data}} = \frac{(\bar{x}_1 - \bar{x}_2)}{\sqrt{\frac{s_1^2}{n_1} + \frac{s_2^2}{n_2}}} = \frac{(29-21)}{\sqrt{\frac{59^2}{100} + \frac{52^2}{100}}} \approx 1.017.$$

Since $t_{data} \approx 1.017$ is not ≥ 1.662, we do not reject H_0. There is insufficient evidence at the α = 0.05 level of significance that the population mean coached SAT improvement is greater than the population mean noncoached improvement.

39. **(a)** An increase in the sample size will result in an increase in the degrees of freedom. Depending on the new sample size, the "conservative" $t_{\alpha/2}$ will either stay the same or decrease. Since the sample sizes are in the denominator of $t_{\alpha/2}\sqrt{\frac{s_1^2}{n_1} + \frac{s_2^2}{n_2}}$, this quantity will decrease. Since the width of the confidence interval is $2\, t_{\alpha/2}\sqrt{\frac{s_1^2}{n_1} + \frac{s_2^2}{n_2}}$, the width of the confidence interval will decrease.

(b) If the width of the confidence interval decreases so that 0 is no longer in the confidence interval, then the conclusion will change to reject H_0. Otherwise, the conclusion will remain do not reject H_0.

41. **(a)** If the birth weights of the babies in each group are increased by the same amount c, then $\bar{x}_1$ and $\bar{x}_2$ are increased by the same amount c. Since $(\bar{x}_1 + c) - (\bar{x}_2 + c) = \bar{x}_1 - \bar{x}_2$, the difference between the sample means $\bar{x}_1 - \bar{x}_2$ will remain the same.

(b) From (a), $\bar{x}_1 - \bar{x}_2$ will remain the same. An increase in the birth weights of the babies in each group will not affect the sample sizes n_1 and n_2 or the population variances σ_1^2 and σ_2^2. Therefore, $Z_{\text{data}} = \frac{\bar{x}_1 - \bar{x}_2}{\sqrt{\frac{\sigma_1^2}{n_1} + \frac{\sigma_2^2}{n_2}}}$ will remain the same.

(c) From (b), Z_{data} will remain the same. Since the p-value $= 2P(Z > Z_{\text{data}})$, it will remain the same.

(d) From (c), the p-value will remain the same. Therefore, the conclusion will remain the same.

43. $H_0: \mu_1 = \mu_2$ versus $H_a: \mu_1 \neq \mu_2$

Reject H_0 if the p-value $\leq \alpha = 0.01$.

TI-83/84: $Z_{\text{data}} \approx -5.0267$.

p-value $= 2P(Z > |Z_{data}|) = 2P(Z > |-5.0267|) = 2P(Z > 5.0267) = 0.0002583121504$ (TI-83/84: 0.000092880601)

The p-value $= 0.0002583121504$ is $\leq \alpha = 0.01$, so we reject H_0. There is evidence at level of significance $\alpha = 0.01$ that the population mean effect sizes of zooplankton differs from that of the phytoplankton.

45. $(\bar{x}_1 - \bar{x}_2) \pm t_{\alpha/2}\sqrt{\frac{s_1^2}{n_1} + \frac{s_2^2}{n_2}} = (16.7 - 13.9) \pm 1.987\sqrt{\frac{5^2}{100} + \frac{5^2}{100}} \approx 2.8 \pm 1.4050 =$ $(1.3950, 4.2050)$

47. $H_0: \mu_1 = \mu_2$ versus $H_a: \mu_1 \neq \mu_2$

Reject H_0 if the p-value $\leq \alpha = 0.05$.

$$t_{\text{data}} = \frac{(\bar{x}_1 - \bar{x}_2)}{\sqrt{\frac{s_1^2}{n_1} + \frac{s_2^2}{n_2}}} = \frac{(16.7 - 13.9)}{\sqrt{\frac{5^2}{100} + \frac{5^2}{100}}} \approx 3.9598$$

The degrees of freedom is the smaller of $n_1 - 1 = 100 - 1 = 99$ and $n_2 - 1 = 100 - 1 = 99$. Thus, df = 99. p-value $= 2P(t > |t_{\text{data}}|) = 2P(t > |3.9598|) = P(t > 3.9598) =$ 0.000141495588 (TI-84: 0.0001045746)

The p-value $= 0.000141495588$ is $\leq \alpha = 0.05$, so we reject H_0. There is evidence at level of significance $\alpha = 0.05$ that the population mean study time of biology majors differs from that of psychology majors.

Yes.

49. $H_0: \mu_1 = \mu_2$ versus $H_a: \mu_1 < \mu_2$

Reject H_0 if the p-value $\leq \alpha = 0.05$.

$$t_{\text{data}} = \frac{(\bar{x}_1 - \bar{x}_2)}{\sqrt{\frac{s_1^2}{n_1} + \frac{s_2^2}{n_2}}} = \frac{(56 - 73)}{\sqrt{\frac{12^2}{19} + \frac{12^2}{19}}} \approx -4.3665$$

The degrees of freedom is the smaller of $n_1 - 1 = 19 - 1 = 18$ and $n_2 - 1 = 19 - 1 = 18$. Thus, df = 18. p-value $= P(t < t_{\text{data}}) = P(t < -4.3665) = 0.0001860220303$ (TI-84: 0.00005106366)

The p-value $= 0.0001860220303$ is $\leq \alpha = 0.05$, so we reject H_0. There is evidence at level of significance $\alpha = 0.05$ that the population mean score for the multitasker's neighbors is less than that of the multitasker's non-neighbors.

51. The degrees of freedom is the smaller of $n_1 - 1 = 25 - 1 = 24$ and $n_2 - 1 = 24 - 1 = 23$. Thus, df = 23. $t_{\alpha/2} = 2.069$

$(\bar{x}_1 - \bar{x}_2) \pm t_{\alpha/2}\sqrt{\frac{s_1^2}{n_1} + \frac{s_2^2}{n_2}} = (2.613 - 2.236) \pm 2.069\sqrt{\frac{0.533^2}{25} + \frac{0.543^2}{24}} \approx 0.377 \pm 0.3182 =$ $(0.0588, 0.6952)$ {TI-84: $(0.0676, 0.6864)$}

53. $H_0: \mu_1 = \mu_2$ versus $H_a: \mu_1 \neq \mu_2$

Reject H_0 if the p-value $\leq \alpha = 0.05$.

$$t_{\text{data}} = \frac{(\bar{x}_1 - \bar{x}_2)}{\sqrt{\frac{s_1^2}{n_1} + \frac{s_2^2}{n_2}}} = \frac{(2.613 - 2.236)}{\sqrt{\frac{0.533^2}{25} + \frac{0.543^2}{24}}} \approx 2.4515$$

The degrees of freedom is the smaller of $n_1 - 1 = 25 - 1 = 24$ and $n_2 - 1 = 24 - 1 = 23$. Thus, df = 23. p-value $= 2P(t > t_{\text{data}}) = 2P(t > 2.4515) = 0.0222425869$ (TI-84: 0.0180119956)

The p-value $= 0.0222425869$ is $\leq \alpha = 0.05$, so we reject H_0. There is evidence at level of significance $\alpha = 0.05$ that the population mean student evaluation rating for the charismatic group differs from that of the punitive group.

55. **(a)** $s^2_{\text{pooled}} = \frac{(n_1-1)s_1^2+(n_2-1)s_2^2}{n_1+n_2-2} = \frac{(25-1)0.533^2+(24-1)0.543^2}{25+24-2} = 0.2893545319$

(b) $H_0: \mu_1 = \mu_2$ versus $H_a: \mu_1 \neq \mu_2$

Reject H_0 if the p-value $\leq \alpha = 0.05$.

$$t_{\text{data}} = \frac{\bar{x}_1-\bar{x}_2}{\sqrt{s^2_{\text{pooled}}\left(\frac{1}{n_1}+\frac{1}{n_2}\right)}} = \frac{2.613-2.236}{\sqrt{0.2893545319\left(\frac{1}{25}+\frac{1}{24}\right)}} \approx 2.4525$$

The degrees of freedom is df $= n_1 + n_2 - 2 = 25 + 24 - 2 = 47$. p-value $= 2P(t > t_{\text{data}}) = 2P(t > 2.4525) = 0.0179558037$.

The p-value $= 0.0179558037$ is $\leq \alpha = 0.05$, so we reject H_0. There is evidence at level of significance $\alpha = 0.05$ that the population mean student evaluation rating for the charismatic group differs from that of the punitive group.

(c) The value of t_{data} in (b) is a little bit larger than the value of t_{data} in 53, so the p-value in (b) is a little smaller than the p-value in 53. The conclusions of both tests are to reject H_0.

57. $H_0: \mu_1 = \mu_2$ versus $H_a: \mu_1 \neq \mu_2$

Reject H_0 if the p-value $\leq \alpha = 0.05$.

$$Z_{\text{data}} = \frac{\bar{x}_1-\bar{x}_2}{\sqrt{\frac{\sigma_1^2}{n_1}+\frac{\sigma_2^2}{n_2}}} = \frac{2.613-2.236}{\sqrt{\frac{0.533^2}{25}+\frac{0.543^2}{24}}} \approx 2.4515$$

p-value $= 2P(Z > Z_{\text{data}}) = 2P(Z > 2.4515) \approx 0.0142254108$

The p-value $= 0.0142254108$ is $\leq \alpha = 0.05$, so we reject H_0. There is evidence at level of significance $\alpha = 0.05$ that the population mean student evaluation rating for the charismatic group differs from that of the punitive group.

The value of Z_{data} in this problem equals the value of t_{data} in Problem 53 but it is a little bit smaller than the value of t_{data} in Problem 55 (b). The p-value is smaller than the other p-values. All three hypothesis tests have the same conclusion of reject H_0.

59. Answers will vary.

61. $H_0: \mu_1 = \mu_2$ versus $H_a: \mu_1 \neq \mu_2$

Rest of answer will vary.

Section 10.3

1. $\hat{p}_1$ and $\hat{p}_2$

3. Z_{data} measures the distance between sample proportions. Extreme values of Z_{data} indicate evidence against the null hypothesis.

5. **(a)** $H_0: p_1 = p_2$ versus $H_a: p_1 \neq p_2$

$\alpha = 0.10$, $Z_{crit} = 1.645$

Reject H_0 if $Z_{data} \leq -1.645$ or $Z_{data} \geq 1.645$.

(b) $\hat{p}_{\text{pooled}} = \frac{x_1+x_2}{n_1+n_2} = \frac{80+30}{100+40} = \frac{110}{140} \approx 0.7857.$

(c) $Z_{\text{data}} = \frac{(\hat{p}_1-\hat{p}_2)}{\sqrt{\hat{p}_{\text{pooled}}(1-\hat{p}_{\text{pooled}})\left(\frac{1}{n_1}+\frac{1}{n_2}\right)}} \approx \frac{(0.8-0.75)}{\sqrt{0.7857(1-0.7857)\left(\frac{1}{100}+\frac{1}{40}\right)}} \approx 0.65.$

(d) Since $Z_{data} = 0.65$ is not ≤ -1.645 and not ≥ 1.645, we do not reject H_0. There is insufficient evidence that the population proportion from Population 1 is different than the population proportion from Population 2.

7. **(a)** $H_0 : p_1 = p_2$ *versus* $H_a : p_1 > p_2$

$Z_{\text{crit}} = 2.33$

Reject H_0 if $Z_{data} \geq 2.33$.

(b) $\hat{p}_{\text{pooled}} = \frac{x_1+x_2}{n_1+n_2} = \frac{60+40}{200+250} = \frac{100}{450} \approx 0.2222.$

(c) $\hat{p}_1 = \frac{x_1}{n_1} = \frac{60}{200} = 0.3$ and $\hat{p}_2 = \frac{x_2}{n_2} = \frac{40}{250} = 0.16.$

$Z_{\text{data}} = \frac{(\hat{p}_1-\hat{p}_2)}{\sqrt{\hat{p}_{\text{pooled}}(1-\hat{p}_{\text{pooled}})\left(\frac{1}{n_1}+\frac{1}{n_2}\right)}} \approx \frac{(0.3-0.16)}{\sqrt{0.2222(1-0.2222)\left(\frac{1}{200}+\frac{1}{250}\right)}} \approx 3.550.$

(d) Since $Z_{data} \approx 3.550$, which is ≥ 2.33, we reject H_0. There is evidence at the $\alpha = 0.01$ level of significance that the population proportion of Population 1 is greater than the population proportion of Population 2.

9. **(a)** $H_0 : p_1 = p_2$ *versus* $H_a : p_1 > p_2$. Reject H_0 if p-value ≤ 0.05.

(b) $\hat{p}_{\text{pooled}} = \frac{x_1+x_2}{n_1+n_2} = \frac{250+200}{400+400} = \frac{450}{800} = 0.5625.$

(c) $\hat{p}_1 = \frac{x_1}{n_1} = \frac{250}{400} = 0.625$ and $\hat{p}_2 = \frac{x_2}{n_2} = \frac{200}{400} = 0.5.$

$Z_{\text{data}} = \frac{(\hat{p}_1-\hat{p}_2)}{\sqrt{\hat{p}_{\text{pooled}}(1-\hat{p}_{\text{pooled}})\left(\frac{1}{n_1}+\frac{1}{n_2}\right)}} \approx \frac{(0.625-0.5)}{\sqrt{0.5625(1-0.5625)\left(\frac{1}{400}+\frac{1}{400}\right)}} \approx 3.563.$

(d) p-value $= P(Z > Z_{data}) = P(Z > 3.563) \approx 0.0002$. **(e)** Since p-value ≈ 0.0002, which is ≤ 0.05, we reject H_0. There is evidence at the $\alpha = 0.05$ level of significance that the population proportion of Population 1 is greater than the population proportion of Population 2.

11. **(a)** $H_0 : p_1 = p_2$ *versus* $H_a : p_1 \neq p_2$. *Reject* H_0 if p-value ≤ 0.10.

(b) $\hat{p}_{\text{pooled}} = \frac{x_1+x_2}{n_1+n_2} = \frac{412+498}{527+613} = \frac{910}{1140} \approx 0.7982.$

(c) $\hat{p}_1 = \frac{x_1}{n_1} = \frac{412}{527} \approx 0.7818$ and $\hat{p}_2 = \frac{x_2}{n_2} = \frac{498}{613} \approx 0.8124.$

$Z_{\text{data}} = \frac{(\hat{p}_1-\hat{p}_2)}{\sqrt{\hat{p}_{\text{pooled}}(1-\hat{p}_{\text{pooled}})\left(\frac{1}{n_1}+\frac{1}{n_2}\right)}} \approx \frac{(0.7818-0.8124)}{\sqrt{0.7982(1-0.7982)\left(\frac{1}{527}+\frac{1}{613}\right)}} \approx -1.283.$

TI-83/84: $Z_{data} \approx -1.284$.

(d) p-value = $2P(Z < Z_{data}) = 2P(Z < -1.284) \approx 0.1991$. **(e)** Since p-value ≈ 0.1991, which is not ≤ 0.10, we do not reject H_0. There is insufficient evidence at the $\alpha = 0.10$ level of significance that the population proportion of Population 1 is different than the population proportion of Population 2.

13. **(a)** We have $x_1 = 80 \geq 5$, $n_1 - x_1 = 20 \geq 5$, $x_2 = 30 \geq 5$, and $n_2 - x_2 = 10 \geq 5$, so it is appropriate to construct a 95% confidence interval for $p_1 - p_2$.

(b) $\hat{p}_1 = \frac{x_1}{n_1} = \frac{80}{100} = 0.80$ and $\hat{p}_2 = \frac{x_2}{n_2} = \frac{30}{40} = 0.75$, so $\hat{p}_1 - \hat{p}_2 = 0.80 - 0.75 = 0.05$

(c) $E = Z_{\alpha/2}\sqrt{\frac{\hat{p}_1 \cdot \hat{q}_1}{n_1} + \frac{\hat{p}_2 \cdot \hat{q}_2}{n_2}} = 1.96\sqrt{\frac{(0.80)(0.20)}{100} + \frac{(0.75)(0.25)}{40}} \approx 0.1554.$

The point estimate $\hat{p}_1 - \hat{p}_2 = 0.05$ will lie within 0.1554 of the difference in population proportions $p_1 - p_2$ 95% of the time.

(d) $(\hat{p}_1 - \hat{p}_2) \pm E = 0.05 \pm 0.1554 = (-0.1054, 0.2054)$. We are 95% confident that the difference in population proportions lies between −0.1054 and 0.2054.

15. **(a)** We have $x_1 = 60 \geq 5$, $n_1 - x_1 = 140 \geq 5$, $x_2 = 40 \geq 5$, and $n_2 - x_2 = 210 \geq 5$, so it is appropriate to construct a 95% confidence interval for $p_1 - p_2$.

(b) $\hat{p}_1 = \frac{x_1}{n_1} = \frac{60}{200} = 0.30$ and $\hat{p}_2 = \frac{x_2}{n_2} = \frac{40}{250} = 0.16$, so $\hat{p}_1 - \hat{p}_2 = 0.30 - 0.16 = 0.14$

(c) $E = Z_{\alpha/2}\sqrt{\frac{\hat{p}_1 \cdot \hat{q}_1}{n_1} + \frac{\hat{p}_2 \cdot \hat{q}_2}{n_2}} \approx 1.96\sqrt{\frac{(0.30)(0.70)}{200} + \frac{(0.16)(0.84)}{250}} \approx 0.078.$

The point estimate $\hat{p}_1 - \hat{p}_2 = 0.14$ will lie within 0.078 of the difference in population proportions $p_1 - p_2$ 95% of the time.

(d) $(\hat{p}_1 - \hat{p}_2) \pm E = 0.14 \pm 0.078 = (0.062, 0.218)$. We are 95% confident that the difference in population proportions lies between 0.062 and 0.218.

17. **(a)** We have $x_1 = 490 \geq 5$, $n_1 - x_1 = 510 \geq 5$, $x_2 = 620 \geq 5$, and $n_2 - x_2 = 380 \geq 5$, so it is appropriate to construct a 95% confidence interval for $p_1 - p_2$.

(b) $\hat{p}_1 = \frac{x_1}{n_1} = \frac{490}{1000} = 0.49$ and $\hat{p}_2 = \frac{x_2}{n_2} = \frac{620}{1000} = 0.62$, so $\hat{p}_1 - \hat{p}_2 = 0.49 - 0.62 = -0.13$

(c) $E = Z_{\alpha/2}\sqrt{\frac{\hat{p}_1 \cdot \hat{q}_1}{n_1} + \frac{\hat{p}_2 \cdot \hat{q}_2}{n_2}} \approx 1.96\sqrt{\frac{(0.49)(0.51)}{1000} + \frac{(0.62)(0.38)}{1000}} \approx 0.0432.$

The point estimate $\hat{p}_1 - \hat{p}_2 = -0.13$ will lie within 0.0432 of the difference in population proportions $p_1 - p_2$ 95% of the time.

(d) $(\hat{p}_1 - \hat{p}_2) \pm E = -0.13 \pm 0.0432 = (-0.1732, -0.0868)$. We are 95% confident that the difference in population proportions lies between −0.1732 and −0.0868

19. **(a)** $H_0 : p_1 - p_2 = 0$ *versus* $H_a : p_1 - p_2 \neq 0$. The hypothesized value of 0 lies outside the interval (0.5, 0.6), so we reject H_0 at the $\alpha = 0.05$ level of significance.

(b) $H_0 : p_1 - p_2 = 0.1$ *versus* $H_a : p_1 - p_2 \neq 0.1$. The hypothesized value of 0.1 lies outside of the interval (0.5, 0.6), so we reject H_0 at the $\alpha = 0.05$ level of significance.

(c) $H_0 : p_1 - p_2 = 0.57$ *versus* $H_a : p_1 - p_2 \neq 0.57$. The hypothesized value of 0.57 lies inside the interval (0.5, 0.6), so we do not reject H_0 at the $\alpha = 0.05$ level of significance.

21. **(a)** $H_0 : p_1 - p_2 = 0.151$ *versus* $H_a : p_1 - p_2 \neq 0.151$. The hypothesized value of 0.151 lies outside of the interval (0.1, 0.11), so we reject H_0 at the $\alpha = 0.10$ level of significance.

(b) $H_0 : p_1 - p_2 = 0.115$ *versus* $H_a : p_1 - p_2 \neq 0.115$. The hypothesized value of 0.115 lies outside of the interval (0.1, 0.11), so we reject H_0 at the $\alpha = 0.10$ level of significance.

(c) $H_0 : p_1 - p_2 = 0.105$ versus $H_a : p_1 - p_2 \neq 0.105$. The hypothesized value of 0.105 lies inside of the interval (0.1, 0.11), so we do not reject H_0 at the $\alpha = 0.10$ level of significance.

23. **(a)** $x_1 = 85 \geq 5, n_1 - x_1 = 144 - 85 = 59 \geq 5, x_2 = 166 \geq 5$, and $n_2 - x_2 = 297 - 166 = 131 \geq 5$. Therefore it is appropriate to perform the Z test for the difference in population proportions.

(b) p_1 refers to the population proportion of people 18–29 years old who agree that technology will lead to a better future and p_2 refers to the population proportion of people 65 years old or older who agree that technology will lead to a better future.

(c) H_0: $p_1 = p_2$ versus H_a: $p_1 \neq p_2$

Reject H_0 if the p-value $\leq \alpha = 0.05$.

$$\hat{p}_{\text{pooled}} = \frac{x_1+x_2}{n_1+n_2} = \frac{85+166}{144+297} = \frac{251}{441} \approx 0.5692.$$

$$\hat{p}_1 = \frac{x_1}{n_1} = \frac{85}{144} \approx 0.5903 \text{ and } \hat{p}_2 = \frac{x_2}{n_2} = \frac{166}{297} \approx 0.5589.$$

$$Z_{\text{data}} = \frac{(\hat{p}_1-\hat{p}_2)}{\sqrt{\hat{p}_{pooled}(1-\hat{p}_{pooled})\left(\frac{1}{n_1}+\frac{1}{n_2}\right)}} \approx \frac{(0.5903-0.5589)}{\sqrt{0.5692(1-0.5692)\left(\frac{1}{144}+\frac{1}{297}\right)}} \approx 0.62$$

p-value $= 2P(Z > |Z_{data}|) \approx 2P(Z > 0.62) \approx 0.5352$ (TI-84: 0.5329195849)

The p-value = 0.5352 is not $\leq \alpha = 0.05$, so we do not reject H_0. There is insufficient evidence at level of significance $\alpha = 0.05$ that the population proportions agreeing that technology will lead to a better future differ between the younger group and the older group.

25. $x_1 = 34 \geq 5$, $(n_1 - x_1) = 100 - 34 = 66 \geq 5$, $x_2 = 64 \geq 5$, and $(n_2 - x_2) = 200 - 64 = 136 \geq 5$. Therefore, it is appropriate to perform the Z-test for the difference in population proportions. p_1 is the population proportion of Ohio businesses that are owned by women, and p_2 is the population proportion of New Jersey businesses that are owned by women.

Critical value method:

$H_0 : p_1 = p_2$ *versus* $H_a : p_1 > p_2$

$Z_{crit} = 1.28$

Reject H_0 if $Z_{data} \geq 1.28$.

$$\hat{p}_{\text{pooled}} = \frac{x_1+x_2}{n_1+n_2} = \frac{34+64}{100+200} = \frac{98}{300} \approx 0.3267.$$

$$\hat{p}_1 = \frac{x_1}{n_1} = \frac{34}{100} = 0.34 \text{ and } \hat{p}_2 = \frac{x_2}{n_2} = \frac{64}{200} = 0.32$$

$$Z_{\text{data}} = \frac{(\hat{p}_1-\hat{p}_2)}{\sqrt{\hat{p}_{\text{pooled}}(1-\hat{p}_{\text{pooled}})\left(\frac{1}{n_1}+\frac{1}{n_2}\right)}} \approx \frac{(0.34-0.32)}{\sqrt{0.3267(1-0.3267)\left(\frac{1}{100}+\frac{1}{200}\right)}} \approx 0.348$$

Since $Z_{data} \approx 0.348$ is not ≥ 1.28, we do not reject H_0. There is insufficient evidence at the $\alpha =$ 0.10 level of significance that the population proportion of Ohio businesses that are owned by women is greater than the population proportion of New Jersey businesses that are owned by women.

p-value method:

$H_0 : p_1 = p_2$ *versus* $H_a : p_1 > p_2$

Reject H_0 if the p-value ≤ 0.10.

$\hat{p}_{\text{pooled}} = \frac{x_1+x_2}{n_1+n_2} = \frac{34+64}{100+200} = \frac{98}{300} \approx 0.3267.$

$\hat{p}_1 = \frac{x_1}{n_1} = \frac{34}{100} = 0.34$ and $\hat{p}_2 = \frac{x_2}{n_2} = \frac{64}{200} = 0.32$

$$Z_{\text{data}} = \frac{(\hat{p}_1-\hat{p}_2)}{\sqrt{\hat{p}_{\text{pooled}}(1-\hat{p}_{\text{pooled}})\left(\frac{1}{n_1}+\frac{1}{n_2}\right)}} \approx \frac{(0.34-0.32)}{\sqrt{0.3267(1-0.3267)\left(\frac{1}{100}+\frac{1}{200}\right)}} \approx 0.348.$$

p-value $= P(Z > Z_{data}) = P(Z > 0.348) \approx 0.3639$

Since the p-value ≈ 0.3639 is not ≤ 0.10, we do not reject H_0. There is insufficient evidence at the $\alpha = 0.10$ level of significance that the population proportion of Ohio businesses that are owned by women is greater than the population proportion of New Jersey businesses that are owned by women.

27. **(a)** $\hat{p}_1 = \frac{x_1}{n_1} = \frac{85}{144} \approx 0.5903$ and $\hat{p}_2 = \frac{x_2}{n_2} = \frac{166}{297} \approx 0.5589.$

$(\hat{p}_1 - \hat{p}_2) \pm Z_{\alpha/2}\sqrt{\frac{\hat{p}_1 \cdot \hat{q}_1}{n_1} + \frac{\hat{p}_2 \cdot \hat{q}_2}{n_2}} \approx (0.5903 - 0.5589) \pm 1.96\sqrt{\frac{(0.5903)(0.4097)}{144} + \frac{(0.5589)(0.4411)}{297}} \approx$ $0.0314 \pm 0.0982 \approx (-0.0668, 0.1296)$. TI-83/84 $(-0.0668, 0.1295)$. We are 95% confident that the difference in the population proportions people 18–29 years old and people age 65 years or older who agree that technology will lead to a better future lies between -0.0668 and 0.1295.

(b) $H_0\text{: } p_1 = p_2$ versus $H_a\text{: } p_1 \neq p_2$

0 lies in the interval so we do not reject H_0.

(c) Yes.

29. **(a)** $\hat{p}_1 = \frac{x_1}{n_1} = \frac{34}{100} = 0.34$, $\hat{q}_1 = 1 - \hat{p}_1 = 1 - 0.34 = 0.66$,

$\hat{p}_2 = \frac{x_2}{n_2} = \frac{64}{200} = 0.32$, $\hat{q}_2 = 1 - \hat{p}_2 = 1 - 0.32 = 0.68$.

$(\hat{p}_1 - \hat{p}_2) \pm Z_{\alpha/2}\sqrt{\frac{\hat{p}_1 \cdot \hat{q}_1}{n_1} + \frac{\hat{p}_2 \cdot \hat{q}_2}{n_2}} = (0.34 - 0.32) \pm 1.645\sqrt{\frac{(0.34)(0.66)}{100} + \frac{(0.32)(0.68)}{200}} \approx 0.02 \pm$ $0.0950 = (-0.0750, 0.1150)$. TI-83/84 : $(-0.0749, 0.1150)$.

We are 90% confident that the difference of the population proportion of Ohio businesses that are owned by women and the population proportion of New Jersey businesses that are owned by women lies between -0.0750 (-0.0749) and 0.1150.

(b) $H_0 : p_1 = p_2$ *versus* $H_a : p_1 \neq p_2$. Our hypothesized value of 0 lies inside the interval in (a), so we do not reject H_0. There is insufficient evidence that the population proportion of Ohio businesses that are owned by women differs from the population proportion of New Jersey businesses that are owned by women.

(c) No, it is a one-sided test, and confidence intervals can only be used to perform two-sided tests.

31. $H_0 : p_1 = p_2$ versus $H_a : p_1 \neq p_2$

Reject H_0 if p-value ≤ 0.05.

$$\hat{p}_{pooled} = \frac{x_1+x_2}{n_1+n_2} = \frac{236+234}{305+305} = \frac{470}{610} \approx 0.7705.$$

$$\hat{p}_1 = \frac{x_1}{n_1} = \frac{236}{305} \approx 0.774 \text{ and } \hat{p}_2 = \frac{x_2}{n_2} = \frac{234}{305} \approx 0.767.$$

$$Z_{\text{data}} = \frac{(\hat{p}_1-\hat{p}_2)}{\sqrt{\hat{p}_{\text{pooled}}(1-\hat{p}_{\text{pooled}})\left(\frac{1}{n_1}+\frac{1}{n_2}\right)}} \approx \frac{(0.774-0.767)}{\sqrt{0.7705(1-0.7705)\left(\frac{1}{305}+\frac{1}{305}\right)}} \approx 0.21 .$$

From the TI-83/84: p-value $= 2\,P\,(Z > Z_{data}) = 2\,P\,(Z > 0.21) = 0.8336677306$.

From the table: p-value $= 2\,P\,(Z > Z_{data}) = 2\,P\,(Z > 0.21) = 2\,(1 - 0.5832) = 0.8336$.
Since p-value $= 0.8336$ is not ≤ 0.05, we do not reject H_0. There is insufficient evidence that the proportion of the people who wore the ionized bracelets who reported improvement in their maximum pain index is different than the proportion of the people who wore the placebo bracelets who reported improvement in their
maximum pain index.

33. p_1 = the population proportion of 18–24-year-old males who listen to the radio each week and p_2 = the population proportion of males age 65 or older who listen to the radio each week.

35. $$E = Z_{\alpha/2}\sqrt{\frac{\hat{p}_1 \cdot \hat{q}_1}{n_1} + \frac{\hat{p}_2 \cdot \hat{q}_2}{n_2}} \approx 1.96\sqrt{\frac{(0.92)(0.08)}{1000} + \frac{(0.87)(0.13)}{1000}} \approx 0.0268.$$

The point estimate of the difference in the population proportion of 18–24-year-old males who listen to the radio each week and the population proportion of males 65 years and older who listen to the radio each week will lie within 0.0268 of the difference in population proportions $p_1 - p_2$ 95% of the time.

37. **(a)** $H_0 : p_1 - p_2 = 0$ versus $H_a : p_1 - p_2 \neq 0$. The hypothesized value of 0 does not lie in the interval from Exercise 36, so we reject H_0. There is evidence that the difference in the population proportion of 18–24-year-old males who listen to the radio each week and the population proportion of males 65 years and older who listen to the radio each week differs from 0.

(b) $H_0 : p_1 - p_2 = 0.01$ versus $H_a : p_1 - p_2 \neq 0.01$. The hypothesized value of 0.01 does not lie in the interval from Exercise 36, so we reject H_0. There is evidence that the difference in the population proportion of 18–24-year old males who listen to the radio each week and the population proportion of males 65 years and older who listen to the radio each week differs from 0.01.

(c) $H_0 : p_1 - p_2 = 0.05$ versus $H_a : p_1 - p_2 \neq 0.05$. The hypothesized value of 0.05 lies in the interval from Exercise 36, so we do not reject H_0. There is insufficient evidence that the difference in the population proportion of 18–24-year old males who listen to the radio each week and the population proportion of males 65 years and older who listen to the radio each week differs from 0.05.

39. Critical value method:

$H_0: p_1 = p_2$ versus $H_a: p_1 > p_2$

$Z_{crit} = 1.645$

Reject H_0 if $Z_{data} \geq 1.645$.

$x_1 = p_1 n_1 = 0.92(1000) = 920$ and $x_2 = p_2 n_2 = 0.87(1000) = 870$.

$$\hat{p}_{\text{pooled}} = \frac{x_1+x_2}{n_1+n_2} = \frac{920+870}{1000+1000} = \frac{1790}{2000} = 0.895.$$

$$Z_{\text{data}} = \frac{(\hat{p}_1-\hat{p}_2)}{\sqrt{\hat{p}_{\text{pooled}}(1-\hat{p}_{\text{pooled}})\left(\frac{1}{n_1}+\frac{1}{n_2}\right)}} = \frac{(0.92-0.87)}{\sqrt{0.895(1-0.895)\left(\frac{1}{305}+\frac{1}{305}\right)}} \approx 3.647.$$

Since $Z_{\text{data}} = 3.647$, which is ≥ 1.645, we reject H_0. There is evidence at the $\alpha = 0.05$ level of significance that the population proportion of 18–24-year-old males who listen to the radio each week is greater than the population proportion of males 65 years and older who listen to the radio each week.

p-value method:

$H_0: p_1 = p_2$ versus $H_a: p_1 > p_2$

Reject H_0 if the *p*-value ≤ 0.05.

$x_1 = p_1 n_1 = 0.92(1000) = 920$ and $x_2 = p_2 n_2 = 0.87(1000) = 870$.

$$\hat{p}_{\text{pooled}} = \frac{x_1+x_2}{n_1+n_2} = \frac{920+870}{1000+1000} = \frac{1790}{2000} = 0.895.$$

$$Z_{\text{data}} = \frac{(\hat{p}_1-\hat{p}_2)}{\sqrt{\hat{p}_{\text{pooled}}(1-\hat{p}_{\text{pooled}})\left(\frac{1}{n_1}+\frac{1}{n_2}\right)}} = \frac{(0.92-0.87)}{\sqrt{0.895(1-0;895)\left(\frac{1}{1000}+\frac{1}{1000}\right)}} \approx 3.647.$$

p-value $= P(Z > Z_{data}) = P(Z > 3.647) \approx 0.00013$

Since the *p*-value ≈ 0.00013, which is ≤ 0.05, we reject H_0. There is evidence at the $\alpha = 0.05$ level of significance that the population proportion of 18–24-year-old males who listen to the radio each week is greater than the population proportion of males 65 years and older who listen to the radio each week.

41. Answers will vary.

43. H_0: $p_1 = p_2$ versus H_a: $p_1 \neq p_2$

Rest of answer will vary.

Section 10.4

1. Two population standard deviations.

3. True.

5. An *F* distribution with $n_1 - 1$ degrees of freedom in the numerator and $n_2 - 1$ degrees of freedom in the denominator.

7. The *F* value with $df_1 = n_1 - 1$ numerator degrees of freedom and $df_2 = n_2 - 1$ denominator degrees of freedom, with area α to the right of F_{α,n_1-1,n_2-1}.

9. $F_{\alpha,n_1-1,n_2-1} = F_{0.05,10,9} = 3.14.$

11. $F_{\alpha,n_1-1,n_2-1} = F_{0.01,5,4} = 15.52.$

13. $F_{1-\alpha,n_1-1,n_2-1} = F_{0.95,20,3} = \frac{1}{F_{\alpha,n_2-1,n_1-1}} = \frac{1}{F_{0.05,3,20}} = \frac{1}{3.10} \approx 0.3226.$

15. $F_{1-\alpha,n_1-1,n_2-1} = F_{0.95,3,20} = \frac{1}{F_{\alpha,n_2-1,n_1-1}} = \frac{1}{F_{0.05,20,3}} = \frac{1}{8.66} \approx 0.1155.$

17. $F_{\alpha/2,n_1-1,n_2-1} = F_{0.025,15,6} = 5.27, F_{1-\alpha/2,n_1-1,n_2-1} = F_{0.975,15,6} = \frac{1}{F_{\alpha/2,n_2-1,n_1-1}} = \frac{1}{F_{0.025,6,15}} = \frac{1}{3.41} \approx 0.2933.$

19. $F_{\alpha/2,n_1-1,n_2-1} = F_{0.025,6,15} = 3.41, F_{1-\alpha/2,n_1-1,n_2-1} = F_{0.975,6,15} = \frac{1}{F_{\alpha/2,n_2-1,n_1-1}} = \frac{1}{F_{0.025,15,6}} = \frac{1}{5.27} \approx 0.1898.$

21. **(a)** $H_0 : \sigma_1 = \sigma_2$ versus $H_a : \sigma_1 > \sigma_2$.

(b) $F_{crit} = F_{\alpha,n_1-1,n_2-1} = F_{0.05,10,6} = 4.06$. Reject H_0 if $F_{\text{data}} \geq 4.06$.

(c) $F_{data} = \frac{s_1^2}{s_2^2} = \frac{2^2}{1^2} = \frac{4}{1} = 4.$

(d) Since $F_{\text{data}} = 4$ is not ≥ 4.06, we do not reject H_0. There is insufficient evidence at the $\alpha = 0.05$ level of significance that the population standard deviation of Population 1 is greater than the population standard deviation of Population 2.

23. **(a)** $H_0 : \sigma_1 = \sigma_2$ versus $H_a : \sigma_1 < \sigma_2$.

(b) $F_{\text{crit}} = F_{1-\alpha,n_1-1,n_2-1} = F_{0.90,9,20} = \frac{1}{F_{\alpha,n_2-1,n_1-1}} = \frac{1}{F_{0.10,20,9}} = \frac{1}{2.30} \approx 0.4348.$

Reject H_0 if $F_{\text{data}} \leq 0.4348$.

(c) $F_{data} = \frac{s_1^2}{s_2^2} = \frac{2.5^2}{3.5^2} = \frac{6.25}{12.25} \approx 0.5102.$

(d) Since $F_{\text{data}} = 0.5102$, which is not ≤ 0.4348, we do not reject H_0. There is insufficient evidence at the $\alpha = 0.10$ level of significance that the population standard deviation of Population 1 is less than the population standard deviation of Population 2.

25. **(a)** $H_0 : \sigma_1 = \sigma_2$ versus $H_a : \sigma_1 \neq \sigma_2$.

(b) $F_{\alpha/2,n_1-1,n_2-1} = F_{0.025,5,12} = 3.89, F_{1-\alpha/2,n_1-1,n_2-1} = F_{0.975,5,12} = \frac{1}{F_{\alpha/2,n_2-1,n_1-1}} = \frac{1}{F_{0.025,12,5}}.$

Since $\text{df}_1 = 12$ is not in the table we use the conservative $\text{df}_1 = 10$. Thus, we use

$\frac{1}{F_{0.025,10,5}} = \frac{1}{6.62} \approx 0.1511.$ Reject H_0 if $F_{\text{data}} \geq 3.89$ or if $F_{\text{data}} \leq 0.1511$.

(c) $F_{\text{data}} = \frac{s_1^2}{s_2^2} = \frac{3.1^2}{1.3^2} = \frac{9.61}{1.69} \approx 5.686.$

(d) Since $F_{\text{data}} \approx 5.686$, which is ≥ 3.89, we reject H_0. There is evidence at the $\alpha = 0.05$ level of significance that the population standard deviation of Population 1 differs from the population standard deviation of Population 2.

27. **(a)** $H_0 : \sigma_1 = \sigma_2$ versus $H_a : \sigma_1 > \sigma_2$. Reject H_0 if the p-value ≤ 0.01.

(b) $F_{data} = \frac{s_1^2}{s_2^2} = \frac{5^2}{3^2} = \frac{25}{9} \approx 2.778$.

(c) $df_1 = n_1 - 1 = 9 - 1 = 8$, $df_2 = n_2 - 1 = 16 - 1 = 15$. p-value $= P(F > F_{data}) = P(F > 2.778) \approx 0.0420$.

(d) Since the p-value ≈ 0.0420, which is not ≤ 0.01, we do not reject H_0. There is insufficient evidence at the $\alpha = 0.01$ level of significance that the population standard deviation of Population 1 is greater than the population standard deviation of Population 2.

29. **(a)** $H_0 : \sigma_1 = \sigma_2$ versus $H_a : \sigma_1 < \sigma_2$. Reject H_0 if the p-value ≤ 0.05.

(b) $F_{data} = \frac{s_1^2}{s_2^2} = \frac{2.8^2}{5.2^2} = \frac{7.84}{27.04} \approx 0.2899$.

(c) $df_1 = n_1 - 1 = 16 - 1 = 15$, $df_2 = n_2 - 1 = 21 - 1 = 20$. p-value $= P(F < F_{data}) = P(F < 0.2899) \approx 0.0090$.

(d) Since the p-value ≈ 0.0090, which is ≤ 0.05, we reject H_0. There is evidence at the $\alpha = 0.05$ level of significance that the population standard deviation of Population 1 is less than the population standard deviation of Population 2.

31. **(a)** $H_0 : \sigma_1 = \sigma_2$ versus $H_a : \sigma_1 \neq \sigma_2$. Reject H_0 if the p-value ≤ 0.10.

(b) $F_{data} = \frac{s_1^2}{s_2^2} = \frac{150^2}{150^2} = \frac{22{,}500}{22{,}500} = 1$.

(c) $df_1 = n_1 - 1 = 8 - 1 = 7$ $df_2 = n_2 - 1 = 8 - 1 = 7$ The p-value is the smaller of $2P(F > F_{data}) = 2P(F > 1) \approx 1$ and $2P(F < F_{data}) = 2P(F < 1) \approx 1$. Therefore, p-value ≈ 1.

(d) Since the p-value ≈ 1, which is not ≤ 0.10, we do not reject H_0. There is insufficient evidence at the $\alpha = 0.10$ level of significance that the population standard deviation of Population 1 is different than the population standard deviation of Population 2.

33. Critical value method:

$H_0 : \sigma_1 = \sigma_2$ versus $H_a : \sigma_1 < \sigma_2$

$$F_{crit} = F_{1-\alpha,n_1-1,n_2-1} = F_{0.95,1159,1570} = \frac{1}{F_{\alpha,n_2-1,n_1-1}} = \frac{1}{F_{0.05,1570,1159}}.$$

Since $df_1 = 1570$ and $df_2 = 1159$ are not in the table, we use the conservative $df_1 = 1000$ and $df_2 = 1000$. Thus, we use

$$F_{crit} = \frac{1}{F_{0.05,1000,1000}} = \frac{1}{1.11} \approx 0.9009.$$

Reject H_0 if $F_{data} \leq 0.9009$.

$$F_{data} = \frac{s_1^2}{s_2^2} = \frac{3.3^2}{4.3^2} = \frac{10.89}{18.49} \approx 0.5890.$$

Since $F_{data} \approx 0.5890$ is ≤ 0.9009, we reject H_0. There is evidence at the $\alpha = 0.05$ level of significance that the population standard deviation of the males' BMI is less than the population standard deviation of the females' BMI.

p-value method:

$H_0 : \sigma_1 = \sigma_2$ versus $H_a : \sigma_1 < \sigma_2$

Reject H_0 if the p-value ≤ 0.05.

$$F_{\text{data}} = \frac{s_1^2}{s_2^2} = \frac{3.3^2}{4.3^2} = \frac{10.89}{18.49} \approx 0.5890.$$

$\text{df}_1 = n_1 - 1 = 1160 - 1 = 1159$, $\text{df}_2 = n_2 - 1 = 1571 - 1 = 1570$.

p-value $= P(F < F_{\text{data}}) = P(F < 0.5890) \approx 0$

Since the p-value = 0 is ≤ 0.05, we reject H_0. There is evidence at the $\alpha = 0.05$ level of significance that the population standard deviation of the males' BMI is less than the population standard deviation of the females' BMI.

35. Critical value method:

$H_0 : \sigma_1 = \sigma_2$ versus $H_a : \sigma_1 \neq \sigma_2$

$F_{\text{crit}} = F_{\alpha/2, n_1-1, n_2-1} = F_{0.05,84,340}$.

Since $\text{df}_1 = 84$ and $\text{df}_2 = 340$ are not in the table, we use the conservative $\text{df}_1 = 60$ and $\text{df}_2 = 200$. Thus, we use $F_{\text{crit}} = F_{0.05,\,60,\,200} = 1.39$.

$$F_{\text{crit}} = F_{1-\alpha/2, n_1-1, n_2-1} = F_{0.95,84,340} = \frac{1}{F_{\alpha/2, n_2-1, n_1-1}} = \frac{1}{F_{0.05,340,84}}.$$

Since $\text{df}_1 = 340$ and $\text{df}_2 = 84$ are not in the table, we use the conservative $\text{df}_1 = 120$ and $\text{df}_2 = 50$. Thus, we use

$$F_{\text{crit}} = \frac{1}{F_{0.05,120,50}} = \frac{1}{1.51} \approx 0.6623.$$

Reject H_0 if $F_{\text{data}} \leq 0.6623$ or if $F_{\text{data}} \geq 1.39$.

$$F_{\text{data}} = \frac{s_1^2}{s_2^2} = \frac{15.08^2}{16.56^2} = \frac{227.4064}{274.2336} \approx 0.8292.$$

Since $F_{\text{data}} \approx 0.8292$, which is not ≤ 0.6623 and not ≥ 1.39, we do not reject H_0. There is insufficient evidence at the $\alpha = 0.10$ level of significance that the population standard deviation of the IQ scores of children with autism differs from the population standard deviation of the IQ scores of children with Asperger's syndrome.

p-value method:

$H_0 : \sigma_1 = \sigma_2$ versus $H_a : \sigma_1 \neq \sigma_2$

Reject H_0 if the p-value ≤ 0.10.

$$F_{\text{data}} = \frac{s_1^2}{s_2^2} = \frac{15.08^2}{16.56^2} = \frac{227.4064}{274.2336} \approx 0.8292.$$

$\text{df}_1 = n_1 - 1 = 85 - 1 = 84$, $\text{df}_2 = n_2 - 1 = 341 - 1 = 340$

The p-value is the smaller of $2P(F > F_{\text{data}}) = 2P(F > 0.8292) \approx 1.6970$ and $2P(F < F_{\text{data}}) = 2P(F < 0.8292) \approx 0.3030$. Therefore, p-value ≈ 0.3030.

Since the p-value ≈ 0.3030 is not ≤ 0.10, we do not reject H_0. There is insufficient evidence at the $\alpha = 0.10$ level of significance that the population standard deviation of the IQ scores of children with autism differs from the population standard deviation of the IQ scores of children with Asperger's syndrome.

37. **(a)** No. All of the points in both graphs lie between the curved lines and most of the points in both graphs lie near the center line.

(b) Critical value method:

$H_0 : \sigma_1 = \sigma_2$ versus $H_a : \sigma_1 > \sigma_2$

$F_{\text{crit}} = F_{\alpha, n_1 - 1, n_2 - 1} = F_{0.01,13,13} = 4.16$. Reject H_0 if $F_{\text{data}} \geq 4.16$.

Using the TI-83/84, $s_1^2 \approx 3173.62408$ and $s_2^2 \approx 1.164810989$.

$$F_{\text{data}} = \frac{s_1^2}{s_2^2} = \frac{3173.62408}{1.164810989} \approx 2724.5829.$$

Since $F_{\text{data}} \approx 2724.5829$ is ≥ 4.16, we reject H_0. There is evidence at the $\alpha = 0.01$ level of significance that the population standard deviation of gold prices is greater than the population standard deviation of silver prices.

p-value method:

$H_0 : \sigma_1 = \sigma_2$ versus $H_a : \sigma_1 > \sigma_2$

Reject H_0 if the p-value ≤ 0.01.

Using the TI-83/84, $s_1^2 \approx 3173.62408$ and $s_2^2 \approx 1.164810989$.

$$F_{\text{data}} = \frac{s_1^2}{s_2^2} = \frac{3173.62408}{1.164810989} \approx 2724.5829.$$

$\text{df}_1 = n_1 - 1 = 14 - 1 = 13$, $\text{df}_2 = n_2 - 1 = 14 - 1 = 13$

p-value $= P(F > F_{\text{data}}) = P(F > 2724.5829) \approx 0$

Since the p-value ≈ 0, which is ≤ 0.01, we reject H_0. There is evidence at the $\alpha = 0.01$ level of significance that the population standard deviation of gold prices is greater than the population standard deviation of silver prices.

Chapter 10 Review Exercises

1.

Subject	**1**	**2**	**3**	**4**	**5**	**6**	**7**	**8**
Sample 1	100.7	110.2	105.3	107.1	95.6	109.9	112.3	94.7
Sample 2	104.4	112.5	105.9	111.4	99.8	109.9	115.7	97.7
Differences (Sample 1 minus Sample 2)	-3.7	-2.3	-0.6	-4.3	-4.2	0.0	-3.4	-3.0

(a) From the TI-83/84, $\bar{x}_d = -2.6875$ and $s_d \approx 1.6146$.

(b) df $= n - 1 = 8 - 1 = 7$

$$\bar{x}_d \pm t_{\alpha/2}\left(\frac{s_d}{\sqrt{n}}\right) = -2.6875 \pm 2.365\left(\frac{1.6146}{\sqrt{8}}\right) = -2.6875 \pm 1.3501 \approx (-4.0376, -1.3374).$$

3. $H_0 : \mu_d = 0$ versus $H_a : \mu_d < 0$

Reject H_0 if p-value ≤ 0.05.

$$t_{\text{data}} = \frac{\bar{x}_d}{s_d/\sqrt{n}} \approx \frac{-2.6875}{1.6146/\sqrt{8}} \approx -4.708.$$

df = $n - 1 = 8 - 1 = 7$, p-value = $P(t < t_{data}) = P(t < -4.708) = 0.0010939869$
Since p-value = 0.0010939869 is ≤ 0.05, we reject H_0. There is evidence that the population mean of the differences is less than 0.

5. $\bar{x}_1 - \bar{x}_2 = 14.4 - 14.3 = 0.1.$

7. $(\bar{x}_1 - \bar{x}_2) \pm t_{\alpha/2}\sqrt{\frac{s_1^2}{n_1} + \frac{s_2^2}{n_2}} = (14.4 - 14.3) \pm 2.030\sqrt{\frac{0.01^2}{36} + \frac{0.02^2}{81}} \approx 0.1 \pm 0.0056 =$ $(0.0944,\ 0.1056)$.

We are 95% confident that the interval (0.0944, 0.1056) captures the difference in population means.

9. **(a)** $\hat{p}_1 = \frac{x_1}{n_1} = \frac{641}{823} \approx 0.7789$ and $\hat{p}_2 = \frac{x_2}{n_2} = \frac{658}{824} \approx 0.7985.$

$\hat{q}_1 = 1 - \hat{p}_1 \approx 1 - 0.7789 = 0.2211$ and $\hat{q}_2 = 1 - \hat{p}_2 \approx 1 - 0.7985 = 0.2015$

$(\hat{p}_1 - \hat{p}_2) \pm Z_{\alpha/2}\sqrt{\frac{\hat{p}_1 \cdot \hat{q}_1}{n_1} + \frac{\hat{p}_2 \cdot \hat{q}_2}{n_2}} =$

$(07789 - 0.7985) \pm 2.576\sqrt{\frac{(0.7789)(0.2211)}{823} + \frac{(0.7985)(0.2015)}{824}} \approx -0.0196 \pm 0.0518 =$ $(-0.0714, 0.322)$.

(b) $H_0 : p_1 = p_2$ versus $H_a : p_1 \neq p_2$
Reject H_0 if p-value ≤ 0.01.

$\hat{p}_{\text{pooled}} = \frac{x_1 + x_2}{n_1 + n_2} = \frac{641+658}{823+824} = \frac{1299}{1647} \approx 0.7887.$

$\hat{p}_1 = \frac{x_1}{n_1} = \frac{641}{823} \approx 0.7789$ and $\hat{p}_2 = \frac{x_2}{n_2} = \frac{658}{824} \approx 0.7985.$

$$Z_{\text{data}} = \frac{(\hat{p}_1 - \hat{p}_2)}{\sqrt{\hat{p}_{pooled}(1 - \hat{p}_{pooled})\left(\frac{1}{n_1} + \frac{1}{n_2}\right)}} \approx \frac{(0.7789 - 0.7985)}{\sqrt{0.7887(1 - 0.7887)\left(\frac{1}{823} + \frac{1}{824}\right)}} \approx -0.97\ .$$

From the TI-83/84, p-value = $2\,P(Z < Z_{data}) = 2\,P(Z < -0.97) = 0.3320464796$.

From the table, p-value = $2\,P(Z < Z_{data}) = 2\,P(Z < -0.97) = 2\,(0.1660) = 0.3320$.
Since p-value = 0.3320 is not ≤ 0.01, we do not reject H_0. There is insufficient evidence that the population proportion of mothers living in Florida who took their babies in for a checkup within 1 week of delivery is different than the population proportion of mothers living in North Carolina who took their babies in for a checkup within 1 week of delivery.

11. $H_0 : \sigma_1 = \sigma_2$ *versus* $H_a : \sigma_1 < \sigma_2$

$$F_{\text{crit}} = F_{1-\alpha, n_1 - 1, n_2 - 1} = F_{0.95,124,26} = \frac{1}{F_{\alpha, n_2 - 1, n_1 - 1}} = \frac{1}{F_{0.05,26,124}}.$$

Since $df_1 = 26$ and $df_2 = 124$ are not in the table, we use the conservative $df_1 = 25$ and $df_2 = 100$. Thus, we use

$$F_{\text{crit}} = \frac{1}{F_{0.05,25,100}} = \frac{1}{1.62} \approx 0.6173.$$

Reject H_0 if $F_{data} \leq 0.6173$.

$$F_{\text{data}} = \frac{s_1^2}{s_2^2} = \frac{5.7^2}{10.2^2} = \frac{32.49}{104.04} \approx 0.3123.$$

Since $F_{data} \approx 0.3123$, which is ≤ 0.6173, we reject H_0. There is evidence at the $\alpha = 0.05$ level of significance that the population standard deviation of Population 1 is less than the population standard deviation of Population 2.

13. $H_0 : \sigma_1 = \sigma_2$ *versus* $H_a : \sigma_1 < \sigma_2$

Reject H_0 if the p-value ≤ 0.01.

$$F_{\text{data}} = \frac{s_1^2}{s_2^2} = \frac{25.1^2}{35.8^2} = \frac{630.01}{1281.64} \approx 0.4916.$$

$\text{df}_1 = n_1 - 1 = 150 - 1 = 149$, $\text{df}_2 = n_2 - 1 = 80 - 1 = 79$

p-value $= P(F < F_{data}) = P(F < 0.4916) \approx 0.0001$

Since *the* p-value ≈ 0.0001, which is ≤ 0.01, we reject H_0. There is evidence at the $\alpha = 0.01$ level of significance that the population standard deviation of Population 1 is less than the population standard deviation of Population 2.

Chapter 10 Quiz

1. True.

3. False. The test statistic Z_{data} measures the distance between the sample proportions.

5. Margin of error.

7. $\mu_1 - \mu_2$

9. No difference.

11. **(a)** Critical value method:

$H_0 : \mu_1 = \mu_2$ versus $H_a : \mu_1 < \mu_2$

The degrees of freedom is the smaller of $n_1 - 1 = 40 - 1 = 39$ and $n_2 - 1 = 36 - 1 = 35$. Thus, df = 35. This is a left-tailed test with $\alpha = 0.05$, so $t_{crit} = -1.690$. Reject H_0 if $t_{data} \leq -1.690$.

$$t_{\text{data}} = \frac{(\bar{x}_1 - \bar{x}_2)}{\sqrt{\frac{s_1^2}{n_1} + \frac{s_2^2}{n_2}}} = \frac{(50{,}000 - 65{,}000)}{\sqrt{\frac{15{,}000^2}{40} + \frac{20{,}000^2}{36}}} \approx -3.667.$$

Since $t_{data} \approx -3.667$ is ≤ -1.690, we reject H_0. There is evidence at the $\alpha = 0.05$ level of significance that the population mean income in Suburb A is less than the population mean income in Suburb B.

p-value method:

$H_0 : \mu_1 = \mu_2$ versus $H_a : \mu_1 < \mu_2$

Reject H_0 if p-value ≤ 0.05.

$$t_{\text{data}} = \frac{(\bar{x}_1 - \bar{x}_2)}{\sqrt{\frac{s_1^2}{n_1} + \frac{s_2^2}{n_2}}} = \frac{(50{,}000 - 65{,}000)}{\sqrt{\frac{15{,}000^2}{40} + \frac{20{,}000^2}{36}}} \approx -3.667.$$

The degrees of freedom is the smaller of $n_1 - 1 = 40 - 1 = 39$ and $n_2 - 1 = 36 - 1 = 35$. Thus, df = 35.

p-value = $P(t < t_{data}) = P(t < -3.667) \approx 0.0004$

Since p-value ≈ 0.0004, which is ≤ 0.05, we reject H_0. There is evidence at the $\alpha = 0.05$ level of significance that the population mean income in Suburb A is less than the population mean income in Suburb B.

(b) $n_1 - 1 = 40 - 1 = 39$ and $n_2 = 36 - 1 = 35$, so df = 35.

$(\bar{x}_1 - \bar{x}_2) \pm t_{\alpha/2}\sqrt{\frac{s_1^2}{n_1} + \frac{s_2^2}{n_2}} = (50{,}000 - 65{,}000) \pm 2.030\sqrt{1\frac{5{,}000^2}{40} + \frac{20{,}000^2}{36}} \approx -15{,}000 \pm 8{,}304.69 = (-23{,}304.69,\ -6{,}695.31)$.

We are 95% confident that the interval (–23,304.69, –6,695.31) captures the difference of the population mean income of Suburb A and the population mean income of Suburb B.

13. **(a)** $H_0 : \mu_1 = \mu_2$ versus $H_a : \mu_1 > \mu_2$

$t_{crit} = 1.662$. Reject H_0 if $t_{data} \geq 1.662$, $t_{data} = 2.561$.

$$t_{\text{data}} = \frac{(\bar{x}_1 - \bar{x}_2)}{\sqrt{\frac{s_1^2}{n_1} + \frac{s_2^2}{n_2}}} = \frac{(200 - 190)}{\sqrt{\frac{30^2}{100} + \frac{25^2}{100}}} \approx 2.561.$$

Since $t_{data} = 2.561$ is ≥ 1.662, we reject H_0. There is evidence that the population mean number of bottles processed by the updated machine is greater than the mean number of bottles processed by the nonupdated machine.

(b) Since confidence intervals can only be used to perform two-tail tests and the hypothesis test in (a) is a one-tail test, the confidence interval in Exercise 12 cannot be used to perform the hypothesis test in (a).

Chapter 11: Categorical Data Analysis

Section 11.1

1. A random variable is *multinomial* if it satisfies each of the following conditions:

- Each independent trial of the experiment has k possible outcomes, $k = 2, 3, 4, \ldots$
- The ith outcome (category) occurs with probability p_i, where $i = 1, 2, \ldots, k$.
- $\sum_{i=1}^{k} p_i = 1$

3. It is the long run mean of that random variable after an arbitrarily large number of trials.

5. Multinomial.

7. Multinomial.

9. **(a)** $E_1 = n \cdot p_1 = (100)\,(0.50) = 50$, $E_2 = n \cdot p_2 = (100)\,(0.25) = 25$, $E_3 = n \cdot p_3 = (100)\,(0.25) = 25$

(b) Since none of the expected frequencies is less than 1 and none of the expected frequencies is less than 5, the conditions for performing the χ^2 goodness of fit test are met.

11. **(a)** $E_1 = n \cdot p_1 = (100)\,(0.9) = 90$, $E_2 = n \cdot p_2 = (100)\,(0.05) = 5$, $E_3 = n \cdot p_3 = (100)\,(0.04) = 4$, $E_4 = n \cdot p_4 = (100)\,(0.01) = 1$

(b) Since 50% of the expected frequencies are less than 5, the conditions for performing the χ^2 goodness of fit test are not met.

13.

O_i	E_i	$O_i - E_i$	$(O_i - E_i)^2$	$\frac{(O_i - E_i)^2}{E_i}$
10	12	–2	4	4/12 = 0.3333333333
12	12	0	0	0/12 = 0
14	12	2	4	4/12 = 0.3333333333
				χ^2_{data} = 0.6666666666

Thus, $\chi^2_{\text{data}} = \sum \frac{(O_i - E_i)^2}{E_i} = 0.6666666666 \approx 0.667.$

15.

O_i	E_i	$O_i - E_i$	$(O_i - E_i)^2$	$\frac{(O_i - E_i)^2}{E_i}$
20	25	–5	25	25/25 = 1
30	25	5	25	25/25 = 1
40	30	10	100	100/30 = 3.333333333
40	50	–10	100	100/50 = 2
				χ^2_{data} = 7.333333333

Thus, $\chi^2_{\text{data}} = \sum \frac{(O_i - E_i)^2}{E_i} = 7.333333333 \approx 7.333.$

17.

O_i	E_i	$O_i - E_i$	$(O_i - E_i)^2$	$\frac{(O_i - E_i)^2}{E_i}$
1	6	–5	25	25/6 = 4.166666667
10	6	4	16	16/6 = 2.666666667
8	6	2	4	4/6 = 0.666666667
0	6	–6	36	36/6 = 6
11	6	5	25	25/6 = 4.166666667
				$\chi^2_{data} = 17.666666667$

Thus, $\chi^2_{\text{data}} = \sum \frac{(O_i - E_i)^2}{E_i} = 17.666666667 \approx 17.667.$

19. **(a)** $n = O_1 + O_2 + O_3 = 50 + 25 + 25 = 100.$

$E_1 = n \cdot p_1 = (100)\,(0.4) = 40$, $E_2 = n \cdot p_2 = (100)\,(0.3) = 30$,

$E_3 = n \cdot p_3 = (100)\,(0.3) = 30.$

Since none of the expected frequencies is less than 1 and none of the expected frequencies is less than 5, the conditions for performing the χ^2 goodness of fit test are met.

(b) df = $k - 1 = 3 - 1 = 2$, $\alpha = 0.05$, $\chi^2_{crit} = \chi^2_{0.05} = 5.991.$

Reject H_0 if $\chi^2_{\text{data}} \geq 5.991.$

(c)

O_i	E_i	$O_i - E_i$	$(O_i - E_i)^2$	$\frac{(O_i - E_i)^2}{E_i}$
50	40	10	100	100/40 = 2.5
25	30	–5	25	25/30 = 0.8333333333
25	30	–5	25	25/30 = 0.8333333333
				$\chi^2_{data} = 4.1666666666$

Thus, $\chi^2_{\text{data}} = \sum \frac{(O_i - E_i)^2}{E_i} = 4.166666666 \approx 4.167.$

(d) Since χ^2_{data} is not $\geq$ 5.991, we do not reject H_0. There is insufficient evidence that the random variable does not follow the distribution specified in H_0.

21. **(a)** $n = O_1 + O_2 + O_3 + O_4 + O_5 = 90 + 75 + 15 + 15 + 5 = 200.$

$E_1 = n \cdot p_1 = (200)\,(0.4) = 80$, $E_2 = n \cdot p_2 = (200)\,(0.35) = 70$,

$E_3 = n \cdot p_3 = (200)\,(0.10) = 20$, $E_4 = n \cdot p_4 = (200)\,(0.10) = 20$,

$E_5 = n \cdot p_5 = (200)\,(0.05) = 10.$

Since none of the expected frequencies is less than 1 and none of the expected frequencies is less than 5, the conditions for performing the χ^2 goodness of fit test are met.

(b) df = $k - 1 = 5 - 1 = 4$, $\alpha = 0.10$, $\chi^2_{\text{crit}} = \chi^2_{0.10} = 7.779.$

Reject H_0 if $\chi^2_{\text{data}} \geq 7.779$.

(c)

O_i	E_i	$O_i - E_i$	$(O_i - E_i)^2$	$\frac{(O_i - E_i)^2}{E_i}$
90	80	10	100	100/80 = 1.25
75	70	5	25	25/70 = 0.3571428571
15	20	–5	25	25/20 = 1.25
15	20	–5	25	25/20 = 1.25
5	10	–5	25	25/10 = 2.5
				$\chi^2_{data} = 6.607142857$

Thus, $\chi^2_{\text{data}} = \sum \frac{(O_i - E_i)^2}{E_i} = 6.607142857 \approx 6.607$.

(d) Since χ^2_{data} is not ≥ 7.779, we do not reject H_0. There is insufficient evidence that the random variable does not follow the distribution specified in H_0.

23. **(a)** Reject H_0 if p-value ≤ 0.05. $n = O_1 + O_2 = 40 + 60 = 100$.

$E_1 = n \cdot p_1 = (100)\ (0.50) = 50$, $E_2 = n \cdot p_2 = (100)\ (0.50) = 50$.

Since none of the expected frequencies is less than 1 and none of the expected frequencies is less than 5, the conditions for performing the χ^2 goodness of fit test are met.

(b)

O_i	E_i	$O_i - E_i$	$(O_i - E_i)^2$	$\frac{(O_i - E_i)^2}{E_i}$
40	50	–10	100	100/50 = 2
60	50	10	100	100/50 = 2
				$\chi^2_{\text{data}} = 4$

Thus, $\chi^2_{\text{data}} = \sum \frac{(O_i - E_i)^2}{E_i} = 4$.

(c) df = $k - 1 = 2 - 1 = 1$. The p-value $= P(\chi^2 > \chi^2_{\text{data}}) = P(\chi^2 > 4) \approx 0.0455$.

(d) Since p-value is ≤ 0.05, we reject H_0. There is evidence that the random variable does not follow the distribution specified in H_0.

25. **(a)** Reject H_0 if p-value ≤ 0.10.

$n = O_1 + O_2 + O_3 + O_4 = 90 + 55 + 40 + 15 = 200$.

$E_1 = n \cdot p_1 = (200)\ (0.5) = 100$, $E_2 = n \cdot p_2 = (200)\ (0.25) = 50$,

$E_3 = n$ **v** $p_3 = (200)\ (0.15) = 30$, $E_4 = n \cdot p_4 = (200)\ (0.1) = 20$.

Since none of the expected frequencies is less than 1 and none of the expected frequencies is less than 5, the conditions for performing the χ^2 goodness of fit test are met.

(b)

O_i	E_i	$O_i - E_i$	$(O_i - E_i)^2$	$\frac{(O_i - E_i)^2}{E_i}$
90	100	−10	100	100/100 = 1
55	50	5	25	25/50 = 0.5
40	30	10	100	100/30 = 3.333333333
15	20	−5	25	25/20 = 1.25
				X^2_{data} = 6.083333333

Thus, $\chi^2_{\text{data}} = \sum \frac{(O_i - E_i)^2}{E_i} = 6.083333333 \approx 6.083.$

(c) df = $k - 1 = 4 - 1 = 3$. The p-value $= P(\chi^2 > \chi^2_{\text{data}}) \approx P(\chi^2 > 6.083) \approx 0.1076.$

(d) Since p-value is not ≤ 0.10, we do not reject H_0. There is insufficient evidence that the random variable does not follow the distribution specified in H_0.

27. H_0: $p_{\text{Barnes and Noble}} = 0.23,\ p_{\text{Amazon}} = 0.20,\ p_{\text{Others}} = 0.57$

H_a: The random variable does not follow the distribution specified in H_0.

Checking the conditions, the expected frequencies are

$E_{\text{Barnes and Noble}} = n \cdot p_{\text{Barnes and Noble}} = 1000(0.23) = 230,$

$E_{\text{Amazon}} = n \cdot p_{\text{Amazon}} = 1000(0.2) = 200,$

$E_{\text{Others}} = n \cdot p_{\text{Others}} = 1000(0.57) = 570.$

Because none of these expected frequencies is less than 1, and none of the expected frequencies is less than 5, the conditions for performing the goodness of fit test are met.

df= $k - 1 = 3 - 1 = 2$. $\chi^2_{\text{crit}} = \chi^2_{0.05} = 5.991$. Reject H_0 if $\chi^2_{\text{data}} \geq 5.991$.

O_i	E_i	$O_i - E_i$	$(O_i - E_i)^2$	$\frac{(O_i - E_i)^2}{E_i}$
200	230	−30	900	900/230 ≈ 3.913043478
220	200	20	400	400/200 = 2
580	570	10	100	100/570 ≈ 0.1754385965
				X^2_{data} = 6.088482075

Thus, $\chi^2_{\text{data}} = \sum \frac{(O_i - E_i)^2}{E_i} = 6.088482075 \approx 6.088.$

Compare χ^2_{data} with χ^2_{crit}. $\chi^2_{\text{data}} \approx 6.088$ is $\geq \chi^2_{\text{crit}} = 5.991$. Therefore, we reject H_0. There is evidence that the variable *seller* does not follow the distribution specified in H_0. In other words, there is evidence that the market share for sales of children's books has changed.

29. H_0: $p_{\text{Windows 7}} = 0.51,\ p_{\text{Windows XP}} = 0.25,\ p_{\text{Windows 8}} = 0.13,\ p_{\text{Others}} = 0.11$

H_a: The random variable does not follow the distribution specified in H_0.

Checking the conditions, the expected frequencies are

$E_{\text{Windows 7}} = n \cdot p_{\text{Windows 7}} = 10{,}000(0.51) = 5{,}100,$

$E_{\text{Windows XP}} = n \cdot p_{\text{Windows XP}} = 10{,}000(0.25) = 2{,}500,$

$E_{\text{Windows 8}} = n \cdot p_{\text{Windows 8}} = 10{,}000(0.13) = 1{,}300,$

$E_{\text{Others}} = n \cdot p_{\text{Others}} = 10{,}000(0.11) = 1{,}100.$

Because none of these expected frequencies is less than 1, and none of the expected frequencies is less than 5, the conditions for performing the goodness of fit test are met.

df $= k - 1 = 4 - 1 = 3$. $\chi^2_{\text{crit}} = \chi^2_{0.05} = 7.815$. Reject H_0 if $\chi^2_{\text{data}} \geq 7.815$.

O_i	E_i	$O_i - E_i$	$(O_i - E_i)^2$	$\frac{(O_i - E_i)^2}{E_i}$
4,500	5,100	–600	360,000	360,000/5,100 ≈ 70.588235294118
1,500	2,500	–1,000	1,000,000	1,000,000/2,500 = 400
2,000	1,300	700	490,000	490,000/1,300 ≈ 376.92307692308
2,000	1,100	900	810,000	810,000/1,100 ≈736.36363636364
				X^2_{data} = 1,583.874949

Thus, $\chi^2_{\text{data}} = \sum \frac{(O_i - E_i)^2}{E_i} = 1583.874949 \approx 1583.875.$

Compare χ^2_{data} with χ^2_{crit}. $\chi^2_{\text{data}} \approx 1583.875$ is $\geq \chi^2_{\text{crit}} = 7.815$. Therefore, we reject H_0. There is evidence that the variable *operating system* does not follow the distribution specified in H_0. In other words, there is evidence that the desktop operating systems market share has changed.

31. $H_0 : p_{\textit{fast food}} = 0.30, p_{\textit{food courts}} = 0.46, p_{\textit{restaurants}} = 0.24$

H_a : The random variable does not follow the distribution specified in H_0.

$E_{\textit{fast food}} = n \cdot p_{\textit{fast food}} = (100)\ (0.30) = 30,$

$E_{\textit{food courts}} = n \cdot p_{\textit{food courts}} = (100)\ (0.46) = 46,$

$E_{\textit{restaurants}} = n \cdot p_{\textit{restaurants}} = (100)\ (0.24) = 24.$

Since none of the expected frequencies is less than 1 and none of the expected frequencies is less than 5, the conditions for performing the χ^2 goodness of fit test are met.

df $= k - 1 = 3 - 1 = 2$, $\alpha = 0.10$, $\chi^2_{\text{crit}} = \chi^2_{0.10} = 4.605$. Reject H_0 if $\chi^2_{\text{data}} \geq 4.605$.

$O_{\textit{restaurants}} = n - O_{\textit{fast food}} - O_{\textit{food courts}} = 100 - 32 - 49 = 19.$

O_i	E_i	$O_i - E_i$	$(O_i - E_i)^2$	$\frac{(O_i - E_i)^2}{E_i}$
32	30	2	4	4/30 = 0.1333333333
49	46	3	9	9/46 = 0.1956521739
19	24	–5	25	25/24 = 1.041666667
				χ^2_{data} = 1.370652174

Thus, $\chi^2_{\text{data}} = \sum \frac{(O_i - E_i)^2}{E_i} = 1.370652174 \approx 1.371.$

Since χ^2_{data} is not ≥ 4.605, we do not reject H_0. There is insufficient evidence that the random variable does not follow the distribution specified in H_0.

33. $H_0 : p_{\text{pizza}} = 0.25, p_{\text{cheeseburger}} = 0.25, p_{\text{quiche}} = 0.25, p_{\text{sushi}} = 0.25$

H_a : The random variable does not follow the distribution specified in H_0.

$E_{\text{pizza}} = n \cdot p_{\text{pizza}} = (500)\ (0.25) = 125,$

$E_{\text{cheeseburger}} = n \cdot p_{\text{cheeseburger}} = (500)\ (0.25) = 125,$

$E_{\text{quiche}} = n \cdot p_{\text{quiche}} = (500)\ (0.25) = 125,$

$E_{\text{sushi}} = n \cdot p_{\text{sushi}} = (500)\ (0.25) = 125.$

Since none of the expected frequencies is less than 1 and none of the expected frequencies is less than 5, the conditions for performing the χ^2 goodness of fit test are met.

df = $k - 1 = 4 - 1 = 3$, $\alpha = 0.01$, $\chi^2_{\text{crit}} = \chi^2_{0.01} = 11.345$. Reject H_0 if $\chi^2_{\text{data}} \geq 11.345$.

O_i	E_i	$O_i - E_i$	$(O_i - E_i)^2$	$\frac{(O_i - E_i)^2}{E_i}$
250	125	125	15,625	15,625/125 = 125
215	125	90	8,100	8,100/125 = 64.8
30	125	–95	9,025	9,025/125 = 72.2
5	125	–120	14,400	14,400/125 = 115.2
				$\chi^2_{\text{data}} = 377.2$

Thus, $\chi^2_{\text{data}} = \sum \frac{(O_i - E_i)^2}{E_i} = 377.2.$

Since χ^2_{data} is ≥ 11.345, we reject H_0. There is evidence that the random variable does not follow the distribution specified in H_0.

35. **(a)** The hypotheses are not dependent on the observed frequencies so changes in the observed frequencies would not affect the hypotheses.

(b) The expectant frequencies are not dependent on the observed frequencies so changes in the observed frequencies would not affect the expected frequencies.

(c) Let c be the amount by which the observed frequency for Comcast is increased and the observed frequency for Others is decreased.

O_i	E_i	$O_i - E_i$	$(O_i - E_i)^2$	$\frac{(O_i - E_i)^2}{E_i}$
$900 + c$	700	$200 + c$	$(200 + c)^2 = 40{,}000+400c+c^2$	$\frac{40{,}000 + 400c + c^2}{700} \approx 57.142857142857 + \frac{4c}{7} + \frac{c^2}{700}$
800	700	100	10,000	$\frac{10{,}000}{700} \approx 14{,}285714285714$
600	500	100	10,000	$\frac{10{,}000}{500} = 20$
700	500	200	40,000	$\frac{40{,}000}{500} = 80$
$7600 - c$	7600	$-600 - c$	$(-600 - c)^2 = 360{,}000 + 1{,}200c + c^2$	$\frac{360{,}000 + 1{,}200c + c^2}{7600} \approx 47.368421052632 + \frac{3c}{19} + \frac{c^2}{7600}$
				$\chi^2_{data} \approx 218.7969925 + \frac{4c}{7} + \frac{c^2}{700} + \frac{3c}{19} + \frac{c^2}{7600}$

Thus, $\chi^2_{\text{data}} = \sum \frac{(O_i - E_i)^2}{E_i} \approx 218.7969925 + \frac{4c}{7} + \frac{c^2}{700} + \frac{3c}{19} + \frac{c^2}{7600}$. Since c is positive, $\frac{4c}{7} + \frac{c^2}{700} + \frac{3c}{19} + \frac{c^2}{7600}$ is positive. Therefore χ^2_{data} will increase.

(d) The p-value $= P(\chi^2 > \chi^2_{data})$. From part (c), χ^2_{data} will increase. Therefore, the p-value will decrease.

(e) The p-value is already $\leq \alpha = 0.01$. From (d), the p-value will decrease. After a decrease in the p-value it will still be $\leq \alpha = 0.01$. Therefore the conclusion will still be reject H_0.

Section 11.2

1. A contingency table is a tabular summary of the relationship between two categorical variables.

3. The two-sample Z test for the difference in proportions from Chapter 10 is for comparing proportions of two independent populations and the χ^2 test for homogeneity of proportions is for comparing proportions of k independent populations.

5. Table of observed frequencies:

	A1	**A2**	**Total**
B1	10	20	30
B2	12	18	30
Total	22	38	60

Table of expected frequencies:

	A1	A2	Total
B1	$\frac{(22)(30)}{60} = 11$	$\frac{(38)(30)}{60} = 19$	30
B2	$\frac{(22)(30)}{60} = 11$	$\frac{(38)(30)}{60} = 19$	30
Total	22	38	60

7. Table of observed frequencies:

	E1	E2	E3	Total
F1	30	20	10	60
F2	35	24	8	67
Total	65	44	18	127

Table of expected frequencies:

	E1	E2	E3	Total
F1	$\frac{(60)(65)}{127} \approx 30.71$	$\frac{(60)(44)}{127} \approx 20.79$	$\frac{(60)(18)}{127} \approx 8.50$	60
F2	$\frac{(67)(65)}{127} \approx 34.29$	$\frac{(67)(44)}{127} \approx 23.21$	$\frac{(67)(18)}{127} \approx 9.50$	67

9. Table of observed frequencies:

	I1	I2	I3	Total
J1	100	90	105	295
J2	50	60	55	165
J3	25	15	20	60
Total	175	165	180	520

Table of expected frequencies:

	I1	I2	I3	Total
J1	$\frac{(295)(175)}{520} \approx 99.28$	$\frac{(295)(165)}{520} \approx 93.61$	$\frac{(295)(180)}{520} = 102.12$	295.01
J2	$\frac{(165)(175)}{520} \approx 55.53$	$\frac{(165)(165)}{520} \approx 52.36$	$\frac{(165)(180)}{520} = 57.12$	165.01
J3	$\frac{(60)(175)}{520} \approx 20.19$	$\frac{(60)(165)}{520} \approx 19.04$	$\frac{(60)(180)}{520} = 20.77$	60
Total	175	165.01	180.01	520.02

11. **(a)** H_0 : Variable A and variable B are independent.

H_a : Variable A and variable B are dependent.

(b)

	A1	A2	Total
B1	$\frac{(22)(30)}{60} = 11$	$\frac{(38)(30)}{60} = 19$	30
B2	$\frac{(22)(30)}{60} = 11$	$\frac{(38)(30)}{60} = 19$	30
Total	22	38	60

Since none of the expected frequencies is less than 1 and none of the expected frequencies is less than 5, the conditions for performing the χ^2 test for independence are met.

(c) df = $(r - 1)(c - 1) = (2 - 1)(2 - 1) = 1$, $\alpha = 0.05$, $\chi^2_{\text{crit}} = \chi^2_{0.05} = 3.841$. Reject H_0 if $\chi^2_{\text{data}} \geq 3.841$.

(d) $\chi^2_{\text{data}} = \sum \frac{(O_i - E_i)^2}{E_i} = \frac{(10-11)^2}{11} + \frac{(20-19)^2}{19} + \frac{(12-11)^2}{11} + \frac{(18-19)^2}{19} \approx 0.287$

(e) Since χ^2_{data} is not ≥ 3.841, we do not reject H_0. There is insufficient evidence that variable A and variable B are dependent.

13. **(a)** H_0 : Variable I and variable J are independent.

H_a : Variable I and variable J are dependent.

(b)

	I1	**I2**	**I3**	**Total**
J1	$\frac{(295)(175)}{520} \approx 99.28$	$\frac{(295)(165)}{520} \approx 93.61$	$\frac{(295)(180)}{520} \approx 102.12$	295.01
J2	$\frac{(165)(175)}{520} \approx 55.53$	$\frac{(165)(165)}{520} \approx 52.36$	$\frac{(165)(180)}{520} \approx 57.12$	165.01
J3	$\frac{(60)(175)}{520} \approx 20.19$	$\frac{(60)(165)}{520} \approx 19.04$	$\frac{(60)(180)}{520} \approx 20.77$	60
Total	175	165.01	180.01	520.02

Since none of the expected frequencies is less than 1 and none of the expected frequencies is less than 5, the conditions for performing the χ^2 test for independence are met.

(c) df = $(r - 1)(c - 1) = (3 - 1)(3 - 1) = 4$, $\alpha = 0.01$, $\chi^2_{\text{crit}} = \chi^2_{0.01} = 13.277$. Reject H_0 if $\chi^2_{\text{data}} \geq 13.277$.

(d)

$$\chi^2_{\text{data}} = \sum \frac{(O_i - E_i)^2}{E_i} = \frac{(100-99.28)^2}{99.28} + \frac{(90-93.61)^2}{93.61} + \frac{(105-102.12)^2}{102.12} + \frac{(50-55.53)^2}{55.53}$$
$$+ \frac{(60-52.36)^2}{52.36} + \frac{(55-57.12)^2}{57.12} + \frac{(25-20.19)^2}{20.19} + \frac{(15-19.04)^2}{19.04} + \frac{(20-20.77)^2}{20.77} \approx 4.002$$

[TI-83/84: $\chi^2_{\text{data}} \approx 4.000$.]

(e) Since χ^2_{data} is not ≥ 13.277, we do not reject H_0. There is insufficient evidence that variable I and variable J are dependent.

15. **(a)** H_0 : Variable C and variable D are independent.

H_a : Variable C and variable D are dependent. Reject H_0 if the p-value ≤ 0.05.

	C1	**C2**	**Total**
D1	$\frac{(110)(150)}{300} = 55$	$\frac{(190)(150)}{300} = 95$	150
D2	$\frac{(110)(150)}{300} = 55$	$\frac{(190)(150)}{300} = 95$	150
Total	110	190	300

Since none of the expected frequencies is less than 1 and none of the expected frequencies is less than 5, the conditions for performing the χ^2 test for independence are met.

(b) $\chi^2_{\text{data}} = \sum \frac{(O_i - E_i)^2}{E_i} = \frac{(50-55)^2}{55} + \frac{(100-95)^2}{95} + \frac{(60-55)^2}{55} + \frac{(90-95)^2}{95} \approx 1.435$

(c) df = $(r - 1)(c - 1) = (2 - 1)(2 - 1) = 1$. p-value = 0.2309.

(d) Since the p-value is not ≤ 0.05, we do not reject H_0. There is insufficient evidence that variable C and variable D are dependent.

17. **(a)** H_0 : Variable K and variable L are independent.

H_a : Variable K and variable L are dependent.

Reject H_0 if the p-value ≤ 0.01.

	K1	K2	K3	K4	Total
L1	$\frac{(300)(90)}{720} = 37.5$	$\frac{(300)(175)}{720} \approx 72.92$	$\frac{(300)(215)}{720} \approx 89.58$	$\frac{(300)(240)}{720} = 100$	300
L2	$\frac{(190)(90)}{720} = 23.75$	$\frac{(190)(175)}{720} \approx 46.18$	$\frac{(190)(215)}{720} \approx 56.74$	$\frac{(190)(240)}{720} \approx 63.33$	190
L3	$\frac{(230)(90)}{720} = 28.75$	$\frac{(230)(175)}{720} \approx 55.90$	$\frac{(230)(215)}{720} \approx 68.68$	$\frac{(230)(240)}{720} \approx 76.67$	230
Total	90	175	215	240	720

Since none of the expected frequencies is less than 1 and none of the expected frequencies is less than 5, the conditions for performing the χ^2 test for independence are met.

(b) $\chi^2_{\text{data}} = \sum \frac{(O_i - E_i)^2}{E_i} = \frac{(40-37.5)^2}{37.5} + \frac{(70-72.92)^2}{72.92} + \frac{(90-89.58)^2}{89.58} + \frac{(100-100)^2}{100} + \frac{(20-23.75)^2}{23.75} + \frac{(40-46.18)^2}{46.18} + \frac{(60-56.74)^2}{56.74} + \frac{(70-63.33)^2}{63.33} + \frac{(30-28.75)^2}{28.75.} + \frac{(65-55.90)^2}{55.90} + \frac{(65-68.68)^2}{68.68} + \frac{(70-76.67)^2}{76.67} \approx$ 4.908. **[TI-83/84:** $\chi^2_{\text{data}} \approx 4.906$.]

(c) df = $(r-1)(c-1) = (3-1)(4-1) = 6$. p-value ≈ 0.5557.

[TI-83/84: p-value $= 0.5560$.]

(d) Since the p-value is not ≤ 0.01, we do not reject H_0. There is insufficient evidence that variable K and variable L are dependent.

19. **(a)** $H_0 : p_1 = p_2 = p_3$

H_a : Not all the proportions in H_0 are equal.

(b) Observed frequencies:

	Sample 1	Sample 2	Sample 3	Total
Successes	10	20	30	60
Failures	20	45	62	127
Total	30	65	92	187

Expected frequencies:

	Sample 1	Sample 2	Sample 3	Total
Successes	$\frac{(60)(30)}{187} \approx 9.63$	$\frac{(60)(65)}{187} \approx 20.86$	$\frac{(60)(92)}{187} \approx 29.52$	60.01
Failures	$\frac{(127)(30)}{187} \approx 20.37$	$\frac{(127)(65)}{187} \approx 44.14$	$\frac{(127)(92)}{187} \approx 62.48$	126.99
Total	30	65	92	187

Since none of the expected frequencies is less than 1 and none of the expected frequencies is less than 5, the conditions for performing the χ^2 test for homogeneity of proportions are met.

(c) df = $(r-1)(c-1) = (2-1)(3-1) = 2$, $\alpha = 0.05$, $\chi^2_\alpha = \chi^2_{0.05} = 5.991$. Reject H_0 if $\chi^2_{\text{data}} \geq$ 5.991.

(d) $\chi^2_{\text{data}} = \sum \frac{(O_i - E_i)^2}{E_i} = \frac{(10-9.63)^2}{9.63} + \frac{(20-20.86)^2}{20.86} + \frac{(30-29.52)^2}{29.52} + \frac{(20-20.37)^2}{20.37} + \frac{(45-44.14)^2}{44.14}$ $+ \frac{(62-62.48)^2}{62.48} \approx 0.085.$

(e) Since χ^2_{data} is not ≥ 5.991, we do not reject H_0. There is insufficient evidence that not all of the proportions in H_0 are equal.

21. **(a)** $H_0 : p_1 = p_2 = p_3 = p_4$

H_a : Not all the proportions in H_0 are equal.

(b) Observed frequencies:

	Sample 1	Sample 2	Sample 3	Sample 4	Total
Successes	10	15	20	25	70
Failures	15	24	32	40	111
Total	25	39	52	65	181

Expected frequencies:

	Sample 1	Sample 2	Sample 3	Sample 4	Total
Successes	$\frac{(70)(25)}{181} \approx 9.67$	$\frac{(70)(39)}{181} \approx 15.08$	$\frac{(70)(52)}{181} \approx 20.11$	$\frac{(70)(65)}{181} \approx 25.14$	70
Failures	$\frac{(111)(25)}{181} \approx 15.33$	$\frac{(111)(39)}{181} \approx 23.92$	$\frac{(111)(52)}{181} \approx 31.89$	$\frac{(111)(65)}{181} \approx 39.86$	111
Total	25	39	52	65	181

Since none of the expected frequencies is less than 1 and none of the expected frequencies is less than 5, the conditions for performing the χ^2 test for homogeneity of proportions are met.

(c) df = $(r-1)(c-1) = (2-1)(4-1) = 3$, $\alpha = 0.05$, $\chi^2_{\text{crit}} = \chi^2_{0.05} = 7.815$.

Reject H_0 if $\chi^2_{\text{data}} \geq 7.815$.

(d) $\chi^2_{\text{data}} = \sum \frac{(O_i - E_i)^2}{E_i} = \frac{(10-9.67)^2}{9.67} + \frac{(15-15.08)^2}{15.08} + \frac{(20-20.11)^2}{20.11} + \frac{(25-25.14)^2}{25.14} + \frac{(15-15.33)^2}{15.33}$ $+ \frac{(24-23.92)^2}{23.92} + \frac{(32-31.89)^2}{31.89} + \frac{(40-39.86)^2}{39.86} \approx 0.0213.$ **[TI-83/84:** $\chi^2_{\text{data}} \approx 0.0215$.]

(e) Since χ^2_{data} is not ≥ 7.815, we do not reject H_0. There is insufficient evidence that not all of the proportions in H_0 are equal.

23. **(a)** $H_0 : p_1 = p_2 = p_3$

H_a : Not all the proportions in H_0 are equal.

Reject H_0 if the p-value ≤ 0.05.

Observed frequencies:

	Sample 1	**Sample 2**	**Sample 3**	**Total**
Successes	30	60	90	180
Failures	10	25	50	85
Total	40	85	140	265

Expected frequencies:

	Sample 1	**Sample 2**	**Sample 3**	**Total**
Successes	$\frac{(180)(40)}{265} \approx 27.17$	$\frac{(180)(85)}{265} \approx 57.74$	$\frac{(180)(140)}{265} \approx 95.09$	180
Failures	$\frac{(85)(40)}{265} \approx 12.83$	$\frac{(85)(85)}{265} \approx 27.26$	$\frac{(85)(140)}{265} \approx 44.91$	85
Total	40	85	140	265

Since none of the expected frequencies is less than 1 and none of the expected frequencies is less than 5, the conditions for performing the χ^2 test for homogeneity of proportions are met.

(b) $\chi^2_{\text{data}} = \sum \frac{(O_i - E_i)^2}{E_i} = \frac{(30-27.17)^2}{27.17} + \frac{(60-57.74)^2}{57.74} + \frac{(90-95.09)^2}{95.09} + \frac{(10-12.83)^2}{12.83} + \frac{(25-27.26)^2}{27.26}$ $+ \frac{(50-44.91)^2}{44.91} \approx 2.044$. **[TI-83/84:** $\chi^2_{\text{data}} \approx 2.047$].

(c) df = $(r - 1)(c - 1) = (2 - 1)(3 - 1) = 2$. p-value ≈ 0.3599.

[TI-83/84: p-value $\approx$ **0.3594]**

(d) Since the p-value is not ≤ 0.05, we do not reject H_0. There is insufficient evidence that not all of the proportions in H_0 are equal.

25. **(a)** $H_0 : p_1 = p_2 = p_3 = p_4$

H_a : Not all the proportions in H_0 are equal.

Reject H_0 if the p-value ≤ 0.05.

Observed frequencies:

	Sample 1	**Sample 2**	**Sample 3**	**Sample 4**	**Total**
Successes	10	12	24	32	78
Failures	6	10	15	30	61
Total	16	22	39	62	139

Expected frequencies:

	Sample 1	Sample 2	Sample 3	Sample 4	Total
Successes	$\frac{(78)(16)}{139} \approx 8.98$	$\frac{(78)(22)}{139} \approx 12.35$	$\frac{(78)(39)}{139} \approx 21.88$	$\frac{(78)(62)}{139} \approx 34.79$	78
Failures	$\frac{(61)(16)}{139} \approx 7.02$	$\frac{(61)(22)}{139} \approx 9.65$	$\frac{(61)(39)}{139} \approx 17.12$	$\frac{(61)(62)}{139} \approx 27.21$	61
Total	16	22	39	62	139

Since none of the expected frequencies is less than 1 and none of the expected frequencies is less than 5, the conditions for performing the χ^2 test for homogeneity of proportions are met.

(b)

$$\chi^2_{\text{data}} = \sum\frac{(O_i - E_i)^2}{E_i} = \frac{(10-8.98)^2}{8.98} + \frac{(12-12.35)^2}{12.35} + \frac{(24-21.88)^2}{21.88} + \frac{(32-34.79)^2}{34.79} + \frac{(6-7.02)^2}{7.02} + \frac{(10-9.65)^2}{9.65} + \frac{(15-17.12)^2}{17.12} + \frac{(30-27.21)^2}{27.21} \approx 1.264$$

[TI-83/84: $\chi^2_{\text{data}} \approx 1.263$].

(c) df = $(r - 1)(c - 1) = (2 - 1)(4 - 1) = 3$. p-value ≈ 0.7377.

[TI-83/84: p-value ≈ 0.7379].

(d) Since the p-value is not ≤ 0.05, we do not reject H_0. There is insufficient evidence that not all of the proportions in H_0 are equal.

27. $H_0 : p_{Edit} = p_{Arrange}$

H_a : Not all the proportions in H_0 are equal.

Reject H_0 if the p-value ≤ 0.05.

Observed frequencies:

	By email	By phone or in person	Total
Edit or review documents	670	330	1000
Arrange meetings or appointments	630	370	1000
Total	1300	700	2000

Expected frequencies:

	By e-mail	By phone or in person	Total
Edit or review documents	$\frac{(1000)(1300)}{2000} = 650$	$\frac{(1000)(700)}{2000} = 350$	1000
Arrange meetings or appointments	$\frac{(1000)(1300)}{2000} = 650$	$\frac{(1000)(700)}{2000} = 350$	1000
Total	1300	700	2000

Since none of the expected frequencies is less than 1 and none of the expected frequencies is less than 5, the conditions for performing the χ^2 test for homogeneity of proportions are met.

$$\chi^2_{\text{data}} = \sum\frac{(O_i - E_i)^2}{E_i} = \frac{(670-650)^2}{650} + \frac{(330-350)^2}{350} + \frac{(630-650)^2}{650} + \frac{(370-350)^2}{350} \approx 3.516$$

df = $(r - 1)(c - 1) = (2 - 1)(2 - 1) = 1$. p-value = $P(\chi^2 > 3.516) \approx 0.0608$. Since the p-value is not ≤ 0.05, we do not reject H_0. There is insufficient evidence that not all of the proportions in H_0 are equal.

29. H_0: *Cause of death* and *age group* are independent.

H_a: *Cause of death* and *age group* are dependent.

Chi-Square Test for Association: Age group, Worksheet columns

```
Rows: Age group   Columns: Worksheet columns

               Heat-related  Cold-related  Floods/Storms/Lightening   All

15 to 24                106           286                        97   489

                      142.7         310.4                      35.9

75 to 84                490          1010                        53  1553

                      453.3         985.6                     114.1

All                     596          1296                       150  2042

Cell Contents:          Count

                        Expected count

Pearson Chi-Square = 151.500, DF = 2, P-Value = 0.000

Likelihood Ratio Chi-Square = 127.302, DF = 2, P-Value = 0.000
```

From the Minitab output above, none of these expected frequencies is less than 1, and none of the expected frequencies is less than 5. Therefore, the conditions for performing the χ^2 test for independence are met.

Reject H_0 if the p-value $\leq \alpha = 0.05$.

$\chi^2_{\text{data}} = 151.500$

p-value ≈ 0.

The p-value ≈ 0 is less than or equal to $\alpha = 0.05$. Therefore, we reject H_0. Evidence exists, at level of significance $\alpha = 0.05$, that the variables *Cause of death* and *age group* are dependent.

31. $H_0: p_{Work} = p_{Personal}$

H_a: Not all the proportions in H_0 are equal.

Reject H_0 if the p-value ≤ 0.01.

Observed frequencies:

	None	**Some**	**A lot**	**Total**
Work email	53	36	11	100
Personal email	22	48	30	100
Total	75	84	41	200

Expected frequencies:

	None	**Some**	**A lot**	**Total**
Work email	$\frac{(100)(75)}{200} = 37.5$	$\frac{(100)(84)}{200} = 42$	$\frac{(100)(41)}{200} = 20.5$	100
Personal email	$\frac{(100)(75)}{200} = 37.5$	$\frac{(100)(84)}{200} = 42$	$\frac{(100)(41)}{200} = 20.5$	100
Total	75	84	41	200

Since none of the expected frequencies is less than 1 and none of the expected frequencies is less than 5, the conditions for performing the χ^2 test for homogeneity of proportions are met.

$\chi^2_{\text{data}} = \sum \frac{(O_i-E_i)^2}{E_i} = \frac{(53-37.5)^2}{37.5} + \frac{(36-42)^2}{42} + \frac{(11-20.5)^2}{20.5} + \frac{(22-37.5)^2}{37.5} + \frac{(48-42)^2}{42} + \frac{(30-20.5)^2}{20.5} \approx$ 23.332.

df = $(r-1)(c-1) = (2-1)(3-1) = 2$. p-value = $P(\chi^2 > 23.332) \approx 0$. Since the p-value is ≤ 0.01, we reject H_0. There is evidence that the proportions who report "a lot of spam" are not the same for work email and personal email.

33. $H_0 : p_{Urban} = p_{Suburban} = p_{Rural}$

H_a : Not all the proportions in H_0 are equal.

Reject H_0 if the p-value ≤ 0.05.

Observed frequencies:

	Use online dating	**Don't use online dating**	**Total**
Urban	(0.13) (1000) = 130	1000 – 130 = 870	1000
Suburban	(0.10) (1000) = 100	1000 – 100 = 900	1000
Rural	(0.09) (1000) = 90	1000 – 90 = 910	1000
Total	320	2680	3000

Expected Frequencies:

	Use online dating	**Don't use online dating**	**Total**
Urban	$\frac{(1000)(320)}{3000} \approx 106.67$	$\frac{(1000)(2680)}{3000} \approx 893.33$	1000
Suburban	$\frac{(1000)(320)}{3000} \approx 106.67$	$\frac{(1000)(2680)}{3000} \approx 893.33$	1000
Rural	$\frac{(1000)(320)}{3000} \approx 106.67$	$\frac{(1000)(2680)}{3000} \approx 893.33$	1000
Total	320.01	2679.99	3000

Since none of the expected frequencies is less than 1 and none of the expected frequencies is less than 5, the conditions for performing the χ^2 test for homogeneity of proportions are met.

$$\chi^2_{\text{data}} = \sum \frac{(O_i - E_i)^2}{E_i} = \frac{(130 - 106.67)^2}{106.67} + \frac{(870 - 893.33)^2}{893.33} + \frac{(100 - 106.67)^2}{106.67} + \frac{(900 - 893.33)^2}{893.33} + \frac{(90 - 106.67)^2}{106.67} + \frac{(910 - 893.33)^2}{893.33} \approx 9.095$$

df = $(r-1)(c-1) = (2-1)(3-1) = 2$. p-value = $P(\chi^2_{\text{data}} > 9.095) \approx 0.0106$. Since the p-value is ≤ 0.05, we reject H_0. There is evidence that not all of the proportions of people who use online dating in urban, suburban, and rural areas are equal.

35. **(a)** Dependent.

(b) H_0 : *Gender* and *goals* are independent.

H_a : *Gender* and *goals* are dependent.

Reject H_0 if p-value ≤ 0.05.

Tabulated statistics: GENDER, GOALS

```
Rows: GENDER   Columns: GOALS

        Grades  Popular  Sports  All

boy        117       50      60  227
girl       130       91      30  251
All        247      141      90  478

Cell Contents:       Count

Pearson Chi-Square = 21.455, DF = 2, P-Value = 0.000
Likelihood Ratio Chi-Square = 21.769, DF = 2, P-Value = 0.000
```

Since p-value ≈ 0, p-value is ≤ 0.05. Thus, we reject H_0. There is evidence that *gender* and *goals* are dependent.

37. **(a)** Dependent.

(b) H_0 : *Goals* and *urb_rur* are independent.

H_a : *Goals* and *urb_rur* are dependent.

Reject H_0 if p-value ≤ 0.10.

Tabulated statistics: URB_RUR, GOALS

```
Rows: URB_RUR   Columns: GOALS

             Grades  Popular  Sports  All

Rural            57       50      42  149
Suburban         87       42      22  151
Urban           103       49      26  178
All             247      141      90  478

Cell Contents:       Count


Pearson Chi-Square = 18.828, DF = 4, P-Value = 0.001
Likelihood Ratio Chi-Square = 18.571, DF = 4, P-Value = 0.001
```

Since p-value 0.001, p-value is ≤ 0.10. Thus, we reject H_0. There is evidence that *urb_rural* and *goals* are dependent.

39. $H_0 : p_{Jan} = p_{Feb} = p_{Mar} = p_{Apr} = p_{May} = p_{June} = p_{July} = p_{Aug} = p_{Sept} = p_{Oct} = p_{Nov} = p_{Dec}$

H_a : Not all the proportions in H_0 are equal.

Reject H_0 if the p-value ≤ 0.01.

Expected frequencies:

Month	**Dates not drafted**	**Dates drafted**	**All**
Jan.	$\frac{(31)(171)}{366} \approx 14.48$	$\frac{(31)(195)}{366} \approx 16.52$	31
Feb.	$\frac{(29)(171)}{366} \approx 13.55$	$\frac{(29)(195)}{366} \approx 15.45$	29
Mar.	$\frac{(31)(171)}{366} \approx 14.48$	$\frac{(31)(195)}{366} \approx 16.52$	31
Apr.	$\frac{(30)(171)}{366} \approx 14.02$	$\frac{(30)(195)}{366} \approx 15.98$	30
May	$\frac{(31)(171)}{366} \approx 14.48$	$\frac{(31)(195)}{366} \approx 16.52$	31
June	$\frac{(30)(171)}{366} \approx 14.02$	$\frac{(30)(195)}{366} \approx 15.98$	30
July	$\frac{(31)(171)}{366} \approx 14.48$	$\frac{(31)(195)}{366} \approx 16.52$	31
Aug.	$\frac{(31)(171)}{366} \approx 14.48$	$\frac{(31)(195)}{366} \approx 16.52$	31
Sept.	$\frac{(30)(171)}{366} \approx 14.02$	$\frac{(30)(195)}{366} \approx 15.98$	30
Oct.	$\frac{(31)(171)}{366} \approx 14.48$	$\frac{(31)(195)}{366} \approx 16.52$	31
Nov.	$\frac{(30)(171)}{366} \approx 14.02$	$\frac{(30)(195)}{366} \approx 15.98$	30
Dec.	$\frac{(31)(171)}{366} \approx 14.48$	$\frac{(31)(195)}{366} \approx 16.52$	31
All	170.99	195.01	366

$$\chi^2_{\text{data}} = \sum\frac{(O_i - E_i)^2}{E_i} = \frac{(17-14.48)^2}{14.48} + \frac{(14-16.52)^2}{16.52} + \frac{(16-13.55)^2}{13.55}$$
$$+ \frac{(13-15.45)^2}{15.45} + \frac{(21-14.48)^2}{14.48} + \frac{(10-16.52)^2}{16.52}$$
$$+ \frac{(18-14.02)^2}{14.02} + \frac{(12-15.98)^2}{15.98} + \frac{(17-14.48)^2}{14.48}$$
$$+ \frac{(14-16.52)^2}{16.52} + \frac{(16-14.02)^2}{14.02} + \frac{(14-15.98)^2}{15.98}$$
$$+ \frac{(13-14.48)^2}{14.48} + \frac{(18-16.52)^2}{16.52} + \frac{(12-14.48)^2}{14.48}$$
$$+ \frac{(19-16.52)^2}{16.52} + \frac{(11-14.02)^2}{14.02} + \frac{(19-15.98)^2}{15.98}$$
$$+ \frac{(17-14.48)^2}{14.48} + \frac{(14-16.52)^2}{16.52} + \frac{(8-14.02)^2}{14.02}$$
$$+ \frac{(22-15.98)^2}{15.98} + \frac{(5-14.48)^2}{14.48} + \frac{(26-16.52)^2}{16.52} \approx 30.257$$

[TI-83/84: $\chi^2_{\text{data}} \approx 30.254$.]

df = $(r - 1)(c - 1) = (12 - 1)(2 - 1) = 11$. p-value = $P(\chi^2 > 30.254) \approx 0.0014$.

Since the p-value is ≤ 0.01, we reject H_0. There is evidence that not all of the proportions of dates drafted for all 12 months are equal.

Chapter 11 Review Exercises

1. $H_0 : p_{USCan} = 0.32, p_{USMex} = 0.22, p_{CanUS} = 0.31, p_{MexUS} = 0.15$

H_a : The random variable does not follow the distribution specified in H_0.

$n = O_{USCan} + O_{USMex} + O_{CanUS} + O_{MexUS}$ = \$25 billion + \$15 billion + \$20 billion + \$10 billion = \$70 billion

$E_{USCan} = n \cdot p_{USCan}$ = (\$70 billion)(0.32) = \$22.4 billion

$E_{USMex} = n \cdot p_{USMex}$ = (\$70 billion) (0.22) = \$15.4 billion

$E_{CanUS} = n \cdot p_{CanUS}$ = (\$70 billion) (0.31) = \$21.7 billion

$E_{MexUS} = n \cdot p_{MexUS}$ = (\$70 billion) (0.15) = \$10.5 billion

Since none of the expected frequencies is less than 1 and none of the expected frequencies is less than 5, the conditions for performing the χ^2 goodness of fit test are met. Reject H_0 if the p-value ≤ 0.05.

O_i	E_i	$O_i - E_i$	$(O_i - E_i)^2$	$\frac{(O_i-E_i)^2}{E_i}$
25	22.4	2.6	6.76	6.76/22.4 = 0.3017857143
15	15.4	-0.4	0.16	0.16/15.4 = 0.0103896104
20	21.7	-1.7	2.89	2.89/21.7 = 0.1331797235
10	10.5	-0.5	0.25	0.25/10.5 = 0.0238095238
				$\chi^2_{\text{data}} = 0.469164572$

Thus, $\chi^2_{\text{data}} = \sum \frac{(O_i-E_i)^2}{E_i} = 0.469164572 \approx 0.469$.

df = $k - 1 = 4 - 1 = 3$, p-value = $P(\chi^2 > \chi^2_{\text{data}}) = P(\chi^2 > 0.469) = 0.9256472412$.

Since the p-value is not ≤ 0.05, we do not reject H_0. There is insufficient evidence that the random variable does not follow the distribution specified in H_0.

3. $H_0 : p_0 = 0.10, p_1 = 0.10, p_2 = 0.10, p_3 = 0.10, p_4 = 0.10, p_5 = 0.10, p_6 = 0.10, p_7 = 0.10, p_8 = 0.10, p_9 = 0.10$

H_a : Not all the proportions in H_0 are equal.

Reject H_0 if the p-value ≤ 0.05.

$E_0 = n \cdot p_0 = (218)(0.10) = 21.8$

$E_1 = n \cdot p_1 = (218)(0.10) = 21.8$

$E_2 = n \cdot p_2 = (218)(0.10) = 21.8$

$E_3 = n \cdot p_3 = (218)(0.10) = 21.8$

$E_4 = n \cdot p_4 = (218)\,(0.10) = 21.8$

$E_5 = n \cdot p_5 = (218)\,(0.10) = 21.8$

$E_6 = n \cdot p_6 = (218)\,(0.10) = 21.8$

$E_7 = n \cdot p_7 = (218)\,(0.10) = 21.8$

$E_8 = n \cdot p_8 = (218)\,(0.10) = 21.8$

$E_9 = n \cdot p_9 = (218)\,(0.10) = 21.8$

Since none of the expected frequencies is less than 1 and none of the expected frequencies is less than 5, the conditions for performing the χ^2 goodness of fit test are met.

O_i	E_i	$O_i - E_i$	$(O_i - E_i)^2$	$\frac{(O_i-E_i)^2}{E_i}$
26	21.8	4.2	17.64	17.64/21.8 = 0.8091743119
12	21.8	-9.8	96.04	96.04/21.8 = 4.405504587
26	21.8	4.2	17.64	17.64/21.8 = 0.8091743119
18	21.8	-3.8	14.44	14.44/21.8 = 0.6623853211
23	21.8	1.2	1.44	1.44/21.8 = 0.0660550459
19	21.8	-2.8	7.84	7.84/21.8 = 0.3596330275
18	21.8	-3.8	14.44	14.44/21.8 = 0.6623853211
27	21.8	5.2	27.04	27.04/21.8 = 1.240366972
30	21.8	8.2	67.24	67.24/21.8 = 3.08440367
19	21.8	-2.8	7.84	7.84/21.8 = 0.3596330275
				X^2_{data} = 12.4587156

Thus, $\chi^2_{\text{data}} = \sum \frac{(O_i-E_i)^2}{E_i} = 12.4587156 \approx 12.459.$

df = $k - 1 = 10 - 1 = 9$. p-value = $P\,(\chi^2 > 12.459) = 0.1886519376$.

Since the p-value is not ≤ 0.05, we do not reject H_0. There is insufficient evidence that the random variable does not follow the distribution specified in H_0.

5. **(a)** A higher proportion of the females with high GPAs takes the SAT exam than the proportion of the females with lower GPAs.

(b) $H_0 : p_{A+} = p_A = p_{A-} = p_B = p_C = p_{D/F}$

H_a : Not all the proportions in H_0 are equal.

Reject H_0 if the p-value ≤ 0.05.

Observed frequencies:

	High school grade point average						
Gender	**A+**	**A**	**A–**	**B**	**C**	**D–F**	**Total**
Female	60	62	59	53	43	43	320
Male	40	38	41	47	57	57	280
Total	100	100	100	100	100	100	600

Expected frequencies:

	High school grade point average						
Gender	**A+**	**A**	**A–**	**B**	**C**	**D–F**	**Total**
Female	$\frac{(320)(100)}{600} \approx 53.33$	$\frac{(320)(100)}{600} \approx 53.33$	$\frac{(320)(100)}{600} \approx 53.33$	$\frac{(320)(100)}{600} \approx 53.33$	$\frac{(320)(100)}{600} \approx 53.33$	$\frac{(320)(100)}{600} \approx 53.33$	319.98
Male	$\frac{(280)(100)}{600} \approx 46.67$	$\frac{(280)(100)}{600} \approx 46.67$	$\frac{(280)(100)}{600} \approx 46.67$	$\frac{(280)(100)}{600} \approx 46.67$	$\frac{(280)(100)}{600} \approx 46.67$	$\frac{(280)(100)}{600} \approx 46.67$	280.02
Total	100	100	100	100	100	100	600

Since none of the expected frequencies is less than 1 and none of the expected frequencies is less than 5, the conditions for performing the χ^2 test for homogeneity of proportions are met.

$$\chi^2_{\text{data}} = \sum \frac{(O_i - E_i)^2}{E_i} = \frac{(60-53.33)^2}{53.33} + \frac{(62-53.33)^2}{53.33} + \frac{(59-53.33)^2}{53.33} + \frac{(53-53.33)^2}{100} + \frac{(43-53.33)^2}{53.33} + \frac{(43-53.33)^2}{53.33} + \frac{(40-46.67)^2}{46.67} + \frac{(38-46.67)^2}{46.67} + \frac{(41-46.67)^2}{46.67.} + \frac{(47-46.67)^2}{46.67} + \frac{(57-46.67)^2}{46.67} + \frac{(57-46.67)^2}{46.67} \approx 14.678.$$

[TI-83/84: $\chi^2_{\text{data}} \approx 14.679$]

df = $(r-1)(c-1) = (2-1)(6-1) = 5$. p-value $= P(\chi^2 > 14.678) = 0.0118$. Since the p-value is ≤ 0.05, we reject H_0. There is evidence that not all of the proportions in H_0 are equal.

7. H_0 : Age and radio station type are independent.

H_a : Age and radio station type are dependent.

Reject H_0 if the p-value ≤ 0.05.

Observed frequencies:

	Pop contemporary hit radio stations	**Alternative radio stations**	**Total**
12–17 years old	240	170	410
18–24 years old	250	260	510
Total	490	430	920

Expected frequencies:

	Pop contemporary hit radio stations	**Alternative radio stations**	**Total**
12–17 years old	$\frac{(410)(490)}{920} \approx 218.37$	$\frac{(410)(430)}{920} \approx 191.63$	410
18–24 years old	$\frac{(510)(490)}{920} \approx 271.63$	$\frac{(510)(430)}{920} \approx 238.37$	510
Total	490	430	920

Since none of the expected frequencies is less than 1 and none of the expected frequencies is less than 5, the conditions for performing the χ^2 test for independence are met.

$$\chi^2_{\text{data}} = \sum \frac{(O_i - E_i)^2}{E_i} = \frac{(240-218.37)^2}{218.37} + \frac{(170-191.63)^2}{191.63} + \frac{(250-271.63)^2}{271.63} + \frac{(260-238.37)^2}{238.37} \approx 8.269.$$

df = $(r-1)(c-1) = (2-1)(2-1) = 1$. p-value $= P(\chi^2 > 8.269) = 0.004$. Since the p-value is

≤ 0.05, we reject H_0. There is evidence that age and radio station type are dependent.

Chapter 11 Quiz

1. True.

3. False. In the test for homogeneity of proportions, the alternative hypothesis states that the random variable does not follow the distribution specified in H_0.

5. Equal.

7. The critical value method and the p-value method.

9. Degrees of freedom = $(r - 1)(c - 1)$, where r = the number of categories in the row variable and c = the number of categories in the column variable.

11. $n = O_1 + O_2 + O_3 + O_4 + O_5 + O_6 = 50 + 40 + 30 + 20 + 10 + 10 = 160.$

$E_1 = n \cdot p_1 = (160)(0.3) = 48$, $E_2 = n \cdot p_2 = (160)(0.25) = 40$,

$E_3 = n \cdot p_3 = (160)(0.2) = 32$, $E_4 = n \cdot p_4 = (160)(0.15) = 24$,

$E_5 = n \cdot p_5 = (160)(0.06) = 9.6$, $E_6 = n \cdot p_6 = (160)(0.04) = 6.4$.

Since none of the expected frequencies is less than 1 and none of the expected frequencies is less than 5, the conditions for performing the χ^2 goodness of fit test are met.

df $= k - 1 = 6 - 1 = 5$, $\alpha = 0.05$, $\chi^2_{\text{crit}} = \chi^2_{0.05} = 11.071$. Reject H_0 if $\chi^2_{\text{data}} \geq 11.071$.

O_i	E_i	$O_i - E_i$	$(O_i - E_i)^2$	$\frac{(O_i-E_i)^2}{E_i}$
50	48	2	4	4/48 = 0.0833333333
40	40	0	0	0/40 = 0
30	32	–2	4	4/32 = 0.125
20	24	–4	16	16/24 = 0.6666666667
10	9.6	0.4	0.16	0.16/9.6 = 0.0166666667
10	6.4	3.6	12.96	12.96/6.4 = 2.025
				X^2_{data} = 2.916666667

Thus, $\chi^2_{\text{data}} = \sum \frac{(O_i-E_i)^2}{E_i} = 2.916666667 \approx 2.917$.

Since χ^2_{data} is not ≥ 11.071, we do not reject H_0. There is insufficient evidence that the random variable does not follow the distribution specified in H_0.

13. $n = O_1 + O_2 + O_3 + O_4 + O_5 + O_6 = 65 + 55 + 30 + 25 + 15 + 10 = 200.$

$E_1 = n \cdot p_1 = (200)(0.3) = 60$, $E_2 = n \cdot p_2 = (200)(0.25) = 50$,

$E_3 = n \cdot p_3 = (200)(0.20) = 40$, $E_4 = n \cdot p_4 = (200)(0.15) = 30$,

$E_5 = n \cdot p_5 = (200)(0.06) = 12$, $E_6 = n \cdot p_6 = (200)(0.04) = 8$.

Since none of the expected frequencies is less than 1 and none of the expected frequencies is less than 5, the conditions for performing the χ^2 goodness of fit test are met.

df $= k - 1 = 6 - 1 = 5$, $\alpha = 0.05$, $\chi^2_{\text{crit}} = \chi^2_{0.05} = 11.071$. Reject H_0 if $\chi^2_{\text{data}} \geq 11.071$.

O_i	E_i	$O_i - E_i$	$(O_i - E_i)^2$	$\frac{(O_i-E_i)^2}{E_i}$
65	60	5	25	25/60 = 0.4166666667
55	50	5	25	25/50 = 0.5
30	40	−10	100	100/40 = 2.5
25	30	−5	25	25/30 = 0.8333333333
15	12	3	9	9/12 = 0.75
10	8	2	4	4/8 = 0.5
				$\chi^2_{data} = 5.5$

Thus, $\chi^2_{\text{data}} = \sum \frac{(O_i-E_i)^2}{E_i} = 5.5.$

Since χ^2_{data} is not ≥ 11.071, we do not reject H_0. There is insufficient evidence that the random variable does not follow the distribution specified in H_0.

15. H_0 : Gender and sport preference are independent.

H_a : Gender and sport preference are dependent.

Reject H_0 if the *p*-value ≤ 0.05.

Expected frequencies:

		Sport reference		
Gender	**Basketball**	**Soccer**	**Swimming**	**Total**
Female	$\frac{(100)(80)}{200} = 40$	$\frac{(100)(50)}{200} = 25$	$\frac{(100)(70)}{200} = 35$	100
Male	$\frac{(100)(80)}{200} = 40$	$\frac{(100)(50)}{200} = 25$	$\frac{(100)(70)}{200} = 35$	100
Total	80	50	70	200

Since none of the expected frequencies is less than 1 and none of the expected frequencies is less than 5, the conditions for performing the χ^2 test for independence are met.

$$\chi^2_{\text{data}} = \sum \frac{(O_i-E_i)^2}{E_i} = \frac{(30-40)^2}{40} + \frac{(20-25)^2}{25} + \frac{(50-35)^2}{35} + \frac{(50-40)^2}{40} + \frac{(30-25)^2}{25} + \frac{(20-35)^2}{35} \approx 19.857.$$

df = $(r-1)(c-1) = (2-1)(3-1) = 2$. *p*-value = $P(\chi^2 > 19.857) \approx 0.00004876$. Since the *p*-value is ≤ 0.05, we reject H_0. There is evidence that gender and sport preference are dependent.

Chapter 12: Analysis of Variance

Section 12.1

1. No. If the sample sizes are not all the same, then we need to calculate the overall sample mean by calculating the weighted mean of the sample means where the weights are the sample sizes.

3. ANOVA works by comparing the variability in the sample means with the variability within each sample. When the variability in the sample means is larger than the variability within each sample, this is evidence that the population means are not all equal and that we should reject the null hypothesis.

5. Against.

7. **(a)**

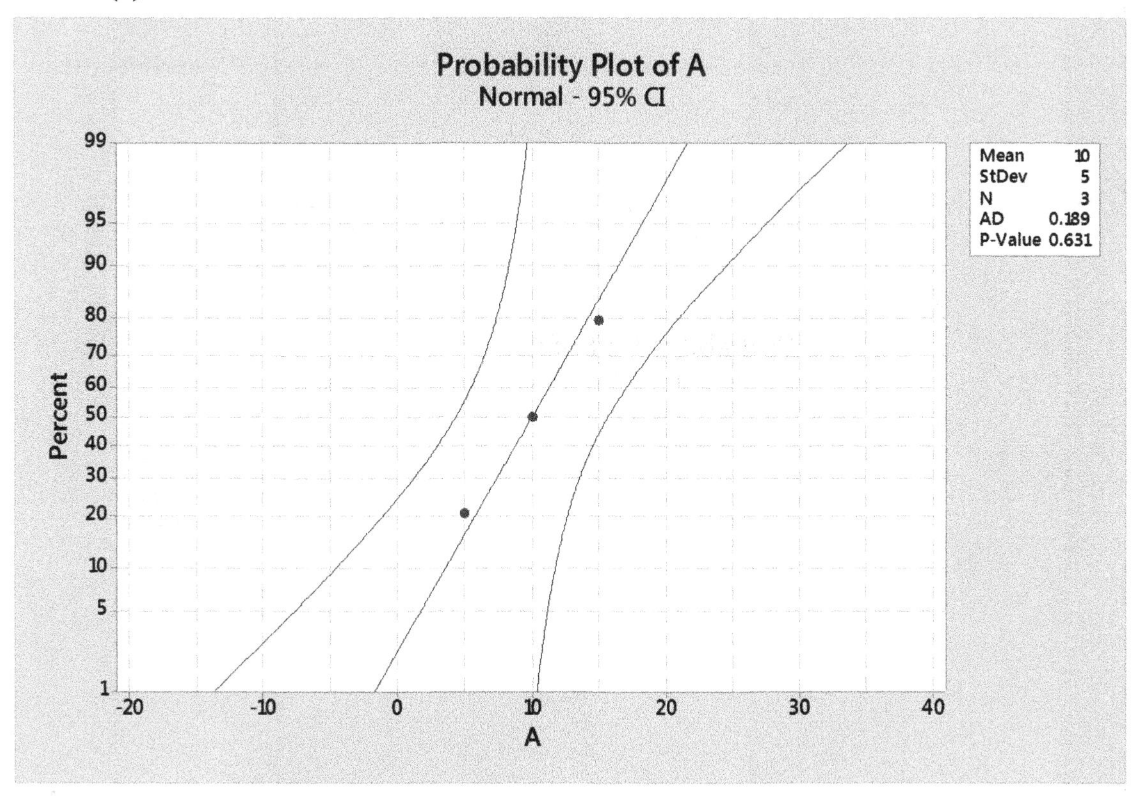

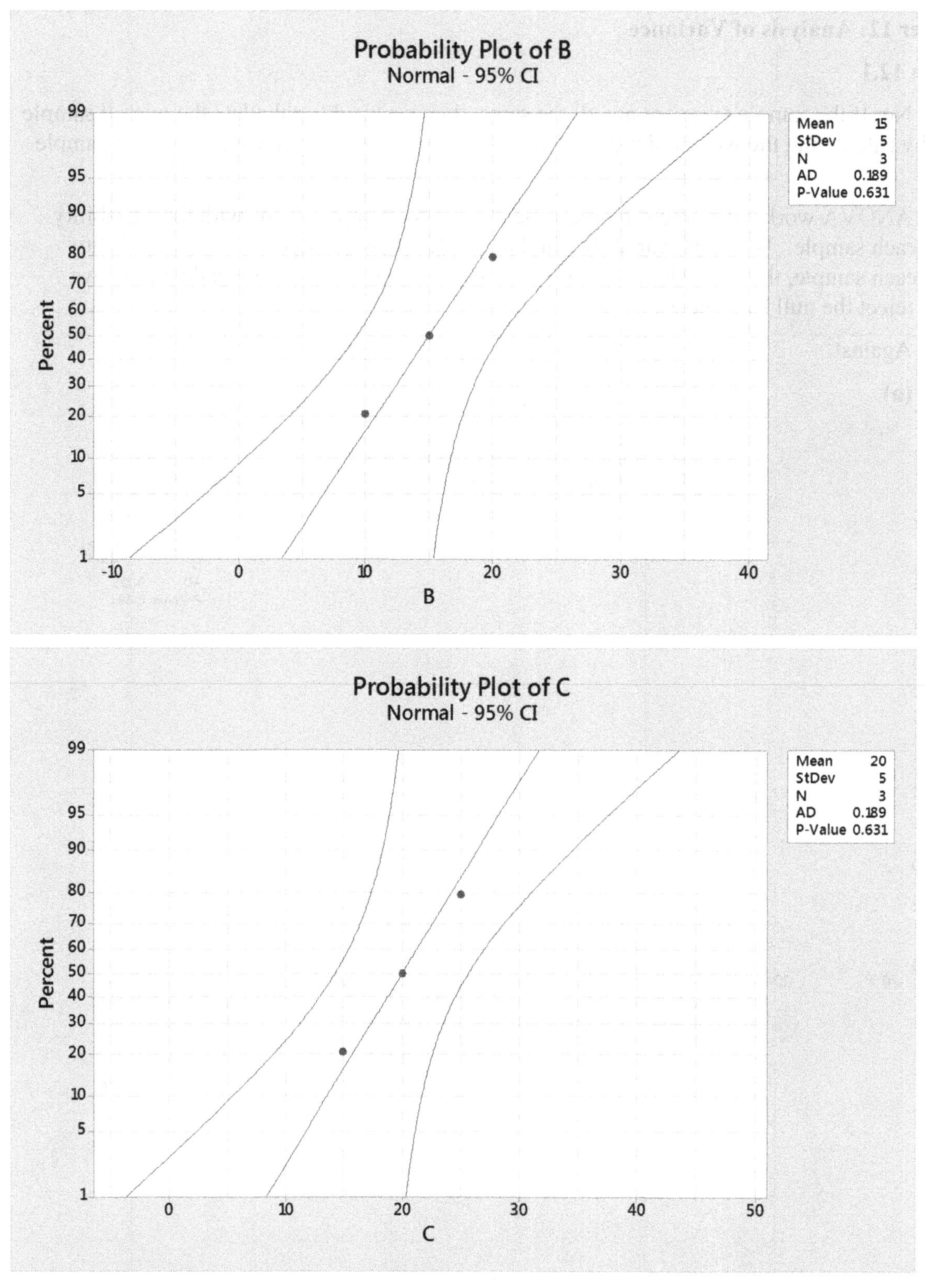

All three graphs show acceptable normality.

(b) `Means`

```
Factor  N   Mean  StDev
A       3  10.00   5.00
B       3  15.00   5.00
C       3  20.00   5.00
```

All three standard deviations are equal to 5, so all of them are less than $2(5.00) = 10.00$. Therefore, the equal variance requirement is satisfied.

9. **(a)**

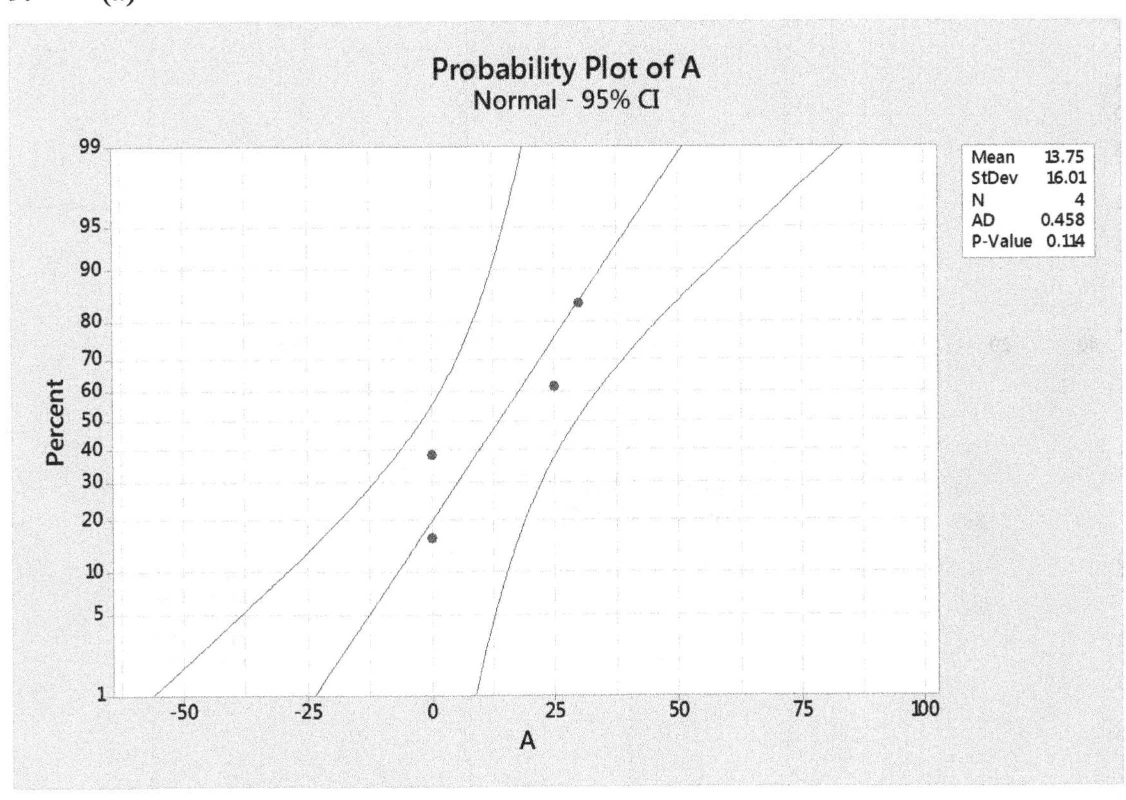

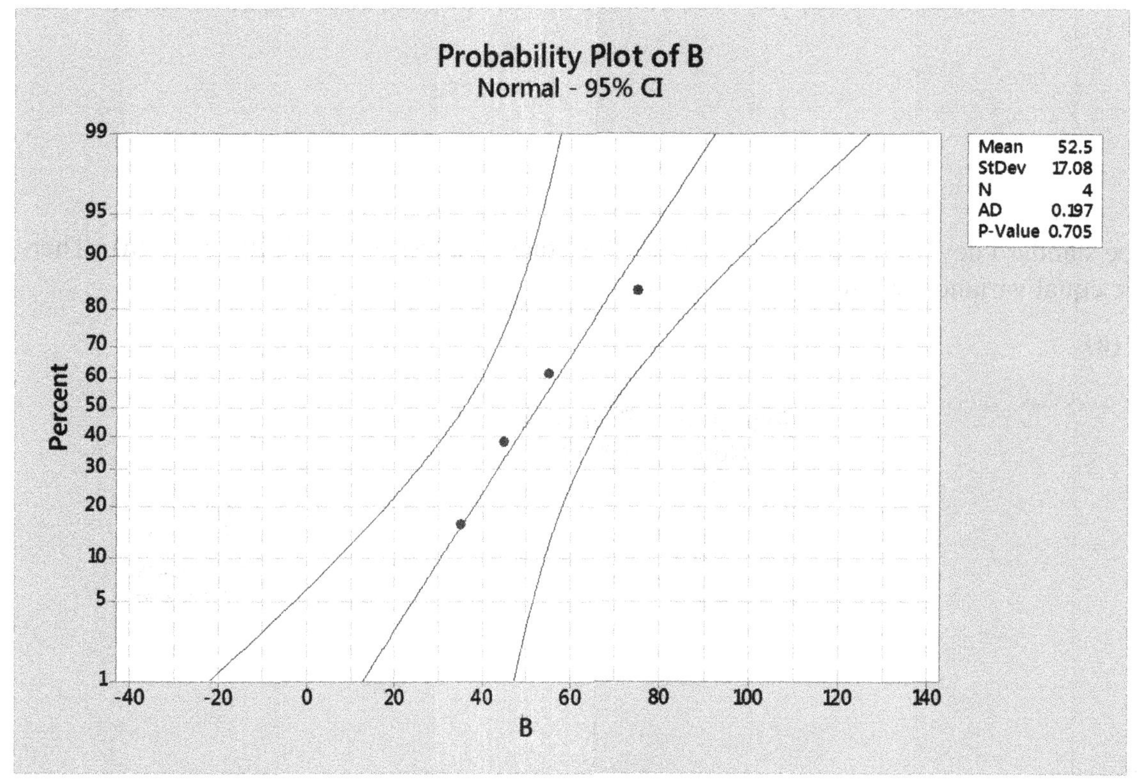
Probability Plot of B
Normal - 95% CI
Mean 52.5
StDev 17.08
N 4
AD 0.197
P-Value 0.705
Percent
99
95
90
80
70
60
50
40
30
20
10
5
1
-40
-20
0
20
40
60
80
100
120
140
B

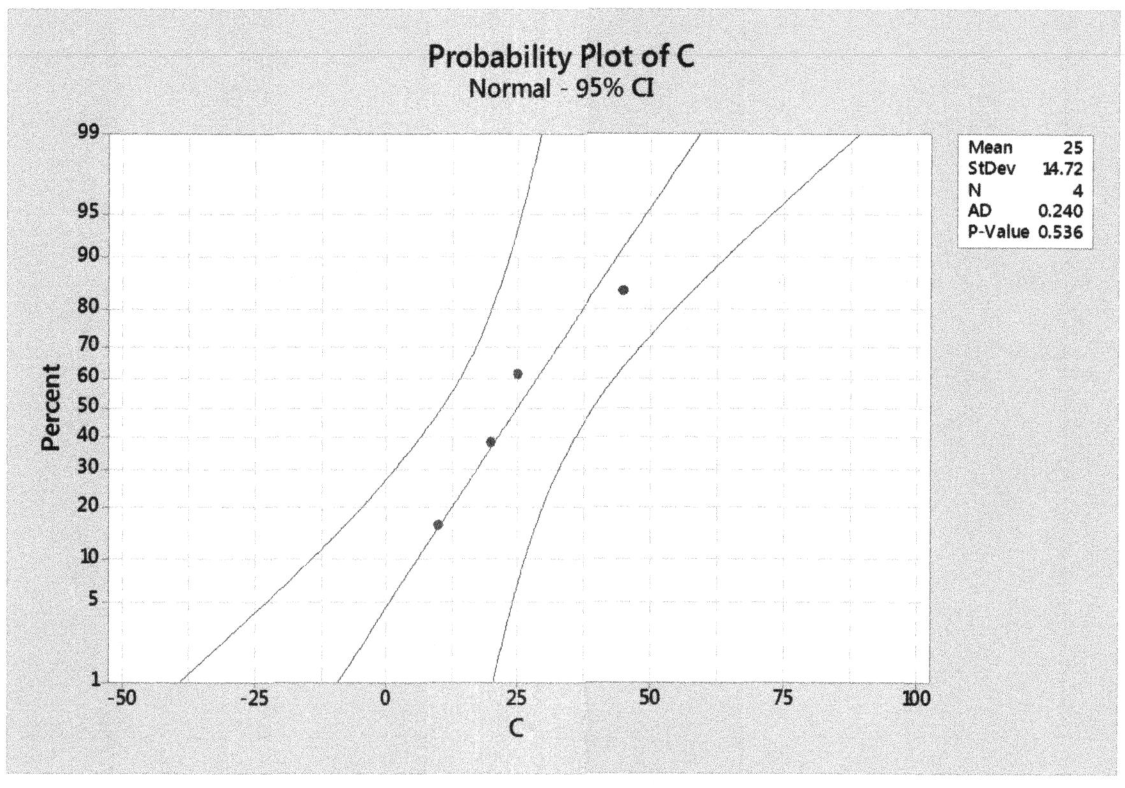
Probability Plot of C
Normal - 95% CI
Mean 25
StDev 14.72
N 4
AD 0.240
P-Value 0.536
Percent
99
95
90
80
70
60
50
40
30
20
10
5
1
-50
-25
0
25
50
75
100
C

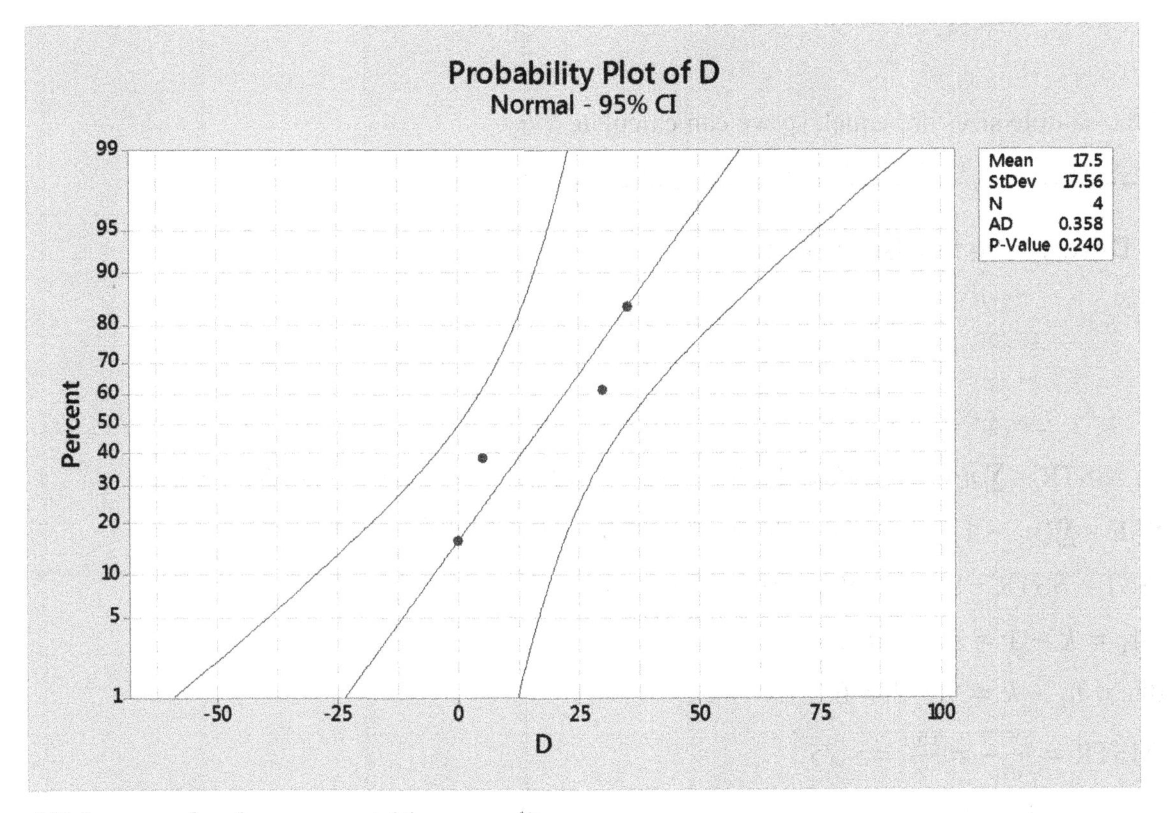

All four graphs show acceptable normality.

(b) Means

Factor	N	Mean	StDev
A	4	13.75	16.01
B	4	52.50	17.08
C	4	25.00	14.72
D	4	17.50	17.56

The smallest standard deviation is 14.72, which is the standard deviation for group C. Twice 14.72 is 29.44. All other standard deviations are less than 29.44 so the equal variance requirement is satisfied.

11. Means

Factor	N	Mean	StDev
A	3	10.00	5.00
B	3	15.00	5.00
C	3	20.00	5.00

All of the sample sizes are equal, so we can calculate $\bar{\bar{x}}$ by

$$\bar{\bar{x}} = \frac{\bar{x}_A + \bar{x}_B + \bar{x}_C}{k} = \frac{10+15+20}{3} = \frac{45}{3} = 15.$$

13. Means

Factor	N	Mean	StDev
A	4	13.75	16.01
B	4	52.50	17.08

C	4	25.00	14.72
D	4	17.50	17.56

All of the sample sizes are equal, so we can calculate $\bar{\bar{x}}$ by

$$\bar{\bar{x}} = \frac{\bar{x}_A+\bar{x}_B+\bar{x}_C+\bar{x}_D}{k} = \frac{13.75+52.50+25.00+17.50}{4} = \frac{108.75}{4} = 27.1875.$$

15. Descriptive Statistics: A, B, C

Variable	N	Mean	Variance
A	3	10.00	25.00
B	3	15.00	25.00
C	3	20.00	25.00

(a) (i) SSTR $= \sum n_i(\bar{x}_i - \bar{\bar{x}})^2 = 3(10-15)^2 + 3(15-15)^2 + 3(20-15)^2 = 150$

(ii) SSE $= \sum(n_i - 1)s_i^2 = (3-1)5^2 + (3-1)5^2 + (3-1)5^2 = 150$

(iii) SST = SSTR + SSE = 150 + 150 = 300

(iv) $\text{df}_1 = k - 1 = 3 - 1 = 2$

(v) $\text{df}_2 = n_t - k = 9 - 3 = 6$

(vi) MSTR $= \frac{\text{SSTR}}{\text{df}_1} = \frac{150}{2} = = 75$

(vii) MSE $= \frac{\text{SSE}}{\text{df}_2} = \frac{150}{6} = 25$

(b) Analysis of Variance

Source	DF	Adj SS	Adj MS	F-Value	P-Value
Factor	2	150.0	75.00	3.00	0.125
Error	6	150.0	25.00		
Total	8	300.0			

Source of variation	Sum of squares	Degrees of freedom	Mean square	F
Treatments	150	2	75	$F = \frac{\text{MSTR}}{\text{MSE}} = \frac{75}{25} = 3$
Error	150	6	25	
Total	300			

17. Descriptive Statistics: A, B, C, D

Variable	Total Count	Mean	Variance
A	4	13.75	256.25
B	4	52.50	291.67
C	4	25.00	216.67
D	4	17.50	308.33

(a) (i) SSTR $= \sum n_i(\bar{x}_i - \bar{\bar{x}})^2 = 4(13.75 - 27.1875)^2 + 4(52.50 - 27.1875)^2 + 4(25 - 27.1875)^2 + 4(17.50 - 27.1875)^2 \approx 3680$

(ii) SSE $= \sum(n_i - 1)s_i^2 = (4-1)256.25 + (4-1)291.67 + (4-1)216.67 + (4-1)308.33 \approx 3219$

(iii) SST = SSTR + SSE ≈ 3680 + 3219 = 6898

(iv) $\text{df}_1 = k - 1 = 4 - 1 = 3$

(v) $\text{df}_2 = n_t - k = 16 - 4 = 12$

(vi) MSTR $= \frac{\text{SSTR}}{\text{df}_1} = \frac{3680}{3} \approx 1226.6$

(vii) MSE $= \frac{\text{SSE}}{\text{df}_2} = \frac{3219}{12} \approx 268.2$

(b)

```
Analysis of Variance
Source   DF   Adj SS   Adj MS   F-Value   P-Value
Factor    3     3680   1226.6      4.57     0.023
Error    12     3219    268.2
Total    15     6898
```

Source of variation	Sum of squares	Degrees of freedom	Mean square	F
Treatments	3680	3	1226.6	$F = \frac{\text{MSTR}}{\text{MSE}} \approx \frac{1226.6}{268.2} \approx 4.57$
Error	3219	12	268.2	
Total	6898			

19 **(a)** **(i)** All of the sample sizes are equal, so we can calculate $\bar{\bar{x}}$ by

$\bar{\bar{x}} = \frac{\bar{x}_A + \bar{x}_B + \bar{x}_C}{k} = \frac{10+12+8}{3} = \frac{30}{3} = 10.$

SSTR $= \sum n_i(\bar{x}_i - \bar{\bar{x}})^2 = 5(10-10)^2 + 5(12-10)^2 + 5(8-10)^2 = 40$

(ii) SSE $= \sum(n_i - 1)s_i^2 = (5-1)1^2 + (5-1)1^2 + (5-1)1^2 = 12$

(iii) SST = SSTR + SSE = 40 + 12 = 52

(iv) $\text{df}_1 = k - 1 = 3 - 1 = 2$

(v) $\text{df}_2 = n_t - k = 15 - 3 = 12$

(vi) MSTR $= \frac{\text{SSTR}}{\text{df}_1} = \frac{40}{2} = 20$

(vii) MSE $= \frac{\text{SSE}}{\text{df}_2} = \frac{12}{12} = 1$

(b)

Source of variation	Sum of squares	Degrees of freedom	Mean square	F
Treatments	40	2	20	$F = \frac{\text{MSTR}}{\text{MSE}} = \frac{20}{1} = 20$
Error	12	12	1	
Total	52			

21. **(a)** **(i)** $n_t = n_A + n_B + n_C + n_D = 100 + 150 + 200 + 150 = 600$

$\bar{\bar{x}} = \frac{n_A\bar{x}_A + n_B\bar{x}_B + n_C\bar{x}_C + n_D\bar{x}_D}{n_t} = \frac{100(50)+150(75)+200(100)+150(125)}{600} = \frac{55{,}000}{600} \approx 91.66666667.$

SSTR $= \sum n_i(\bar{x}_i - \bar{\bar{x}})^2 = 100(50 - 91.66666667)^2 + 150(75 - 91.66666667)^2 + 200(100 - 91.66666667)^2 + 150(125 - 91.66666667)^2 = 395{,}833.3333$

(ii) SSE $= \sum(n_i - 1)s_i^2 = (100 - 1)5^2 + (150 - 1)4^2 + (200 - 1)6^2 + (150 - 1)5^2 = 15{,}748$

(iii) SST = SSTR + SSE = 395,833.3333 + 15,748 = 411,581.3333

(iv) $\text{df}_1 = k - 1 = 4 - 1 = 3$

(v) $\text{df}_2 = n_t - k = 600 - 4 = 596$

(vi) MSTR $= \frac{\text{SSTR}}{\text{df}_1} = \frac{395{,}833.3333}{3} \approx 131{,}944.4444$

(vii) MSE $= \frac{\text{SSE}}{\text{df}_2} = \frac{15{,}748}{596} \approx 26.4228$

(b)

Source of variation	Sum of squares	Degrees of freedom	Mean square	F
Treatments	395,833.3333	3	131,944.4444	$F = \frac{\text{MSTR}}{\text{MSE}} = \frac{131{,}944.4444}{26.4228} =$ 4993.5794
Error	15,748	596	26.4228	
Total	411,581.3333			

23. H_0: $\mu_A = \mu_B = \mu_C$ versus H_a: Not all the population means are equal.

Reject H_0 if the p-value $\leq \alpha = 0.05$.

$F_{\text{data}} = 3.00$

p-value = 0.125

The p-value = 0.125 is not $\leq \alpha = 0.05$. Therefore we do not reject H_0. There is not enough evidence to conclude at level of significance $\alpha = 0.05$ that the population means are not all equal.

25. H_0: $\mu_A = \mu_B = \mu_C = \mu_D$ versus H_a: Not all the population means are equal.

Reject H_0 if the p-value $\leq \alpha = 0.05$.

$F_{\text{data}} = 4.57$

p-value = 0.023

The p-value = 0.023 is $\leq \alpha = 0.05$. Therefore we reject H_0. There is evidence at level of significance $\alpha = 0.05$ that the population means are not all equal.

27. H_0: $\mu_A = \mu_B = \mu_C$

H_a: Not all the population means are equal.

Reject H_0 if the p-value $\leq \alpha = 0.05$.

$F_{\text{data}} = 20$

p-value =0.0002

The p-value = 0.0002 is $\leq \alpha = 0.05$. Therefore we reject H_0. There is evidence at level of significance $\alpha = 0.05$ that the population means are not all equal.

29. H_0: $\mu_A = \mu_B = \mu_C = \mu_D$

H_a: Not all the population means are equal.

$F_{\text{crit}} = 3.815$ Reject H_0 if $F_{data} \geq 3.815$

$F_{\text{data}} = 4993.5794$
$F_{\text{data}} = 4793.5794$ is ≥ 3.815. Therefore we reject H_0. There is evidence at level of significance $\alpha = 0.01$ that the population means are not all equal.

31. **(a)**

Means

Factor	N	Mean	StDev
Wraps	5	14.20	6.61
Muffins	5	5.60	3.65
Chips	5	9.00	4.47

The smallest standard deviation is 3.65, which is the standard deviation for muffins. Twice 3.65 is 7.30. All other standard deviations are less than 7.30, so the equal variance requirement is satisfied.

(b)

Descriptive Statistics: Wraps, Muffins, Chips

Variable	Total Count	Mean	Variance
Wraps	5	14.20	43.70
Muffins	5	5.60	13.30
Chips	5	9.00	20.00

(i) $\text{df}_1 = k - 1 = 3 - 1 = 2$, $\text{df}_2 = n_t - k = 15 - 3 = 12$

(ii) All of the sample sizes are equal, so we can calculate $\bar{\bar{x}}$ by

$$\bar{\bar{x}} = \frac{\bar{x}_A + \bar{x}_B + \bar{x}_C}{k} = \frac{14.20+5.60+9}{3} = \frac{28.80}{3} = 9.6.$$

(iii) $\text{SSTR} = \sum n_i(\bar{x}_i - \bar{\bar{x}})^2 = 5(14.20 - 9.6)^2 + 5(5.60 - 9.6)^2 + 5(9 - 9.6)^2 = 187.6$

(iv) $\text{SSE} = \sum(n_i - 1)s_i^2 = (5-1)43.70 + (5-1)13.30 + (5-1)20 = 308$

(v) SST = SSTR + SSE = 187.6 + 308 = 495.6

(vi) $\text{MSTR} = \frac{\text{SSTR}}{\text{df}_1} = \frac{187.6}{2} = 93.80$

(vii) $\text{MSE} = \frac{\text{SSE}}{\text{df}_2} = \frac{308}{12} \approx 25.67$

(viii) $F_{\text{data}} = \frac{\text{MSTR}}{\text{MSE}} \approx \frac{93.80}{25.67} \approx 3.65$

(c) Analysis of Variance

Source	DF	Adj SS	Adj MS	F-Value	P-Value
Factor	2	187.6	93.80	3.65	0.058
Error	12	308.0	25.67		
Total	14	495.6			

Source of variation	Sum of squares	Degrees of freedom	Mean square	F
Treatments	187.6	2	93.80	3.6!
Error	308	12	25.67	
Total	495.6			

(d) H_0: $\mu_{\text{Wraps}} = \mu_{\text{Muffins}} = \mu_{\text{Chips}}$

H_a: Not all the population means are equal.

Reject H_0 if the p-value $\leq \alpha = 0.05$.

$F_{\text{data}} = 3.65$

p-value = 0.058

The p-value = 0.058 is not $\leq \alpha = 0.05$. Therefore we do not reject H_0. There is not enough evidence to conclude at level of significance $\alpha = 0.05$ that the population means of the number of different types of food items are not all equal.

33. **(a)** Means

Factor	N	Mean	StDev
Younger	5	132.00	13.91
Middle	5	114.26	17.14
Older	5	135.24	25.9

The smallest standard deviation is 13.91, which is the standard deviation for the younger. Twice 13.91 is 27.82. All other standard deviations are less than 27.82, so the equal variance requirement is satisfied.

(b) Descriptive Statistics: Younger, Middle, Older

Variable	Total Count	Mean	Variance
Younger	5	132.00	193.43
Middle	5	114.26	293.91
Older	5	135.24	672.7

(i) $\text{df}_1 = k - 1 = 3 - 1 = 2$, $\text{df}_2 = n_t - k = 15 - 3 = 12$

(ii) $\bar{\bar{x}} = \frac{132+114.26+135.24}{3} = \frac{381.5}{3} \approx 127.1666667 \approx 127.17$

(iii) $\text{SSTR} = \sum n_i(\bar{x}_i - \bar{\bar{x}})^2 = 5(132 - 127.1666667)^2 + 5(114.26 - 127.1666667)^2 + 5(135.24 - 127.1666667)^2 \approx 1276$

(iv) $\text{SSE} = \sum(n_i - 1)s_i^2 \approx (5-1)193.43 + (5-1)293.91 + (5-1)672.7 \approx 4640$

(v) SST = SSTR + SSE ≈ 1276 + 4640 = 5916

(vi) $\text{MSTR} = \frac{\text{SSTR}}{\text{df}_1} \approx \frac{1276}{2} = 638$ (Minitab: 637.8)

(vii) $\text{MSE} = \frac{\text{SSE}}{\text{df}_2} \approx \frac{4640}{12} \approx 386.7$

(viii) $F_{\text{data}} = \frac{\text{MSTR}}{\text{MSE}} \approx \frac{637.8}{386.7} \approx 1.65$

(c) Analysis of Variance

Source	DF	Adj SS	Adj MS	F-Value	P-Value
Factor	2	1276	637.8	1.65	0.233
Error	12	4640	386.7		
Total	14	5916			

Source of variation	Sum of squares	Degrees of freedom	Mean square	*F*
Treatments	1276	2	637.8	1.6.
Error	4640	12	386.7	
Total	5916			

(d) H_0: $\mu_{\text{Younger}} = \mu_{\text{Middle}} = \mu_{\text{Older}}$

H_a: Not all the population means are equal.

Reject H_0 if the p-value $\leq \alpha = 0.05$.

$F_{\text{data}} = 1.65$

p-value = 0.233

The p-value = 0.233 is not $\leq \alpha = 0.05$. Therefore we do not reject H_0. There is not enough evidence to conclude at level of significance $\alpha = 0.05$ that the population mean weights of the different age groups are not all equal.

35. **(a)** SST = SSTR + SSE = 120 + 315 = 435, $\text{df}_1 = k - 1 = 7 - 1 = 6$,

$n_t = 7(10) = 70$, $\text{df}_2 = n_t - k = 70 - 7 = 63$, $\text{MSTR} = \frac{\text{SSTR}}{\text{df}_1} = \frac{120}{6} = 20$.

$\text{MSE} = \frac{\text{SSE}}{\text{df}_2} = \frac{315}{63} = 5$.

$F_{data} = \frac{\text{MSTR}}{\text{MSE}} = \frac{20}{5} = 4$.

Missing values are bold.

Source of variation	Sum of squares	Degrees of freedom	Mean square	F-test statistic
Treatment	SSTR = 120	**$df_1 = 6$**	**MSTR = 20**	**$F_{data} = 4$**
Error	SSE = 315	**$df_2 = 63$**	**MSE = 5**	
Total	**SST = 435**			

(b) $H_0 : \mu_1 = \mu_2 = \mu_3 = \mu_4 = \mu_5 = \mu_6 = \mu_7$

H_a : Not all of the population means are equal.

Reject H_0 if the p-value ≤ 0.05. $F_{data} = 4$. Since $df_1 = 6$, $df_2 = 63$, and p-value $= P(F > F_{data}) = P(F > 4) \approx 0.0019$. Since the p-value ≈ 0.0019, which is ≤ 0.05, we reject H_0. There is evidence that not all of the population means are equal.

37. **(a)** $F_{\text{data}} = \frac{\text{MSTR}}{\text{MSE}} = \frac{10}{\text{MSE}} = 1.0$, so MSE = 10.

$\text{MSTR} = \frac{\text{SSTR}}{\text{df}_1} = \frac{\text{SSTR}}{4} = 10$, so SSTR = 4(10) = 40.

SSE = SST − SSTR = 440 − 40 = 400.

$\text{MSE} = \frac{\text{SSE}}{\text{df}_2} = \frac{400}{\text{df}_2} = 10$, so $\text{df}_2 = \frac{400}{10} = 40$.

Missing values are bold.

Source of variation	Sum of squares	Degrees of freedom	Mean square	F-test statistic
Treatment	**SSTR = 40**	$df_1 = 4$	MSTR = 10	$F_{data} = 1.0$
Error	**SSE = 400**	**$df_2 = 40$**	**MSE = 10**	
Total	SST = 440			

(b) $H_0 : \mu_1 = \mu_2 = \mu_3 = \mu_4 = \mu_5$

H_a : Not all of the population means are equal.

$df_1 = 4$, $df_2 = 40$, $\alpha = 0.10$, $F_{crit} = 2.09$. Reject H_0 if $F_{data} \geq 2.09$.

$F_{data} = 1.0$. Since $F_{data} = 1.0$ is not ≥ 2.09, we do not reject H_0. There is insufficient evidence that not all of the population means are equal.

39. **(a)** $H_0 : \mu_{Females} = \mu_{Males}$

H_a : Not all of the population means are equal.

$\mu_{Females}$ = the population mean heart rate for females.

μ_{Males} = the population mean heart rate for males.

Reject H_0 if the p-value ≤ 0.05.

The largest sample standard deviation $s_{\text{females}} = 0.743$ is not more than twice the smallest sample standard deviation $2 \cdot s_{\text{male}} = 2(0.699) = 1.398$. $n_t = 65 + 65 = 130$, $df_1 = k - 1 = 2 - 1 = 1$, $df_2 = n_t - k = 130 - 2 = 128$.

$\bar{\bar{x}} = \frac{98.384+98.104}{2} = \frac{196.488}{2} = 98.244.$

$\text{SSTR} = \sum n_i(\bar{x}_i - \bar{\bar{x}})^2 = 65(98.384 - 98.244)^2 + 65(98.104 - 98.244)^2 \approx 2.548.$

SSE = $\sum(n_i - 1)s_i^2 = (65 - 1)0.743^2 + (65 - 1)0.699^2 = 66.6016$

SST = SSTR + SSE = 2.548 + 66.6016 = 69.1496

$\text{MSTR} = \frac{\text{SSTR}}{\text{df}_1} = \frac{2.548}{1} = 2.548$

$\text{MSE} = \frac{\text{SSE}}{\text{df}_2} = \frac{66.6016}{128} = 0.520325$

$F_{\text{data}} = \frac{\text{MSTR}}{\text{MSE}} \approx \frac{2.548}{0.520325} = 4.896939413 \approx 4.90$

Source of variation	Sum of squares	Degrees of freedom	Mean square	F
Treatment	2.548	1	2.548	4.896939413
Error	66.6016	128	0.520325	
Total	69.1496			

Since $\text{df}_1 = 1$, $\text{df}_2 = 128$, the *p*-value = $P(F > F_{\text{data}}) = P(F > 4.8969) \approx 0.0287$. Since the *p*-value ≈ 0.0287, which is ≤ 0.05, we reject H_0. There is evidence that the population mean heart rates are not equal.

(b) Inference for Two Independent Means, Section 10.2.

41. $H_0 : \mu_{Sopranos} = \mu_{Altos} = \mu_{Tenors} = \mu_{Basses}$

H_a : Not all of the population means are equal.

Reject H_0 if *p*-value ≤ 0.01.

The largest sample standard deviation is $s_{\text{Tenors}} = 3.216$ is not more than twice the smallest sample standard deviation $2 \cdot s_{\text{Sopranos}} = 2(1.873) = 3.746$. $n_t = 36 + 35 + 20 + 39 = 130$, $\text{df}_1 = k - 1 = 4 - 1 = 3$, $\text{df}_2 = n_t - k = 130 - 4 = 126$.

$\bar{\bar{x}} = \frac{36(64.250)+35(64.886)+20(69.150)+39(70.718)}{130} = \frac{8725.012}{130} = 67.11547692$

$\text{SSTR} = \sum n_i(\bar{x}_i - \bar{\bar{x}})^2 = 36(64.250 - 67.11547692)^2 + 35(64.886 - 67.11547692)^2 +$
$= \sum n_i(\bar{x}_i - \bar{\bar{x}})^2 = 36(64.250 - 67.11547692)^2 + 35(64.886 - 67.11547692)^2 + 20(69.150 - 67.11547692)^2 + 39(70.718 - 67.11547692)^2 \approx 1058.498756$

SSE = $\sum(n_i - 1)s_i^2 = (36 - 1)1.873^2 + (35 - 1)2.795^2 + (20 - 1)3.216^2 + (39 - 1)2.361^2 = 796.728027$

SST = SSTR + SSE = 1058.498756 + 796.728027 = 1855.226783

$\text{MSTR} = \frac{\text{SSTR}}{\text{df}_1} = \frac{1058.498756}{3} = 352.8329187$

$\text{MSE} = \frac{\text{SSE}}{\text{df}_2} = \frac{796.728027}{126} = 6.32323831$

$F_{\text{data}} = \frac{\text{MSTR}}{\text{MSE}} = \frac{352.8329187}{6.32323831} = 55.79940236 \approx 55.799$

Source of variation	Sum of squares	Degrees of freedom	Mean square	F
Treatment	1058.498756	3	352.8329187	55.79940236
Error	796.728027	126	6.32323831	
Total	1855.226783			

$\text{df}_1 = 3$, $\text{df}_2 = 126$, p-value $= P(F > F_{data}) = P(F > 55.7994) \approx 0$.

Since the p-value ≈ 0, which is ≤ 0.01, we reject H_0. There is evidence that not all of the population mean heights are equal.

43. Since $s_{Quaker\ Oats} = 10.603$ is not more than $11.018 = 2 \cdot s_{Nabisco}$, the equal variance requirement is satisfied.

45. $H_0 : \mu_{General\ Mills} = \mu_{Kelloggs} = \mu_{Nabisco} = \mu_{Post} = \mu_{Quaker\ Oats} = \mu_{Ralston\ Purina}$

H_a : Not all of the population means are equal.

Reject H_0 if the p-value ≤ 0.05. $F_{data} = 15.71$. p-value ≈ 0.

Since the p-value ≈ 0, which is ≤ 0.05, we reject H_0. There is evidence that the population mean nutritional ratings of the cereals made by the different manufacturers are not all equal.

47. **(a)** **(i)** $\text{df}_1 = k - 1$ and $\text{df}_2 = n_t - k$ where k is the number of samples and n_t is the total sample size (sum of the k sample sizes). Adding 3 points to each value in the DJIA would not affect the number of samples k or the total sample size n_t. Therefore, df_1 and df_2 would stay the same.

(ii) Adding 3 points to each value in the DJIA would increase $\bar{x}_{\text{DJIA}}$ by 3 points. Then the new $\bar{\bar{x}} = \frac{\bar{x}_{\text{Pros}}+\bar{x}_{\text{Darts}}+\bar{x}_{\text{DJIA}}+3}{3} = \frac{\bar{x}_{\text{Pros}}+\bar{x}_{\text{Darts}}+\bar{x}_{\text{DJIA}}}{3} + \frac{3}{3} = \frac{\bar{x}_{\text{Pros}}+\bar{x}_{\text{Darts}}+\bar{x}_{\text{DJIA}}}{3} + 1 =$ the old $\bar{\bar{x}} + 1$. Therefore, $\bar{\bar{x}}$ would increase.

(iii) The sample mean for the DJIA would increase by 3 points. This would bring it closer to the other two sample means, so SSTR would decrease.

(iv) Each sample size and variance would remain the same, so SSE would stay the same.

(v) SST= SSTR + SSE. Since SSTR would decrease and SSE would stay the same, SST would decrease.

(vi) $\text{MSTR} = \frac{\text{SSTR}}{\text{df}_1}$. Since SSTR would decrease and df_1 would stay the same, MSTR would decrease.

(vii) $\text{MSE} = \frac{\text{SSE}}{\text{df}_2}$, Since SSE and df_2 would stay the same, MSE would stay the same.

(viii) $F_{\text{data}} = \frac{\text{MSTR}}{\text{MSE}}$. Since MSTR would decrease MSE would stay the same, and F_{data} would decrease.

(ix) The p-value $= P(F > F_{\text{data}})$. Since F_{data} would decrease, the p-value would increase.

(b) F_{data} would decrease and the p-value would increase. Therefore, the conclusion would still be do not reject H_0.

49. **(a)** **(i)** $\text{df} = n_1 - 1 = 68 - 1 = 67$,

$$\bar{x}_1 \pm t_{\alpha/2}\left(\frac{s_1}{\sqrt{n_1}}\right) = 27.603 \pm 2.660\left(\frac{6.58}{\sqrt{68}}\right) \approx 27.603 \pm 2.1225 = (25.4805, 29.7255)$$

(ii) $\text{df} = n_2 - 1 = 79 - 1 = 78$,

$$\bar{x}_2 \pm t_{\alpha/2}\left(\frac{s_2}{\sqrt{n_2}}\right) = 30.451 \pm 2.648\left(\frac{6.09}{\sqrt{79}}\right) \approx 30.451 \pm 1.8144 = (28.6366, 32.2654)$$

(iii) df = $n_3 - 1 = 245 - 1 = 244$,

$$\bar{x}_3 \pm t_{\alpha/2}\left(\frac{s_3}{\sqrt{n_3}}\right) = 20.033 \pm 2.626\left(\frac{6.440}{\sqrt{245}}\right) \approx 20.033 \pm 1.0804 = (18.9526, 21.1134)$$

(b) The confidence interval for the population mean gas mileage of American cars does not overlap the other two confidence intervals. This is evidence against the null hypothesis that all of the population means are equal.

51. From Minitab:

One-way ANOVA: HEAD_INJ versus SIZE2

```
Method
Null hypothesis          All means are equal
Alternative hypothesis   At least one mean is different
Significance level       α = 0.05
Rows unused              14
Equal variances were assumed for the analysis.
Factor Information
Factor  Levels  Values
SIZE2        8  1, 2, 3, 4, 5, 6, 7, 8
Analysis of Variance
Source   DF    Adj SS   Adj MS  F-Value  P-Value
SIZE2     7  10879518  1554217     8.27    0.000
Error   330  62003922   187891
Total   337  72883441
Model Summary
      S    R-sq  R-sq(adj)  R-sq(pred)
433.464  14.93%     13.12%      10.16%
Means
SIZE2   N    Mean  StDev        95% CI
1      83   784.9  339.6  ( 691.3,  878.5)
2      73   826.2  375.6  ( 726.4,  926.0)
3      58   740.3  427.8  ( 628.3,  852.2)
4      14     838    490  (   610,   1066)
5      14   853.6  366.4  ( 625.7, 1081.5)
6      30    1244    686  (  1088,   1399)
7      36  1050.5  460.6  ( 908.4, 1192.6)
8      30  1239.7  456.4  (1084.0, 1395.4)
Pooled StDev = 433.464
```

$H_0 : \mu_1 = \mu_2 = \mu_3 = \mu_4 = \mu_5 = \mu_6 = \mu_7 = \mu_8$

H_a : Not all of the population means are equal.

Reject H_0 if the p-value ≤ 0.05.

$F_{data} = 8.27$

p-value ≈ 0

Since the p-value $\approx$ 0, which is ≤ 0.05, we reject H_0.

There is evidence that not all of the population means are equal.

53. From Minitab:

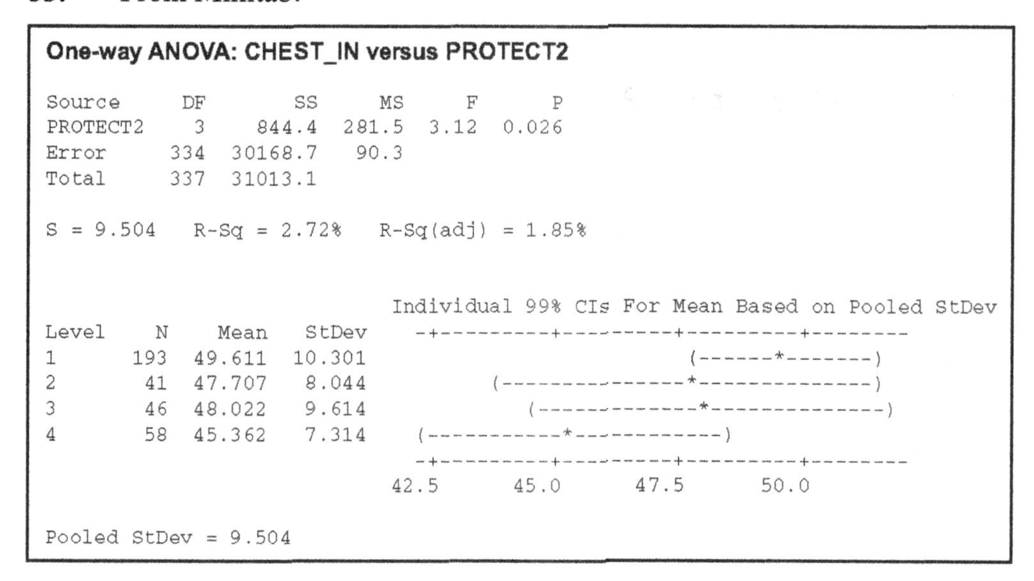

One-way ANOVA: CHEST_IN versus PROTECT2

```
Source     DF       SS     MS     F      P
PROTECT2    3    844.4  281.5  3.12  0.026
Error     334  30168.7   90.3
Total     337  31013.1

S = 9.504   R-Sq = 2.72%   R-Sq(adj) = 1.85%


                                Individual 99% CIs For Mean Based on Pooled StDev
Level    N    Mean  StDev   -+---------+---------+---------+--------
1      193  49.611 10.301                        (------*-------)
2       41  47.707  8.044            (--------------*--------------)
3       46  48.022  9.614               (-------------*--------------)
4       58  45.362  7.314   (------------*------------)
                            -+---------+---------+---------+--------
                          42.5      45.0      47.5      50.0

Pooled StDev = 9.504
```

$H_0 : \mu_1 = \mu_2 = \mu_3 = \mu_4$

H_a : Not all of the population means are equal.

Reject H_0 if the p-value ≤ 0.01.

$F_{data} = 3.12$

p-value = 0.026

Since the p-value $\approx$ 0.026, which is not ≤ 0.01, we do not reject H_0.

There is insufficient evidence that not all of the population means are equal.

55. (a)

Original screen:

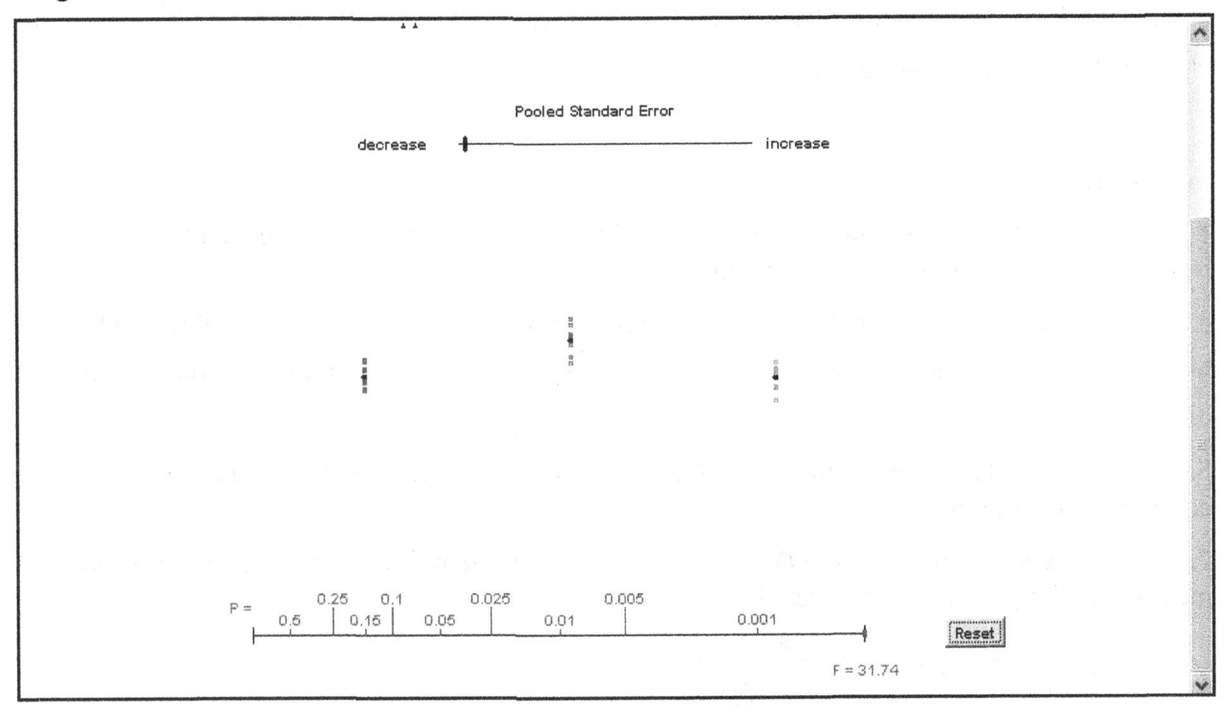

Screen after black dots are made almost level horizontally:

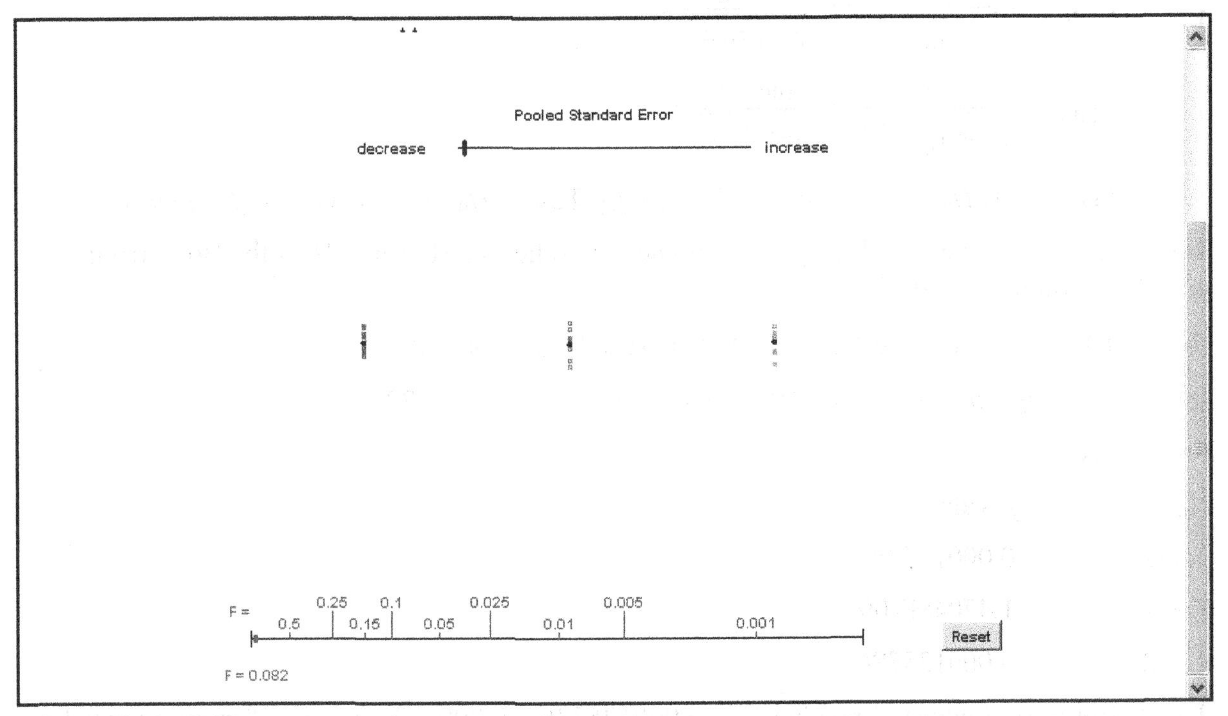

F decreases.

(b) If all of the sample means are about the same, then the between-sample variability is small and all of the sample means are close to $\bar{\bar{x}}$. Therefore, $\text{SSTR} = \sum n_i(\bar{x}_i - \bar{\bar{x}})^2$ decreases. Since the number of observations in each sample remains the same, the total number of observations remains the same. The number of treatments also remains the same. Therefore, $\text{df}_1 = k - 1$ remains the same, so $\text{MSTR} = \frac{\text{SSTR}}{\text{df}_1}$ decreases. Since the within-sample variability does not change, SSE does not change. Therefore, $\text{df}_2 = n_t - k$ remains the same, so $\text{MSE} = \frac{\text{SSE}}{\text{df}_2}$ does not change. Thus, $F_{\text{data}} = \frac{\text{MSTR}}{\text{MSE}}$ decreases.

Section 12.2

1. ANOVA only tests for whether or not all of the population means are equal. It does not test for which pairs of means are not equal.

3. It is performing a hypothesis test to determine whether two means are different. The number of pairwise comparisons is $c = {}_kC_2 = \frac{k!}{2!(k-2)!}$ where k is the number of population means and $k \geq 3$.

5. By multiplying the p-value of each pairwise hypothesis test by the number of comparisons being made.

7. The requirements for ANOVA have been met and the null hypothesis that the population means are all equal has been rejected.

9. $t_{\text{data}} = \frac{\bar{x}_1 - \bar{x}_2}{\sqrt{\text{MSE} \cdot \left(\frac{1}{n_1} + \frac{1}{n_2}\right)}} = \frac{85-75}{\sqrt{25 \cdot \left(\frac{1}{10} + \frac{1}{10}\right)}} \approx 4.472$

11. $t_{\text{data}} = \frac{\bar{x}_2 - \bar{x}_3}{\sqrt{\text{MSE} \cdot \left(\frac{1}{n_2} + \frac{1}{n_3}\right)}} = \frac{75-65}{\sqrt{25 \cdot \left(\frac{1}{10} + \frac{1}{10}\right)}} \approx 4.472$

13. $t_{\text{data}} = \frac{\bar{x}_1 - \bar{x}_3}{\sqrt{\text{MSE} \cdot \left(\frac{1}{n_1} + \frac{1}{n_3}\right)}} = \frac{100-85}{\sqrt{100 \cdot \left(\frac{1}{8} + \frac{1}{8}\right)}} = 3$

15. **(a)** Test 1: $H_0 : \mu_1 = \mu_2$ versus $H_a : \mu_1 \neq \mu_2$. Test 2: $H_0 : \mu_1 = \mu_3$ versus $H_a : \mu_1 \neq \mu_3$. Test 3: $H_0 : \mu_2 = \mu_3$ versus $H_a : \mu_2 \neq \mu_3$. For each hypothesis test, reject H_0 if the Bonferroni-adjusted p-value $\leq \alpha = 0.05$.

(b) Test 1: $t_{\text{data}} \approx 4.472$. Test 2: $t_{\text{data}} \approx 8.944$. Test 3: $t_{\text{data}} \approx 4.472$.

(c) $n_T = n_1 + n_2 + n_3 = 10 + 10 + 10 = 30$. $\text{df} = n_T - k = 30 - 3 = 27$.

From Excel:

t_{data}	p-value
4.472	0.000125786
8.944	1.47048E-09
4.472	0.000125786

Test 1: Bonferroni-adjusted p-value $= 3 \cdot 0.000125786 \approx 0.0004$. Test 2: Bonferroni-adjusted p-value $= 3 \cdot 1.47048 \times 10^{-9} \approx 0$. Test 3: Bonferroni-adjusted p-value $= 3 \cdot 0.000125786 \approx 0.0004$.

(d) Test 1: The adjusted p-value ≈ 0.0004, which is ≤ 0.05. Therefore, we reject H_0. There is evidence at the $\alpha = 0.05$ level of significance that the population mean of Population 1 differs from the population mean of Population 2. Test 2: The adjusted p-value ≈ 0, which is ≤ 0.05.

Therefore, we reject H_0. There is evidence at the $\alpha = 0.05$ level of significance that the population mean of Population 1 differs from the population mean of Population 3. Test 3: The adjusted p-value ≈ 0.0004, which is ≤ 0.05. Therefore, we reject H_0. There is evidence at the $\alpha = 0.05$ level of significance that the population mean of Population 2 differs from the population mean of Population 3.

17. Test 1: $H_0 : \mu_A = \mu_B$ versus $H_a : \mu_A \neq \mu_B$.

Test 2: $H_0 : \mu_A = \mu_C$ versus $H_a : \mu_A \neq \mu_C$.

Test 3: $H_0 : \mu_B = \mu_C$ versus $H_a : \mu_B \neq \mu_C$. For each hypothesis test, reject H_0 if the Bonferroni-adjusted p-value $\leq \alpha = 0.01$.

Test 1: $t_{\text{data}} = \dfrac{\bar{x}_A - \bar{x}_B}{\sqrt{\text{MSE} \cdot \left(\frac{1}{n_A} + \frac{1}{n_B}\right)}} = \dfrac{10-12}{\sqrt{1 \cdot \left(\frac{1}{5} + \frac{1}{5}\right)}} \approx -3.162$

Test 2: $t_{\text{data}} = \dfrac{\bar{x}_A - \bar{x}_C}{\sqrt{\text{MSE} \cdot \left(\frac{1}{n_A} + \frac{1}{n_C}\right)}} = \dfrac{10-8}{\sqrt{1 \cdot \left(\frac{1}{5} + \frac{1}{5}\right)}} \approx 3.162$

Test 3: $t_{\text{data}} = \dfrac{\bar{x}_B - \bar{x}_C}{\sqrt{\text{MSE} \cdot \left(\frac{1}{n_B} + \frac{1}{n_C}\right)}} = \dfrac{12-8}{\sqrt{1 \cdot \left(\frac{1}{5} + \frac{1}{5}\right)}} \approx 6.325$

$n_T = n_A + n_B + n_C = 5 + 5 + 5 = 15$.

$k = 3$.

$\text{df} = n_T - k = 15 - 3 = 12$.

From Excel:

t_{data}	p-value
–3.162	0.008190378
3.162	0.008190378
6.325	3.80321E-05

Test 1: Bonferroni-adjusted p-value $= 3 \cdot 0.008190378 \approx 0.0246$.

Test 2: Bonferroni-adjusted p-value $= 3 \cdot 0.008190378 \approx 0.0246$.

Test 3: Bonferroni-adjusted p-value $= 3 \cdot 0.0000380321 \approx 0.0001$.

Test 1: The adjusted p-value ≈ 0.0246, which is not ≤ 0.01. Therefore, we do not reject H_0. There is insufficient evidence at the $\alpha = 0.01$ level of significance that the population mean of Population A differs from the population mean of Population B.

Test 2: The adjusted p-value ≈ 0.0246, which is not ≤ 0.01. Therefore, we do not reject H_0. There is insufficient evidence at the $\alpha = 0.01$ level of significance that the population mean of Population A differs from the population mean of Population C.

Test 3: The adjusted p-value ≈ 0.0001, which is ≤ 0.01. Therefore, we reject H_0. There is evidence at the $\alpha = 0.01$ level of significance that the population mean of Population B differs from the population mean of Population C.

19. $q_{\text{data}} = \dfrac{\bar{x}_2 - \bar{x}_1}{\sqrt{\frac{\text{MSE}}{2} \cdot \left(\frac{1}{n_2} + \frac{1}{n_1}\right)}} = \dfrac{61-60}{\sqrt{\frac{100}{2} \cdot \left(\frac{1}{50} + \frac{1}{50}\right)}} \approx 0.707$

21. $q_{\text{data}} = \dfrac{\bar{x}_3 - \bar{x}_2}{\sqrt{\frac{\text{MSE}}{2} \cdot \left(\frac{1}{n_3} + \frac{1}{n_2}\right)}} = \dfrac{65-61}{\sqrt{\frac{100}{2} \cdot \left(\frac{1}{50} + \frac{1}{50}\right)}} \approx 2.828$

23. $q_{\text{data}} = \dfrac{\bar{x}_1 - \bar{x}_3}{\sqrt{\frac{\text{MSE}}{2} \cdot \left(\frac{1}{n_1} + \frac{1}{n_3}\right)}} = \dfrac{220-210}{\sqrt{\frac{430}{2} \cdot \left(\frac{1}{25} + \frac{1}{25}\right)}} \approx 2.357$

25. $n_T = n_1 + n_2 + n_3 = 50 + 50 + 50 = 150$. Calculating the degrees of freedom gives df $= n_T - k = 150 - 3 = 147$, which is not in Table H. Therefore, we use the conservative df $= 120$ to get $q_{\text{crit}} = 3.356$.

27. **(a)** Test 1: $H_0 : \mu_1 = \mu_2$ versus $H_a : \mu_1 \neq \mu_2$. Test 2: $H_0 : \mu_1 = \mu_3$ versus $H_a : \mu_1 \neq \mu_3$. Test 3: $H_0 : \mu_2 = \mu_3$ versus $H_a : \mu_2 \neq \mu_3$.

(b) $q_{\text{crit}} = 3.356$. Reject H_0 if $q_{\text{data}} \geq 3.356$.

(c) Test 1: $q_{\text{data}} \approx 0.707$. Test 2: $q_{\text{data}} \approx 3.536$. Test 3: $q_{\text{data}} \approx 2.828$.

(d) Test 1: $q_{\text{data}} \approx 0.707$, which is not ≥ 3.356. Therefore, we do not reject H_0. There is insufficient evidence at the $\alpha = 0.05$ level of significance that the population mean of Population 1 differs from the population mean of Population 2. Test 2: $q_{\text{data}} \approx 3.536$, which is ≥ 3.356. Therefore, we reject H_0. There is evidence at the $\alpha = 0.05$ level of significance that the population mean of Population 1 differs from the population mean of Population 3. Test 3: $q_{\text{data}} \approx 2.828$, which is not ≥ 3.356. Therefore, we do not reject H_0. There is insufficient evidence at the $\alpha = 0.05$ level of significance that the population mean of Population 2 differs from the population mean of Population 3.

29. Test 1: $H_0 : \mu_A = \mu_B$ versus $H_a : \mu_A \neq \mu_B$.

Test 2: $H_0 : \mu_A = \mu_C$ versus $H_a : \mu_A \neq \mu_C$.

Test 3: $H_0 : \mu_A = \mu_D$ versus $H_a : \mu_A \neq \mu_D$.

Test 4: $H_0 : \mu_B = \mu_C$ versus $H_a : \mu_B \neq \mu_C$.

Test 5: $H_0 : \mu_B = \mu_D$ versus $H_a : \mu_B \neq \mu_D$.

Test 6: $H_0 : \mu_C = \mu_D$ versus $H_a : \mu_C \neq \mu_D$.

$n_T = n_A + n_B + n_C + n_D = 100 + 150 + 200 + 150 = 600$.

$k = 4$.

Calculating the degrees of freedom gives df $= n_T - k = 600 - 4 = 596$, which is not in Table H. Therefore, we use the conservative df $= 120$ to get $q_{\text{crit}} = 3.685$. For each hypothesis test, reject H_0 if $q_{\text{data}} \geq 3.685$.

Test 1: $q_{\text{data}} = \dfrac{\bar{x}_B - \bar{x}_A}{\sqrt{\frac{\text{MSE}}{2} \cdot \left(\frac{1}{n_B} + \frac{1}{n_A}\right)}} \approx \dfrac{75-50}{\sqrt{\frac{26.4228}{2} \cdot \left(\frac{1}{150} + \frac{1}{100}\right)}} \approx 53.277$

Test 2: $q_{\text{data}} = \dfrac{\bar{x}_C - \bar{x}_A}{\sqrt{\frac{\text{MSE}}{2} \cdot \left(\frac{1}{n_C} + \frac{1}{n_A}\right)}} \approx \dfrac{100-50}{\sqrt{\frac{26.4228}{2} \cdot \left(\frac{1}{200} + \frac{1}{100}\right)}} \approx 112.318$

Test 3: $q_{\text{data}} = \dfrac{\bar{x}_D - \bar{x}_A}{\sqrt{\frac{\text{MSE}}{2} \cdot \left(\frac{1}{n_D} + \frac{1}{n_A}\right)}} \approx \dfrac{125-50}{\sqrt{\frac{26.4228}{2} \cdot \left(\frac{1}{150} + \frac{1}{100}\right)}} \approx 159.832$

Test 4: $q_{\text{data}} = \dfrac{\bar{x}_C - \bar{x}_B}{\sqrt{\frac{\text{MSE}}{2} \cdot \left(\frac{1}{n_C} + \frac{1}{n_B}\right)}} \approx \dfrac{100-75}{\sqrt{\frac{26.4228}{2} \cdot \left(\frac{1}{200} + \frac{1}{150}\right)}} \approx 63.678$

Test 5: $q_{\text{data}} = \dfrac{\bar{x}_D - \bar{x}_B}{\sqrt{\frac{\text{MSE}}{2} \cdot \left(\frac{1}{n_D} + \frac{1}{n_B}\right)}} \approx \dfrac{125-75}{\sqrt{\frac{26.4228}{2} \cdot \left(\frac{1}{150} + \frac{1}{150}\right)}} \approx 119.131$

Test 6: $q_{\text{data}} = \dfrac{\bar{x}_D - \bar{x}_C}{\sqrt{\frac{\text{MSE}}{2} \cdot \left(\frac{1}{n_D} + \frac{1}{n_C}\right)}} \approx \dfrac{125-100}{\sqrt{\frac{26.4228}{2} \cdot \left(\frac{1}{150} + \frac{1}{200}\right)}} \approx 63.678$

Test 1: $q_{\text{data}} \approx 53.277$, which is ≥ 3.685. Therefore, we reject H_0. There is evidence at the $\alpha = 0.05$ level of significance that the population mean of Population A differs from the population mean of Population B.

Test 2: $q_{\text{data}} \approx 112.318$, which is ≥ 3.685. Therefore, we reject H_0. There is evidence at the $\alpha =$ 0.05 level of significance that the population mean of Population A differs from the population mean of Population C.

Test 3: $q_{\text{data}} \approx 159.832$, which is ≥ 3.685. Therefore, we reject H_0. There is evidence at the $\alpha =$ 0.05 level of significance that the population mean of Population A differs from the population mean of Population D.

Test 4: $q_{\text{data}} \approx 63.678$ which is ≥ 3.685. Therefore, we reject H_0. There is evidence at the $\alpha = 0.05$ level of significance that the population mean of Population B differs from the population mean of Population C.

Test 5: $q_{\text{data}} \approx 119.131$, which is ≥ 3.685. Therefore, we reject H_0. There is evidence at the $\alpha =$ 0.05 level of significance that the population mean of Population B differs from the population mean of Population D.

Test 6: $q_{\text{data}} \approx 63.678$ which is ≥ 3.685. Therefore, we reject H_0. There is evidence at the $\alpha = 0.05$ level of significance that the population mean of Population C differs from the population mean of Population D.

31. Test 1: $H_0 : \mu_A = \mu_B$ versus $H_a : \mu_A \neq \mu_B$. 95% confidence interval for $\mu_B - \mu_A$ is (–5.642, 0.309), which does contain zero, so we do not reject $H_0 : \mu_A = \mu_B$ for level of significance $\alpha =$ 0.05.

Test 2: $H_0 : \mu_A = \mu_C$ versus $H_a : \mu_A \neq \mu_C$. 95% confidence interval for $\mu_C - \mu_A$ is (–7.642, –1.691), which does not contain zero, so we reject $H_0 : \mu_A = \mu_C$ for level of significance $\alpha = 0.05$.

Test 3: $H_0 : \mu_B = \mu_C$ versus $H_a : \mu_B \neq \mu_C$. 95% confidence interval for $\mu_C - \mu_B$ is (–4.976, 0.976), which does contain zero, so we do not reject $H_0 : \mu_B = \mu_C$ for level of significance $\alpha = 0.05$.

33. Since the conclusion of the ANOVA is do not reject H_0, it is not appropriate to perform multiple comparisons.

35. Test 1: $H_0 : \mu_{Sopranos} = \mu_{Altos}$ versus $H_a : \mu_{Sopranos} \neq \mu_{Altos}$.

Test 2: $H_0 : \mu_{Sopranos} = \mu_{Tenors}$ versus $H_a : \mu_{Sopranos} \neq \mu_{Tenors}$.

Test 3: $H_0 : \mu_{Sopranos} = \mu_{Basses}$ versus $H_a : \mu_{Sopranos} \neq \mu_{Basses}$.

Test 4: $H_0 : \mu_{Altos} = \mu_{Tenors}$ versus $H_a : \mu_{Altos} \neq \mu_{Tenors}$.

Test 5: $H_0 : \mu_{Altos} = \mu_{Basses}$ versus $H_a : \mu_{Altos} \neq \mu_{Basses}$.

Test 6: $H_0 : \mu_{Tenors} = \mu_{Basses}$ versus $H_a : \mu_{Tenors} \neq \mu_{Basses}$.

For each hypothesis test, reject H_0 if the Bonferroni-adjusted p-value $\leq \alpha = 0.01$.

Test 1: $$t_{\text{data}} = \frac{\bar{x}_{\text{Sopranos}} - \bar{x}_{\text{Altos}}}{\sqrt{\text{MSE} \cdot \left(\frac{1}{n_{\text{Sopranos}}} + \frac{1}{n_{\text{Altos}}}\right)}} \approx \frac{64.250 - 64.886}{\sqrt{6.32323831 \cdot \left(\frac{1}{36} + \frac{1}{35}\right)}} \approx -1.065$$

Test 2: $$t_{\text{data}} = \frac{\bar{x}_{\text{Sopranos}} - \bar{x}_{\text{Tenors}}}{\sqrt{\text{MSE} \cdot \left(\frac{1}{n_{\text{Sopranos}}} + \frac{1}{n_{\text{Tenors}}}\right)}} \approx \frac{64.250 - 69.150}{\sqrt{6.32323831 \cdot \left(\frac{1}{36} + \frac{1}{20}\right)}} \approx -6.987$$

Test 3: $$t_{\text{data}} = \frac{\bar{x}_{\text{Sopranos}} - \bar{x}_{\text{Basses}}}{\sqrt{\text{MSE} \cdot \left(\frac{1}{n_{\text{Sopranos}}} + \frac{1}{n_{\text{Basses}}}\right)}} \approx \frac{64.250 - 70.718}{\sqrt{6.32323831 \cdot \left(\frac{1}{36} + \frac{1}{39}\right)}} \approx -11.129$$

Test 4: $$t_{\text{data}} = \frac{\bar{x}_{\text{Altos}} - \bar{x}_{\text{Tenors}}}{\sqrt{\text{MSE} \cdot \left(\frac{1}{n_{\text{Altos}}} + \frac{1}{n_{\text{Tenors}}}\right)}} \approx \frac{64.886 - 69.150}{\sqrt{6.32323831 \cdot \left(\frac{1}{35} + \frac{1}{20}\right)}} \approx -6.049$$

Test 5: $$t_{\text{data}} = \frac{\bar{x}_{\text{Altos}} - \bar{x}_{\text{Basses}}}{\sqrt{\text{MSE} \cdot \left(\frac{1}{n_{\text{Altos}}} + \frac{1}{n_{\text{Basses}}}\right)}} \approx \frac{64.886 - 70.718}{\sqrt{6.32323831 \cdot \left(\frac{1}{35} + \frac{1}{39}\right)}} \approx -9.961$$

Test 6: $$t_{\text{data}} = \frac{\bar{x}_{\text{Tenors}} - \bar{x}_{\text{Basses}}}{\sqrt{\text{MSE} \cdot \left(\frac{1}{n_{\text{Tenors}}} + \frac{1}{n_{\text{Basses}}}\right)}} \approx \frac{69.150 - 70.718}{\sqrt{6.32323831 \cdot \left(\frac{1}{20} + \frac{1}{39}\right)}} \approx -2.267$$

$n_T = n_{Sopranos} + n_{Altos} + n_{Tenors} + n_{Basses} = 36 + 35 + 20 + 39 = 130.$

$k = 4$.

df $= n_T - k = 130 - 4 = 126$.

From Excel:

t_{data}	p-value
–1.065	0.288912544
–6.987	1.44519E-10
–11.129	1.85751E-20
–6.049	1.54215E-08
–9.961	1.36022E-17
–2.267	0.025094383

Test 1: Bonferroni-adjusted p-value $= 6 \cdot 0.288912544 \approx 1.7335$, which is greater than 1. Therefore, use Bonferroni-adjusted p-value = 1.

Test 2: Bonferroni-adjusted p-value $= 6 \cdot 1.44519 \times 10^{-10} \approx 0$.

Test 3: Bonferroni-adjusted p-value $= 6 \cdot 1.85751 \times 10^{-20} \approx 0$.

Test 4: Bonferroni-adjusted p-value $= 6 \cdot 1.54215 \times 10^{-8} \approx 0$.

Test 5: Bonferroni-adjusted p-value $= 6 \cdot 1.36022 \times 10^{-17} \approx 0$.

Test 6: Bonferroni-adjusted p-value $= 6 \cdot 0.025094383 \approx 0.1506$.

Test 1: The adjusted p-value = 1, which is not ≤ 0.01. Therefore, we do not reject H_0. There is insufficient evidence at the $\alpha = 0.01$ level of significance that the population mean height of sopranos differs from the population mean height of altos.

Test 2: The adjusted p-value ≈ 0, which is ≤ 0.01. Therefore, we reject H_0. There is evidence at the $\alpha = 0.01$ level of significance that the population mean height of sopranos differs from the population mean height of tenors.

Test 3: The adjusted p-value ≈ 0, which is ≤ 0.01. Therefore, we reject H_0. There is evidence at the $\alpha = 0.01$ level of significance that the population mean height of sopranos differs from the population mean height of basses.

Test 4: The adjusted p-value ≈ 0, which is ≤ 0.01. Therefore, we reject H_0. There is evidence at the $\alpha = 0.01$ level of significance that the population mean height of altos differs from the population mean height of tenors.

Test 5: The adjusted p-value ≈ 0, which is ≤ 0.01. Therefore, we reject H_0. There is evidence at the 0.01 level of significance that the population mean height of altos differs from the population mean height of basses.

Test 6: The adjusted p-value ≈ 0.1506, which is not ≤ 0.01. Therefore, we do not reject H_0. There is insufficient evidence at the $\alpha = 0.01$ level of significance that the population mean height of tenors differs from the population mean height of basses.

37. Since the conclusion of the ANOVA is do not reject H_0, it is not appropriate to perform multiple comparisons.

39.

One-way ANOVA: Nutritional Rating versus Manufacturer

Source	DF	SS	MS	F	P
Manufacturer	5	5717.7	1143.5	15.71	0.000
Error	62	4511.7	72.8		
Total	67	10229.4			

```
S = 8.531      R-Sq = 55.89%      R-Sq(adj) = 52.34%

                                  Individual 95% CIs for Mean Based on
                            Pooled StDev
Level  N    Mean     StDev  ----+---------+---------+---------+-----
G      22   34.486   8.947  (--*--)
K      17   41.761   8.102         (---*--)
N       6   67.969   5.509                      (-----*----)
P       9   41.706  10.048         (----*---)
Q       6   50.569  10.603               (-----*-----)
R       8   41.543   6.081      (----*----)
                            ----+---------+---------+---------+-----
```

```
                                  36      48      60      72
Pooled StDev = 8.531
Grouping Information Using Tukey Method
Manufacturer  N    Mean  Grouping
N             6    67.969    A
Q             6    50.569    B
K             17   41.761    B    C
P             9    41.706    B    C
R             8    41.543    B    C
G             22   34.486         C
Means that do not share a letter are significantly different.
Tukey 95% Simultaneous Confidence Intervals
All Pairwise Comparisons among Levels of Manufacturer
Individual confidence level = 99.54%
Manufacturer = G subtracted from:
Manufacturer  Lower    Center Upper  ------+---------+---------+---------+---
K             –0.828   7.275  15.378              (--*--)
N             21.926   33.483 45.040                    (---*----)
P             –2.709   7.220  17.149              (---*---)
Q             4.526    16.083 27.640                 (---*----)
R             –3.303   7.057  17.417              (---*---)
                                     ------+---------+---------+---------+---
                                         -25       0        25       50
Manufacturer = K subtracted from:
Manufacturer  Lower    Center   Upper  ------+---------+---------+---------+---
N             14.292   26.207   38.123                   (---*----)
P             -10.400  -0.056   10.289            (---*---)
Q             -3.108   8.808    20.723               (----*---)
R             -10.977  -0.218   10.540            (---*---)
                                       ------+---------+---------+---------+---
                                           -25       0        25       50
Manufacturer = N subtracted from:
Manufacturer  Lower    Center   Upper
```

```
P             -39.488      -26.263      -13.038
Q             -31.887      -17.400       -2.912
R             -39.977      -26.426      -12.874

Manufacturer  ------+---------+---------+---------+---
P             (----*-----)
Q                (-----*-----)
R             (----*-----)
              ------+---------+---------+---------+---
                   -25         0        25        50

Manufacturer = P subtracted from:

Manufacturer  Lower        Center       Upper   ------+---------+---------+---------+---
Q             -4.362       8.863        22.088              (-----*----)
R             -12.356      -0.163       12.030          (----*----)
                                                ------+---------+---------+---------+---
                                                     -25         0        25        50

Manufacturer = Q subtracted from:

Manufacturer  Lower        Center       Upper   ------+---------+---------+---------+---
R             -22.578      -9.026       4.526           (----*-----)
                                                ------+---------+---------+---------+---
                                                     –25         0        25        50
```

Test 1: $H_0 : \mu_{General\ Mills} = \mu_{Kellogg's}$ versus $H_a : \mu_{General\ Mills} \neq \mu_{Kellogg's}$. 95% confidence interval for $\mu_{Kellogg's} - \mu_{General\ Mills}$ is (–0.828, 16.378), which does contain zero, so we do not reject H_0 : $\mu_{General\ Mills} = \mu_{Kellogg's}$ for level of significance $\alpha = 0.05$.

Test 2: $H_0 : \mu_{General\ Mills} = \mu_{Nabisco}$ versus $H_a : \mu_{General\ Mills} \neq \mu_{Nabisco}$. 95% confidence interval for $\mu_{Nabisco} - \mu_{General\ Mills}$ is (21.926, 45.040), which does not contain zero, so we reject $H_0 : \mu_{General\ Mills} = \mu_{Nabisco}$ for level of significance $\alpha = 0.05$.

Test 3: $H_0 : \mu_{General\ Mills} = \mu_{Post}$ versus $H_a : \mu_{General\ Mills} \neq \mu_{Post}$. 95% confidence interval for $\mu_{Post} - \mu_{General\ Mills}$ is (–2.709, 17.149), which does contain zero, so we do not reject $H_0 : \mu_{General\ Mills} = \mu_{Post}$ for level of significance $\alpha = 0.05$.

Test 4: $H_0 : \mu_{General\ Mills} = \mu_{Quaker\ Oats}$ versus $H_a : \mu_{General\ Mills} \neq \mu_{Quaker\ Oats}$. 95% confidence interval for $\mu_{Quaker\ Oats} - \mu_{General\ Mills}$ is (4.526, 27.640), which does not contain zero, so we reject H_0 : $\mu_{General\ Mills} = \mu_{Quaker\ Oats}$ for level of significance $\alpha = 0.05$.

Test 5: $H_0 : \mu_{General\ Mills} = \mu_{Ralston\ Purina}$ versus $H_a : \mu_{General\ Mills} \neq \mu_{Ralston\ Purina}$. 95% confidence interval for $\mu_{Ralston\ Purina} - \mu_{General\ Mills}$ is (–3.303, 17.417), which does contain zero, so we do not reject $H_0 : \mu_{General\ Mills} = \mu_{Ralston\ Purina}$ for level of significance $\alpha = 0.05$.

Test 6: $H_0 : \mu_{Kellogg's} = \mu_{Nabisco}$ versus $H_a : \mu_{Kellogg's} \neq \mu_{Nabisco}$. 95% confidence interval for $\mu_{Nabisco} -$

$\mu_{Kellogg's}$ is (14.292, 38.123), which does not contain zero, so we reject $H_0 : \mu_{Kellogg's} = \mu_{Nabisco}$ for level of significance $\alpha = 0.05$.

Test 7: $H_0 : \mu_{Kellogg's} = \mu_{Post}$ versus $H_a : \mu_{Kellogg's} \neq \mu_{Post}$. 95% confidence interval for $\mu_{Post} - \mu_{Kellogg's}$ is (–10.400, 10.289), which does contain zero, so we do not reject $H_0 : \mu_{Kellogg's} = \mu_{Post}$ for level of significance $\alpha = 0.05$.

Test 8: $H_0 : \mu_{Kellogg's} = \mu_{Quaker\ Oats}$ versus $H_a : \mu_{Kellogg's} \neq \mu_{Quaker\ Oats}$. 95% confidence interval for $\mu_{Quaker\ Oats} - \mu_{Kellogg's}$ is (–3.108, 20.723), which does contain zero, so we do not reject $H_0 : \mu_{Kellogg's} = \mu_{Quaker\ Oats}$ for level of significance $\alpha = 0.05$.

Test 9: $H_0 : \mu_{Kellogg's} = \mu_{Ralston\ Purina}$ versus $H_a : \mu_{Kellogg's} \neq \mu_{Ralston\ Purina}$. 95% confidence interval for $\mu_{Ralston\ Purina} \neq \mu_{Kellogg's}$ is (–10.977, 10.540), which does contain zero, so we do not reject $H_0 : \mu_{Kellogg's} = \mu_{Ralston\ Purina}$ for level of significance $\alpha = 0.05$.

Test 10: $H_0 : \mu_{Nabisco} = \mu_{Post}$ versus $H_a : \mu_{Nabisco} \neq \mu_{Post}$. 95% confidence interval for $\mu_{Post} - \mu_{Nabisco}$ is (–39.488, –13.038), which does not contain zero, so we reject $H_0 : \mu_{Nabisco} = \mu_{Post}$ for level of significance $\alpha = 0.05$.

Test 11: $H_0 : \mu_{Nabisco} = \mu_{Quaker\ Oats}$ versus $H_a : \mu_{Nabisco} \neq \mu_{Quaker\ Oats}$. 95% confidence interval for $\mu_{Quaker\ Oats} - \mu_{Nabisco}$ is (–31.887, –2.912), which does not contain zero, so we reject $H_0 : \mu_{Nabisco} = \mu_{Quaker\ Oats}$ for level of significance $\alpha = 0.05$.

Test 12: $H_0 : \mu_{Nabisco} = \mu_{Ralston\ Purina}$ versus $H_a : \mu_{Nabisco} \neq \mu_{Ralston\ Purina}$. 95% confidence interval for $\mu_{Ralston\ Purina} - \mu_{Nabisco}$ is (–39.977, –12.874), which does not contain zero, so we reject $H_0 : \mu_{Nabisco} = \mu_{Ralston\ Purina}$ for level of significance $\alpha = 0.05$.

Test 13: $H_0 : \mu_{Post} = \mu_{Quaker\ Oats}$ versus $H_a : \mu_{Post} \neq \mu_{Quaker\ Oats}$. 95% confidence interval for $\mu_{Quaker\ Oats} - \mu_{Post}$ is (–4.362, 22.088), which does contain zero, so we do not reject $H_0 : \mu_{Post} = \mu_{Quaker\ Oats}$ for level of significance $\alpha = 0.05$.

Test 14: $H_0 : \mu_{Post} = \mu_{Ralston\ Purina}$ versus $H_a : \mu_{Post} \neq \mu_{Ralston\ Purina}$. 95% confidence interval for $\mu_{Ralston\ Purina} - \mu_{Post}$ is (–12.356, 12.030), which does contain zero, so we do not reject $H_0 : \mu_{Post} = \mu_{Ralston\ Purina}$ for level of significance $\alpha = 0.05$.

Test 15: $H_0 : \mu_{Post} = \mu_{Ralston\ Purina}$ versus $H_a : \mu_{Post} \neq \mu_{Ralston\ Purina}$. 95% confidence interval for $\mu_{Ralston\ Purina} - \mu_{Post}$ is (–22.578, 4.526), which does contain zero, so we do not reject $H_0 : \mu_{Post} = \mu_{Ralston\ Purina}$ for level of significance $\alpha = 0.05$.

Section 12.3

1. The variable of interest is the variable we are interested in studying and the blocking factor is a variable that is not of primary interest to the researcher, but is included in the ANOVA in order to improve the ability of the ANOVA to find significant differences among the treatment means. Treatment. Nuisance factor.

3. Our F statistic from the ANOVA table is $F = \frac{\text{MSTR}}{\text{MSE}} = \frac{\text{MSTR}}{\text{SSE}/\text{df}_2}$. We reject H_0 when the p-value is small. Also note that:

- The p-value is small only when F is large.
- F is large when SSE is small.

The smaller SSE is, the larger F is; therefore, the smaller the p-value is.

5. **(a)** $H_0 : \mu_1 = \mu_2 = \mu_3 = \mu_4$ versus H_a : Not all of the population means are equal. Reject H_0 if p-value ≤ 0.05.

(b) The p-value ≈ 0.010, which is ≤ 0.05. Therefore, we reject H_0. There is evidence at level of significance $\alpha = 0.05$ that the population means are not all equal.

7. **(a)** $H_0 : \mu_1 = \mu_2 = \mu_3 = \mu_4$ versus H_a : Not all of the population means are equal. Reject H_0 if p-value ≤ 0.05.

(b) The p-value ≈ 0.009, which is ≤ 0.05. Therefore, we reject H_0. There is evidence at level of significance $\alpha = 0.05$ that the population means are not all equal.

9. **(a)** $H_0 : \mu_1 = \mu_2 = \mu_3$ versus H_a : Not all of the population means are equal. Reject H_0 if p-value ≤ 0.05.

(b) SSE = (Degrees of freedom for error)(MSE) = (7.5)(8) = 60. SST = SSTR + SSB + SSE = 90 + 50 + 60 = 200. The degrees of freedom for treatments is $\frac{\text{SSTR}}{\text{MSTR}} = \frac{90}{45} = 2$. The degrees of freedom for blocks is 15 – 2 – 8 = 5. $\text{MSB} = \frac{\text{SSB}}{\text{Degrees of freedom for blocks}} = \frac{50}{5} = 10.$

$F = \frac{\text{MSTR}}{\text{MSE}} = \frac{45}{7.5} = 6.$

The complete ANOVA table is given below.

Answers to missing values are in bold.

Source	**Sum of squares**	**Degrees of freedom**	**MeansSquare**	***F***
Treatments	90	**2**	45	**6**
Blocks	50	**5**	**10**	
Error	**60**	8	7.5	
Total	**200**	15		

(c) The p-value = $P(F > F_{\text{data}}) = P(F > 6) \approx 0.0256$, which is ≤ 0.05. Therefore, we reject H_0. There is evidence at level of significance $\alpha = 0.05$ that the population means are not all equal.

11. **(a)** To get SSE for the one-way ANOVA, we add SSB and SSE from the RBD. Thus, SSE = 50 + 60 = 110. To get the degrees of freedom for error in the one-way ANOVA we add the degrees of freedom for blocks and error in the RBD. Thus, the degrees of freedom for error in the one-way ANOVA is 5 + 8 = 13.

$\text{MSE} = \frac{\text{SSE}}{\text{Degrees of freedom for error}} = \frac{110}{13} \approx 8.4615.$

$F = \frac{\text{MSTR}}{\text{MSE}} \approx \frac{45}{8.4615} \approx 5.3182.$

The one-way ANOVA table is given below.

Source	**Sum of squares**	**Degrees of freedom**	**Mean square**	***F***
Treatments	90	2	45	5.3182
Error	110	13	8.4615	
Total	200	15		

(b) $H_0 : \mu_1 = \mu_2 = \mu_3$ versus H_a : Not all of the population means are equal. Reject H_0 if p-value $\leq$

0.05. The p-value = $P(F > F_{data}) = P(F > 5.3182) \approx 0.0205$, which is ≤ 0.05. Therefore, we reject H_0. There is evidence at level of significance $\alpha = 0.05$ that the population means are not all equal.

(c) The conclusions are the same for both tests. No.

13. $b - 1 + (k - 1)(b - 1) = b - 1 + k \cdot b - k - b + 1 = k \cdot b - k = n_T - k$

15. Analysis of Variance

Source	DF	Adj SS	Adj MS	F-Value	P-Value
Year	2	519.6	259.8	0.96	0.423
Country	4	24201.1	6050.3	22.35	0.000
Error	8	2165.7	270.7		
Total	14	26886.4			

H_0: $\mu_{1995} = \mu_{2007} = \mu_{2011}$ versus H_a: Not all the population means are equal.

Reject H_0 if the p-value $\leq \alpha = 0.10$.

$F_{data} = 0.96$

p-value = 0.423

The p-value = 0.423 is not $\leq \alpha = 0.10$. Therefore we do not reject H_0. There is not enough evidence to conclude at level of significance $\alpha = 0.10$ that not all of the population mean math scores are the same for all 3 years.

17. **(a)** Analysis of Variance

Source	DF	Adj SS	Adj MS	F-Value	P-Value
Year	2	0.08448	0.04224	0.11	0.892
Error	15	5.51523	0.36768		
Total	17	5.59971			

H_0: $\mu_{2012} = \mu_{2013} = \mu_{2014}$ versus H_a: Not all the population means are equal.

Reject H_0 if the p-value $\leq \alpha = 0.05$.

$F_{data} = 0.11$

p-value = 0.892

The p-value = 0.892 is not $\leq \alpha = 0.05$. Therefore we do not reject H_0. There is not enough evidence to conclude at level of significance $\alpha = 0.05$ that not all of the population mean prices of the vegetables are the same for all 3 years.

(b) Yes.

19. **(a)** Analysis of Variance

Source	DF	Adj SS	Adj MS	F-Value	P-Value
Year	2	3137	1568	0.05	0.948
Error	15	442861	29524		
Total	17	445998			

H_0: $\mu_{2010} = \mu_{2011} = \mu_{2012}$ versus H_a: Not all the population means are equal.

Reject H_0 if the p-value $\leq \alpha = 0.10$.

$F_{\text{data}} = 0.05$

p-value = 0.948

The p-value = 0.948 is not $\leq \alpha = 0.10$. Therefore we do not reject H_0. There is not enough evidence to conclude at level of significance $\alpha = 0.10$ that not all of the population mean number of children not covered by health insurance are the same for all 3 years.

(b) No.

Section 12.4

1. We are interested in testing for the significance of both factors and we need to test for interaction of the two factors.

3. Factor interaction

5. It is a graphical representation of the cell means for each cell in the contingency table. It is used to investigate the presence of interaction between the two factors.

7. Test for interaction between the factors, test for Factor A effect, and test for Factor B effect.

9. The lines intersect and are not parallel at all. Therefore, there is significant interaction between the factors.

11. The lines are nearly parallel. Therefore, there is some interaction between the factors.

13.

The lines are parallel, so there is no interaction between Factor A and Factor B.

15.

The lines are nearly parallel, so there is no interaction between Factors A and B.

17.

Source	DF	SS	MS	F	P
Factor A	2	2.00	1.000	8.00	0.020
Factor B	1	3.00	3.000	24.00	0.003
Interaction	2	0.00	0.000	0.00	1.000
Error	6	0.75	0.125		
Total	11	5.75			

(a) H_0 : There is no interaction between carrier (Factor A) and type (Factor B). H_a : There is interaction between carrier (Factor A) and type (Factor B). Reject H_0 if the p-value ≤ 0.05. The p-value = 1.00, which is not ≤ 0.05. Therefore, we do not reject H_0. There is insufficient evidence of interaction between carrier (Factor A) and type (Factor B) at level of significance α = 0.05. This result agrees with the interaction plot in Exercise 13.

(b) H_0 : There is no carrier (Factor A) effect. That is, the population means do not differ by carrier. H_a : There is a carrier (Factor A) effect. That is, the population means do differ by carrier. Reject H_0 if the p-value ≤ 0.05. The p-value ≈ 0.020, which is ≤ 0.05. Therefore, we reject H_0. There is evidence for a carrier (Factor A). In other words, there is evidence at level of significance $\alpha = 0.05$ that the population means do differ by carrier.

(c) H_0 : There is no type (Factor B) effect. That is, the population means do not differ by type. H_a : There is a type (Factor B) effect. That is, the population means do differ by type. Reject H_0 if the p-value ≤ 0.05. The p-value ≈ 0.003, which is ≤ 0.05. Therefore, we reject H_0. There is evidence for a type (Factor B) effect. In other words, there is evidence at level of significance α = 0.05 that the population means do differ by type.

19.

Source	DF	SS	MS	F	P
Factor A	2	228.167	114.083	114.08	0.000
Factor B	1	48.000	48.000	48.00	0.000
Interaction	2	3.500	1.750	1.75	0.252
Error	6	6.000	1.000		
Total	11	285.667			

(a) H_0 : There is no interaction between carrier (Factor A) and type (Factor B). H_a : There is interaction between carrier (Factor A) and type (Factor B). Reject H_0 if the p-value ≤ 0.05. The p-value ≈ 0.252, which is not ≤ 0.05. Therefore, we do not reject H_0. There is insufficient evidence of interaction between carrier (Factor A) and type (Factor B) at level of significance α = 0.05. This result agrees with the interaction plot in Exercise 15.

(b) H_0 : There is no carrier (Factor A) effect. That is, the population means do not differ by carrier. H_a : There is a carrier (Factor A) effect. That is, the population means do differ by carrier. Reject H_0 if the p-value ≤ 0.05. The p-value ≈ 0, which is ≤ 0.05. Therefore, we reject H_0. There is evidence for a carrier (Factor A) effect. In other words, there is evidence at level of significance $\alpha = 0.05$ that the population means do differ by carrier.

(c) H_0 : There is no type (Factor B) effect. That is, the population means do not differ by type. H_a: There is a type (Factor B) effect. That is, the population means do differ by type. Reject H_0 if the p-value ≤ 0.05. The p-value ≈ 0, which is ≤ 0.05. Therefore, we reject H_0. There is evidence for a type (Factor B) effect. In other words, there is evidence at level of significance $\alpha = 0.05$ that the population means do differ by type.

21.

Source	DF	SS	MS	F	P
Hypertension	1	8.6528	8.6528	17.76	0.014
Gender	1	2.1218	2.1218	4.35	0.105
Interaction	1	0.3528	0.3528	0.72	0.443
Error	4	1.9492	0.4873		
Total	7	13.0766			

(a) H_0 : There is no interaction between hypertension (Factor A) and gender (Factor B). H_a : There is interaction between hypertension (Factor A) and gender (Factor B). Reject H_0 if the p-value ≤ 0.05. The p-value ≈ 0.443, which is not ≤ 0.05. Therefore, we do not reject H_0. There is insufficient evidence of interaction between hypertension (Factor A) and gender (Factor B) at level of significance $\alpha = 0.05$.

(b) H_0 : There is no hypertension (Factor A) effect. That is, the population means do not differ by whether or not the person has hypertension. H_a : There is a hypertension (Factor A) effect. That is, the population means do not differ by whether or not the person has hypertension. Reject H_0 if the p-value ≤ 0.05. The p-value ≈ 0.014, which is ≤ 0.05. Therefore, we reject H_0. There is evidence for a hypertension (Factor A) effect. Thus, we can conclude at level of significance $\alpha = 0.05$ that there is a significant difference in mean cardiac output between patients who have hypertension and patients who don't.

(c) H_0 : There is no gender (Factor B) effect. That is, the population means do not differ by gender. H_a : There is a gender (Factor B) effect, that is, the population means do differ by gender. Reject H_0 if the p-value ≤ 0.05. The p-value ≈ 0.105, which is not ≤ 0.05. Therefore, we do not reject H_0. There is insufficient evidence for a gender (Factor B) effect.

23.

Source	DF	SS	MS	F	P
Brand	1	387.81	387.81	8.31	0.045
Fiber	1	1267.56	1267.56	27.15	0.006
Interaction	1	1.05	1.05	0.02	0.888
Error	4	186.73	46.68		
Total	7	1843.15			

H_0 : There is no interaction between brand (Factor A) and fiber (Factor B). H_a: There is interaction between brand (Factor A) and fiber (Factor B). Reject H_0 if the p-value ≤ 0.05. The p-value ≈ 0.888, which is not ≤ 0.05. Therefore, we do not reject H_0. There is insufficient evidence of interaction between brand (Factor A) and fiber (Factor B) at level of significance $\alpha = 0.05$.

H_0: There is no brand (Factor A) effect. That is, the population means do not differ by brand. H_a: There is a brand (Factor A) effect. That is, the population means do differ by brand. Reject H_0 if the p-value ≤ 0.05. The p-value ≈ 0.045, which is ≤ 0.05. Therefore, we reject H_0. There is evidence for a brand (Factor A) effect. Thus, we can conclude at level of significance $\alpha = 0.05$ that there is a significant difference in mean nutritional ratings between Kellogg's cereals and General Mills cereals. H_0: There is no fiber (Factor B) effect. That is, the population means do not differ by whether or not the cereal has fiber. H_a: There is a fiber (Factor B) effect. That is, the population means do differ by whether or not the cereal has fiber. Reject H_0 if the p-value $\leq$ 0.05. The p-value ≈ 0.006, which is ≤ 0.05. Therefore, we reject H_0. There is evidence for a fiber (Factor B) effect. Thus, we can conclude at level of significance $\alpha = 0.05$ that there is a significant difference in mean nutritional ratings between cereals with a high fiber content and cereals with a low fiber content.

Chapter 12 Review Exercises

1. **(a)** $\text{df}_1 = k - 1 = 4 - 1 = 3$, $n_t = n_A + n_B + n_C + n_D = 50 + 100 + 50 + 100 = 300$, $\text{df}_2 = n_t - k = 300 - 4 = 296$.

(b) $\bar{\bar{x}} = \frac{50(0)+100(10)+50(20)+100(10)}{300} = \frac{3000}{300} = 10.$

(c) $\text{SSTR} = \sum n_i(\bar{x}_i - \bar{\bar{x}})^2 = 50(0-10)^2 + 100(10-10)^2 + 50(20-10)^2 + 100(10-10)^2 = 10{,}000$

(d) $\text{SSE} = \sum(n_i - 1)s_i^2 = (50 - 1)1.5^2 + (100 - 1)2.25^2 + (50 - 1)1.75^2 + (100 - 1)2.0^2 = 1157.5$

(e) SST = SSTR + SSE = 10,000 + 1157.5 = 11,157.5.

(f) $\text{MSTR} = \frac{\text{SSTR}}{\text{df}_1} = \frac{10{,}000}{3} = 3333.333333.$

(g) $\text{MSE} = \frac{\text{SSE}}{\text{df}_2} = \frac{1157.5}{296} = 3.910472973.$

(h) $F_{\text{data}} = \frac{\text{MSTR}}{\text{MSE}} = \frac{3{,}333.333333}{3.910472973} = 852.4117985 \approx 852.4118$

3. $H_0: \mu_1 = \mu_2 = \mu_3$

H_a: Not all of the population means are equal.

where μ_1 = population mean level of satisfaction for medical treatment 1

μ_2 = population mean level of satisfaction for medical treatment 2

μ_3 = population mean level of satisfaction for medical treatment 3

$\text{df}_1 = 2$, $\text{df}_2 = 18$, $\alpha = 0.05$, $F_{crit} = 3.35$

Reject H_0 if $F_{data} \geq 3.35$.

From Minitab:

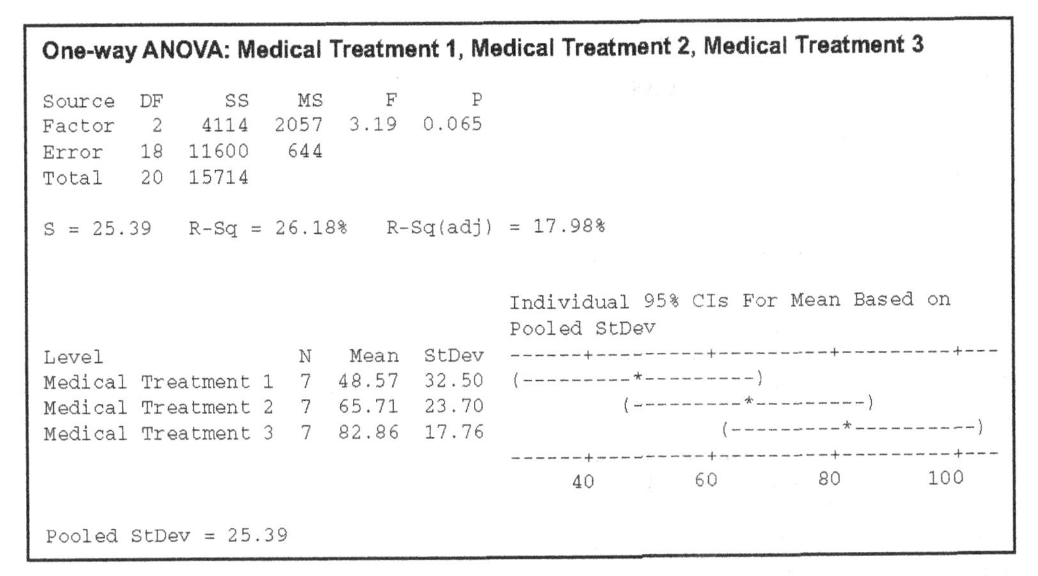

One-way ANOVA: Medical Treatment 1, Medical Treatment 2, Medical Treatment 3

```
Source  DF     SS    MS     F      P
Factor   2   4114  2057  3.19  0.065
Error   18  11600   644
Total   20  15714

S = 25.39   R-Sq = 26.18%   R-Sq(adj) = 17.98%

                                       Individual 95% CIs For Mean Based on
                                       Pooled StDev
Level                N   Mean  StDev  ------+---------+---------+---------+---
Medical Treatment 1  7  48.57  32.50  (---------*---------)
Medical Treatment 2  7  65.71  23.70            (---------*---------)
Medical Treatment 3  7  82.86  17.76                     (---------*---------)
                                      ------+---------+---------+---------+---
                                           40        60        80       100

Pooled StDev = 25.39
```

From the Minitab output, $F_{data} = 3.19$. Since F_{data} is not ≥ 3.55, we do not reject H_0.

There is insufficient evidence that not all of the population means are equal.

Source	DF	SS	MS	F	P
Factor	2	4114	2057	3.19	0.065
Error	18	11600	644		
Total	20	15714			

Or we can calculate F_{data} using the formulas.

$\bar{x}_1 = 48.5714$, $s_1 = 32.4954$, $n_1 = 7$

$\bar{x}_2 = 65.7143$, $s_2 = 23.7045$, $n_2 = 7$

$\bar{x}_3 = 82.8571$, $s_3 = 17.7616$, $n_3 = 7$

(i) $\text{df}_1 = k - 1 = 3 - 1 = 2$, $n_t = 7 + 7 + 7 = 21$

$\text{df}_2 = n_t - k = 21 - 3 = 18$

(ii) $\bar{\bar{x}} = \frac{48.5714+65.7143+82.8571}{3} = \frac{197.1428}{3} \approx 65.7142667$

(iii) SSTR $= \sum n_i(\bar{x}_i - \bar{\bar{x}})^2 = 7(48.5714 - 65.7142667)^2 + 7(65.7143 - 65.7142667)^2$

$+7(82.8571 - 65.7142667)^2 = 4114.282286$

(iv) $\text{SSE} = \sum(n_i - 1)s_i^2 = (7 - 1)32.4954^2 + (7 - 1)23.7045^2 + (7 - 1)17.7616^2 =$
$11{,}599.97266$

(v) $\text{SST} = \text{SSTR} + \text{SSE} = 4114.282286 + 11{,}599.97266 = 15{,}714.25494$

(vi) MSTR $= \frac{\text{SSTR}}{\text{df}_1} = \frac{4114.282286}{2} = 2057.141143$

(vii) $\text{MSE} = \frac{\text{SSE}}{\text{df}_2} = \frac{11{,}599.97266}{18} = 644.4429256$

(viii) $F_{data} = \frac{\text{MSTR}}{\text{MSE}} = \frac{2057.141143}{644.4429256} \approx 3.1921$

Source	**df**	**SS**	**MS**	***F***	***P***
Factor	2	4114	2057	3.19	0.065
Error	18	11600	644		
Total	20	15714			

5. $t_{\text{data}} = \dfrac{\bar{x}_1 - \bar{x}_2}{\sqrt{\text{MSE} \cdot \left(\frac{1}{n_1} + \frac{1}{n_2}\right)}} = \dfrac{50 - 75}{\sqrt{1250 \cdot \left(\frac{1}{25} + \frac{1}{25}\right)}} = -2.5$

7. $t_{\text{data}} = \dfrac{\bar{x}_2 - \bar{x}_3}{\sqrt{\text{MSE} \cdot \left(\frac{1}{n_2} + \frac{1}{n_3}\right)}} = \dfrac{75 - 65}{\sqrt{1250 \cdot \left(\frac{1}{25} + \frac{1}{25}\right)}} = 1$

9. $q_{\text{data}} = \dfrac{\bar{x}_2 - \bar{x}_1}{\sqrt{\frac{\text{MSE}}{2} \cdot \left(\frac{1}{n_2} + \frac{1}{n_1}\right)}} = \dfrac{224 - 200}{\sqrt{\frac{14{,}400}{2} \cdot \left(\frac{1}{100} + \frac{1}{100}\right)}} = 2$

11. $q_{\text{data}} = \dfrac{\bar{x}_3 - \bar{x}_2}{\sqrt{\frac{\text{MSE}}{2} \cdot \left(\frac{1}{n_3} + \frac{1}{n_2}\right)}} = \dfrac{248 - 224}{\sqrt{\frac{14{,}400}{2} \cdot \left(\frac{1}{100} + \frac{1}{100}\right)}} = 2$

13. **(a)** Test 1: $H_0 : \mu_1 = \mu_2$ versus $H_a : \mu_1 \neq \mu_2$. Test 2: $H_0 : \mu_1 = \mu_3$ versus $H_a : \mu_1 \neq \mu_3$. Test 3: $H_0 : \mu_2 = \mu_3$ versus $H_a : \mu_2 \neq \mu_3$.

(b) $n_T = n_1 + n_2 + n_3 = 100 + 100 + 100 = 300$.

$k = 3$.

Calculate degrees of freedom df $= n_T - k = 300 - 3 = 297$, which is not in Table H. Therefore, we use the conservative df $= 120$ to get $q_{\text{crit}} = 3.356$. Reject H_0 if $q_{\text{data}} \geq 3.356$.

(c) Test 1: $q_{\text{data}} = 2$. Test 2: $q_{\text{data}} = 4$. Test 3: $q_{\text{data}} = 2$.

(d) Test 1: $q_{\text{data}} = 2$, which is not ≥ 3.356. Therefore, we do not reject H_0. There is insufficient evidence at the $\alpha = 0.05$ level of significance that the population mean of Population 1 differs from the population mean of Population 2. Test 2: $q_{\text{data}} = 4$, which is ≥ 3.356. Therefore, we reject H_0. There is evidence at the $\alpha = 0.05$ level of significance that the population mean of Population 1 differs from the population mean of Population 3. Test 3: $q_{\text{data}} = 2$, which is not $\geq$ 3.356. Therefore, we do not reject H_0. There is insufficient evidence at the $\alpha = 0.05$ level of significance that the population mean of Population 2 differs from the population mean of Population 3.

15. $H_0 : \mu_{women} = \mu_{men}$ versus H_a : Not all of the population means are equal. Reject H_0 if p-value ≤ 0.05.

Source	DF	SS	MS	F	P
Gender of owner	1	0.0	0.0000	0.00	1.000
Industry	3	125.5	41.8333	1.74	0.330
Error	3	72.0	24.0000		
Total	7	197.5			

The p-value = 1.000, which is not ≤ 0.05. Therefore, we do not reject H_0. There is insufficient evidence at level of significance $\alpha = 0.05$ that the population mean number of small businesses owned by women differs from the population mean number of small businesses owned by men.

The lines are not parallel, so there is some interaction between Factor A and Factor B.

17.

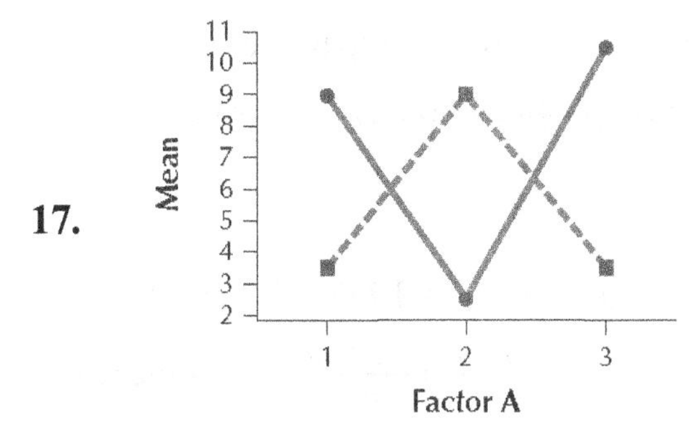

There is significant interaction between Factor A and Factor B.

19.

Source	DF	SS	MS	F	P
Factor A	2	3.167	1.5833	0.47	0.643
Factor B	1	12.000	12.0000	3.60	0.107
Interaction	2	109.500	54.7500	16.42	0.004
Error	6	20.000	3.3333		
Total	11	144.667			

(a) H_0: There is no interaction between carrier (Factor A) and type (Factor B). H_a: There is interaction between carrier (Factor A) and type (Factor B). Reject H_0 if the p-value is ≤ 0.05. The p-value ≈ 0.004, which is ≤ 0.05. Therefore, we reject H_0. There is evidence of interaction between (carrier) Factor A and type (Factor B) at level of significance $\alpha = 0.05$. This result agrees with the interaction plot in Exercise 17.

(b) Not appropriate.

(c) Not appropriate.

Chapter 12 Quiz

1. False.

3. False. If we reject the null hypothesis in an ANOVA, we conclude that not all of the population means are different. This means that at least one of the population means is different from the rest.

5. Mean square treatment.

7. $\bar{\bar{x}}$.

9. F_{data}

11. $H_0: \mu_{\text{Married}} = \mu_{\text{Widowed}} = \mu_{\text{Divorced}} = \mu_{\text{Separated}} = \mu_{\text{Nevermarried}}$

H_a: Not all of the population means are equal.

μ_{Married} = the population mean number of hours worked by people who are married

μ_{Widowed} = the population mean number of hours worked by people who are widowed

μ_{Divorced} = the population mean number of hours worked by people who are divorced

$\mu_{\text{Separated}}$ = the population mean number of hours worked by people who are separated

$\mu_{\text{Nevermarried}}$ = the population mean number of hours worked by people who have never been married

Reject H_0 if p-value ≤ 0.05.

$n_t = 964 + 72 + 342 + 79 + 478 = 1935$, $df_1 = k - 1 = 5 - 1 = 4$, $df_2 = n_t - k = 1935 - 5 = 1930$.

$$\bar{\bar{x}} = \frac{n_1\bar{x}_1 + n_2\bar{x}_2 + n_3\bar{x}_3 + n_4\bar{x}_4 + n_5\bar{x}_5}{n_t} = \frac{964(42.76)+72(40.13)+342(43.69)+79(41.66)+478(41.03)}{1935} = \frac{81{,}955.46}{1935} = 42.35424289.$$

$$\text{SSTR} = \sum n_i(\bar{x}_i - \bar{\bar{x}})^2 = 964(42.76 - 42.35424289)^2 + 72(40.13 - 42.35424289)^2 + 342(43.69 - 42.35424289)^2 + 79(41.66 - 42.35424289)^2 + 478(41.03 - 42.35424289)^2 = 2001.432666$$

$$\text{SSE} = \sum(n_i - 1)s_i^2 = (964 - 1)\,14.08^2 + (72 - 1)14.28^2 + (342 - 1)13.93^2 + (79 - 1)15.71^2 + (478 - 1)14.03^2 = 384{,}702.6296$$

SST = SSTR + SSE = 2001.432666 + 384,702.6296 = 386,704.0623

$$\text{MSTR} = \frac{\text{SSTR}}{df_1} = \frac{2001.432666}{4} = 500.3581665$$

$$\text{MSE} = \frac{\text{SSE}}{df_2} = \frac{384{,}702.6296}{1930} = 199.3277874$$

$$F_{\text{data}} = \frac{\text{MSTR}}{\text{MSE}} = \frac{500.3581665}{199.3277874} = 2.510227867$$

Source of variation	**Sum of squares**	**Degrees of freedom**	**Mean square**	F
Treatment	2,001.4326666	4	500.3581665	2.510227867
Error	384,702.6296	1930	199.3277874	
Total	386,704.0623			

p-value = $P(F > 2.5102) = 0.040092846$.

Since the p-value ≤ 0.05, we reject H_0. There is evidence that not all of the population mean numbers of hours worked are equal.

Chapter 13: Inference in Regression

Note to instructors and students: Some answers may differ by 1 or 2 in the last digit or two depending on whether you round for intermediate steps or wait until you get the final answer to round. Also, different software and different forms of technology may give slightly different answers.

Section 13.1

1. The regression equation is calculated from a sample and is only valid for values of x in the range of the sample data. The population regression equation may be used to approximate the relationship between the predictor variable x and the response variable y for the entire population of (x, y) pairs.

3. We construct a scatterplot of the residuals against the fitted values and a normal probability plot of the residuals. We must make sure that the scatter plot contains no strong evidence of any unhealthy patterns and that the normal probability plot indicates no evidence of departures from normality in residuals.

5. It means that there is no relationship between x and y.

7. **(a) and (b)**

x	y	**Predicted value** $\hat{y} = 2.5x + 13.5$	**Residual** $(y - \hat{y})$
1	15	16	-1
2	20	18.5	1.5
3	20	21	-1
4	25	23.5	1.5
5	25	26	-1

(c)

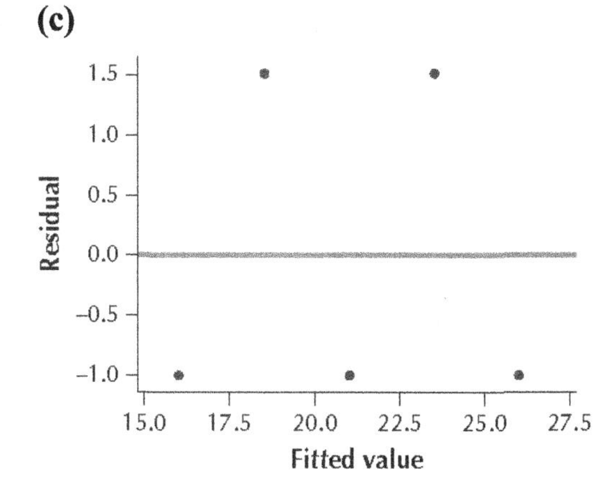

(d)

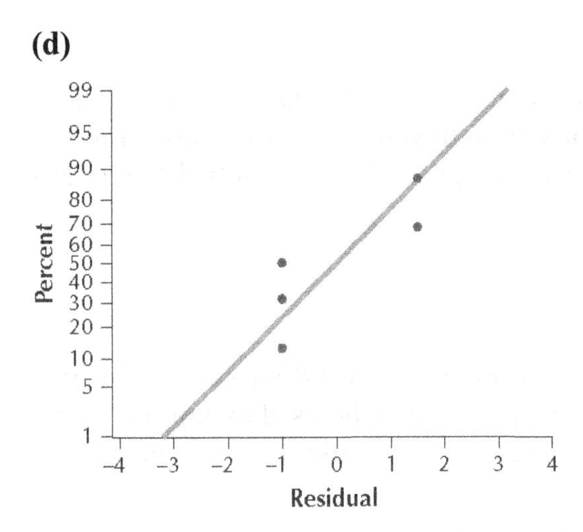

(e) The scatterplot of the residuals contains an unhealthy pattern, so the regression assumptions are not verified.

9. **(a)** and **(b)**

x	y	**Predicted value $\hat{y} = 4x + 21.6$**	**Residual** $(y - \hat{y})$
-5	0	1.6	-1.6
-4	8	5.6	2.4
-3	8	9.6	-1.6
-2	16	13.6	2.4
-1	16	17.6	-1.6

(c)

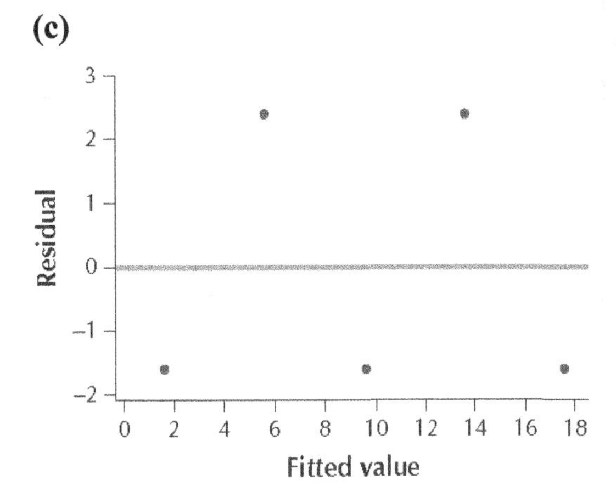

(d)

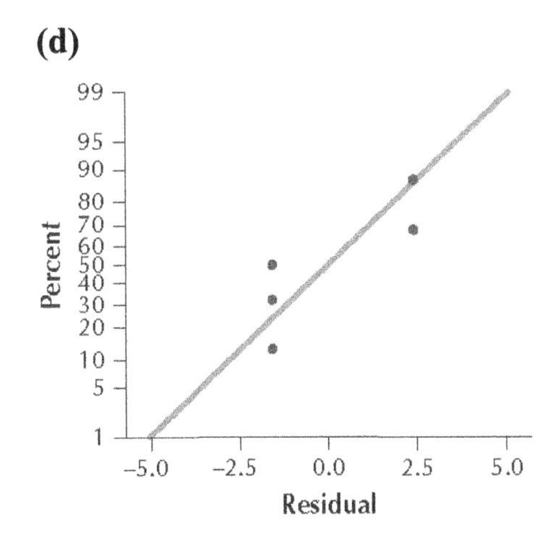

(e) The scatterplot of the residuals contains an unhealthy pattern, so the regression assumptions are not verified.

11. **(a)** and **(b)**

x	y	**Predicted value** $\hat{y} = -0.5x + 104$	**Residual** $(y - \hat{y})$
10	100	99	1
20	95	94	1
30	85	89	-4
40	85	84	1
50	80	79	1

(c)

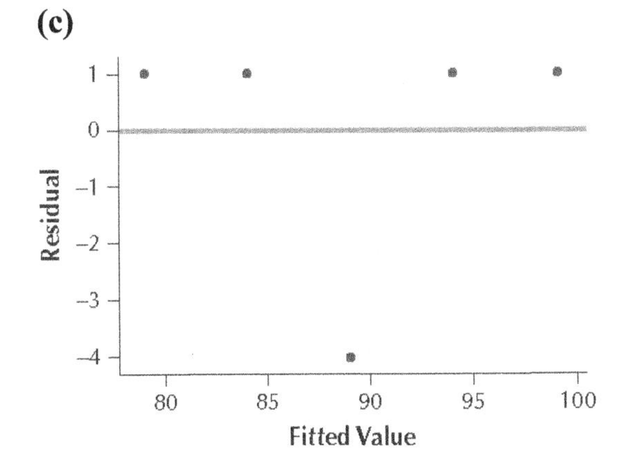

(d)

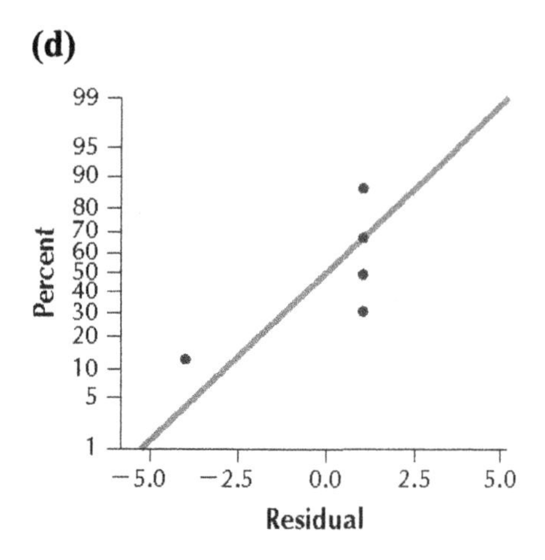

(e) The scatterplot of the residuals contains an unhealthy pattern, so the regression assumptions are not verified.

13. **(a)** and **(b)**

x	y	$\hat{y} = 0.6x + 0.2$	$y - \hat{y}$
1	1	0.8	0.2
2	1	1.4	-0.4
3	2	2	0
4	3	2.6	0.4
5	3	3.2	-0.2

(c)

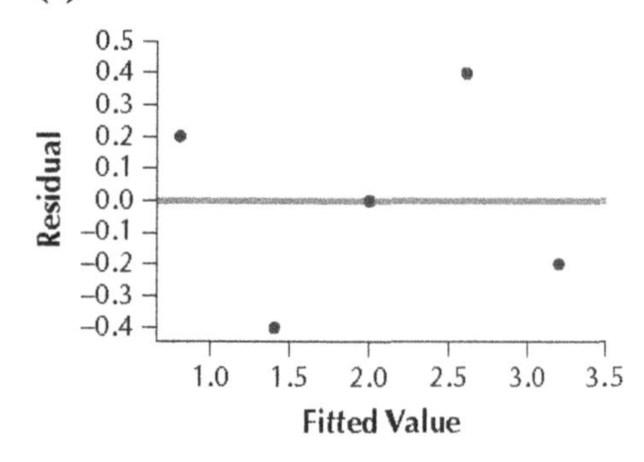

(d)

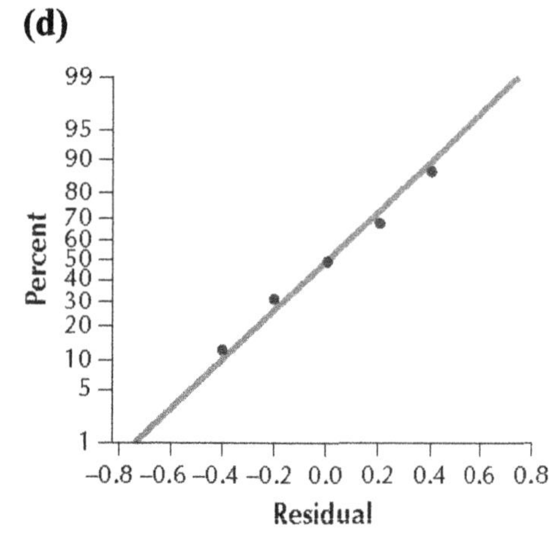

(e) The scatterplot in (c) of the residuals versus fitted values shows no strong evidence of unhealthy patterns. Thus, the independence assumption, the constant variance assumption, and the zero-mean assumption are verified. Also, the normal probability plot of the residuals in (d) indicates no evidence of departure from normality of the residuals. Therefore, we conclude that the regression assumptions are verified.

15. **(a)** df = $n - 2 = 5 - 2 = 3$, $t_{crit} = 3.182$.

(b)

$y - \hat{y}$	$(y - \hat{y})^2$
–1	1
1.5	2.25
–1	1
1.5	2.25
–1	1

SSE = $\sum(y - \hat{y})^2 = 7.5$. $s = \sqrt{\frac{\text{SSE}}{n-2}} = \sqrt{\frac{7.5}{5-2}} \approx 1.58113883$.

(c) From the TI-83/84: $s_x^2 = 2.5$. Therefore,

$\sum(x - \bar{x})^2 = (n - 1)s_x^2 = (5 - 1)2.5 = 10$.

(d) $t_{\text{data}} = \frac{b_1}{s/\sqrt{\sum(x-\bar{x})^2}} \approx \frac{2.5}{1.58113883/\sqrt{10}} = 5$.

(e) $H_0 : \beta_1 = 0$: There is no linear relationship between x and y.

$H_a : \beta_1 \neq 0$: There is a linear relationship between x and y.

Reject H_0 if $t_{\text{data}} \geq 3.182$ or $t_{\text{data}} \leq -3.182$.

Since $t_{\text{data}} = 5$, which is ≥ 3.182, we reject H_0. There is evidence at level of significance $\alpha = 0.05$ that $\beta_1 \neq 0$ and that there is a linear relationship between x and y.

17. **(a)** df = $n - 2 = 5 - 2 = 3$, $t_{crit} = 3.182$

(b)

$y - \hat{y}$	$(y - \hat{y})^2$
–1.6	2.56
2.4	5.76
–1.6	2.56
2.4	5.76
–1.6	2.56

SSE = $\sum(y - \hat{y})^2 = 19.2$. $s = \sqrt{\frac{\text{SSE}}{n-2}} = \sqrt{\frac{19.2}{5-2}} \approx 2.529822128$.

(c) From the TI-83/84: $s_x^2 = 2.5$. Therefore,

$\sum(x - \bar{x})^2 = (n - 1)s_x^2 = (5 - 1)2.5 = 10$.

(d) $t_{\text{data}} = \frac{b_1}{s/\sqrt{\sum(x-\bar{x})^2}} \approx \frac{4}{2.529822128/\sqrt{10}} = 5$.

(e) $H_0 : \beta_1 = 0$: There is no linear relationship between x and y.

$H_a : \beta_1 \neq 0$: There is a linear relationship between x and y.

Reject H_0 if $t_{\text{data}} \geq 3.182$ or $t_{\text{data}} \leq -3.182$.

Since $t_{\text{data}} = 5$, which is ≥ 3.182, we reject H_0. There is evidence at level of significance $\alpha = 0.05$ that $\beta_1 \neq 0$ and that there is a linear relationship between x and y.

19. (a)

$y - \hat{y}$	$(y - \hat{y})^2$
1	1
1	1
−4	16
1	1
1	1

$\text{SSE} = \sum(y - \hat{y})^2 = 20.$ $s = \sqrt{\frac{\text{SSE}}{n-2}} = \sqrt{\frac{20}{5-2}} \approx 2.581988897.$

(b) From the TI-83/84: $s_x^2 = 250$. Therefore,

$\sum(x - \bar{x})^2 = (n-1)s_x^2 = (5-1)250 = 1000.$

(c) $t_{\text{data}} = \frac{b_1}{s/\sqrt{\sum(x-\bar{x})^2}} \approx \frac{-0.5}{2.581988897/\sqrt{1000}} \approx -6.1237.$

(d) df = $n - 2 = 5 - 2 = 3$. p-value $= 2P(t > |t_{\text{data}}|) = 2P(t > 6.1237) \approx 0.0088$

(e) $H_0 : \beta_1 = 0$: There is no linear relationship between x and y.

$H_a : \beta_1 \neq 0$: There is a linear relationship between x and y.

Reject H_0 if p-value ≤ 0.05.

Since p-value ≈ 0.0088, which is ≤ 0.05, we reject H_0. There is evidence at level of significance $\alpha = 0.05$ that $\beta_1 \neq 0$ and that there is a linear relationship between x and y.

21. (a)

$y - \hat{y}$	$(y - \hat{y})^2$
0.2	0.04
−0.4	0.16
0	0
0.4	0.16
−0.2	0.04

$\text{SSE} = \sum(y - \hat{y})^2 = 0.4.$ $s = \sqrt{\frac{\text{SSE}}{n-2}} = \sqrt{\frac{0.4}{5-2}} \approx 0.3651483717.$

(b) From the TI-83/84: $s_x^2 = 2.5$. Therefore,

$\sum(x - \bar{x})^2 = (n-1)s_x^2 = (5-1)2.5 = 10.$

(c) $t_{\text{data}} = \frac{b_1}{s/\sqrt{\sum(x-\bar{x})^2}} \approx \frac{0.6}{0.3651483717/\sqrt{10}} \approx 5.1962.$

(d) df = $n - 2 = 5 - 2 = 3$. p-value $= 2P(t > |t_{\text{data}}|) = 2P(t > 5.1962) \approx 0.0138.$

(e) $H_0 : \beta_1 = 0$: There is no linear relationship between x and y.

$H_a : \beta_1 \neq 0$: There is a linear relationship between x and y.

Reject H_0 if p-value ≤ 0.05.

Since p-value ≈ 0.0138, which is ≤ 0.05, we reject H_0. There is evidence at level of significance $\alpha = 0.05$ that $\beta_1 \neq 0$ and that there is a linear relationship between x and y.

23. **(a)** df $= n - 2 = 5 - 2 = 3$, $t_{\alpha/2} = 3.182$.

(b) From Exercise 15, $s \approx 1.58113883$ and $\sum(x-\bar{x})^2 = 10$. Therefore,

$$E = t_{\alpha/2} \cdot \frac{s}{\sqrt{\sum(x-\bar{x})^2}} = (3.182)\left(\frac{1.58113883}{\sqrt{10}}\right) \approx 1.591.$$

(c) $b_1 \pm t_{\alpha/2} \cdot \frac{s}{\sqrt{\sum(x-\bar{x})^2}} = 2.5 \pm (3.182)\left(\frac{1.58113883}{\sqrt{10}}\right) \approx 2.5 \pm 1.591 = (0.909, 4.091)$.

25. **(a)** df $= n - 2 = 5 - 2 = 3$, $t_{\alpha/2} = 3.182$.

(b) From Exercise 17, $s \approx 2.529822128$ and $\sum(x-\bar{x})^2 = 10$. Therefore,

$$E = t_{\alpha/2} \cdot \frac{s}{\sqrt{\sum(x-\bar{x})^2}} = (3.182)\left(\frac{2.529822128}{\sqrt{10}}\right) \approx 2.5456.$$

(c) $b_1 \pm t_{\alpha/2} \cdot \frac{s}{\sqrt{\sum(x-\bar{x})^2}} = 4 \pm (3.182)\left(\frac{2.529822128}{\sqrt{10}}\right) \approx 4 \pm 2.5456 = (1.4544, 6.5456)$.

27. **(a)** df $= n - 2 = 5 - 2 = 3$, $t_{\alpha/2} = 3.182$.

(b) From Exercise 19, $s \approx 2.581988897$ and $\sum(x-\bar{x})^2 = 1000$. Therefore,

$$E = t_{\alpha/2} \cdot \frac{s}{\sqrt{\sum(x-\bar{x})^2}} = (3.182)\left(\frac{2.581988897}{\sqrt{1000}}\right) \approx 0.2598.$$

(c)

$b_1 \pm t_{\alpha/2} \cdot \frac{s}{\sqrt{\sum(x-\bar{x})^2}} = -0.5 \pm (3.182)\left(\frac{2.581988897}{\sqrt{1000}}\right) \approx -0.5 \pm 0.2598 = (-0.7598, -0.2402)$.

29. **(a)** df $= n - 2 = 5 - 2 = 3$, $t_{\alpha/2} = 3.182$.

(b) From Exercise 21, $s \approx 0.3651483717$ and $\sum(x-\bar{x})^2 = 10$. Therefore,

$$E = t_{\alpha/2} \cdot \frac{s}{\sqrt{\sum(x-\bar{x})^2}} = (3.182)\left(\frac{0.3651483717}{\sqrt{10}}\right) \approx 0.3674.$$

(c) $b_1 \pm t_{\alpha/2} \cdot \frac{s}{\sqrt{\sum(x-\bar{x})^2}} = 0.6 \pm (3.182)\left(\frac{0.3651483717}{\sqrt{10}}\right) \approx 0.6 \pm 0.3674 = (0.2326, 0.9674)$.

TI-83/84: (0.2325, 0.9675).

31. $H_0 : \beta_1 = 0$: There is no linear relationship between x and y.

$H_a : \beta_1 \neq 0$: There is a linear relationship between x and y.

Since the confidence interval from Exercise 23 (c) does not contain zero, we may conclude that $\beta_1 \neq 0$ and that a linear relationship exists between x and y, at level of significance $\alpha = 0.05$.

33. $H_0 : \beta_1 = 0$: There is no linear relationship between x and y.

$H_a : \beta_1 \neq 0$: There is a linear relationship between x and y.

Since the confidence interval from Exercise 25 (c) does not contain zero, we may conclude that $\beta_1 \neq 0$ and that a linear relationship exists between x and y, at level of significance $\alpha = 0.05$.

35. $H_0 : \beta_1 = 0$: There is no linear relationship between x and y.

$H_a : \beta_1 \neq 0$: There is a linear relationship between x and y.

Since the confidence interval from Exercise 27 (c) does not contain zero, we may conclude that $\beta_1 \neq 0$ and that a linear relationship exists between x and y, at level of significance $\alpha = 0.05$.

37. $H_0 : \beta_1 = 0$: There is no linear relationship between x and y.

$H_a : \beta_1 \neq 0$: There is a linear relationship between x and y.

Since the confidence interval from Exercise 29 (c) does not contain zero, we may conclude that $\beta_1 \neq 0$ and that a linear relationship exists between x and y, at level of significance $\alpha = 0.05$

39. Minitab results.

Regression Analysis: Weight (y) versus Volume (x)

Analysis of Variance

Source	DF	Adj SS	Adj MS	F-Value	P-Value
Regression	1	409.600	409.600	236.31	0.001
Volume (x)	1	409.600	409.600	236.31	0.001
Error	3	5.200	1.733		
Total	4	414.800			

Model Summary

S	R-sq	R-sq(adj)	R-sq(pred)
1.31656	98.75%	98.33%	96.64%

Coefficients

Term	Coef	SE Coef	T-Value	P-Value	VIF
Constant	4.00	1.38	2.90	0.063	
Volume (x)	1.600	0.104	15.37	0.001	1.00

Regression Equation

Weight (y) = 4.00 + 1.600 Volume (x)

Critical-value method:

$H_0 : \beta_1 = 0$. There is no relationship between *volume* (x) and *weight* (y).

$H_a : \beta_1 \neq 0$. There is a linear relationship between *volume* (x) and *weight* (y).

$\text{df} = n - 2 = 5 - 2 = 3$. $t_{\text{crit}} = 3.182$

Reject H_0 if $t_{\text{data}} \leq -3.182$ or $t_{\text{data}} \geq 3.182$

TI-83/84: $t_{\text{data}} = 15.37230277$ (From Minitab: $t_{\text{data}} = 15.37$)

Since $t_{\text{data}} = 15.37230277$ is ≥ 3.182, we reject H_0. There is evidence for a linear relationship between *volume* (x) and *weight* (y).

p-value method:

$H_0 : \beta_1 = 0$. There is no relationship between *volume* (x) and *weight* (y).

$H_a : \beta_1 \neq 0$. There is a linear relationship between *volume* (x) and *weight* (y).

Reject H_0 if p-value ≤ 0.05.

df $= n - 2 = 5 - 2 = 3$.

TI-83/84: p-value = 0.00059796674 (From Minitab: p-value = 0.001).

Since p-value = 0.00059796674 is ≤ 0.05, we reject H_0. There is evidence for a linear relationship between *volume* (x) and *weight* (y).

41. Minitab results:

Regression Analysis: High (y) versus Low (x)

The regression equation is

High (y) = 11.9 + 1.05 Low (x)

Predictor	Coef	SE Coef	T	P
Constant	11.947	3.898	3.07	0.037
Low (x)	1.04658	0.08654	12.09	0.000

S = 4.17671 R-Sq = 97.3% R-Sq(adj) = 96.7%

Critical-value method:

df $= n - 2 = 6 - 2 = 4$. $t_{data} = 2.776$.

$H_0 : \beta_1 = 0$. There is no relationship between *low* (x) and *high* (y).

$H_a : \beta_1 \neq 0$. There is a linear relationship between *low* (x) and *high* (y).

Reject H_0 if $t_{data} \leq -2.776$ or $t_{data} \geq 2.776$.

TI-83/84: $t_{data} = 12.09393244$ (Minitab: $t_{data} = 12.09$).

Since $t_{data} = 12.09393244$ is ≥ 2.776, we reject H_0. There is evidence for a linear relationship between *low* (x) and *high* (y).

p-value method:

$H_0 : \beta_1 = 0$. There is no relationship between *low* (x) and *high* (y).

$H_a : \beta_1 \neq 0$. There is a linear relationship between *low* (x) and *high* (y).

Reject H_0 if p-value ≤ 0.05.

df $= n - 2 = 6 - 2 = 4$.

TI-83/84: p-value = 0.00026812718 (Minitab: p-value = 0.000)

Since p-value = 0.00026812718 is ≤ 0.05, we reject H_0. There is evidence for a linear relationship between *low* (x) and *high* (y).

43. Minitab Results

Regression Analysis: Darts versus DJIA

Analysis of Variance

Source	DF	Adj SS	Adj MS	F-Value	P-Value
Regression	1	1613	1613.0	4.73	0.073
DJIA	1	1613	1613.0	4.73	0.073
Error	6	2046	341.0		
Total	7	3659			

Model Summary

S	R-sq	R-sq(adj)	R-sq(pred)
18.4665	44.08%	34.76%	1.09%

Coefficients

Term	Coef	SE Coef	T-Value	P-Value	VIF
Constant	4.39	6.93	0.63	0.549	
DJIA	1.414	0.650	2.17	0.073	1.00

Regression Equation

Darts = 4.39 + 1.414 DJIA

H_0: $\beta_1 = 0$ No linear relationship exists between DJIA and darts.

H_a: $\beta_1 \neq 0$ A linear relationship exists between DJIA and darts.

Reject H_0 if the p-value $\leq \alpha = 0.05$.

$t_{\text{data}} \approx 2.1749$.

p –value = 0.0725703888.

The p –value = 0.0725703888 is not $\leq \alpha = 0.05$, so we do not reject H_0. Insufficient evidence exists, at level of significance $\alpha = 0.05$, for a linear relationship between DJIA and darts.

45. Minitab results

Regression Analysis: Shots versus Age

Analysis of Variance

Source	DF	Adj SS	Adj MS	F-Value	P-Value
Regression	1	0.03125	0.03125	0.03	0.860
Age	1	0.03125	0.03125	0.03	0.860
Error	8	7.56875	0.94609		
Lack-of-Fit	6	5.06875	0.84479	0.68	0.700
Pure Error	2	2.50000	1.25000		
Total	9	7.60000			

Model Summary

S	R-sq	R-sq(adj)	R-sq(pred)
0.972674	0.41%	0.00%	0.00%

Coefficients

Term	Coef	SE Coef	T-Value	P-Value	VIF
Constant	1.93	1.49	1.30	0.231	

```
Age        0.0156   0.0860     0.18    0.860  1.00
Regression Equation
Shots = 1.93 + 0.0156 Age
Fits and Diagnostics for Unusual Observations
Obs  Shots    Fit   Resid  Std Resid
  7  2.000  2.325  -0.325      -0.53  X
X  Unusual X
```

H_0: $\beta_1 = 0$ No linear relationship exists between age and shots.

H_a: $\beta_1 \neq 0$ A linear relationship exists between age and shots.

Reject H_0 if the p-value $\leq \alpha = 0.05$.

$t_{\text{data}} \approx 0.1817$.

p –value = 0.8603048707.

The p-value = 0.8603048707 is not $\leq \alpha = 0.05$, so we do not reject H_0. Insufficient evidence exists, at level of significance $\alpha = 0.05$, for a linear relationship between a patient's age and the number of shots taken by the patient.

47. **(a)** df = $n - 2 = 5 - 2 = 3$, $t_{\alpha/2} = 3.182$. TI-83/84: $s \approx 1.316561177$. TI-83/84: $s_x^2 \approx 40$. Therefore,

$\Sigma(x - \bar{x})^2 = (n - 1)s_x^2 = (5 - 1)40 = 160.$

$E = t_{\alpha/2} \cdot \frac{s}{\sqrt{\Sigma(x-\bar{x})^2}} = (3.182)\left(\frac{1.31656117}{\sqrt{160}}\right) \approx 0.3312.$

(b) TI-83/84: $b_1 = 1.6$.

$b_1 \pm t_{\alpha/2} \cdot \frac{s}{\sqrt{\Sigma(x-\bar{x})^2}} = 1.6 \pm (3.182)\left(\frac{1.31656117}{\sqrt{160}}\right) \approx 0.6 \pm 0.3312 = (1.2688, 1.9312).$

(c) We are 95% confident that the interval (1.2688, 1.9312) captures the population slope β_1 of the relationship between *volume* and *weight.*

49. **(a)** df = $n - 2 = 6 - 2 = 4$, $t_{\alpha/2} = 2.776$. TI-83/84: $s \approx 4.17671465$. TI-83/84: $s_x^2 \approx 465.9$. Therefore,

$\Sigma(x - \bar{x})^2 = (n - 1)s_x^2 = (6 - 1)465.9 = 2329.5.$

$E = t_{\alpha/2} \cdot \frac{s}{\sqrt{\Sigma(x-\bar{x})^2}} = (2.776)\left(\frac{4.17671465}{\sqrt{2329.5}}\right) \approx 0.2402.$

(b) TI-83/84: $b_1 \approx 1.0466$.

$b_1 \pm t_{\alpha/2} \cdot \frac{s}{\sqrt{\Sigma(x-\bar{x})^2}} = 1.0466 \pm (2.776)\left(\frac{4.17671465}{\sqrt{2329.5}}\right) \approx 1.0466 \pm 0.2402 = (0.8064, 1.2868).$

(c) We are 95% confident that the interval (0.8064, 1.2868) captures the population slope β_1 of the relationship between *low* and *high.*

51. **(a)** df = $n - 2 = 8 - 2 = 6$, $t_{\alpha/2} = 2.447$. TI-83/84: $s \approx 18.46647035$. TI-83/84: $s_x^2 \approx$ 115.2392857. Therefore,

$$\sum (x - \bar{x})^2 = (n-1)s_x^2 = (8-1)115.2392857 = 806.675.$$

$E = t_{\alpha/2} \cdot \frac{s}{\sqrt{\sum(x-\bar{x})^2}} = (2.447)\left(\frac{18.46647035}{\sqrt{806.675}}\right) \approx 1.5910.$

(b) TI-83/84: $b_1 \approx 1.4141$.

$b_1 \pm t_{\alpha/2} \cdot \frac{s}{\sqrt{\sum(x-\bar{x})^2}} \approx 1.4141.\pm(2.447)\left(\frac{18.46647035}{\sqrt{806.675}}\right) \approx 1.4141.\pm 1.5910 =$

$(-0.1769, 3.0051)$. TI-83/84: $(-0.1769, 3.005)$

(c) We are 95% that the interval $(-0.1769,\ 3.005)$ captures the slope β_1 of the population regression line. That is, we are 95% confident that for each additional \$1 the stocks in the DJIA gain in one day, the daily change in the stocks in the portfolio predicted by the darts lies between $-\$0.1769$ and \$3.005.

53. **(a)** df $= n - 2 = 10 - 2 = 8$, $t_{\alpha/2} = 2.306$. TI-83/84: $s \approx 0.9726735064$. TI-83/84: $s_X^2 \approx 14.22222222$. Therefore,

$\sum(x - \bar{x})^2 = (n-1)s_X^2 = (10-1)14.22222222 = 128.$

$E = t_{\alpha/2} \cdot \frac{s}{\sqrt{\sum(x-\bar{x})^2}} = (2.306)\left(\frac{0.9726735064}{\sqrt{128}}\right) \approx 0.1983$ shot.

(b) TI-83/84: $b_1 = 0.015625$.

$b_1 \pm t_{\alpha/2} \cdot \frac{s}{\sqrt{\sum(x-\bar{x})^2}} \approx 0.015625 \pm (2.306)\left(\frac{0.9726735064}{\sqrt{128}}\right) \approx 0.015625 \pm 0.1983 \approx$

$(-0.1827, 0.2139)$. TI-83/84: $(-0.1826, 0.21388)$

(c) We are 95% confident that the interval $(-0.1826, 0.21388)$ captures the slope β_1 of the population regression line. That is, we are 95% confident that for each additional year in a patient's age, the change in the number of shots taken by the patient lies between -0.1826 shot and 0.21388 shot.

55. **(a)** TI-83/84: $(-6.647, 49.817)$

(b) 0 lies in the interval, so we do not reject H_0.

57.

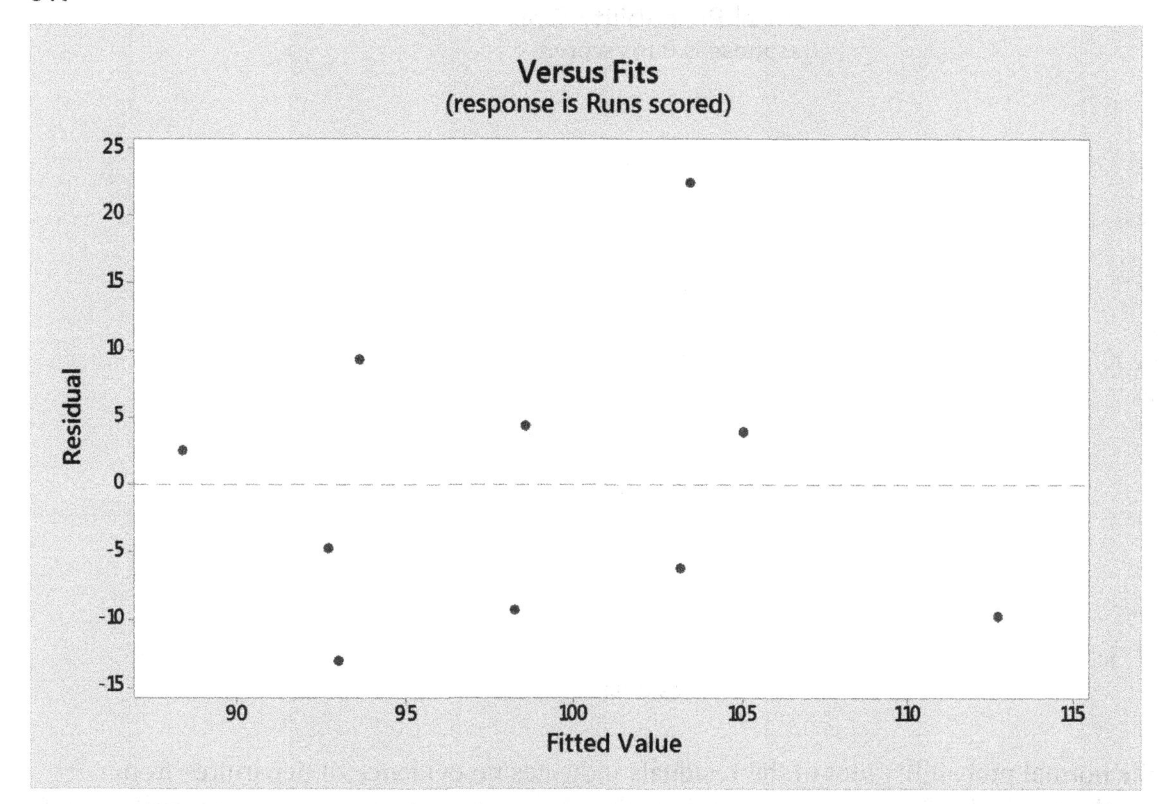

The scatterplot of the residuals versus the fitted values shows no evidence of the unhealthy patterns shown in Figure 4. Thus, the independence assumption, the constant variance assumption, and the zero-mean assumption are verified.

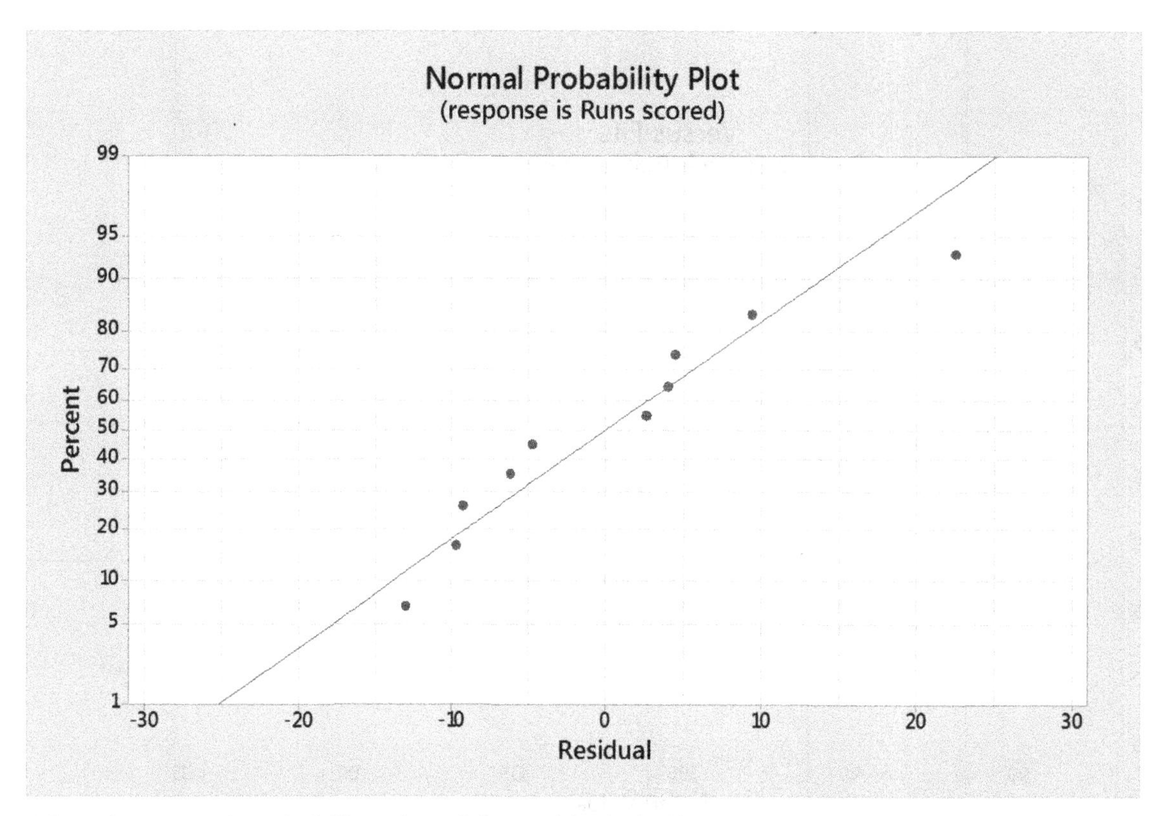

Also, the normal probability plot of the residuals indicates no evidence of departures from normality in the residuals. Therefore, we conclude that the regression assumptions are verified and it is okay to proceed with the regression.

59. 22.51 runs; by far the highest number of runs scored but the third highest batting average.

61. There is evidence against the null hypothesis that no linear relationship exists. The points appear to lie near a line with a positive slope.

63. $H_0 : \beta_1 = 0$. There is no linear relationship between *SAT Reading score* (x) and *SAT Math score* (y). $H_a : \beta_1 \neq 0$. There is a linear relationship between *SAT Reading score* (x) and *SAT Math score* (y).

df $= n - 2 = 5 - 2 = 3$, $t_{\text{crit}} = 2.353$.

Reject H_0 if $t_{\text{data}} \geq 2.353$, or $t_{\text{data}} \leq -2.353$.

TI-83/84: $t_{\text{data}} \approx 3.1963$.

Since $t_{\text{data}} \approx 3.1963$, which is ≥ 2.353, we reject H_0. There is evidence at level of significance $\alpha = 0.10$ that $\beta_1 \neq 0$ and that there is a linear relationship between *SAT Reading score* (x) and *SAT Math score* (y).

65. Yes. Since the confidence interval from Exercise 64 does not contain zero, we may conclude that $\beta_1 \neq 0$ and that a linear relationship exists between *SAT Reading score* (x) and *SAT Math score* (y), at level of significance $\alpha = 0.10$.

67. **(a)** Use part (c) and the identities $t_{data} = \frac{b_1}{s/\sqrt{\sum(x-\bar{x})^2}}$ and $b_1 = \frac{\sum(x-\bar{x})(y-\bar{y})}{\sum(x-\bar{x})^2}$. Since the five added points all have coordinates$(\bar{x}, \bar{y})$, the quantities $(x - \bar{x})$, $(x - \bar{x})^2$, and $(y - \bar{y})$ are all 0 for those five points and remain the same for the original points. Thus, b_1 and $\sqrt{\sum(x-\bar{x})^2}$ remain the same. From part (c) s decreases. Thus, t_{data} increases if b_1 is positive and decreases if b_1 is negative.

(b) Use 66 (c), (d), and the identity $r^2 = \frac{\text{SSR}}{\text{SST}}$. Since SSR and SST remain the same, r^2 remains the same.

(c) Use 66 (a), (b), and the identity $= \sqrt{\frac{\text{SSE}}{n-2}}$. Since SSE remains the same and n increases, s decreases.

(d) Use (a) and the identity p-value $= 2 \cdot P(t > |t_{\text{data}}|)$. From (a), $|t_{\text{data}}|$ increases. Thus, the p-value decreases.

(e) Since we don't know what the new p-value will be, we don't know the p-value will decrease enough to change the conclusion from do not reject H_0 to reject H_0.

69. (a) $t_{\text{data}} = \frac{b_1}{s/\sqrt{\sum(x-\bar{x})^2}}$. Since the 10 points that are added are on the regression line, b_1 remains the same. Since all of the data points that were added have different x-values, s_x^2 increases. Since $\sum(x-\bar{x})^2 = (n-1)\,s_x^2$ and n increases by 10, $\sum(x-\bar{x})^2$ increases. Since $s = \sqrt{\frac{\text{SSE}}{n-2}}$, $t_{\text{data}} = \frac{b_1}{s/\sqrt{\sum(x-\bar{x})^2}} = \frac{b_1}{\sqrt{\frac{\text{SSE}}{n-2}}/\sqrt{\sum(x-\bar{x})^2}} = \frac{b_1}{\sqrt{\text{SSE}}/\sqrt{(n-2)\sum(x-\bar{x})^2}}$. From 68 (b) SSE remains the same, so the denominator in t_{data} decreases. Therefore, t_{data} decreases if b_1 is negative and increases if b_1 is positive.

(b) Use 68 (b), (c), and the identities $r^2 = \frac{\text{SSR}}{\text{SST}}$ and SSR = SST − SSE. Combining the identities gives us $r^2 = \frac{\text{SSR}}{\text{SST}} = \frac{\text{SST}-\text{SSE}}{\text{SST}} = 1 - \frac{\text{SSE}}{\text{SST}}$. From 68 (b), SSE remains the same. From (c), SST increases. Thus, $\frac{\text{SSE}}{\text{SST}}$ decreases, so r^2 increases.

(c) Use the identity $s = \sqrt{\frac{\text{SSE}}{n-2}}$. Since SSE remains the same and n increases, s decreases.

(d) Use (a) and the identity p-value $= 2 \cdot P(t > |t_{\text{data}}|)$. From (a), $|t_{\text{data}}|$ increases. Thus, the p-value decreases.

(e) Since the conclusion was already reject H_0 and the p-value decreases from (d), the conclusion will still be reject H_0.

71. (a) Here is an example:

Original data set: {(1, 11), (2, 12), (3, 13), (4, 14), (5, 14)}

New data set with 3 added to the largest value of x, which is 5:

:{(1, 11), (2, 12), (3, 13), (4, 14), (8, 14)}

Minitab results of regression analysis on original data set.

Regression Analysis: y versus x

```
Analysis of Variance

Source      DF  Adj SS  Adj MS  F-Value  P-Value
```

```
Regression    1  6.4000  6.4000    48.00    0.006
  x           1  6.4000  6.4000    48.00    0.006
Error         3  0.4000  0.1333
Total         4  6.8000
Model Summary
       S    R-sq  R-sq(adj)  R-sq(pred)
0.365148  94.12%     92.16%      75.90%
Coefficients
Term         Coef  SE Coef  T-Value  P-Value   VIF
Constant   10.400    0.383    27.16    0.000
x           0.800    0.115     6.93    0.006  1.00
Regression Equation
y = 10.400 + 0.800 x
```

Minitab results of regression analysis on new data set.

Regression Analysis: y versus x1

```
Analysis of Variance
Source       DF  Adj SS  Adj MS  F-Value  P-Value
Regression    1   4.608  4.6082     6.31    0.087
  x1          1   4.608  4.6082     6.31    0.087
Error         3   2.192  0.7306
Total         4   6.800
Model Summary
       S    R-sq  R-sq(adj)  R-sq(pred)
0.854748  67.77%     57.02%       0.00%
Coefficients
Term         Coef  SE Coef  T-Value  P-Value   VIF
Constant   11.370    0.686    16.58    0.000
x1          0.397    0.158     2.51    0.087  1.00
Regression Equation
y = 11.370 + 0.397 x1
```

Scatterplot of both data sets with corresponding regression lines.

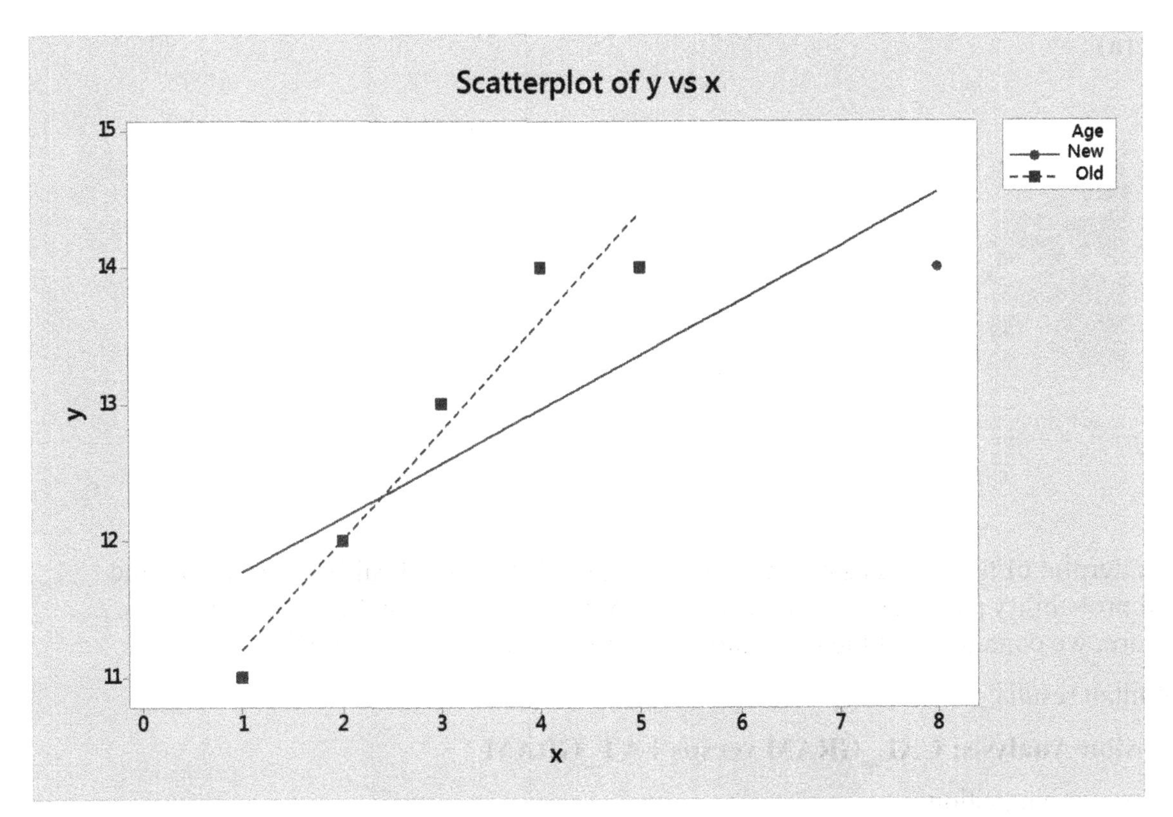

$t_{\text{data}} = \frac{b_1}{s/\sqrt{\Sigma(x-\bar{x})^2}}$.. The graph above shows that b_1 would decrease but remain positive. The graph above shows that increasing the *x*-value of the biggest value of *x* makes the data points more spread out about the regression line. Therefore, the variability in b_1, $\frac{s}{\sqrt{\Sigma(x-\bar{x})^2}}$, increases. Thus, t_{data} decreases.

(b) Use 70 (c), (d), and the identity $r^2 = \frac{\text{SSR}}{\text{SST}}$. From 70 (d), SSR decreases. From 70 (c), SST remains the same. Thus, r^2 decreases.

(c) Use 70 (a), (b), and the identity $s = \sqrt{\frac{\text{SSE}}{n-2}}$. Since SSE increases and *n* remains the same, *s* increases.

(d) Use (a) and the identities *p*-value = $2 \cdot P(t > |t_{\text{data}}|)$ and $t_{\text{data}} = \frac{b_1}{s/\sqrt{\Sigma(x-\bar{x})^2}}$. Since b_1, *s*, and $\sqrt{\Sigma(x-\bar{x})^2}$ are both positive, t_{data} is positive. From (a), $t_{\text{data}} = |t_{\text{data}}|$ decreases. Thus, the *p*-value increases.

(e) It depends on the new *p*-value.

73. **(a)**

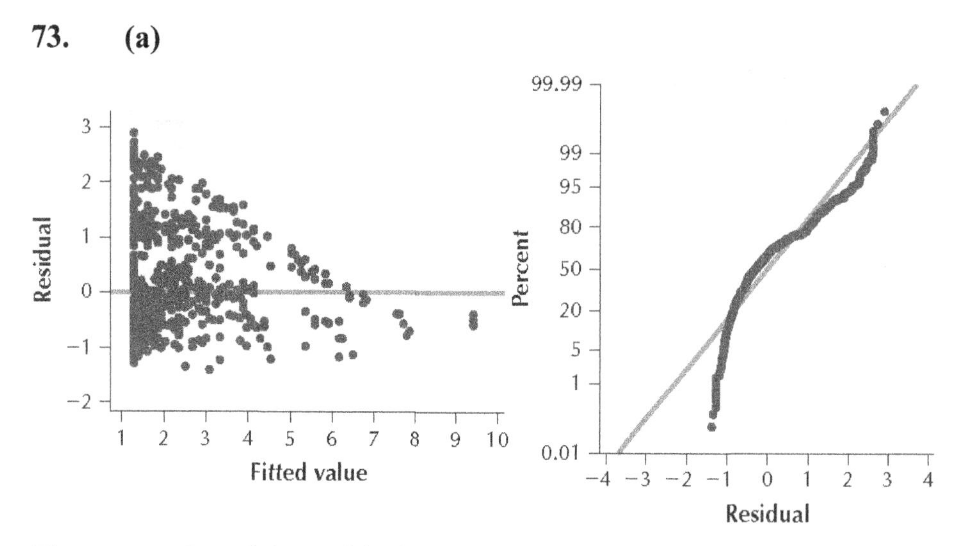

The scatterplot of the residuals contains no strong evidence of unhealthy patterns but the normal probability plot indicates evidence of departures from normality in the residuals. Therefore, we conclude that the regression assumptions are not fully verified.

(b) Minitab results:

Regression Analysis: CAL_GRAM versus FAT_GRAM

The regression equation is

CAL_GRAM = 1.28 + 8.13 FAT_GRAM

Predictor	Coef	SE Coef	T	P
Constant	1.28238	0.03711	34.55	0.000
FAT_GRAM	8.1290	0.1568	51.84	0.000

S = 0.993259 R-Sq = 73.7%R-Sq(adj) = 73.7%

Descriptive Statistics: FAT_GRAM

	Total	
Variable	Count	Variance
FAT_GRAM	961	0.04179

From Minitab, $b_1 = 8.1290$, $s = 0.993259$ and df $= n - 2 = 961 - 2 = 959$, and $s^2 \approx 0.04179$. Therefore,

$\sum(x-\bar{x})^2 = (n-1)s^2 = (961-1)0.04179 = 40.1184.$

$t_{\alpha/2} = 1.984.$

Therefore the 95% confidence interval for β_1 is

$$b_1 \pm t_{\alpha/2} \cdot \frac{s}{\sqrt{\sum(x-\bar{x})^2}} \approx 8.1290 \pm (1.984)\left(\frac{0.993259}{\sqrt{40.1184}}\right) \approx 8.1290 \pm 0.3111 \approx (7.8179, 8.4401).$$

Thus, we are 95% confident that the interval (7.8179, 8.4401) captures the slope β_1 of the regression line. That is, we are 95% confident that, for each additional gram of fat per gram of food, the increase in the number of calories per gram lies between 7.8179 and 8.4401 calories per gram.

(c) Since 0 does not lie in the confidence interval, we would expect to reject the null hypothesis that $\beta_1 = 0$.

(d) $H_0 : \beta_1 = 0$. There is no relationship between *fat per gram* (x) and *calories per gram* (y).

$H_a : \beta_1 \neq 0$. There is a linear relationship between *fat per gram* (x) and *calories per gram* (y).

Reject H_0 if p-value ≤ 0.05. From the Minitab results, the p-value = 0.000. Since p-value = 0.000 is ≤ 0.05, we reject H_0. There is evidence for a linear relationship between *fat per gram* (x) and *calories per gram* (y).

75. **(a)** No.

(b) Positive relationship.

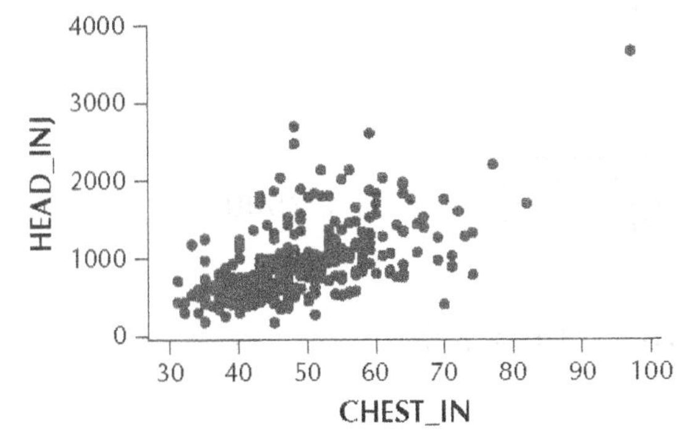

(c) Unclear. It is possible to have a head injury without having a chest injury or to have a chest injury without having a head injury, so it is unclear which variable should be the response variable and which variable should be the predictor variable.

77. **(a)** No.

(b) No apparent relationship between the variables.

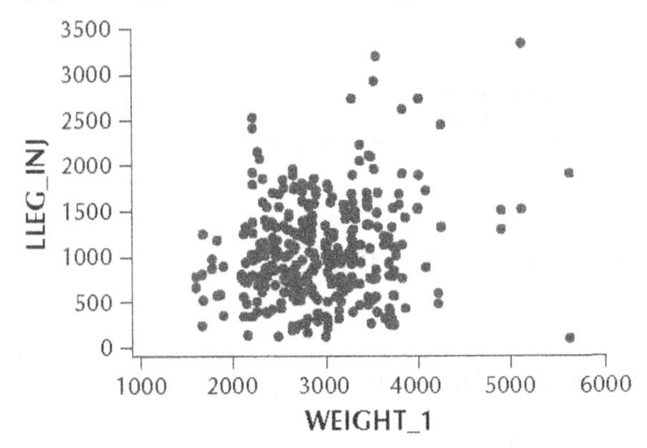

(c) Weight of the vehicles should be the predictor variable and the severity of the left leg injuries should be the response variable. It may be possible that the weight of the vehicle affects the severity of the left leg injury but the severity of the left leg injury will not affect the weight of the vehicle.

Section 13.2

1. The confidence interval is for the mean value of y for a given x and the prediction interval is for a randomly selected value of y for a given x.

3. $x^* = 3, \hat{y} = 2.5x + 13.5 = 2.5(3) + 13.5 = 7.5 + 13.5 = 21.$

df = $n - 2 = 5 - 2 = 3$, $t_{\alpha/2} = 3.182$.

TI-83/84: $\bar{x} = 3$, $s \approx 1.58113883$, and $s^2 = 2.5$.

Thus $\sum(x - \bar{x})^2 = (n - 1)s^2 = (5 - 1)2.5 = 10.$

Lower bound:

$\hat{y} - t_{\alpha/2} \cdot s\sqrt{\frac{1}{n} + \frac{(x^* - \bar{x})^2}{\sum(x - \bar{x})^2}} \approx 21 - (3.182)(1.58113883)\sqrt{\frac{1}{5} + \frac{(3-3)^2}{10}} \approx 21 - 2.2500 \approx$ 18.7500 [Minitab: 18.7497].

Upper bound:

$\hat{y} + t_{\alpha/2} \cdot s\sqrt{\frac{1}{n} + \frac{(x^* - \bar{x})^2}{\sum(x - \bar{x})^2}} \approx 21 + (3.182)(1.58113883)\sqrt{\frac{1}{5} + \frac{(3-3)^2}{10}} \approx 21 + 2.2500 \approx$ 23.2500 [Minitab: 23.2500].

5. $x^* = -4, \hat{y} = 4x + 21.6 = 4(-4) + 21.6 = -16 + 21.6 = 5.6.$

df = $n - 2 = 5 - 2 = 3$, $t_{\alpha/2} = 3.182$.

TI-83/84: $\bar{x} = -3$. $s \approx 2.529822128$. $s^2 = 2.5$, so $\sum(x - \bar{x})^2 = (n - 1)s^2 = (5 - 1)2.5 = 10.$

Lower bound:

$\hat{y} - t_{\alpha/2} \cdot s\sqrt{\frac{1}{n} + \frac{(x^* - \bar{x})^2}{\sum(x - \bar{x})^2}} \approx 5.6 - (3.182)(2.529822128)\sqrt{\frac{1}{5} + \frac{(-4-(-3))^2}{10}} \approx 5.6 - 4.4091 =$ 1.1909 [Minitab: 1.19027].

Upper bound:

$\hat{y} + t_{\alpha/2} \cdot s\sqrt{\frac{1}{n} + \frac{(x^* - \bar{x})^2}{\sum(x - \bar{x})^2}} \approx 5.6 + (3.182)(2.529822128)\sqrt{\frac{1}{5} + \frac{(-4-(-3))^2}{10}} \approx 5.6 + 4.4091 =$ 10.0091 [Minitab: 10.097].

7. $x^* = 10, \hat{y} = -0.5x + 104 = -0.5(10) + 104 = -5 + 104 = 99.$

$df = n - 2 = 5 - 2 = 3$, $t_{\alpha/2} = 3.182$.

TI-83/84: $\bar{x} = 30$. $s \approx 2.581988897$. $s^2 = 250$, so $\sum(x - \bar{x})^2 = (n - 1)s^2 = (5 - 1)250 =$ 1000.

Lower bound:

$\hat{y} - t_{\alpha/2} \cdot s\sqrt{\frac{1}{n} + \frac{(x^* - \bar{x})^2}{\sum(x - \bar{x})^2}} \approx 99 - (3.182)(2.581988897)\sqrt{\frac{1}{5} + \frac{(10-30)^2}{1000}} \approx 99 - 6.3640 =$ 92.6360 [Minitab: 92.6351].

Upper bound:

$\hat{y} + t_{\alpha/2} \cdot s\sqrt{\frac{1}{n} + \frac{(x^*-\bar{x})^2}{\Sigma(x-\bar{x})^2}} \approx 99 + (3.182)(2.581988897)\sqrt{\frac{1}{5} + \frac{(10-30)^2}{1000}} \approx 99 + 6.3640 =$ 105.3640 [Minitab: 105.365].

9. Lower bound:

$\hat{y} - t_{\alpha/2} \cdot s\sqrt{1 + \frac{1}{n} + \frac{(x^*-\bar{x})^2}{\Sigma(x-\bar{x})^2}} \approx 21 - (3.182)(1.58113883)\sqrt{1 + \frac{1}{5} + \frac{(3-3)^2}{10}} \approx 21 - 5.5114 =$ 15.4886 [Minitab: 15.4878].

Upper bound:

$\hat{y} + t_{\alpha/2} \cdot s\sqrt{1 + \frac{1}{n} + \frac{(x^*-\bar{x})^2}{\Sigma(x-\bar{x})^2}} \approx 21 + (3.182)(1.58113883)\sqrt{1 + \frac{1}{5} + \frac{(3-3)^2}{10}} \approx 21 + 5.5114 =$ 26.5114 [Minitab: 26.5122].

11. Lower bound:

$\hat{y} - t_{\alpha/2} \cdot s\sqrt{1 + \frac{1}{n} + \frac{(x^*-\bar{x})^2}{\Sigma(x-\bar{x})^2}} \approx 5.6 - (3.182)(2.529822128)\sqrt{1 + \frac{1}{5} + \frac{(-4-(-3))^2}{10}} \approx 5.6 -$ $9.1783 = -3.5783$ [Minitab: -3.57958].

Upper bound:

$\hat{y} + t_{\alpha/2} \cdot s\sqrt{1 + \frac{1}{n} + \frac{(x^*-\bar{x})^2}{\Sigma(x-\bar{x})^2}} \approx 5.6 + (3.182)(2.529822128)\sqrt{1 + \frac{1}{5} + \frac{(-4-(-3))^2}{10}} \approx 5.6 +$ $9.1783 = 14.7783$ [Minitab: 14.7796].

13. Lower bound:

$\hat{y} - t_{\alpha/2} \cdot s\sqrt{1 + \frac{1}{n} + \frac{(x^*-\bar{x})^2}{\Sigma(x-\bar{x})^2}} \approx 99 - (3.182)(2.581988897)\sqrt{1 + \frac{1}{5} + \frac{(10-30)^2}{1000}} \approx 99 -$ $10.3924 = 88.6076$ [Minitab: 88.6062].

Upper bound:

$\hat{y} + t_{\alpha/2} \cdot s\sqrt{1 + \frac{1}{n} + \frac{(x^*-\bar{x})^2}{\Sigma(x-\bar{x})^2}} \approx 99 + (3.182)(2.581988897)\sqrt{1 + \frac{1}{5} + \frac{(10-30)^2}{1000}} \approx 99 +$ $10.3924 = 109.3924$ [Minitab: 109.394].

15. **(a)** $x^* = 4$ cubic meters, $\hat{y} = 1.6x + 4 = 1.6(4) + 4 = 6.4 + 4 = 10.4$ kilograms.

(b) df $= n - 2 = 5 - 2 = 3$, $t_{\alpha/2} = 3.182$. TI-83/84: $\bar{x} = 12$. TI-83/84: $s \approx 1.316561177$. TI-83/84: $s^2 = 40$, so $\sum(x - \bar{x})^2 = (n - 1)s^2 = (5 - 1)40 = 160$.

Lower bound:

$\hat{y} - t_{\alpha/2} \cdot s\sqrt{\frac{1}{n} + \frac{(x^*-\bar{x})^2}{\Sigma(x-\bar{x})^2}} \approx 10.4 - (3.182)(1.316561177)\sqrt{\frac{1}{5} + \frac{(4-12)^2}{160}} \approx 10.4 - 3.2450 =$ 7.1550 kilograms [Minitab: 7.15453 kilograms].

Upper bound:

$\hat{y} + t_{\alpha/2} \cdot s\sqrt{\frac{1}{n} + \frac{(x^*-\bar{x})^2}{\Sigma(x-\bar{x})^2}} \approx 10.4 + (3.182)(1.316561177)13.6450\sqrt{\frac{1}{5} + \frac{(4-12)^2}{160}} \approx 10.4 +$

3.2450 = kilograms [Minitab: 13.6455 kilograms].

17. **(a)** $x^* = 30$ degrees, $\hat{y} \approx 1.046576519x + 11.94698433 = 1.046576519(30) + 11.94698433 = 31.39729557 + 11.94698433 \approx 43.3443$ degrees.

(b) df $= n - 2 = 6 - 2 = 4$, $t_{\alpha/2} = 4.604$. TI-83/84: $\bar{x} = 40.5$. TI-83/84: $s \approx 4.17671465$. TI-83/84: $s^2 = 465.9$, so $\sum(x - \bar{x})^2 = (n-1)s^2 = (6-1)465.9 = 2329.5$.

Lower bound:

$\hat{y} - t_{\alpha/2} \cdot s\sqrt{\frac{1}{n} + \frac{(x^*-\bar{x})^2}{\Sigma(x-\bar{x})^2}} \approx 43.3443 - (4.604)(4.17671465)\sqrt{\frac{1}{6} + \frac{(30-40.5)^2}{2329.5}} \approx 43.3443 - 8.8955 = 34.4488$ degrees [Minitab: 34.4486 degrees].

Upper bound:

$\hat{y} + t_{\alpha/2} \cdot s\sqrt{\frac{1}{n} + \frac{(x^*-\bar{x})^2}{\Sigma(x-\bar{x})^2}} \approx 43.3443 + (4.604)(4.17671465)\sqrt{\frac{1}{6} + \frac{(30-40.5)^2}{2329.5}} \approx 43.3443 + 8.8955 = 52.2298$ degrees [Minitab: 52.2400 degrees].

19. **(a)** Lower bound:

$\hat{y} - t_{\alpha/2} \cdot s\sqrt{1 + \frac{1}{n} + \frac{(x^*-\bar{x})^2}{\Sigma(x-\bar{x})^2}} \approx 10.4 - (3.182)(1.316561177)\sqrt{1 + \frac{1}{5} + \frac{(4-12)^2}{160}} \approx 10.4 - 5.2991 = 5.1009$ kilograms [Minitab: 5.1009 kilograms].

Upper bound:

$\hat{y} + t_{\alpha/2} \cdot s\sqrt{1 + \frac{1}{n} + \frac{(x^*-\bar{x})^2}{\Sigma(x-\bar{x})^2}} \approx 10.4 + (3.182)(1.316561177)\sqrt{1 + \frac{1}{5} + \frac{(4-12)^2}{160}} \approx 10.4 + 5.2991 = 15.6991$ kilograms [Minitab: 15.6998 kilograms].

(b) The interval in Exercise 19 (a) is wider. Individual values are more variable than their mean. The interval in Exercise 15 (b) is a 95% confidence interval for the mean value of y for the given value $x = 4$ cubic meters and the interval in Exercise 19 (a) is a prediction interval for a randomly chosen value of y for the given value $x = 4$ cubic meters.

21 **(a)** Lower bound:

$\hat{y} - t_{\alpha/2} \cdot s\sqrt{\frac{1}{n} + \frac{(x^*-\bar{x})^2}{\Sigma(x-\bar{x})^2}} \approx 43.3443 - (4.604)(4.17671465)\sqrt{1 + \frac{1}{6} + \frac{(30-40.5)^2}{2329.5}} \approx 43.3443 - 21.1874 = 22.1569$ degrees [Minitab: 22.1564 degrees].

Upper bound:

$\hat{y} + t_{\alpha/2} \cdot s\sqrt{\frac{1}{n} + \frac{(x^*-\bar{x})^2}{\Sigma(x-\bar{x})^2}} \approx 43.3443 + (4.604)(4.17671465)\sqrt{1 + \frac{1}{6} + \frac{(30-40.5)^2}{2329.5}} \approx 43.3443 + 21.1874 = 64.5317$ degrees [Minitab: 64.5322 degrees].

(b) The range of the low temperatures in the data set is from 7°C to 70°C, inclusive. Therefore, a low temperature of 0°C is outside of the range of our data set, and any predictions or estimates using the regression equation for 0°C represent extrapolation. Extrapolation should be avoided if possible because the relationship between the variables may no longer be linear outside the range of x.

23.

New Obs	Fit	SE Fit	95% CI	95% PI
1	98.337	0.092	(98.153, 98.521)	(96.891, 99.783)

(a) (98.153, 98.521)

(b) (96.891, 99.783)

Section 13.3

1. $\hat{y} = b_0 + b_1x_1 + b_2x_2 + b_3x_3$.

3. The F test for the overall significance of the multiple regression.

5. The F test is for the overall significance of the multiple regression and the t test is for testing whether a particular x-variable has a significant relationship with the response variable y.

7. The coefficient of a dummy variable can be interpreted as the estimated increase in y for those observations with the value of the dummy variable equal to 1, as compared to those with the value of the dummy variable equal to 0 when all of the other x-variables are held constant.

9. For each increase in one unit of the variable x_1 the estimated value of y increases by five units when the value of x_2 is held constant.

11. The estimated value of y when $x_1 = 0$ and $x_2 = 0$ is $b_0 = 10$.

$b_1 = 5$ means that for each increase of one unit of the variable x_1, the estimated value of y increases by five units when the value of x_2 is held constant.

$b_2 = 8$ means that for each increase of one unit of the variable x_2, the estimated value of y increases by eight units when the value of x_1 is held constant.

13. For each increase in one unit of the variable x_1 the estimated value of y decreases by 0.1 unit when the value of x_2 is held constant.

15. The estimated value of y when $x_1 = 0$ and $x_2 = 0$ is $b_0 = 0.5$.

$b_1 = -0.1$ means that for each increase of one unit of the variable x_1, the estimated value of y decreases by 0.1 unit when the value of x_2 is held constant.

$b_2 = 0.9$ means that for each increase of one unit of the variable x_2, the estimated value of y increases by 0.9 unit when the value of x_1 is held constant.

17. 50% of the variability in y is accounted for by this multiple regression equation.

19. 75% of the variability in y is accounted for by this multiple regression equation.

21. **Regression Analysis: y versus x1, x2, x3**

Analysis of Variance

Source	DF	Adj SS	Adj MS	F-Value	P-Value
Regression	3	157.376	52.459	**14.37**	**0.004**
x1	1	50.489	50.489	13.83	0.010
x2	1	27.512	27.512	7.54	0.034

```
  x3          1    2.765   2.765      0.76    0.418
Error         6   21.905   3.651
Total         9  179.281
Model Summary
      S    R-sq  R-sq(adj)  R-sq(pred)
1.91071  87.78%     81.67%      68.14%
Coefficients
Term        Coef  SE Coef  T-Value  P-Value    VIF
Constant  -37.8      14.0    -2.71    0.035
x1          4.50     1.21     3.72    0.010  33.04
x2          3.37     1.23     2.75    0.034  33.10
x3         0.306    0.352     0.87    0.418   1.02
Regression Equation
y = -37.8 + 4.50 x1 + 3.37 x2 + 0.306 x3
```

From the Minitab output above the regression equation is

$\hat{y} = -37.8 + 4.50x_1 + 3.37x_2 + 0.306x_3$.

23. **(a)** Test 1

$H_0 : \beta_1 = 0$: There is no linear relationship between y and x_1.

$H_a : \beta_1 \neq 0$: There is a linear relationship between y and x_1.

Reject H_0 if the p-value $\leq \alpha = 0.05$.

Test 2

$H_0 : \beta_2 = 0$: There is no linear relationship between y and x_2.

$H_a : \beta_2 \neq 0$: There is a linear relationship between y and x_2.

Reject H_0 if the p-value $\leq \alpha = 0.05$.

Test 3

$H_0 : \beta_3 = 0$: There is no linear relationship between y and x_3.

$H_a : \beta_3 \neq 0$: There is a linear relationship between y and x_3.

Reject H_0 if the p-value $\leq \alpha = 0.05$.

(b)

Predictor	Coef	SE Coef	T	P
Constant	—37.83	13.98	–2.71	0.035
x1	4.497	1.209	3.72	0.010
x2	3.374	1.229	2.75	0.034
x3	0.3059	0.3515	0.87	0.418

Test 1: $t = 3.72$, with p-value = 0.010

Test 2: $t = 2.75$, with p-value = 0.034

Test 3: $t = 0.87$, with p-value = 0.418

(c) Test 1: The p-value = 0.010, which is $\leq \alpha = 0.05$. Therefore, we reject H_0. There is evidence of a linear relationship between y and x_1.

Test 2: The p-value = 0.034, which is $\leq \alpha = 0.05$. Therefore, we reject H_0. There is evidence of a linear relationship between y and x_2.

Test 3: The p-value = 0.418, which is not $\leq \alpha = 0.05$. Therefore, we do not reject H_0. There is insufficient evidence of a linear relationship between y and x_3.

25.

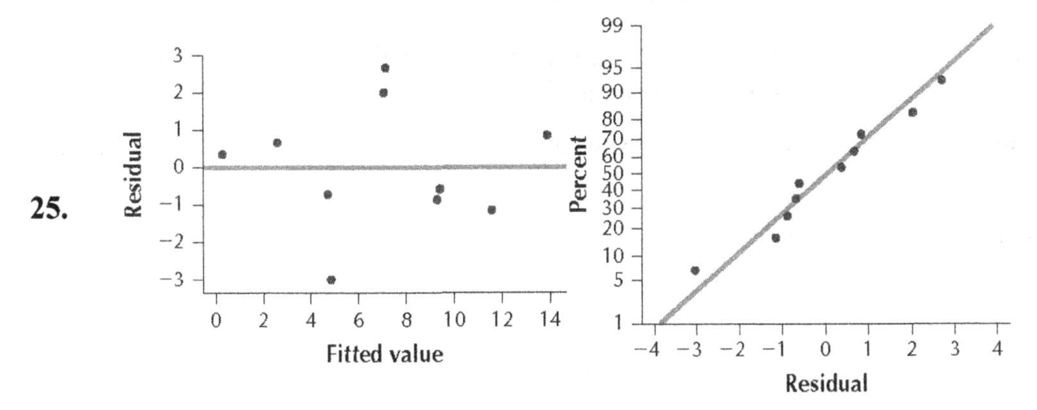

The scatterplot above of the residuals versus fitted values shows no strong evidence of unhealthy patterns. Thus, the independence assumption, the constant variance assumption, and the zero-mean assumption are verified. Also, the normal probability plot of the residuals above indicates no evidence of departure from normality of the residuals. Therefore, we conclude that the regression assumptions are verified.

27. Regression Analysis: y versus x1, x2, x3

Analysis of Variance

Source	DF	Adj SS	Adj MS	F-Value	P-Value
Regression	3	508.433	169.478	185.68	0.000
x1	1	395.199	395.199	432.99	0.000
x2	1	0.963	0.963	1.06	0.344
x3	1	30.295	30.295	33.19	0.001
Error	6	5.476	0.913		
Total	9	513.909			

Model Summary

S	R-sq	R-sq(adj)	R-sq(pred)
0.955362	98.93%	98.40%	96.33%

Coefficients

Term	Coef	SE Coef	T-Value	P-Value	VIF
Constant	-2.981	0.698	-4.27	0.005	
x1	1.1349	0.0545	20.81	0.000	1.08
x2	-0.175	0.170	-1.03	0.344	1.06

x3	3.550	0.616	5.76	0.001	1.04

Regression Equation

y = -2.981 + 1.1349 x1 - 0.175 x2 + 3.550 x3

Fits and Diagnostics for Unusual Observations

Obs	y	Fit	Resid	Std Resid	
9	18.800	17.272	1.528	2.17	R

R Large residual

The regression equation is $\hat{y} = -2.981 + 1.1349x_1 - 0.175x_2 + 3.550x_3$.

29. For each increase in one unit of the variable x_3, the estimated value of y increases by 3.55 units when the values of x_1 and x_2 are held constant.

31. The p-value for x_2 is the only p-value greater than $\alpha = 0.01$, so we eliminate x_2 from the regression equation.

Regression Analysis: y versus x1, x3

Analysis of Variance

Source	DF	Adj SS	Adj MS	F-Value	P-Value
Regression	2	507.469	253.735	275.82	0.000
x1	1	420.444	420.444	457.04	0.000
x3	1	31.549	31.549	34.30	0.001
Error	7	6.439	0.920		
Total	9	513.909			

Model Summary

S	R-sq	R-sq(adj)	R-sq(pred)
0.959129	98.75%	98.39%	97.15%

Coefficients

Term	Coef	SE Coef	T-Value	P-Value	VIF
Constant	-3.122	0.687	-4.55	0.003	
x1	1.1462	0.0536	21.38	0.000	1.03
x3	3.607	0.616	5.86	0.001	1.03

Regression Equation

y = -3.122 + 1.1462 x1 + 3.607 x3

The new regression equation is $\hat{y} = -3.122 + 1.1462x_1 + 3.607x_3$.

(a) Test 1

$H_0 : \beta_1 = 0$: There is no linear relationship between y and x_1.

$H_a : \beta_1 \neq 0$: There is a linear relationship between y and x_1.

Reject H_0 if the p-value $\leq \alpha = 0.01$.

Test 2

$H_0 : \beta_3 = 0$: There is no linear relationship between y and x_3.

$H_a : \beta_3 \neq 0$: There is a linear relationship between y and x_3.

Reject H_0 if the p-value $\leq \alpha = 0.01$.

(b)

Predictor	Coef	SE Coef	T	P
Constant	–3.1225	0.6866	–4.55	0.003
x1	1.14625	0.05362	21.38	0.000
x3	3.6075	0.6160	5.86	0.001

Test 1: $t = 21.38$, with p-value = 0.000

Test 2: $t = 5.86$, with p-value = 0.001

(c)

Test 1: The p-value = 0.000, which is $\leq \alpha = 0.01$. Therefore, we reject H_0. There is evidence of a linear relationship between y and x_1.

Test 2: The p-value = 0.001, which is $\leq \alpha = 0.01$. Therefore, we reject H_0. There is evidence of a linear relationship between y and x_3.

Since all of the variables are significant, we have our final multiple regression equation.

33. **(a)** The final multiple regression equation is $\hat{y} = -3.12 + 1.15x_1 + 3.61x_3$.

For $x_3 = 0$, the regression equation is $\hat{y} = -3.12 + 1.15x_1$.

For $x_3 = 1$, the regression equation is $\hat{y} = 0.49 + 1.15x_1$.

(b) For each increase in one unit of the variable x_1, the estimated value of y increases by 1.15 units.

The estimated increase in y for those observations with $x_3 = 1$, as compared to those with $x_3 = 0$, when x_1 is held constant, is 3.61.

(c) S = 0.959129 R-Sq = 98.7% R-Sq(adj) = 98.4%

Using the multiple regression equation in (a), the size of the typical prediction error will be about 0.959129. 98.4% of the variability in y is accounted for by this multiple regression equation.

35. **(a)** The regression equation is

Easiness Score = 99.1 + 0.0473 Starting a Business – 0.0942 Employing Workers – 0.101 Paying Taxes

S = 1.96021 R-Sq = 78.5% R-Sq(adj) = 70.4%

Analysis of Variance

Source	DF	SS	MS	F	P
Regression	3	112.261	37.420	9.74	0.005
Residual Error	8	30.739	3.842		

Total	11	143.000

$H_0 : \beta_1 = \beta_2 = \beta_3 = 0$: There is no linear relationship between *Easiness Score* and the set *Starting a Business*, *Employing Workers*, and *Paying Taxes.* The overall multiple regression is not significant.

H_a : At least one of the βs $\neq 0$: There is a linear relationship between *Easiness Score* and the set *Starting a Business, Employing Workers,* and *Paying Taxes.* The overall multiple regression is significant.

Reject H_0 if the p-value $\leq \alpha = 0.05$.

$F = 9.74$. p-value = 0.005.

The p-value = 0.005 $\leq \alpha = 0.05$, so we reject H_0. There is evidence at level of significance α = 0.05 for a linear relationship between *Easiness Score* and the set *Starting a Business, Employing Workers,* and *Paying Taxes.* The overall multiple regression is significant.

(b) Test 1

$H_0 : \beta_1 = 0$: There is no linear relationship between *Easiness Score* and *Starting a Business.*

$H_0 : \beta_1 \neq 0$: There is a linear relationship between *Easiness Score* and *Starting a Business*.

Reject H_0 if the p-value $\leq \alpha = 0.05$.

Test 2

$H_0 : \beta_2 = 0$: There is no linear relationship between *Easiness Score* and *Employing Workers.*

$H_a : \beta_2 \neq 0$: There is a linear relationship between *Easiness Score* and *Employing Workers*.

Reject H_0 if the p-value $\leq \alpha = 0.05$.

Test 3

$H_0 : \beta_3 = 0$: There is no linear relationship between *Easiness Score* and *Paying Taxes*.

$H_a : \beta_3 \neq 0$: There is a linear relationship between *Easiness Score* and *Paying Taxes*.

Reject H_0 if the p-value $\leq \alpha = 0.05$.

Predictor	Coef	SE Coef	T	P
Constant	99.133	1.031	96.17	0.000
Starting a Business	0.04725	0.05545	0.85	0.419
Employing Workers	–0.09423	0.02500	–3.77	0.005
Paying Taxes	–0.10125	0.03138	–3.23	0.012

Test 1: $t = 0.85$, with p-value = 0.419.

Test 2: $t = -3.77$, with p-value = 0.005.

Test 3: $t = -3.23$, with p-value = 0.012.

Test 1: The p-value = 0.419, which is not $\leq \alpha = 0.05$. Therefore, we do not reject H_0. There is insufficient evidence of a linear relationship between *Easiness Score* and *Starting a Business*.

Test 2: The p-value = 0.005, which is $\leq \alpha = 0.05$. Therefore, we reject H_0. There is evidence of a linear relationship between *Easiness Score* and *Employing Workers*.

Test 3: The p-value = 0.012, which is $\leq \alpha = 0.05$. Therefore, we reject H_0. There is evidence of a linear relationship between *Easiness Score* and *Paying Taxes*.

The p-value for Starting a Business is the only p-value greater than $\alpha = 0.05$, so we eliminate Starting a Business from our equation.

The regression equation is

Easiness Score = 98.9 – 0.0826 Employing Workers – 0.0804 Paying Taxes

Predictor	Coef	SE Coef	T	P
Constant	98.9455	0.9916	99.78	0.000
Employing Workers	–0.08259	0.02062	–4.01	0.003
Paying Taxes	–0.08038	0.01933	–4.16	0.002

S = 1.93016 R-Sq = 76.6% R-Sq(adj) = 71.3%

Analysis of Variance

Source	DF	SS	MS	F	P
Regression	2	109.470	54.735	14.69	0.001
Residual Error	9	33.530	3.726		
Total	11	143.000			

Test 1

$H_0 : \beta_2 = 0$: There is no linear relationship between *Easiness Score* and *Employing Workers*.

$H_a : \beta_2 \neq 0$: There is a linear relationship between *Easiness Score* and *Employing Workers*.

Reject H_0 if the p-value $\leq \alpha = 0.05$.

Test 2

$H_0 : \beta_3 = 0$: There is no linear relationship between *Easiness Score* and *Paying Taxes*.

$H_a : \beta_3 \neq 0$: There is a linear relationship between *Easiness Score* and *Paying Taxes*.

Reject H_0 if the p-value $\leq \alpha = 0.05$.

Predictor	Coef	SE Coef	T	P
Constant	98.9455	0.9916	99.78	0.000
Employing Workers	–0.08259	0.02062	–4.01	0.003
Paying Taxes	–0.08038	0.01933	–4.16	0.002

Test 1: $t = -4.01$, with p-value = 0.003.

Test 2: $t = -4.16$, with p-value = 0.002.

Test 1: The p-value = 0.003, which is $\leq \alpha = 0.05$. Therefore, we reject H_0. There is evidence of a linear relationship between *Easiness Score* and *Employing Workers*.

Test 2: The p-value = 0.002, which is $\leq \alpha = 0.05$. Therefore, we reject H_0. There is evidence of a linear relationship between *Easiness Score* and *Paying Taxes*.

Since all variables are significant, we have our final model.

(c)

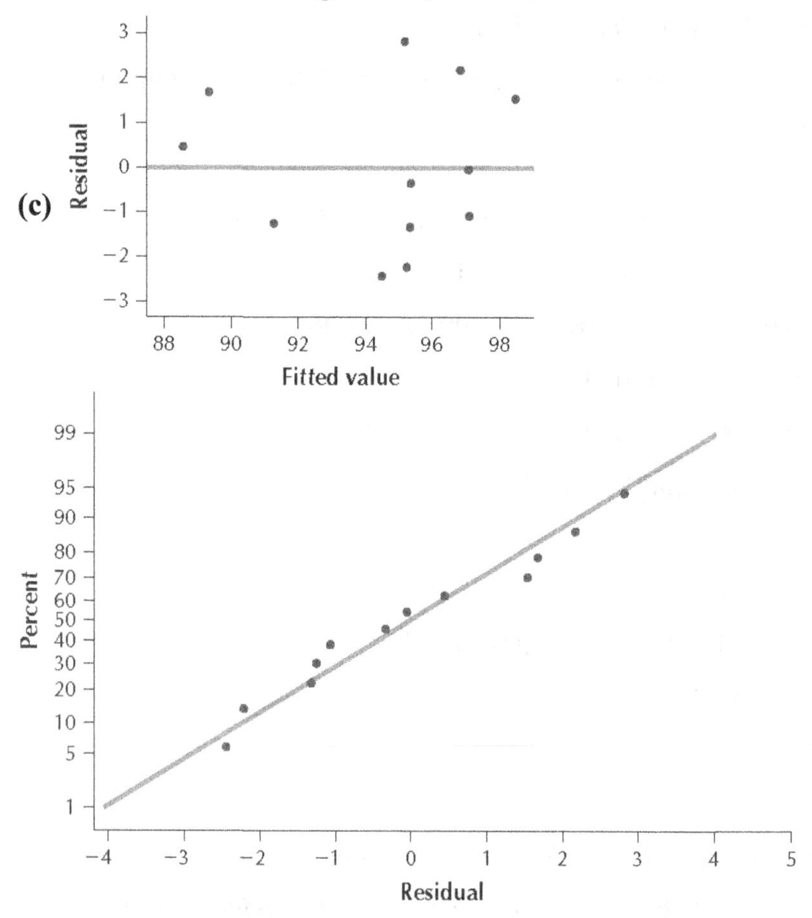

The scatterplot above of the residuals versus fitted values shows no strong evidence of unhealthy patterns. Thus, the independence assumption, the constant variance assumption, and the zero-mean assumption are verified. Also, the normal probability plot of the residuals above indicates no evidence of departure from normality of the residuals. Therefore, we conclude that the regression assumptions are verified.

(d) The regression equation is

Easiness Score = 98.9 – 0.0826 Employing Workers – 0.0804 Paying Taxes

The estimated value of *Easiness Score* when *Employing Workers* = 0 and *Paying Taxes* = 0 is 98.9.

For each increase in one unit of the variable *Employing Workers*, the estimated value of *Easiness Score* decreases by 0.0826 when the value of *Paying Taxes* is held constant.

For each increase in one unit of the variable *Paying Taxes*, the estimated value of *Easiness Score* decreases by 0.0804 unit when the value of *Employing Workers* is held constant.

S = 1.93016 R-Sq = 76.6% R-Sq(adj) = 71.3%

Using the multiple regression equation above, the size of the typical prediction error will be about 1.93016. 71.3% of the variability in *Easiness Score* is accounted for by this multiple regression equation.

37. **(a)** The regression equation is

Persons Covered = 906 + 11.5 Adults Not Covered – 35.1 Children Not Covered

S = 1403.22 R-Sq = 69.3% R-Sq(adj) = 65.7%

Analysis of Variance

Source	DF	SS	MS	F	P
Regression	2	75654326	37827163	19.21	0.000
Residual Error	17	33473280	1969016		
Total	19	109127606			

$H_0 : \beta_1 = \beta_2 = 0$: There is no linear relationship between *Persons Covered* and the set *Adults Not Covered* and *Children Not Covered.* The overall multiple regression is not significant.

H_a : At least one of the βs $\neq 0$: There is a linear relationship between *Persons Covered* and the set *Adults Not Covered* and *Children Not Covered.* The overall multiple regression is significant.

Reject H_0 if the p-value $\leq \alpha = 0.05$.

$F = 19.21$. p-value = 0.000

The p-value = 0.000, which is $\leq \alpha = 0.05$, so we reject H_0. There is evidence at level of significance $\alpha = 0.05$ for a linear relationship between *Persons Covered* (y) and the set *Adults Not Covered* and *Children Not Covered.* The overall multiple regression is significant.

(b) Test 1

$H_0 : \beta_1 = 0$: There is no linear relationship between *Persons Covered* and *Adults Not Covered.*

$H_a : \beta_1 \neq 0$: There is a linear relationship between *Persons Covered* and *Adults Not Covered.*

Reject H_0 if the p-value $\leq \alpha = 0.05$

Test 2

$H_0 : \beta_2 = 0$: There is no linear relationship between *Persons Covered* and *Children Not Covered.*

$H_a : \beta_2 \neq 0$: There is a linear relationship between *Persons Covered* and *Children Not Covered.*

Reject H_0 if the p-value $\leq \alpha = 0.05$.

Predictor	Coef	SE Coef	T	P
Constant	906.1	932.0	0.97	0.345
Adults Not Covered	11.458	2.338	4.90	0.000
Children Not Covered	–35.15	10.82	–3.25	0.005

Test 1: $t = 4.90$, with p-value = 0.000.

Test 2: $t = -3.25$, with p-value = 0.005.

Test 1: The p-value = 0.000, which is $\leq \alpha = 0.05$. Therefore, we reject H_0. There is evidence of a linear relationship between *Persons Covered* and *Adults Not Covered.*

Test 2: The p-value = 0.005, which is $\leq \alpha = 0.05$. Therefore, we reject H_0. There is evidence of a linear relationship between *Persons Covered* and *Children Not Covered.*

Since all of the variables are significant, we have our final multiple regression equation.

(c)

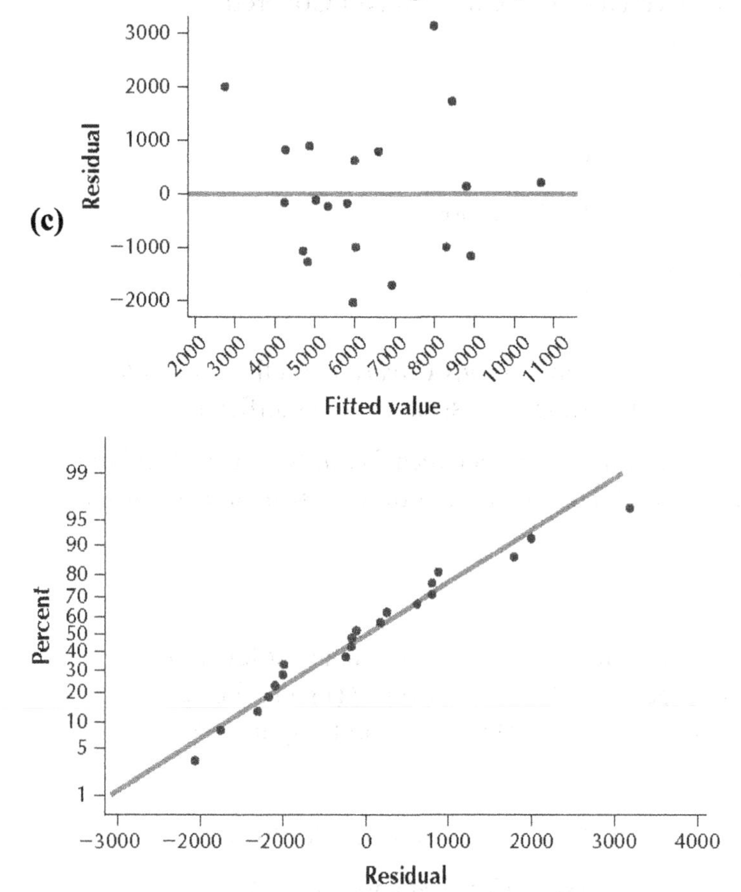

The scatterplot above of the residuals versus fitted values shows no strong evidence of unhealthy patterns. Thus, the independence assumption, the constant variance assumption, and the zero-mean assumption are verified. Also, the normal probability plot of the residuals above indicates no evidence of departure from normality of the residuals. Therefore, we conclude that the regression assumptions are verified.

(d)

The regression equation is

Persons Covered = 906 + 11.5 Adults Not Covered – 35.1 Children Not Covered

The estimated value of *Persons Covered* when *Adults Not Covered* = 0 and *Children Not Covered* = 0 is 906.

For each increase in one unit of the variable *Adults Not Covered,* the estimated value of *Persons Covered* increases by 11.5 when the value of *Children Not Covered* is held constant.

For each increase in one unit of the variable *Children Not Covered,* the estimated value of *Persons Covered* decreases by 35.1 when the value of *Adults Not Covered* is held constant.

S = 1403.22 R-Sq = 69.3% R-Sq(adj) = 65.7%

Using the multiple regression equation above, the size of the typical prediction error will be about 1403.22. 65.7% of the variability in *Persons Covered* is accounted for by this multiple regression equation.

39. The regression equation is

systolic blood pressure = 70.6 – 1.12 gender + 4.34 married + 10.0 smoke
– 5.21 exercise + 0.0872 age – 0.050 weight
+ 0.602 height + 1.26 body mass index + 2.03 stress
– 0.02 salt + 2.49 income – 0.41 educatn

Predictor	Coef	SE Coef	T	P
Constant	70.64	47.00	1.50	0.133
Gender	–1.119	2.564	–0.44	0.663
Married	4.343	2.376	1.83	0.068
Smoke	10.004	2.389	4.19	0.000
Exercise	–5.209	1.398	–3.73	0.000
Age	0.08717	0.08874	0.98	0.326
Weight	–-0.0504	0.1352	–0.37	0.709
Height	0.6019	0.7097	0.85	0.397
body mass index	1.2646	0.8007	1.58	0.115
stress	2.030	1.473	1.38	0.169
salt	–0.017	1.436	–0.01	0.991
income	2.490	1.458	1.71	0.088
educatn	–0.407	1.443	–0.28	0.778

S = 26.2530 R-Sq = 14.2% R-Sq(adj) = 12.1%

Analysis of Variance

Source	DF	SS	MS	F	P
Regression	12	55425.7	4618.8	6.70	0.000
Residual Error	487	335649.1	689.2		
Total	499	391074.8			

$H_0 : \beta_1 = \beta_2 = \beta_3 = \beta_4 = \beta_5 = \beta_6 = \beta_7 = \beta_8 = \beta_9 = \beta_{10} = \beta_{11} = \beta_{12} = 0$: There is no linear relationship between *Systolic Blood Pressure* and the set *gender, married, smoke, exercise, age, weight, height, body mass index, stress, salt, income,* and *education.* The overall multiple regression is not significant.

H_a : At least one of the βs $\neq 0$: There is a linear relationship between *Systolic Blood Pressure* and the set *gender, married, smoke, exercise, age, weight, height, body mass index, stress, salt, income,* and *education.* The overall multiple regression is significant.

Reject H_0 if the p-value $\leq \alpha = 0.05$.

$F = 6.70$. p-value = 0.000.

The p-value = 0.000, which is $\leq \alpha = 0.05$, so we reject H_0. There is evidence at level of significance $\alpha = 0.05$ for a linear relationship between *Systolic Blood Pressure* and the set *gender, married, smoke, exercise, age, weight, height, body mass index, stress, salt, income*, and *education*. The overall multiple regression is significant.

The largest p-value that is not greater than $\alpha = 0.05$ is the p-value for *Salt*, so we eliminate *Salt* from the multiple regression equation.

The regression equation is

systolic blood pressure = 70.6 – 1.12 gender + 4.34 married + 10.0 smoke
– 5.21 exercise + 0.0871 age – 0.050 weight
+ 0.601 height + 1.26 body mass index + 2.03 stress
+ 2.49 income – 0.41 educatn

Predictor	Coef	SE Coef	T	P
Constant	70.63	46.94	1.50	0.133
gender	–1.118	2.560	–0.44	0.663
married	4.344	2.370	1.83	0.067
smoke	10.006	2.384	4.20	0.000
exercise	–5.210	1.396	–3.73	0.000
age	0.08711	0.08854	0.98	0.326
weight	–0.0504	0.1350	–0.37	0.709
height	0.6014	0.7079	0.85	0.396
body mass index	1.2643	0.7995	1.58	0.114
stress	2.031	1.470	1.38	0.168
income	2.491	1.456	1.71	0.088
educatn	–0.405	1.436	–0.28	0.778

S = 26.2261 R-Sq = 14.2% R-Sq(adj) = 12.2%

Analysis of Variance

Source	DF	SS	MS	F	P
Regression	11	55425.6	5038.7	7.33	0.000
Residual Error	488	335649.2	687.8		
Total	499	391074.8			

The variable with the largest p-value greater than $\alpha = 0.05$ is the p-value for *Education*, so we eliminate *Education* from the multiple regression equation.

The regression equation is

systolic blood pressure = 70.5 – 1.03 gender + 4.32 married + 10.0 smoke
– 5.20 exercise + 0.0875 age – 0.049 weight
+ 0.590 height + 1.26 body mass index + 2.03 stress
+ 2.50 income

Predictor	Coef	SE Coef	T	P
Constant	70.51	46.90	1.50	0.133
Gender	–1.027	2.538	–0.40	0.686
married	4.324	2.367	1.83	0.068
smoke	10.014	2.382	4.20	0.000
exercise	–5.197	1.394	–3.73	0.000
age	0.08754	0.08844	0.99	0.323
weight	–0.0491	0.1348	–0.36	0.716
height	0.5898	0.7061	0.84	0.404
body mass index	1.2554	0.7981	1.57	0.116
stress	2.034	1.469	1.38	0.167
income	2.500	1.454	1.72	0.086

S = 26.2014 R-Sq = 14.2% R-Sq(adj) = 12.4%

Analysis of Variance

Source	DF	SS	MS	F	P
Regression	10	55370.8	5537.1	8.07	0.000
Residual Error	489	335704.0	686.5		
Total	499	391074.8			

The variable with the largest *p*-value greater than $\alpha = 0.05$ is the *p*-value for *Weight*, so we eliminate *Weight* from the multiple regression equation.

The regression equation is

systolic blood pressure = 85.9 – 1.11 gender + 4.31 married + 10.0 smoke
– 5.23 exercise + 0.0883 age + 0.350 height
+ 0.972 body mass index + 2.01 stress + 2.53 income

Predictor	Coef	SE Coef	T	P
Constant	85.94	20.19	4.26	0.000
gender	–1.107	2.526	–0.44	0.661
married	4.315	2.364	1.82	0.069

smoke	10.013	2.380	4.21	0.000
exercise	–5.235	1.389	–.77	0.000
age	0.08826	0.08834	1.00	0.318
height	0.3497	0.2539	1.38	0.169
body mass index	0.9716	0.1762	5.51	0.000
stress	2.006	1.465	1.37	0.172
income	2.534	1.450	1.75	0.081

S = 26.1782 R-Sq = 14.1% R-Sq(adj) = 12.6%

Analysis of Variance

Source	DF	SS	MS	F	P
Regression	9	55279.6	6142.2	8.96	0.000
Residual Error	490	335795.3	685.3		
Total	499	391074.8			

The variable with the largest p-value greater than $\alpha = 0.05$ is the p-value for *Gender*, so we eliminate *Gender* from the multiple regression equation.

The regression equation is

systolic blood pressure = 88.5 + 4.34 married + 10.1 smoke – 5.20 exercise
+ 0.0880 age + 0.309 height + 0.954 body mass index
+ 2.00 stress + 2.51 income

Predictor	Coef	SE Coef	T	P
Constant	88.50	19.31	4.58	0.000
married	4.343	2.362	1.84	0.067
smoke	10.053	2.376	4.23	0.000
exercise	–5.203	1.386	–3.75	0.000
age	0.08801	0.08827	1.00	0.319
height	0.3092	0.2363	1.31	0.191
body mass index	0.9541	0.1715	5.56	0.000
stress	2.005	1.464	1.37	0.172
income	2.515	1.448	1.74	0.083

S = 26.1566 R-Sq = 14.1% R-Sq(adj) = 12.7%

Analysis of Variance

Source	DF	SS	MS	F	P
Regression	8	55147.9	6893.5	10.08	0.000

Residual Error	491	335926.9	684.2
Total	499	391074.8	

The variable with the largest *p*-value greater than $\alpha = 0.05$ is the *p*-value for *Age*, so we eliminate *Age* from the multiple regression equation.

The regression equation is

systolic blood pressure = 91.8 + 4.32 married + 10.0 smoke – 5.14 exercise
+ 0.308 height + 0.954 body mass index + 2.06 stress
+ 2.56 income

Predictor	Coef	SE Coef	T	P
Constant	91.84	19.02	4.83	0.000
married	4.316	2.361	1.83	0.068
smoke	10.041	2.376	4.23	0.000
exercise	–5.140	1.384	–3.71	0.000
height	0.3077	0.2363	1.30	0.193
body mass index	0.9538	0.1715	5.56	0.000
stress	2.062	1.463	1.41	0.159
income	2.556	1.448	1.77	0.078

S = 26.1565 R-Sq = 13.9% R-Sq(adj) = 12.7%

Analysis of Variance

Source	DF	SS	MS	F	P
Regression	7	54467.8	7781.1	11.37	0.000
Residual Error	492	336607.1	684.2		
Total	499	391074.8			

The variable with the largest *p*-value greater than $\alpha = 0.05$ is the *p*-value for *Height*, so we eliminate *Height* from the multiple regression equation.

The regression equation is

systolic blood pressure = 115 + 4.23 married + 10.0 smoke – 5.07 exercise
+ 0.821 body mass index + 2.20 stress + 2.63 income

Predictor	Coef	SE Coef	T	P
Constant	115.100	6.541	17.60	0.000
Married	4.230	2.362	1.79	0.074
Smoke	10.019	2.378	4.21	0.000
Exercise	–5.068	1.384	–3.66	0.000

body mass index	0.8214	0.1382	5.94	0.000
stress	2.199	1.460	1.51	0.133
income	2.627	1.448	1.81	0.070

S = 26.1749   R-Sq = 13.6%   R-Sq(adj) = 12.6%

Analysis of Variance

Source	DF	SS	MS	F	P
Regression	6	53307.6	8884.6	12.97	0.000
Residual Error	493	337767.2	685.1		
Total	499	391074.8			

The variable with the largest *p*-value greater than $\alpha = 0.05$ is the *p*-value for *Stress*, so we eliminate *Stress* from the multiple regression equation.

The regression equation is

systolic blood pressure = 120 + 3.94 married + 10.1 smoke – 5.12 exercise
+ 0.820 body mass index + 2.67 income

Predictor	Coef	SE Coef	T	P
Constant	119.735	5.779	20.72	0.000
married	3.940	2.357	1.67	0.095
smoke	10.143	2.379	4.26	0.000
exercise	–5.117	1.386	–3.69	0.000
body mass index	0.8198	0.1384	5.93	0.000
income	2.667	1.449	1.84	0.066

S = 26.2085   R-Sq = 13.2%   R-Sq(adj) = 12.4%

Analysis of Variance

Source	DF	SS	MS	F	P
Regression	5	51753	10351	15.07	0.000
Residual Error	494	339322	687		
Total	499	391075			

The variable with the largest *p*-value greater than $\alpha = 0.05$ is the *p*-value for *Married*, so we eliminate *Married* from the multiple regression equation.

The regression equation is

systolic blood pressure = 122 + 10.3 smoke – 5.21 exercise
+ 0.801 body mass index + 2.64 income

Predictor	Coef	SE Coef	T	P

Constant	122.292	5.583	21.90	0.000
smoke	10.304	2.382	4.33	0.000
exercise	–5.206	1.387	–3.75	0.000
body mass index	0.8008	0.1381	5.80	0.000
income	2.635	1.452	1.82	0.070

S = 26.2560   R-Sq = 12.7%   R-Sq(adj) = 12.0%

Analysis of Variance

Source	DF	SS	MS	F	P
Regression	4	49834	12459	18.07	0.000
Residual Error	495	341241	689		
Total	499	391075			

The variable with the largest p-value greater than $\alpha = 0.05$ is the p-value for *Income*, so we eliminate *Income* from the multiple regression equation.

The regression equation is

systolic blood pressure = 127 + 9.90 smoke – 4.98 exercise
+ 0.802 body mass index

Predictor	Coef	SE Coef	T	P
Constant	127.202	4.895	25.98	0.000
Smoke	9.900	2.377	4.17	0.000
Exercise	–4.981	1.385	–3.60	0.000
body mass index	0.8022	0.1385	5.79	0.000

S = 26.3166   R-Sq = 12.2%   R-Sq(adj) = 11.6%

Analysis of Variance

Source	DF	SS	MS	F	P
Regression	3	47563	15854	22.89	0.000
Residual Error	496	343512	693		
Total	499	391075			

Since all of the variables are significant, we have our final multiple regression equation.

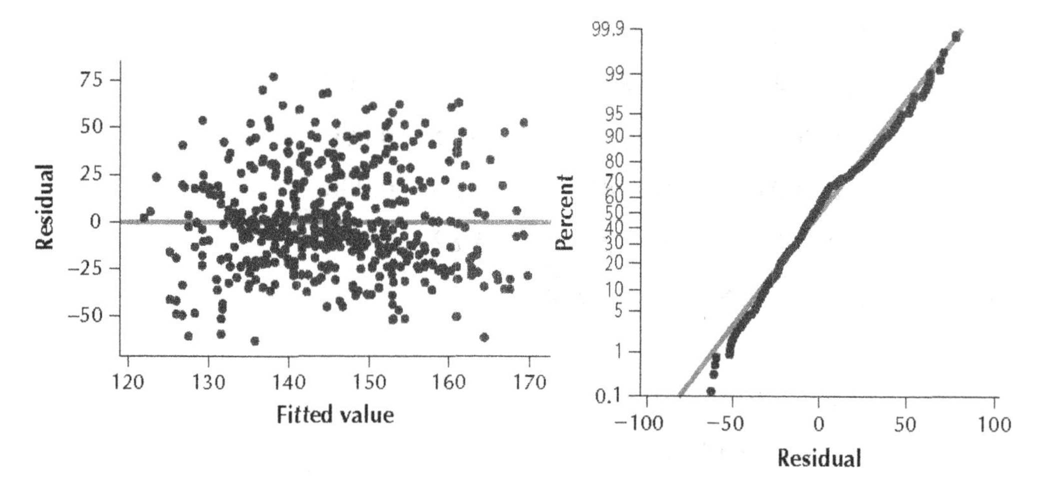

The scatterplot above of the residuals versus fitted values shows no strong evidence of unhealthy patterns. Thus, the independence assumption, the constant variance assumption, and the zero-mean assumption are verified. Also, the normal probability plot of the residuals above indicates no evidence of departure from normality of the residuals. Therefore, we conclude that the regression assumptions are verified.

The regression equation is

Systolic blood pressure = 127 + 9.90 Smoke – 4.98 Exercise
+ 0.802 Body Mass Index

The estimated value of *Systolic Blood Pressure* when *Smoke* = 0, *Exercise* = 0, and *Body Mass Index* = 0 is 127.

For each increase in one unit of the variable *Smoke*, the estimated value of *Systolic Blood Pressure* increases by 9.90 when the values of *Exercise* and *Body Mass Index* are held constant.

For each increase in one unit of the variable *Exercise*, the estimated value of *Systolic Blood Pressure* decreases by 4.98 when the value of *Smoke* and *Body Mass Index* are held constant.

For each increase in one unit of the variable *Body Mass Index*, the estimated value of *Systolic Blood Pressure* increases by 0.802 when the value of *Smoke* and *Exercise* are held constant.

S = 26.3166 R-Sq = 12.2% R-Sq(adj) = 11.6%

Using the multiple regression equation above, the size of the typical prediction error will be about 26.3166. 11.6% of the variability in *Systolic Blood Pressure* is accounted for by this multiple regression equation.

41. Regression Analysis: CALORIES versus PROTEIN, FAT, SAT_FAT, CHOLEST, CARBO, CALCIUM, ...

Analysis of Variance

Source	DF	Adj SS	Adj MS	F-Value	P-Value
Regression	13	282704802	21746523	77981.55	0.000
PROTEIN	1	367727	367727	1318.65	0.000
FAT	1	22696651	22696651	81388.65	0.000
SAT_FAT	1	187	187	0.67	0.413

CHOLEST	1	1082	1082	3.88	0.049
CARBO	1	23097255	23097255	82825.19	0.000
CALCIUM	1	7698	7698	27.60	0.000
PHOSPHOR	1	10809	10809	38.76	0.000
IRON	1	13204	13204	47.35	0.000
POTASS	1	19354	19354	69.40	0.000
SODIUM	1	2649	2649	9.50	0.002
THIAMIN	1	11540	11540	41.38	0.000
NIACIN	1	726	726	2.60	0.107
ASCORBIC	1	412	412	1.48	0.224
Error	947	264088	279		
Lack-of-Fit	927	263971	285	48.82	0.000
Pure Error	20	117	6		
Total	960	282968889			

Model Summary

S	R-sq	R-sq(adj)	R-sq(pred)
16.6993	99.91%	99.91%	99.89%

Coefficients

Term	Coef	SE Coef	T-Value	P-Value	VIF
Constant	2.078	0.724	2.87	0.004	
PROTEIN	4.547	0.125	36.31	0.000	5.53
FAT	8.8003	0.0308	285.29	0.000	3.60
SAT_FAT	-0.092	0.113	-0.82	0.413	5.05
CHOLEST	0.01444	0.00733	1.97	0.049	2.66
CARBO	3.8859	0.0135	287.79	0.000	3.87
CALCIUM	0.03008	0.00573	5.25	0.000	3.07
PHOSPHOR	-0.04277	0.00687	-6.23	0.000	6.80
IRON	-2.340	0.340	-6.88	0.000	3.92
POTASS	-0.01958	0.00235	-8.33	0.000	2.77
SODIUM	0.00360	0.00117	3.08	0.002	1.84
THIAMIN	24.13	3.75	6.43	0.000	4.57
NIACIN	0.554	0.343	1.61	0.107	4.11
ASCORBIC	-0.0245	0.0202	-1.22	0.224	1.43

Regression Equation

CALORIES = 2.078 + 4.547 PROTEIN + 8.8003 FAT - 0.092 SAT_FAT + 0.01444 CHOLEST
+ 3.8859 CARBO + 0.03008 CALCIUM - 0.04277 PHOSPHOR - 2.340 IRON -
0.01958 POTASS
+ 0.00360 SODIUM + 24.13 THIAMIN + 0.554 NIACIN - 0.0245 ASCORBIC

The p-value = 0.000 for the F test. This is significant, so a linear relationship exists between y = calories and at least one of the x variables.

The p-value = 0.413 for saturated fat. This is greater than $\alpha = 0.05$, so we eliminate saturated fat from the model.

Regression Analysis: CALORIES versus PROTEIN, FAT, CHOLEST, CARBO, CALCIUM, PHOSPHOR, ...

Analysis of Variance

Source	DF	Adj SS	Adj MS	F-Value	P-Value
Regression	12	282704614	23558718	84509.25	0.000
PROTEIN	1	380867	380867	1366.24	0.000
FAT	1	49876286	49876286	178914.97	0.000
CHOLEST	1	897	897	3.22	0.073
CARBO	1	23143857	23143857	83021.07	0.000
CALCIUM	1	7513	7513	26.95	0.000
PHOSPHOR	1	11487	11487	41.21	0.000
IRON	1	13016	13016	46.69	0.000
POTASS	1	19397	19397	69.58	0.000
SODIUM	1	2614	2614	9.38	0.002
THIAMIN	1	12005	12005	43.06	0.000
NIACIN	1	667	667	2.39	0.122
ASCORBIC	1	376	376	1.35	0.246
Error	948	264275	279		
Lack-of-Fit	916	264158	288	79.10	0.000
Pure Error	32	117	4		
Total	960	282968889			

Model Summary

S	R-sq	R-sq(adj)	R-sq(pred)
16.6964	99.91%	99.91%	99.90%

Coefficients

Term	Coef	SE Coef	T-Value	P-Value	VIF
Constant	2.022	0.721	2.80	0.005	
PROTEIN	4.564	0.123	36.96	0.000	5.38
FAT	8.7817	0.0208	422.98	0.000	1.63
CHOLEST	0.01169	0.00651	1.79	0.073	2.10
CARBO	3.8853	0.0135	288.13	0.000	3.86
CALCIUM	0.02944	0.00567	5.19	0.000	3.01
PHOSPHOR	-0.04360	0.00679	-6.42	0.000	6.65
IRON	-2.308	0.338	-6.83	0.000	3.87
POTASS	-0.01960	0.00235	-8.34	0.000	2.77
SODIUM	0.00358	0.00117	3.06	0.002	1.84
THIAMIN	24.46	3.73	6.56	0.000	4.52

NIACIN	0.528	0.342	1.55	0.122	4.08
ASCORBIC	-0.0234	0.0201	-1.16	0.246	1.43

Regression Equation

```
CALORIES = 2.022 + 4.564 PROTEIN + 8.7817 FAT + 0.01169 CHOLEST + 3.8853 CARBO
         + 0.02944 CALCIUM - 0.04360 PHOSPHOR - 2.308 IRON - 0.01960 POTASS
         + 0.00358 SODIUM + 24.46 THIAMIN + 0.528 NIACIN - 0.0234 ASCORBIC
```

The p-value = 0.246 for ascorbic acid. This is greater than $\alpha = 0.05$, so we eliminate ascorbic acid from the model.

Regression Analysis: CALORIES versus PROTEIN, FAT, CHOLEST, CARBO, CALCIUM, PHOSPHOR, ...

Analysis of Variance

Source	DF	Adj SS	Adj MS	F-Value	P-Value
Regression	11	282704239	25700385	92158.01	0.000
PROTEIN	1	385118	385118	1380.98	0.000
FAT	1	50010977	50010977	179332.42	0.000
CHOLEST	1	801	801	2.87	0.090
CARBO	1	23165354	23165354	83067.75	0.000
CALCIUM	1	7630	7630	27.36	0.000
PHOSPHOR	1	11115	11115	39.86	0.000
IRON	1	13018	13018	46.68	0.000
POTASS	1	28748	28748	103.09	0.000
SODIUM	1	2747	2747	9.85	0.002
THIAMIN	1	11727	11727	42.05	0.000
NIACIN	1	685	685	2.46	0.117
Error	949	264651	279		
Lack-of-Fit	914	264534	289	86.83	0.000
Pure Error	35	117	3		
Total	960	282968889			

Model Summary

S	R-sq	R-sq(adj)	R-sq(pred)
16.6995	99.91%	99.91%	99.90%

Coefficients

Term	Coef	SE Coef	T-Value	P-Value	VIF
Constant	1.907	0.714	2.67	0.008	
PROTEIN	4.576	0.123	37.16	0.000	5.35
FAT	8.7828	0.0207	423.48	0.000	1.63
CHOLEST	0.01100	0.00649	1.69	0.090	2.09
CARBO	3.8848	0.0135	288.21	0.000	3.85
CALCIUM	0.02965	0.00567	5.23	0.000	3.01

PHOSPHOR	-0.04232	0.00670	-6.31	0.000	6.48
IRON	-2.308	0.338	-6.83	0.000	3.87
POTASS	-0.02091	0.00206	-10.15	0.000	2.13
SODIUM	0.00366	0.00117	3.14	0.002	1.83
THIAMIN	24.09	3.71	6.48	0.000	4.48
NIACIN	0.535	0.342	1.57	0.117	4.08

Regression Equation

```
CALORIES = 1.907 + 4.576 PROTEIN + 8.7828 FAT + 0.01100 CHOLEST + 3.8848 CARBO
         + 0.02965 CALCIUM - 0.04232 PHOSPHOR - 2.308 IRON - 0.02091 POTASS
         + 0.00366 SODIUM + 24.09 THIAMIN + 0.535 NIACIN
```

The p-value = 0.117 for niacin. This is greater than $\alpha = 0.05$, so we eliminate niacin from the model.

Regression Analysis: CALORIES versus PROTEIN, FAT, CHOLEST, CARBO, CALCIUM, PHOSPHOR, ...

Analysis of Variance

Source	DF	Adj SS	Adj MS	F-Value	P-Value
Regression	10	282703554	28270355	101218.54	0.000
PROTEIN	1	499076	499076	1786.88	0.000
FAT	1	50125869	50125869	179469.53	0.000
CHOLEST	1	631	631	2.26	0.133
CARBO	1	23166870	23166870	82946.14	0.000
CALCIUM	1	6946	6946	24.87	0.000
PHOSPHOR	1	11212	11212	40.14	0.000
IRON	1	12363	12363	44.27	0.000
POTASS	1	28580	28580	102.33	0.000
SODIUM	1	2743	2743	9.82	0.002
THIAMIN	1	18162	18162	65.03	0.000
Error	950	265335	279		
Lack-of-Fit	915	265218	290	86.96	0.000
Pure Error	35	117	3		
Total	960	282968889			

Model Summary

S	R-sq	R-sq(adj)	R-sq(pred)
16.7123	99.91%	99.91%	99.90%

Coefficients

Term	Coef	SE Coef	T-Value	P-Value	VIF
Constant	1.976	0.713	2.77	0.006	
PROTEIN	4.662	0.110	42.27	0.000	4.28
FAT	8.7812	0.0207	423.64	0.000	1.62

CHOLEST	0.00968	0.00644	1.50	0.133	2.05
CARBO	3.8849	0.0135	288.00	0.000	3.85
CALCIUM	0.02694	0.00540	4.99	0.000	2.73
PHOSPHOR	-0.04250	0.00671	-6.34	0.000	6.47
IRON	-2.211	0.332	-6.65	0.000	3.74
POTASS	-0.02085	0.00206	-10.12	0.000	2.13
SODIUM	0.00366	0.00117	3.13	0.002	1.83
THIAMIN	26.72	3.31	8.06	0.000	3.56

Regression Equation

```
CALORIES = 1.976 + 4.662 PROTEIN + 8.7812 FAT + 0.00968 CHOLEST + 3.8849 CARBO
           + 0.02694 CALCIUM - 0.04250 PHOSPHOR - 2.211 IRON - 0.02085 POTASS
           + 0.00366 SODIUM + 26.72 THIAMIN
```

The p-value = 0.133 for cholesterol. This is greater than $\alpha = 0.05$, so we eliminate cholesterol from the model.

Regression Analysis: CALORIES versus PROTEIN, FAT, CARBO, CALCIUM, PHOSPHOR, IRON, ...

Analysis of Variance

Source	DF	Adj SS	Adj MS	F-Value	P-Value
Regression	9	282702923	31411436	112316.16	0.000
PROTEIN	1	582434	582434	2082.58	0.000
FAT	1	56324642	56324642	201396.95	0.000
CARBO	1	25983381	25983381	92907.36	0.000
CALCIUM	1	7462	7462	26.68	0.000
PHOSPHOR	1	11708	11708	41.86	0.000
IRON	1	12143	12143	43.42	0.000
POTASS	1	30371	30371	108.60	0.000
SODIUM	1	2664	2664	9.53	0.002
THIAMIN	1	17669	17669	63.18	0.000
Error	951	265966	280		
Lack-of-Fit	913	265049	290	12.03	0.000
Pure Error	38	917	24		
Total	960	282968889			

Model Summary

S	R-sq	R-sq(adj)	R-sq(pred)
16.7233	99.91%	99.91%	99.90%

Coefficients

Term	Coef	SE Coef	T-Value	P-Value	VIF
Constant	1.880	0.711	2.64	0.008	
PROTEIN	4.720	0.103	45.64	0.000	3.76

```
FAT           8.7914    0.0196    448.77    0.000  1.45
CARBO         3.8915    0.0128    304.81    0.000  3.45
CALCIUM      0.02778   0.00538      5.17    0.000  2.70
PHOSPHOR    -0.04329   0.00669     -6.47    0.000  6.43
IRON          -2.190     0.332     -6.59    0.000  3.73
POTASS      -0.02128   0.00204    -10.42    0.000  2.09
SODIUM       0.00360   0.00117      3.09    0.002  1.83
THIAMIN        25.38      3.19      7.95    0.000  3.30
Regression Equation
CALORIES = 1.880 + 4.720 PROTEIN + 8.7914 FAT + 3.8915 CARBO + 0.02778 CALCIUM
           - 0.04329 PHOSPHOR - 2.190 IRON - 0.02128 POTASS + 0.00360 SODIUM
+ 25.38 THIAMIN
```

All p-values are less than or equal to $\alpha = 0.05$, so no more variables need to be eliminated from the model.

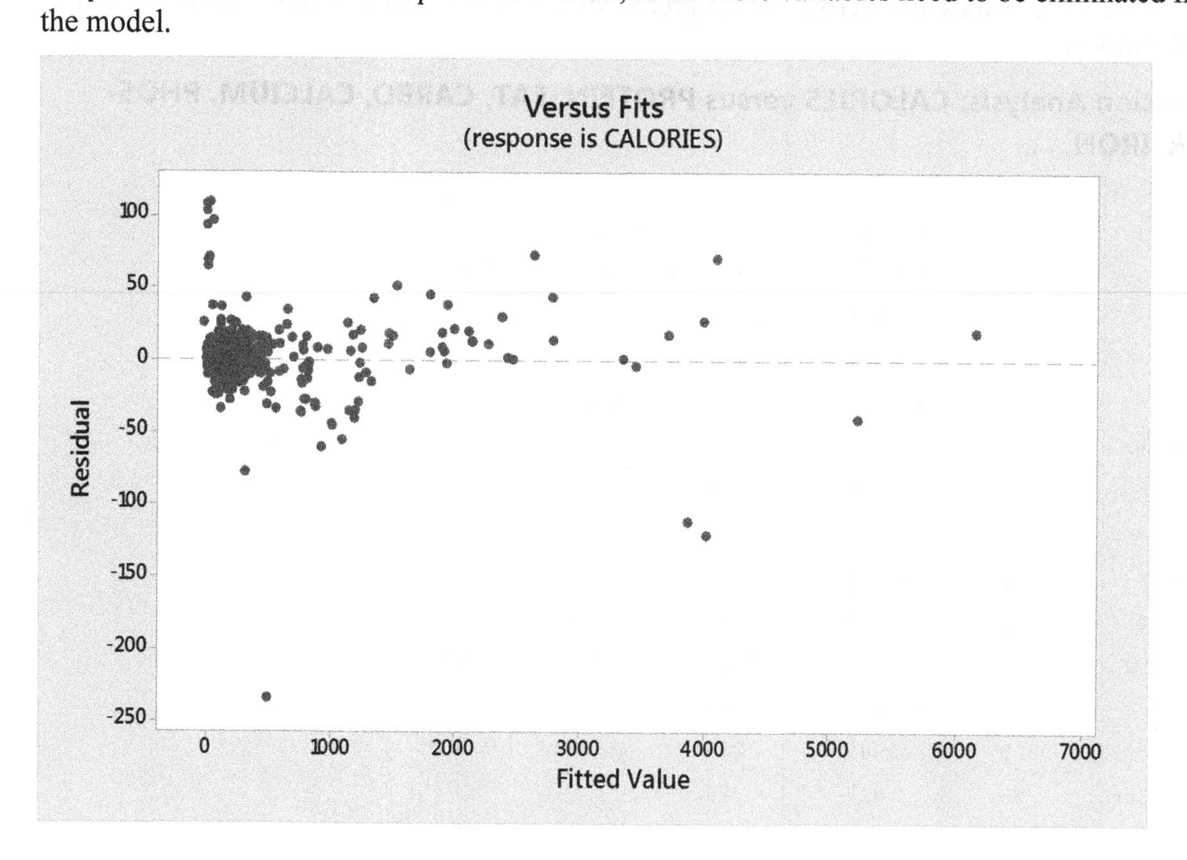

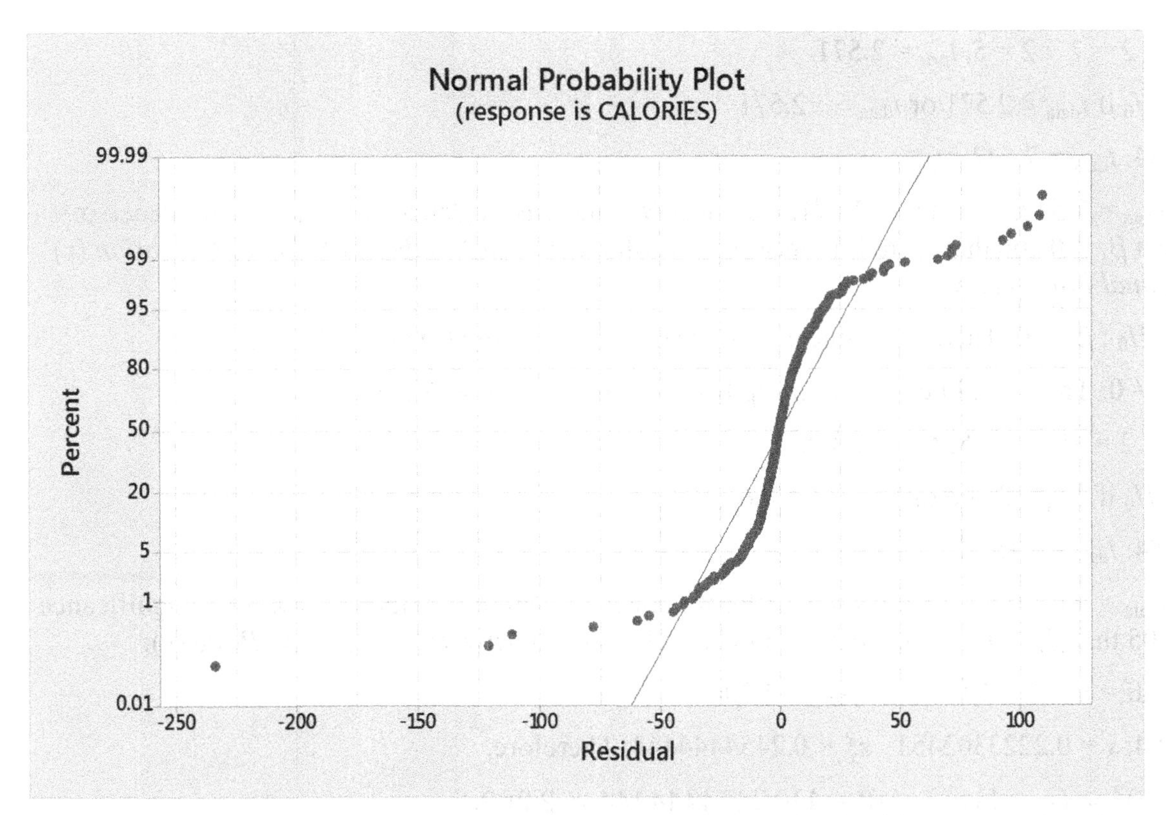

The scatterplot of the residuals contains no strong evidence of unhealthy patterns but the normal probability plot indicates evidence of departures from normality in the residuals. Therefore, we conclude that the regression assumptions are not fully verified.

```
Regression Equation

CALORIES = 1.880 + 4.720 PROTEIN + 8.7914 FAT + 3.8915 CARBO + 0.02778 CALCIUM

           - 0.04329 PHOSPHOR - 2.190 IRON - 0.02128 POTASS + 0.00360 SODIUM
+ 25.38 THIAMIN

Model Summary

      S    R-sq  R-sq(adj)  R-sq(pred)

16.7233  99.91%     99.91%      99.90%
```

The standard error in the estimate for the final model is $s = 16.7233$. That is, using the multiple regression equation given above, the size of the typical prediction error will be about 16.7233 calories.

The adjusted coefficient of variation is $R^2_{\text{adj}} = 99.91\%$. In other words, 99.91% of the variation in calories is accounted for by this multiple regression equation.

Chapter 13 Review Exercises

1. $H_0 : \beta_1 = 0$: There is no linear relationship between *Number of Years of Education* (x) and *Annual Earnings* (y).

$H_a : \beta_1 \neq 0$: There is a linear relationship between *Number of Years of Education* (x) and *Annual Earnings* (y).

$df = n - 2 = 7 - 2 = 5$, $t_{crit} = 2.571$

Reject H_0 if $t_{data} \geq 2.571$ or $t_{data} \leq -2.571$

TI-83/84: $t_{data} \approx 7.542$.

Since $t_{data} \approx 7.542$, which is ≥ 2.571, we reject H_0. There is evidence at level of significance $\alpha = 0.05$ that $\beta_1 \neq 0$ and that there is a linear relationship between *Number of Years of Education* (x) and *Annual Earnings* (y).

3. $H_0 : \beta_1 = 0$: There is no linear relationship between *Age* (x) and *Price* (y).

$H_a : \beta_1 \neq 0$: There is a linear relationship between *Age* (x) and *Price* (y).

df $= n - 2 = 10 - 2 = 8$, $t_{crit} = 2.306$.

Reject H_0 if $t_{data} \geq 2.306$ or $t_{data} \leq -2.306$.

TI-83/84: $t_{data} \approx -15.01$.

Since $t_{data} \approx -15.0124$, which is ≤ -2.306, we reject H_0. There is evidence at level of significance $\alpha = 0.05$ that $\beta_1 \neq 0$ and that there is a linear relationship between *Age* (x) and *Price* (y).

5. df $= n - 2 = 10 - 2 = 8$, $t_{\alpha/2} = 2.306$.

TI-83/84: $s \approx 0.2223302451$. $s_x^2 \approx 0.2454444444$. Therefore,

$\sum(x - \bar{x})^2 = (n-1)s_x^2 \approx (10-1)0.2454444444 = 2.209$.

TI-83/84: $b_1 \approx 0.6840$.

$b_1 \pm t_{\alpha/2} \cdot \frac{s}{\sqrt{\sum(x-\bar{x})^2}} \approx 0.6840 \pm (2.306)\left(\frac{0.2223302451}{\sqrt{2.209}}\right) \approx 0.6840 \pm 0.3450 =$ $(0.3390, 1.0290)$ TI-83/84: $(0.3391, 1.0290)$.

7. **(a)** $x^* = 10$ years, $\hat{y} \approx 4.468571429x - 23.76285714 = 4.468571429(10) - 23.76285714$ $= 44.68571429 - 23.76285714 \approx 20.9229$ thousand dollars.

(b)

New Obs	Fit	SE Fit	95% CI	95% PI
1	20.92	2.58	(14.28, 27.56)	(6.55, 35.29)

df $= n - 2 = 7 - 2 = 5$, $t_{\alpha/2} = 2.571$

TI-83/84: $\bar{x} = 13$

TI-83/84: $s \approx 4.957118979$

TI-83/84: $s_x^2 \approx 11.66666667$, so $\sum(x - \bar{x})^2 = (n-1)\, s_x^2 \approx (7-1)11.66666667 = 70$

Lower bound:

$$\hat{y} - t_{\alpha/2} \cdot s\sqrt{\frac{1}{n} + \frac{(x^* - \bar{x})^2}{\sum(x_i - \bar{x})^2}} \approx 20.9229 - (2.571)(4.957118979)\sqrt{\frac{1}{7} + \frac{(10-13)^2}{70}}$$

$$\approx 20.9229 - 6.6399 = 14.2830$$

Upper bound:

$$\hat{y} + t_{\alpha/2} \cdot s\sqrt{\frac{1}{n} + \frac{(x^* - \bar{x})^2}{\sum(x_i - \bar{x})^2}} \approx 20.9229 + (2.571)(4.957118979)\sqrt{\frac{1}{7} + \frac{(10 - 13)^2}{70}}$$
$$\approx 20.9229 + 6.6399 = 27.5628$$

(14.2830, 27.5630). Minitab: (14.28, 27.56). We are 95% confident that the mean annual salary for people with 10 years of education lies between 14.28 thousand dollars and 27.56 thousand dollars.

(c) Lower bound:

$$\hat{y} - t_{\alpha/2} \cdot s\sqrt{1 + \frac{1}{n} + \frac{(x^* - \bar{x})^2}{\sum(x_i - \bar{x})^2}} \approx 20.9229 - (2.571)(4.957118979)\sqrt{1 + \frac{1}{7} + \frac{(10 - 13)^2}{70}}$$
$$\approx 20.9229 - 14.3707 = 6.5522$$

Upper bound:

$$\hat{y} + t_{\alpha/2} \cdot s\sqrt{1 + \frac{1}{n} + \frac{(x^* - \bar{x})^2}{\sum(x_i - \bar{x})^2}} \approx 20.9229 + (2.571)(4.957118979)\sqrt{1 + \frac{1}{7} + \frac{(10 - 13)^2}{70}}$$
$$\approx 20.9229 + 14.3707 = 35.2936$$

(6.5522, 35.2936). Minitab: (6.55, 35.29). We are 95% confident that the annual salary for a randomly selected person with 10 years of education lies between 6.55 thousand dollars and 35.29 thousand dollars.

9. **(a)** $x^* = 8$ years old, $\hat{y} \approx -1.623529412x + 19.80588235 = -1.623529412(8) + 19.80588235 = -12.9882353 + 19.80588235 \approx 6.8176$ thousand dollars.

(b) Predicted Values for New Observations

New Obs	Fit	SE Fit	95% CI	95% PI
1	6.818	0.439	(5.805, 7.831)	(4.902, 8.733)

$df = n - 2 = 10 - 2 = 8$, $t_{\alpha/2} = 2.306$.

TI-83/84: $\bar{x} = 4.5$.

TI-83/84: $s \approx 0.7050239879$.

TI-83/84: $s_x^2 \approx 4.722222222$, so

$\sum(x - \bar{x})^2 = (n - 1)\, s_x^2 \approx (10 - 1)4.722222222 = 42.5$.

Lower bound:

$$\hat{y} - t_{\alpha/2} \cdot s\sqrt{\frac{1}{n} + \frac{(x^* - \bar{x})^2}{\sum(x_i - \bar{x})^2}} \approx 6.8176 - (2.306)(0.70502398789)\sqrt{\frac{1}{10} + \frac{(8 - 4.5)^2}{42.5}}$$
$$\approx 6.8176 - 1.0130 = 5.8046.$$

Upper bound:

$$\hat{y} + t_{\alpha/2} \cdot s\sqrt{\frac{1}{n} + \frac{(x^* - \bar{x})^2}{\sum(x_i - \bar{x})^2}} \approx 6.8176 + (2.306)(0.70502398789)\sqrt{\frac{1}{10} + \frac{(8 - 4.5)^2}{42.5}}$$
$$\approx 6.8176 + 1.0130 = 7.8306.$$

(5.8046, 7.8306). Minitab: (5.805, 7.831). We are 95% confident that the mean price for 8-year-old cars of this make and model lies between 5.805 thousand dollars and 7.831 thousand dollars.

(c) Lower bound:

$$\hat{y} - t_{\alpha/2} \cdot s\sqrt{1 + \frac{1}{n} + \frac{(x^* - \bar{x})^2}{\sum(x_i - \bar{x})^2}} \approx 6.8176 - (2.306)(0.70502398789)\sqrt{1 + \frac{1}{10} + \frac{(8 - 4.5)^2}{42.5}}$$
$$\approx 6.8176 - 1.9156 = 4.902.$$

Upper bound:

$$\hat{y} + t_{\alpha/2} \cdot s\sqrt{1 + \frac{1}{n} + \frac{(x^* - \bar{x})^2}{\sum(x_i - \bar{x})^2}} \approx 6.8176 + (2.306)(0.70502398789)\sqrt{1 + \frac{1}{10} + \frac{(8 - 4.5)^2}{42.5}}$$
$$\approx 6.8176 + 1.9156 = 8.7332.$$

(4.902, 8.7332). Minitab: (4.902, 8.733). We are 95% confident that the price for a randomly selected 8-year-old car of this make and model lies between 4.902 thousand dollars and 8.733 thousand dollars.

11. **(a)** Test 1

$H_0 : \beta_1 = 0$: There is no linear relationship between y and x_1.

$H_a : \beta_1 \neq 0$: There is a linear relationship between y and x_1.

Reject H_0 if the p-value $\leq \alpha = 0.05$.

Test 2

$H_0 : \beta_2 = 0$: There is no linear relationship between y and x_2.

$H_a : \beta_2 \neq 0$: There is a linear relationship between y and x_2.

Reject H_0 if the p-value $\leq \alpha = 0.05$.

Test 3

$H_0 : \beta_3 = 0$: There is no linear relationship between y and x_3.

$H_a : \beta_3 \neq 0$: There is a linear relationship between y and x_3.

Reject H_0 if the p-value $\leq \alpha = 0.05$.

(b)

Predictor	Coef	SE Coef	T	P
Constant	21.87	10.11	2.16	0.074
x1	0.6467	0.3478	1.86	0.112
x2	–0.01968	0.07337	–0.27	0.798
x3	–0.6531	0.6111	–1.07	0.326

Test 1: $t = 1.86$, with p-value = 0.112.

Test 2: $t = -0.27$, with p-value = 0.798.

Test 3: $t = -1.07$, with p-value = 0.326.

(c) Test 1: The p-value = 0.112, which is not $\leq \alpha = 0.05$. Therefore, we do not reject H_0. There is insufficient evidence of a linear relationship between y and x_1.

Test 2: The p-value = 0.798, which is not $\leq \alpha = 0.05$. Therefore, we do not reject H_0. There is insufficient evidence of a linear relationship between y and x_2.

Test 3: The p-value = 0.326, which is not $\leq \alpha = 0.05$. Therefore, we do not reject H_0. There is insufficient evidence of a linear relationship between y and x_3.

13.

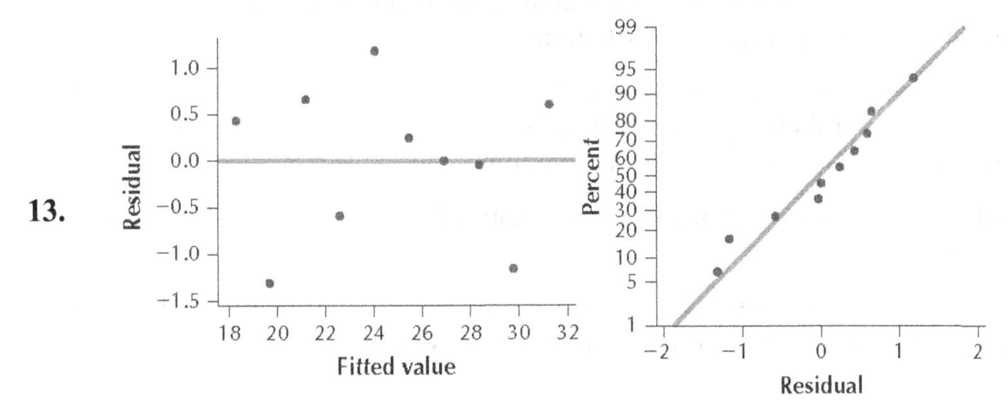

The scatterplot above of the residuals versus fitted values shows no strong evidence of unhealthy patterns. Thus, the independence assumption, the constant variance assumption, and the zero-mean assumption are verified. Also, the normal probability plot of the residuals above indicates no evidence of departure from normality of the residuals. Therefore, we conclude that the regression assumptions are verified.

Chapter 13 Quiz

1. False. We also need to use a normal probability plot of the residuals.

3. True

5. β_1, b_1

7. Zero mean assumption $E(\varepsilon) = 0$, constant variance assumption, independence assumption, normality assumption

9. R^2_{adj} is preferable to R^2 as a measure of the goodness of a regression equation, because R^2_{adj} will decrease if an unhelpful x-variable is added to the regression equation.

11. **(a)** No. Since H_0 was rejected at level of significance $\alpha = 0.01$ in the hypothesis test in Exercise 10, the conclusion is that $\beta_1 \neq 0$.

(b) TI-83/84: (4.3998, 10.17). We are 99% confident that the interval (4.3998, 10.17) captures the slope β_1 of the regression line. That is, we are 99% confident that, for each additional inch in height of a male student, the increase in the weight lies between 4.3998 and 10.17.

13. **(a)** The regression equation is $\hat{y} = 16.8 + 0.718x_1$

(b) The estimated value of y when $x_1 = 0$ is 16.8. For each increase in one unit of the variable x_1, the estimated value of y increases by 0.718.

(c) S = 0.854028 R-Sq = 96.7% R-Sq(adj) = 96.3%

Using the multiple regression equation above, the size of the typical prediction error will be about $s \approx 0.854028$. 96.3% of the variability in y is accounted for by this multiple regression equation.

Chapter 14: Nonparametric Statistics

Section 14.1

1. Population characteristic; used to test claims about population parameters and often require the population to follow a particular distribution

3. Parametric hypothesis tests are used to test claims about a population parameter and they often require the population to follow a particular distribution. Nonparametric hypothesis tests do not require the population to follow a particular distribution.

5. May be used for a greater variety of data, can be applied to categorical data, the manual computations tend to be easier.

7. The ratio of the sample size required for the corresponding parametric test to the sample size required for the nonparametric test in order to achieve the same result.

Section 14.2

1. One-sample t test for the population mean μ

3. True.

5. False. The test statistic for the large sample case is $Z_{\text{data}} = \frac{(S_{\text{data}}+0.5)-\frac{n}{2}}{\frac{\sqrt{n}}{2}}$.

7. Paired-sample t test

9. **(a)** n = 10 plus signs + 10 minus signs = 20 ≤ 25, $\alpha = 0.05$, one-tailed test, so $S_{crit} = 5$.

(b) Reject H_0 if $S_{data} \leq 5$.

(c) Since this is a right-tailed test, S_{data} = the number of minus signs = 10.

(d) Since $S_{data} = 10$, which is not ≤ to 5, the conclusion is do not reject H_0. There is insufficient evidence that the population median is greater than 10.

11. **(a)** n = 0 plus signs and 8 minus signs = 8 ≤ 25, $\alpha = 0.10$, two-tailed test, so $S_{crit} = 1$.

(b) Reject H_0 if $S_{data} \leq 1$.

(c) Since this is a two-tailed test, S_{data} = the number of minus signs or plus signs, whichever is smaller. Therefore $S_{\text{data}} = 0$.

(d) Since $S_{data} = 0$, which is ≤ 1, the conclusion is reject H_0. There is evidence that the population median is different than 0.

13. **(a)** $H_0 : M = 10$ versus $H_a : M < 10$.

Since n = the number of values in the data set not equal to 10 = 11 is ≤ 25, we use the small sample case. For a one-tailed test with $n = 11$ and $\alpha = 0.05$, $S_{crit} = 2$.

(b) Reject H_0 if $S_{data} \leq 2$.

(c)

Number	10	8	9	5	11	10	6	9	3	12	1	7	2
Sign	None	–	–	–	+	None	–	–	–	+	–	–	–

Since this is a left-tailed test, S_{data} = the number of plus signs = 2.

(d) Since $S_{data} = 2$, which is ≤ 2, reject H_0. There is evidence that the population median is less than 10.

15. **(a)** $H_0 : M = 400$ versus $H_a : M < 400$. Since n = the number of values in the data set not equal to 400 = 11 is ≤ 25, we use the small sample case. For a one-tailed test with $n = 11$ and $\alpha = 0.05$, $S_{crit} = 2$.

(b) Reject H_0 if $S_{data} \leq 2$.

(c)

Number	105	219	100	136	345	996	100	400	102	100	229	331
Sign	–	–	–	–	–	+	–	None	–	–	–	–

Since this is a left-tailed test, S_{data} = the number of plus signs = 1.

(d) Since $S_{data} = 1$, which is ≤ 2, reject H_0. There is evidence that the population median is less than 400.

17. **(a)** $H_0 : M = 3.14$ versus $H_a : M > 3.14$. Since n = the number of pluses + the number of minuses =100 + 10 =110 is > 25, we use the large sample case.

(b) This is a one-tailed test with $\alpha = 0.05$. Therefore, $Z_{crit} = -1.645$. Reject H_0 if $Z_{data} \leq -1.645$.

(c) Since this is a right-tailed test, S_{data} = the number of minus signs = 10. Thus,

$$Z_{\text{data}} = \frac{(S_{\text{data}}+0.5)-\frac{n}{2}}{\frac{\sqrt{n}}{2}} = \frac{(10+0.5)-\frac{110}{2}}{\frac{\sqrt{110}}{2}} \approx -8.49$$

(d) Since $Z_{\text{data}} \approx -8.49$ is ≤ -1.645, we reject H_0. There is evidence that the population median is greater than 3.14.

19. **(a)** $H_0 : M = -0.25$ versus $H_a : M \neq -0.25$. Since n = the number of plusses + the number of minuses = 225 + 5 =230 is > 25, we use the large sample case.

(b) This is a two-tailed test with $\alpha = 0.10$. Therefore, $Z_{crit} = -1.645$. Reject H_0 if $Z_{data} \leq -1.645$.

(c) Since this is a two-tailed test, S_{data} = the number of minus signs or plus signs, whichever is smaller = 5. Thus,

$$Z_{\text{data}} = \frac{(S_{\text{data}}+0.5)-\frac{n}{2}}{\frac{\sqrt{n}}{2}} = \frac{(5+0.5)-\frac{230}{2}}{\frac{\sqrt{230}}{2}} \approx -14.44$$

(d) Since $Z_{\text{data}} \approx -14.44$ is ≤ -1.645, we reject H_0. There is evidence that the population median is different than -0.25.

21. **(a)** $H_0 : M_d = 0$ versus $H_a : M_d > 0$.

(b) Since $n =$ the number of differences not equal to 0 = 4 is ≤ 25, we use the small sample case. Table I indicates that it is not possible to get a value in the critical region for a one-tailed test with $n = 4$ and $\alpha = 0.05$. Therefore, we would not reject H_0.

(c)

Subject	1	2	3	4	5
Sample 1	3.0	2.5	3.5	3.0	4.0
Sample 2	2.5	2.5	2.0	2.0	1.5
Difference (Sample 1 – Sample 2)	0.5	0	1.5	1.0	2.5
Sign	+	None	+	+	+

Since this is a right-tailed test, S_{data} = the number of minus signs = 0.

(d) Do not reject H_0. There is insufficient evidence that the population median of the differences is greater than 0.

23. **(a)** $H_0 : M_d = 0$ versus $H_a : M_d > 0$.

(b) Since n = the number of differences not equal to 0 = 7 is ≤ 25, we use the small sample case. For a one-tailed test with $n = 7$ and $\alpha = 0.05$, $S_{crit} = 0$. Reject H_0 if $S_{data} \leq 0$.

(c)

Subject	1	2	3	4	5	6	7
Sample 1	20	25	15	10	20	30	15
Sample 2	30	30	20	20	25	35	25
Difference (Sample 1 – Sample 2)	–10	–5	–5	–10	–5	–5	–10
Sign	–	–	–	–	–	–	–

Since this is a right-tailed test, S_{data} = the number of minus signs = 7.

(d) Since $S_{data} = 7$, which is not ≤ 0, do not reject H_0. There is insufficient evidence that the population median of the differences is greater than 0.

25. $H_0 : p = 0.5$ versus $H_a : p < 0.5$. Since $n =$ the sample size = 200, which is > 25, we use the large sample case. This is a one-tailed test with $\alpha = 0.01$, so $Z_{crit} = -2.326$. Reject H_0 if $Z_{data} \leq -2.326$. Since this is a left-tailed test, S_{data} = the number of plus signs = the number of successes = 75.

$$Z_{\text{data}} = \frac{(S_{\text{data}}+0.5)-\frac{n}{2}}{\frac{\sqrt{n}}{2}} = \frac{(75+0.5)-\frac{200}{2}}{\frac{\sqrt{200}}{2}} \approx -3.46$$

Since $Z_{data} \approx -3.46$, which is ≤ -2.326, we reject H_0. There is evidence that the population proportion of successes is less than 0.5.

27. $H_0 : M = 90$ versus $H_a : M > 90$. Since n = the number of values in the data set not equal to 90 = 5 is ≤ 25, we use the small sample case. For a one-tailed test with $n = 5$ and $\alpha = 0.05$, $S_{crit} = 0$. Reject H_0 if $S_{data} \leq 0$.

Mileage (MPGe) **Sign**

89	–
99	+
105	+
112	+
119	+

Since this is a right-tailed test, S_{data} = the number of minus signs = 1

Since $S_{data} = 1$ is not ≤ 0, do not reject H_0. There is insufficient evidence that the population median mileage is greater than 90 MPGe.

29. $H_0 : M = 500{,}000$ versus $H_a : M > 500{,}000$. Since n = the number of values in the data set not equal to 500,000 = 4 is ≤ 25, we use the small sample case. For a one-tailed test with n = 4 and $\alpha = 0.01$, Table I indicates that it is not possible to get a value in the critical region for this sample size and level of significance. Therefore, we do not reject H_0.

Cleanup costs ($ millions)	Sign
0.85	+
0.70	+
0.50	None
1.15	+
0.50	None
1.35	+

Since this is a right-tailed test, S_{data} = the number of minus signs = 0

Table I indicates that it is not possible to get a value in the critical region for this sample size and level of significance. Therefore, we do not reject H_0. There is insufficient evidence that the population median cleanup cost exceeds $500,000.

31. $H_0 : M_d = 0$ versus $H_a : M_d \neq 0$. Since n = the number of differences not equal to 0 =10 , which is ≤ 25, we use the small sample case. For a two-tailed test with n = 10 and $\alpha = 0.05$, $S_{crit} = 1$. Reject H_0 if $S_{data} \leq 1$.

High (°C)	Low (°C)	Difference (High – Low) (°C)	Sign
9.4	0.8	8.6	+
6.1	−8.9	15	+
5.9	−1.3	7.2	+
29.1	19.3	9.8	+
11.9	6.7	5.2	+
30.6	21.5	9.1	+
23.1	10.5	12.6	+
33.1	18.7	14.4	+
14.8	7.4	7.4	+
0.1	−9.9	10.0	+

Since this is a two-tailed test, S_{data} = the number of plus signs or minus signs, whichever is smaller. Therefore $S_{data} = 0$.

Since $S_{data} = 0$ is ≤ 1, reject H_0. There is evidence that the population median of the difference in temperature differs from zero.

33. $H_0 : p = 0.5$ versus $H_a : p \neq 0.5$. Since n = the sample size = 1000 is > 25, we use the large sample case. $Z_{crit} = -1.645$. Reject H_0 if $Z_{data} \leq -1.645$. Since this is a two-tailed test, S_{data} = the number of plus signs or minus signs, whichever is smaller = the number of Internet browsers in Finland using Firefox = 472.

$$Z_{\text{data}} = \frac{(S_{\text{data}}+0.5)-\frac{n}{2}}{\frac{\sqrt{n}}{2}} = \frac{(472+0.5)-\frac{1000}{2}}{\frac{\sqrt{1000}}{2}} \approx -1.7393$$

Since $Z_{data} \approx -1.7393$, which is ≤ -1.645, reject H_0. There is evidence that the population proportion of Internet browsers in Finland using Firefox differs from 0.5.

Section 14.3

1. The requirements for the Wilcoxon signed rank test are that the sample data should be randomly selected and that the distribution of the differences should be symmetric.

3. The boxplot has whiskers of approximately equal length and the median is situated approximately in the center of the box.

5. **(a)** For each data value, find the difference d between the data values for each matched pair for the population median of the differences for matched-pair data from two dependent samples or the data point and the hypothesized median for the population median of a single population. Omit observations where $d = 0$.

(b) Find the absolute value of the differences.

(c) Rank the absolute values of the differences from smallest to largest. If two or more data values are tied with the same rank, assign to each the mean value of their ranks had they not been tied.

(d.) Attach to each rank the sign of its corresponding value of d.

7. For a right-tailed test, $T_{data} = |T_-|$, where T_- is the sum of the negative-signed ranks. For a left-tailed test, $T_{data} = T_+$ where T_+ is the sum of the positive-signed ranks. For a two-tailed test, $T_{data} = T_+$ or $|T_-|$, whichever is smaller.

9. Small-sample case: Reject H_0 if $T_{data} \leq T_{crit}$. Large-sample case: Reject H_0 if $Z_{data} \leq Z_{crit}$.

11. The left whisker is shorter than the right whisker and the median is situated closer to Q1 than to Q3. Therefore, the distribution is not symmetric, so it is not appropriate to perform the Wilcoxon signed rank test.

13. The left whisker and the right whisker are approximately the same length and the median is situated approximately in the center of the box. Therefore, the distribution is symmetric, so it is appropriate to perform the Wilcoxon signed rank test.

15. Since $n = 12$ is less than or equal to 30, we use the Wilcoxon signed rank test for the small sample case. Since this is a right-tailed test with $\alpha = 0.05$, $T_{crit} = 17$. Reject H_0 if $T_{data} \leq 17$. Since this is a right-tailed test, $T_{data} = |T_-| = 20$. Since $T_{data} = 20$, which is not $\leq T_{crit} = 17$, we do not reject H_0. There is insufficient evidence that the population median difference is greater than 0.

17. Since $n = 18$ is less than or equal to 30, we use the Wilcoxon signed rank test for the small sample case. Since this is a left-tailed test with $\alpha = 0.05$, $T_{crit} = 47$. Reject H_0 if $T_{data} \leq 47$. Since this is a left-tailed test, $T_{data} = T_+ = 45$. Since $T_{data} = 45$, which is $\leq T_{crit} = 47$, we reject H_0. There is evidence that the population median difference is less than 0.

19. Since $n = 20$ is less than or equal to 30, we use the Wilcoxon signed rank test for the small sample case. Since this is a right-tailed test with $\alpha = 0.05$, $T_{crit} = 60$. Reject H_0 if $T_{data} \leq 60$. Since this is a right-tailed test, $T_{data} = |T_-| = 50$. Since $T_{data} = 50$, which is $\leq T_{crit} = 60$, we reject H_0. There is evidence that the population median is greater than 10.

21. Since $n = 15$ is less than or equal to 30, we use the Wilcoxon signed rank test for the small sample case. Since this is a left-tailed test with $\alpha = 0.05$, $T_{crit} = 30$. Reject H_0 if $T_{data} \leq 30$. Since this is a left-tailed test, $T_{data} = T_+ = 20$. Since $T_{data} = 20$, which is $\leq T_{crit} = 30$, we reject H_0. There is evidence that the population median is less than 45.

23. The left whisker and the right whisker are approximately the same length and the median is situated approximately in the center of the box. Therefore, the distribution is symmetric so it is appropriate to perform the Wilcoxon signed rank test. $H_0: M = 98.6$ versus $H_a: M \neq 98.6$. Since $M_0 = 98.6$, n equals the number of data values that are not equal to 98.6. Since none of the data values is equal to 98.6, $n = 7$. Since $n = 7$ is less than or equal to 30, we use the Wilcoxon signed rank test for the small sample case. Since this is a two-tailed test with $\alpha = 0.05$, $T_{crit} = 2$. Reject H_0 if $T_{data} \leq 2$.

Temperature	**Temperature – M_0 =**	**$\lvert d \rvert$**	**Rank of $\lvert d \rvert$**	**Signed rank**
97.2	97.2 – 98.6 = –1.4	1.4	7	–7
97.8	97.8 – 98.6 = –0.8	0.8	6	–6
98.1	98.1 – 98.6 = –0.5	0.5	4	–4
98.3	98.3 – 98.6 = –0.3	0.3	3	–3
98.7	98.7 – 98.6 = 0.1	0.1	1	1
98.8	98.8 – 98.6 = 0.2	0.2	2	2
99.3	99.3 – 98.6 = 0.7	0.7	5	5

$T_+ = 1 + 2 + 5 = 8$ and $|T_-| = |(-7) + (-6) + (-4) + (-3)| = 20$.

Since this is a two-tailed test, $T_{data} = T_+$ or $|T_-|$, whichever is smaller. Thus $T_{data} = 8$. Since $T_{data} = 8$, which is not $\leq T_{crit} = 2$, we do not reject H_0. There is insufficient evidence that the population median of women's body temperatures differs from 98.6 degrees Fahrenheit.

25. Since perfect symmetry is rare in real-world data, especially with small data sets, we will interpret the boxplot as revealing acceptable symmetry. Therefore, it is appropriate to perform the Wilcoxon signed rank test. $H_0: M = 2500$ versus $H_a: M > 2500$. Since $M_0 = 2500$, n equals the number of data values that are not equal to 2500. Since none of the data values is equal to 2500, $n = 8$. Since $n = 8$ is less than or equal to 30, we use the Wilcoxon signed rank test for the small sample case. Since this is a right-tailed test with $\alpha = 0.01$, $T_{crit} = 2$. Reject H_0 if $T_{data} \leq 2$.

Number of businesses	Number – M_0 =	$\|d\|$	Rank of $\|d\|$	Signed rank
7923	7923 – 2500 = 5423	5423	8	8
3642	3642 – 2500 = 1142	1142	3	3
6909	6909 – 2500 = 4409	4409	7	7
6331	6331 – 2500 = 3831	3831	6	6
4311	4311 – 2500 = 1811	1811	4	4
5578	5578 – 2500 = 3078	3078	5	5
2828	2828 – 2500 = 328	328	2	2
2781	2781 – 2500 = 281	281	1	1

Since this is a right-tailed test, $T_{data} = |T_-| = 0$. Since $T_{data} = 0$, which is $\leq T_{crit} = 2$, we reject H_0. There is evidence that the population median number of businesses per city is greater than 2500.

27. Since perfect symmetry is rare in real-world data, especially with small data sets, we will interpret the boxplot as revealing acceptable symmetry. Therefore, it is appropriate to perform the Wilcoxon signed rank test. $H_0 : M_d = 0$ versus $H_a : M_d < 0$. Since the hypothesized value of M_d is 0, n is the number of the differences not equal to 0. Since 3 of the differences are equal to 0, $n = 13 - 3 = 10$. Since this is a left-tailed test with $\alpha = 0.05$, $T_{crit} = 11$. Reject H_0 if $T_{data} \leq 11$.

Pain after	Pain before	d = (after – before)	$\|d\|$	Rank of $\|d\|$	Signed rank
3	6	3– 6 = –3	3	7	–7
1	2	1 – 2 = –1	1	2	–2
0	2	0 – 2 = –2	2	4.5	–4.5
0	3	0 – 3 = –3	3	7	–7
2	3	2 – 3 = –1	1	2	–2
1	4	1 – 4 = –3	3	7	–7
2	2	2 – 2 = 0	—	—	—
1	5	1 – 5 = –4	4	9	–9
0	1	0 – 1 = –1	1	2	–2
4	6	4 – 6 = –2	2	4.5	–4.5
1	6	1 – 6 = –5	5	10	–10
4	4	4 – 4 = 0	—	—	—
8	8	8 – 8 = 0	—	—	—

Since this is a left-tailed test, $T_{data} = T_+ = 0$. Since $T_{data} = 0$, which is $\leq T_{crit} = 11$, we reject H_0. There is evidence that population median of the differences in the pain levels reported before and after Reiki touch therapy is less than 0. Therefore, there is evidence that Reiki touch therapy was useful in the reduction of chronic pain.

Section 14.4

1. The subjects selected for the first sample do not determine the subjects in the second sample.

3. True.

5. If the null hypothesis is true, we expect R_1 to not be very different from R_2.

7.

Combined data	**2**	**2**	**3**	**3**	3	3	3	3	**4**	**4**	4
Rank	**1.5**	**1.5**	**5.5**	**5.5**	5.5	5.5	5.5	5.5	**10**	**10**	10
Combined data	**5**	**5**	5	5	6	**7**	7	8	**9**	**9**	10
Rank	**13.5**	**13.5**	13.5	13.5	16	**17.5**	17.5	19	**20.5**	**20.5**	22
Combined data	11										
Rank	23										

The values in **bold** are from Sample 1. Therefore, $R_1 = 1.5 + 1.5 + 5.5 + 5.5 + 10 + 10 + 13.5 + 13.5 + 17.5 + 20.5 + 20.5 = 119.5$.

9.

Combined data	79	79	**80**	80	**81**	**81**	81	82	83	**84**	**84**
Rank	1.5	1.5	**3.5**	3.5	**6**	**6**	6	8	9	**10.5**	**10.5**
Combined data	85	85	86	**88**	**88**	**89**	90	**94**	**96**	96	96
Rank	12.5	12.5	14	**15.5**	**15.5**	**17**	18	**19**	**21**	21	21
Combined data	**97**	97	97	**98**							
Rank	**24**	24	24	**26**							

The values in **bold** are from Sample 1. Therefore, $R_1 = 3.5 + 6 + 6 + 10.5 + 10.5 + 15.5 + 15.5 + 17 + 19 + 21 + 24 + 26 = 174.5$.

11. **(a)** $H_0 : M_1 = M_2$ versus $H_a : M_1 \neq M_2$.

(b) Since this is a two-tailed test with $\alpha = 0.05$, $Z_{crit} = 1.96$.

Reject H_0 if $Z_{data} \leq -1.96$ or $Z_{data} \geq 1.96$.

(c) $\mu_R = \frac{n_1(n_1+n_2+1)}{2} = \frac{11(11+12+1)}{2} = 132.$

$$\sigma_R = \sqrt{\frac{n_1 n_2(n_1+n_2+1)}{12}} = \sqrt{\frac{(11)(12)(11+12+1)}{12}} \approx 16.24807681\cdot$$

$$Z_{\text{data}} = \frac{R_1-\mu_R}{\sigma_R} = \frac{119.5-132}{16.24807681} \approx -0.77.$$

(d) Since $Z_{data} \approx -0.77$ is neither ≤ -1.96 nor ≥ 1.96, we do not reject H_0. There is insufficient evidence that the population median of Population 1 differs from the population median of Population 2.

13. **(a)** $H_0 : M_1 = M_2$ versus $H_a : M_1 \neq M_2$.

(b) Since this is a two-tailed test with $\alpha = 0.10$, $Z_{crit} = 1.645$.

Reject H_0 if $Z_{data} \leq -1.645$ or $Z_{data} \geq 1.645$.

(c) $\mu_R = \frac{n_1(n_1+n_2+1)}{2} = \frac{12(12+14+1)}{2} = 162.$

$$\sigma_R = \sqrt{\frac{n_1 n_2(n_1+n_2+1)}{12}} = \sqrt{\frac{(12)(14)(12+14+1)}{12}} \approx 19.4422221\cdot$$

$$Z_{\text{data}} = \frac{R_1-\mu_R}{\sigma_R} = \frac{174.5-162}{19.4422221} \approx 0.64.$$

(d) Since $Z_{data} \approx 0.64$ is neither ≤ -1.645 nor ≥ 1.645, we do not reject H_0. There is insufficient evidence that the population median of Population 1 differs from the population median of Population 2.

15. $H_0 : M_1 = M_2$ versus $H_a : M_1 \neq M_2$. Since this is a two-tailed test with $\alpha = 0.10$, $Z_{crit} = 1.645$. Reject H_0 if $Z_{data} \leq -1.645$ or $Z_{data} \geq 1.645$.

Combined data	146	**167**	**191**	**197**	**227**	**248**	363	377	379	**394**
Rank	1	**2**	**3**	**4**	**5**	**6**	7	8	9	**10**
Combined data	**399**	**459**	**497**	545	546	567	**568**	**640**	807	855
Rank	**11**	**12**	**13**	14	15	16	**17**	**18**	19	20
Combined data	981	1244	1445							
Rank	21	22	23							

The values in **bold** are from Sample 1. Therefore, $R_1 = 2 + 3 + 4 + 5 + 6 + 10 + 11 + 12 + 13 + 17 + 18 = 101$.

$n_1 = 11, n_2 = 12$.

$$\mu_R = \frac{n_1(n_1+n_2+1)}{2} = \frac{11(11+12+1)}{2} = 132.$$

$$\sigma_R = \sqrt{\frac{n_1 n_2(n_1+n_2+1)}{12}} = \sqrt{\frac{(11)(12)(11+12+1)}{12}} \approx 16.24807681\cdot$$

$$Z_{\text{data}} = \frac{R_1-\mu_R}{\sigma_R} = \frac{101-132}{16.24807681} \approx -1.91.$$

Since $Z_{data} \approx -1.91$, which is ≤ -1.645, we reject H_0. There is evidence that the population median of fans of Facebook pages for games differs from the population median of fans of Facebook pages for television shows.

17. $H_0 : M_1 = M_2$ versus $H_a : M_1 \neq M_2$. Since this is a two-tailed test with $\alpha = 0.05$, $Z_{crit} = 1.96$. Reject H_0 if $Z_{data} \leq -1.96$ or $Z_{data} \geq 1.96$.

Combined data	120	123	129	138	143	147	154	164	**165**
Rank	1	2	3	4	5	6	7	8	**9**
Combined data	176	**177**	**180**	201	206	207	**225**	**245**	**268**
Rank	10	**11**	**12**	13	14	15	**16**	**17**	**18**
Combined data	**270**	**270**	**285**	**289**	**291**	**292**	**298**	**315**	**400**
Rank	**19.5**	**19.5**	**21**	**22**	**23**	**24**	**25**	**26**	**27**

The values in **bold** are from Sample 1. Therefore, $R_1 = 9 + 11 + 12 + 16 + 17 + 18 + 19.5 + 19.5 + 21 + 22 + 23 + 24 + 25 + 26 + 27 = 290$.

$n_1 = 15, n_2 = 12$.

$$\mu_R = \frac{n_1(n_1+n_2+1)}{2} = \frac{15(15+12+1)}{2} = 210.$$

$$\sigma_R = \sqrt{\frac{n_1 n_2(n_1+n_2+1)}{12}} = \sqrt{\frac{(15)(12)(15+12+1)}{12}} \approx 20.49390153\cdot$$

$$Z_{\text{data}} = \frac{R_1-\mu_R}{\sigma_R} = \frac{290-210}{20.49390153} \approx 3.90.$$

Since $Z_{data} \approx 3.90$ is ≥ 1.96 we reject H_0. There is evidence that the population median property tax in Ohio differs from that in North Carolina.

19. $H_0 : M_1 = M_2$ versus $H_a : M_1 \neq M_2$. Since this is a two-tailed test with $\alpha = 0.05$, $Z_{crit} = 1.96$. Reject H_0 if $Z_{data} \leq -1.96$ or $Z_{data} \geq 1.96$.

Combined data	**0**	**0**	**7**	7	**14**	17	18	31	**31**	**34**	**36**
Rank	**1.5**	**1.5**	**3.5**	3.5	**5**	6	7	8.5	**8.5**	**10**	**11**
Combined data	50	**54**	**56**	62	**66**	66	**71**	**89**	**97**	103	106
Rank	12	**13**	**14**	15	**16.5**	16.5	**18**	**19**	**20**	21	22
Combined data	**110**	**117**	**121**	**129**	**131**	132	**140**	**142**	**143**	**222**	237
Rank	**23**	**24**	**25**	**26**	**27**	28	**29**	**30**	**31**	**32**	33
Combined data	265	271	278	292	298	302	336	387	**395**	**424**	627
Rank	34	35	36	37	38	39	40	41	**42**	**43**	44
Combined data	**927**										
Rank	**45**										

The values in **bold** are from Sample 1. Therefore, $R_1 = 1.5 + 1.5 + 3.5 + 5 + 8.5 + 10 + 11 + 13 + 14 + 16.5 + 18 + 19 + 20 + 23 + 24 + 25 + 26 + 27 + 29 + 30 + 31 + 32 + 42 + 43 + 45 = 518.5$.

$n_1 = 25$, $n_2 = 20$.

$$\mu_R = \frac{n_1(n_1+n_2+1)}{2} = \frac{25(25+20+1)}{2} = 575.$$

$$\sigma_R = \sqrt{\frac{n_1 n_2(n_1+n_2+1)}{12}} = \sqrt{\frac{(25)(20)(25+20+1)}{12}} \approx 43.77975179.$$

$$Z_{\text{data}} = \frac{R_1 - \mu_R}{\sigma_R} = \frac{518.5-575}{43.77875179} \approx -1.29.$$

Since $Z_{data} \approx -1.29$, which is neither ≤ -1.96 nor ≥ 1.96, we do not reject H_0. There is insufficient evidence that the population median amount of phosphorus differs from the population median amount of potassium.

Section 14.5

1. The Wilcoxon rank sum test tests whether population medians of two independent random samples are equal by temporarily combining the two samples and calculating the ranks of the combined data values. Then the ranks of the data values in Sample 1 are summed and the sum is used to calculate the test statistic for this test. The Kruskal-Wallis test extends this method from two populations to three or more populations by temporarily combining all of the samples and calculating the ranks of the combined data values. Then the ranks are summed separately for each sample and the sums are used to calculate the test statistic for this test.

3. True.

5. Total number of data values in all of the samples combined

7.

Combined data	1	**2**	2	**3**	3	3	**4**	**5**	**5**	5	*6*	*7*	*9*	*9*
Rank	1	**2.5**	2.5	**5**	5	5	**7**	**9**	**9**	9	*11*	*12*	*13.5*	*13.5*
Combined data	*10*													
Rank	*15*													

Sample 1 is in **bold**, Sample 2 is in *italics*, and Sample 3 is neither. $R_1 = 2.5 + 5 + 7 + 9 + 9 = 32.5$, $R_2 = 11 + 12 + 13.5 + 13.5 + 15 = 65$, $R_3 = 1 + 2.5 + 5 + 5 + 9 = 22.5$, $n_1 = 5$, $n_2 = 5$, $n_3 = 5$, $N = n_1 + n_2 + n_3 = 5 + 5 + 5 = 15$.

9.

Combined data	*112*	***112***	114	***127***	129	***133***	137	**143**	***144***	*145*
Rank	*1.5*	***1.5***	3	***4***	5	***6***	7	**8**	***9***	*10*
Combined data	***149***	***150***	152	**152**	*155*	158	162	**164**	**168**	172
Rank	***11***	***12***	13.5	**13.5**	*15*	16	17	**18**	**19**	21
Combined data	172	172	*182*	**183**	**184**	187	*193*	193		
Rank	21	21	*23*	**24**	**25**	26	*27.5*	27.5		

Sample 1 is in **bold**, Sample 2 is in *italics*, Sample 3 is in ***bold italics***, Sample 4 is underlined, and Sample 5 is none of these. $R_1 = 8 + 13.5 + 18 + 19 + 24 + 25 = 107.5$, $R_2 = 1.5 + 10 + 15 + 23 + 27.5 = 77$, $R_3 = 1.5 + 4 + 6 + 9 + 11 + 12 = 43.5$, $R_4 = 5 + 17 + 21 + 21 + 26 + 27.5 = 117.5$, $R_5 = 3 + 7 + 13.5 + 16 + 21 = 60.5$, $n_1 = 6$, $n_2 = 5$, $n_3 = 6$, $n_4 = 6$, $n_5 = 5$, $N = n_1 + n_2 + n_3 + n_4 + n_5 = 6 + 5 + 6 + 6 + 5 = 28$.

11. $\chi^2_{\text{data}} = \frac{12}{N(N+1)}\left(\frac{R_1^2}{n_1} + \frac{R_2^2}{n_2} + \frac{R_3^2}{n_3}\right) - 3(N+1) = \frac{12}{15(15+1)}\left(\frac{32.5^2}{5} + \frac{65^2}{5} + \frac{22.5^2}{5}\right) - 3(15 + 1) = 9.875$

13. $\chi^2_{\text{data}} = \frac{12}{N(N+1)}\left(\frac{R_1^2}{n_1} + \frac{R_2^2}{n_2} + \frac{R}{n_3} + \frac{R_4^2}{n_4} + \frac{R_5^2}{n_5}\right) - 3(N+1) = \frac{12}{28(28+1)}\left(\frac{107.5^2}{6} + \frac{77^2}{5} + \frac{43.5^2}{6} + \frac{117.5^2}{6} + \frac{60.5^2}{5}\right) - 3(28 + 1) \approx 8.472536946$

15. Each sample is independent and randomly selected and there are at least 5 data values in each sample. Thus, the conditions for the Kruskal-Wallis test are met, and we may proceed with the hypothesis test.

(a) H_0 : The population medians are all equal. H_a : Not all the population medians are equal.

(b) $\alpha = 0.05$, df $= k - 1 = 3 - 1 = 2$, $X^2_{crit} = 5.991$. Reject H_0 if $X^2_{data} \geq 5.991$.

(c) $X^2_{data} = 9.875$.

(d) Since $X^2_{data} = 9.875$ is $\geq X^2_{crit} = 5.991$, we reject H_0. There is evidence that not all of the population medians are equal.

17. Each sample is independent and randomly selected and there are at least 5 data values in each sample. Thus, the conditions for the Kruskal-Wallis test are met, and we may proceed with the hypothesis test.

(a) H_0 : The population medians are all equal. H_a : Not all the population medians are equal.

(b) $\alpha = 0.01$, df $= k - 1 = 5 - 1 = 4$, $X^2_{crit} = 13.277$. Reject H_0 if $X^2_{data} \geq 13.277$.

(c) $X^2_{data} \approx 8.472536946$.

(d) Since $X^2_{data} \approx 8.472536946$ is not $\geq X^2_{crit} = 13.277$, we do not reject H_0. There is insufficient evidence that not all of the population medians are equal.

19. Each sample is independent and randomly selected and there are at least 5 data values in each sample. Thus, the conditions for the Kruskal-Wallis test are met, and we may proceed with the hypothesis test. H_0 : The population medians are all equal. H_a : Not all the population medians are equal. $\alpha = 0.05$, df $= k - 1 = 3 - 1 = 2$, $X^2_{crit} = 5.991$. Reject H_0 if $X^2_{data} \geq 5.991$.

Combined data	*1*	*3*	4	**5**	*6*	7	*8*	8	*10*	10	**12**	**13**	16
Rank	*1*	*2*	3	**4**	*5*	6	*7.5*	7.5	*9.5*	9.5	**11**	**12**	13
Combined data	**19**	**22**											
Rank	**14**	**15**											

Number of wraps is in **bold**, number of muffins sold is in *italics*, and number of chips sold is neither. $R_1 = 4 + 11 + 12 + 14 + 15 = 56$, $R_2 = 1 + 2 + 5 + 7.5 + 9.5 = 25$, $R_3 = 3 + 6 + 7.5 + 9.5 + 13 = 39$, $n_1 = 5$, $n_2 = 5$, $n_3 = 5$, $N = n_1 + n_2 + n_3 = 5 + 5 + 5 = 15$.

$$\chi^2_{\text{data}} = \frac{12}{N(N+1)}\left(\frac{R_1^2}{n_1} + \frac{R_2^2}{n_2} + \frac{R_3^2}{n_3}\right) - 3(N+1) = \frac{12}{15(15+1)}\left(\frac{56^2}{5} + \frac{25^2}{5} + \frac{39^2}{5}\right) - 3(15+1) = 4.82.$$

Since $X^2_{data} = 4.82$ is not $\geq X^2_{crit} = 5.991$, we do not reject H_0. There is insufficient evidence that not all of the population median numbers of items sold are the same in all three groups.

21. Each sample is independent and randomly selected and there are at least 5 data values in each sample. Thus, the conditions for the Kruskal-Wallis test are met, and we may proceed with the hypothesis test. H_0 : The population medians are all equal. H_a : Not all the population medians are equal. $\alpha = 0.05$, df $= k - 1 = 3 - 1 = 2$, $X^2_{crit} =$ 5.991 Reject H_0 if $X^2_{data} \geq$ 5.991

Combined data	*98.8*	*104.3*	107.4	*110.2*	*115.1*	**119.0**	121.3	122.8	**124.8**	**130.1**	**130.7**	*142.9*
Rank	*1*	*2*	3	*4*	*5*	**6**	7	8	**9**	**10**	**11**	*12*
Combined data	**155.4**	155.4	169.3									
Rank	**13.5**	13.5	15									

Younger is in **bold**, middle is in *italics*, and older is in neither. $R_1 = 6 + 9 + 10 + 11 + 13.5 = 49.5$, $R_2 = 1 + 2 + 4 + 5 + 12 = 24$, $R_3 = 3 + 7 + 8 + 13.5 + 15 = 46.5$, $n_1 = 5$, $n_2 = 5$, $n_3 = 5$, $N = n_1 + n_2 + n_3 = 5 + 5 + 5 = 15$.

$$\chi^2_{\text{data}} = \frac{12}{N(N+1)}\left(\frac{R_1^2}{n_1} + \frac{R_2^2}{n_2} + \frac{R_3^2}{n_3}\right) - 3(N+1) = \frac{12}{15(15+1)}\left(\frac{49.5^2}{5} + \frac{24^2}{5} + \frac{46.5^2}{5}\right) - 3(15+1) = 3.885.$$

Since $X^2_{data} = 3.885$ is not $\geq X^2_{crit} = 5.991$, we do not reject H_0. There is insufficient evidence that not all of the population median women's weights are the same in all three groups.

23. Each sample is independent and randomly selected and there are at least 5 data values in each sample. Thus, the conditions for the Kruskal-Wallis test are met, and we may proceed with the hypothesis test. H_0 : The population medians are all equal. H_a : Not all the population medians are equal. $\alpha = 0.01$, df $= k - 1 = 3 - 1 = 2$, $X^2_{crit} = 9.210$. Reject H_0 if $X^2_{data} \geq 9.210$.

Combined data	*4.4*	*4.8*	*5.5*	*5.6*	**5.8**	**6.4**	**7.1**	**7.2**	*7.5*	**7.7**	**8.4**	**8.5**
Rank	*1*	*2*	*3*	*4*	**5**	**6**	**7**	**8**	*9*	**10**	**11**	**12**
Combined data	22.6	23.4	25.3	26.0	27.0	28.0						
Rank	13	14	15	16	17	18						

U.S. is in **bold**, Canada is in *italics*, and Mexico is neither.

$R_1 = 5 + 6 + 7 + 8 + 10 + 11 + 12 = 59$, $R_2 = 1 + 2 + 3 + 4 + 9 = 19$, $R_3 = 13 + 14 + 15 + 16 + 17 + 18 = 93$, $n_1 = 7$, $n_2 = 5$, $n_3 = 6$, $N = n_1 + n_2 + n_3 = 7 + 5 + 6 = 18$,

$$\chi^2_{\text{data}} = \frac{12}{N(N+1)}\left(\frac{R_1^2}{n_1} + \frac{R_2^2}{n_2} + \frac{R_3^2}{n_3}\right) - 3(N+1) = \frac{12}{18(18+1)}\left(\frac{59^2}{7} + \frac{19^2}{5} + \frac{93^2}{6}\right) - 3(18+1) \approx 13.56090226.$$

Since $X^2_{data} \approx 13.56090226$ is $\geq X^2_{crit} = 9.210$, we reject H_0. There is evidence that not all of the population median infant mortality rates are the same for the United States, Canada, and Mexico.

25. Each sample is independent and randomly selected and there are at least 5 data values in each sample. Thus, the conditions for the Kruskal-Wallis test are met, and we may proceed with the hypothesis test. H_0 : The population medians are all equal. H_a : Not all the population medians are equal. $\alpha = 0.05$, df $= k - 1 = 3 - 1 = 2$, $X^2_{crit} = 5.991$. Reject H_0 if $X^2_{data} \geq 5.991$.

Combined data	19.8236	22.3965	31.0722	**31.4360**	**33.1741**	*34.3848*
Rank	1	2	3	**4**	**5**	*6*
Combined data	**36.4715**	**36.5237**	39.1062	39.2411	**39.2592**	39.7034
Rank	**7**	**8**	9	10	**11**	12
Combined data	**40.5602**	*41.4450*	**41.5035**	*41.9989*	*44.3309*	*49.7874*
Rank	**13**	*14*	**15**	*16*	*17*	*18*

Kellogg's is in **bold**, Ralston-Purina is in *italics*, and General Mills is neither. $R_1 = 4 + 5 + 7 + 8 + 11 + 13 + 15 = 63$, $R_2 = 6 + 14 + 16 + 17 + 18 = 71$, $R_3 = 1 + 2 + 3 + 9 + 10 + 12 = 37$

$n_1 = 7$, $n_2 = 5$, $n_3 = 6$, $N = n_1 + n_2 + n_3 = 7 + 5 + 6 = 18$.

$$\chi^2_{\text{data}} = \frac{12}{N(N+1)}\left(\frac{R_1^2}{n_1} + \frac{R_2^2}{n_2} + \frac{R_3^2}{n_3}\right) - 3(N+1) = \frac{12}{18(18+1)}\left(\frac{63^2}{7} + \frac{71^2}{5} + \frac{37^2}{6}\right) - 3(18+1) \approx 6.276023392.$$

Since $X^2_{data} \approx 6.276023392$ is $\geq X^2_{crit} = 5.991$, we reject H_0. There is evidence that the population median nutritional ratings differ by manufacturer.

Section 14.6

1. To investigate whether two variables are related; may also be used to detect a nonlinear relationship between two variables.

3. Dependent samples; test statistic is based on the squares of the paired differences in ranks.

5. H_0 : There is no rank correlation between the two variables. H_a : There is a rank correlation between the two variables.

7. $\sum d^2 = 0$, $r_{data} = 1$.

9. **(a)** and **(b)**

Sample 1	Sample 2	Rank of Sample 1	Rank of Sample 2	Difference d	d^2
0	9	1	4.5	–3.5	12.25
1	1	2.5	1	1.5	2.25
1	6	2.5	2.5	0	0
3	10	4	6	–2	4
4	9	5	4.5	0.5	0.25
7	6	6	2.5	3.5	12.25
					$\sum d^2 = 31$

(c) $r_{\text{data}} = 1 - \frac{6\sum d^2}{n(n^2-1)} = 1 - \frac{6(31)}{6(6^2-1)} \approx 0.1142857143.$

11. **(a)** and **(b)**

Sample 1	Sample 2	Rank of Sample 1	Rank of Sample 2	Difference d	d^2
19	64	1	4	–3	9
20	69	2	6.5	–4.5	20.25
21	62	3	3	0	0
25	60	4.5	2	2.5	6.25
25	69	4.5	6.5	–2	4
27	58	6	1	5	25
28	65	7.5	5	2.5	6.25
28	70	7.5	8	–0.5	0.25
					$\sum d^2 = 71$

(c) $r_{\text{data}} = 1 - \frac{6\sum d^2}{n(n^2-1)} = 1 - \frac{6(71)}{8(8^2-1)} \approx 0.1547619048$

13. $n = 6 \leq 30$, $\alpha = 0.01$. From Table K, it is not possible to get a value in the critical region for this sample size and level of significance.

15. $n = 8 \leq 30$, $\alpha = 0.10$, $r_{crit} = 0.643$. Reject H_0 if $r_{data} \leq -0.643$ or $r_{data} \geq 0.643$.

17. **(a)** H_0 : There is no rank correlation between the two variables. H_a : There is a rank correlation between the two variables.

(b) $n = 6 \leq 30$, $\alpha = 0.01$. From Table K, it is not possible to get a value in the critical region for this sample size and level of significance.

(c) $r_{data} \approx 0.1142857143$.

(d) From Table K, it is not possible to get a value in the critical region for this sample size and level of significance. Thus, the conclusion is do not reject H_0. There is insufficient evidence that there is a rank correlation between the two variables.

19. **(a)** H_0 : There is no rank correlation between the two variables. H_a : There is a rank correlation between the two variables.

(b) $n = 8 \leq 30$, $\alpha = 0.10$, $r_{crit} = 0.643$. Reject H_0 if $r_{data} \leq -0.643$ or $r_{data} \geq 0.643$.

(c) $r_{data} \approx 0.1547619048$.

(d) Since $r_{\text{data}} \approx 0.1547619048$ is not ≤ -0.643 and not ≥ 0.643, we do not reject H_0. There is insufficient evidence that there is a rank correlation between the two variables.

21. H_0 : There is no rank correlation between the two variables. H_a : There is a rank correlation between the two variables. $n = 5 \leq 30$, $\alpha = 0.10$, $r_{crit} = 0.900$. Reject H_0 if $r_{data} \leq -0.900$ or $r_{data} \geq 0.900$.

President	Liberal rank	Conservative rank	Difference d	d^2
Abraham Lincoln	1	1	0	0
George Washington	3	2	1	1
Franklin Roosevelt	2	3	−1	1
Thomas Jefferson	4	4	0	0
Theodore Roosevelt	5	5	0	0
				$\sum d^2 = 2$

$$r_{\text{data}} = 1 - \frac{6\sum d^2}{n(n^2-1)} = 1 - \frac{6(2)}{5(5^2-1)} = 0.9.$$

Since $r_{data} = 0.9$, which is ≥ 0.900, the conclusion is reject H_0. There is evidence that a rank correlation exists between the liberal ranks and the conservative ranks.

23. H_0 : There is no rank correlation between the two variables. H_a : There is a rank correlation between the two variables. $n = 25 \leq 30$, $\alpha = 0.10$, $r_{crit} = 0.337$. Reject H_0 if $r_{data} \leq -0.337$ or $r_{data} \geq 0.337$.

College	**AP poll**	***USA Today* poll**	**AP rank**	***USA Today* rank**	**Difference *d***	d^2
Florida State	1500	1475	1	1	0	0
Auburn	1428	1388	2	2	0	0
Michigan State	1385	1375	3	3	0	0
South Carolina	1247	1219	4	4	0	0
Missouri	1236	1200	5	5	0	0
Oklahoma	1205	1189	6	6	0	0
Alabama	1114	1086	7	8	−1	1
Clemson	1078	1091	8	7	1	1
Oregon	974	975	9	9	0	0
UCF	959	865	10	12	−2	4
Stanford	936	872	11	10.5	0.5	0.25
Ohio State	816	872	12	10.5	1.5	2.25
Baylor	778	796	13	13	0	0
LSU	717	719	14	14	0	0
Louisville	693	703	15	15	0	0
UCLA	632	597	16	16	0	0
Oklahoma State	598	587	17	17	0	0
Texas A&M	459	443	18	18	0	0
USC	299	313	19	19	0	0
Notre Dame	256	125	20	24	-4	16
Arizona State	255	302	21	20	1	1
Wisconsin	245	266	22	21	1	1
Duke	190	202	23	22	1	1
Vanderbilt	117	180	24	23	1	1
						$\sum d^2 = 28.5$

$$r_{\text{data}} = 1 - \frac{6\sum d^2}{n(n^2-1)} = 1 - \frac{6(28.5)}{24(24^2-1)} \approx 0.9876086957.$$

Since $r_{data} \approx 0.9876086957$ is ≥ 0.337, the conclusion is reject H_0. There is evidence that a rank correlation exists between the AP poll and the *USA Today* poll.

25. H_0 : There is no rank correlation between the two variables. H_a : There is a rank correlation between the two variables. $n = 6 \leq 30$, $\alpha = 0.10$, $r_{crit} = 0.829$. Reject H_0 if $r_{data} \leq -0.829$ or $r_{data} \geq 0.829$.

Game	Playstation 3 mean reviewer score	Xbox 360 mean reviewer score	Playstation 3 rank	Xbox 360 rank	Difference *d*	d^2
Grand Theft Auto IV	0.9373	0.9656	3	1	2	4
BioShock	0.9403	0.9525	1	3	−2	4
Call of Duty 4: Modern Warfare	0.9378	0.9416	2	4	−2	4
Rock Band	0.9119	0.9225	4	5	−1	1
The Orange Box	0.8838	0.9624	5	2	3	9
Guitar Hero III: Legends of Rock	0.8390	0.8622	6	6	0	0
						$\sum d^2 = 22$

$r_{\text{data}} = 1 - \frac{6\sum d^2}{n(n^2-1)} = 1 - \frac{6(22)}{6(6^2-1)} \approx 0.3714285714.$

Since $r_{data} \approx 0.3714285714$, which is not ≤ -0.829 and not ≥ 0.829, the conclusion is do not reject H_0. There is insufficient evidence that a rank correlation exists between the average reviewer score for the Playstation 3 platform and the average reviewer score for the Xbox 360 platform.

27. H_0 : There is no rank correlation between the two variables. H_a : There is a rank correlation between the two variables. $n = 30 \leq 30$, $\alpha = 0.05$, $r_{crit} = 0.362$. Reject H_0 if $r_{data} \leq -0.362$ or $r_{data} \geq 0.362$.

Community college	Rank of overall quality	Tuition and fees	Rank of tuition and fees	Differenc e d	d^2
Atlanta Technical College, GA	1	$1362	6	–5	25
Cascadia Community College, WA	2	$2642	17	–15	225
Southern Univ. at Shreveport, LA	3	$2252	14	–11	121
Southwestern CC, NC	4	$1171	2	2	4
Hazard CC, KY	5	$2616	16	–11	121
North Florida Community College, FL	6	$1910	9	–3	9
Indianhead College, WI	7	$2912	21	–14	196
Southeast Kentucky CC, KY	8	$2760	19	–11	121
Zane State College, OH	9	$3849	25	–16	256
Baldwin College, GA	10	$2098	12	–2	4
Texas State Technical College, Marshall, TX	11	$3930	26	–15	225
Lake City CC, FL	12	$2979	22	–10	100
Itasca CC, MN	13	$4590	29	–16	256
South Piedmont CC, NC	14	$1319	4	10	100
Vermillion CC, MN	15	$4366	28	–13	169
Hawaii CC, HI	16	$1478	7	9	81
Ellsworth CC, IA	17	$3108	24	–7	49
Chipola College, FL	18	$2137	13	5	25
Martin CC, NC	19	$1302	3	16	256
Texas State Technical College, TX	20	$3105	23	–3	9
South Texas College, TX	21	$1966	10	11	121
Skagit Valley College, WA	22	$2712	18	4	16
Valencia CC, FL	23	$2091	11	12	144
MiraCosta College, CA	24	$590	1	23	529
Florida CC at Jacksonville, FL	25	$1714	8	17	289
New Hampshire CC, NH	26	$5464	30	–4	16
Frank Phillips College, TX	27	$2766	20	7	49
Mesabi Range, CC MN	28	$4174	27	1	1
Northwest Vista College, TX	29	$2292	15	14	196
New Mexico University Grants, NM	30	$1320	5	25	625
					$\sum d^2 =$ 4338

$$r_{\text{data}} = 1 - \frac{6\sum d^2}{n(n^2-1)} = 1 - \frac{6(4338)}{30(30^2-1)} \approx 0.0349276974.$$

Since $r_{data} \approx 0.0349276974$, which is not ≤ -0.362 and is not ≥ 0.362, the conclusion is do not reject H_0. There is insufficient evidence that a rank correlation exists between overall quality and tuition and fees.

29. H_0 : There is no rank correlation between the two variables. H_a : There is a rank correlation between the two variables. $n = 10 \le 30$, $\alpha = 0.05$, $r_{crit} = 0.648$. Reject H_0 if $r_{data} \le -0.648$ or $r_{data} \ge 0.648$.

State	Cigarettes per capita (100s)	Deaths from bladder cancer per 100,000 people	Cigarettes rank	Deaths rank	Difference d	d^2
Kansas	21.84	2.91	7	10	−3	9
Washington	21.17	4.04	9	6	3	9
Oklahoma	23.44	2.93	6	9	−3	9
Maryland	25.91	5.21	5	1	4	16
Texas	20.08	2.94	10	8	2	4
Louisiana	21.58	4.65	8	4	4	16
Massachusetts	26.92	4.69	4	3	1	1
Rhode Island	29.18	4.99	2	2	0	0
Florida	28.27	4.46	3	5	−2	4
Alaska	30.34	3.46	1	7	−6	36
						$\sum d^2 =$ 104

$$r_{\text{data}} = 1 - \frac{6\sum d^2}{n(n^2-1)} = 1 - \frac{6(104)}{10(10^2-1)} \approx 0.3696969697.$$

Since $r_{data} \approx 0.3696969697$, which is not ≤ -0.648 and is not ≥ 0.648, the conclusion is do not reject H_0. There is insufficient evidence that a rank correlation exists between the number of deaths from bladder cancer and the per capita number of cigarettes smoked.

Section 14.7

1. An ordered data set.

3. A sequence of observations sharing the same value (of two possible values), preceded or followed by data having the other possible value or no data at all.

5. **(a)** n_1 = the number of Ms = 11.

(b) n_2 = the number of Fs = 6.

(c) $n = n_1 + n_2 = 11 + 6 = 17$.

(d) $\underbrace{MM}_{1}\ \underbrace{FFF}_{2}\ \underbrace{M}_{3}\ \underbrace{F}_{4}\ \underbrace{M}_{5}\ \underbrace{F}_{6}\ \underbrace{MMMM}_{7}\ \underbrace{F}_{8}\ \underbrace{MMM}_{9}$. **So,** $G = 9$.

7. **(a)** n_1 = the number of Ts = 21.

(b) n_2 = the number of Fs = 21.

(c) $n = n_1 + n_2 = 21 + 21 = 42$.

(d) $\underbrace{T}_{1}\underbrace{F}_{2}\underbrace{T}_{3}\underbrace{F}_{4}\underbrace{T}_{5}\underbrace{FF}_{6}\underbrace{TT}_{7}\underbrace{F}_{8}\underbrace{T}_{9}\underbrace{F}_{10}\underbrace{T}_{11}\underbrace{F}_{12}\underbrace{TT}_{13}\underbrace{F}_{14}\underbrace{T}_{15}\underbrace{F}_{16}\underbrace{T}_{17}\underbrace{F}_{18}\underbrace{T}_{19}\underbrace{FFF}_{20}\underbrace{T}_{21}\underbrace{F}_{22}\underbrace{T}_{23}\underbrace{F}_{24}\underbrace{T}_{25}\underbrace{F}_{26}\underbrace{T}_{27}\underbrace{FF}_{28}\underbrace{T}_{29}\underbrace{F}_{30}\underbrace{T}_{31}\underbrace{F}_{32}\underbrace{TTT}_{33}\underbrace{F}_{34}$. So, $G = 34$

9. $n_1 = 11 \le 20$, $n_2 = 6 \le 20$, $\alpha = 0.05$. From Table K, $G_{crit,lower} = 4$ and $G_{crit,upper} = 13$.

11. $n_1 = n_2 = 21 > 20$, $\alpha = 0.01$, $Z_{crit} = 2.58$.

13. $G = 9$

15. $n_1 = 21$, $n_2 = 21$, $G = 34$.

(a) $\mu_G = \frac{2n_1n_2}{n_1+n_2} + 1 = \frac{2(21)(21)}{21+21} + 1 = 22.$

(b) $\sigma_G = \sqrt{\frac{(2n_1n_2)(2n_1n_2-n_1-n_2)}{(n_1+n_2)^2(n_1+n_2-1)}} = \sqrt{\frac{(2(21)(21))(2(21)(21)-21-21)}{(21+21)^2(21+21-1)}} \approx 3.200609698.$

(c) $Z_{data} = \frac{G-\mu_G}{\sigma_G} \approx \frac{34-22}{3.200609698} \approx 3.749285646 \approx 3.7493.$

17. **(a)** H_0 : The sequence of data is random. H_a : The sequence of data is not random.

(b) $n_1 = 11 \le 20$ and $n_2 = 6 \le 20$ and $\alpha = 0.05$. From Table L, $G_{crit,lower} = 4$ and $G_{crit,upper} = 13$. Reject H_0 if $G \le 4$ or if $G \ge 13$.

(c) $G = 9$.

(d) Since $G = 9$ is not ≤ 4 and is not ≥ 13, do not reject H_0. There is insufficient evidence that the sequence of data is not random.

19. **(a)** H_0 : The sequence of data is random. H_a : The sequence of data is not random.

(b) $n_1 = n_2 = 21 > 20$ so use the Z test, $\alpha = 0.01$, $Z_{crit} = 2.58$. Reject H_0 if $Z_{data} \le -2.58$ or if $Z_{data} \ge 2.58$.

(c) $Z_{data} \approx 3.7493$.

(d) Since $Z_{data} \approx 3.7493$ is ≥ 2.58, reject H_0. There is evidence that the sequence of data is not random.

21. H_0 : The sequence of data is random. H_a : The sequence of data is not random. We have n_1 = the number of As = $19 \le 20$ and n_2 = the number of Bs = $3 \le 20$ and $\alpha = 0.05$. From Table L, $G_{crit,lower} = 3$ and $G_{crit,upper} = 8$.

Reject H_0 if $G \le 3$ or if $G \ge 8$.

$$\underbrace{\text{AAAAAAAAAAAAAA}}_{1}\ \underbrace{\text{BB}}_{2}\ \underbrace{\text{AAAA}}_{3}\ \underbrace{\text{B}}_{4}\ \underbrace{\text{A}}_{5}.$$

So, $G = 5$. Since $G = 5$ is not ≤ 3 or ≥ 8, do not reject H_0. There is insufficient evidence that the A/B sequence is not random.

23. H_0 : The sequence of data is random. H_a : The sequence of data is not random. n_1 = the number of As = 25 which is > 20 and n_2 = the number of Ns = $5 \le 20$ and $\alpha = 0.05$. Since n_1 = the number of As = 25 is > 20, we use a normal approximation, so $Z_{crit} = 1.96$. Reject H_0 if $Z_{data} \le -1.96$ or if $Z_{data} \ge 1.96$.

$$\underbrace{\text{AAAAA}}_{1}\ \underbrace{\text{O}}_{2}\ \underbrace{\text{AA}}_{3}\ \underbrace{\text{OO}}_{4}\ \underbrace{\text{AAAAAA}}_{5}\ \underbrace{\text{O}}_{6}\ \underbrace{\text{AAA}}_{7}\ \underbrace{\text{O}}_{8}\ \underbrace{\text{AAAAAAAAA}}_{9}.$$

So, $G = 9$.

$\mu_G = \frac{2n_1n_2}{n_1+n_2} + 1 = \frac{2(25)(5)}{25+5} + 1 \approx 9.333333333.$

$$\sigma_G = \sqrt{\frac{(2n_1n_2)(2n_1n_2-n_1-n_2)}{(n_1+n_2)^2(n_1+n_2-1)}} = \sqrt{\frac{(2(25)(5))(2(25)(5)-25-5)}{(25+5)^2(25+5-1)}} \approx 1.451647234.$$

$$Z_{\text{data}} = \frac{G-\mu_G}{\sigma_G} \approx \frac{9-9.333333333}{1.451647234} \approx -0.2296241988 \approx -0.2296.$$

Since $Z_{data} \approx -0.2296$ is not ≤ -1.96 and not ≥ 1.96, the conclusion is do not reject H_0. There is insufficient evidence that the sequence is not random.

25. H_0 : The sequence of residuals is random. H_a : The sequence of residuals is not random. BBBBBBAAAAAAAABBBBB. n_1 = the number of Bs = 11 $\leq$ 20 and n_2 = the number of As = 8 $\leq$ 20 and $\alpha = 0.05$. From Table L, $G_{crit,lower} = 5$ and $G_{crit,upper} = 15$. Reject H_0 if $G \leq 5$ or if $G \geq 15$.

$$\underbrace{\text{BBBBBB}}_{1}\ \underbrace{\text{AAAAAAAA}}_{2}\ \underbrace{\text{BBBBB}}_{3}.$$

So, $G = 3$. Since $G = 3$ is $\leq G_{crit,\text{lower}} = 5$, the conclusion is reject H_0.

There is evidence that the sequence of residuals is not random.

Chapter 14 Review Exercises

1. Nonparametric tests do not require the population to follow a particular distribution. Because nonparametric hypothesis tests sometimes do involve claims about a population parameter, such as the population median, it may be more accurate to call them distribution-free hypothesis tests.

3. There is no parametric test for randomness.

5. M_d is the population median of the differences.

7. **(a)** $n = 9 + 0 = 9 \leq 25$, $\alpha = 0.10$, two-tailed test, so $S_{crit} = 1$.

(b) Reject H_0 if $S_{\text{data}} \leq 1$.

(c) Since this is a two-tailed test, S_{data} = the number of minus signs or plus signs, whichever is smaller. Thus $S_{\text{data}} = 0$.

(d) Since $S_{\text{data}} = 0$ is $\leq$ 1, reject H_0. There is evidence that the population median of the differences differs from 500.

9. $H_0 : p = 0.5$ versus $H_a : p < 0.5$. Since $n =$ the sample size = 100, which is > 25, we use the large sample case. This is a one-tailed test with $\alpha = 0.05$, so $Z_{crit} = -1.645$. Reject H_0 if $Z_{data} \leq -1.645$. Since this is a left-tailed test, S_{data} = the number of plus signs = the number of successes = 41.

$$Z_{\text{data}} = \frac{(S_{\text{data}}+0.5)-\frac{n}{2}}{\frac{\sqrt{n}}{2}} = \frac{(41+0.5)-\frac{100}{2}}{\frac{\sqrt{100}}{2}} = -1.7$$

Since $Z_{data} = -1.7$ is ≤ -1.645, we reject H_0. There is evidence that the population proportion of successes is less than 0.5.

11. Since $n = 20$ is $\leq$ 30, we use the Wilcoxon signed rank test for the small sample case. Since this is a right-tailed test with $\alpha = 0.05$, $T_{crit} = 60$. Reject H_0 if $T_{\text{data}} \leq 60$. Since this is a right-tailed test, $T_{\text{data}} = |T_-| = 70$. Since $T_{\text{data}} = 70$, which is not $\leq$ $T_{crit} = 60$, we do not reject H_0. There is insufficient evidence that the population median is greater than 10.

13. Since $n = 18$ is ≤ 30, we use the Wilcoxon signed rank test for the small sample case. Since this is a left-tailed test with $\alpha = 0.05$, $T_{crit} = 47$. Reject H_0 if $T_{\text{data}} \leq$ 47. Since this is a left-tailed test, $T_{\text{data}} = T_+ = 49$. Since $T_{\text{data}} = 49$, which is not $\leq$ $T_{crit} = 47$, we do not reject H_0. There is insufficient evidence that the population median difference is less than 0.

15.

45 50 55 60 65

Annual precipitation (inches)

Since perfect symmetry is rare in real-world data, especially with small data sets, we will interpret the boxplot as revealing acceptable symmetry. Therefore, it is appropriate to perform the Wilcoxon signed rank test. $H_0: M = 50$ versus $H_a: M \neq 50$. Since $M_0 = 50$, n equals the number of data values that are not equal to 50. Since none of the data values is equal to 50, $n = 6$. Since $n = 6$ is less than or equal to 30, we use the Wilcoxon signed rank test for the small sample case. Since this is a two-tailed test with $\alpha = 0.05$, $T_{crit} = 1$. Reject H_0 if $T_{\text{data}} \leq 1$.

Precipitation	**Precipitation – M_0 =**	**$\|d\|$**	**Rank of $\|d\|$**	**Signed rank**
49.56	49.56 – 50 = –0.44	0.44	1	–1
51.88	51.88 – 50 = 1.88	1.88	3	3
58.53	58.53 – 50 = 8.53	8.53	5	5
44.77	44.77 – 50 = –5.23	5.23	4	–4
64.19	64.19 – 50 = 14.19	14.19	6	6
48.35	48.35 – 50 = –1.65	1.65	2	–2

$T_+ = 3 + 5 + 6 = 14$ and $|T_-| = |(-1) + (-4) + (-2)| = 7$. Since this is a two-tailed test, $T_{\text{data}} = |T_+|$ or $|T_-|$, whichever is smaller. Thus $T_{\text{data}} = 7$. Since $T_{\text{data}} = 7$, which is not $\leq T_{crit} = 1$, we do not reject H_0. There is insufficient evidence that the population median precipitation in Florida differs from 50 inches.

17. $H_0: M_1 = M_2$ versus $H_a: M_1 \neq M_2$. Since this is a two-tailed test with $\alpha = 0.01$, $Z_{crit} =$ 2.58. Reject H_0 if $Z_{data} \leq -2.58$ or $Z_{data} \geq 2.58$.

Combined data	**20**	**22**	**22**	23	**24**	**24**	25	**26**	**27**	27
Rank	**1**	**2.5**	**2.5**	4	**5.5**	**5.5**	7	**8**	**9.5**	9.5
Combined data	**28**	28	29	29	**32**	**32**	32	**33**	**33**	**34**
Rank	**11.5**	11.5	13.5	13.5	**16**	**16**	16	**18.5**	**18.5**	**21**
Combined data	**34**	34	36	**38**	38	38	**40**			
Rank	**21**	21	23	**25**	25	25	**27**			

The values in **bold** are from Sample 1. Therefore, $R_1 = 1 + 2.5 + 2.5 + 5.5 + 5.5 + 8 + 9.5 + 11.5 + 16 + 16 + 18.5 + 18.5 + 21 + 21 + 25 + 27 = 209$. $n_1 = 16$, $n_2 = 11$.

$\mu_R = \frac{n_1(n_1+n_2+1)}{2} = \frac{16(16+11+1)}{2} = 224.$

$$\sigma_R = \sqrt{\frac{n_1 n_2(n_1+n_2+1)}{12}} = \sqrt{\frac{(16)(11)(16+11+1)}{12}} \approx 20.2649122.$$

$$Z_{\text{data}} = \frac{R_1 - \mu_R}{\sigma_R} = \frac{209-224}{20.2649122} \approx -0.7402.$$

Since $Z_{data} \approx -0.7402$, which is not ≤ -2.58 and not ≥ 2.58, the conclusion is do not reject H_0. There is insufficient evidence that the population median of Sample 1 differs from the population median of Sample 2.

19. $H_0 : M_1 = M_2$ versus $H_a : M_1 \neq M_2$. Since this is a two-tailed test with $\alpha = 0.05$, $Z_{crit} = 1.96$. Reject H_0 if $Z_{data} \leq -1.96$ or $Z_{data} \geq 1.96$.

Combined data	2.8	3.4	4.2	4.4	4.6	5.3	**5.4**	**6.4**	**6.6**	7.3
Rank	1	2	3	4	5	6	**7**	**8**	**9**	10
Combined data	**7.5**	**7.6**	**7.9**							
Rank	**11**	**12**	**13**							

The values in **bold** are from Ohio. Therefore, $R_1 = 7 + 8 + 9 + 11 + 12 + 13 = 60$. $n_1 = 6$, $n_2 = 7$.

$$\mu_R = \frac{n_1(n_1+n_2+1)}{2} = \frac{6(6+7+1)}{2} = 42.$$

$$\sigma_R = \sqrt{\frac{n_1 n_2(n_1+n_2+1)}{12}} = \sqrt{\frac{(6)(7)(6+7+1)}{12}} = 7.$$

$$Z_{\text{data}} = \frac{R_1 - \mu_R}{\sigma_R} = \frac{60-42}{7} \approx 2.5714.$$

Since $Z_{data} \approx 2.5714$, which is ≥ 1.96, the conclusion is reject H_0. There is evidence that the population median unemployment rate differs between cities in Ohio and Virginia.

21. Each sample is independent and randomly selected, and there are at least 5 data values in each sample. Thus, the conditions for the Kruskal-Wallis test are met, and we may proceed with the hypothesis test. H_0 : The population medians are all equal. H_a : Not all the population medians are equal. $\alpha = 0.05$, df $= k - 1 = 4 - 1 = 3$, $X^2_{crit} = 7.815$. Reject H_0 if $X^2_{data} \geq 7.815$.

Combined data	*11*	***11***	***11***	**12**	**12**	**12**	*12*	***12***	**13**	*14*	**15**	**15**	*15*
Rank	*2*	***2***	***2***	**6**	**6**	**6**	*6*	***6***	**9**	*10*	**12**	**12**	*12*
Combined data	**16**	*17*	*17*	***17***	***17***	***17***	17	*18*	**19**	***19***	19	*20*	20
Rank	**14**	*17.5*	*17.5*	***17.5***	***17.5***	***17.5***	17.5	*21*	**23**	***23***	23	*25.5*	25.5
Combined data	21	22	23										
Rank	27	28	29										

Sample 1 is in **bold**, Sample 2 is in *italics*, Sample 3 is in ***bold italics***, and Sample 4 is none of these.

$R_1 = 6 + 6 + 6 + 9 + 12 + 12 + 14 + 23 = 88$, $R_2 = 2 + 6 + 10 + 12 + 17.5 + 17.5 + 21 + 25.5 = 111.5$, $R_3 = 2 + 2 + 6 + 17.5 + 17.5 + 17.5 + 23 = 85.5$, $R_4 = 17.5 + 23 + 25.5 + 27 + 28 + 29 = 150$. $n_1 = 8$, $n_2 = 8$, $n_3 = 7$, $n_4 = 6$, $N = n_1 + n_2 + n_3 + n_4 = 8 + 8 + 7 + 6 = 29$.

$$\chi^2_{\text{data}} = \frac{12}{N(N+1)}\left(\frac{R_1^2}{n_1} + \frac{R_2^2}{n_2} + \frac{R_3^2}{n_3} + \frac{R_4^2}{n_4}\right) - 3(N+1) = \frac{12}{29(29+1)}\left(\frac{88^2}{8} + \frac{111.5^2}{8} + \frac{85.5^2}{7} + \frac{150^2}{6}\right) - 3(29+1) \approx 10.91520936.$$

Since $X^2_{data} \approx 10.91520936$, which is $\geq X^2_{crit} = 7.815$, we reject H_0. There is evidence that the population medians are not all the same.

23. H_0 : There is no rank correlation between the two variables. H_a : There is a rank correlation between the two variables. $n = 6 \leq 30$, $\alpha = 0.05$, $r_{crit} = 0.886$. Reject H_0 if $r_{data} \leq -0.886$ or $r_{data} \geq 0.886$.

Sample 1	Sample 2	Rank of Sample 1	Rank of Sample 2	Difference *d*	d^2
10	9	1	6	–5	25
12	6	2	4	–2	4
13	7	3	5	–2	4
15	1	4.5	1.5	3	9
15	3	4.5	3	1.5	2.25
18	1	6	1.5	4.5	20.25
					$\sum d^2 = 64.5$

$r_{\text{data}} = 1 - \frac{6\sum d^2}{n(n^2-1)} = 1 - \frac{6(64.5)}{6(6^2-1)} \approx -0.8428571429.$

Since $r_{data} \approx -0.8428571429$, which is not ≤ -0.886 and is not ≥ 0.886, the conclusion is do not reject H_0. There is insufficient evidence that a rank correlation exists between Sample 1 and Sample 2.

25. H_0 : There is no rank correlation between the two variables. H_a : There is a rank correlation between the two variables. $n = 13 \leq 30$, $\alpha = 0.05$, $r_{crit} = 0.560$. Reject H_0 if $r_{data} \leq -0.560$ or $r_{data} \geq 0.560$.

California city	Heating degree-days	Cooling degree-days	Heating rank	Cooling rank	Difference *d*	d^2
Arcadia	1295	1575	11.5	1.5	10	100
Burlingame	2720	184	1	13	–12	144
Simi Valley	1822	1485	4	3	1	1
Azusa	1727	1191	6.5	5.5	1	1
Palo Alto	2584	452	2	11	–9	81
Lake Forest	1465	1183	9	8	1	1
Santee	1313	1261	10	4	6	36
Torrance	1526	742	8	9	–1	1
Whittier	1295	1575	11.5	1.5	10	100
Dana Point	1756	666	5	10	–5	25
Camarillo	1961	389	3	12	–9	81
Glendora	1727	1191	6.5	5.5	1	1
Bellflower	1211	1186	13	7	6	36
						$\sum d^2 = 608$

$r_{\text{data}} = 1 - \frac{6\sum d^2}{n(n^2-1)} = 1 - \frac{6(608)}{13(13^2-1)} \approx -0.67032967.$

Since $r_{data} \approx -0.67032967$, which is ≤ -0.560, the conclusion is reject H_0. There is evidence that a rank correlation exists between heating degree-days and cooling degree-days in California.

27. H_0 : The sequence of data is random. H_a : The sequence of data is not random. n_1 = the number of Ms = $11 \leq 20$ and n_2 = the number of Fs = $11 \leq 20$ and $\alpha = 0.05$. From Table L, $G_{crit,lower} = 7$ and $G_{crit,upper} = 17$. Reject H_0 if $G \leq 7$ or if $G \geq 17$.

$\underbrace{M}_{1}\ \underbrace{FF}_{2}\ \underbrace{MMM}_{3}\ \underbrace{FF}_{4}\ \underbrace{M}_{5}\ \underbrace{FF}_{6}\ \underbrace{M}_{7}\ \underbrace{F}_{8}\ \underbrace{MMM}_{9}\ \underbrace{FFF}_{10}\ \underbrace{M}_{11}\ \underbrace{F}_{12}\ \underbrace{M}_{13}.$

So, $G = 13$. Since $G = 13$ is not $\leq G_{crit,lower} = 7$ and is not $\geq G_{crit,upper} = 17$, the conclusion is do not reject H_0. There is insufficient evidence that the sequence is not random.

Chapter 14 Quiz

1. True.

3. True.

5. 30

7. The differences equal to 0 need to be eliminated.

9. The conditions for performing the Kruskal-Wallis test are that there are 3 or more independent samples and that each of the sample sizes must be at least 5.

11. Boxplot of the differences of the carbon emissions of various countries:

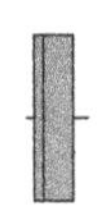

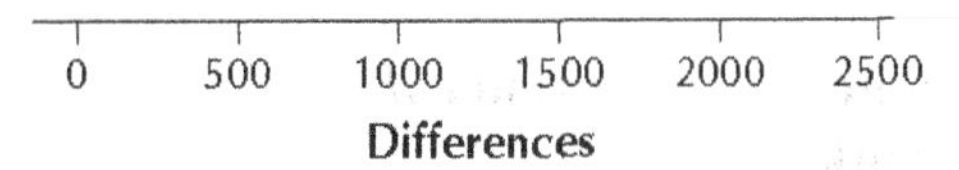

Since the boxplot indicates that the median is closer to Q1 to Q3 and the left whisker is shorter that the right whisker, the distribution of the differences is not symmetric. Therefore, the signed-rank test is not appropriate. Since the sample was randomly selected, perform the sign test for matched-pair data from two-dependent samples. $H_0 : M_d = 0$ versus $H_a : M_d > 0$.

Nation	Carbon emissions in 2000 (millions of metric tons)	Carbon emissions in 2005 (millions of metric tons)	Difference (2005 – 2000)	Sign
Brazil	342.1	360.6	18.5	+
Canada	558.4	631.3	72.9	+
China	2912.6	5322.7	2410.1	+
France	399.0	415.3	16.3	+
India	994.1	1165.7	171.6	+
Ireland	40.4	44.1	3.7	+
South Africa	383.4	423.8	40.4	+
Thailand	160.6	234.2	73.6	+
Vietnam	47.39	80.4	33.01	+
United States	5823.5	5957.0	133.5	+

$n = 10 \le 25$, $\alpha = 0.05$, one-tailed test, so $S_{\text{crit}} = 1$. Reject H_0 if $S_{\text{data}} \le 1$. Since this is a right-tailed test, S_{data} = the number of minus signs = 0. Since $S_{\text{data}} = 0$, which is ≤ 1, the conclusion is reject H_0. There is evidence that the population median of the difference in carbon dioxide emissions is greater than 0.

13. $H_0 : M_1 = M_2$ versus $H_a : M_1 \neq M_2$. Since this is a two-tailed test with $\alpha = 0.05$, $Z_{crit} = 1.96$. Reject H_0 if $Z_{data} \le -1.96$ or $Z_{data} \ge 1.96$.

Combined data	–256,207	–82,760	**–44,513**	–25,230	**–21,436**	–20,948
Rank	1	2	**3**	4	**5**	6
Combined data	–14,300	**–14,140**	–12,918	–11,968	–7,775	**–7,497**
Rank	7	**8**	9	10	11	**12**
Combined data	**–6,629**	**–4,256**	–2,976	–2,325	**–2,133**	**–1,168**
Rank	**13**	**14**	15	16	**17**	**18**
Combined data	–61	**475**	**918**	**10,009**	**14,560**	
Rank	19	**20**	**21**	**22**	**23**	

The values in **bold** are from European countries. Therefore, $R_1 = 3 + 5 + 8 + 12 + 13 + 14 + 17 + 18 + 20 + 21 + 22 + 23 = 176$. $n_1 = 12$, $n_2 = 11$.

$$\mu_R = \frac{n_1(n_1+n_2+1)}{2} = \frac{12(12+11+1)}{2} = 144.$$

$$\sigma_R = \sqrt{\frac{n_1 n_2(n_1+n_2+1)}{12}} = \sqrt{\frac{(12)(11)(12+11+1)}{12}} \approx 16.24807681.$$

$$Z_{\text{data}} = \frac{R_1-\mu_R}{\sigma_R} \approx \frac{176-144}{16.24807681} \approx 1.9695.$$

Since $Z_{data} \approx 1.9695$, which is ≥ 1.96, we reject H_0. There is evidence that the population median trade balance with European countries differs from the population median trade balance with Asian countries.